AF293973

Universitext

Springer Science+Business Media, LLC

Universitext

Editors (North America): S. Axler, F.W. Gehring, and K.A. Ribet

Aksoy/Khamsi: Nonstandard Methods in Fixed Point Theory
Andersson: Topics in Complex Analysis
Aupetit: A Primer on Spectral Theory
Berberian: Fundamentals of Real Analysis
Booss/Bleecker: Topology and Analysis
Borkar: Probability Theory: An Advanced Course
Böttcher/Sibermann: Introduction to Large Truncated Toeplitz Matrices
Carleson/Gamelin: Complex Dynamics
Cecil: Lie Sphere Geometry: With Applications to Submanifolds
Chae: Lebesgue Integration (2nd ed.)
Charlap: Bieberbach Groups and Flat Manifolds
Chern: Complex Manifolds Without Potential Theory
Cohn: A Classical Invitation to Algebraic Numbers and Class Fields
Curtis: Abstract Linear Algebra
Curtis: Matrix Groups
DiBenedetto: Degenerate Parabolic Equations
Dimca: Singularities and Topology of Hypersurfaces
Edwards: A Formal Background to Mathematics I a/b
Edwards: A Formal Background to Mathematics II a/b
Foulds: Graph Theory Applications
Friedman: Algebraic Surfaces and Holomorphic Vector Bundles
Fuhrmann: A Polynomial Approach to Linear Algebra
Gardiner: A First Course in Group Theory
Gårding/Tambour: Algebra for Computer Science
Goldblatt: Orthogonality and Spacetime Geometry
Gustafson/Rao: Numerical Range: The Field of Values of Linear Operators and Matrices
Hahn: Quadratic Algebras, Clifford Algebras, and Arithmetic Witt Groups
Holmgren: A First Course in Discrete Dynamical Systems
Howe/Tan: Non-Abelian Harmonic Analysis: Applications of $SL(2, R)$
Howes: Modern Analysis and Topology
Hsieh/Sibuya: Basic Theory of Ordinary Differential Equations
Humi/Miller: Second Course in Ordinary Differential Equations
Hurwitz/Kritikos: Lectures on Number Theory
Jennings: Modern Geometry with Applications
Jones/Morris/Pearson: Abstract Algebra and Famous Impossibilities
Kannan/Krueger: Advanced Analysis
Kelly/Matthews: The Non-Euclidean Hyperbolic Plane
Kostrikin: Introduction to Algebra
Luecking/Rubel: Complex Analysis: A Functional Analysis Approach
MacLane/Moerdijk: Sheaves in Geometry and Logic
Marcus: Number Fields
McCarthy: Introduction to Arithmetical Functions
Meyer: Essential Mathematics for Applied Fields
Mines/Richman/Ruitenburg: A Course in Constructive Algebra

(continued after index)

Po-Fang Hsieh   Yasutaka Sibuya

Basic Theory of Ordinary Differential Equations

With 114 Illustrations

Springer

Po-Fang Hsieh
Department of Mathematics
  and Statistics
Western Michigan University
Kalamazoo, MI 49008
USA
philip.hsieh@wmich.edu

Yasutaka Sibuya
School of Mathematics
University of Minnesota
206 Church Street SE
Minneapolis, MN 55455
USA
sibuya@math.umn.edu

Mathematics Subject Classification (1991): 34-01

Library of Congress Cataloging-in-Publication Data
Hsieh, Po-Fang.
    Basic theory of ordinary differential equations / Po-Fang Hsieh,
Yasutaka Sibuya.
        p.   cm. — (Universitext)
        Includes bibliographical references and index.
        ISBN 978-1-4612-7171-0   ISBN 978-1-4612-1506-6 (eBook)
        DOI 10.1007/978-1-4612-1506-6
        1. Differential equations.   I. Sibuya, Yasutaka, 1930–
    II. Title.   III. Series.
    QA372.H84   1999
    515′.35—dc21   99-18392

Printed on acid-free paper.

Production managed by MaryAnn Cottone; manufacturing supervised by Jeffrey Taub.
Photocomposed copy prepared from the authors' $\mathcal{A}_{\mathcal{M}}\mathcal{S}$-TₑX 2.1 files.

9 8 7 6 5 4 3 2 1

ISBN 978-1-4612-7171-0

To Emmy and Yasuko

PREFACE

This graduate level textbook is developed from courses in ordinary differential equations taught by the authors in several universities in the past 40 years or so. Prerequisite of this book is a knowledge of elementary linear algebra, real multivariable calculus, and elementary manipulation with power series in several complex variables. It is hoped that this book would provide the reader with the very basic knowledge necessary to begin research on ordinary differential equations. To this purpose, materials are selected so that this book would provide the reader with methods and results which are applicable to many problems in various fields. In order to accomplish this purpose, the book *Theory of Differential Equations* by E. A. Coddington and Norman Levinson is used as a role model. Also, the teaching of Masuo Hukuhara and Mitio Nagumo can be found either explicitly or in spirit in many chapters. This book is useful for both pure mathematician and user of mathematics.

This book may be divided into four parts. The first part consists of Chapters I, II, and III and covers fundamental existence, uniqueness, smoothness with respect to data, and nonuniqueness. The second part consists of Chapters IV, VI, and VII and covers the basic results concerning linear differential equations. The third part consists of Chapters VIII, IX, and X and covers nonlinear differential equations. Finally, Chapters V, XI, XII, and XIII cover the basic results concerning power series solutions.

The particular contents of each chapter are as follows. The fundamental existence and uniqueness theorems and smoothness in data of an initial problem are explained in Chapters I and II, whereas the results concerning nonuniqueness are explained in Chapter III. Topics in Chapter III include the Kneser theorem and maximal and minimal solutions. Also, utilizing comparison theorems, some sufficient conditions for uniqueness are studied. In Chapter IV, the basic theorems concerning linear differential equations are explained. In particular, systems with constant or periodic coefficients are treated in detail. In this study, the S-N decomposition of a matrix is used instead of the Jordan canonical form. The S-N decomposition is equivalent to the block-diagonalization separating distinct eigenvalues. Computation of the S-N decomposition is easier than that of the Jordan canonical form. A detailed explanation of linear Hamiltonian systems with constant or periodic coefficients is also given. In Chapter V, formal power series solutions and their convergence are explained. The main topic is singularities of the first kind. The convergence of formal power series solutions is proven for nonlinear systems. Also, the transformation of a linear system to a standard form at a singular point of the first kind is explained as the S-N decomposition of a linear differential operator. The main idea is originally due to R. Gérard and A. H. M. Levelt. The Gérard-Levelt theorem is presented as the S-N decomposition of a matrix of infinite order. At the end of Chapter V, the classification of the singularities of linear differential equations is given. In Chapter VI, the main topics are the basic results concerning boundary-value problems of the second-order linear differential equations. The comparison theorems, oscillation and nonoscillation of solutions, eigenvalue problems for the Sturm-Liouville boundary conditions, scattering problems (in the case of reflectionless potentials), and periodic potentials are studied. The authors learned much about the scattering problems from the book by S. Tanaka and E. Date [TD]. In Chapter VII, asymptotic behaviors of solutions of linear systems as the independent variable approaches infinity are treated. Topics include the Liapounoff numbers and the Levinson theorem together with its various improvements. In Chapter VIII, some fundamental theorems concerning stability, asymptotic stability, and perturbations of 2×2 linear systems are explained, whereas in Chapter IX, results on autonomous systems which include the LaSalle-Lefschetz theorem concerning behavior of solutions (or orbits) as the independent variable tends to infinity, the basic properties of limit-invariant sets including the Poincaré-Bendixson theorem, and applications of indices of simple closed curves are studied. Those theorems are applied to some nonlinear oscillation problems in Chapter X. In particular, the van der Pol equation is treated as both a problem of regular perturbations and a problem of singular perturbations. In Chapters XII and XIII, asymptotic solutions of nonlinear differential equations as a parameter or the independent variable tends to its singularity are explained. In these chapters, the asymptotic expansions in the sense of Poincaré are used most of time. However, asymptotic solutions in the sense of the Gevrey asymptotics are explained briefly. The basic properties of asymptotic expansions in the sense of Poincaré as well as of the Gevrey asymptotics are explained in Chapter XI.

At the beginning of each chapter, the contents and their history are discussed briefly. Also, at the end of each chapter, many problems are given as exercises. The purposes of the exercises are (i) to help the reader to understand the materials in each chapter, (ii) to encourage the reader to read research papers, and (iii) to help the reader to develop his (or her) ability to do research. Hints and comments for many exercises are provided.

The authors are indebted to many colleagues and former students for their valuable suggestions, corrections, and assistance at the various stages of writing this book. In particular, the authors express their sincere gratitude to Mrs. Susan Coddington and Mrs. Zipporah Levinson for allowing the authors to use the materials in the book *Theory of Differential Equations* by E. A. Coddington and Norman Levinson.

Finally, the authors could not have carried out their work all these years without the support of their wives and children. Their contribution is immeasurable. We thank them wholeheartedly.

PFH
YS

March, 1999

CHAPTER I

FUNDAMENTAL THEOREMS OF
ORDINARY DIFFERENTIAL EQUATIONS

In this chapter, we explain the fundamental problems of the existence and uniqueness of the initial-value problem

$$\text{(P)} \qquad \frac{d\vec{y}}{dt} = \vec{f}(t, \vec{y}), \qquad \vec{y}(t_0) = \vec{c}_0$$

in the case when the entries of $\vec{f}(t, \vec{y})$ are real-valued and continuous in the variable $(t, \vec{y})$, where t is a real independent variable and $\vec{y}$ is an unknown quantity in $\mathbb{R}^n$. Here, $\mathbb{R}$ is the real line and $\mathbb{R}^n$ is the set of all n-column vectors with real entries. In §I-1, we treat the problem when $\vec{f}(t, \vec{y})$ satisfies the Lipschitz condition in $\vec{y}$. The main tools are successive approximations and Gronwall's inequality (Lemma I-1-5). In §I-2, we treat the problem without the Lipschitz condition. In this case, approximating $\vec{f}(t, \vec{y})$ by smooth functions, ϵ-approximate solutions are constructed. In order to find a convergent sequence of approximate solutions, we use Arzelà-Ascoli's lemma concerning a bounded and equicontinuous set of functions (Lemma I-2-3). The existence Theorem I-2-5 is due to A. L. Cauchy and G. Peano [Pea1] and the existence and uniqueness Theorem I-1-4 is due to É. Picard [Pi] and E. Lindelöf [Lind1, Lind2]. The extension of these local solutions to a larger interval is explained in §I-3, assuming some basic requirements for such an extension. In §I-4, using successive approximations, we explain the power series expansion of a solution in the case when $\vec{f}(t, \vec{y})$ is analytic in $(t, \vec{y})$. In each section, examples and remarks are given for the benefit of the reader. In particular, remarks concerning other methods of proving these fundamental theorems are given at the end of §I-2.

I-1. Existence and uniqueness with the Lipschitz condition

We shall use the following notations throughout this book:

$$\text{(I.1.1)} \qquad \vec{y} = \begin{bmatrix} y_1 \\ y_2 \\ \vdots \\ y_n \end{bmatrix}, \qquad \vec{f} = \begin{bmatrix} f_1 \\ f_2 \\ \vdots \\ f_n \end{bmatrix}, \qquad |\vec{y}| = \max\{|y_1|, |y_2|, \ldots, |y_n|\}.$$

In §§I-1, I-2, and I-3 we consider problem (P) under the following assumption.

Assumption 1. *The entries of $\vec{f}(t, \vec{y})$ are real-valued and continuous on a rectangular region:*

$$\mathcal{R} = \mathcal{R}((t_0, \vec{c}_0), a, b) = \{(t, \vec{y}) : |t - t_0| \leq a, \ |\vec{y} - \vec{c}_0| \leq b\},$$

where a and b are two positive numbers.

1

Since $\vec{f}(t,\vec{y})$ is continuous on $\mathcal{R}$, $|\vec{f}(t,\vec{y})|$ is bounded. Let us denote by M the maximum value of $|\vec{f}(t,\vec{y})|$ on $\mathcal{R}$, i.e., $M = \max_{\mathcal{R}}\left|\vec{f}(t,\vec{y})\right|$. We define a positive number α by

$$\alpha = \begin{cases} a & \text{if } M = 0, \\ \min\left(a, \dfrac{b}{M}\right) & \text{if } M > 0. \end{cases}$$

To locate a solution in a neighborhood of its initial point, the number α plays an important role as shown in the following lemma.

Lemma I-1-1. *If $\vec{y} = \vec{\phi}(t)$ is a solution of problem (P) in an interval $|t - t_0| < \tilde{\alpha} \le \alpha$, then $|\vec{\phi}(t) - \vec{c}_0| < b$ in $|t - t_0| < \tilde{\alpha}$, i.e., $(t, \vec{\phi}(t)) \in \mathcal{R}((t_0, \vec{c}_0), \tilde{\alpha}, b)$ for $|t - t_0| < \tilde{\alpha}$.*

Proof.

Assume that Lemma I-1-1 is not true. Then, by the continuity of $\vec{\phi}(t)$, there exists a positive number β such that

(i) $\qquad\qquad \beta < \tilde{\alpha},$

(ii)

$$\begin{cases} |\vec{\phi}(t) - \vec{c}_0| < b & \text{for} & |t - t_0| < \beta, \\ |\vec{\phi}(t_0 + \beta) - \vec{c}_0| = b & \text{or} & |\vec{\phi}(t_0 - \beta) - \vec{c}_0| = b. \end{cases}$$

The assumption $\tilde{\alpha} \le \alpha \le a$ implies $\beta < a$. Hence, $(t, \vec{\phi}(t)) \in \mathcal{R}$ for $|t - t_0| \le \beta$. Therefore, $|\vec{f}(t, \vec{\phi}(t))| \le M$ for $|t - t_0| \le \beta$. From $\vec{\phi}'(t) = \vec{f}(t, \vec{\phi}(t))$ and $\vec{\phi}(t_0) = \vec{c}_0$, it follows that

$$(\text{I.1.2}) \qquad \vec{\phi}(t) = \vec{c}_0 + \int_{t_0}^{t} \vec{f}(s, \vec{\phi}(s))\,ds \qquad \text{for } |t - t_0| \le \beta.$$

Hence,

$$(\text{I.1.3}) \qquad |\vec{\phi}(t) - \vec{c}_0| = \left|\int_{t_0}^{t} \vec{f}(s, \vec{\phi}(s))\,ds\right| \le M|t - t_0| \qquad \text{for } |t - t_0| \le \beta.$$

This implies that

$$(\text{I.1.4}) \qquad |\vec{\phi}(t) - \vec{c}_0| \le M\beta < M\tilde{\alpha} \le M\alpha \le M\frac{b}{M} = b \qquad \text{for } |t - t_0| \le \beta.$$

In particular, $|\vec{\phi}(t_0 \pm \beta) - \vec{c}_0| < b$. This contradicts assumption (ii). $\quad\square$

Similarly, we obtain the following result.

Lemma I-1-2. *If $\vec{y} = \vec{\phi}(t)$ is a solution of problem (P) in an interval $|t - t_0| < \tilde{\alpha} \le \alpha$, then $|\vec{\phi}(t_1) - \vec{\phi}(t_2)| \le M|t_1 - t_2|$ whenever t_1 and t_2 are in the interval $|t - t_0| < \tilde{\alpha}$.*

Remark I-1-3. *Using Lemma I-1-2, it is concluded that if $\vec{y} = \vec{\phi}(t)$ is a solution of problem (P) on an interval $|t - t_0| < \tilde{\alpha} \leq \alpha$, then $\vec{\phi}(t_0 + \tilde{\alpha} - 0)$ and $\vec{\phi}(t_0 - \tilde{\alpha} + 0)$ exist.*

Now, consider problem (P) with Assumption 1 and the following assumption.

Assumption 2. *The function $\vec{f}$ satisfies a Lipschitz condition on $\mathcal{R}$: i.e., there exists a positive constant L such that*

$$|\vec{f}(t, \vec{y}_1) - \vec{f}(t, \vec{y}_2)| \leq L|\vec{y}_1 - \vec{y}_2|$$

whenever $(t, \vec{y}_1)$ and $(t, \vec{y}_2)$ are on the region $\mathcal{R}$ (cf. [Lip]).

Note that L is dependent on the region $\mathcal{R}$. The constant L is called the *Lipschitz constant* of $\vec{f}$ with respect to $\mathcal{R}$.

The following theorem is the main conclusion of this section.

Theorem I-1-4. *Under Assumptions 1 and 2, there exists a unique solution of problem (P) on the interval $|t - t_0| < \alpha$.*

In order to prove this theorem, the following lemma is useful.

Lemma I-1-5 (T. H. Gronwall [Gr]). *If*
  (i) $g(t)$ is continuous on $t_0 \leq t \leq t_1$,
  (ii) $g(t)$ satisfies the inequality

$$(\text{I.1.5}) \qquad 0 \leq g(t) \leq K + L \int_{t_0}^{t} g(s)ds \quad on \;\; t_0 \leq t \leq t_1,$$

then

$$(\text{I.1.6}) \qquad 0 \leq g(t) \leq K \exp[L(t - t_0)]$$

on $t_0 \leq t \leq t_1$.

Proof.

Set $v(t) = \displaystyle\int_{t_0}^{t} g(s)ds$. Then, by (I.1.5), $\dfrac{dv(t)}{dt} \leq K + Lv(t)$ and $v(t_0) = 0$. Hence,

$$(\text{I.1.7}) \qquad \frac{d}{dt}\{\exp[-L(t - t_0)]v(t)\} \leq K \exp[-L(t - t_0)]$$

and, consequently,

$$(\text{I.1.8}) \qquad \exp[-L(t - t_0)]v(t) \leq \frac{K}{L}\{1 - \exp[-L(t - t_0)]\}.$$

This, in turn, implies that $Lv(t) \leq K\{\exp[L(t - t_0)] - 1\}$. Therefore, $g(t) \leq K + Lv(t) \leq K \exp[L(t - t_0)]$. $\square$

Proof of Theorem I-1-4.

We shall prove this theorem in six steps by using successive approximations which are defined as follows:

$$(\text{I.1.9}) \qquad \begin{cases} \vec{\phi}_0(t) = \vec{c}_0, \\[2mm] \vec{\phi}_{k+1}(t) = \vec{c}_0 + \displaystyle\int_{t_0}^{t} \vec{f}(s, \vec{\phi}_k(s))ds, \qquad k = 0, 1, 2, \ldots. \end{cases}$$

Step 1. Each function $\vec{\phi}_k$ is well defined and $(t, \vec{\phi}_k(t)) \in \mathcal{R}$ for $|t - t_0| \leq \alpha$ $(k = 0, 1, 2, \dots)$.

Proof.

We use mathematical induction in Steps 1 and 2. The statement above is evident for $k = 0$. Assume that $(t, \vec{\phi}_k(t)) \in \mathcal{R}$ on $|t - t_0| \leq \alpha$. Then, $|\vec{f}(t, \vec{\phi}_k(t))| \leq M$ on $|t - t_0| \leq \alpha$. Hence,

$$|\vec{\phi}_{k+1}(t) - \vec{c}_0| = \left| \int_{t_0}^{t} \vec{f}(s, \vec{\phi}_k(s)) ds \right| \leq M|t - t_0| \leq M\alpha \leq M\frac{b}{M} = b.$$

Thus, $(t, \vec{\phi}_{k+1}(t)) \in \mathcal{R}$ for $|t - t_0| \leq \alpha$.

Step 2. The successive approximations given by (I.1.9) satisfy the estimates

$$|\vec{\phi}_{k+1}(t) - \vec{\phi}_k(t)| \leq \frac{ML^k}{(k+1)!}|t - t_0|^{k+1} \qquad \text{for } |t - t_0| \leq \alpha, \quad k = 0, 1, 2, \dots .$$

Proof.

For $k = 0$, we have $|\vec{\phi}_1(t) - \vec{\phi}_0(t)| = \left| \int_{t_0}^{t} \vec{f}(s, \vec{c}_0) ds \right| \leq M|t - t_0|$. Assume that $|\vec{\phi}_k(t) - \vec{\phi}_{k-1}(t)| \leq \dfrac{ML^{k-1}}{k!}|t - t_0|^k$ for $|t - t_0| \leq \alpha$. Then,

$$|\vec{\phi}_{k+1}(t) - \vec{\phi}_k(t)| = \left| \int_{t_0}^{t} \left\{ \vec{f}(s, \vec{\phi}_k(s)) - \vec{f}(s, \vec{\phi}_{k-1}(s)) \right\} ds \right|$$

$$\leq L \left| \int_{t_0}^{t} |\vec{\phi}_k(t) - \vec{\phi}_{k-1}(t)| ds \right|$$

$$\leq \frac{ML^k}{k!} \left| \int_{t_0}^{t} |t - t_0|^k ds \right| = \frac{ML^k}{(k+1)!}|t - t_0|^{k+1}.$$

Step 3. The series $\displaystyle\sum_{k=1}^{+\infty} \left(\vec{\phi}_k(t) - \vec{\phi}_{k-1}(t) \right)$ is uniformly convergent on the interval $|t - t_0| \leq \alpha$.

Proof.

By virtue of the result of Step 2, for a given $\epsilon > 0$, there exists a positive integer N such that

$$\sum_{k=N}^{\infty} |\vec{\phi}_k(t) - \vec{\phi}_{k-1}(t)| \leq \frac{M}{L} \sum_{k=N}^{\infty} \frac{(L|t - t_0|)^k}{k!} \leq \frac{M}{L} \sum_{k=N}^{\infty} \frac{(L\alpha)^k}{k!} < \epsilon.$$

Step 4. The sequence $\{\vec{\phi}_k(t) : k = 0, 1, 2, \dots\}$ converges to

$$(\mathrm{I.1.10}) \qquad \vec{\phi}(t) = \vec{\phi}_0(t) + \sum_{k=1}^{\infty} \left(\vec{\phi}_k(t) - \vec{\phi}_{k-1}(t) \right)$$

uniformly on $|t - t_0| \le \alpha$ as $k \to +\infty$.

Proof.

Use the identity $\vec{\phi}_N(t) = \vec{\phi}_0(t) + \sum_{k=1}^{N} \left(\vec{\phi}_k(t) - \vec{\phi}_{k-1}(t) \right)$ and the result of Step 3.

Step 5. $\vec{y} = \vec{\phi}(t)$ given by (I.1.10) is a solution of problem (P) on $|t - t_0| \le \alpha$.

Proof.

The definition

$$\vec{\phi}_{k+1}(t) = \vec{c}_0 + \int_{t_0}^{t} \vec{f}(s, \vec{\phi}_k(s)) ds,$$

the continuity of $\vec{f}$, and the results of Steps 3 and 4 imply that

$$\vec{\phi}(t) = \vec{c}_0 + \int_{t_0}^{t} \vec{f}(s, \vec{\phi}(s)) ds.$$

Step 6. $\vec{y} = \vec{\phi}(t)$ is the unique solution of problem (P) on $|t - t_0| < \alpha$.

Proof.

Suppose that $\vec{y} = \vec{\psi}(t)$ is another solution of problem (P) on an interval $|t - t_0| < \tilde{\alpha} \le \alpha$ for some $\tilde{\alpha}$. Lemma I-1-1 and Remark I-1-3 imply that $(t, \vec{\psi}(t)) \in \mathcal{R}$ on $|t - t_0| \le \tilde{\alpha}$. Note that

$$\vec{\psi}(t) = \vec{c}_0 + \int_{t_0}^{t} \vec{f}(s, \vec{\psi}(s)) ds$$

on $|t - t_0| \le \tilde{\alpha}$. Hence,

$$|\vec{\phi}(t) - \vec{\psi}(t)| = \left| \int_{t_0}^{t} \left\{ \vec{f}(s, \vec{\phi}(s)) - \vec{f}(s, \vec{\psi}(s)) \right\} ds \right| \le L \left| \int_{t_0}^{t} |\vec{\phi}(s) - \vec{\psi}(s)| ds \right|$$

on $|t - t_0| \le \tilde{\alpha}$. Now, by applying Lemma I-1-5 with $K = 0$ to the inequality

$$|\vec{\phi}(t) - \vec{\psi}(t)| \le L \int_{t_0}^{t} |\vec{\phi}(s) - \vec{\psi}(s)| ds$$

on $t_0 \le t \le t_0 + \tilde{\alpha}$, we conclude that $|\vec{\phi}(t) - \vec{\psi}(t)| = 0$ on the interval $t_0 \le t \le t_0 + \tilde{\alpha}$. Similarly, the same conclusion is obtained on the interval $t_0 - \tilde{\alpha} \le t \le t_0$. Steps 1 through 6 complete the proof of Theorem I-1-4. $\square$

The following lemma gives a sufficient condition that $\vec{f}(t, \vec{y})$ satisfies Assumption 2 (i.e., the Lipschitz condition).

Lemma I-1-6. *If $\dfrac{\partial \vec{f}(t,\vec{y})}{\partial y_j}$ $(j = 1,2,\ldots,n)$ exist, are continuous, and satisfy the condition:* $\left| \dfrac{\partial \vec{f}(t,\vec{y})}{\partial y_j} \right| \leq L_0$ $(j = 1,2,\ldots,n)$ *for $(t,\vec{y}) \in \mathcal{I} \times \mathcal{D}$, where $\mathcal{I}$ is an interval, $\mathcal{D}$ is a region, and L_0 is a constant, and if $\mathcal{D}$ is a convex open set, then $|\vec{f}(t,\vec{y}_1) - \vec{f}(t,\vec{y}_2)| \leq nL_0|\vec{y}_1 - \vec{y}_2|$ for $t \in \mathcal{I}$, $\vec{y}_1 \in \mathcal{D}$, and $\vec{y}_2 \in \mathcal{D}$.*

Proof.

Fix $t \in \mathcal{I}$, and $\vec{y}_1$ and $\vec{y}_2$ in $\mathcal{D}$. Then, $\theta \vec{y}_1 + (1-\theta)\vec{y}_2 \in \mathcal{D}$ for $0 \leq \theta \leq 1$ since $\mathcal{D}$ is convex. Setting $\vec{g}(\theta) = \vec{f}(t, \theta\vec{y}_1 + (1-\theta)\vec{y}_2)$, we derive

$$\begin{cases} \vec{g}(0) = \vec{f}(t,\vec{y}_2), \qquad \vec{g}(1) = \vec{f}(t,\vec{y}_1), \\ \dfrac{d\vec{g}}{d\theta} = \sum_{j=1}^{n} (y_{1,j} - y_{2,j})\dfrac{\partial \vec{f}}{\partial y_j}(t, \theta\vec{y}_1 + (1-\theta)\vec{y}_2), \end{cases}$$

where $y_{h,1}, y_{h,2}, \ldots, y_{h,n}$ are entries of $\vec{y}_h$ $(h = 1,2)$. Hence,

$$|\vec{f}(t,\vec{y}_1) - \vec{f}(t,\vec{y}_2)| = \left| \int_0^1 \frac{d\vec{g}}{d\theta}(\theta)d\theta \right| \leq nL_0|\vec{y}_1 - \vec{y}_2|.$$

This is true for all $t \in \mathcal{I}$, $\vec{y}_1 \in \mathcal{D}$, and $\vec{y}_2 \in \mathcal{D}$. Thus, the lemma is proved. $\square$

The following two simple problems illustrate Theorem I-1-4.

Problem 1. Using Theorem I-1-4, show that on the interval $|t - 1| < \sqrt{2}$, the initial-value problem

$$\frac{dy}{dt} = \frac{1}{(3 - (t-1)^2)(9 - (y-5)^2)}, \qquad y(1) = 5$$

has one and only one solution.

Answer.

The standard strategy is to find a suitable region

$$\mathcal{R} = \{(t,y) : |t-1| \leq a, \quad |y-5| \leq b\}$$

and find the maximum value M of $|f(t,y)|$ in $\mathcal{R}$. Then, $\alpha = \min\left(a, \dfrac{b}{M}\right)$. If $\alpha \geq \sqrt{2}$, the problem is solved. In other words, the main idea is to trap the solution curve in a suitable region $\mathcal{R}$. To do this, for example, use the fact that the function

$$f(t,y) = \frac{1}{(3 - (t-1)^2)(9 - (y-5)^2)}$$

is continuous and satisfies the Lipschitz condition

$$|f(t,y_1) - f(t,y_2)| \leq \frac{|y_1 - y_2|}{5}$$

on the region $\mathcal{R} = \{(t, y) : |t-1| \leq \sqrt{2}, |y-5| \leq 2\}$ (cf. Exercise I-2). Furthermore,

$$|f(t,y)| \leq \frac{1}{(3 - (\sqrt{2})^2)(9 - 2^2)} = \frac{1}{5} \quad \text{on} \ \mathcal{R}.$$

Set $a = \sqrt{2}$, $b = 2$, and $M = \frac{1}{5}$. Then,

$$\alpha = \min\left(a, \frac{b}{M}\right) = \min\left(\sqrt{2}, \frac{2}{\frac{1}{5}}\right) = \min(\sqrt{2}, 10) = \sqrt{2}.$$

Therefore, the given initial-value problem has one and only one solution on the interval $|t - 1| < \sqrt{2}$.

Problem 2. Using Theorem I-1-4, show that the initial-value problem

$$(\tilde{P}) \qquad \frac{dy}{dt} = \frac{1}{(1 + (t - 4)^2)(5 + (y - 3)^2)}, \qquad y(4) = 3$$

has one and only one solution on the interval $-\infty < t < +\infty$.

Answer.

The function

$$f(t, y) = \frac{1}{(1 + (t - 4)^2)(5 + (y - 3)^2)}$$

is continuous and satisfies the Lipschitz condition everywhere in the (t, y)-plane. Also, $|f(t, y)| \leq \frac{1}{5}$ everywhere. This implies that $\alpha = \min(a, 5b)$, if Problem $(\tilde{P})$ is considered in a region

$$\mathcal{R} = \{(t, y) : |t - 4| \leq a, \quad |y - 3| \leq b\}.$$

Hence, if $a \leq 5b$, Problem $(\tilde{P})$ has one and only one solution on the interval $|t - 4| < a$. Now, this solution is independent of a due to the uniqueness. Therefore, we can conclude that $(\tilde{P})$ has one and only one solution on the interval $-\infty < t < +\infty$.

Remark I-1-7. It is usually claimed that the *general solution* of the system $\frac{d\vec{y}}{dt} = \vec{f}(t, \vec{y})$ depends on n independent arbitrary constants. Theorem I-1-4 verifies this claim if we define the *general solution* very carefully, since the initial data $(t_0, \vec{c})$ represent n arbitrary constants for a fixed t_0. However, if the reader looks into this definition a little deeper, he will find various complicated situations. It is important for the reader to know that the number of independent arbitrary constants contained in the *general solution* depends strongly on the space of functions to which the *general solution* should belong. For example, if the differential equation

$$(E) \qquad t\frac{dy}{dt} = 1$$

is considered on the set $\mathcal{J} = \{t \in \mathbb{R} : t \neq 0\}$, the general solution is $y(t) = c_1 + \ln|t|$ for $t > 0$ and $y(t) = c_2 + \ln|t|$ for $t < 0$, where c_1 and c_2 are independent arbitrary constants. If the Heaviside function

$$H(t) = \begin{cases} 1 & \text{if} \quad t > 0, \\ -1 & \text{if} \quad t < 0 \end{cases}$$

is used, this general solution is given by

$$y(t) = \frac{c_1 - c_2}{2} H(t) + \frac{c_1 + c_2}{2} + \ln|t|.$$

Equation (E) may be looked at in terms of the generalized functions (or the distributions of L. Schwartz [Sc]). The transformation $y = u + \ln|t|$, changes equation (E) to

$$(E') \qquad\qquad t\frac{du}{dt} = 0.$$

As a differential equation on the generalized functions, (E') is reduced to

$$(E'') \qquad\qquad \frac{du}{dt} = c\delta(t),$$

where c is an arbitrary constant and $\delta(t)$ is the Dirac delta function. Integrating (E''), we obtain $u(t) = cH(t) + \tilde{c}$, where $\tilde{c}$ is another arbitrary constant. Therefore, the general solution of (E) is $y(t) = cH(t) + \tilde{c} + \ln|t|$. For more information on differential equations on the generalized functions, see [Ko] for examples.

The following example might be a little strange. Let us construct solutions of the differential equation

$$(\tilde{E}) \qquad\qquad \frac{dy}{dt} + 2y(\cos t + \cos(2t)) = 0$$

as a formal Fourier series

$$(S) \qquad\qquad y(t) = \sum_{m=-\infty}^{+\infty} a_m\, e^{imt}.$$

Inserting (S) into ($\tilde{E}$) we obtain

$$(R) \qquad ima_m + a_{m-1} + a_{m+1} + a_{m-2} + a_{m+2} = 0 \qquad (m \in \mathbb{Z}),$$

where $\mathbb{Z}$ is the set of all integers. Equation (R) is a fourth-order difference equation on a_m. Therefore, solution (S) contains four independent arbitrary constants. Of course, if (S) is restricted to be a convergent series, three of these four constants should be eliminated. A similar phenomenon may be observed if solutions of a certain differential equation are expanded in positive and negative powers of independent variables such as $\sum_{m=-\infty}^{+\infty} a_m x^m$. For such an example, see [Dw].

I-2. Existence without the Lipschitz condition

To treat the initial-value problem (P) without the Lipschitz condition, we need some preparation.

Definition I-2-1. *A set $\mathcal{F}$ of functions $\vec{f}(t)$ defined on an interval $\mathcal{I}$ is said to be* equicontinuous *on $\mathcal{I}$ if for every $\epsilon > 0$, there exists a $\delta(\epsilon) > 0$ such that $|\vec{f}(t_1) - \vec{f}(t_2)| \leq \epsilon$ for every $\vec{f} \in \mathcal{F}$ whenever $|t_1 - t_2| \leq \delta(\epsilon)$, $t_1 \in \mathcal{I}$ and $t_2 \in \mathcal{I}$.*

For example, if the set $\mathcal{F}$ consists of all functions $\vec{f}$ such that $\vec{f}$ is continuous on a closed interval $|t - t_0| \leq a$, $\vec{f}'$ is continuous in $|t - t_0| < a$, and $|\vec{f}'(t)| \leq L$ in $|t - t_0| < a$, then $\mathcal{F}$ is equicontinuous on the interval $\mathcal{I} = \{t : |t - t_0| \leq a\}$. In fact, since $\vec{f}(t_1) - \vec{f}(t_2) = \displaystyle\int_{t_2}^{t_1} \vec{f}'(s)ds$ for $t_1 \in \mathcal{I}$ and $t_2 \in \mathcal{I}$, we have $|\vec{f}(t_1) - \vec{f}(t_2)| \leq \epsilon$ whenever $|t_1 - t_2| \leq \delta(\epsilon) = \dfrac{\epsilon}{L}$.

Definition I-2-2. *A set $\mathcal{F}$ of functions is said to be* bounded *on the interval $\mathcal{I}$, if for every $t \in \mathcal{I}$ there exists a non-negative number $M(t)$ such that $|\vec{f}(t)| \leq M(t)$ for every $t \in \mathcal{I}$ and $\vec{f} \in \mathcal{F}$.*

Lemma I-2-3 (C. Arzelà [Ar]-G. Ascoli [As]). *On a bounded interval $\mathcal{I}$, let $\mathcal{F}$ be an infinite, bounded, and equicontinuous set of functions. Then, $\mathcal{F}$ contains an infinite sequence which is uniformly convergent on $\mathcal{I}$.*

Proof.

Let $\mathcal{Q}$ be the set of all rational numbers contained in the interval $\mathcal{I}$. Then,
(1) $\mathcal{Q}$ is dense in $\mathcal{I}$; i.e., for every $\delta > 0$ and every $t \in \mathcal{I}$, there exists a rational number $r(t, \delta) \in \mathcal{Q}$ such that $|t - r(t, \delta)| \leq \delta$,
(2) $\mathcal{Q}$ is an enumerable set; i.e., $\mathcal{Q} = \{r_j : j = 1, 2, 3, \dots\}$.
Note that the choice of the rational number $r(t, \delta)$ is not unique. We prove this lemma in two steps.

Step 1. The set $\mathcal{F}$ contains an infinite sequence $\{\vec{f}_h : h = 1, 2, \dots\}$ such that $\displaystyle\lim_{h \to +\infty} \vec{f}_h(r)$ exists for every rational number $r \in \mathcal{Q}$.

Proof.

Choose from $\mathcal{F}$ a sequence of subsequences $\mathcal{F}_j = \{\vec{f}_{j,\ell} : \ell = 1, 2, \dots\}$, $j = 1, 2, \dots$, such that
(i) $\mathcal{F}_{j+1} \subset \mathcal{F}_j$ for every j,
(ii) $\displaystyle\lim_{\ell \to +\infty} \vec{f}_{j,\ell}(r_j) = c_j$ exists for every j.
To do this, first look at functions in $\mathcal{F}$ at $t = r_1$. Using the boundedness of the set $\{\vec{f}(r_1) : \vec{f} \in \mathcal{F}\}$, we choose an infinite subsequence $\mathcal{F}_1 = \{\vec{f}_{1,\ell} : \ell = 1, 2, \dots\}$ from $\mathcal{F}$ so that $\displaystyle\lim_{\ell \to +\infty} \vec{f}_{1,\ell}(r_1) = c_1$ exists. Next, look at the sequence $\mathcal{F}_1$ at $t = r_2$. Using the boundedness of the set $\{\vec{f}_{1,\ell}(r_2) : \ell = 1, 2, \dots\}$, we choose an infinite subsequence $\mathcal{F}_2 = \{\vec{f}_{2,\ell} : \ell = 1, 2, \dots\}$ from $\mathcal{F}_1$ so that $\displaystyle\lim_{\ell \to +\infty} \vec{f}_{2,\ell}(r_2) = c_2$ exists. Continuing this way, we can choose subsequences $\mathcal{F}_{j+1}$ from $\mathcal{F}_j$ successively.

Set $\vec{f}_h = \vec{f}_{h,h}$ $(h = 1, 2, 3, \dots)$. Then, $\vec{f}_h \in \mathcal{F}_j$ for every $h \geq j$ since $\vec{f}_h = \vec{f}_{h,h} \in \mathcal{F}_h \subset \mathcal{F}_j$ if $h \geq j$. Therefore, $\displaystyle\lim_{h \to +\infty} \vec{f}_h(r_j) = c_j$ exists for every j, i.e., $\displaystyle\lim_{h \to +\infty} \vec{f}_h(r)$ exists for every rational number $r \in \mathcal{Q}$.

Step 2. The sequence $\{\vec{f_h} : h = 1, 2, \ldots\}$ of Step 1 converges uniformly on the interval $\mathcal{I}$.

Proof.

For a given positive number ϵ and a rational number $r \in \mathcal{Q}$, there exist a positive number $\delta(\epsilon)$ and a positive integer $N(r, \epsilon)$ such that

$$\begin{cases} |\vec{f_h}(t) - \vec{f_h}(r)| \leq \epsilon & \text{whenever} \quad |t - r| \leq \delta(\epsilon), \\ |\vec{f_h}(r) - \vec{f_\ell}(r)| \leq \epsilon & \text{whenever} \quad h \geq N(r, \epsilon) \quad \text{and} \quad \ell \geq N(r, \epsilon), \\ |\vec{f_\ell}(r) - \vec{f_\ell}(t)| \leq \epsilon & \text{whenever} \quad |t - r| \leq \delta(\epsilon). \end{cases}$$

Now, observe that

$$|\vec{f_h}(t) - \vec{f_\ell}(t)| \leq |\vec{f_h}(t) - \vec{f_h}(r)| + |\vec{f_h}(r) - \vec{f_\ell}(r)| + |\vec{f_\ell}(r) - \vec{f_\ell}(t)|,$$

where $t \in \mathcal{I}$ and $r \in \mathcal{Q}$. Therefore, by choosing $r = r\left(t, \delta\left(\dfrac{\epsilon}{3}\right)\right)$, we obtain

$$(I.2.1) \qquad\qquad |\vec{f_h}(t) - \vec{f_\ell}(t)| \leq \epsilon$$

whenever $h \geq N\left(r\left(t, \delta\left(\dfrac{\epsilon}{3}bigg)\right), \dfrac{\epsilon}{3}\right)\right.$ and $\ell \geq N\left(r\left(t, \delta\left(\dfrac{\epsilon}{3}\right)\right), \dfrac{\epsilon}{3}\right)$.

Since the interval $\mathcal{I}$ is bounded, we can cover $\mathcal{I}$ by a finite number of open intervals $\mathcal{I}_1, \mathcal{I}_2, \ldots, \mathcal{I}_{p(\epsilon)}$ in such a way that the length of each interval $\mathcal{I}_j$ is not greater than $\delta\left(\dfrac{\epsilon}{3}\right)$, where $p(\epsilon)$ is a positive integer depending on ϵ. For every ℓ, choose a rational number $r_\ell(\epsilon) \in \mathcal{I}_\ell \cap \mathcal{Q}$. For all $t \in \mathcal{I}_\ell$, set $r\left(t, \delta\left(\dfrac{\epsilon}{3}\right)\right) = r_\ell(\epsilon)$ and set

$$N(\epsilon) = \max_{1 \leq \ell \leq p(\epsilon)} N\left(r_\ell(\epsilon), \dfrac{\epsilon}{3}\right).$$

Then, (I.2.1) is satisfied whenever $h \geq N(\epsilon)$ and $\ell \geq N(\epsilon)$. $\square$

To construct a solution of the initial-value problem (P) without the Lipschitz condition, approximate the given differential equation by another equation which satisfies the Lipschitz condition. The unique solution of such an approximate problem is called an *ϵ-approximate solution*. Using the following lemma, an ϵ-approximate solution is found.

Lemma I-2-4. *Suppose that $\vec{f}(t, \vec{y})$ is continuous on a rectangular region*

$$\mathcal{R} = \{(t, \vec{y}) : |t - t_0| \leq a, \ |\vec{y} - \vec{c_0}| \leq b\}.$$

Then, for every positive number ϵ, there exists a function $\vec{F}_\epsilon(t, \vec{y})$ such that
  (i) $\vec{F}_\epsilon$ is continuous for $|t - t_0| \leq a$ and all $\vec{y}$,
  (ii) $\vec{F}_\epsilon$ has continuous partial derivatives of all orders with respect to $y_1, y_2, \ldots, y_n$ for $|t - t_0| \leq a$ and all $\vec{y}$,
  (iii) $|\vec{F}_\epsilon(t, \vec{y})| \leq \max\limits_{(\tau, \vec{\eta}) \in \mathcal{R}} |\vec{f}(\tau, \vec{\eta})|$ for $|t - t_0| \leq a$ and all $\vec{y}$,
  (iv) $|\vec{F}_\epsilon(t, \vec{y}) - \vec{f}(t, \vec{y})| \leq \epsilon$ on $\mathcal{R}$.

Proof.

We prove this lemma in three steps.

Step 1. There exists a function $\vec{F}(t, \vec{y})$ such that
  (1) $\vec{F}$ is continuous for $|t - t_0| \leq a$ and all $\vec{y}$,
  (2) $|\vec{F}(t, \vec{y})| \leq \max\limits_{(\tau, \vec{\eta}) \in \mathcal{R}} |\vec{f}(\tau, \vec{\eta})|$ for $|t - t_0| \leq a$ and all $\vec{y}$,
  (3)

$$\vec{F}(t, \vec{y}) = \begin{cases} \vec{f}(t, \vec{y}) & \text{on} \quad \mathcal{R}, \\ 0 & \text{for} \quad |t - t_0| \leq a \quad \text{and} \quad |\vec{y} - \vec{c}_0| \geq b + 1. \end{cases}$$

The construction of such an $\vec{F}$ is left to the reader as an exercise. Since $\vec{f}$ is uniformly continuous on $\mathcal{R}$, properties (1), (2), and (3) of $\vec{F}$ imply that
  (4) for every positive number ϵ, there exists a positive number $\delta(\epsilon)$ such that

$$|\vec{F}(t, \vec{y}_1) - \vec{F}(t, \vec{y}_2)| \leq \epsilon \qquad \text{whenever} \quad |t - t_0| \leq a, \quad |\vec{y}_1 - \vec{y}_2| \leq \delta(\epsilon).$$

Step 2. For every positive number δ, there exists a real-valued function $\mu(\xi, \delta)$ of a real variable ξ such that
  (a) μ has continuous derivatives of all orders with respect to ξ for $-\infty < \xi < +\infty$,
  (b) $\mu(\xi, \delta) \geq 0$ for $-\infty < \xi < +\infty$,
  (c) $\mu(\xi, \delta) = 0$ for $|\xi| \geq \delta$,
  (d) $\displaystyle\int_{-\infty}^{+\infty} \mu(\xi, \delta) d\xi = 1$.
The construction of such a function μ is left to the reader as an exercise.

  Set $\Phi(\vec{y}, \delta) = \mu(y_1, \delta)\mu(y_2, \delta) \cdots \mu(y_n, \delta)$. Then
  (a') Φ has continuous partial derivatives of all orders with respect to $y_1, y_2, \ldots, y_n$
      for all $\vec{y}$,
  (b') $\Phi(\vec{y}, \delta) \geq 0$ for all $\vec{y}$,
  (c') $\Phi(\vec{y}, \delta) = 0$ for $|\vec{y}| \geq \delta$,
  (d') $\displaystyle\int_{-\infty}^{+\infty} \cdots \int_{-\infty}^{+\infty} \Phi(\vec{y}, \delta) dy_1 dy_2 \cdots dy_n = 1$.

Step 3. Set

$$\vec{F}_\epsilon(t, \vec{y}) = \int_{-\infty}^{+\infty} \cdots \int_{-\infty}^{+\infty} \Phi(\vec{y} - \vec{\eta},\ \delta(\epsilon)) \vec{F}(t, \vec{\eta}) d\eta_1 \cdots d\eta_n,$$

where $\delta(\epsilon)$ is defined in (4) of Step 1. Then, $\vec{F}_\epsilon(t, \vec{y})$ satisfies all of the requirements (i), (ii), (iii), and (iv) of Lemma I-2-4.

Proof.

  Properties (i) and (ii) are evident. Property (iii) follows from the estimate

$$|\vec{F}_\epsilon(t, \vec{y})| \leq \max\left\{ |\vec{F}(t, \vec{\eta})| : \ \text{all } \vec{\eta} \right\} \int_{-\infty}^{+\infty} \cdots \int_{-\infty}^{+\infty} \Phi(\vec{y} - \vec{\eta}, \delta(\epsilon)) d\eta_1 \cdots d\eta_n$$

$$\leq \max\limits_{(\tau, \vec{\eta}) \in \mathcal{R}} |\vec{f}(\tau, \vec{\eta})|.$$

Property (iv) follows from the estimate

$$|\vec{F}_\epsilon(t,\vec{y}) - \vec{F}(t,\vec{y})|$$

$$= \left| \int_{-\infty}^{+\infty} \cdots \int_{-\infty}^{+\infty} \Phi(\vec{y}-\vec{\eta},\delta(\epsilon)) \left\{ \vec{F}(t,\vec{\eta}) - \vec{F}(t,\vec{y}) \right\} d\eta_1 \cdots d\eta_n \right|$$

$$= \left| \int \cdots \int_{|\vec{y}-\vec{\eta}| \leq \delta(\epsilon)} \Phi(\vec{y}-\vec{\eta},\delta(\epsilon)) \left\{ \vec{F}(t,\vec{\eta}) - \vec{F}(t,\vec{y}) \right\} d\eta_1 \cdots d\eta_n \right|$$

$$\leq \epsilon \left| \int \cdots \int_{|\vec{y}-\vec{\eta}| \leq \delta(\epsilon)} \Phi(\vec{y}-\vec{\eta},\delta(\epsilon)) d\eta_1 \cdots d\eta_n \right| = \epsilon.$$

This completes the proof of Lemma I-2-4. $\quad\square$

Now, we prove the fundamental existence theorem without the Lipschitz condition.

Theorem I-2-5. *If $\vec{f}(t,\vec{y})$ is continuous on a region*

$$\mathcal{R} = \{(t,\vec{y}) : |t-t_0| \leq a, \quad |\vec{y}-\vec{c}_0| \leq b\}$$

and $M = \max\limits_{\mathcal{R}} |\vec{f}(t,\vec{y})|$, there exists a function $\vec{\phi}(t)$ such that

(i) $\vec{\phi}(t_0) = \vec{c}_0$,
(ii) $\vec{\phi}'(t) = \vec{f}(t,\vec{\phi}(t))$ on $|t-t_0| < \alpha$,
where

$$\alpha = \begin{cases} a & \text{if } M = 0, \\ \min\left(a, \dfrac{b}{M}\right) & \text{if } M > 0. \end{cases}$$

Proof.

We prove this theorem in four steps.

Step 1. For every positive number ϵ, construct a function $\vec{F}_\epsilon(t,\vec{y})$ which satisfies all of the requirements given in Lemma I-2-4. Using Theorem I-1-4, we construct a function $\vec{\phi}_\epsilon(t)$ such that

(1) $\vec{\phi}_\epsilon(t_0) = \vec{c}_0$,
(2) $\vec{\phi}_\epsilon'(t) = \vec{F}_\epsilon(t,\vec{\phi}_\epsilon(t))$ on $|t-t_0| < \alpha$.
Furthermore,
(3) $(t,\vec{\phi}_\epsilon(t)) \in \mathcal{R}$ on $|t-t_0| \leq \alpha$ (cf. Lemma I-1-1 and Remark I-1-3).

Step 2. The set $\mathcal{F} = \{\vec{\phi}_\epsilon : \epsilon > 0\}$ is bounded and equicontinuous on $|t-t_0| \leq \alpha$.

Proof.

Property (3) of functions $\vec{\phi}_\epsilon$ implies that $\mathcal{F}$ is bounded on $|t-t_0| \leq \alpha$ and that $|\vec{F}_\epsilon(t,\vec{\phi}_\epsilon(t))| \leq M$ on $|t-t_0| \leq \alpha$. Hence, property (2) of $\vec{\phi}$ implies that

$$|\vec{\phi}_\epsilon(t_1) - \vec{\phi}_\epsilon(t_2)| \leq \left| \int_{t_2}^{t_1} \vec{\phi}_\epsilon'(s)ds \right| \leq M|t_1 - t_2|$$

if $|t_1 - t_0| \leq \alpha$ and $|t_2 - t_0| \leq \alpha$ (cf. Lemma I-1-2). Therefore, for a given positive number μ, we have $|\vec{\phi}_\epsilon(t_1) - \vec{\phi}_\epsilon(t_2)| \leq \mu$ whenever $|t_1 - t_0| \leq \alpha$, $|t_2 - t_0| \leq \alpha$, and $|t_1 - t_2| \leq \dfrac{\mu}{M}$. This shows that $\mathcal{F}$ is equicontinuous on $|t-t_0| \leq \alpha$.

Step 3. Using Lemma I-2-3 (Arzelà-Ascoli) choose a sequence $\{\epsilon_\ell : \ell = 1, 2, \ldots\}$ of positive numbers such that $\lim\limits_{\ell \to +\infty} \epsilon_\ell = 0$ and that the sequence $\{\vec{\phi}_{\epsilon_\ell} : \ell = 1, 2, \ldots\}$ converges uniformly on $|t - t_0| \leq \alpha$ as $\ell \to +\infty$. Then, set

$$\vec{\phi}(t) \;=\; \lim_{\ell \to +\infty} \vec{\phi}_{\epsilon_\ell}(t) \qquad \text{on} \quad |t - t_0| \leq \alpha.$$

Step 4. Observe that

$$\vec{\phi}_\epsilon(t) = \vec{c}_0 + \int_{t_0}^t \vec{F}_\epsilon(s, \vec{\phi}_\epsilon(s))ds$$

$$= \vec{c}_0 + \int_{t_0}^t \vec{f}(s, \vec{\phi}_\epsilon(s))ds + \int_{t_0}^t \left\{ \vec{F}_\epsilon(s, \vec{\phi}_\epsilon(s)) - \vec{f}(s, \vec{\phi}_\epsilon(s)) \right\} ds$$

and that

$$\left| \int_{t_0}^t \left| \vec{F}_\epsilon(s, \vec{\phi}_\epsilon(s)) - \vec{f}(s, \vec{\phi}_\epsilon(s)) \right| ds \right| \leq \epsilon |t - t_0| \qquad \text{on} \quad |t - t_0| \leq \alpha.$$

This is true for all $\epsilon \geq 0$. Letting $\epsilon \to 0$, we obtain

$$\vec{\phi}(t) \;=\; \vec{c}_0 + \int_{t_0}^t \vec{f}(s, \vec{\phi}(s))ds.$$

This completes the proof of Theorem I-2-5. $\square$

Example I-2-6. Theorem I-2-5 applies to the initial-value problem

$$(\tilde{\mathrm{P}}) \qquad\qquad \frac{dy}{dx} \;=\; xy^{1/5}, \qquad y(3) \;=\; 0.$$

However, Theorem I-1-4 does not apply to $(\tilde{\mathrm{P}})$. In fact $y(x) = 0$ is a solution of Problem $(\tilde{\mathrm{P}})$ and also

$$y(x) \;=\; \begin{cases} 0 & \text{for} \quad x < 3\,, \\[2mm] \left(\dfrac{2(x^2 - 9)}{5} \right)^{5/4} & \text{for} \quad x \geq 3 \end{cases}$$

is a solution of Problem $(\tilde{\mathrm{P}})$. Note that the right-hand side of the differential equation $(\tilde{\mathrm{P}})$ does not satisfy the Lipschitz condition on any y-interval containing $y = 0$.

Two other methods of proving Theorem I-2-5 are summarized in the following two remarks.

Remark I-2-7. Let every entry of an n-dimensional vector $\vec{f}(t, \vec{y})$ be a real-valued continuous function of $n + 1$ independent variables t and $\vec{y} = (y_1, \ldots, y_n)$ on a rectangular region $\mathcal{R} = \{(t, \vec{y}) : |t - t_0| \leq a, |\vec{y} - \vec{c}_0| \leq b\}$. Assume that $|\vec{f}(t, \vec{y})| \leq M$

on $\mathcal{R}$, where M is a positive number. Set $\alpha = \min\left(a, \dfrac{b}{M}\right)$. Since $\vec{f}(t, \vec{y})$ is uniformly continuous on $\mathcal{R}$, there exists a positive number $\rho(\epsilon)$ for every given positive number ϵ such that $|\vec{f}(t_1, \vec{y}_1) - \vec{f}(t_2, \vec{y}_2)| \leq \epsilon$ whenever $|t_1 - t_2| \leq \rho(\epsilon)$ and $|\vec{y}_1 - \vec{y}_2| \leq \rho(\epsilon)$ for $(x_1, \vec{y}_1) \in \mathcal{R}$ and $(x_2, \vec{y}_2) \in \mathcal{R}$. Let $\Delta(\epsilon) : t_0 = \tau_0 < \tau_1 < \cdots < \tau_{n(\epsilon)} = t_0 + \alpha$ be a subdivision of the interval $t_0 \leq t \leq t_0 + \alpha$ such that $\displaystyle\max_{j=0}^{n(\epsilon)-1}(|\tau_j - \tau_{j+1}|) \leq \min\left(\rho(\epsilon), \dfrac{\rho(\epsilon)}{M}\right)$. Set

$$
\vec{y}_\epsilon(t) = \begin{cases}
\vec{c}_0 & \text{for} \quad t = t_0, \\
\vec{y}_\epsilon(\tau_{j-1}) + \vec{f}(\tau_{j-1}, \vec{y}_\epsilon(\tau_{j-1}))(t - \tau_{j-1}) & \text{for} \quad \tau_{j-1} < t \leq \tau_j, \\
& \qquad (j = 1, 2, \ldots, n(\epsilon)).
\end{cases}
$$

Then, we can show that

(i) every entry of $\vec{y}_\epsilon(t)$ is piecewise linear and continuous in t and $(t, \vec{y}_\epsilon(t)) \in \mathcal{R}$ on the interval $t_0 \leq t \leq t_0 + \alpha$,

(ii) the set $\{\vec{y}_\epsilon : \epsilon > 0\}$ is bounded and equicontinuous on the interval $t_0 \leq t \leq t_0 + \alpha$,

(iii) if we choose a decreasing sequence $\{\epsilon_j : j = 1, 2, \ldots\}$ such that $\displaystyle\lim_{j \to +\infty} \epsilon_j = 0$ in a suitable way, the sequence $\{\vec{y}_{\epsilon_j}(x) : j = 1, 2, \ldots\}$ converges to a solution of the initial-value problem

$$
(\mathrm{I.2.2}) \qquad\qquad \frac{d\vec{y}}{dt} = \vec{f}(t, \vec{y}), \qquad \vec{y}(t_0) = \vec{c}_0
$$

uniformly on $t_0 \leq t \leq t_0 + \alpha$ as $j \to +\infty$.

For more details, see [CL, pp. 3-5].

Remark I-2-8. We use the same assumption, M, and α as in Remark I-2-7. Also, let $\vec{\eta}(t)$ be a continuously differentiable function of t such that $\vec{\eta}(t_0) = \vec{c}_0$ and $(t, \vec{\eta}(t)) \in \mathcal{R}$ on the interval $t_0 - \tau \leq t \leq t_0$, where τ is a positive number. For every ϵ such that $0 < \epsilon \leq \tau$, define

$$
\vec{y}_\epsilon(t) = \begin{cases}
\vec{\eta}(t) & \text{for} \quad t_0 - \tau \leq t \leq t_0, \\
\vec{c}_0 + \displaystyle\int_{t_0}^{t} \vec{f}(s, \vec{y}_\epsilon(s - \epsilon))\, ds & \text{for} \quad t_0 \leq t \leq t_0 + \alpha.
\end{cases}
$$

Then, we can show that

(i) every entry of $\vec{y}_\epsilon(t)$ is continuous in t and $(t, \vec{y}_\epsilon(t)) \in \mathcal{R}$ on the interval $t_0 - \tau \leq t \leq t_0 + \alpha$,

(ii) every entry of $\vec{y}_\epsilon'(t)$ is continuous in t on $t_0 - \tau \leq t \leq t_0 + \alpha$ except a jump at $x = t_0$,

(iii) the set $\{\vec{y}_\epsilon : 0 < \epsilon \leq \tau\}$ is bounded and equicontinuous on the interval $t_0 - \tau \leq t \leq t_0 + \alpha$,

(iv) if we choose a decreasing sequence $\{\epsilon_j : j = 1, 2, \ldots\}$ such that $\displaystyle\lim_{j \to +\infty} \epsilon_j = 0$ in a suitable way, the sequence $\{\vec{y}_{\epsilon_j}(x) : j = 1, 2, \ldots\}$ converges to a solution of the initial-value problem (I.2.2) uniformly on $x_0 \leq x \leq x_0 + \alpha$ as $j \to +\infty$.

For more details, see [CL, pp. 43-44] and [Har2, pp. 10-11].

Remark I-2-9. If $\vec{f}(x, \vec{y})$ is assumed to be measurable for each fixed $\vec{y}$, continuous in $\vec{y}$ for each fixed t, and $|\vec{f}(t, \vec{y})|$ is bounded by a Lebesgue-integrable function when $(t, \vec{y}) \in \mathcal{R}$, a result similar to Theorem I-2-5 (Carathéodory's existence theorem) [Ca, pp. 665-688] is obtained. Similarly, if, in addition, it satisfies an inequality similar to Lipschitz condition with the Lipschitz constant replaced by a Lebesgue-integrable function $L(t)$ when $(t, \vec{y}) \in \mathcal{R}$, we obtain also a result similar to Theorem I-1-4. (See [CL, p. 43], [Har2, p. 10], and [SC, p. 15].)

I-3. Some global properties of solutions

We can construct a solution to an initial-value problem (P) in a neighborhood of the initial point by using Theorem I-1-4 or I-2-5. To extend such a local solution to a larger interval of the independent variable t, the following lemma is useful.

Lemma I-3-1. *Assume that*
(i) a function $\vec{f}(t, \vec{y})$ is continuous in an open set $\mathcal{D}$ in the $(t, \vec{y})$-space,
(ii) a function $\vec{\phi}(t)$ satisfies the condition $\vec{\phi}'(t) = \vec{f}(t, \vec{\phi}(t))$, and $(t, \vec{\phi}(t)) \in \mathcal{D}$, in an open interval $\mathcal{I} = \{t : \tau_1 < t < \tau_2\}$.
Under this assumption, if $\lim\limits_{j \to +\infty} (t_j, \vec{\phi}(t_j)) = (\tau_1, \vec{\eta}) \in \mathcal{D}$ for some sequence $\{t_j : j = 1, 2, \ldots\}$ of points in the interval $\mathcal{I}$, then $\lim\limits_{t \to \tau_1} (t, \vec{\phi}(t)) = (\tau_1, \vec{\eta})$. Similarly, if $\lim\limits_{j \to +\infty} (t_j, \vec{\phi}(t_j)) = (\tau_2, \vec{\eta}) \in \mathcal{D}$ for some sequence $\{t_j : j = 1, 2, \ldots\}$ of points in the interval $\mathcal{I}$, then $\lim\limits_{t \to \tau_2} (t, \vec{\phi}(t)) = (\tau_2, \vec{\eta})$.

Proof.

We shall prove the part of the lemma which concerns the left endpoint τ_1 of the interval $\mathcal{I}$. The other part can be proven similarly.

Let $\mathcal{U}$ be an open neighborhood of $(\tau_1, \vec{\eta})$. We show that $(t, \vec{\phi}(t)) \in \mathcal{U}$ in an interval $\tau_1 < t < \tau(\mathcal{U})$ for some $\tau(\mathcal{U})$ which is determined by $\mathcal{U}$. Assume that the closure of $\mathcal{U}$ is contained in $\mathcal{D}$ and that $|\vec{f}(t, \vec{y})| \leq M$ in $\mathcal{U}$ for some positive number M. For every positive integer j and every positive number ϵ, consider a rectangular region $\mathcal{R}_j(\epsilon) = \{(t, \vec{y}) : |t - t_j| \leq \epsilon, \ |\vec{y} - \vec{\phi}(t_j)| \leq M\epsilon\}$. Then, there exist an $\epsilon > 0$ and a j such that $(\tau_1, \vec{\eta}) \in \mathcal{R}_j(\epsilon) \subset \mathcal{U}$. Note that $\epsilon = \min\left(\epsilon, \dfrac{M\epsilon}{M}\right)$ and $t_j - \epsilon \leq \tau_1$.

Applying Lemma I-1-1 to the solution $\vec{y} = \vec{\phi}(t)$ of the initial-value problem $\dfrac{d\vec{y}}{dt} = \vec{f}(t, \vec{y})$, $\vec{y}(t_j) = \vec{\phi}(t_j)$, we obtain $(t, \vec{\phi}(t)) \in \mathcal{R}_j(\epsilon) \subset \mathcal{U}$ on the interval $\tau_1 < t \leq t_j$. Since $\mathcal{U}$ is an arbitrary open neighborhood of $(\tau_1, \vec{\eta})$, we conclude that $\lim\limits_{j \to +\infty} (t_j, \vec{\phi}(t_j)) = (\tau_1, \vec{\eta}) \in \mathcal{D}$. $\square$

From Lemma I-3-1, we derive the following result concerning the maximal interval on which a solution can be extended.

Theorem I-3-2. *If*
(i) a function $\vec{f}(t, \vec{y})$ is continuous in an open set $\mathcal{D}$ in the $(t, \vec{y})$-space,

(ii) a function $\vec{\phi}(t)$ satisfies the condition $\vec{\phi}'(t) = \vec{f}(t, \vec{\phi}(t))$ and $(t, \vec{\phi}(t)) \in \mathcal{D}$, in an open interval $\mathcal{I} = \{t : \tau_1 < t < \tau_2\}$,

(iii) $\vec{\phi}(t)$ cannot be extended to the left of τ_1(or, respectively, to the right of τ_2) with property (ii),

(iv) $\lim\limits_{j \to +\infty}(t_j, \vec{\phi}(t_j)) = (\tau_1, \vec{\eta})$ (or, respectively, $(\tau_2, \vec{\eta})$) exists for some sequence $\{t_j : j = 1, 2, \dots\}$ of points in the interval $\mathcal{I}$,

then the limit point $(\tau_1, \vec{\eta})$ (or, respectively, $(\tau_2, \vec{\eta})$) must be on the boundary of $\mathcal{D}$.

Proof.

We shall prove this result by deriving a contradiction from the assumption that $(\tau_1, \vec{\eta}) \in \mathcal{D}$ (or, respectively, $(\tau_2, \vec{\eta}) \in \mathcal{D}$). Applying Lemma I-3-1 to this situation, we obtain $\lim\limits_{t \to \tau_1}(t, \vec{\phi}(t)) = (\tau_1, \vec{\eta})$ (or, respectively, $(\tau_2, \vec{\eta})$). Hence, by applying Theorem I-2-5 to the initial-value problem

$$\frac{d\vec{y}}{dt} = \vec{f}(t, \vec{y}), \qquad \vec{y}(\tau_1) = \vec{\eta} \quad (\text{or, respectively, } \vec{y}(\tau_2) = \vec{\eta}),$$

the solution $\vec{\phi}(t)$ can be extended to the left of τ_1 (or, respectively, to the right of τ_2). This is a contradiction. $\square$

The following example illustrates how to use Lemma I-3-1.

Problem I-3-3. Show that the solution of the initial-value problem

$$\frac{d^2y}{dx^2} - 2xy\frac{dy}{dx} = \frac{1}{y^3} - y^2, \qquad y(x_0) = \eta, \quad y'(x_0) = \zeta,$$

exists at least on the interval $0 \leq x \leq x_0$ if $x_0 > 0$, $\eta > 0$, and ζ is any real number.

Answer.

Assume that the solution $y = \phi(x)$ of the given initial-value problem exists on an interval $\mathcal{I} = \{x : \xi < x \leq x_0\}$ for some positive number ξ. Observe that

$$\left[\frac{1}{2}(\phi')^2 + \frac{1}{3}\phi^3 + \frac{1}{2}\phi^{-2}\right]' = 2x\phi(\phi')^2 \geq 0$$

on $\mathcal{I}$. Hence,

$$\frac{1}{2}(\phi'(x))^2 + \frac{1}{3}\phi(x)^3 + \frac{1}{2}\phi(x)^{-2} \leq \frac{1}{2}(\phi'(x_0))^2 + \frac{1}{3}\phi(x_0)^3 + \frac{1}{2}\phi(x_0)^{-2} \; (= M > 0)$$

on $\mathcal{I}$. Therefore, we have

$$(\phi'(x))^2 \leq 2M, \qquad \phi(x)^3 \leq 3M, \qquad \phi(x)^2 \geq \frac{1}{2M}.$$

Now, apply Lemma I-3-1.

The proof of the following result is left to the reader as an exercise.

Corollary I-3-4. *Assume that $\vec{f}(t,\vec{y})$ is continuous for $t_1 < t < t_2$ and all $\vec{y} \in \mathbb{R}^n$. Assume also that a function $\vec{\phi}(t)$ satisfies the following conditions:*

(a) $\vec{\phi}$ and $\vec{\phi}'$ are continuous in a subinterval $\mathcal{I}$ of the interval $t_1 < t < t_2$,

(b) $\vec{\phi}'(t) = \vec{f}(t,\vec{\phi}(t))$ in $\mathcal{I}$.

Then, either

(i) $\vec{\phi}(t)$ can be extended to the entire interval $t_1 < t < t_2$ as a solution of the differential equation $\dfrac{d\vec{y}}{dt} = \vec{f}(t,\vec{y})$, or

(ii) $\lim\limits_{t \to \tau} \vec{\phi}(t) = \infty$ for some τ in the interval $t_1 < \tau < t_2$.

Using Corollary I-3-4, we obtain the following important result concerning a linear nonhomogeneous differential equation

$$(\text{I.3.1}) \qquad\qquad \frac{d\vec{y}}{dt} = A(t)\vec{y} + \vec{b}(t),$$

where the entries of the $n \times n$ matrix $A(t)$ and the entries of the $\mathbb{R}^n$-valued function $\vec{b}(t)$ are continuous in an open interval $\mathcal{I} = \{t : t_1 < t < t_2\}$.

Theorem I-3-5. *Every solution of differential equation (I.3.1) which is defined in a subinterval of the interval $\mathcal{I}$ can be extended uniquely to the entire interval $\mathcal{I}$ as a solution of (I.3.1).*

Proof.

Suppose that a solution $\vec{y} = \vec{\phi}(t)$ of (I.3.1) exists in a subinterval $\mathcal{I}' = \{t : \tau_1 < t < \tau_2\}$ of the interval $\mathcal{I}$ such that $t_1 < \tau_1 < \tau_2 < t_2$. Then,

$$|\vec{\phi}(t)| \leq |\vec{\phi}(t_0)| + \left| \int_{t_0}^{t} \left\{ A(s)\vec{\phi}(s) + \vec{b}(s) \right\} ds \right|$$

in $\mathcal{I}'$, where t_0 is any point in the interval $\mathcal{I}'$. Setting

$$\begin{cases} K = |\vec{\phi}(t_0)| + (\tau_2 - \tau_1) \max\limits_{\tau_1 \leq t \leq \tau_2} |\vec{b}(t)|, \\ L = \max\limits_{\tau_1 \leq t \leq \tau_2} |A(t)|, \end{cases}$$

we obtain

$$|\vec{\phi}(t)| \leq K + L \left| \int_{t_0}^{t} |\vec{\phi}(s)| ds \right| \qquad \text{in} \quad \mathcal{I}'.$$

Thus, the estimate

$$|\vec{\phi}(t)| \leq K \exp[L|t - t_0|] \leq K \exp[L(\tau_2 - \tau_1)] < \infty \qquad \text{in} \quad \mathcal{I}'$$

is derived by using Lemma I-1-5. Hence, case (ii) of Corollary I-3-4 is eliminated. Since the right-hand side of (I.3.1) satisfies the Lipschitz condition, the extension of the solution $\vec{\phi}(t)$ is unique. $\square$

For nonlinear ordinary differential equations, we cannot expect to obtain the same result as in Theorem I-3-5. The following example shows the local blowup of solutions of the differential equation of Problem I-3-3.

Problem I-3-6. Show that for any fixed $\xi > 0$, there exists a real-valued function $\phi(x)$ such that

(1) $\phi(x)$ is continuous on an interval $0 \leq \xi - x \leq \delta$ for some small positive number δ which depends on ξ,

(2) $\dfrac{d\phi(x)}{dx}$ is continuous on an interval $0 < \xi - x < \delta$,

(3) the function

$$\text{(Sol)} \qquad y(x) \; = \; \frac{1 + (\xi - x)\phi(x)}{\xi(\xi - x)}$$

satisfies the differential equation

$$\text{(Eq)} \qquad \frac{d^2 y}{dx^2} - 2xy\frac{dy}{dx} \; = \; \frac{1}{y^3} - y^2$$

on $0 < \xi - x < \delta$.

Answer.

Step 1. If we set $t = \xi - x$ and $y \; = \; \dfrac{u}{t}$, the given differential equation is changed to

$$(1) \qquad t^2 u'' - 2tu' + 2u + 2\xi(tu' - u)u \; = \; 2t(tu' - u)u + t^6 u^{-3} - tu^2,$$

where $f' = \dfrac{df}{dt}$.

Step 2. If we set $u = \dfrac{1 + w_1}{\xi}$ and $tu' = \dfrac{w_2}{\xi}$, differential equation (1) becomes

$$(2) \qquad \begin{cases} t\dfrac{dw_1}{dt} \; = \; w_2, \\[2mm] t\dfrac{dw_2}{dt} \; = \; -2(1 + w_1) + 3w_2 - 2(w_2 - (1 + w_1))(1 + w_1) \\[2mm] \qquad\qquad + 2\xi^{-1}t(w_2 - (1 + w_1))(1 + w_1) \\[2mm] \qquad\qquad + t^6 \xi^4 (1 + w_1)^{-3} - t\xi^{-1}(1 + w_1)^2. \end{cases}$$

If we set $t = 0$, $w_1 = 0$, and $w_2 = 0$, the right-hand side of (2) vanishes. Now, (2) can be written in the form

$$(3) \qquad \begin{cases} t\dfrac{dw_1}{dt} \; = \; w_2, \\[2mm] t\dfrac{dw_2}{dt} \; = \; 2w_1 + w_2 - 2w_1(w_2 - w_1) \\[2mm] \qquad\qquad + 2\xi^{-1}t(w_2 - (1 + w_1))(1 + w_1) \\[2mm] \qquad\qquad + t^6 \xi^4 (1 + w_1)^{-3} - t\xi^{-1}(1 + w_1)^2. \end{cases}$$

Step 3. Write (3) in the following form:

$$(4) \qquad t\vec{w}' \; = \; \Omega\vec{w} + \vec{F}(\vec{w}) + t\vec{G}(t,\vec{w}),$$

where

$$\vec{w} = \begin{bmatrix} w_1 \\ w_2 \end{bmatrix}, \quad \Omega = \begin{bmatrix} 0 & 1 \\ 2 & 1 \end{bmatrix}, \quad \vec{F}(\vec{w}) = -\begin{bmatrix} 0 \\ 2w_1(w_2 - w_1) \end{bmatrix}, \quad \vec{G} = \begin{bmatrix} 0 \\ G(t,\vec{w}) \end{bmatrix}$$

with

$$(5) \qquad \begin{aligned} G(t,\vec{w}) = {} & 2\xi^{-1}(w_2 - (1 + w_1))(1 + w_1) \\ & + t^5\xi^4(1 + w_1)^{-3} - \xi^{-1}(1 + w_1)^2. \end{aligned}$$

Two eigenvalues of Ω are -1 and 2, which are distinct. Let us diagonalize Ω by the transformation

$$\vec{w} \; = \; P\vec{v}, \qquad \text{where} \quad P = \begin{bmatrix} 1 & 1 \\ -1 & 2 \end{bmatrix}.$$

Then, (4) becomes

$$(6) \qquad t\vec{v}' \; = \; \Lambda\vec{v} + P^{-1}\vec{F}(P\vec{v}) + tP^{-1}\vec{G}(t, P\vec{v}), \qquad \Lambda = \begin{bmatrix} -1 & 0 \\ 0 & 2 \end{bmatrix}.$$

Set $\vec{v} = t\vec{z}$. Then, $t\vec{v}' = t\vec{z} + t^2\vec{z}'$ and $\vec{F}(P\vec{v}) = t^2\vec{F}(P\vec{z})$. Therefore, (6) becomes

$$(7) \qquad t\vec{z}' \; = \; \Lambda_0\vec{z} + \vec{f}(t, \vec{z}),$$

where $\Lambda_0 = \begin{bmatrix} -2 & 0 \\ 0 & 1 \end{bmatrix}$ and $\vec{f}(t, \vec{z}) = tP^{-1}\vec{F}(P\vec{z}) + P^{-1}\vec{G}(t, tP\vec{z})$. If $t \geq 0$ and if $t|\vec{z}|$ and $t|\vec{\zeta}|$ are small, there exist two positive constants K_0 and L such that $|\vec{f}(t, \vec{z}) - \vec{f}(t, \vec{\zeta})| \leq tL|\vec{z} - \vec{\zeta}|$ and $|\vec{f}(t, \vec{z})| \leq K_0 + tL|\vec{z}|$.

Step 4. We change (7) to the following system of integral equations:

$$(8) \qquad \begin{cases} z_1(t) = t^{-2} \displaystyle\int_0^t \tau f_1(\tau, \vec{z}(\tau))d\tau \;, \\[2ex] z_2(t) = ct + t \displaystyle\int_{t_0}^t \tau^{-2} f_2(\tau, \vec{z}(\tau))d\tau, \end{cases}$$

where t_0 is a sufficiently small positive number and c is an arbitrary constant, and $\vec{z} = \begin{bmatrix} z_1 \\ z_2 \end{bmatrix}$ and $\vec{f} = \begin{bmatrix} f_1 \\ f_2 \end{bmatrix}$. We want to construct a bounded solution of (8) on

$0 \leq t \leq t_0$. To do this, first note that, if $|\vec{z}(t)| \leq K$ on $0 \leq t \leq t_0$, we obtain

$$(9) \quad \begin{cases} |z_1(t)| \leq t^{-2} \displaystyle\int_0^t \tau |f_1(\tau, \vec{z}(\tau))| d\tau \leq t^{-2} \displaystyle\int_0^t \tau (K_0 + \tau LK) d\tau \\[2mm] \qquad = \dfrac{K_0}{2} + \dfrac{t}{3} LK \;\leq\; \dfrac{K_0}{2} + \dfrac{t_0}{3} LK, \\[4mm] |z_2(t)| \leq |c|t + t \displaystyle\int_t^{t_0} \tau^{-2} |f_2(\tau, \vec{z}(\tau))| d\tau \\[2mm] \qquad \leq |c|t + t \displaystyle\int_t^{t_0} \tau^{-2} (K_0 + \tau LK) d\tau \\[2mm] \qquad = |c|t + t \left(\dfrac{1}{t} - \dfrac{1}{t_0} \right) K_0 + t \left(\ln \left(\dfrac{t_0}{t} \right) \right) LK \\[2mm] \qquad \leq |c|t_0 + K_0 + e^{-1} t_0 LK. \end{cases}$$

For a positive number c_0, fix another positive number K so that $c_0 + K_0 < \dfrac{K}{2}$ and then fix $t_0 > 0$ so that $t_0 < 1$, $t_0 L < 1$, and $t_0 K |P| < 1$. Here, $|P| = \max\{|P_{jk}| : 1 \leq j, k \leq 2\}$, where $P = [P_{jk}]_{j,k=1}^2$. Define successive approximations by

$$\begin{cases} \vec{\Phi}_0(t, c) = \begin{bmatrix} 0 \\ ct \end{bmatrix}, \\[4mm] \vec{\Phi}_{m+1}(t, c) = \begin{bmatrix} t^{-2} \displaystyle\int_0^t \tau f_1(\tau, \vec{\Phi}_m(\tau, c)) d\tau \\[2mm] ct + t \displaystyle\int_{t_0}^t \tau^{-2} f_2(\tau, \vec{\Phi}_m(\tau, c)) d\tau \end{bmatrix}, \quad m \geq 0. \end{cases}$$

It can be shown that the successive approximations converges uniformly for $|c| \leq c_0$ and $0 \leq t \leq t_0$. This establishes the existence of a bounded solution of the integral equation (8). Thus, we can complete the construction of solution (Sol) of equation (Eq).

I-4. Analytic differential equations

So far, it has been assumed that the independent variable t and the unknown quantity $\vec{y}$ are real. Since a complex number is a two-dimensional real vector, every result in the previous sections can be extended to the case when the entries of $\vec{y}$ and of the function $\vec{f}(t, \vec{y})$ are complex. Furthermore, if the function $\vec{f}$ is analytic with respect to $(t, \vec{y})$, we can construct analytic solutions by means of successive approximations. Denote by $\mathbb{C}$ the set of all complex numbers. The set of all n-column vectors with complex entries is denoted by $\mathbb{C}^n$.

Theorem I-4-1. *Assume that each entry of a $\mathbb{C}^n$-valued function $\vec{f}(z, \vec{y})$ is given by a power series in $(z, y_1, y_2, \ldots, y_n)$ which converges in a polydisk:*

$$\mathcal{D} = \{(z, \vec{y}) : |z - z_0| < a, \; |\vec{y} - \vec{c}_0| < b\},$$

where $z, y_1, y_2, \ldots, y_n$ are complex quantities. Assume also that there exist positive numbers M and L such that

$$\begin{cases} |\vec{f}(z, \vec{y})| \leq M & \text{in } \; \mathcal{D}, \\ |\vec{f}(z, \vec{y}_1) - \vec{f}(z, \vec{y}_2)| \leq L|\vec{y}_1 - \vec{y}_2| & \text{whenever } \; (z, \vec{y}_j) \in \mathcal{D} \; (j = 1, 2). \end{cases}$$

Define successive approximations by

$$\begin{cases} \vec{\phi}_0(z) = \vec{c}_0, \\ \vec{\phi}_m(z) = \vec{c}_0 + \displaystyle\int_{z_0}^{z} \vec{f}(\zeta, \vec{\phi}_{m-1}(\zeta)) d\zeta, & m \geq 1, \end{cases}$$

where the path of integration may be taken to be the line segment $\overline{z_0 z}$ joining z_0 to z in the complex z-plane. Set $\alpha = \min\left\{a, \dfrac{b}{M}\right\}$. Then,

(i) For every m, the entries of the $\mathbb{C}^n$-valued function $\vec{\phi}_m(z)$ are power series in $z - z_0$ which are convergent in the open disk $\Delta = \{z : |z - z_0| < \alpha\}$,

(ii) the sequence $\{\vec{\phi}_m(z) : m = 0, 1, \ldots\}$ satisfies the estimates

$$|\vec{\phi}_{m+1}(z) - \vec{\phi}_m(z)| \leq \frac{ML^m}{(m+1)!}|z - z_0|^{m+1} \quad \text{on } \Delta, \quad m = 0, 1, \ldots,$$

(iii) the sequence $\{\vec{\phi}_m(z) : m = 0, 1, \ldots\}$ converges to

$$\vec{\phi}(z) = \vec{\phi}_0(z) + \sum_{k=1}^{\infty} \left(\vec{\phi}_k(z) - \vec{\phi}_{k-1}(z)\right)$$

uniformly in Δ as $m \to +\infty$,

(iv) the entries of the $\mathbb{C}^n$-valued function $\vec{\phi}(z)$ are power series in $z - z_0$ which are convergent in Δ,

(v) the function $\vec{\phi}(z)$ is the unique solution to the initial-value problem

$$\frac{d\vec{y}}{dz} = \vec{f}(z, \vec{y}), \qquad \vec{y}(z_0) = \vec{c}_0.$$

The proof of this theorem is left to the reader as an exercise. The reader must notice that even if a sequence of analytic functions $f_m(t)$ of a real variable t converges to a function $f(t)$ uniformly on an interval $\mathcal{I}$, the limit function $f(t)$ may not be analytic at all. For example, any continuous function can be uniformly approximated by polynomials on a bounded closed interval. In order to derive analyticity of $f(t)$, uniform convergence must be proved on a domain in the complex t-plane.

Observation I-4-2. The estimates

$$|\vec{\phi}_{m+1}(z) - \vec{\phi}_m(z)| \leq \frac{ML^m}{(m+1)!}|z - z_0|^{m+1} \quad \text{in } \Delta, \quad m = 0, 1, \ldots,$$

imply that there exists a power series $\displaystyle\sum_{k=0}^{+\infty}\vec{c}_k(z-z_0)^k$ such that

$$\vec{\phi}_m(z) \ = \ \sum_{k=0}^{m}\vec{c}_k(z-z_0)^k + O(|z-z_0|^{m+1}), \qquad m = 0,1,2,\ldots\ .$$

Observation I-4-3. Since

$$\vec{\phi}_m(z) = \vec{\phi}_0(z) + \sum_{k=1}^{m}\left(\vec{\phi}_k(z) - \vec{\phi}_{k-1}(z)\right),$$

we obtain $\vec{\phi}(z) - \vec{\phi}_m(z) = O(|z-z_0|^{m+1})$ and, hence, $\vec{\phi}(z) = \displaystyle\sum_{k=0}^{\infty}\vec{c}_k(z-z_0)^k$.

Observation I-4-4. Set

$$\vec{\phi}_m(z) = \sum_{k=0}^{m}\vec{c}_k(z-z_0)^k + \sum_{k=m+1}^{\infty}\vec{c}_{m,k}(z-z_0)^k.$$

Since $|\vec{\phi}_m(z) - \vec{c}_0| \leq b$ in the disk Δ, it follows that

$$|\vec{c}_{m,k}| \leq \frac{b}{\alpha^k} \qquad (k = m+1,\, m+2,\ldots).$$

Note that

$$\vec{c}_{m,k} \ = \ \frac{1}{2\pi i}\int_{|z-z_0|=\rho}\frac{\vec{\phi}(z)-\vec{c}_0}{(z-z_0)^{k+1}}dz$$

for any positive number $\rho < \alpha$. This, in turn, implies that

$$\left|\vec{\phi}_m(z) - \sum_{k=0}^{m}\vec{c}_k(z-z_0)^k\right| \leq b\sum_{k=m+1}^{\infty}\left|\frac{z-z_0}{\alpha}\right|^k$$

$$= b\left|\frac{z-z_0}{\alpha}\right|^{m+1}\frac{1}{1-\left|\dfrac{z-z_0}{\alpha}\right|}$$

in the disk Δ.

Example I-4-5. In the case of the initial-value problem $\dfrac{dy}{dz} = (1-z)y^2$, $y(0) = 2$, we obtain

$$\phi(z) \ = \ 2\sum_{k=0}^{\infty}(k+1)z^k = \frac{2}{(1-z)^2}$$

by using the successive approximations

$$
\begin{cases}
\phi_0(z) = 2 , \\
\phi_1(z) = 2 + 4z - 2z^2, \\
\phi_2(z) = 2 + 4z + 6z^2 - \dfrac{8z^3}{3} - 6z^4 + 4z^5 - \dfrac{2z^6}{3}, \\
\phi_3(z) = 2 + 4z + 6z^2 + 8z^3 - \dfrac{2z^4}{3} - \dfrac{28z^5}{3} - \dfrac{82z^6}{9} \\
\qquad\qquad + \dfrac{256z^7}{63} + \dfrac{124z^8}{9} - \dfrac{68z^9}{9} - \dfrac{46z^{10}}{9} + \dfrac{56z^{11}}{9} \\
\qquad\qquad - \dfrac{22z^{12}}{9} + \dfrac{4z^{13}}{9} - \dfrac{2z^{14}}{63}, \\
\qquad \vdots
\end{cases}
$$

Combining Theorem I-3-5 and Theorem I-4-1, we obtain the following theorem:

Theorem I-4-6. *If all the entries of an $n \times n$ matrix $A(t)$ and $\mathbb{C}^n$-valued function $\vec{b}(t)$ are analytic on an interval $\mathcal{I}$, then every solution of linear differential equation (I.3.1) is analytic on $\mathcal{I}$.*

EXERCISES I

I-1. Solve the initial-value problem

$$
\frac{d^2y}{dt^2} + 2y\frac{dy}{dt} = 0, \qquad y(\tau) = \eta_0, \quad y'(\tau) = \eta_1.
$$

The reader must consider various cases concerning the initial data (τ, η_0, η_1).

I-2. Show that the function $f(x,y) = \dfrac{1}{(3 - (x-1)^2)(9 - (y-5)^2)}$ satisfies the Lipschitz condition

$$
|f(x, y_1) - f(x, y_2)| \leq \frac{|y_1 - y_2|}{5}
$$

if

$$
|x - 1| \leq \sqrt{2}, \qquad |y_1 - 5| \leq 2, \qquad |y_2 - 5| \leq 2.
$$

I-3. Show that the initial-value problem

$$
\text{(P)} \qquad \frac{dy}{dx} = \sin\left(\frac{x^3 + 3x + 1}{\sqrt{101 - y^2}}\right), \qquad y(5) = 3,
$$

has one and only one solution on the interval $|x - 5| < 7$.

I-4. Suppose that $u(x)$ is continuous and satisfies the integral equation

$$u(x) \; = \; \int_0^x \sin(u(t))u(t)^p dt$$

on the interval $0 \le x \le 1$. Show that $u(x) = 0$ on this interval if $p \ge 0$. What would happen if $p < 0$?

Hint. In case $p < 0$, the problem is reduced to the initial-value problem $\dfrac{u^{-p}}{\sin(u)}du = dx$, $u(0) = 0$. Thus, the solution $u(x)$ is not identically zero and has a branch point at $x = 0$.

I-5. Show that if real-valued continuous functions $f(x)$, $g(x)$, and $h(x)$ satisfy the inequalities

$$f(x) \; \ge \; 0, \qquad g(x) \; \le \; h(x) + \int_0^x f(\xi)g(\xi)d\xi$$

on an interval $0 \le x \le x_0$, then

$$g(x) \; \le \; h(x) + \int_0^x \left\{ f(\xi)h(\xi) \exp\left[\int_\xi^x f(\eta)d\eta \right] \right\} d\xi$$

on $0 \le x \le x_0$.

Hint. If we put $y(x) = \displaystyle\int_0^x f(\xi)g(\xi)d\xi$, then $\dfrac{dy}{dx} \le f(x)h(x) + f(x)y$ and $y(0) = 0$.

I-6. Let $a(t)$ be a real-valued function which is continuous on the interval $\mathcal{I} = \{t : 0 \le t \le 1\}$. For every positive integer m, set

$$f_m(t) = 1 + ta(0) \qquad \text{for} \quad 0 \le t \le \frac{1}{m}$$

and

$$f_m(t) = \left[\prod_{h=0}^{k-1} \left(1 + \frac{1}{m}a\left(\frac{h}{m}\right) \right) \right] \left(1 + \left(t - \frac{k}{m} \right) a\left(\frac{k}{m}\right) \right)$$

for $\dfrac{k}{m} \le t \le \dfrac{k+1}{m}$ and $k = 1, 2, \dots, m - 1$. Show that the sequence $\{f_m : m = 1, 2, \dots\}$ converges uniformly on the interval $\mathcal{I}$. Also, find $\displaystyle\lim_{m \to \infty} f_m$.

Hint. For large m, we have $1 + \dfrac{1}{m}a\left(\dfrac{h}{m}\right) > 0$ and

$$\exp\left[\frac{1}{m}a\left(\frac{h}{m}\right) - \frac{K}{m^2} \right] \le 1 + \frac{1}{m}a\left(\frac{h}{m}\right) \le \exp\left[\frac{1}{m}a\left(\frac{h}{m}\right) + \frac{K}{m^2} \right].$$

for some positive number K. Also

$$\left| \int_0^t a(\tau)d\tau - \sum_{m=0}^{k-1} \frac{1}{m}a\left(\frac{h}{m}\right) - \left(t - \frac{k}{m} \right) a\left(\frac{k}{m}\right) \right| \le \epsilon_m$$

for some positive ϵ_m such that $\displaystyle\lim_{m \to +\infty} \epsilon_m = 0$.

I-7. Let $f(t, y)$ be a real-valued function of two independent variables t and y which is continuous on a region:

$$\mathcal{D} = \{(t, y) : \ g_1(t) \leq y \leq g_2(t), \ \ \tau_1 \leq t \leq \tau_2\}.$$

Suppose that
  (i) g_1, g_1', g_2, and g_2' are continuous on $\tau_1 \leq t \leq \tau_2$,
  (ii) $g_1(t) < g_2(t)$ on $\tau_1 \leq t \leq \tau_2$,
  (iii)

$$\begin{cases} g_1'(t) < f(t, g_1(t)), \\ g_2'(t) > f(t, g_2(t)), \end{cases} \qquad \tau_1 \leq t \leq \tau_2,$$

  (iv) $\tau_1 < t_0 < \tau_2$ and $g_1(t_0) < c_0 < g_2(t_0)$.

Show that there exists a function $\phi(t)$ such that
  (1) ϕ and ϕ' are continuous on $t_0 \leq t \leq \tau_2$ and $(t, \phi(t)) \in \mathcal{D}$ on $t_0 \leq t \leq \tau_2$,
  (2) $\phi'(t) = f(t, \phi(t))$ on $t_0 \leq t \leq \tau_2$ and $\phi(t_0) = c_0$.

I-8. Set

$$f(y) = \begin{cases} 0 & \text{for} & -\infty < y \leq 0, \\ \sqrt{y} & \text{for} & 0 \leq y < +\infty. \end{cases}$$

Also, let $\eta(x)$ be an arbitrary continuous function of x on an interval $0 \leq x \leq x_0$, where x_0 is a positive number. Define successive approximations by

$$y_m(x) = \begin{cases} \eta(x) & \text{for} & m = 0, \\ \displaystyle\int_0^x f(y_{m-1}(t))dt & \text{for} & m = 1,\ 2,\ \ldots. \end{cases}$$

Show that $\displaystyle\lim_{m \to +\infty} y_m(x)$ exists uniformly on the interval $0 \leq x \leq x_0$.

Hint. For this problem, see [Na3, pp. 143-160; in particular Bemerkung 2].

I-9. Let $\mathcal{B}$ be a Banach space over the field $\mathbb{R}$ of real numbers. Also, let $f(t, x)$ be a $\mathcal{B}$-valued function of the variables $(t, x) \in \mathbb{R} \times \mathcal{B}$. Consider an initial-value problem

(IP) $$\qquad\qquad \frac{dx}{dt} = f(t, x), \qquad x(\tau) = \xi,$$

where (τ, ξ) is a given point in $\mathbb{R} \times \mathcal{B}$. Assume that f satisfies the following conditions:
  (i) $f(t, x)$ is continuous on a region $\mathcal{R} = \{(t, x) \in \mathbb{R} \times \mathcal{B} : |t - \tau| \leq a, \ \|x - \xi\| \leq b\}$, where a and b are positive numbers and $\| \ \|$ is the norm of $\mathcal{B}$,
  (ii) $f(t, x)$ also satisfies the Lipschitz condition $\|f(t, x_1) - f(t, x_2)\| \leq L\|x_1 - x_2\|$ whenever $(t, x_1) \in \mathcal{R}$ and $(t, x_2) \in \mathcal{R}$, where L is a positive constant.
Show that
  (1) there exists a positive number M such that $\|f(t, x)\| \leq M$ for $(t, x) \in \mathcal{R}$,
  (2) problem (IP) has one and only one solution on the interval $|t - \tau| < \alpha$, where

$$\alpha = \min\left(a, \frac{b}{M}\right).$$

Hint. See [Huk7].

I-10. Let $\mathcal{B}$ be a Banach space over the field $\mathbb{R}$ of real numbers and let $\mathcal{C}$ be a nonempty compact subset of $\mathcal{B}$. Also, on a bounded interval $\mathcal{I}$, let $\mathcal{F}$ be an infinite and equicontinuous set of $\mathcal{C}$-valued functions. Show that $\mathcal{F}$ contains an infinite sequence which is uniformly convergent on $\mathcal{I}$.

I-11. Let $\mathcal{B}$ be a Banach space over the field $\mathbb{R}$ of real numbers and let $\mathcal{C}$ be a nonempty compact and convex subset of $\mathcal{B}$. Assume that a $\mathcal{C}$-valued function $f(t,x)$ is continuous on a region

$$\mathcal{R} = \{(t,x) \in \mathbb{R} \times \mathcal{B} : |t - \tau| \leq a, \|x - \xi\| \leq b\},$$

where $\tau \in \mathbb{R}$, $\xi \in \mathcal{B}$, $a > 0$, $b > 0$, and $\|\ \|$ is the norm of $\mathcal{B}$. Show that
  (1) there exists a positive number M such that $\|f(t,x)\| \leq M$ for $(t,x) \in \mathcal{R}$,
  (2) the initial-value problem $\dfrac{dx}{dt} = f(t,x)$, $x(\tau) = \xi$ has a solution on the interval

$$|t - \tau| < \alpha, \text{ where } \alpha = \min\left(a, \frac{b}{M}\right).$$

Hint. Use a method similar to that of Remark I-2-7. See, also, [Huk7].

I-12. Let $\vec{f}(t,\vec{x})$ be $\mathbb{R}^n$-valued and continuous for $(t,\vec{x}) \in \mathbb{R}^{n+1}$. Assume also that $\mathcal{E}$ is a nonempty closed set in $\mathbb{R}^{n+1}$. Show that in order that for every $(\tau,\vec{\xi}) \in \mathcal{E}$, there exists a real number $\sigma(\tau,\ \vec{\xi})$ depending on $(\tau,\ \vec{\xi})$ such that $\tau < \sigma(\tau,\vec{\xi})$ and that the intial-value problem $\dfrac{d\vec{x}}{dt} = \vec{f}(t,\vec{x})$, $\vec{x}(\tau) = \vec{\xi}$ has a solution $\vec{x} = \vec{\phi}(t,\tau,\vec{\xi})$ satisfying the condition $(t, \vec{\phi}(t,\tau,\vec{\xi})) \in \mathcal{E}$ on an interval $\tau \leq t \leq \sigma(\tau)$, it is necessary and sufficient that for every $(\tau,\vec{\xi}) \in \mathcal{E}$ and every positive number ϵ, there exists a point $(\alpha,\vec{\beta}) \in \mathcal{E}$ such that

$$0 < \alpha - \tau < \epsilon \quad \text{and} \quad \left|\frac{\vec{\beta} - \vec{\xi}}{\alpha - \tau} - \vec{f}(\tau,\vec{\xi})\right| < \epsilon,$$

where $(\alpha,\vec{\beta})$ generaly depends on $(\tau,\vec{\xi},\epsilon)$.

Hint. The necessity is easy to prove. For sufficiency, use an idea similar to that of Remark I-2-7. See, also, [Huk6].

I-13. Assume that $f(x, y_1, y_2, \ldots, y_n)$ is real-valued, continuous, and continuously differentiable function of $(x, y_1, \ldots, y_n)$ on a region $\mathcal{R} = \{(x, y_1, \ldots, y_n) : a < x < b, |y_j| < r\ (j = 1, \ldots, n)\}$, where a and b are real numbers such that $a < b$, and r is a positive number. Assume, also, that (1) c is a real number such that $a < c < b$, (2) $\phi(x)$ is a real-valued solution of $\dfrac{d^n y}{dx^n} = f\left(x, y, \dfrac{dy}{dx}, \ldots, \dfrac{d^{n-1}y}{dx^{n-1}}\right)$ on an interval $|x - c| < \rho$, where ρ is a positive number such that $a \leq c - \rho < c + \rho \leq b$, (3) $\left|\dfrac{d^k \phi}{dx^k}(x)\right| < r\ (k = 0, 1, \ldots, n-1)$, and (4) $f(x, 0, \ldots, 0) = 0$ on $a < x < b$. Show that if there eixsts a sequence $\{c_h : h = 1, 2, \ldots\}$ of real numbers such that $\lim\limits_{h \to +\infty} c_h = c$ and that $\phi(c_h) = 0\ (h = 1, 2, \ldots)$, then $\phi(x) = 0$ identically on $a < x < b$.

I-14. Let $u(x,y)$ be a continuously differentiable function of two independent variables (x,y) in a domain $\mathcal{D} = \{(x,y) : y > 0, \ a + Ay \le x \le b - Ay\}$, where A is a positive number and the two quantities a and b are real numbers such that $a < b$. Assume that $u(x,y)$ is continuous on the closure $\bar{\mathcal{D}}$ of $\mathcal{D}$ and satisfies the condition

$$\left|\frac{\partial u}{\partial y}\right| \le A\left|\frac{\partial u}{\partial x}\right| + B|u| + C$$

on $\mathcal{D}$, where B and C are positive numbers. Set $M = \max\{|u(x,0)| : \ a \le x \le b\}$. Show that

$$|u(x,y)| \le Me^{By} + \frac{C}{B}(e^{By} - 1)$$

on $\bar{\mathcal{D}}$.

Hint. The last inequality is the Haar inequality. (See [Haa], [Na5, pp. 51-56], and [Har2, pp. 140-141]).

I-15. Let $H(u,t,x,q)$ be a function defined on an open set in $\mathbb{R}^4$ containing the point $(u,t,x,q) = (0,0,0,0)$ and satisfy a uniform Lipschitz condition with respect to (u,q). Let $\phi(x)$ be a function of class C^1 satisfying $\phi(0) = 0$ and $\phi_x(0) = 0$. Show that the initial-value problem

$$u_t + H(u,t,x,u_x) = 0, \qquad u(x,0) = \phi(x)$$

has at most one solution of class C^1 in a neighborhood of $x = 0$.

Hint. This is Haar's uniqueness theorem of the given partial differential equation (cf. the references in Exercise I-14). Upon applying Exercise I-14, with $M = C = 0$, to the difference $\hat{u}(t,x) = u_1(t,x) - u_2(t,x)$ of two solutions $u_1(t,x)$ and $u_2(t,x)$, the proof follows immediately.

CHAPTER II

DEPENDENCE ON DATA

The initial-value problem

$$\text{(P)} \qquad \frac{d\vec{y}}{dt} = \vec{f}(t, \vec{y}), \qquad \vec{y}(t_0) = \vec{c}_0$$

is equivalent to the integral equation

$$\text{(E)} \qquad \vec{y}(t) \;=\; \vec{c}_0 \;+\; \int_{t_0}^{t} \vec{f}(s, \vec{y}(s))\,ds.$$

The right-hand side of (E) is regarded as a function of $(t, t_0, \vec{c}_0, \vec{f}, \vec{y})$. In this chapter, we explain some basic properties of solutions with respect to $(t, t_0, \vec{c}_0, \vec{f})$. A parameter ϵ is used to represent the variable $\vec{f}$. In §II-1, we explain the continuity of the solution of (P) with respect to $(t, t_0, \vec{c}_0, \epsilon)$ at $(t, \tau, \vec{c}, \epsilon_0)$ when the solution of (P) is unique for the initial data $(\tau, \vec{c}, \epsilon_0)$ (Theorem II-1-2). Theorem II-1-2 was first proved by I. Bendixson [Ben1] for the scalar equation and later by G. Peano [Pea2] and E. Lindelöf [Lind2] for a system of equations. In §II-2, we explain the differentiability with respect to initial data $(t_0, \vec{c}_0)$ (Theorem II-2-1) and with respect to a parameter ϵ (Theorem II-2-2). The discussion of these topics can be found in [CL, pp. 22-28, 57-60] and [Har2, pp. 94-100].

II-1. Continuity with respect to initial data and parameters

Let ϵ be a variable which takes values in a set with a separable topology. We consider $\mathbb{R}^n$-valued functions $\vec{f}(t, \vec{y}, \epsilon)$ and $\vec{\phi}(t)$ under the following assumption.

Assumption 1. *For the initial-value problem (P), assume that*
  (i) $\vec{f}(t, \vec{y}, \epsilon)$ is continuous in $(t, \vec{y}, \epsilon)$ in an open set $\mathcal{D}_0$ in the $(t, \vec{y}, \epsilon)$-space,
  (ii) $\vec{\phi}'(t) = \vec{f}(t, \vec{\phi}(t), \epsilon_0)$ and $(t, \vec{\phi}(t), \epsilon_0) \in \mathcal{D}_0$ on an interval $\mathcal{I}_0 = \{t : t_1 \le t \le t_2\}$, where ϵ_0 is a fixed value of the variable ϵ,
  (iii) $\vec{y} = \vec{\phi}(t)$ is unique in the sense that if $\vec{\psi}'(t) = \vec{f}(t, \vec{\psi}(t), \epsilon_0)$ and $(t, \vec{\psi}(t), \epsilon_0) \in \mathcal{D}_0$ on a subinterval $\mathcal{I}$ of the interval $\mathcal{I}_0$ and if $\vec{\psi}(\tau) = \vec{\phi}(\tau)$ at a point τ of the interval $\mathcal{I}$, then $\vec{\psi}(t) = \vec{\phi}(t)$ identically on the interval $\mathcal{I}$.

We start our discussion with the following lemma.

Lemma II-1-1. *Conditions (i) and (ii) of Assumption 1 imply that there exist an open set Δ in the $(t, \vec{y})$-space and an open neighborhood $\mathcal{U}_0$ of ϵ_0 in the ϵ-space such that*
  (a) $\Delta \times \mathcal{U}_0 \subset \mathcal{D}_0$,
  (b) $(t, \vec{\phi}(t)) \in \Delta$ on the interval $\mathcal{I}_0$,

(c) $\left|\vec{f}(t,\vec{y},\epsilon)\right| \leq M$ *on* $\Delta \times \mathcal{U}_0$ *for some positive number* M.

Proof.

Let $\tau \in \mathcal{I}_0$. Since $\vec{f}$ is continuous in the open set $\mathcal{D}_0$ and $(\tau, \vec{\phi}(\tau), \epsilon_0) \in \mathcal{D}_0$, there exist an open neighborhood $\Delta(\tau)$ of $(\tau, \vec{\phi}(\tau))$ in the $(t, \vec{y})$-space and an open neighborhood $\mathcal{U}(\tau)$ of ϵ_0 in the ϵ-space such that $\Delta(\tau) \times \mathcal{U}(\tau) \subset \mathcal{D}_0$ and $\left|\vec{f}(t,\vec{y},\epsilon) - \vec{f}(\tau, \vec{\phi}(\tau), \epsilon_0)\right| \leq 1$ on $\Delta(\tau) \times \mathcal{U}(\tau)$. This implies that $\left|\vec{f}(t,\vec{y},\epsilon)\right| \leq 1 + M_0$ on $\bigcup_{\tau \in \mathcal{I}_0} \Delta(\tau) \times \mathcal{U}(\tau)$, where $M_0 = \max_{\tau \in \mathcal{I}_0} \left|\vec{f}(\tau, \vec{\phi}(\tau), \epsilon_0)\right|$. Since $(t, \vec{\phi}(t)) \in \bigcup_{\tau \in \mathcal{I}_0} \Delta(\tau)$ on the interval $\mathcal{I}_0$ and the interval $\mathcal{I}_0$ is compact, there exists a finite set $\{\tau_j : j = 1, 2, \ldots, N\}$ of points on the interval $\mathcal{I}_0$ such that $(t, \vec{\phi}(t)) \in \bigcup_{N \geq j \geq 1} \Delta(\tau_j)$ for all $t \in \mathcal{I}_0$. Set

$$\Delta = \bigcup_{j=1}^{N} \Delta(\tau_j), \qquad \mathcal{U}_0 = \bigcap_{j=1}^{N} \mathcal{U}(\tau_j), \qquad M = 1 + M_0.$$

Then, all the requirements of Lemma II-1-1 are satisfied by Δ, $\mathcal{U}_0$, and M. $\square$

The main purpose of this section is to prove the following theorem under Assumption 1.

Theorem II-1-2. *Let $\mathcal{E}$ be an open neighborhood of ϵ_0 and let $\mathcal{I} = \{t : s_1 \leq t \leq s_2\}$ be a subinterval of $\mathcal{I}_0$. Assume that*

$$\text{(II.1.1)} \qquad \vec{\psi}'(t,\epsilon) = \vec{f}(t, \vec{\psi}(t,\epsilon), \epsilon) \quad \text{and} \quad (t, \vec{\psi}(t,\epsilon), \epsilon) \in \mathcal{D}_0 \quad \text{on} \quad \mathcal{I} \times \mathcal{E}.$$

Assume also that there exists a real-valued function $t(\epsilon)$ such that
(1) $t(\epsilon) \in \mathcal{I}$ *for* $\epsilon \in \mathcal{E}$,
(2) $\lim_{\epsilon \to \epsilon_0} t(\epsilon) = t_0$,
(3) $\lim_{\epsilon \to \epsilon_0} \vec{\psi}(t(\epsilon), \epsilon) = \vec{\phi}(t_0)$.
Then,

$$\lim_{\epsilon \to \epsilon_0} \vec{\psi}(t,\epsilon) = \vec{\phi}(t)$$

uniformly on the interval $\mathcal{I}$.

Proof.

We shall prove this theorem in two steps. In Step 1, we prove Theorem II-1-2 assuming that there exists an open neighborhood $\mathcal{U}$ of ϵ_0 such that $(t, \vec{\psi}(t,\epsilon)) \in \Delta$ on $\mathcal{I} \times (\mathcal{E} \cap \mathcal{U})$, where Δ is the open set in the $(t, \vec{y})$-space which was determined by Lemma II-1-1. The existence of such an open neighborhood $\mathcal{U}$ of ϵ_0 is proved in Step 2.

Step 1. We can assume without any loss of generality that $\mathcal{U}$ is contained in the neighborhood $\mathcal{U}_0$ of ϵ_0 which was determined by Lemma II-1-1. Furthermore, using condition (3) of Theorem II-1-2, we choose $\mathcal{U}$ in such a way that $\left|\vec{\psi}(t(\epsilon), \epsilon)\right|$ is bounded on $\mathcal{U}$.

Since

$$(\text{II.1.2}) \qquad \vec{\psi}(t,\epsilon) \; = \; \vec{\psi}(t(\epsilon),\epsilon) \; + \; \int_{t(\epsilon)}^{t} \vec{f}(s,\vec{\psi}(s,\epsilon),\epsilon)ds \qquad \text{on} \;\; \mathcal{I} \times \mathcal{E},$$

we obtain $\left|\vec{\psi}(t,\epsilon)\right| \leq \left|\vec{\psi}(t(\epsilon),\epsilon)\right| + M|t - t(\epsilon)|$ on $\mathcal{I} \times (\mathcal{E} \cap \mathcal{U})$ and $\left|\vec{\psi}(t,\epsilon) - \vec{\psi}(\tau,\epsilon)\right| \leq$ $M|t - \tau|$ if (t,ϵ) and $(\tau,\epsilon) \in \mathcal{I} \times (\mathcal{E} \cap \mathcal{U})$, where M is the positive constant which was determined by Lemma II-1-1. Therefore, the set $\mathcal{F} = \{\vec{\psi}(\cdot,\epsilon) : \epsilon \in \mathcal{E} \cap \mathcal{U}\}$ is bounded and equicontinuous on the interval $\mathcal{I}$.

Let us derive a contradiction from the assumption that $\vec{\psi}(t,\epsilon)$ does not converge to $\vec{\phi}(t)$ uniformly on the interval $\mathcal{I}$ as $\epsilon \to \epsilon_0$. This assumption means that $\max_{t \in \mathcal{I}} \left|\vec{\psi}(t,\epsilon) - \vec{\phi}(t)\right|$ does not tend to zero as $\epsilon \to \epsilon_0$; i.e., there exists a positive number ρ and a sequence $\{\epsilon_j : j = 1, 2, \ldots\}$ such that

$$(\text{II.1.3}) \qquad \begin{cases} \epsilon_j \in \mathcal{E} \cap \mathcal{U} \quad \text{and} \quad \lim_{j \to +\infty} \epsilon_j \; = \; \epsilon_0, \\[2mm] \max_{t \in \mathcal{I}} \left|\vec{\psi}(t,\epsilon_j) - \vec{\phi}(t)\right| \; \geq \; \rho. \end{cases}$$

Here, use was made of the assumption that the topology of the ϵ-space is separable.

Observe that the sequence $\{\vec{\psi}(\cdot,\epsilon_j) : j = 1, 2, \ldots\}$ is a subset of $\mathcal{F}$. Hence, this sequence is bounded and equicontinuous on $\mathcal{I}$. Therefore, by Lemma I-2-3 (Arzelà-Ascoli), there exists a subsequence $\{\vec{\psi}(\cdot,\epsilon_{j_\nu}) : \nu = 1, 2, \ldots\}$ such that $j_\nu \to +\infty$ as $\nu \to +\infty$ and that $\vec{\psi}(t,\epsilon_{j_\nu})$ converges uniformly on $\mathcal{I}$ as $\nu \to +\infty$. Set $\vec{\psi}(t) = \lim_{\nu \to +\infty} \vec{\psi}(t,\epsilon_{j_\nu})$ on $\mathcal{I}$. Since (II.1.2) holds for $\epsilon = \epsilon_{j_\nu}$ for all ν, we obtain

$$\vec{\psi}(t) \; = \; \vec{\phi}(t_0) \; + \; \int_{t_0}^{t} \vec{f}(s,\vec{\psi}(s),\epsilon_0)ds \qquad \text{on} \;\; \mathcal{I}.$$

Hence, $\vec{\psi}'(t) = \vec{f}(t,\vec{\psi}(t),\epsilon_0)$ on $\mathcal{I}$ and $\vec{\psi}(t_0) = \vec{\phi}(t_0)$. Now, condition (iii) of Assumption 1 implies that $\vec{\psi}(t) = \vec{\phi}(t)$ identically on the interval $\mathcal{I}$. This, in turn, implies that $\lim_{\nu \to +\infty} \max_{t \in \mathcal{I}} \left|\vec{\phi}(t,\epsilon_{j_\nu}) - \vec{\phi}(t)\right| = 0$. This contradicts (II.1.3).

Step 2. We shall prove that there exists an open neighborhood $\mathcal{U}$ of ϵ_0 such that $\mathcal{U} \subset \mathcal{U}_0$ and $(t,\vec{\psi}(t,\epsilon)) \in \Delta$ on $\mathcal{I} \times (\mathcal{E} \cap \mathcal{U})$.

Choose $a_0 > 0$ so that the rectangular region

$$\mathcal{R}_0(\tau) \; = \; \{(t,\vec{y}) : \; |t - \tau| \leq a_0, \;\; |\vec{y} - \vec{\phi}(\tau)| \leq M a_0\}$$

is contained in Δ for every $\tau \in \mathcal{I}$ (cf. Figure 1).

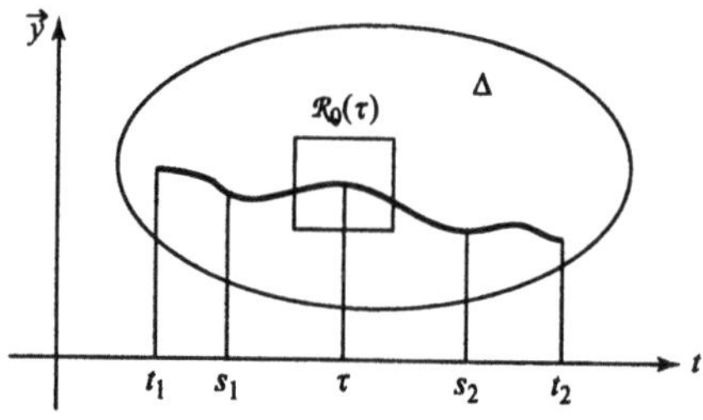

FIGURE 1.

For every positive number a less than a_0, set

$$\mathcal{R}((\hat{\tau}, \vec{\eta}); a) = \{(t, \vec{y}) : |t - \hat{\tau}| \leq a, \quad |\vec{y} - \vec{\eta}| \leq Ma\}.$$

Then, $\mathcal{R}((\hat{\tau}, \vec{\eta}); a) \subset \mathcal{R}_0(\tau)$ if $|\hat{\tau} - \tau| \leq a_0 - a$, $|\vec{\eta} - \vec{\phi}(\tau)| \leq M(a_0 - a)$. Fixing a positive number a less than a_0, choose $\tau_1, \tau_2, \ldots, \tau_N$ so that

$$\begin{cases} s_1 = \tau_1 < \tau_2 < \cdots < \tau_N = s_2, \\ t_0 = \tau_{j_0} \qquad \text{for some} \quad j_0, \\ 0 < \tau_{j+1} - \tau_j < a \qquad (j = 1, 2, \ldots, N - 1). \end{cases}$$

Since $\lim_{\epsilon \to \epsilon_0} t(\epsilon) = t_0$ and $\lim_{\epsilon \to \epsilon_0} \vec{\psi}(t(\epsilon), \epsilon) = \vec{\phi}(t_0)$, there exists a neighborhood $\mathcal{U}_1$ of ϵ_0 such that $|t(\epsilon) - t_0| \leq a_0 - a$ and $\left|\vec{\psi}(t(\epsilon), \epsilon) - \vec{\phi}(t_0)\right| \leq M(a_0 - a)$ for $\epsilon \in \mathcal{E} \cap \mathcal{U}_1$. Hence, $\mathcal{R}((t(\epsilon), \vec{\psi}(t(\epsilon), \epsilon)); a) \subset \mathcal{R}_0(t_0) \subset \Delta$ for $\epsilon \in \mathcal{E} \cap \mathcal{U}_1$. Now, using Lemma I-1-1, we can verify that $(t, \vec{\psi}(t, \epsilon)) \in \mathcal{R}((t(\epsilon), \vec{\psi}(t(\epsilon), \epsilon)); a) \subset \Delta$ for $|t - t(\epsilon)| \leq a$ and $\epsilon \in \mathcal{E} \cap \mathcal{U}_1$. Also, since $\lim_{\epsilon \to \epsilon_0} t(\epsilon) = t_0 = \tau_{j_0}$, we have $|t - t(\epsilon)| \leq a$ for $\tau_{j_0-1} \leq t \leq \tau_{j_0+1}$ and $\epsilon \in \mathcal{E} \cap \mathcal{U}_1$, if the neighborhood $\mathcal{U}_1$ of ϵ_0 is sufficiently small.

Using the same argument as in Step 1 for $\vec{\psi}(t, \epsilon)$ on the interval $\tau_{j_0-1} \leq t \leq \tau_{j_0+1}$ and $\epsilon \in \mathcal{E} \cap \mathcal{U}_1$, we obtain $\lim_{\epsilon \to \epsilon_0} \vec{\psi}(t, \epsilon) = \vec{\phi}(t)$ uniformly on $\tau_{j_0-1} \leq t \leq \tau_{j_0+1}$. In particular,

$$\lim_{\epsilon \to \epsilon_0} \vec{\psi}(\tau_{j_0-1}, \epsilon) = \vec{\phi}(\tau_{j_0-1}) \qquad \text{and} \qquad \lim_{\epsilon \to \epsilon_0} \vec{\psi}(\tau_{j_0+1}, \epsilon) = \vec{\phi}(\tau_{j_0+1}).$$

Using the same method in the neighborhoods of $(\tau_{j_0-1}, \vec{\phi}(\tau_{j_0-1}))$ and $(\tau_{j_0+1}, \vec{\phi}(\tau_{j_0+1}))$, we obtain $\lim_{\epsilon \to \epsilon_0} \vec{\psi}(t, \epsilon) = \vec{\phi}(t)$ uniformly on $\tau_{j_0-2} \leq t \leq \tau_{j_0}$ and $\tau_{j_0} \leq t \leq \tau_{j_0+2}$. Hence, $\lim_{\epsilon \to \epsilon_0} \vec{\psi}(t, \epsilon) = \vec{\phi}(t)$ uniformly on $\tau_{j_0-2} \leq t \leq \tau_{j_0+2}$. Continuing this process, we find an open neighborhood $\mathcal{U}$ of ϵ_0 such that $\mathcal{U} \subset \mathcal{U}_0$ and $(t, \vec{\psi}(t, \epsilon)) \in \Delta$ on $\mathcal{I} \times (\mathcal{E} \cap \mathcal{U})$ in a finite number of steps. $\square$

Remark II-1-3. If $\vec{f}(t, \vec{y}, \epsilon)$ is independent of ϵ, Theorem II-1-2 yields continuity of solutions of the initial-value problem $\dfrac{d\vec{y}}{dt} = \vec{f}(t, \vec{y})$, $\vec{y}(t_0) = \vec{c}_0$ with respect to the initial data $(t_0, \vec{c}_0)$.

In Theorem II-1-2, it was assumed that $(t, \vec{\psi}(t, \epsilon), \epsilon) \in \mathcal{D}_0$ on $\mathcal{I} \times \mathcal{E}$ (cf. (II.1.1)). With regard to this assumption, we can improve Theorem II-1-2 as follows.

Theorem II-1-4. *Assume that Assumption 1 is satisfied. Let $\mathcal{E}$ be an open neighborhood of ϵ_0. Assume further that there exist functions $t(\epsilon)$ and $\vec{c}(\epsilon)$ defined on $\mathcal{E}$ such that*

(i) $t(\epsilon) \in \mathcal{I}_0$ and $(t(\epsilon), \vec{c}(\epsilon), \epsilon) \in \mathcal{D}_0$ for $\epsilon \in \mathcal{E}$,

(ii) $\lim_{\epsilon \to \epsilon_0} t(\epsilon) = t_0$ and $\lim_{\epsilon \to \epsilon_0} \vec{c}(\epsilon) = \vec{\phi}(t_0)$.

Then, we can choose a neighborhood $\mathcal{U}$ of ϵ_0 sufficiently small so that the initial-value problem

$$\frac{d\vec{y}}{dt} = \vec{f}(t, \vec{y}, \epsilon), \qquad \vec{y}(t(\epsilon)) = \vec{c}(\epsilon)$$

has on the region $\mathcal{I}_0 \times (\mathcal{E} \cap \mathcal{U})$ *a solution* $\vec{y} = \vec{\psi}(t, \epsilon)$ *such that* $(t, \vec{\psi}(t, \epsilon), \epsilon) \in \mathcal{D}_0$.

Theorem II-1-4 can be proved by using an argument similar to Step 2 of the proof of Theorem II-1-2 and the local existence theorem (cf. Theorem I-2-5). Details are left to the reader as an exercise.

II-2. Differentiability

In this section, we explain the differentiability of solutions of an initial-value problem with respect to the initial data under the following assumption.

Assumption 2. *For the initial-value problem (P), assume that*

(i) $\vec{f}(t, \vec{y})$ *and* $\dfrac{\partial \vec{f}}{\partial y_j}(t, \vec{y})$ $(j = 1, 2, \ldots, n)$ *are continuous in* $(t, \vec{y})$ *on an open set* Δ *in the* $(t, \vec{y})$*-space,*

(ii) $\vec{\phi}'(t) = \vec{f}(t, \vec{\phi}(t))$ *and* $(t, \vec{\phi}(t)) \in \Delta$ *on an interval* $\mathcal{I}_0 = \{t : t_1 \leq t \leq t_2\}$.

Using Lemma I-1-6, Theorems I-1-4, II-1-2, II-1-4, and Remark II-1-3, we can show that there exists a positive number ρ such that the initial-value problem

$$(\text{II.2.1}) \qquad\qquad \frac{d\vec{y}}{dt} = \vec{f}(t, \vec{y}), \qquad \vec{y}(\tau) = \vec{\eta}$$

has a unique solution $\vec{y} = \vec{\phi}(t, \tau, \vec{\eta})$ on the interval $\mathcal{I}_0$ if $\tau \in \mathcal{I}_0$ and $|\vec{\eta} - \vec{\phi}(\tau)| \leq \rho$. The solution $\vec{y} = \vec{\phi}(t, \tau, \vec{\eta})$ is continuous in $(t, \tau, \vec{\eta})$ and satisfies the condition $(t, \vec{\phi}(t, \tau, \vec{\eta})) \in \Delta$ if $t \in \mathcal{I}_0$, $\tau \in \mathcal{I}_0$, and $|\vec{\eta} - \vec{\phi}(\tau)| \leq \rho$. The solution $\vec{\phi}(t, \tau, \vec{\eta})$ satisfies the integral equation

$$\vec{\phi}(t, \tau, \vec{\eta}) = \vec{\eta} + \int_{\tau}^{t} \vec{f}(s, \vec{\phi}(s, \tau, \vec{\eta}))ds.$$

If $\vec{\phi}(t, \tau, \vec{\eta})$ is differentiable with respect to $(\tau, \vec{\eta})$, then

$$\begin{cases} \dfrac{\partial \vec{\phi}(t, \tau, \vec{\eta})}{\partial \tau} = -\vec{f}(\tau, \vec{\eta}) + \displaystyle\int_{\tau}^{t} \dfrac{\partial \vec{f}}{\partial \vec{y}}(s, \vec{\phi}(s, \tau, \vec{\eta})) \dfrac{\partial \vec{\phi}(s, \tau, \vec{\eta})}{\partial \tau}ds, \\[4mm] \dfrac{\partial \vec{\phi}(t, \tau, \vec{\eta})}{\partial \eta_j} = \vec{e}_j + \displaystyle\int_{\tau}^{t} \dfrac{\partial \vec{f}}{\partial \vec{y}}(s, \vec{\phi}(s, \tau, \vec{\eta})) \dfrac{\partial \vec{\phi}(s, \tau, \vec{\eta})}{\partial \eta_j}ds, \end{cases}$$

where η_j is the j-th entry of $\vec{\eta}$, $\vec{e}_j$ is the vector in $\mathbb{R}^n$ with entries 0 except for 1 at the j-th entry, and $\dfrac{\partial \vec{f}}{\partial \vec{y}}$ is the $n \times n$ matrix whose j-th column is $\dfrac{\partial \vec{f}}{\partial y_j}$. From this speculation, we obtain the following result.

Theorem II-2-1. *Partial derivatives* $\dfrac{\partial \vec{\phi}}{\partial \tau}$ *and* $\dfrac{\partial \vec{\phi}}{\partial \eta_j}$ $(j = 1, 2, \ldots, n)$ *exist and are continuous in* $(t, \tau, \vec{\eta})$ *in the domain*

$$(\text{II.2.2}) \qquad \mathcal{D} = \{(t, \tau, \vec{\eta}) : t \in \mathcal{I}_0, \; t_1 < \tau < t_2, \; |\vec{\eta} - \vec{\phi}(\tau)| < \rho\}.$$

Furthermore, $\dfrac{\partial \vec{\phi}}{\partial \tau}$ $\left(\text{respectively } \dfrac{\partial \vec{\phi}}{\partial \eta_j}\right)$ *is the unique solution of the initial-value problem*

$$(\text{II.2.3}) \qquad \begin{cases} \dfrac{d\vec{z}}{dt} = \dfrac{\partial \vec{f}}{\partial \vec{y}}(t, \vec{\phi}(t, \tau, \vec{\eta})) \; \vec{z}, \\[2mm] \vec{z}(\tau) = -\vec{f}(\tau, \vec{\eta}) \qquad (\text{ respectively } \vec{e}_j). \end{cases}$$

Proof.

We prove the existence and continuity of $\dfrac{\partial \vec{\phi}}{\partial \tau}$ by showing that they are the unique solution of initial-value problem (II.2.3) with the initial-value $\vec{z}(\tau) = -\vec{f}(\tau, \vec{\eta})$. The existence and continuity of $\dfrac{d\vec{\phi}}{d\eta_j}$ can be proved similarly.

Let $\vec{\psi}(t, \tau, \vec{\eta})$ be the unique solution of (II.2.3) with the initial condition $\vec{z}(\tau) = -\vec{f}(\tau, \vec{\eta})$. From Theorems I-3-5 and II-1-2 and Remark II-1-3, it follows that $\vec{\psi}(t, \tau, \vec{\eta})$ exists and is continuous in $(t, \tau, \vec{\eta})$ on the closure $\overline{\mathcal{D}}$ of domain (II.2.2). Furthermore,

$$(\text{II.2.4}) \qquad \vec{\psi}(t, \tau, \vec{\eta}) = -\vec{f}(\tau, \vec{\eta}) + \int_\tau^t \frac{\partial \vec{f}}{\partial \vec{y}}(s, \vec{\phi}(s, \tau, \vec{\eta})) \; \vec{\psi}(s, \tau, \vec{\eta})ds$$

on $\overline{\mathcal{D}}$. To show that $\dfrac{\partial \vec{\phi}}{\partial \tau} = \vec{\psi}(t, \tau, \vec{\eta})$, we first derive

$$(\text{II.2.5}) \qquad \begin{cases} \vec{\phi}(t, \tau + h, \vec{\eta}) = \vec{\eta} + \displaystyle\int_{\tau+h}^t \vec{f}(s, \vec{\phi}(s, \tau + h, \vec{\eta}))ds, \\[3mm] \vec{\phi}(t, \tau, \vec{\eta}) = \vec{\eta} + \displaystyle\int_\tau^t \vec{f}(s, \vec{\phi}(s, \tau, \vec{\eta}))ds, \end{cases}$$

from (II.2.1) on $\mathcal{D}$ assuming that $h \neq 0$ is sufficiently small. Using (II.2.4) and (II.2.5), we compute

$$(\text{II.2.6}) \qquad \vec{g}(t, \tau, \vec{\eta}, h) = \frac{1}{h}\left[\vec{\phi}(t, \tau + h, \vec{\eta}) - \vec{\phi}(t, \tau, \vec{\eta})\right] - \vec{\psi}(t, \tau, \vec{\eta})$$

as follows.

Observation 1. For a fixed $(t, \tau, \vec{\eta}) \in \mathcal{D}$, we have

$$\vec{\phi}(t, \tau + h, \vec{\eta}) - \vec{\phi}(t, \tau, \vec{\eta}) = \int_{\tau}^{t} \left[\vec{f}(s, \vec{\phi}(s, \tau + h, \vec{\eta})) - \vec{f}(s, \vec{\phi}(s, \tau, \vec{\eta})) \right] ds$$
$$- \int_{\tau}^{\tau + h} \vec{f}(s, \vec{\phi}(s, \tau + h, \vec{\eta})) ds$$

if $h \neq 0$ is sufficiently small.

Observation 2. Using an idea similar to the proof of Lemma I-1-6, for a sufficiently small h, we obtain

$$\vec{f}(s, \vec{\phi}(s, \tau + h, \vec{\eta})) - \vec{f}(s, \vec{\phi}(s, \tau, \vec{\eta}))$$
$$= \left[\int_{0}^{1} \frac{\partial \vec{f}}{\partial \vec{y}}(s, \theta \vec{\phi}(s, \tau + h, \vec{\eta}) + (1 - \theta) \vec{\phi}(s, \tau, \vec{\eta})) d\theta \right] \left(\vec{\phi}(t, \tau + h, \vec{\eta}) - \vec{\phi}(t, \tau, \vec{\eta}) \right)$$

for $s \in \mathcal{I}_0$ and a fixed $(\tau, \vec{\eta})$ such that $(s, \tau, \vec{\eta}) \in \mathcal{D}$. Note that $(s, \theta \vec{\phi}(s, \tau + h, \vec{\eta}) + (1 - \theta) \vec{\phi}(s, \tau, \vec{\eta})) \in \Delta$ for $s \in \mathcal{I}_0$ and a fixed $(\tau, \vec{\eta})$ such that $(s, \tau, \vec{\eta}) \in \mathcal{D}$, if $h \neq 0$ is sufficiently small.

Finally, applying Observations 1 and 2 to (II.2.4), (II.2.5), and (II.2.6), we obtain

$$\vec{g}(t, \tau, \vec{\eta}, h)$$
$$= - \left[\frac{1}{h} \int_{\tau}^{\tau + h} \vec{f}(s, \vec{\phi}(s, \tau + h, \vec{\eta})) ds - \vec{f}(\tau, \vec{\eta}) \right]$$
$$\text{(II.2.7)} \qquad + \int_{\tau}^{t} \left[\int_{0}^{1} \frac{\partial \vec{f}}{\partial \vec{y}}(s, \theta \vec{\phi}(s, \tau + h, \vec{\eta}) + (1 - \theta) \vec{\phi}(s, \tau, \vec{\eta})) \, d\theta \right] \vec{g}(s, \tau, \vec{\eta}, h) ds$$
$$+ \int_{\tau}^{t} \left[\int_{0}^{1} \frac{\partial \vec{f}}{\partial \vec{y}}(s, \theta \vec{\phi}(s, \tau + h, \vec{\eta}) + (1 - \theta) \vec{\phi}(s, \tau, \vec{\eta})) d\theta \right.$$
$$\left. - \frac{\partial \vec{f}}{\partial \vec{y}}(s, \vec{\phi}(s, \tau, \vec{\eta})) \right] \vec{\psi}(s, \tau, \vec{\eta}) ds.$$

Observation 3. The first and the third terms of the right-hand side of (II.2.7) tend to zero as $h \to 0$. Also, since $\vec{\phi}(\tau, \tau, \vec{\eta}) = \vec{\eta}$ and $\dfrac{\partial \vec{f}}{\partial \vec{y}}(s, \vec{\phi}(s, \tau, \vec{\eta}))$ is bounded on $\overline{\mathcal{D}}$, there exist a positive constant L and a non-negative function $K(h)$ such that $\lim_{h \to 0} K(h) = 0$ and that

$$|\vec{g}(t, \tau, \vec{\eta}, h)| \leq K(h) + L \left| \int_{\tau}^{t} |\vec{g}(s, \tau, \vec{\eta}, h)| ds \right| \qquad \text{for} \quad t \in \mathcal{I}_0.$$

Now, using Lemma I-1-5, we obtain the estimate

$$|\vec{g}(t, \tau, \vec{\eta}, h)| \leq K(h) \exp[L|t - \tau|] \qquad \text{on} \quad \mathcal{I}_0.$$

Thus, we conclude that $\vec{g}(t, \tau, \vec{\eta}, h) \to 0$ as $h \to 0$, i.e., $\dfrac{\partial \vec{\phi}}{\partial \tau}(t, \tau, \vec{\eta}) = \vec{\psi}(t, \tau, \vec{\eta})$. $\square$

Upon applying Theorem II-2-1 to the initial-value problem

$$\frac{d\vec{y}}{dt} = \vec{f}(t, \vec{y}, u), \qquad \frac{du}{dt} = 0, \qquad \vec{y}(\tau) = \vec{\eta}, \qquad u(\tau) = \epsilon,$$

we can prove the following theorem.

Theorem II-2-2. *Let ϵ be a real variable. Assume that $\vec{f}(t, \vec{y}, \epsilon)$, $\dfrac{\partial \vec{f}}{\partial y_j}(t, \vec{y}, \epsilon)$ $(j = 1, 2, \ldots, n)$, and $\dfrac{\partial \vec{f}}{\partial \epsilon}(t, \vec{y}, \epsilon)$ are continuous on an open set $\mathcal{D}_0$ in the $(t, \vec{y}, \epsilon)$-space. Let $\vec{y} = \vec{\phi}(t, \tau, \vec{\eta}, \epsilon)$ be the unique solution of the initial-value problem*

$$\frac{d\vec{y}}{dt} = \vec{f}(t, \vec{y}, \epsilon), \qquad \vec{y}(\tau) = \vec{\eta}.$$

Then, there exists an open set Ω in the $(t, \tau, \vec{\eta}, \epsilon)$-space such that the function $\vec{\phi}(t, \tau, \vec{\eta}, \epsilon)$ and $\dfrac{\partial \vec{\phi}}{\partial \epsilon}$ are continuous on Ω. Furthermore, $\vec{z} = \dfrac{\partial \vec{\phi}}{\partial \epsilon}(t, \tau, \vec{\eta}, \epsilon)$ is the unique solution of the initial-value problem

$$\frac{d\vec{z}}{dt} = \frac{\partial \vec{f}}{\partial \vec{y}}(t, \vec{\phi}(t, \tau, \vec{\eta}, \epsilon), \epsilon)\vec{z} + \frac{\partial \vec{f}}{\partial \epsilon}(t, \vec{\phi}(t, \tau, \vec{\eta}, \epsilon), \epsilon), \qquad \vec{z}(\tau) = 0.$$

EXERCISES II

II-1. Given an interval $\mathcal{I} = \{x : a \leq x \leq b\}$, show that if $|\epsilon|$ is sufficiently small, the initial-value problem

$$\frac{dy}{dx} = \cos(x(y - x)), \qquad y(a) = a + \epsilon$$

has the unique solution $y = \phi(x, \epsilon)$ on $\mathcal{I}$. Show also that $\lim\limits_{\epsilon \to 0} \phi(x, \epsilon) = x$ uniformly on $\mathcal{I}$.

Hint. The function $\cos(x(y - x))$ is continuously differentiable on the entire (x, y)-plane. Furthermore, $|\cos(x(y - x))| \leq 1$. This shows the existence of the unique solution. Note that $y = x$ is the unique solution in case $\epsilon = 0$. Complete the proof by using Theorem II-1-2.

II-2. Let $y = \phi_1(x, \lambda)$ and $y = \phi_2(x, \lambda)$ be two solutions of the differential equation

$$\frac{d^2y}{dx^2} + \lambda(1 + x^2)y = 0$$

determined respectively by the initial conditions

$$\phi_1(0, \lambda) = 1, \quad \frac{\partial\phi_1}{\partial x}(0, \lambda) = 0 \quad \text{and} \quad \phi_2(0, \lambda) = 0, \quad \frac{\partial\phi_2}{\partial x}(0, \lambda) = 1.$$

Show that

   (i) ϕ_1 and ϕ_2 are analytic in (x, λ) everywhere in the complex (x, λ)-space (i.e., in $\mathbb{C}^2$),

  (ii) $\dfrac{\partial\phi_2}{\partial\lambda}(1, \lambda_0) \neq 0$, if $\phi_2(1, \lambda_0) = 0$.

Hint. Theorem II-2-2 implies that $z = \dfrac{\partial\phi_2}{\partial\lambda}(x, \lambda)$ is the solution of the initial-value problem

$$\frac{d^2z}{dx^2} + \lambda(1 + x^2)z = -(1 + x^2)\phi_2(x, \lambda), \qquad z(0) = 0, \quad z'(0) = 0.$$

The method of variation of parameters (cf. (IV.7.9) in Remark IV-7-2 of Chapter IV; also see, for example, [Cod, pp. 67-68, 122-123] or [Rab, pp. 241-244]) yields

$$\frac{\partial\phi_2}{\partial\lambda}(x, \lambda)$$
$$= \phi_1(x, \lambda)\int_0^x \phi_2(\xi, \lambda)^2(1 + \xi^2)d\xi - \phi_2(x, \lambda)\int_0^x \phi_1(\xi, \lambda)\phi_2(\xi, \lambda)(1 + \xi^2)d\xi.$$

Hence, we obtain

$$\frac{\partial\phi_2}{\partial\lambda}(1, \lambda_0) = \phi_1(1, \lambda_0)\int_0^1 \phi_2(\xi, \lambda_0)^2(1 + \xi^2)d\xi \neq 0.$$

Note that ϕ_1 and ϕ_2 are linearly independent and hence $\phi_1(1, \lambda_0) \neq 0$.

II-3. Show that if $|\epsilon|$ is small, the following boundary-value problem has the unique solution which is continuous in ϵ:

$$\frac{d^2y}{dx^2} = \epsilon \sin\left(\frac{x}{100 - y^2}\right), \qquad y(-1) = 0, \quad y(1) = 0.$$

Hint. Determine $\phi(x, c, \epsilon)$ by the initial-value problem

$$\frac{d^2y}{dx^2} = \epsilon \sin\left(\frac{x}{100 - y^2}\right), \qquad y(-1) = 0, \quad y'(-1) = c.$$

Then, the given boundary-value problem is reduced to the equation

(E) $$\phi(1, c, \epsilon) = 0.$$

First of all, $\phi(1, c, \epsilon)$ is analytic for small $|\epsilon|$ and $|c|$. Also, $\phi(1, c, 0) = 2c$ and, hence, $\dfrac{\partial\phi(1, c, 0)}{\partial c} = 2 \neq 0$. Therefore, equation (E) has the unique solution $c = c(\epsilon)$ such that $c(0) = 0$. Furthermore, $c(\epsilon)$ is analytic for small $|\epsilon|$.

II-4. Let $\vec{f}(x, \vec{y})$ be an $\mathbb{R}^n$-valued function whose entries are continuously differentiable with respect to $(x, \vec{y})$ in a domain $\mathcal{D} \subset \mathbb{R}^{n+1}$. Also, let $\vec{y} = \vec{\phi}(x, \vec{\eta})$ be a solution of the system $\dfrac{d\vec{y}}{dx} = \vec{f}(x, \vec{y})$ such that $\vec{\phi}(0) = \vec{\eta}$ and that $(x, \vec{\phi}(x, \vec{\eta})) \in \mathcal{D}$ for $0 \leq x \leq 1$ and $\vec{\eta} \in \Delta_0$, where Δ_0 is a domain in $\mathbb{R}^n$. Note that we must have $(0, \vec{\eta}) \in \mathcal{D}$ for $\vec{\eta} \in \Delta_0$. Set $\Delta_1 = \{\vec{\phi}(1, \vec{\eta}) : \vec{\eta} \in \Delta_0\}$. Show that
  (a) Δ_1 is open in $\mathbb{R}^n$,
  (b) the mapping $\vec{\phi}(1, \vec{\eta}) : \Delta_0 \to \Delta_1$ is one-to-one, onto, and differentiable with respect to $\vec{\eta}$ in Δ_0,
  (c) if we denote the inverse of the mapping $\vec{\phi}(1, \vec{\eta})$ by $\vec{\psi}(\zeta) : \Delta_1 \to \Delta_0$, then $\vec{\psi}(\zeta)$ is also differentiable with respect to $\vec{\zeta}$.

II-5. For the differential equation

$$(1 + x^2)\frac{dy}{dx} - 2xy = 2xy^2,$$

find
  (i) the unique solution $y = \phi(x, \xi, \eta)$ satisfying the initial-condition $y(\xi) = \eta$;
  (ii) the partial derivative $\dfrac{\partial\phi(x, \xi, \eta)}{\partial\xi}$;
  (iii) the partial derivative $\dfrac{\partial\phi(x, \xi, \eta)}{\partial\eta}$;
  (iv) $\dfrac{\partial\phi(x, \xi, \eta)}{\partial\xi}$ at $x = \xi$;
  (v) $\dfrac{\partial\phi(x, \xi, \eta)}{\partial\eta}$ at $x = \xi$.

Show also that $z = \dfrac{\partial\phi(x, \xi, \eta)}{\partial\xi}$ and $z = \dfrac{\partial\phi(x, \xi, \eta)}{\partial\eta}$ both satisfy the differential equation:

$$(1 + x^2)\frac{dz}{dx} - 2xz = 4x\phi(x, \xi, \eta)z .$$

II-6. Assume that $f(t, x_1, x_2, \dots, x_n)$ is a real-valued, continuous, and continuously differentiable function of $(t, x_1, \dots, x_n)$ on an open set Δ in the $(t, x_1, \dots, x_n)$-space. Assume also that $\phi_0(t)$ is a real-valued solution of the n-th order differential equation $x^{(n)} = f(t, x, x', \dots, x^{(n-1)})$ and $(t, \phi_0(t), \phi_0'(t), \dots, \phi_0^{(n-1)}(t)) \in \Delta$ on an interval $\mathcal{I}_0 = \{t : t_1 \leq t \leq t_2\}$. Show that there exists a positive number ρ such that
  (i) the initial-value problem

$$x^{(n)} = f(t, x, x', \dots, x^{(n-1)}), \quad x(\tau) = \xi_1, x'(\tau) = \xi_2, \dots, x^{(n-1)}(\tau) = \xi_n$$

has a unique solution $x = \phi(t, \tau, \xi_1, \dots, \xi_n)$ on the interval $\mathcal{I}_0$ if $\tau \in \mathcal{I}_0$ and $|\xi_j - \phi_0^{(j-1)}(\tau)| \leq \rho \ (j = 1, 2, \dots, n)$;
  (ii) the solution $x = \phi(t, \tau, \xi_1, \dots, \xi_n)$ is continuous in $(t, \tau, \xi_1, \dots, \xi_n)$ and satisfies the condition $(t, \phi(t, \tau, \xi_1, \dots, \xi_n), \phi'(t, \tau, \xi_1, \dots, \xi_n), \dots, \phi^{(n-1)}(t, \tau, \xi_1, \dots, \xi_n)) \in \Delta$ if $t \in \mathcal{I}_0$, $\tau \in \mathcal{I}_0$, and $|\xi_j - \phi_0^{(j-1)}(\tau)| \leq \rho \ (j = 1, 2, \dots, n)$;

(iii) the partial derivative $\dfrac{\partial^{n+1}\phi(t,\tau,\xi_1,\ldots,\xi_n)}{\partial\tau\partial t^n}$ exists and is continuous in $(t,\tau,\xi_1,$
$\ldots,\xi_n)$ on the domain $\mathcal{D} = \{(t,\tau,\xi_1,\ldots,\xi_n) : t \in \mathcal{I}_0,\ t_1 < \tau < t_2,\ |\xi_j - \phi_0^{(j-1)}(\tau)| < \rho\ \ (j=1,2,\ldots,n)\}$;

(iv) $u = \dfrac{\partial\phi}{\partial\tau}(t,\tau,\xi_1\ldots,\xi_n)$ is the unique solution of the initial-value problem

$$\begin{cases} \dfrac{d^n u}{dt^n} = \sum_{j=1}^n \dfrac{\partial f}{\partial x_j}(t,\phi(t,\tau,\xi_1,\ldots,\xi_n),\ldots,\phi^{(n-1)}(t,\tau,\xi_1,\ldots,\xi_n))\dfrac{d^{j-1}u}{dt^{j-1}}, \\[2mm] u(\tau) = -\xi_2,\ u'(\tau) = -\xi_3,\ \ldots,\ u^{(n-2)}(\tau) = -\xi_n,\ u^{(n-1)} = -f(\tau,\xi_1,\ldots,\xi_n). \end{cases}$$

II-7. Assume that $f(t,x_1,x_2)$ is a real-valued, continuous, and continuously differentiable function of (t,x_1,x_2) on an open set Δ in the (t,x_1,x_2)-space. Assume also that $\phi_0(t)$ is a real-valued solution of the second-order differential equation $x'' = f(t,x,x')$ and $(t,\phi_0(t),\phi_0'(t)) \in \Delta$ on the interval $\mathcal{I}_0 = \{t : 0 \le t \le 1\}$. Set $\phi_0(0) = a$ and $\phi_0'(0) = b$. Denote by $\phi(t,\beta)$ the unique solution of the initial value-problem $x'' = f(t,x,x')$, $x(0) = a$, $x'(0) = \beta$, where $|b - \beta|$ is sufficiently small. Show that $\dfrac{\partial\phi}{\partial\beta}(1,b) > 0$ if $\dfrac{\partial f}{\partial x}(t,\phi_0(t),\phi_0'(t)) > 0$ for $t \in \mathcal{I}_0$.

II-8. Let $g(t)$ be a real-valued and continuous function of t on the interval $0 \le t \le 1$. Also, let λ be a real parameter.

(1) Show that, if $\phi(t,\lambda)$ is a real-valued solution of the boundary-value problem

$$\frac{d^2 u}{dt^2} + (g(t) + \lambda)u = 0, \quad u(0) = 0, \quad u(1) = 0,$$

and if $\dfrac{\partial^3\phi}{\partial\lambda\partial t^2}(t,\lambda)$ is continuous on the region $\Delta = \{(t,\lambda) : 0 \le t \le 1, a < \lambda < b\}$, then $\phi(t,\lambda)$ is identically equal to zero on Δ, where a and b are real numbers.

(2) Does the same conclusion hold if $\phi(t,\lambda)$ is merely continuous on Δ?

II-9. Let $a(x,y)$ and $b(x,y)$ be two continuously differentiable functions of two variables (x,y) in a domain $\Delta_0 = \{(x,y) : |x| < \alpha,\ |y| < \beta\}$ and let $F(x,y,z)$ be a continuously differentiable function of three variables (x,y,z) in a domain $\mathcal{D}_0 = \{(x,y,z) : |x| < \alpha,\ |y| < \beta,\ |z| < \gamma\}$. Also, let $x = f(t,\eta)$, $y = g(t,\eta)$, and $z = h(t,\eta)$ be the unique solution of the initial-value problem

$$\begin{cases} \dfrac{dx}{dt} = a(x,y), \quad & \dfrac{dy}{dt} = b(x,y), \quad & \dfrac{dz}{dt} = F(x,y,z), \\[2mm] x(0) = 0, & y(0) = \eta, & z(0) = c(\eta), \end{cases}$$

where $c(\eta)$ is a differentiable function of η in the domain $\mathcal{U}_0 = \{\eta : |\eta| < \beta\}$. Assume that $(t,\eta) = (\phi(x,y),\psi(x,y))$ is the inverse of the relation $(x,y) = (f(t,\eta),g(t,\eta))$, where we assume that $\phi(x,y)$ and $\psi(x,y)$ are continuously differentiable with respect to (x,y) in a domain $\Delta_1 = \{(x,y) : |x| < r,\ |y| < \rho\}$. Set $H(x,y) = h(\phi(x,y),\psi(x,y))$. Show that the function $H(x,y)$ satisfies the partial differential equation

$$a(x,y)\frac{\partial H}{\partial x} + b(x,y)\frac{\partial H}{\partial y} = F(x,y,H)$$

and the initial-condition $H(0,y) = c(y)$.

Hint. Differentiate $H(x, y)$ with respect to (x, y).

II-10. Let $F(x, y, z, p, q)$ be a twice continuously differentiable function of (x, y, u, p, q) for $(x, y, u, p, q) \in \mathbb{R}^5$. Also, let

$$x = x(t, s), \qquad y = y(t, s), \qquad z = z(t, s), \qquad p = p(t, s), \qquad q = q(t, s)$$

be the solution of the following system:

$$\begin{cases} \dfrac{dx}{dt} = \dfrac{\partial F}{\partial p}(x, y, z, p, q), \\[2mm] \dfrac{dy}{dt} = \dfrac{\partial F}{\partial q}(x, y, z, p, q), \\[2mm] \dfrac{dz}{dt} = p\dfrac{\partial F}{\partial p}(x, y, z, p, q) + q\dfrac{\partial F}{\partial q}(x, y, z, p, q), \\[2mm] \dfrac{dp}{dt} = -\dfrac{\partial F}{\partial x}(x, y, z, p, q) - p\dfrac{\partial F}{\partial z}(x, y, z, p, q), \\[2mm] \dfrac{dq}{dt} = -\dfrac{\partial F}{\partial y}(x, y, z, p, q) - q\dfrac{\partial F}{\partial z}(x, y, z, p, q) \end{cases}$$

satisfying the initial condition

$$x(0, s) = x_0(s), \qquad y(0, s) = y_0(s), \qquad z(0, s) = z_0(s),$$
$$p(0, s) = p_0(s), \qquad q(0, s) = q_0(s),$$

where $x_0(s)$, $y_0(s)$, $z_0(s)$, $p_0(s)$, and $q_0(s)$ are differentiable functions of s on $\mathbb{R}$ such that

$$F(x_0(s), y_0(s), z_0(s), p_0(s), q_0(s)) = 0, \qquad \frac{dz_0}{ds}(s) = p_0(s)\frac{dx_0}{ds}(s) + q_0(s)\frac{dy_0}{ds}(s)$$

on $\mathbb{R}$. Show that

$$\begin{cases} F(x(t, s), y(t, s), z(t, s), p(t, s), q(t, s)) = 0, \\[2mm] \dfrac{\partial z}{\partial t}(t, s) = p(t, s)\dfrac{\partial x}{\partial t}(t, s) + q(t, s)\dfrac{\partial y}{\partial t}(t, s), \\[2mm] \dfrac{\partial z}{\partial s}(t, s) = p(t, s)\dfrac{\partial x}{\partial s}(t, s) + q(t, s)\dfrac{\partial y}{\partial s}(t, s) \end{cases}$$

as long as the solution (x, y, z, p, q) exists.

Comment. This is a traditional way of solving the partial differential equation $F(x, y, z, p, q) = 0$, where $p = \dfrac{\partial z}{\partial x}$ and $q = \dfrac{\partial z}{\partial y}$. For more details, see [Har2, pp. 131-143].

II-11. Let $H(t, x, y, p, q)$ be a twice continuously differentiable function of (t, x, y, p, q) in $\mathbb{R}^5$. Show that we can solve the partial differential equation

$$\frac{\partial z}{\partial t} + H\left(t, x, y, \frac{\partial z}{\partial x}, \frac{\partial z}{\partial y}\right) = 0$$

by using the system of ordinary differential equations

$$\begin{cases} \dfrac{dx}{dt} = \dfrac{\partial H}{\partial p}, & \dfrac{dy}{dt} = \dfrac{\partial H}{\partial q}, \\[2mm] \dfrac{dp}{dt} = -\dfrac{\partial H}{\partial x}, & \dfrac{dq}{dt} = -\dfrac{\partial H}{\partial y}, \\[2mm] \dfrac{dz}{dt} = u + p\dfrac{\partial H}{\partial p} + q\dfrac{\partial H}{\partial q}, & \dfrac{du}{dt} = -\dfrac{\partial H}{\partial t}. \end{cases}$$

CHAPTER III

NONUNIQUENESS

We consider, in this chapter, an initial-value problem

$$(P) \qquad \frac{d\vec{y}}{dt} = \vec{f}(t, \vec{y}), \qquad \vec{y}(\tau) = \vec{\eta}$$

without assuming the uniqueness of solutions. Some examples of nonuniqueness are given in §III-1. Topological properties of a set covered by solution curves of problem (P) are explained in §§III-2 and III-3. The main result is the Kneser theorem (Theorem III-2-4, cf. [Kn]). In §III-4, we explain maximal and minimal solutions and their continuity with respect to data. In §§III-5 and III-6, using differential inequalities, we derive a comparison theorem to estimate solutions of (P) and also some sufficient conditions for the uniqueness of solutions of (P). An application of the Kneser theorem to a second-order nonlinear boundary-value problem will be given in Chapter X (cf. §X-1).

III-1. Examples

In this section, four examples are given to illustrate the nonuniqueness of solutions of initial-value problems. As already known, problem (P) has the unique solution if $\vec{f}(t, \vec{y})$ satisfies a Lipschitz condition (cf. Theorem I-1-4). Therefore, in order to create *nonuniqueness*, $\vec{f}(t, \vec{y})$ must be chosen so that the Lipschitz condition is not satisfied.

Example III-1-1. The initial-value problem

$$(III.1.1) \qquad \frac{dy}{dt} = y^{1/3}, \qquad y(t_0) = 0$$

has at least three solutions

$$(S.1.1) \qquad y(t) = 0 \qquad (-\infty < t < +\infty),$$

$$(S.1.2) \qquad y(t) = \begin{cases} \left[\dfrac{2}{3}(t - t_0)\right]^{3/2}, & t \geq t_0, \\ 0, & t \leq t_0, \end{cases}$$

and

$$(S.1.3) \qquad y(t) = \begin{cases} -\left[\dfrac{2}{3}(t - t_0)\right]^{3/2}, & t \geq t_0, \\ 0, & t \leq t_0 \end{cases}$$

(cf. Figure 1). Actually, the region bounded by two *solution curves* (S.1.2) and
(S.1.3) is covered by solution curves of problem (III.1.1). Note that, in this case,
solution (S.1.1) is the unique solution of problem (P) for $t \leq t_0$. Solutions are not
unique only for $t \geq t_0$.

Example III-1-2. Consider a curve defined by

$$(\text{III.1.2}) \qquad\qquad y = \sin t \qquad (-\infty < t < +\infty)$$

and translate (III.1.2) along a straight line of slope 1. In other words, consider a
family of curves

$$(\text{III.1.3}) \qquad\qquad y = \sin(t - c) + c,$$

where c is a real parameter. By eliminating c from the relations

$$\frac{dy}{dt} = \cos(t - c), \qquad y - t = \sin(t - c) - (t - c),$$

we can derive the differential equation for family (III.1.3). In fact, since $\sin u - u$
is strictly decreasing, the relation $v = \sin u - u$ can be solved with respect to u
to obtain $u = G(v) - v$, where $G(v)$ is continuous and periodic of period 2π in v,
$G(2n\pi) = 0$ for every integer n, and $G(v)$ is differentiable except at $v = 2n\pi$ for
every integer n. The differential equation for family (III.1.3) is given by

$$(\text{III.1.4}) \qquad\qquad \frac{dy}{dt} = \cos[G(y - t) - (y - t)].$$

Since $G(2n\pi) = 0$ and $\cos(-2n\pi) = 1$ for every integer n, differential equation
(III.1.4) has *singular solutions* $y = t + 2n\pi$, where n is an arbitrary integer. These
lines are envelopes of family (III.1.3) (cf. Figure 2).

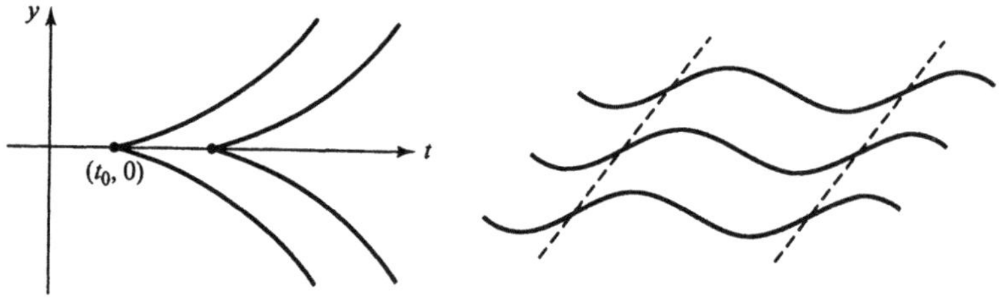

FIGURE 1.   FIGURE 2.

Example III-1-3. The initial-value problem

$$(\text{III.1.5}) \qquad\qquad \frac{dy}{dt} = \sqrt{|y|}, \qquad y(t_0) = 0$$

has at least two solutions

$$(\text{S.3.1}) \qquad\qquad y(t) = 0 \qquad (-\infty < t < +\infty),$$

and

(S.3.2)
$$y(t) = \begin{cases} \dfrac{1}{4}(t - t_0)^2, & t \geq t_0 , \\[2mm] -\dfrac{1}{4}(t - t_0)^2, & t \leq t_0 \end{cases}$$

(cf. Figure 3). The region bounded by two solution curves (S.3.1) and (S.3.2) is covered by solution curves of problem (III.1.5).

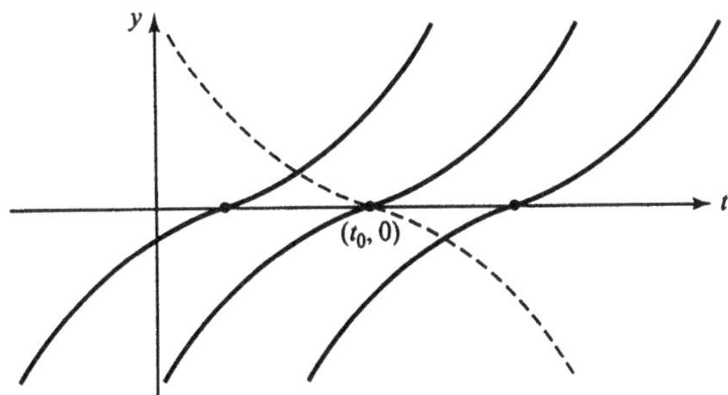

FIGURE 3.

Consider the following two perturbations of problem (III.1.5):

(III.1.6)
$$\frac{dy}{dt} = \sqrt{|y|} + \epsilon, \qquad y(t_0) = 0$$

and

(III.1.7)
$$\frac{dy}{dt} = \frac{y^2}{y^2 + \epsilon^2} \sqrt{|y|}, \qquad y(t_0) = 0,$$

where ϵ is a real positive parameter. Each of these two differential equations satisfies the Lipschitz condition. In particular, the unique solution of problem (III.1.6) is given by

(S.3.3)
$$y(t) = \begin{cases} \dfrac{1}{4}(t - t_0 + 2\sqrt{\epsilon})^2 - \epsilon, & t \geq t_0 , \\[2mm] -\dfrac{1}{4}(t - t_0 - 2\sqrt{\epsilon})^2 + \epsilon, & t \leq t_0 \end{cases}$$

(cf. Figure 4). On the other hand, (S.3.1) is the unique solution of problem (III.1.7). Figure 5 shows shapes of solution curves of differential equation (III.1.7). Note that nontrivial solution of (III.1.7) is an increasing function of t, but it does not reach $y = 0$ due to the uniqueness.

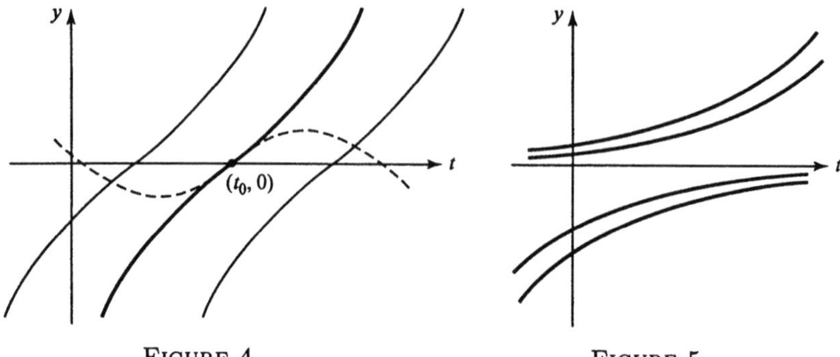

FIGURE 4.   FIGURE 5.

Generally speaking, starting from a differential equation which does not satisfy any uniqueness condition, we can create two drastically different families of curves by utilizing two different smooth perturbations. In other words, a differential equation without uniqueness condition can be regarded as a *branch point* in the space of differential equations (cf. [KS]).

Example III-1-4. The general solution of the differential equation

$$(III.1.8) \qquad \left(\frac{dy}{dt}\right)^2 + y^2 = 1$$

is given by

$$(III.1.9) \qquad y = \sin(t + c),$$

where c is a real arbitrary constant. Also, $y = 1$ and $y = -1$ are two singular solutions. Two solution curves (III.1.9) with two different values of c intersect each other. Hence, the uniqueness of solutions is violated (cf. Figure 6).

This phenomenon may be explained by observing that (III.1.8) actually consists of two differential equations:

$$(III.1.10) \qquad \frac{dy}{dt} = \sqrt{1 - y^2} \quad \text{and} \quad \frac{dy}{dt} = -\sqrt{1 - y^2}.$$

Each of these two differential equations satisfies the Lipschitz condition for $|y| < 1$. Figures 6-A and 6-B show solution curves of these two differential equations, respectively.

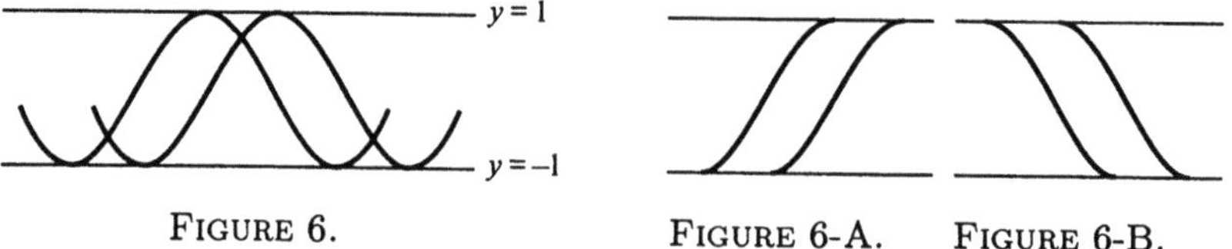

FIGURE 6.   FIGURE 6-A.   FIGURE 6-B.

Observe that each of these two pictures gives only a partial information of the complete picture (Figure 6).

We can regard differential equation (III.1.8) as a differential equation

$$(\text{III.1.11}) \qquad \frac{dy}{dt} = w$$

on the circle

$$(\text{III.1.12}) \qquad w^2 + y^2 = 1.$$

If circle (III.1.12) is parameterized as $y = \sin u$, $w = \cos u$, differential equation (III.1.11) becomes

$$(\text{III.1.13}) \qquad \frac{du}{dt} = 1 \quad \text{or} \quad \cos u = 0 \,.$$

Solution curves $u = t$ of (III.1.13) can be regarded as a curve on the cylinder

$$\{(t, y, w) : y = \sin u, \ w = \cos u, \ -\infty < u < +\infty \ (\text{mod } 2\pi), \ -\infty < t < +\infty\}$$

(cf. Figure 7). Figure 6 is the projection of this curve onto (t, y)-plane.

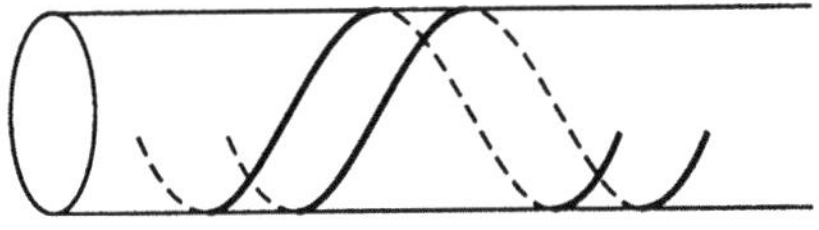

FIGURE 7.

In a case such as this example, a differential equation on a manifold would give a better explanation. To study a differential equation on a manifold, we generally use a covering of the manifold by open sets. We first study the differential equation on each open set (locally). Putting those local informations together, we obtain a global result. Each of Figures 6-A and 6-B is a local picture. If these two pictures are put together, the complete picture (Figure 6) is obtained.

III-2. The Kneser theorem

We consider a differential equation

$$(\text{III.2.1}) \qquad \frac{d\vec{y}}{dt} = \vec{f}(t, \vec{y})$$

under the assumption that the $\mathbb{R}^n$-valued function $\vec{f}$ is continuous and bounded on a region

$$(\text{III.2.2}) \qquad \Omega = \{(t, \vec{y}) : a \le t \le b \,, \ |\vec{y}| < +\infty \,\}.$$

Under this assumption, every solution of differential equation (III.2.1) exists on the interval $\mathcal{I}_0 = \{t : a \le t \le b\}$ if $(t_0, \vec{y}(t_0)) \in \Omega$ for some $t_0 \in \mathcal{I}_0$ (cf. Theorem I-3-2 and Corollary I-3-4). The main concern in this section is to investigate topological properties of a set which is covered by solution curves of differential equation (III.2.1).

Definition III-2-1. *For a subset $\mathcal{A}$ of the region Ω, we denote by $\mathcal{R}(\mathcal{A})$ the set of all points $(t, \vec{y}) \in \Omega$ such that $\vec{y} = \vec{\phi}(t)$ for some solution $\vec{\phi}$ of differential equation (III.2.1) which satisfies the condition $(t_0, \vec{\phi}(t_0)) \in \mathcal{A}$ for some $t_0 \in \mathcal{I}_0$. Also, we set $\mathcal{S}_c(\mathcal{A}) = \{\vec{y} : (c, \vec{y}) \in \mathcal{R}(\mathcal{A})\}$ for $c \in \mathcal{I}_0$ (cf. Figure 8).*

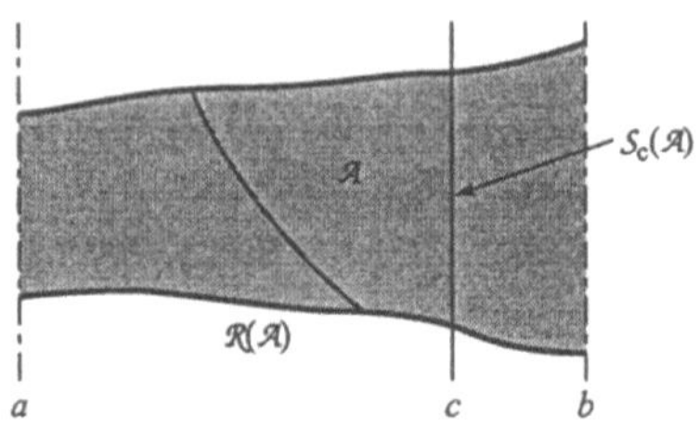

FIGURE 8.

Theorem III-2-2. *If the set $\mathcal{A}$ is closed, then $\mathcal{R}(\mathcal{A})$ is also closed.*

Proof.

Supposing that $\lim_{k \to +\infty} (t_k, \vec{\xi}_k) = (\tau, \vec{\xi})$ for some $(t_k, \vec{\xi}_k) \in \mathcal{R}(\mathcal{A})$ $(k = 1, 2, \ldots)$, we prove that $(\tau, \vec{\xi}) \in \mathcal{R}(\mathcal{A})$. To do this, we must find a solution $\vec{\phi}$ of (III.2.1) such that $\vec{\phi}(\tau) = \vec{\xi}$ and $(\tau_0, \vec{\phi}(\tau_0)) \in \mathcal{A}$ for some $\tau_0 \in \mathcal{I}_0$.

Since $(t_k, \vec{\xi}_k) \in \mathcal{R}(\mathcal{A})$, there exist solutions $\vec{\phi}_k$ of (III.2.1) and points $(\tau_k, \vec{\eta}_k) \in \mathcal{A}$ such that $\vec{\phi}_k(\tau_k) = \vec{\eta}_k$ and $\vec{\phi}_k(t_k) = \vec{\xi}_k$, where $k = 1, 2, \ldots$. It is easy to show that the family $\{\vec{\phi}_k : k = 1, 2, \ldots\}$ is bounded and equicontinuous on the interval $\mathcal{I}_0$ (cf. Definition I-2-1), since the sequence $\{(t_k, \vec{\xi}_k) : k = 1, 2, \ldots\}$ is bounded and the function $\vec{f}$ is also bounded on the region Ω. Hence, there exists a subsequence $\{\vec{\phi}_{k_j} : j = 1, 2, \ldots\}$ (cf. Lemma I-2-3) such that (i) $\lim_{j \to +\infty} k_j = +\infty$, (ii) $\lim_{j \to +\infty} (\tau_{k_j}, \vec{\eta}_{k_j}) = (\tau_0, \vec{\xi}_0)$ exists, and (iii) $\lim_{j \to +\infty} \vec{\phi}_{k_j}(t) = \vec{\phi}(t)$ exists uniformly on $\mathcal{I}_0$. Observe that $(\tau_0, \vec{\xi}_0) \in \mathcal{A}$ since $\mathcal{A}$ is closed and that

$$\vec{\phi}_k(t) = \vec{\xi}_k + \int_{t_k}^t \vec{f}(s, \vec{\phi}_k(s))ds = \vec{\eta}_k + \int_{\tau_k}^t \vec{f}(s, \vec{\phi}_k(s))ds.$$

Therefore,

$$\vec{\phi}(t) = \vec{\xi} + \int_{\tau}^t \vec{f}(s, \vec{\phi}(s))ds = \vec{\xi}_0 + \int_{\tau_0}^t \vec{f}(s, \vec{\phi}(s))ds.$$

Hence, $\vec{\phi}$ is a solution of (III.2.1) such that $\vec{\phi}(\tau) = \vec{\xi}$ and $\vec{\phi}(\tau_0) = \vec{\xi}_0$. $\quad\square$

From Definition III-2-1, the following result is easily derived.

Corollary III-2-3. *If $\mathcal{A}$ is closed, then $\mathcal{S}_c(\mathcal{A})$ is also closed for every $c \in \mathcal{I}_0$.*

Also, if $\mathcal{A}$ is bounded, then $\mathcal{R}(\mathcal{A})$ is bounded, and, hence, $\mathcal{S}_c(\mathcal{A})$ is bounded. This implies that if $\mathcal{A}$ is compact, then $\mathcal{R}(\mathcal{A})$ and $\mathcal{S}_c(\mathcal{A})$ are also compact. It is easy to show that if $\mathcal{A}$ is connected, then $\mathcal{R}(\mathcal{A})$ is connected. The connectedness of the set $\mathcal{S}_c(\mathcal{A})$ is the main claim of the following theorem due to H. Kneser.

Theorem III-2-4 ([Kn]).   *If $\mathcal{A}$ is compact and connected, then $\mathcal{S}_c(\mathcal{A})$ is also compact and connected for every $c \in \mathcal{I}_0$.*

Proof.

The compactness of $\mathcal{S}_c(\mathcal{A})$ was already explained. So, we prove the connectedness only.

Case 1. Suppose that $\mathcal{A}$ consists of a point $(\tau, \vec{\xi})$, where we assume without any loss of generality that $\tau < c$. A contradiction will be derived from the assumption that there exist two nonempty compact sets $\mathcal{F}_1$ and $\mathcal{F}_2$ such that

$$(\text{III.2.3}) \qquad \mathcal{S}_c(\mathcal{A}) = \mathcal{F}_1 \cup \mathcal{F}_2, \qquad \mathcal{F}_1 \cap \mathcal{F}_2 = \emptyset.$$

If $\vec{\xi}_1 \in \mathcal{F}_1$ and $\vec{\xi}_2 \in \mathcal{F}_2$, there exist two solutions $\vec{\phi}_1$ and $\vec{\phi}_2$ of (III.2.1) such that $\vec{\phi}_1(\tau) = \vec{\xi}$, $\vec{\phi}_1(c) = \vec{\xi}_1$, and $\vec{\phi}_2(\tau) = \vec{\xi}$, $\vec{\phi}_2(c) = \vec{\xi}_2$, (cf. Figure 9).
Set

$$\vec{h}(\mu) = \begin{cases} \vec{\phi}_1(\tau + \mu) & \text{for} \quad 0 \leq \mu \leq c - \tau, \\ \vec{\phi}_2(\tau + |\mu|) & \text{for} \quad -(c - \tau) \leq \mu \leq 0. \end{cases}$$

Let $\{\vec{f}_k(t, \vec{y}) : k = 1, 2, \ldots\}$ be a sequence of $\mathbb{R}^n$-valued functions such that
  (a) the functions $\vec{f}_k$ $(k = 1, 2, \ldots)$ are continuously differentiable on Ω,
  (b) $|\vec{f}_k(t, \vec{y})| \leq M$ for $(t, \vec{y}) \in \Omega$, where M is a positive number independent of $(t, \vec{y})$ and k,
  (c) $\displaystyle\lim_{k \to +\infty} \vec{f}_k = \vec{f}$ uniformly on each compact set in Ω

(cf. Lemma I-2-4). For each k, let $\vec{\psi}_k(t, \mu)$ be the unique solution of the initial-value problem $\dfrac{d\vec{y}}{dt} = \vec{f}_k(t, \vec{y})$, $\vec{y}(\tau + |\mu|) = \vec{h}(\mu)$. The solution $\vec{\psi}_k(t, \mu)$ is continuous for $t \in \mathcal{I}_0$ and $|\mu| \leq c - \tau$. It is easy to show that the family $\{\vec{\psi}_k(\cdot, \mu) : k = 1, 2, \ldots ; |\mu| \leq c - \tau\}$ is bounded and equicontinuous on the interval $\mathcal{I}_0$.

Note that the functions $\vec{\psi}_k(c, \mu)$ $(k = 1, 2, \ldots)$ are continuous in μ for $|\mu| \leq c - \tau$ and that $\vec{\psi}_k(c, c - \tau) = \vec{\phi}_1(c) \in \mathcal{F}_1$ and $\vec{\psi}_k(c, -(c - \tau)) = \vec{\phi}_2(c) \in \mathcal{F}_2$. Let d be the distance between two compact sets $\mathcal{F}_1$ and $\mathcal{F}_2$. Since $\vec{\psi}_k(c, \mu)$ is continuous in μ, there exists, for each k, a real number μ_k such that $|\mu_k| < c - \tau$ and $\text{distance}(\vec{\psi}_k(c, \mu_k), \mathcal{F}_1) = \dfrac{d}{2}$. Since the family $\{\vec{\psi}(\cdot, \mu_k) : k = 1, 2, \ldots\}$ is bounded and equicontinuous on the interval $\mathcal{I}_0$, there exists a subsequence $\{\vec{\psi}_{k_j}(\cdot, \mu_{k_j}) : j = 1, 2, \ldots\}$ such that (i) $\displaystyle\lim_{j \to +\infty} k_j = +\infty$, (ii) $\displaystyle\lim_{j \to +\infty} \mu_{k_j} = \mu_0$ exists, and (iii) $\displaystyle\lim_{j \to +\infty} \vec{\psi}_{k_j}(t, \mu_{k_j}) = \vec{\phi}(t)$ exists uniformly on $\mathcal{I}_0$. It is easy to show that

$$\begin{cases} \vec{\phi}(t) = \vec{h}(\mu_0) + \displaystyle\int_{\tau + |\mu_0|}^{t} \vec{f}(s, \vec{\phi}(s))ds, \\[2mm] |\mu_0| \leq c - \tau, \qquad \text{distance}(\vec{\phi}(c), \mathcal{F}_1) = \dfrac{d}{2}. \end{cases}$$

Hence, $\vec{\phi}(c) \in \mathcal{S}_c(\mathcal{A})$ but $\vec{\phi}(c) \notin \mathcal{F}_1 \cup \mathcal{F}_2$. This is a contradiction (cf. Figure 10).

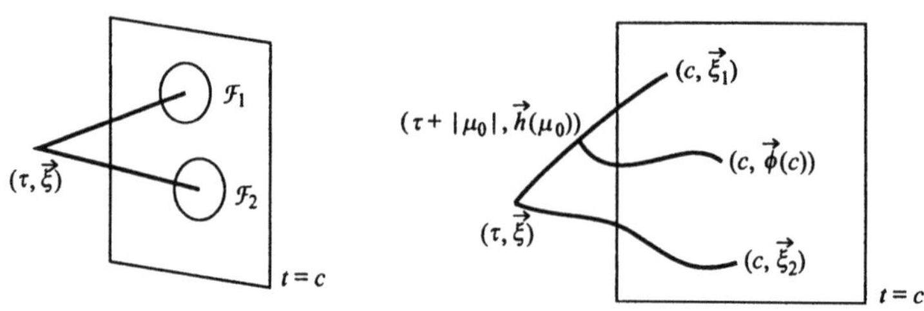

FIGURE 9.   FIGURE 10.

Case 2 (general case). Assume (III.2.3) as in Case 1, set $\mathcal{A}_1 = \mathcal{A} \cap \mathcal{R}(\{c\} \times \mathcal{F}_1)$ and $\mathcal{A}_2 = \mathcal{A} \cap \mathcal{R}(\{c\} \times \mathcal{F}_2)$, where $\{c\} \times \mathcal{F}_j = \{(c, \vec{y}) : \vec{y} \in \mathcal{F}_j\}$ $(j = 1, 2)$. Then, the two sets $\mathcal{A}_j$ $(j = 1, 2)$ are compact and not empty. Note that $\mathcal{A} = \mathcal{A}_1 \cup \mathcal{A}_2$. Since $\mathcal{A}$ is connected, we must have $\mathcal{A}_1 \cap \mathcal{A}_2 \neq \emptyset$. Choose a point $(\tau, \vec{\xi}) \in \mathcal{A}_1 \cap \mathcal{A}_2$. Then, $\mathcal{S}_c((\tau, \vec{\xi})) = \left\{\mathcal{S}_c((\tau, \vec{\xi})) \cap \mathcal{F}_1\right\} \cup \left\{\mathcal{S}_c((\tau, \vec{\xi})) \cap \mathcal{F}_2\right\}$ and $\mathcal{S}_c((\tau, \vec{\xi})) \cap \mathcal{F}_j \neq \emptyset$ $(j = 1, 2)$. This is a contradiction (cf. Case 1). $\square$

In order to apply the Kneser theorem (i.e., Theorem III-2-4), it is desirable to remove the boundedness of $\vec{f}$ from the assumption. To obtain such a refinement of Theorem III-2-4, consider differential equation (III.2.1) under the following assumptions.

Assumption 1. *A set $\mathcal{A}_0$ is a compact and connected subset of the region Ω such that if $\vec{\phi}(t)$ is a solution of (III.2.1) satisfying*

$$(III.2.4) \qquad\qquad (t_0, \vec{\phi}(t_0)) \in \mathcal{A}_0 \quad \text{for some} \quad t_0 \in \mathcal{I}_0,$$

then $\vec{\phi}(t)$ exists on $\mathcal{I}_0$.

As in Definition III-2-1, denote by $\mathcal{R}_0$ the set of all points $(t, \vec{y}) \in \Omega$ such that $\vec{y} = \vec{\phi}(t)$ for some solution $\vec{\phi}$ of differential equation (III.2.1) satisfying condition (III.2.4). Also, set $\mathcal{S}_c = \{\vec{y} : (c, \vec{y}) \in \mathcal{R}_0\}$ for $c \in \mathcal{I}_0$.

Assumption 2. *The set $\mathcal{R}_0$ is bounded.*

Now, we prove the following theorem.

Theorem III-2-5. *If a compact and connected subset $\mathcal{A}_0$ of Ω satisfies Assumptions 1 and 2, then the set $\mathcal{S}_c$ is also compact and connected for every $c \in \mathcal{I}_0$.*

Proof.

Since $\mathcal{R}_0$ is bounded, there exists a positive number M such that

$$\mathcal{R}_0 \subset \{(t, \vec{y}) : t \in \mathcal{I}_0, \ |\vec{y}| \leq M\}.$$

Set

$$\vec{g}(t, \vec{y}) = \begin{cases} \vec{f}(t, \vec{y}) & \text{if} \quad t \in \mathcal{I}_0, \ |\vec{y}| \leq 2M, \\[2mm] \vec{f}\left(t, \dfrac{2M}{|\vec{y}|}\vec{y}\right) & \text{if} \quad t \in \mathcal{I}_0, \ |\vec{y}| \geq 2M. \end{cases}$$

Then, $\vec{g}(t, \vec{y})$ is continuous and bounded on Ω. Using the differential equation $\dfrac{d\vec{y}}{dt} = \vec{g}(t, \vec{y})$, define $\mathcal{R}(\mathcal{A}_0)$ and $\mathcal{S}_c(\mathcal{A}_0)$ by Definition III-2-1. Then, we must have $\mathcal{R}_0 = \mathcal{R}(\mathcal{A}_0)$. Otherwise, there would exist a solution $\vec{\phi}_0(t)$ of differential equation (III.2.1) such that $(t_0, \vec{\phi}_0(t_0)) \in \mathcal{A}_0$ for $t_0 \in \mathcal{I}_0$ and $(t_1, \vec{\phi}_0(t_1)) \notin \mathcal{R}_0$ for $t_1 \in \mathcal{I}_0$. This is a contradiction. The theorem follows from Theorem III-2-4. $\quad\square$

Remark III-2-6. Any kind of shapes can be made by a wire. Therefore, the set $S_c(A)$ needs not to be a convex set even if A is convex.

Using Theorem III-2-4, we can prove the following theorem, which is a refinement of Theorem II-1-4.

Theorem III-2-7. *Assume that the entries of an $\mathbb{R}^n$-valued function $\vec{f}(t, \vec{y})$ are continuous in $(t, \vec{y})$ in a domain $\mathcal{D}_0 \in \mathbb{R}^{n+1}$. Also, let $\vec{\phi}(t)$ be a solution of system (III.2.1) on an interval $a \leq t \leq b$. Assume that $(t, \vec{\phi}(t)) \in \mathcal{D}_0$ on the interval $a \leq t \leq b$. Then, for any given positive number ϵ, there exists another solution $\vec{\psi}(t)$ of (III.2.1) such that*

(i) $(t, \vec{\psi}(t)) \in \mathcal{D}_0$ on the interval $a \leq t \leq b$,

(ii) $|\vec{\phi}(t) - \vec{\psi}(t)| \leq \epsilon$ on the interval $a \leq t \leq b$,

(iii) $\vec{\psi}(\tau) \neq \vec{\phi}(\tau)$ for some τ on the interval $a \leq t \leq b$.

Proof.

Using an idea similar to Step 2 of the proof of Theorem II-1-2, the local existence theorem (Theorem I-2-5), and the connectedness of $S_c(A)$, we can complete the proof of Theorem III-2-7. In fact, subdivide the interval $a \leq t \leq b$ (i.e., $a = t_0 < t_1 < \cdots < t_k = b$) in such a way that $|\vec{y}(t) - \vec{\phi}(t)| \leq \epsilon$ for $t_j \leq t \leq t_{j+1}$ if $\vec{y}(t)$ satisfies (III.2.1) and $|\vec{y}(t_j) - \vec{\phi}(t_j)| \leq \delta$. Set $A_j = \{\vec{c} \in \mathbb{R}^n : |\vec{c} - \vec{\phi}(t_j)| \leq \delta\}$ and $B_1 = \{\vec{y}(t_1) : \vec{y}(t_0) \in A_0\}$, where $\vec{y}(t)$ denotes any solution of (III.2.1) with initial-value $\vec{y}(t_0)$. Since B_1 is connected and contains $\vec{\phi}(t_1)$, we must have $B_1 = \{\vec{\phi}(t_1)\}$ if $B_1 \cap A_1$ contains only $\vec{\phi}(t_1)$. In such a case, the proof is finished.

If $B_1 \cap A_1$ contains more than one point, then $B_1 \cap A_1$ contains a connected set S_1 containing $\phi(t_1)$ and more than one point. Set $B_2 = \{\vec{y}(t_2) : \vec{y}(t_1) \in S_1\}$. Since B_2 is connected and contains $\vec{\phi}(t_2)$, we must have $B_2 = \{\vec{\phi}(t_2)\}$ if $B_2 \cap A_2$ contains only $\vec{\phi}(t_2)$. In such a case, the proof is finished. In this way, the proof is completed in a finite number of steps. $\quad\square$

III-3. Solution curves on the boundary of $\mathcal{R}(\mathcal{A})$

We still consider a differential equation

$$(\text{III.3.1}) \qquad\qquad \frac{d\vec{y}}{dt} = \vec{f}(t, \vec{y})$$

under the assumption that the entries of the $\mathbb{R}^n$-valued function $\vec{f}$ are continuous and bounded on a region

$$(\text{III.3.2}) \qquad\qquad \Omega = \{(t, \vec{y}) : a \leq t \leq b\, , \ |\vec{y}| < +\infty\}.$$

As mentioned in §III-2, under this assumption, every solution of differential equation (III.3.1) exists on the interval $\mathcal{I}_0 = \{t : a \leq t \leq b\}$ if $(t_0, \vec{y}(t_0)) \in \Omega$ for some $t_0 \in \mathcal{I}_0$. Define the sets $\mathcal{R}(\mathcal{A})$ and $\mathcal{S}_c(\mathcal{A})$ for a subset $\mathcal{A}$ of Ω by Definition III-2-1. The main concern of this section is to prove the existence of solution curves on the boundary of $\mathcal{R}(\mathcal{A})$. We start with the following basic lemma.

Lemma III-3-1. *Suppose that (i) the set $\mathcal{A}$ consists of one point $(c_1, \vec{\xi})$ (i.e., $\mathcal{A} = \{(c_1, \vec{\xi})\}$), (ii) $a \leq c_1 < c_0 < c_2 \leq b$, and (iii) $\vec{\eta}$ is on the boundary of $\mathcal{S}_{c_2}(\mathcal{A})$. Let $\mathcal{B} = \{(c_2, \vec{\eta})\}$. Then, $\mathcal{S}_{c_0}(\mathcal{B})$ contains at least a boundary point of $\mathcal{S}_{c_0}(\mathcal{A})$.*

Proof.

A contradiction will be derived from the assumption that $\mathcal{S}_{c_0}(\mathcal{B})$ does not contain any boundary points of $\mathcal{S}_{c_0}(\mathcal{A})$. Note that $\mathcal{S}_{c_0}(\mathcal{B}) \cap \mathcal{S}_{c_0}(\mathcal{A}) \neq \emptyset$. Set $\mathcal{S}_1 = \mathcal{S}_{c_0}(\mathcal{B}) \cap \mathcal{S}_{c_0}(\mathcal{A})$, $\mathcal{S}_2 = \{\vec{y} : \vec{y} \in \mathcal{S}_{c_0}(\mathcal{B}) \text{ and } \vec{y} \notin \mathcal{S}_{c_0}(\mathcal{A})\}$. Then, $\mathcal{S}_{c_0}(\mathcal{B}) = \mathcal{S}_1 \cup \mathcal{S}_2$ and $\mathcal{S}_1 \cap \mathcal{S}_2 = \emptyset$. It is known that $\mathcal{S}_1$ is a nonempty compact set. Also, $\mathcal{S}_2$ is closed and bounded, since $\mathcal{S}_{c_0}(\mathcal{B})$ is compact and does not contain any boundary points of $\mathcal{S}_{c_0}(\mathcal{A})$. If $\mathcal{S}_2$ is not empty, the set $\mathcal{S}_{c_0}(\mathcal{B})$ contains a point in $\mathcal{S}_{c_0}(\mathcal{A})$ and another point which does not belong to $\mathcal{S}_{c_0}(\mathcal{A})$. Then, $\mathcal{S}_{c_0}(\mathcal{B})$ contains a boundary point of $\mathcal{S}_{c_0}(\mathcal{A})$, since $\mathcal{S}_{c_0}(\mathcal{B})$ is connected. This is a contradiction. Therefore, if it is proved that $\mathcal{S}_2$ is not empty, the proof of Lemma III-3-1 will be completed (cf. Figure 11).

To prove that $\mathcal{S}_2$ is not empty, observe first that $\mathcal{S}_{c_0}(\{(c_2, \vec{\zeta})\}) \cap \mathcal{S}_{c_0}(\mathcal{A}) = \emptyset$ if $\vec{\zeta} \notin \mathcal{S}_{c_2}(\mathcal{A})$ (cf. Figure 12).

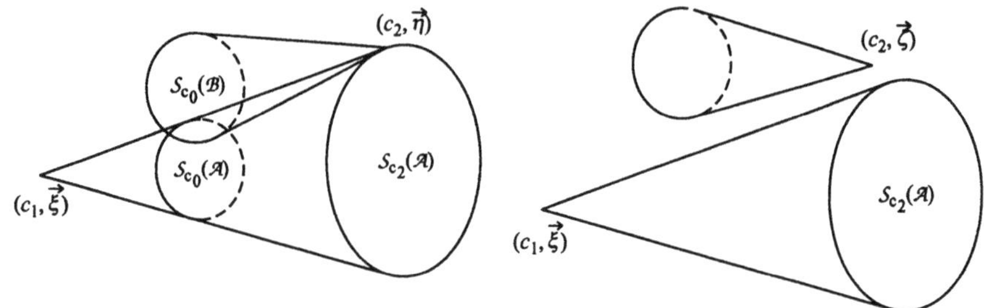

FIGURE 11.   FIGURE 12.

Let $\vec{\eta}$ be on the boundary of $\mathcal{S}_{c_2}(\mathcal{A})$. Then, there exists a sequence of points $\{\vec{\zeta}_k \notin \mathcal{S}_{c_2}(\mathcal{A}) : k = 1, 2, \ldots\}$ and a sequence $\{\vec{\phi}_k : k = 1, 2, \ldots\}$ of solutions of (III.3.1) such that $\lim_{k \to +\infty} \vec{\zeta}_k = \vec{\eta}$, $\vec{\phi}_k(c_2) = \vec{\zeta}_k$, and $\vec{\phi}_k(c_0) \notin \mathcal{S}_{c_0}(\mathcal{A})$. The family $\{\vec{\phi}_k : k = 1, 2, \ldots\}$ is bounded and equicontinuous on the interval $\mathcal{I}_0$. Therefore, there exists a subsequence $\{\vec{\phi}_{k_j} : j = 1, 2, \ldots\}$ such that $\lim_{j \to +\infty} k_j = +\infty$ and $\lim_{j \to +\infty} \vec{\phi}_{k_j}(t) = \vec{\phi}(t)$ exists uniformly on $\mathcal{I}_0$. The limit function $\vec{\phi}$ is a solution of (III.3.1) such that $\vec{\phi}(c_2) = \vec{\eta}$. Hence, $\vec{\phi}(c_0) \in \mathcal{S}_{c_0}(\mathcal{B})$. Since $\mathcal{S}_{c_0}(\mathcal{B})$ does not contain any boundary points of $\mathcal{S}_{c_0}(\mathcal{A})$, we must have $\vec{\phi}(c_0) \notin \mathcal{S}_{c_0}(\mathcal{A})$. This implies that $\mathcal{S}_2$ is not empty. $\square$

The following theorem is the main result in this section which is due to M. Hukuhara [Huk1] (see also [Huk2] and [HN3]).

Theorem III-3-2. *Suppose that*
(1) $a \leq c_1 < c_2 \leq b$,
(2) $\vec{\eta}$ *is on the boundary of* $\mathcal{S}_{c_2}(\{(c_1, \vec{\xi})\})$.
Then, there exists a solution $\vec{\phi}$ of differential equation (III.3.1) such that
(i) $\vec{\phi}(c_1) = \vec{\xi}$, $\vec{\phi}(c_2) = \vec{\eta}$,
(ii) $(t, \vec{\phi}(t))$ *is on the boundary of* $\mathcal{R}(\{(c_1, \vec{\xi})\})$ *for* $c_1 \leq t \leq c_2$.

Proof.

For a subdivision $\Delta : c_1 = \tau_0 < \tau_1 < \cdots < \tau_{m-1} < \tau_m = c_2$ of the interval $c_1 \leq t \leq c_2$, there exists a solution $\vec{\phi}_\Delta$ of (III.3.1) such that
(α) $\vec{\phi}_\Delta(c_1) = \vec{\xi}$, $\vec{\phi}_\Delta(c_2) = \vec{\eta}$,
(β) $\left(\tau_\ell, \vec{\phi}_\Delta(\tau_\ell)\right)$ is on the boundary of $\mathcal{R}(\{(c_1, \vec{\xi})\})$, where $\ell = 1, 2, \ldots, m-1$.
(Cf. Lemma III-3-1 and Figure 13.)

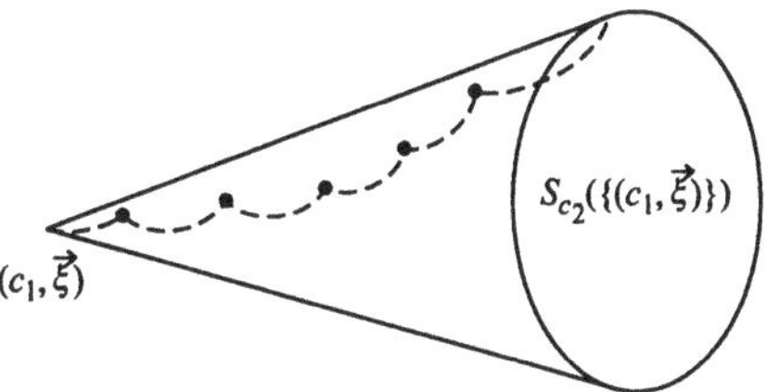

FIGURE 13.

Choose a sequence $\{\Delta_k : k = 1, 2, \ldots\}$ of subdivisions of the interval $c_1 \leq t \leq c_2$ such that

$$\begin{cases} \Delta_k : c_1 = \tau_{k,0} < \tau_{k,1} < \cdots < \tau_{k,2^k - 1} < \tau_{k,2^k} = c_2, \\ \tau_{k,\ell} = c_1 + \dfrac{\ell}{2^k}(c_2 - c_1), \qquad \ell = 0, 1, \ldots, 2^k, \quad k = 1, 2, \ldots. \end{cases}$$

Since $\tau_{k+1,2\ell} = \tau_{k,\ell}$, we have $\{\tau_{k,\ell} : \ell = 0, 1, 2, \ldots, 2^k\} \subset \{\tau_{k+1,\ell} : \ell = 0, 1, 2, \ldots, 2^{k+1}\}$. Set $\vec{\phi}_k = \vec{\phi}_{\Delta_k}$ $(k = 1, 2, \ldots)$. Then,
(α_k) $\vec{\phi}_k(c_1) = \vec{\xi}$, $\vec{\phi}_k(c_2) = \vec{\eta}$,
(β_k) $\left(\tau_{k,\ell}, \vec{\phi}_k(\tau_{k,\ell})\right)$ is on the boundary of $\mathcal{R}(\{(c_1, \vec{\xi})\})$, where $\ell = 1, 2, \ldots, 2^k - 1$.
The family $\{\vec{\phi}_k : k = 1, 2, \ldots\}$ is bounded and equicontinuous on $\mathcal{I}_0$. Hence, there exists a subsequence $\{\vec{\phi}_{k_j} : j = 1, 2, \ldots\}$ such that $\lim_{j \to +\infty} k_j = +\infty$ and that $\lim_{j \to +\infty} \vec{\phi}_{k_j} = \vec{\phi}$ exists uniformly on $\mathcal{I}_0$. Then, $\vec{\phi}$ is a solution of (III.3.1) such that
(α_0) $\vec{\phi}(c_1) = \vec{\xi}$, $\vec{\phi}(c_2) = \vec{\eta}$,
(β_0) $\left(\tau_{k,\ell}, \vec{\phi}_k(\tau_{k,\ell})\right)$ is on the boundary of $\mathcal{R}(\{(c_1, \vec{\xi})\})$, where $\ell = 1, 2, \ldots, 2^k - 1$ and $k = 1, 2, \ldots$.
Since the set $\{\tau_{k,\ell} : \ell = 1, 2, \ldots, 2^k - 1; k = 1, 2, \ldots\}$ is dense on the interval $c_1 \leq t \leq c_2$, the curve $(t, \vec{\phi}(t))$ is on the boundary of $\mathcal{R}(\{(c_1, \vec{\xi})\})$ for $c_1 \leq t \leq c_2$. $\square$

Remark III-3-3.   In general, the conclusion of Theorem III-3-2 does not hold on an interval larger than $c_1 \leq t \leq c_2$.

III-4. Maximal and minimal solutions

Consider a scalar differential equation

$$\text{(III.4.1)} \qquad \frac{dy}{dt} = f(t, y),$$

where t and y are real variables, and assume that f is real-valued, bounded, and continuous on a region $\Omega = \{(t, y) : a < t < b, -\infty < y < +\infty\}$. Choose c in the interval $\mathcal{I} = \{t : a < t < b\}$. Then, $\mathcal{S}_\tau(\{(c, \eta)\})$ is compact and connected for every $\tau \in \mathcal{I}$ and every real number η (cf. Theorem III-2-4). This implies that there exist two numbers $\phi_1(\tau)$ and $\phi_2(\tau)$ such that $\mathcal{S}_\tau(\{(c, \eta)\}) = \{ y \in \mathbb{R} : \phi_1(\tau) \leq y \leq \phi_2(\tau)\}$. Hence, $\mathcal{R}(\{(c, \eta)\}) = \{(t, y) \in \mathbb{R}^2 : \phi_1(t) \leq y \leq \phi_2(t), a < t < b\}$ (cf. Figure 14). The two boundary curves $(t, \phi_1(t))$ and $(t, \phi_2(t))$ of $\mathcal{R}(\{(c, \eta)\})$ are solution curves of differential equation (III.4.1) (cf. Theorem III-3-2). Every solution $\phi(t)$ of (III.4.1) such that $\phi(c) = \eta$ satisfies the inequalities $\phi_1(t) \leq \phi(t) \leq \phi_2(t)$ on $\mathcal{I}$. The solution $\phi_2(t)$ (respectively $\phi_1(t)$) is called the maximal (respectively minimal) solution of the initial-value problem

$$\text{(III.4.2)} \qquad \frac{dy}{dt} = f(t, y), \qquad y(c) = \eta$$

on the interval $\mathcal{I}$. In this section, we explain the basic properties of the maximal and minimal solutions. Before we define the maximal and minimal solutions more precisely, let us make some observations.

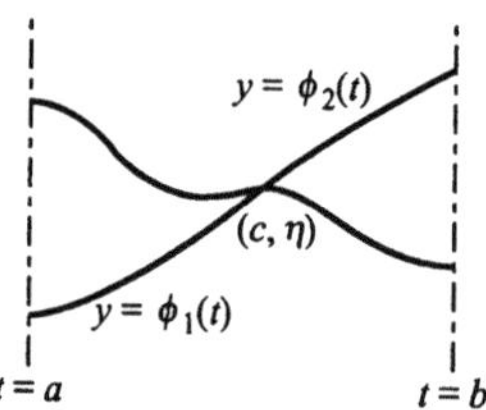

FIGURE 14.

Observation III-4-1.   Assume that $f(t, y)$ is real-valued and continuous on a domain $\mathcal{D}$ in the (t, y)-plane. Set $\mathcal{I}_0 = \{t : a \leq t \leq b\}$. Let $\phi_1(t)$ and $\phi_2(t)$ be two solutions of differential equation (III.4.1) such that $(t, \phi_1(t)) \in \mathcal{D}$ and $(t, \phi_2(t)) \in \mathcal{D}$ for $t \in \mathcal{I}_0$. Note that we do not assume boundedness of f on $\mathcal{D}$. Set $\phi(t) = \max\{\phi_1(t), \phi_2(t)\}$ for $t \in \mathcal{I}_0$. Then, $\phi(t)$ is also a solution of (III.4.1) on the interval $\mathcal{I}_0$.

Proof.

For a fixed $t_0 \in \mathcal{I}_0$, we prove that

$$\text{(III.4.3)} \qquad \lim_{t \to t_0} \frac{\phi(t) - \phi(t_0)}{t - t_0} = f(t_0, \phi(t_0)).$$

If $\phi_1(t_0) > \phi_2(t_0)$, then there exists a positive number δ such that $\phi(t) = \phi_1(t)$ for $|t - t_0| \leq \delta$. Therefore, (III.4.3) holds. Similarly, (III.4.3) holds if $\phi_2(t_0) > \phi_1(t_0)$. Hence, we consider only the case when $\phi(t_0) = \phi_1(t_0) = \phi_2(t_0)$. In this case, for each fixed $t \in \mathcal{I}_0$, we have $\dfrac{\phi(t) - \phi(t_0)}{t - t_0} = \dfrac{\phi_j(t) - \phi_j(t_0)}{t - t_0}$, where $j = 1$ or $j = 2$, depending on t. Hence, by the Mean Value Theorem on $\phi_j(t)$, there exists $\tau \in \mathcal{I}_0$ such that $\tau \to t_0$ as $t \to t_0$ and $\dfrac{\phi(t) - \phi(t_0)}{t - t_0} = f(\tau, \phi_j(\tau))$. Also, $\phi_j(\tau) = \phi_j(t_0) + f(\sigma, \phi_j(\sigma))(\tau - t_0)$ for some $\sigma \in \mathcal{I}_0$ such that $\sigma \to t_0$ as $\tau \to t_0$. Since f is bounded on the two curves $(t, \phi_1(t))$ and $(t, \phi_2(t))$, (III.4.3) follows immediately. $\square$

Observation III-4-2. Let $f(t, y)$, $\mathcal{D}$, and $\mathcal{I}_0$ be the same as in Observation III-4-1. Consider a set $\mathcal{F}$ of solutions of differential equation (III.4.1) such that $(t, \phi(t)) \in \mathcal{D}$ on $\mathcal{I}_0$ for every $\phi \in \mathcal{F}$. Assuming that there exists a real number K such that $\phi(t) \leq K$ on $\mathcal{I}_0$ for every $\phi \in \mathcal{F}$, set $\phi_0(t) = \sup\{\phi(t) : \phi \in \mathcal{F}\}$ for $t \in \mathcal{I}_0$. Assume also that $(t, \phi_0(t)) \in \mathcal{D}$ on $\mathcal{I}_0$. Then, $\phi_0(t)$ is a solution of differential equation (III.4.1) on $\mathcal{I}_0$.

Proof.

As in Observation III-4-1, we prove that

$$\text{(III.4.4)} \qquad \lim_{t \to t_0} \frac{\phi_0(t) - \phi_0(t_0)}{t - t_0} = f(t_0, \phi_0(t_0)).$$

for each fixed $t_0 \in \mathcal{I}_0$. Choose three positive numbers ρ_0, ρ, and δ so that
(i) $\Delta = \{(t, y) : t \in \mathcal{I}_0, |y - \phi_0(t)| \leq \rho_0\} \subset \mathcal{D}$,
(ii) we have $(t, \phi(t)) \in \Delta$ on the interval $|t - t_0| \leq \delta$, if $\phi(t)$ is a solution of (III.4.1) such that $0 \leq \phi_0(\tau) - \phi(\tau) \leq \rho$ for some τ in the interval $|t - t_0| \leq \delta$.
There exists a positive number M such that $|f(t, y)| \leq M$ on Δ.

Let us fix a point τ on the interval $|t - t_0| \leq \delta$. First, we prove the existence of a solution $\psi(t; \tau)$ of (III.4.1) such that

$$\text{(III.4.5)} \qquad \psi(\tau; \tau) = \phi_0(\tau) \quad \text{and} \quad \psi(t; \tau) \leq \phi_0(t) \quad \text{for} \quad |t - t_0| \leq \delta.$$

To do this, select a sequence $\{\phi_k : k = 1, 2, \dots\}$ from the family $\mathcal{F}$ so that $\lim_{k \to +\infty} \phi_k(\tau) = \phi_0(\tau)$. We may assume that $(t, \phi_k(t)) \in \Delta$ on $|t - t_0| \leq \delta$ for all k (cf. (ii) above). Then, the sequence $\{\phi_k : k = 1, 2, \dots\}$ is bounded and equicontinuous on $|t - t_0| \leq \delta$. Hence, we may assume that $\lim_{k \to +\infty} \phi_k(t) = \psi(t; \tau)$ exists uniformly on the interval $|t - t_0| \leq \delta$. It is easy to show that $\psi(t; \tau)$ is a solution of (III.4.1) and that (III.4.5) is satisfied.

Set $\psi(t) = \max\{\psi(t; \tau), \psi(t; t_0)\}$ for $|t - t_0| \leq \delta$. Then, ψ is a solution of (III.4.1) such that (1) $\psi(\tau) = \phi_0(\tau)$ and $\psi(t_0) = \phi_0(t_0)$, (2) $\psi(t) \leq \phi_0(t)$ for $|t - t_0| \leq \delta$, and (3) $(t, \psi(t)) \in \Delta$ for $|t - t_0| \leq \delta$ (cf. Figure 15).

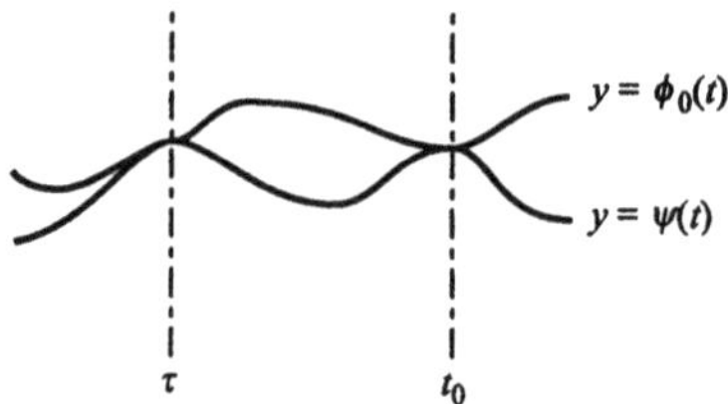

$$\textsc{Figure 15.}$$

Using this solution $\psi(t)$, we obtain $\dfrac{\phi_0(\tau) - \phi_0(t_0)}{\tau - t_0} = \dfrac{\psi(\tau) - \psi(t_0)}{\tau - t_0} = f(\sigma, \psi(\sigma))$ for some σ such that $|\sigma - t_0| \leq \delta$ and that $\sigma \to t_0$ as $\tau \to t_0$. Since $|\psi(\sigma) - \psi(t_0)| \leq M|\sigma - t_0|$, (III.4.4) can be derived immediately. $\square$

Let us define the maximal (respectively minimal) solution of an initial-value problem

$$(\text{III.4.6–}\xi) \qquad\qquad \frac{dy}{dt} = f(t, y), \qquad y(\tau) = \xi,$$

where f is real-valued and continuous on a domain $\mathcal{D}$ in the (t, y)-plane and the initial point (τ, ξ) is fixed in $\mathcal{D}$.

Definition III-4-3. *A solution $\psi(t)$ of problem (III.4.6–ξ) is called the* maximal *(respectively* minimal*) solution of problem (III.4.6–ξ) on an interval $\mathcal{I} = \{t : \tau \leq t \leq \tau'\}$ if*

(a) $\psi(t)$ is defined on $\mathcal{I}$ and $(t, \psi(t)) \in \mathcal{D}$ on $\mathcal{I}$,

(b) if $\phi(t)$ is a solution of problem (III.4.6–ξ) on a subinterval $\tau \leq t \leq \tau''$ of $\mathcal{I}$, then

$$\phi(t) \leq \psi(t) \quad (\textit{respectively } \phi(t) \geq \psi(t)) \qquad \textit{on} \quad \tau \leq t \leq \tau''.$$

The following two theorems are stated in terms of maximal solutions. Similar results can be stated also in terms of minimal solutions. Such details are left to the reader as an exercise. In the first of the two theorems, we consider another initial-value problem

$$(\text{III.4.6–}\eta) \qquad\qquad \frac{dy}{dt} = f(t, y), \qquad y(\tau) = \eta$$

together with problem (III.4.6–ξ).

Theorem III-4-4. *Suppose that the maximal solution ψ_ξ of problem (III.4.6-ξ) exists on an interval $\mathcal{I} = \{t : \tau \leq t \leq \tau'\}$. Then, there exists a positive number δ_0 such that the maximal solution ψ_η of problem (III.4.6-η) exists on $\mathcal{I}$ for $\xi \leq \eta \leq \xi + \delta_0$. Furthermore, $\psi_\xi(t) \leq \psi_\eta(t)$ on $\mathcal{I}$ for $\xi \leq \eta \leq \xi + \delta_0$ and $\lim_{\eta \to \xi} \psi_\eta(t) = \psi_\xi(t)$ uniformly on $\mathcal{I}$.*

Proof.

Set $\Delta(\rho) = \{(t, y) : t \in \mathcal{I}, \psi_\xi(t) \le y \le \psi_\xi(t) + \rho\}$. For a sufficiently small positive number ρ, we have $\Delta(\rho) \subset \mathcal{D}$, and, hence, $|f(t, y)|$ is bounded on $\Delta(\rho)$ for a sufficiently small positive number ρ.

First, we prove that for a given $\rho > 0$, there exists a positive number δ such that, if $\xi \le \eta \le \xi + \delta$, every solution $\phi(t)$ of problem (III.4.6-η) defined on a subinterval $\tau \le t \le \tau''$ of $\mathcal{I}$ satisfies

$$(\text{III.4.7}) \qquad \phi(t) \; < \; \psi_\xi(t) + \rho \quad \text{for} \quad \tau \le t \le \tau''.$$

Otherwise, there exist two sequences $\{\eta_k : k = 1, 2, \dots\}$ and $\{\tau_k : k = 1, 2, \dots\}$ of real numbers such that (1) $\eta_k > \xi$, $\lim\limits_{k \to +\infty} \eta_k = \xi$, and $\tau < \tau_k \le \tau'$, (2) $\phi_k(\tau) = \eta_k$ and $\phi_k(\tau_k) = \psi_\xi(\tau_k) + \rho$, and (3) $\phi_k(t) < \psi_\xi(t) + \rho$ for $\tau \le t < \tau_k$. Furthermore, there exists a real number $\tau(\rho)$ such that $\tau_k \ge \tau(\rho) > \tau$ (cf. Figure 16).

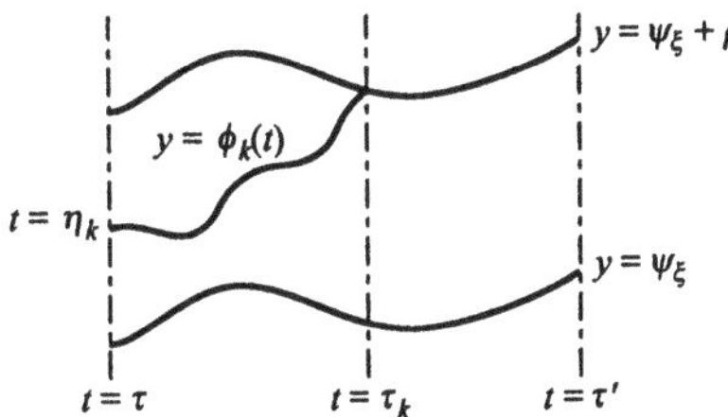

FIGURE 16.

Set
$$\psi_k(t) \;=\; \begin{cases} \max\left(\phi_k(t),\, \psi_\xi(t)\right), & \tau \le t \le \tau_k, \\ \psi_\xi(t) + \rho, & \tau_k \le t \le \tau'. \end{cases}$$

It is easy to show that the sequence $\{\psi_k(t) : k = 1, 2, \dots\}$ is bounded and equicontinuous on the interval $\mathcal{I}$. Hence, we can assume without any loss of generality that

(1) $\lim\limits_{k \to +\infty} \tau_k = \tau_0$ exists,

(2) $\lim\limits_{k \to +\infty} \psi_k(t) = \phi(t)$ exists uniformly on $\mathcal{I}$.

Note that

(3) $\phi(\tau) \;=\; \xi$ and $\phi(\tau_0) = \psi_\xi(\tau_0) + \rho$,

(4) $\tau_0 \ge \tau(\rho) > \tau$.

It is also easy to show that $\phi(t)$ is a solution of (III.4.6-ξ) on the subinterval $\tau \le t \le \tau_0$ of $\mathcal{I}$. Since ψ_ξ is the maximal solution of (III.4.6-ξ) on $\mathcal{I}$, we must have $\phi(t) \le \psi_\xi(t)$ on the interval $\tau \le t \le \tau_0$. This is a contradiction, since $\phi(\tau_0) = \psi_\xi(\tau_0) + \rho$. Thus, (III.4.7) holds.

Now, for a given positive number ρ, there exists another positive number $\delta(\rho)$ such that (III.4.7) holds for any solution $\phi(t)$ of problem (III.4.6-η) if $\xi \le \eta \le \xi + \delta(\rho)$. Hence, using $\max(\phi(t), \psi_\xi(t))$, we can extend $\phi(t)$ on the interval $\mathcal{I}$ in such a way that

$$(\text{III.4.8}) \qquad \psi_\xi(t) \;\le\; \phi(t) \;\le\; \psi_\xi(t) + \rho \qquad \text{on} \quad \mathcal{I}.$$

Let $\mathcal{F}$ be the set of all solutions $\phi(t)$ of (III.4.6-η) which satisfy condition (III.4.8). Set $\psi_\eta(t) = \sup\{\phi(t) : \phi \in \mathcal{F}\}$ for $t \in \mathcal{I}$. Then, ψ_η is the maximal solution of (III.4.6-η) on $\mathcal{I}$. Furthermore, $\psi_\xi(t) \le \psi_\eta(t) \le \psi_\xi(t) + \rho$ if $\xi \le \eta \le \xi + \delta(\rho)$. Letting $\rho \to 0$, we complete the proof of the theorem. $\square$

Assuming again that f is continuous on a domain $\mathcal{D}$ in the (t, y)-plane, consider initial-value problem (III.4.6-ξ) together with another initial-value problem

$$(\text{III.4.9-}\epsilon) \qquad \frac{dy}{dt} = f(t, y) + \epsilon, \qquad y(\tau) = \xi,$$

where $(t, \xi) \in \mathcal{D}$ and ϵ is a positive number. We prove the following theorem.

Theorem III-4-5. *If the maximal solution $\psi(t)$ of problem (III.4.6-ξ) on the interval $\mathcal{I} = \{t : \tau \le t \le \tau'\}$ exists, then, for any positive number ρ, there exists another positive number $\epsilon(\rho)$ such that for $0 < \epsilon \le \epsilon(\rho)$, every solution $\phi(t)$ of (III.4.9-ϵ) exists on $\mathcal{I}$ and $\psi(t) \le \phi(t) \le \psi(t) + \rho$ on the interval $\mathcal{I}$. In particular, for every sufficiently small positive number ϵ, there exists the maximal solution $\psi_\epsilon(t)$ of problem (III.4.9-ϵ) on $\mathcal{I}$ and $\lim_{\epsilon \to 0} \psi_\epsilon = \psi$ uniformly on $\mathcal{I}$.*

Proof.

The maximal solution $\psi(t)$ satisfies the condition $(t, \psi(t)) \in \mathcal{D}$ on $\mathcal{I}$ (cf. Definition III-4-3). Define a function $g(t, y)$ by

$$g(t, y) = \begin{cases} f(t, \psi(t) + \rho), & y \ge \psi(t) + \rho, & t \in \mathcal{I}, \\ f(t, y), & \psi(t) \le y \le \psi(t) + \rho, & t \in \mathcal{I}, \\ f(t, \psi(t)), & y \le \psi(t), & t \in \mathcal{I}. \end{cases}$$

Then, g is continuous and bounded on the domain $\{t \in \mathcal{I}, -\infty < y < +\infty\}$. Hence, every solution $\phi(t, \epsilon)$ of the initial-value problem

$$\frac{dy}{dt} = g(t, y) + \epsilon, \qquad y(\tau) = \xi$$

exists on the interval $\mathcal{I}$.

Note that

$$(\text{III.4.10}) \qquad \frac{d\psi(t)}{dt} = f(t, \psi(t)) = g(t, \psi(t)) < g(t, \psi(t)) + \epsilon$$

on $\mathcal{I}$ (cf. Figure 17). Hence,

$$(\text{III.4.11}) \qquad \psi(t) \le \phi(t, \epsilon) \qquad \text{on} \qquad t \in \mathcal{I}.$$

We prove that for a given positive number ρ, there exists another positive number $\epsilon(\rho)$ such that if $0 < \epsilon \le \epsilon(\rho)$, we have $\phi(t, \epsilon) < \psi(t) + \rho$ on $\mathcal{I}$. Otherwise, there exist a real number $\tau(\rho)$ and two sequences $\{\epsilon_k : k = 1, 2, \ldots\}$ and $\{\tau_k : k = 1, 2, \ldots\}$ of real numbers such that
    (1) $\epsilon_k > 0$ and $\lim_{k \to +\infty} \epsilon_k = 0$,

(2) $\tau' \geq \tau_k \geq \tau(\rho) > \tau$ and $\lim\limits_{k \to +\infty} \tau_k = \tau_0$ exists,

(3) $\phi(\tau_k, \epsilon_k) = \psi(\tau_k) + \rho$,

(4) $\phi(t, \epsilon_k) < \psi(t) + \rho$ for $\tau \leq t < \tau_k$,

(5) $\lim\limits_{k \to +\infty} \phi(t, \epsilon_k) = \phi(t)$ exists uniformly on $\mathcal{I}$.

Then, $\phi(t)$ is a solution of problem (III.4.6-ξ) on the interval $\tau \leq t \leq \tau_0 \leq \tau'$. Since $\phi(\tau_0) = \psi(\tau_0) + \rho$, this is a contradiction. Thus, it was proved that $\psi(t) \leq \phi(t, \epsilon) < \psi(t) + \rho$ on $\mathcal{I}$ for $0 < \epsilon \leq \epsilon(\rho)$. Therefore, Theorem III-4-5 follows immediately. $\square$

In the proof of Theorem III-4-5, (III.4.11) was derived from (III.4.10) (cf. Figure 17). In a similar manner, we can prove the following result.

Lemma III-4-6. *Assume that $f(t, y)$ is continuous on a region $\Omega = \{(t, y) : \omega_-(t) \leq y \leq \omega_+(t), \ t \in \mathcal{I}_0\}$, where $\mathcal{I}_0 = \{t : a \leq t \leq b\}$ and*

(i) ω_+ and ω_- are real-valued, continuous, and differentiable on $\mathcal{I}_0$,

(ii) $\omega_-(t) \leq \omega_+(t)$ on $\mathcal{I}_0$.

Assume also that

$$\begin{cases} \dfrac{d\omega_+(t)}{dt} > f(t, \omega_+(t)) & on \quad \mathcal{I}_0, \\[2mm] \dfrac{d\omega_-(t)}{dt} < f(t, \omega_-(t)) & on \quad \mathcal{I}_0 \end{cases}$$

and $\omega_-(a) \leq \xi \leq \omega_+(a)$. Then, every solution $\phi(t)$ of the initial-value problem

$$(\text{III.4.12}) \qquad \frac{dy}{dt} = f(t, y), \qquad y(a) = \xi$$

exists on $\mathcal{I}_0$ and $\omega_-(t) \leq \phi(t) \leq \omega_+(t)$ on $\mathcal{I}_0$ (i.e. $(t, \phi(t)) \in \Omega$). In particular, the maximal solution $\phi_1(t)$ and the minimal solution $\phi_2(t)$ of problem (III.4.12) on $\mathcal{I}_0$ exist and $\omega_-(t) \leq \phi_2(t) \leq \phi_1(t) \leq \omega_+(t)$ on $\mathcal{I}_0$.

Proof of this lemma is left to the reader as an exercise (cf. Figure 18).

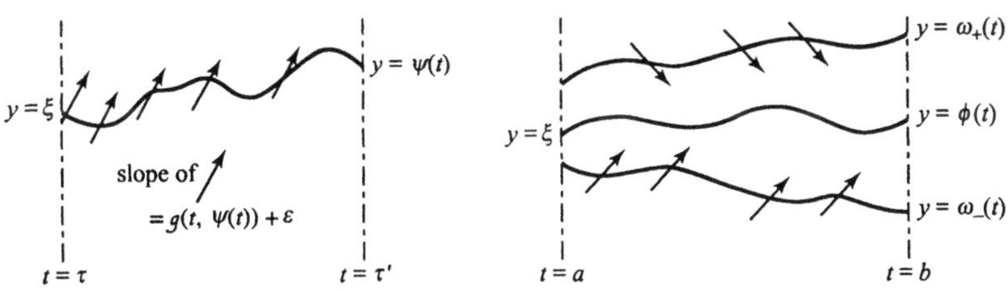

FIGURE 17.   FIGURE 18.

Using Lemma III-4-6, we prove the following theorem due to O. Perron [Per2].

Theorem III-4-7. *Assume that $f(t, y)$ is continuous on a region*

$$\Omega = \{(t, y) : \omega_-(t) \leq y \leq \omega_+(t), \ t \in \mathcal{I}_0\},$$

where $\mathcal{I}_0 = \{t : a \leq t \leq b\}$ and

(i) ω_+ and ω_- are real-valued, continuous, and differentiable on $\mathcal{I}_0$,

(ii) $\omega_-(t) \leq \omega_+(t)$ on $\mathcal{I}_0$.

Assume also that

$$\begin{cases} \dfrac{d\omega_+(t)}{dt} \geq f(t,\omega_+(t)) & \text{on } \mathcal{I}_0, \\[2mm] \dfrac{d\omega_-(t)}{dt} \leq f(t,\omega_-(t)) & \text{on } \mathcal{I}_0. \end{cases}$$

and $\omega_-(a) \leq \xi \leq \omega_+(a)$. Then, there exists a solution $\phi(t)$ of initial-value problem (III.4.12) on $\mathcal{I}_0$ such that

$$\omega_-(t) \leq \phi(t) \leq \omega_+(t) \qquad \text{on } \mathcal{I}_0, \quad \text{i.e.,} \quad (t,\phi(t)) \in \Omega \,.$$

Proof.

Let us define a function $g(t,y)$ by

$$g(t,y) = \begin{cases} f(t,\omega_+(t)), & y \geq \omega_+(t), & t \in \mathcal{I}_0, \\ f(t,y), & \omega_-(t) \leq y \leq \omega_+(t), & t \in \mathcal{I}_0, \\ f(t,\omega_-(t)), & y \leq \omega_-(t), & t \in \mathcal{I}_0. \end{cases}$$

Then, g is bounded and continuous on the region $\{(t,y) : t \in \mathcal{I}_0, -\infty < y < +\infty\}$. Hence, every solution $\phi(t)$ of the initial-value problem

$$\frac{dy}{dt} = g(t,y), \qquad y(a) = \xi$$

exists on $\mathcal{I}_0$.

Set $\omega_+(t,\epsilon) = \omega_+(t) + \epsilon(t-a)$ and $\omega_-(t,\epsilon) = \omega_-(t) - \epsilon(t-a)$, where ϵ is a positive number and $t \in \mathcal{I}_0$. Then,

$$\begin{cases} \dfrac{d\omega_+(t,\epsilon)}{dt} = \dfrac{d\omega_+(t)}{dt} + \epsilon \geq f(t,\omega_+(t)) + \epsilon > g(t,\omega_+(t,\epsilon)), \\[2mm] \dfrac{d\omega_-(t,\epsilon)}{dt} = \dfrac{d\omega_-(t)}{dt} - \epsilon \leq f(t,\omega_-(t)) - \epsilon < g(t,\omega_-(t,\epsilon)). \end{cases}$$

Hence, $\omega_-(t,\epsilon) \leq \phi(t) \leq \omega_+(t,\epsilon)$ on $\mathcal{I}_0$ (cf. Lemma III-4-6). Letting $\epsilon \to 0$, we complete the proof of Theorem III-4-7. $\square$

Comment III-4-8. The maximal and minimal solutions given in this section are also defined and explained in [CL, pp. 45-48] and [Har2, p. 25].

III-5. A comparison theorem

In this section, we derive an estimate for solutions of a differential equation by means of differential inequalities. Let us introduce the basic assumption.

Assumption 1. *Let t and u be two real variables and let $g(t, u)$ be a real-valued and continuous function of (t, u) on a domain $\mathcal{D}$ in the (t, u)-plane. Also let $\psi(t)$ be the maximal solution of the initial-value problem*

$$\frac{du}{dt} = g(t, u), \qquad u(a) = u_0$$

on an interval $\mathcal{I}_0 = \{t : a \leq t \leq b\}$, where $(a, u_0) \in \mathcal{D}$ and $(t, \psi(t)) \in \mathcal{D}$ on $\mathcal{I}_0$.

We first prove the basic theorem given below.

Theorem III-5-1. *Assume that*

(1) $g(t, u) \geq 0$ on $\mathcal{D}$ and $u_0 \geq 0$,

(2) an $\mathbb{R}^n$-valued function $\vec{\phi}(t)$ is continuously differentiable on a subinterval $\mathcal{I} = \{t : a \leq t \leq \tau\}$ of $\mathcal{I}_0$,

(3) $|\vec{\phi}(a)| \leq u_0$,

(4) $(t, |\vec{\phi}(t)|) \in \mathcal{D}$ on $\mathcal{I}$,

(5)

$$\left| \frac{d\vec{\phi}(t)}{dt} \right| \leq g\left(t, \left|\vec{\phi}(t)\right|\right) \qquad on \quad \mathcal{I}.$$

Then,

$$|\vec{\phi}(t)| \leq \psi(t) \qquad on \quad \mathcal{I}.$$

Proof.

Let ϵ be a positive number and let $\psi(t, \epsilon)$ be any solution of the initial-value problem

$$\frac{du}{dt} = g(t, u) + \epsilon, \qquad u(a) = u_0.$$

If $\epsilon > 0$ is sufficiently small, $\psi(t, \epsilon)$ exists on $\mathcal{I}_0$ and $\lim_{\epsilon \to 0} \psi(t, \epsilon) = \psi(t)$ uniformly on $\mathcal{I}_0$ (cf. Theorem III-4-5). Let us make the following observations:

(I) $$|\vec{\phi}(t)| - |\vec{\phi}(s)| \leq \left| \int_s^t \frac{d\vec{\phi}}{d\sigma} \, d\sigma \right| \leq \int_s^t \left| \frac{d\vec{\phi}}{d\sigma} \right| ds \leq \int_s^t g(\sigma, |\vec{\phi}(\sigma)|) d\sigma$$

and

(II) $$\psi(t, \epsilon) - \psi(s, \epsilon) = \int_s^t [g(\sigma, \psi(\sigma, \epsilon)) + \epsilon] d\sigma,$$

where $t \geq s$. Suppose that $|\vec{\phi}(s)| = \psi(s, \epsilon)$ for some $s \in \mathcal{I}$. Then, there exists a positive number $\delta(\epsilon)$ such that

$$|g(\sigma, \psi(\sigma, \epsilon)) - g(\sigma, |\vec{\phi}(\sigma)|)| \leq \frac{\epsilon}{2} \qquad for \quad |\sigma - s| \leq \delta(\epsilon).$$

This implies that

$$\begin{cases} |\vec{\phi}(t)| < \psi(t, \epsilon) & for \quad s < t \leq s + \delta(\epsilon), \\ |\vec{\phi}(t)| > \psi(s, \epsilon) & for \quad s - \delta(\epsilon) \leq t < s \end{cases}$$

(cf. Figure 19).

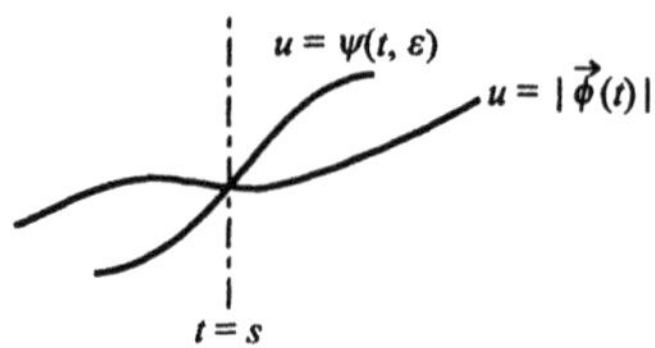

FIGURE 19.

Note that

$$\psi(t,\epsilon) - |\vec{\phi}(t)| \geq \int_s^t [g(\sigma, \psi(\sigma,\epsilon)) - g(\sigma, |\vec{\phi}(\sigma)|)]d\sigma + \epsilon(t - s) \qquad \text{for} \quad t \geq s$$

and

$$\psi(t,\epsilon) - |\vec{\phi}(t)| \leq \int_s^t [g(\sigma, \psi(\sigma,\epsilon)) - g(\sigma, |\vec{\phi}(\sigma)|)]d\sigma - \epsilon(s - t) \qquad \text{for} \quad t \leq s.$$

Thus, it is concluded that $|\vec{\phi}(t)| \leq \psi(t,\epsilon)$ on $\mathcal{I}$. Letting $\epsilon \to 0$, we complete the proof of Theorem III-5-1. $\square$

Example III-5-2. If an $\mathbb{R}^n$-valued function $\vec{\phi}(t)$ satisfies the condition

$$\left| \frac{d\vec{\phi}(t)}{dt} \right| \leq C + M|\vec{\phi}(t)| \quad \text{for} \quad a \leq t \leq b \quad \text{and} \quad |\vec{\phi}(a)| = u_0,$$

we can apply Theorem III-5-1 to $\vec{\phi}(t)$ with $g(u) = C + Mu$. Note that the initial-value problem $\dfrac{du}{dt} = C + Mu, \;\; u(a) = u_0$ has the unique solution

$$\psi(t) = u_0 e^{M(t-a)} + \frac{C}{M}\left(e^{M(t-a)} - 1\right).$$

Hence,

$$|\vec{\phi}(t)| \leq u_0 e^{M(t-a)} + \frac{C}{M}\left(e^{M(t-a)} - 1\right) \qquad \text{for} \quad a \leq t \leq b.$$

We still assume that Assumption 1 is satisfied by the function $g(t,u)$ and $\psi(t)$ and that $u_0 \geq 0$ and $g(t,u) \geq 0$ on $\mathcal{D}$. Consider a differential equation

$$(\text{III.5.1}) \qquad\qquad \frac{d\vec{y}}{dt} = \vec{f}(t, \vec{y})$$

under the following assumption:

(α) the function $\vec{f}(t, \vec{y})$ is continuous on a domain Ω in the $(t, \vec{y})$-space,

(β) $(t, |\vec{y}|) \in \mathcal{D}$ if $(t, \vec{y}) \in \Omega$,

(γ) $|\vec{f}(t,\vec{y})| \leq g(t,|\vec{y}|)$ for $(t,\vec{y}) \in \Omega$.

Then, the following result is a corollary of Theorem III-5-1.

Corollary III-5-3. *If $\vec{\phi}(t)$ is a solution of differential equation (III.5.1) on a subinterval $a \leq t \leq \tau$ of the interval $\mathcal{I}_0$ such that $(t,\vec{\phi}(t)) \in \Omega$ for $a \leq t \leq \tau$ and that $|\vec{\phi}(a)| \leq u_0$, then $|\vec{\phi}(t)| \leq \psi(t)$ for $a \leq t \leq \tau$.*

Proof of this result is left to the reader as an exercise.

III-6. Sufficient conditions for uniqueness

In this section, we derive some sufficient conditions for uniqueness by means of differential inequalities. The basic assumption is given below.

Assumption 1. *Let t and u be two real variables and let $r(t)$, $w(t)$, and $g(t,u)$ be real-valued functions such that*

(1) $r(t)$ and $w(t)$ are continuous on an interval $\mathcal{I} = \{t : a < t \leq b\}$,

(2) $r(t) > 0$ and $w(t) > 0$ on $\mathcal{I}$,

(3) $g(t,u)$ is continuous on the set $\Delta = \{(t,u) : 0 \leq u \leq w(t), t \in \mathcal{I}\}$,

(4) $g(t,u) \geq 0$ on Δ and $g(t,0) = 0$ on $\mathcal{I}$,

(5) the problem

$$
\text{(III.6.1)} \qquad
\begin{cases}
\dfrac{du(t)}{dt} = g(t,u(t)), & (t,u(t)) \in \Delta \qquad on \quad \mathcal{I}, \\[2ex]
\lim\limits_{t \to a} \dfrac{u(t)}{r(t)} = 0
\end{cases}
$$

has only the trivial solution $u = 0$ on $\mathcal{I}$.

Let us consider a problem

$$
\text{(III.6.2)} \qquad
\begin{cases}
\dfrac{d\vec{y}(t)}{dt} = \vec{f}(t,\vec{y}(t)), & (t,\vec{y}(t)) \in \Omega \qquad on \quad \mathcal{I}, \\[2ex]
\lim\limits_{t \to a} \dfrac{|\vec{y}(t)|}{r(t)} = 0,
\end{cases}
$$

where $\Omega = \{(t,\vec{y}) : |\vec{y}| \leq w(t), t \in \mathcal{I}\}$. The main result of this section is the following theorem.

Theorem III-6-1. *If the function $\vec{f}(t,\vec{y})$ is continuous on Ω and if*

$$
|\vec{f}(t,\vec{y})| \leq g(t,|\vec{y}|) \qquad on \quad \Omega,
$$

then problem (III.6.2) has only the trivial solution $\vec{y} = \vec{0}$ on $\mathcal{I}$.

Proof.

Suppose that problem (III.6.2) has a nontrivial solution $\vec{\phi}(t)$ on $\mathcal{I}$. This means that $\vec{\phi}(\alpha) \neq \vec{0}$. for some $\alpha \in \mathcal{I}$. Choose a positive number β so that $\beta < \min\left\{|\vec{\phi}(\alpha)|, \min_{\alpha \leq t \leq b} w(t)\right\}$.

Let us make the following two observations.

Observation 1. Note that the set $\Delta(\beta) = \{(t,u) : \alpha \leq t \leq b, \ 0 \leq u \leq \beta\}$ is a subset of Δ and that $u = 0$ is a solution of the differential equation

$$(\text{III.6.3}) \qquad\qquad \frac{du}{dt} = g(t,u)$$

on the interval $\alpha \leq t \leq b$. Using Theorem III-2-7, we can construct a nontrivial solution $u_0(t)$ of (III.6.3) on the interval $\alpha \leq t \leq b$ so that $(t, u_0(t)) \in \Delta(\beta)$ on $\alpha \leq t \leq b$ (cf. Figure 20).

Observation 2. The nontrivial solution $u_0(t)$ of (III.6.3) which was constructed in Observation 1 can be continued on the interval $a < t \leq \alpha$ so that

$$(\text{III.6.4}) \qquad\qquad 0 \leq u_0(t) \leq |\vec{\phi}(t)| \qquad \text{on} \quad a < t \leq \alpha.$$

To show this, we first remark that for a given positive number ϵ, any solution $\psi(t, \epsilon)$ of the initial-value problem

$$(\text{III.6.5}) \qquad\qquad \frac{du}{dt} = g(t,u) + \epsilon, \qquad u(\alpha) = u_0(\alpha)$$

satisfies the condition $\psi(t, \epsilon) < |\vec{\phi}(t)|$ on any subinterval $\sigma \leq t \leq \alpha$ of $\mathcal{I}$ if it exists on this subinterval (cf. Figure 21).

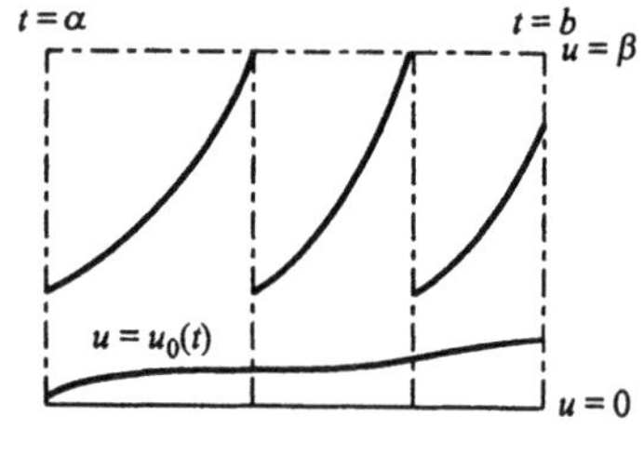

FIGURE 20.

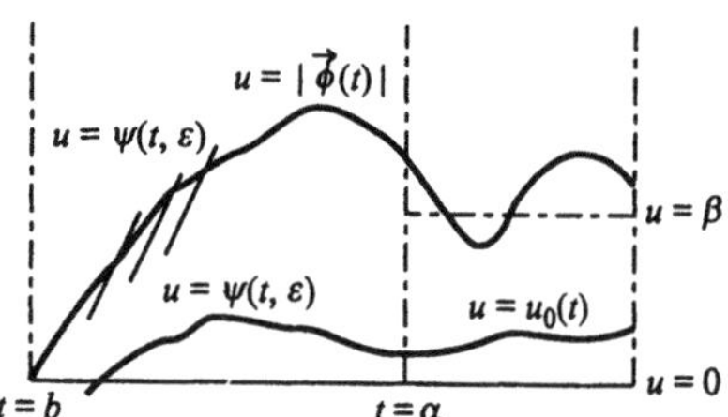

FIGURE 21.

Now, define a real-valued function $G(t, u)$ by

$$G(t,u) = \begin{cases} g(t,u), & u \geq 0, \ a < t \leq \alpha, \\ 0, & u \leq 0, \ a < t \leq \alpha. \end{cases}$$

Then, any solution $\psi(t, \epsilon)$ of problem (III.6.5) can be continued on the interval $a < t \leq \alpha$ as a solution of the differential equation $\dfrac{du}{dt} = G(t,u) + \epsilon$. Since the set

$\{\psi(\cdot,\epsilon) : 0 < \epsilon \leq 1\}$ is bounded and equicontinuous on every closed subinterval of $a < t \leq \alpha$, we can select a sequence $\{\epsilon_k : k = 1, 2, \ldots\}$ of positive numbers such that $\lim_{k\to 0}\epsilon_k = 0$ and $\lim_{k\to 0}\psi(t,\epsilon_k) = \psi(t)$ exists uniformly on each closed subinterval of $a < t \leq \alpha$. It is easy to show that $\psi(t)$ is a solution of the initial-value problem $\dfrac{du}{dt} = g(t,u)$, $u(\alpha) = u_0(\alpha)$ which satisfies condition (III.6.4).

Thus, we constructed a nontrivial solution $u_0(t)$ of differential equation (III.6.3) such that $(t, u_0(t)) \in \Delta$ on $\mathcal{I}$ and $0 \leq u_0(t) \leq |\vec{\phi}(t)|$ on $a < t \leq \alpha$. Since $\lim_{t\to a} \dfrac{|\vec{\phi}(t)|}{r(t)} = 0$, this contradicts condition (5) of Assumption 1. $\square$

From Theorem III-6-1, we derive the following result concerning uniqueness of solutions.

Theorem III-6-2. *Suppose that Assumption 1 is satisfied and that*
(i) an $\mathbb{R}^n$-valued function $\vec{f}(t, \vec{y})$ is continuous on a domain $\mathcal{D}$ in the $(t, \vec{y})$-space,
(ii) $\vec{f}$ satisfies the condition $|\vec{f}(t, \vec{y}_1) - \vec{f}(t, \vec{y}_2)| \leq g(t, |\vec{y}_1 - \vec{y}_2|)$ whenever $t \in \mathcal{I}$,
$\quad |\vec{y}_1 - \vec{y}_2| \leq w(t)$, $(t, \vec{y}_1) \in \mathcal{D}$, and $(t, \vec{y}_2) \in \mathcal{D}$.
Let $\vec{\phi}(t)$ be a solution of the differential equation

$$\frac{d\vec{y}}{dt} = \vec{f}(t, \vec{y}) \qquad on \quad \mathcal{I} = \{t : a < t \leq b\}$$

such that the set $\mathcal{D}_0 = \{(t, \vec{y}) : |\vec{y} - \vec{\phi}(t)| \leq w(t), \ t \in \mathcal{I}\}$ is contained in $\mathcal{D}$. Then, the problem

$$(\text{III.6.6}) \qquad \begin{cases} \dfrac{d\vec{y}(t)}{dt} = \vec{f}(t, \vec{y}(t)), & (t, \vec{y}(t)) \in \mathcal{D}_0 \qquad on \quad \mathcal{I}, \\[2mm] \lim_{t\to a} \dfrac{|\vec{y}(t) - \vec{\phi}(t)|}{r(t)} = 0, \end{cases}$$

has only the solution $\vec{\phi}(t)$ itself.

Proof.

Set $\vec{y}(t) = \vec{z}(t) + \vec{\phi}(t)$. Then, problem (III.6.6) becomes

$$\begin{cases} \dfrac{d\vec{z}(t)}{dt} = \vec{F}(t, \vec{z}(t)) = \vec{f}(t, \vec{z}(t) + \vec{\phi}(t)) - \vec{f}(t, \vec{\phi}(t)) & on \quad \mathcal{I}, \\[2mm] (t, |\vec{z}(t)|) \in \Delta \quad on \quad \mathcal{I}, \quad and \quad \lim_{t\to a} \dfrac{|\vec{z}(t)|}{r(t)} = 0. \end{cases}$$

Since $|\vec{F}(t, \vec{y} - \vec{\phi}(t))| \leq g(t, |\vec{y} - \vec{\phi}(t)|)$ on $\mathcal{D}_0$, Theorem III-6-2 follows from Theorem III.6.1. $\square$

Let us apply Theorem III-6-2 to some initial-value problems.

Example 1 (The Osgood condition (cf. [Os])). Consider a real-valued function $g(t, u) = h(t)p(u)$ in the case when
(1) $h(t)$ is continuous and $h(t) > 0$ on an interval $\mathcal{I} = \{t : a < t \leq b\}$,

(2) $p(u)$ is continuous on an interval $0 \leq u \leq K$, where K is a positive number,

(3) $p(0) = 0$ and $p(u) > 0$ for $0 < u \leq K$,

(4) $\displaystyle\int_{a+} h(t)dt < +\infty$,

(5) $\displaystyle\int_{0+} \frac{du}{p(u)} = +\infty$.

Assume that

(i) an $\mathbb{R}^n$-valued function $\vec{f}(t, \vec{y})$ is continuous on a domain $\mathcal{D}$ in the $(t, \vec{y})$-space,

(ii) $\vec{f}$ satisfies the condition $|\vec{f}(t, \vec{y}_1) - \vec{f}(t, \vec{y}_2)| \leq g(t, |\vec{y}_1 - \vec{y}_2|) = h(t)p(|\vec{y}_1 - \vec{y}_2|)$ whenever $t \in \mathcal{I}$, $|\vec{y}_1 - \vec{y}_2| \leq K$, $(t, \vec{y}_1) \in \mathcal{D}$, and $(t, \vec{y}_2) \in \mathcal{D}$.

Let $\vec{\phi}_1(t)$ and $\vec{\phi}_2(t)$ be two solutions of an initial-value problem $\dfrac{d\vec{y}}{dt} = \vec{f}(t, \vec{y})$, $\vec{y}(a) = \vec{\eta}$ such that

$$(a, \vec{\eta}) \in \mathcal{D} \qquad \text{and} \qquad (t, \vec{\phi}_1(t)) \in \mathcal{D}, \qquad (t, \vec{\phi}_2(t)) \in \mathcal{D}$$

on the interval $\mathcal{I}_0 = \{t : a \leq t \leq b\}$. Then, $\vec{\phi}_1(t) = \vec{\phi}_2(t)$ on $\mathcal{I}_0$.

Proof.

It is sufficient to show that Assumption 1 is satisfied by three functions $r(t) = 1$, $w(t) = K$, and $g(t, u) = h(t)p(u)$, and that $\displaystyle\lim_{t \to a} \frac{|\vec{\phi}_2(t) - \vec{\phi}_1(t)|}{r(t)} = 0$. This limit condition is evidently satisfied, since $\vec{\phi}_2(a) = \vec{\phi}_1(a)$ and $r(t) = 1$. Conditions (1), (2), (3), and (4) of Assumption 1 are also evidently satisfied. Therefore, it suffices to prove that condition (5) is also satisfied.

Let $u(t)$ be a real-valued function such that

(A) $0 < u(t) \leq K$ on a subinterval $\tau < t \leq \sigma$ of $\mathcal{I} = \{t : a < t \leq b\}$,

(B) $\displaystyle\lim_{t \to \tau^+} u(t) = 0$,

(C) $\dfrac{du(t)}{dt} = h(t)p(u(t))$ on $\tau < t \leq \sigma$.

Then,

$$\int_{\tau+}^{\sigma} \frac{1}{p(u(t))} \frac{du(t)}{dt} dt = \int_{\tau+}^{\sigma} h(t)\, dt.$$

This contradicts condition (5). $\square$

Example 2 (The Lipschitz condition). If we choose $h(t) = L$ and $p(u) = u$, where L is a positive constant, then the Osgood condition becomes the Lipschitz condition (cf. Assumption 2 of §I-1).

Example 3 (The Nagumo condition). Define $r(t)$, $w(t)$, and $g(t, u)$ by

$$r(t) = (t - a)^\lambda, \qquad w(t) = K, \qquad g(t, u) = \frac{\lambda}{t - a} u,$$

where λ is a non-negative constant and K is a positive constant. Assume also that $u \geq 0$ and $t > a$. Then, the general solution of the differential equation $\dfrac{du}{dt} = g(t, u)$

is given by $u(t) = c(t-a)^\lambda$, where c is an arbitrary constant. This implies that Assumption 1 is satisfied by $r(t)$, $w(t)$, and $g(t,u)$. In particular, choosing $\lambda = 1$, we derive the following result due to M. Nagumo [Na1 and Na2]. Assume that

(i) $\vec{f}(t,\vec{y})$ is continuous on a domain $\mathcal{D}$ in the $(t,\vec{y})$-space,

(ii) $\vec{f}$ satisfies the condition $|\vec{f}(t,\vec{y}_1) - \vec{f}(t,\vec{y}_2)| \leq \dfrac{|\vec{y}_1 - \vec{y}_2|}{t-a}$ whenever $a < t \leq b$, $|\vec{y}_1 - \vec{y}_2| \leq K$, $(t,\vec{y}_1) \in \mathcal{D}$ and $(t,\vec{y}_2) \in \mathcal{D}$.

Suppose that $\vec{\phi}_1(t)$ and $\vec{\phi}_2(t)$ are two solutions of the initial-value problem $\dfrac{d\vec{y}}{dt} = \vec{f}(t,\vec{y})$, $\vec{y}(a) = \vec{\eta}$ such that

$$(a,\vec{\eta}) \in \mathcal{D} \quad \text{and} \quad (t,\vec{\phi}_1(t)) \in \mathcal{D}, \quad (t,\vec{\phi}_2(t)) \in \mathcal{D}$$

on the interval $\mathcal{I} = \{t : a < t \leq b\}$. Then, $\vec{\phi}_1(t) = \vec{\phi}_2(t)$ on $\mathcal{I}$.

Note that, since $\vec{\phi}_1(a) = \vec{\phi}_2(a) = \vec{\eta}$ and $\dfrac{d\vec{\phi}_1}{dt}(a) = \dfrac{d\vec{\phi}_2}{dt}(a) = \vec{f}(a,\vec{\eta})$, then

$$\lim_{t \to a} \frac{|\vec{\phi}_1(t) - \vec{\phi}_2(t)|}{t-a} = 0.$$

Remark III-6-3. The sufficient conditions of Osgood's [Os] and M. Nagumo's [Na1 and Na2] for uniqueness are also explained in [CL, pp. 48-60] and [Har2, pp. 31-35].

Problem 4. Show that the initial-value problem

$$\frac{dy}{dt} = \begin{cases} t \sin\left(\dfrac{y}{t^2}\right) & \text{if} \quad t \neq 0, \\ 0 & \text{if} \quad t = 0, \end{cases} \qquad y(\tau) = \eta$$

has one and only one solution.

Answer.

Note that $t \sin\left(\dfrac{y_1}{t^2}\right) - t \sin\left(\dfrac{y_2}{t^2}\right) = \dfrac{y_1 - y_2}{t} \cos\left(\dfrac{\tilde{y}}{t^2}\right)$ for some $\tilde{y}$ on the interval between y_1 and y_2. Therefore, we can use the Lipschitz condition for $\tau \neq 0$, whereas we can use the Nagumo condition at $\tau = 0$.

Remark III-6-4. In order to apply Theorem III-6-2 to the initial-value problems in Examples 1 and 3, it was assumed that $\vec{f}(t,\vec{y})$ is continuous at the initial point $(a,\vec{\eta})$. However, the reader must notice that problem (III.6.6) is more general than an initial-value problem. Actually, we can apply Theorem III-6-2 even when $\vec{f}(t,\vec{y})$ is not continuous at $t = a$. For example, the Nagumo condition (ii) of Example 3 is satisfied by the function $\sin\left(\dfrac{y}{t}\right)$ at $a = 0$. Therefore, the problem

$$\left\{\frac{dy(t)}{dt} = \sin\left(\frac{y(t)}{t}\right), \quad \lim_{t \to 0} \frac{y(t)}{t} = 0\right\} \text{ has only the trivial solution } y(t) = 0.$$

EXERCISES III

III-1. Prove the Kneser theorem under the assumption that

(i) $\vec{f}(t,\vec{y})$ is continuous on $\mathcal{R} = \{(t,\vec{y}) : a \le t \le b,\ |\vec{y}| < +\infty\}$,

(ii) $|\vec{f}(t,\vec{y})| \le K + L|\vec{y}|$ on $\mathcal{R}$ for some positive numbers K and L.

Hint. Note that if $|\vec{f}(t,\vec{y})| \le K + L|\vec{y}|$ on $\mathcal{R}$ for some positive numbers K and L, any solution of the system $\dfrac{d\vec{y}}{dt} = \vec{f}(t,\vec{y})$ exists on $a \le t \le b$ and satisfies the estimate $|\vec{y}(t)| \le (|\vec{y}(a)| + K(t-a))e^{L(t-a)}$ for $a \le t \le b$.

III-2. Find the maximal solution and the minimal solution of the initial-value problem $\dfrac{dy}{dt} = \sqrt{|y|},\ \ y(0) = -1$ on the interval $0 \le t \le 10$.

III-3. Set

$$h(u) = \begin{cases} \sqrt{u} & \text{for} \quad u \ge 0, \\ -\sqrt{-u} & \text{for} \quad u < 0, \end{cases}$$

$$H(v) = \begin{cases} \displaystyle\sum_{n=1}^{+\infty} \frac{1}{n\sqrt{n}}\left[h\left(v - \frac{1}{n}\right) - h\left(-\frac{1}{n}\right)\right] & \text{for} \quad v \ge 0, \\ -H(-v) & \text{for} \quad v < 0, \end{cases}$$

and

$$c(t) \begin{cases} = 0 & \text{for} \quad -1 \le t \le -1 + \delta, \\ > 0 & \text{for} \quad -1 + \delta < t < 0, \\ = 0 & \text{for} \quad t = 0, \\ = -c(-t) & \text{for} \quad 0 < t \le 1, \end{cases}$$

where $0 < \delta < 1$. Show that

(a) $v = 0$ and $v = \pm\dfrac{1}{n}$ $(n = 1, 2, \dots)$ are solutions of the differential equation

$$\text{(E)} \qquad \frac{dv}{dt} = \frac{c(t)}{H'(v)}, \qquad \text{where} \quad H'(v) = \frac{dH}{dv};$$

(b) solutions of (E) with the initial-values $v(-1) = 0$ and $v(-1) = \pm\dfrac{1}{n}$ are not unique;

(c) for any positive constant ϵ, the differential equation

$$\frac{dv}{dt} = \frac{v^2 \left|\sin\left(\dfrac{\pi}{v}\right)\right|}{v^2 \left|\sin\left(\dfrac{\pi}{v}\right)\right| + \epsilon}\left[\frac{c(t)}{H'(v)}\right]$$

satisfies the Lipschitz condition in v for $|v| < +\infty$ and $|t| \le 1$.

Also, find the maximal solution and the minimal solution of the solutions of (E) satisfying the initial-condition $v(-1) = 0$ on the interval $-1 \le t \le 1$.

Hint. See [KS].

III-4. Show that if a continuously differentiable $\mathbb{R}^n$–valued function $\vec{\phi}(t)$ satisfies an inequality

$$\left|\frac{d\vec{\phi}(t)}{dt}\right| \leq \sqrt{\left|\vec{\phi}(t)\right|} \quad \text{for} \quad t \geq 0 \text{ and } \vec{\phi}(0) = \vec{0},$$

then $|\vec{\phi}(t)| \leq \dfrac{1}{4}\, t^2$ for $t \geq 0$.

Hint. Use Theorem III-5-1.

III-5. Assuming that an $\mathbb{R}^n$-valued function $\vec{\phi}(t)$ is continuously differentiable for $0 < t \leq 1$, show that if $\vec{\phi}(t)$ satisfies the condition

$$\left|\frac{d\vec{\phi}(t)}{dt}\right| \leq |\vec{\phi}(t)|^2 \quad \text{for} \quad 0 < t \leq 1 \text{ and } \lim_{t \to 0+} |\vec{\phi}(t)| = 0,$$

then $\vec{\phi}(t) = 0$ for $0 < t \leq 1$.

Hint. Use Theorem III-5-1.

III-6. Assuming that an $\mathbb{R}^n$-valued function $\vec{\phi}(t)$ is continuously differentiable for $0 \leq t < +\infty$, show that if $\vec{\phi}(t)$ satisfies the condition

$$\begin{cases} \left|\dfrac{d\vec{\phi}(t)}{dt}\right| \leq 2t|\vec{\phi}(t)| \quad \text{for} \quad 0 \leq t < +\infty, \\[2mm] \lim_{t \to +\infty} |\vec{\phi}(t)|\, \exp[t^2] = 0, \end{cases}$$

then $\vec{\phi}(t) = 0$ for $0 \leq t < +\infty$.

Hint. Use Theorem III-5-1.

III-7. Show that the problem

$$\frac{dy(t)}{dt} = \sin\left(\frac{y(t)}{\sqrt{|t|}}\right), \qquad \lim_{t \to 0} y(t) = 0$$

has only the trivial solution $y(t) = 0$.

Hint. Note that

$$\sin\left(\frac{y_1}{\sqrt{|t|}}\right) - \sin\left(\frac{y_2}{\sqrt{|t|}}\right) = \frac{y_1 - y_2}{\sqrt{|t|}} \cos\left(\frac{\tilde{y}}{\sqrt{|t|}}\right),$$

for some $\tilde{y}$ between y_1 and y_2.

III-8. Let $\vec{f}(t, \vec{x}, \vec{y})$ be an $\mathbb{R}^n$-valued function of $(t, \vec{x}, \vec{y}) \in \mathbb{R} \times \mathbb{R}^n \times \mathbb{R}^m$. Assume that

(1) the entries of $\vec{f}(t, \vec{x}, \vec{y})$ are continuous in the region $\Delta = \{(t, \vec{x}, \vec{y}) : 0 \leq t \leq \alpha, |\vec{x}| \leq a, |\vec{y}| \leq b\}$, where α, a, and b are fixed positive numbers,

(2) there exists a positive number K such that $|\vec{f}(t, \vec{x}_1, \vec{y}) - \vec{f}(t, \vec{x}_2, \vec{y})| \leq K|\vec{x}_1 - \vec{x}_2|$ if $(t, \vec{x}_j, \vec{y}) \in \Delta$ $(j = 1, 2)$.

Let $\mathcal{U}$ denote the set of all $\mathbb{R}^m$-valued functions $\vec{u}(t)$ such that $|\vec{u}(t)| \leq b$ for $0 \leq t \leq \alpha$ and that $|\vec{u}(t) - \vec{u}(\tau)| \leq L|t - \tau|$ if $0 \leq t \leq \alpha$ and $0 \leq \tau \leq \alpha$, where L is a positive constant independent of $\vec{u} \in \mathcal{U}$. Also, let $\vec{\phi}(t; \vec{u})$ denote the unique solution of the initial-value problem $\dfrac{d\vec{x}}{dt} = \vec{f}(t, \vec{x}, \vec{u}(t))$, $\vec{x}(0) = \vec{0}$, where $\vec{u} \in \mathcal{U}$. It is known that there exists a positive number α_0 such that for all $\vec{u} \in \mathcal{U}$, the solution $\vec{\phi}(t, \vec{u})$ exists and $|\vec{\phi}(t, \vec{u})| \leq a$ for $0 \leq t \leq \alpha_0$. Denote by $\mathcal{R}$ the subset of $\mathbb{R}^{n+1}$ which is the union of solution curves $\{(t, \vec{\phi}(t, \vec{u})) : 0 \leq t \leq \alpha_0\}$ for all $\vec{u} \in \mathcal{U}$, i.e., $\mathcal{R} = \{(t, \vec{\phi}(t, \vec{u})) : 0 \leq t \leq \alpha_0, \vec{u} \in \mathcal{U}\}$. Show that $\mathcal{R}$ is a closed set in $\mathbb{R}^{n+1}$.

Hint. See [LM1, Theorem 2, pp. 44-47] and [LM2, Problem 6, pp. 282-283].

III-9. Let $\vec{f}(t, \vec{x}, \vec{y})$ be an $\mathbb{R}^n$-valued function of $(t, \vec{x}, \vec{y}) \in \mathbb{R} \times \mathbb{R}^n \times \mathbb{R}^m$. Assume that

(1) the entries of $\vec{f}(t, \vec{x}, \vec{y})$ are continuous in the region $\Delta = \{(t, \vec{x}, \vec{y}) : 0 \leq t \leq \alpha, |\vec{x}| \leq a, |\vec{y}| \leq b\}$, where α, a, and b are fixed positive numbers,

(2) there exists a positive number K such that $|\vec{f}(t, \vec{x}_1, \vec{y}) - \vec{f}(t, \vec{x}_2, \vec{y})| \leq K|\vec{x}_1 - \vec{x}_2|$ if $(t, \vec{x}_j, \vec{y}) \in \Delta$ $(j = 1, 2)$.

Let $\mathcal{U}$ denote the set of all $\mathbb{R}^m$-valued functions $\vec{u}(t)$ such that $|\vec{u}(t)| \leq b$ for $0 \leq t \leq \alpha$ and that the entries of $\vec{u}$ are piecewise continuous on the interval $0 \leq t \leq \alpha$. Also, let $\vec{\phi}(t; \vec{u})$ denote the unique solution of the initial-value problem $\dfrac{d\vec{x}}{dt} = \vec{f}(t, \vec{x}, \vec{u}(t))$, $\vec{x}(0) = \vec{0}$, where $\vec{u} \in \mathcal{U}$. It is known that there exists a positive number α_0 such that for all $\vec{u} \in \mathcal{U}$, the solution $\vec{\phi}(t, \vec{u})$ exists and $|\vec{\phi}(t, \vec{u})| \leq a$ for $0 \leq t \leq \alpha_0$. Denote by $\mathcal{R}$ the subset of $\mathbb{R}^{n+1}$ which is the union of solution curves $\{(t, \vec{\phi}(t, \vec{u})) : 0 \leq t \leq \alpha_0\}$ for all $\vec{u} \in \mathcal{U}$, i.e., $\mathcal{R} = \{(t, \vec{\phi}(t, \vec{u})) : 0 \leq t \leq \alpha_0, \vec{u} \in \mathcal{U}\}$. Assume that a point $(\tau, \vec{\phi}(\tau, \vec{u}_0))$ is on the boundary of $\mathcal{R}$, where $0 < \tau \leq \alpha_0$ and $\vec{u}_0 \in \mathcal{U}$. Show that the solution curve $\{(t, \vec{\phi}(t, \vec{u}_0)) : 0 \leq t \leq \tau\}$ is also on the boundary of $\mathcal{R}$.

Hint. [LM2, Theorem 3 of Chapter 4 and its remark on pp. 254-257, and Problem 2 on p. 258].

III-10. Let $A(t, \vec{x})$ and $\vec{f}(t, \vec{x})$ be respectively an $n \times n$ matrix-valued and $\mathbb{R}^n$-valued functions whose entries are continuous and bounded in $(t, \vec{x}) \in \mathbb{R}^{n+1}$ on a domain $\Delta = \{(t, \vec{x}) : a < t < b, \vec{x} \in \mathbb{R}^n\}$, where a and b are real numbers. Also, assume that $(\tau, \vec{\xi}) \in \Delta$. Show that every solution of the initial-value problem $\dfrac{d\vec{x}}{dt} = A(t, \vec{x})\vec{x} + \vec{f}(t, \vec{x})$, $\vec{x}(\tau) = \vec{\xi}$ exists on the interval $a < t < b$.

CHAPTER IV

GENERAL THEORY OF LINEAR SYSTEMS

The main topic of this chapter is the structure of solutions of a linear system

$$\text{(LP)} \qquad \frac{d\vec{y}}{dt} = A(t)\vec{y} + \vec{b}(t),$$

where entries of the $n \times n$ matrix $A(t)$ are complex-valued (i.e., $\mathbb{C}$-valued) continuous functions of a real independent variable t, and the $\mathbb{C}^n$-valued function $\vec{b}(t)$ is continuous in t. The existence and uniqueness of solutions of problem (LP) were given by Theorem I-3-5. In §IV-1, we explain some basic results concerning $n \times n$ matrices whose entries are complex numbers. In particular, we explain the S-N decomposition (or the Jordan-Chevalley decomposition) of a matrix (cf. Definition IV-1-12; also see [Bou, Chapter 7], [HirS, Chapter 6], and [Hum, pp. 17-18]). The S-N decomposition is equivalent to the block-diagonalization which separates distinct eigenvalues. It is simpler than the Jordan canonical form. The basic tools for achieving this decomposition are the Cayley-Hamilton theorem (cf. Theorem IV-1-5) and the partial fraction decomposition of reciprocal of the characteristic polynomial. It is relatively easy to obtain this decomposition with an elementary calculation if all eigenvalues of a given matrix are known (cf. Examples IV-1-18 and IV-1-19). In §IV-2, we explain the general aspect of linear homogeneous systems. Homogeneous systems with constant coefficients are treated in §IV-3. More precisely speaking, we define e^{tA} and discuss its properties. In §IV-4, we explain the structure of solutions of a homogeneous system with periodic coefficients. The main result is the Floquet theorem (cf. Theorem IV-4-1 and [Fl]). The Hamiltonian systems with periodic coefficients are the main subject of §IV-5. The Floquet theorem is extended to this case using canonical linear transformations (cf. [Si4] and [Mar]). Also, we go through an elementary part of the theory of symplectic groups. Finally, nonhomogeneous systems and scalar higher-order equations are treated in §IV-6 and §IV-7, respectively. The topics of §§IV-2–IV-4, IV-6, and IV-7 are found also, for example, in [CL, Chapter 3] and [Har2, Chapter IV]. For symplectic groups, see, for example, [Ja, Chapter 6] and [We, Chapters 6 and 8].

IV-1. Some basic results concerning matrices

In this section, we explain the basic results concerning constant square matrices. Let $\mathcal{M}_n(\mathbb{C})$ denote the set of all $n \times n$ matrices whose entries are complex numbers. The set of all invertible matrices with entries in $\mathbb{C}$ is denoted by $\mathrm{GL}(n, \mathbb{C})$, which stands for the *general linear group of order n*. We define a topology in $\mathcal{M}_n(\mathbb{C})$ by the norm $|A| = \max\limits_{1 \leq j,k \leq n} |a_{jk}|$ for $A \in \mathcal{M}_n(\mathbb{C})$, where a_{jk} is the entry of A on the j-th row and the k-th column; i.e., $A = \begin{bmatrix} a_{11} & a_{12} & \cdots & a_{1n} \\ a_{21} & a_{22} & \cdots & a_{2n} \\ \cdots & \cdots & \cdots & \cdots \\ a_{n1} & a_{n2} & \cdots & a_{nn} \end{bmatrix}$. A matrix $A \in \mathcal{M}_n(\mathbb{C})$

69

is said to be *upper-triangular* if $a_{jk} = 0$ for $j > k$. The following lemma is a basic result in the theory of matrices.

Lemma IV-1-1. *For each* $A \in \mathcal{M}_n(\mathbb{C})$, *there exists a matrix* $P \in GL(n, \mathbb{C})$ *such that* $P^{-1}AP$ *is upper-triangular.*

Proof.

Let λ be an eigenvalue of A and $\vec{p}_1$ be an eigenvector of A associated with the eigenvalue λ. Then, $A\vec{p}_1 = \lambda\vec{p}_1$ and $\vec{p}_1 \neq \vec{0}$. Choose $n - 1$ vectors $\vec{p}_j$ $(j = 2, \ldots, n)$ so that $Q = [\vec{p}_1\vec{p}_2 \cdots \vec{p}_n] \in GL(n, \mathbb{C})$, where the $\vec{p}_j$ are column vectors of the matrix Q. Then, the first column vector of $Q^{-1}AQ$ is $\begin{bmatrix} \lambda \\ 0 \\ \vdots \\ 0 \end{bmatrix}$. Hence, we can complete the proof of this lemma by induction on n. $\square$

A matrix $A \in \mathcal{M}_n(\mathbb{C})$ is said to be *diagonal* if $a_{jk} = 0$ for $j \neq k$. We denote by $\mathrm{diag}[d_1, d_2, \ldots, d_n]$ the diagonal matrix with entries $d_1, d_2, \ldots, d_n$ on the main diagonal (i.e., $d_j = a_{jj}$). A matrix $A \in \mathcal{M}_n(\mathbb{C})$ is said to be *diagonalizable* (or *semisimple*) if there exists a matrix $P \in GL(n, \mathbb{C})$ such that $P^{-1}AP$ is diagonal. Denote by $\mathcal{S}_n$ the set of all diagonalizable matrices in $\mathcal{M}_n(\mathbb{C})$. The following lemma is another basic result in the theory of matrices.

Lemma IV-1-2. *A matrix* $A \in \mathcal{M}_n(\mathbb{C})$ *is diagonalizable if and only if* A *has* n *linearly independent eigenvectors* $\vec{p}_1$, $\vec{p}_2$, $\ldots$, $\vec{p}_n$.

Proof.

If A has n linearly independent eigenvectors $\vec{p}_1$, $\vec{p}_2$, $\ldots$, $\vec{p}_n$, set $P = [\vec{p}_1\vec{p}_2 \ldots \vec{p}_n]$ $\in GL(n, \mathbb{C})$. Then, $P^{-1}AP$ is diagonal. Conversely, if $P^{-1}AP$ is diagonal for $P = [\vec{p}_1\vec{p}_2 \cdots \vec{p}_n] \in GL(n, \mathbb{C})$, then $\vec{p}_1$, $\vec{p}_2$, $\ldots$, $\vec{p}_n$ are n linearly independent eigenvectors of A. $\square$

In particular, if a matrix $A \in \mathcal{M}_n(\mathbb{C})$ has n distinct eigenvalues, then n eigenvectors corresponding to these n eigenvalues, respectively, are linearly independent (cf. [Rab, p. 186]). Therefore, we obtain the following corollary of Lemma IV-1-2.

Corollary IV-1-3. *If a matrix* $A \in \mathcal{M}_n(\mathbb{C})$ *has* n *distinct eigenvalues, then* $A \in \mathcal{S}_n$.

The set $\mathcal{M}_n(\mathbb{C})$ is a noncommutative $\mathbb{C}$-algebra. This means that $\mathcal{M}_n(\mathbb{C})$ is a vector space over $\mathbb{C}$ and a noncommutative ring. The set $\mathcal{S}_n$ is not a subalgebra of $\mathcal{M}_n(\mathbb{C})$. However, the following lemma shows an important topological property of $\mathcal{S}_n$ as a subset of $\mathcal{M}_n(\mathbb{C})$.

Lemma IV-1-4. *The set* $\mathcal{S}_n$ *is dense in* $\mathcal{M}_n(\mathbb{C})$.

Proof.

It must be shown that, for each matrix $A \in \mathcal{M}_n(\mathbb{C})$, there exists a sequence $\{B_k : k = 1, 2, \ldots\}$ of matrices in $\mathcal{S}_n$ such that $\lim_{k \to +\infty} B_k = A$. To do this, we may assume without any loss of generality that A is an upper-triangular matrix with the eigenvalues $\lambda_1, \ldots, \lambda_n$ on the main diagonal (cf. Lemma IV-1-1). Set

$B_k = A + \text{diag}[\epsilon_{k,1}, \epsilon_{k,2}, \dots, \epsilon_{k,n}]$, where the quantities $\epsilon_{k,\nu}$ ($\nu = 1, 2, \dots, n$) are chosen in such a way that n numbers $\lambda_1 + \epsilon_{k,1}, \lambda_2 + \epsilon_{k,2}, \dots, \lambda_n + \epsilon_{k,n}$ are distinct and that $\lim_{k \to +\infty} \epsilon_{k,\nu} = 0$ for $\nu = 1, 2, \dots, n$. Then, by Corollary IV-1-3, we obtain $B_k \in \mathcal{S}_n$ and $\lim_{k \to +\infty} B_k = A$. $\square$

For a matrix $A \in \mathcal{M}_n(\mathbb{C})$, denote by $p_A(\lambda)$ the *characteristic polynomial* of A with the expansion

$$\text{(IV.1.1)} \qquad p_A(\lambda) = \det[\lambda I_n - A] = \lambda^n + \sum_{h=1}^{n} p_h(A) \lambda^{n-h},$$

where I_n denotes the $n \times n$ identity matrix. Note that

$$p_A(A) = A^n + \sum_{h=1}^{n} p_h(A) A^{n-h}, \quad A^0 = I_n.$$

Now, let us prove the Cayley-Hamilton theorem (see, for example, [Bel3, pp. 200-201 and 220], [Cu, p. 220], and [Rab, p. 198]).

Theorem IV-1-5 (A. Cayley-W. R. Hamilton). *If $A \in \mathcal{M}_n(\mathbb{C})$, then its characteristic polynomial satisfies $p_A(A) = O$, where O is the zero matrix of appropriate size.*

Remark IV-1-6. The coefficients $p_h(A)$ of $p_A(\lambda)$ are polynomials in entries a_{jk} of the matrix A with integer coefficients.

Proof of Theorem IV-1-5.

Since the entries of $p_A(A)$ are polynomials of entries a_{jk} of the matrix A, they are continuous in the entries of A. Therefore, if $p_A(A) = O$ for $A \in \mathcal{S}_n$, it is also true for every $A \in \mathcal{M}_n(\mathbb{C})$, since $\mathcal{S}_n$ is dense in $\mathcal{M}_n(\mathbb{C})$ (cf. Lemma IV-1-4). Note also that if $B = P^{-1}AP$ for some $P \in \text{GL}(n, \mathbb{C})$, then $p_B(\lambda) = p_A(\lambda)$ and $p_B(B) = P^{-1}p_A(A)P$. Therefore, it suffices to prove Theorem IV-1-5 for diagonal matrices. Set $\Lambda = \text{diag}[\lambda_1, \lambda_2, \dots, \lambda_n]$. Then, $p_\Lambda(\lambda) = (\lambda - \lambda_1)(\lambda - \lambda_2) \cdots (\lambda - \lambda_n)$ and $p_\Lambda(\Lambda) = \text{diag}[p_\Lambda(\lambda_1), p_\Lambda(\lambda_2), \dots, p_\Lambda(\lambda_n)] = 0$. $\square$

It is an important application of Theorem IV-1-5 that an $n \times n$ matrix N satisfies the condition $N^n = O$ if its characteristic polynomial $p_N(\lambda)$ is equal to λ^n. If $N^n = O$, N is said to be *nilpotent*.

Lemma IV-1-7. *A matrix $N \in \mathcal{M}_n(\mathbb{C})$ is nilpotent if and only if all eigenvalues of N are zero.*

Proof.

If $\vec{p}$ is an eigenvector of N associated with an eigenvalue λ of N, then $N^k\vec{p} = \lambda^k\vec{p}$ for every positive integer k. In particular, $N^n\vec{p} = \lambda^n\vec{p}$. Hence, if $N^n = O$, then $\lambda = 0$. On the other hand, if all eigenvalues of N are 0, the characteristic polynomial $p_N(\lambda)$ is equal to λ^n. Hence, N is nilpotent. $\square$

Applying Lemma IV-1-1 to a nilpotent matrix N, we obtain the following result.

Lemma IV-1-8. *A matrix $N \in \mathcal{M}_n(\mathbb{C})$ is nilpotent if and only if there exists a matrix $P \in GL(n, \mathbb{C})$ such that $P^{-1}NP$ is upper-triangular and the entries on the main diagonal of N are all zero. Furthermore, if N is a real matrix, then there exists a real matrix P that satisfies the requirement given above.*

To verify the last statement of this lemma, use a method similar to the proof of Lemma IV-1-1 together with the fact that if an eigenvalue of a real matrix is real, then there exists a real eigenvector associated with this real eigenvalue. Details are left to the reader as an exercise.

The main concern of this section is to explain the S-N decomposition of a matrix $A \in \mathcal{M}_n(\mathbb{C})$ (cf. Theorem IV-1-11). Before introducing the S-N decomposition, we need some preparation.

Let λ_j $(j = 1, 2, \ldots, k)$ be the distinct eigenvalues of A and let m_j $(j = 1, 2, \ldots, k)$ be their respective multiplicities. Then, the characteristic polynomial of the matrix A is given by $p_A(\lambda) = (\lambda - \lambda_1)^{m_1}(\lambda - \lambda_2)^{m_2} \cdots (\lambda - \lambda_k)^{m_k}$. Decompose $\dfrac{1}{p_A(\lambda)}$ into partial fractions $\dfrac{1}{p_A(\lambda)} = \sum_{j=1}^{k} \dfrac{Q_j(\lambda)}{(\lambda - \lambda_j)^{m_j}}$, where, for every j, the quantity Q_j is a nonzero polynomial in λ of degree not greater than $m_j - 1$. Hence, $1 = \sum_{j=1}^{k} Q_j(\lambda) \prod_{h \neq j} (\lambda - \lambda_h)^{m_h}$. Setting

$$(\text{IV.1.2}) \qquad P_j(\lambda) = Q_j(\lambda) \prod_{h \neq j} (\lambda - \lambda_h)^{m_h} ,$$

we obtain

$$(\text{IV.1.3}) \qquad 1 = \sum_{j=1}^{k} P_j(\lambda).$$

Note that this is an identity in λ. Therefore, setting

$$(\text{IV.1.4}) \qquad P_j(A) = Q_j(A) \prod_{h \neq j} (A - \lambda_h I_n)^{m_h} \qquad (j = 1, 2, \ldots, k),$$

we obtain

$$(\text{IV.1.5}) \qquad I_n = \sum_{h=1}^{k} P_h(A).$$

In the following two lemmas, we show that (IV.1.5) is a resolution of the identity in terms of projections $P_h(A)$ onto invariant subspaces of A associated with eigenvalues λ_h, respectively.

Lemma IV-1-9. *The k matrices $P_j(A)$ $(j = 1, 2, \ldots, k)$ given by (IV.1.4) satisfy the following conditions:*
(i) A and $P_j(A)$ $(j = 1, 2, \ldots, k)$ commute,

(ii) $(A - \lambda_j I_n)^{m_j} P_j(A) = O$ $(j = 1, 2, \ldots , k)$,

(iii) $P_j(A)P_h(A) = O$ if $j \neq h$,

(iv) $\sum_{h=1}^{k} P_h(A) = I_n$,

(v) $P_j(A)^2 = P_j(A)$ $(j = 1, 2, \ldots , k)$,

(vi) $P_j(A) \neq O$ $(j = 1, 2, \ldots , k)$.

Proof.

Since $P_j(A)$ is a polynomial of A, we obtain (i). Using Theorem IV-1-5, we derive (ii) and (iii) from (IV.1.4) and (i). Statement (iv) is the same as (IV.1.5). Multiplying the both sides of (IV.1.5) by $P_j(A)$, we obtain

$$(\text{IV.1.6}) \qquad P_j(A) \;=\; \sum_{h=1}^{k} P_j(A)P_h(A).$$

Then, (v) follows from (IV.1.6) and (iii). To prove (vi), let $\vec{p}_j$ be an eigenvector of A associated with the eigenvalue λ_j. Note that (IV.1.2) implies $P_h(\lambda_j) = 0$ if $h \neq j$. Therefore, we derive $P_j(\lambda_j) = 1$ from (IV.1.3). Now, since $P_j(A)\vec{p}_j = P_j(\lambda_j)\vec{p}_j \neq \vec{0}$, we obtain (vi). $\square$

Lemma IV-1-10. *Denote by $\mathcal{V}_j$ the image of the mapping $P_j(A) : \mathbb{C}^n \to \mathbb{C}^n$. Then,*

(1) $\vec{p} \in \mathbb{C}^n$ belongs to $\mathcal{V}_j$ if and only if $P_j(A)\vec{p} = \vec{p}$,

(2) $P_j(A)\vec{p} = \vec{0}$ for all $\vec{p} \in \mathcal{V}_h$ if $j \neq h$,

(3) $\mathbb{C}^n = \mathcal{V}_1 \oplus \mathcal{V}_2 \oplus \cdots \oplus \mathcal{V}_k$ (a direct sum),

(4) for each j, $\mathcal{V}_j$ is an invariant subspace of A,

(5) the restriction of A on $\mathcal{V}_j$ has a coordinates-wise representation:

$$(\text{IV.1.7}) \qquad\qquad A|_{\mathcal{V}_j} \;:\; \lambda_j I_j \;+\; N_j,$$

where I_j is the identity matrix and N_j is a nilpotent matrix,

(6) $\dim_{\mathbb{C}} \mathcal{V}_j = m_j$.

Proof.

Each part of this lemma follows from Lemma IV-1-9 as follows.

(1) A vector $\vec{p} \in \mathcal{V}_j$ if and only if $\vec{p} = P_j(A)\vec{q}$ for some $\vec{q} \in \mathbb{C}^n$. If $\vec{p} = P_j(A)\vec{q}$, we obtain $P_j(A)\vec{p} = P_j(A)^2\vec{q} = P_j(A)\vec{q} = \vec{p}$ from (v) of Lemma IV-1-9.

(2) A vector $\vec{p} \in \mathcal{V}_h$ if and only if $\vec{p} = P_h(A)\vec{q}$ for some $\vec{q} \in \mathbb{C}^n$. Hence, from (iii) of Lemma IV-1-9 we obtain $P_j(A)\vec{p} = P_j(A)P_h(A)\vec{q} = \vec{0}$ if $j \neq h$.

(3) (iv) of Lemma IV-1-9 implies $\vec{p} = P_1(A)\vec{p} + \cdots + P_k(A)\vec{p}$ for every $\vec{p} \in \mathbb{C}^n$, while (1) implies that $P_j(A)\vec{p} \in \mathcal{V}_j$. On the other hand, if $\vec{p} = \vec{p}_1 + \cdots + \vec{p}_k$ for some $\vec{p}_j \in \mathcal{V}_j$ $(j = 1, 2, \ldots , k)$, then, by (1) and (2), we obtain $P_j(A)\vec{p} = P_j(A)\vec{p}_1 + \cdots + P_j(A)\vec{p}_k = \vec{p}_j$.

(4) $A\vec{p} = AP_j(A)\vec{p} = P_j(A)A\vec{p} \in \mathcal{V}_j$ for every $\vec{p} \in \mathcal{V}_j$.

(5) Let n_j be the dimension of the space $\mathcal{V}_j$ over $\mathbb{C}$ and let $\{\vec{p}_{j,\ell} : \ell = 1, 2, \ldots , n_j\}$ be a basis for $\mathcal{V}_j$. Then, there exists an $n_j \times n_j$ matrix N_j, such that

$$(A - \lambda_j I_n)[\vec{p}_{j,1}\vec{p}_{j,2}\ldots\vec{p}_{j,n_j}] = [\vec{p}_{j,1}\vec{p}_{j,2}\ldots\vec{p}_{j,n_j}]N_j$$

as the coordinates-wise representation relative to this basis. This implies that

$$(A - \lambda_j I_n)^\ell P_j(A)[\vec{p}_{j,1}\vec{p}_{j,2}\ldots\vec{p}_{j,n_j}] = (A - \lambda_j I_n)^\ell[\vec{p}_{j,1}\vec{p}_{j,2}\ldots\vec{p}_{j,n_j}]$$
$$= [\vec{p}_{j,1}\vec{p}_{j,2}\ldots\vec{p}_{j,n_j}]N_j^\ell \quad \text{for} \quad (\ell = 1, 2, \ldots).$$

In particular, from (ii) of Lemma IV-1-9, we derive $N_j^{m_j} = O$. Thus, we obtain

$$(\text{IV.1.8}) \qquad A[\vec{p}_{j,1}\vec{p}_{j,2}\ldots\vec{p}_{j,n_j}] = [\vec{p}_{j,1}\vec{p}_{j,2}\ldots\vec{p}_{j,n_j}](\lambda_j I_j + N_j),$$

where I_j is the $n_j \times n_j$ identity matrix. This proves (IV.1.7).

(6) Let $\{\vec{p}_{j,\ell} : \ell = 1, 2, \ldots, n_j\}$ be a basis for $\mathcal{V}_j$ $(j = 1, 2, \ldots, k)$. Set

$$(\text{IV.1.9}) \qquad P_0 = [\vec{p}_{1,1}\ldots\vec{p}_{1,n_1}\vec{p}_{2,1}\ldots\vec{p}_{2,n_2}\ldots\vec{p}_{k,1}\ldots\vec{p}_{k,n_k}].$$

Then, $P_0 \in \mathrm{GL}(n, \mathbb{C})$ and (IV.1.8) implies

$$(\text{IV.1.10}) \qquad P_0^{-1}AP_0 = \mathrm{diag}[\lambda_1 I_1 + N_1, \lambda_2 I_2 + N_2, \ldots, \lambda_k I_k + N_k],$$

where the right-hand side of (IV.1.10) is a matrix in a block-diagonal form with entries $\lambda_1 I_1 + N_1, \lambda_2 I_2 + N_2, \ldots, \lambda_k I_k + N_k$ on the main diagonal blocks. Hence, $p_A(\lambda) = (\lambda - \lambda_1)^{n_1}(\lambda - \lambda_2)^{n_2}\cdots(\lambda - \lambda_k)^{n_k}$. Also, $p_A(\lambda) = (\lambda - \lambda_1)^{m_1}(\lambda - \lambda_2)^{m_2}\cdots(\lambda - \lambda_k)^{m_k}$. Therefore, $\dim_{\mathbb{C}} \mathcal{V}_j = n_j = m_j$ $(j = 1, 2, \ldots, k)$. $\square$

The following theorem defines the *S-N decomposition* of a matrix $A \in \mathcal{M}_n(\mathbb{C})$.

Theorem IV-1-11. *Let A be an $n \times n$ matrix whose entries are complex numbers. Then, there exist two $n \times n$ matrices S and N such that*
(a) S is diagonalizable,
(b) N is nilpotent,
(c) $A = S + N$,
(d) $SN = NS$.
The two matrices S and N are uniquely determined by these four conditions. If A is real, then S and N are also real. Furthermore, they are polynomials in A with coefficients in the smallest field $\mathcal{Q}(a_{jk}, \lambda_h)$ containing the field $\mathcal{Q}$ of rational numbers, the entries a_{jk} of A, and the eigenvalues $\lambda_1, \lambda_2 \ldots, \lambda_k$ of A.

Proof.

We prove this theorem in three steps.

Step 1. *Existence of S and N.* Using the projections $P_j(A)$ given by (IV.1.4), define S and N by

$$S = \lambda_1 P_1(A) + \lambda_2 P_2(A) + \cdots + \lambda_k P_k(A), \qquad N = A - S.$$

If P_0 is given by (IV.1.9), then

$$(\text{IV.1.11}) \qquad P_0^{-1}SP_0 = \mathrm{diag}[\lambda_1 I_1, \lambda_2 I_2, \ldots, \lambda_k I_k]$$

and

(IV.1.12) $$P_0^{-1}NP_0 = \mathrm{diag}[N_1, N_2, \ldots, N_k]$$

from Lemmas IV-1-9 and IV-1-10 and (IV.1.10). Hence, S is diagonalizable and N is nilpotent. Furthermore, $NS = SN$ since S and N are polynomials in A. This shows the existence of S and N satisfying (a), (b), (c), and (d). Moreover, from (IV.1.4), it follows that two matrices S and N are polynomials in A with coefficients in the field $\mathcal{Q}(a_{jk}, \lambda_h)$.

Step 2. *Uniqueness of S and N.* Assume that there exists another pair $(\hat{S}, \hat{N})$ of $n \times n$ matrices satisfying conditions (a), (b), (c), and (d). Then, (c) and (d) imply that $\hat{S}A = A\hat{S}$ and $\hat{N}A = A\hat{N}$. Hence, $S\hat{S} = \hat{S}S$, $N\hat{S} = \hat{S}N$, $S\hat{N} = \hat{N}S$, and $N\hat{N} = \hat{N}N$ since S and N are polynomials in A. This implies that $S - \hat{S}$ is diagonalizable and $N - \hat{N}$ is nilpotent. Therefore, from $S - \hat{S} = \hat{N} - N$, it follows that $S - \hat{S} = \hat{N} - N = O$.

Step 3. *The case when S and N are real.* In case when A is real, let $\overline{S}$ and $\overline{N}$ be the complex conjugates of S and N, respectively. Then, $A = S + N = \overline{S} + \overline{N}$. Hence, the uniqueness of S and N implies that $S = \overline{S}$ and $N = \overline{N}$.

This completes the proof of Theorem IV-1-11. $\square$

Definition IV-1-12. *The decomposition $A = S + N$ of Theorem IV-1-11 is called the S-N decomposition of A.*

Remark IV-1-13. From (IV.1.11), it follows immediately that S and A have the same eigenvalues, counting their multiplicities. Therefore, S is invertible if and only if A is invertible.

Observation IV-1-14. Let A be an $n \times n$ matrix whose distinct eigenvalues are λ_1, λ_2, ..., λ_k. Let $A = S + N$ be the S-N decomposition of A. It can be shown that $n \times n$ matrices P_1, P_2, ..., P_k are uniquely determined by the following three conditions:
  (i) $I_n = P_1 + P_2 + \cdots + P_k$,
  (ii) $P_j P_\ell = O$ if $j \neq \ell$,
  (iii) $S = \lambda_1 P_1 + \lambda_2 P_2 + \cdots + \lambda_k P_k$.

Proof.

Note that

$$\begin{cases} I_n = P_1(A) + P_2(A) + \cdots + P_k(A), \\ P_j(A)P_h(A) = O \quad \text{if} \quad j \neq h, \\ S = \lambda_1 P_1(A) + \lambda_2 P_2(A) + \cdots + \lambda_k P_k(A). \end{cases}$$

First, derive that $P_j S = S P_j = \lambda_j P_j$. Then, this implies that $\lambda_j P_j = \sum_{h=1}^{k} \lambda_h P_j P_h(A)$.

Hence, $\lambda_j P_j P_h(A) = \lambda_h P_j P_h(A)$. Thus, $P_j P_h(A) = O$ whenever $j \neq h$. Therefore, it follows that $P_j = P_j(A) = P_j P_j(A)$. $\square$

Observation IV-1-15. Let $A = S + N$ be the S-N decomposition of an $n \times n$ matrix A. Let T be an $n \times n$ invertible matrix such that if we set $\Lambda = T^{-1}ST$, then $\Lambda = \mathrm{diag}[\lambda_1 I_1, \lambda_2 I_2, \dots, \lambda_k I_k]$, where $\lambda_1, \lambda_2, \dots, \lambda_k$ are distinct eigenvalues of S (and also of A), I_j is the $m_j \times m_j$ identity matrix, and m_j is the multiplicity of the eigenvalue λ_j. It is easy to show that
- (i) if we set $M = T^{-1}NT$, then M is nilpotent, $M\Lambda = \Lambda M$, and $M = \mathrm{diag}[M_1, M_2, \dots, M_k]$, where M_j are $m_j \times m_j$ nilpotent matrices,
- (ii) if we set $\mathcal{P}_j = T\,\mathrm{diag}[E_{j1}, E_{j2}, \dots, E_{jk}]T^{-1}$, where $E_{j\ell} = O$ if $j \neq \ell$, while $E_{jj} = I_j$, we obtain

$$\begin{cases} I &= \mathcal{P}_1 + \mathcal{P}_2 + \cdots + \mathcal{P}_k, \qquad \mathcal{P}_j\mathcal{P}_h = O \qquad (j \neq h), \\ S &= \lambda_1\mathcal{P}_1 + \lambda_2\mathcal{P}_2 + \cdots + \lambda_k\mathcal{P}_k. \end{cases}$$

Therefore, $\mathcal{P}_j = P_j(S) = P_j(A)$ $(j = 1, 2, \dots, k)$ (cf. Observation IV-1-14).

The following two remarks concern real diagonalizable matrices.

Remark IV-1-16. Let A be a real $n \times n$ diagonalizable matrix and let $\lambda_1, \lambda_2 \dots, \lambda_n$ be the eigenvalues of A. Then, there exists a real $n \times n$ invertible matrix P such that
- (1) in the case when all eigenvalues λ_j $(j = 1, 2, \dots, n)$ are real, then $P^{-1}AP$ is a real diagonal matrix whose entries on the main diagonal are $\lambda_1, \lambda_2, \dots, \lambda_n$,
- (2) in the case when all eigenvalues are not real, then n is an even integer $2m$ and $P^{-1}AP = \mathrm{diag}[D_1, D_2, \dots, D_m]$, where $\lambda_{2j-1} = a_j + ib_j$, $\lambda_{2j} = a_j - ib_j$, and
$$D_j = \begin{bmatrix} a_j & b_j \\ -b_j & a_j \end{bmatrix},$$
- (3) in other cases, $P^{-1}AP = \mathrm{diag}[D_1, D_2, \dots, D_h]$, where $\lambda_{2j-1} = a_j + ib_j$, $\lambda_{2j} = a_j - ib_j$, and $D_j = \begin{bmatrix} a_j & b_j \\ -b_j & a_j \end{bmatrix}$ for $j = 1, 2, \dots, h-1$ and D_h is a real diagonal matrix whose entries on the main diagonal are λ_j $(j = 2h - 1, \dots, n)$.

Remark IV-1-17. For any given real $n \times n$ matrix A, there exists a sequence $\{B_k : k = 1, 2, \dots\}$ of real $n \times n$ diagonalizable matrices such that $\lim\limits_{k \to +\infty} B_k = A$. This can be proved in the following way:
- (i) let $A = S + N$ be S-N decomposition of A,
- (ii) using Remark IV-1-16, assume that $S = \mathrm{diag}[D_1, D_2, \dots, D_h]$, as in (3) of Remark IV-1-16,
- (iii) find the form of N by $SN = NS$,
- (iv) triangularize N without changing S,
- (v) use a method similar to the proof of Lemma IV-1-4.

Details of proofs of Remark IV-1-16 and IV-1-17 are left to the reader as exercises.

Now, we give two examples of calculation of the S-N decomposition.

Example IV-1-18. The matrix $A = \begin{bmatrix} 252 & 498 & 4134 & 698 \\ -234 & -465 & -3885 & -656 \\ 15 & 30 & 252 & 42 \\ -10 & -20 & -166 & -25 \end{bmatrix}$ has two distinct eigenvalues 3 and 4, and

$$p_A(\lambda) = (\lambda - 4)^2(\lambda - 3)^2, \qquad \frac{1}{p_A(\lambda)} = \frac{1}{(\lambda - 4)^2} - \frac{2}{\lambda - 4} + \frac{1}{(\lambda - 3)^2} + \frac{2}{\lambda - 3}.$$

Set
$$P_1(\lambda) = (\lambda - 3)^2 - 2(\lambda - 4)(\lambda - 3)^2, \qquad P_2(\lambda) = (\lambda - 4)^2 + 2(\lambda - 3)(\lambda - 4)^2.$$

Then,
$$P_1(A) = \begin{bmatrix} -1 & -2 & 134 & 198 \\ 1 & 2 & -125 & -186 \\ 0 & 0 & 9 & 12 \\ 0 & 0 & -6 & -8 \end{bmatrix}, \qquad P_2(A) = \begin{bmatrix} 2 & 2 & -134 & -198 \\ -1 & -1 & 125 & 186 \\ 0 & 0 & -8 & -12 \\ 0 & 0 & 6 & 9 \end{bmatrix}.$$

Therefore,
$$\begin{cases} S = 4P_1(A) + 3P_2(A) = \begin{bmatrix} 2 & -2 & 134 & 198 \\ 1 & 5 & -125 & -186 \\ 0 & 0 & 12 & 12 \\ 0 & 0 & -6 & -5 \end{bmatrix}, \\[4em] N = A - S = \begin{bmatrix} 250 & 500 & 4000 & 500 \\ -235 & -470 & -3760 & -470 \\ 15 & 30 & 240 & 30 \\ -10 & -20 & -160 & -20 \end{bmatrix}. \end{cases}$$

Example IV-1-19. The matrix $A = \begin{bmatrix} 3 & 4 & 3 \\ 2 & 7 & 4 \\ -4 & 8 & 3 \end{bmatrix}$ has two distinct eigenvalues $\lambda_1 = 11$, $\lambda_2 = 1$, and

$$p_A(\lambda) = (\lambda - 1)^2(\lambda - 11), \qquad \frac{1}{p_A(\lambda)} = \frac{1}{100(\lambda - 11)} - \frac{(\lambda + 9)}{100(\lambda - 1)^2}.$$

Hence,
$$1 = \frac{(\lambda - 1)^2}{100} - \frac{(\lambda + 9)(\lambda - 11)}{100}.$$

Set
$$P_1(\lambda) = \frac{(\lambda - 1)^2}{100}, \qquad P_2(\lambda) = -\frac{(\lambda + 9)(\lambda - 11)}{100}.$$

Then,
$$P_1(A) = \frac{1}{100} \begin{bmatrix} 0 & 56 & 28 \\ 0 & 76 & 38 \\ 0 & 48 & 24 \end{bmatrix}, \qquad P_2(A) = \frac{1}{100} \begin{bmatrix} 100 & -56 & -28 \\ 0 & 24 & -38 \\ 0 & -48 & 76 \end{bmatrix}.$$

Therefore,
$$\begin{cases} S = 11P_1(A) + P_2(A) = \frac{1}{10} \begin{bmatrix} 10 & 56 & 28 \\ 0 & 86 & 38 \\ 0 & 48 & 34 \end{bmatrix}, \\[4em] N = A - S = \frac{1}{10} \begin{bmatrix} 20 & -16 & 2 \\ 20 & -16 & 2 \\ -40 & 32 & -4 \end{bmatrix}. \end{cases}$$

In this case, $SN = NS = N$ and $N^2 = O$. Let $\mathcal{V}_j = \text{Image}\,(P_j(A))$ $(j = 1, 2)$. Then, by virtue of (1) of Lemma IV-1-10, $P_j(A)\vec{p} = \vec{p}$ for all $\vec{p} \in \mathcal{V}_j$ $(j = 1, 2)$. Furthermore, $\mathcal{V}_1$ is spanned by $\begin{bmatrix} 14 \\ 19 \\ 12 \end{bmatrix}$ and $\mathcal{V}_2$ is spanned by $\begin{bmatrix} 1 \\ 0 \\ 0 \end{bmatrix}$ and $\begin{bmatrix} 0 \\ 1 \\ -2 \end{bmatrix}$. Set $P_0 = \begin{bmatrix} 14 & 1 & 0 \\ 19 & 0 & 1 \\ 12 & 0 & -2 \end{bmatrix}$. Then,

$$P_0^{-1}SP_0 = \begin{bmatrix} 11 & 0 & 0 \\ 0 & 1 & 0 \\ 0 & 0 & 1 \end{bmatrix}, \quad P_0^{-1}NP_0 = 2\begin{bmatrix} 0 & 0 & 0 \\ 0 & 1 & -1 \\ 0 & 1 & -1 \end{bmatrix}.$$

It is noteworthy that there is only one linearly independent eigenvector $\vec{x} = \begin{bmatrix} 1 \\ 1 \\ -2 \end{bmatrix}$ for the eigenvalue $\lambda_2 = 1$.

It is not difficult to make a program for calculation of S and N with a computer. For more examples of calculation of S and N, see [HKS].

IV-2. Homogeneous systems of linear differential equations

In this section, we explain the basic results concerning the structure of solutions of a homogeneous system of linear differential equations given by

$$\text{(IV.2.1)} \qquad \frac{d\vec{y}}{dt} = A(t)\vec{y},$$

where the entries of the $n \times n$ matrix $A(t)$ are continuous on an interval $\mathcal{I} = \{t : a \leq t \leq b\}$. Let us prove the following basic theorem.

Theorem IV-2-1. *The solutions of (IV.2.1) forms an n-dimensional vector space over $\mathbb{C}$.*

We break the entire proof into three observations.

Observation IV-2-2. Any linear combination of a finite number of solutions of (IV.2.1) is also a solution of (IV.2.1). We can prove the existence of n linearly independent solutions of (IV.2.1) on the interval $\mathcal{I}$ by using Theorem I-3-5 with n linearly independent initial conditions at $t = t_0$. Notice that each column vector of a solution Y of the differential equation

$$\text{(IV.2.2)} \qquad \frac{dY}{dt} = A(t)Y$$

on an $n \times n$ unknown matrix Y is a solution of system (IV.2.1). Therefore, constructing an invertible solution Y of (IV.2.2), we can construct n linearly independent solutions of (IV.2.1) all at once. If an $n \times n$ matrix $Y(t)$ is a solution of equation (IV.2.2) on an interval $\mathcal{I} = \{t : a \leq t \leq b\}$ and $Y(t) \in \text{GL}(n, \mathbb{C})$ for all $t \in \mathcal{I}$, then $Y(t)$ is called a *fundamental matrix solution* of system (IV.2.1) on $\mathcal{I}$. Furthermore, n columns of a fundamental matrix solution $Y(t)$ of (IV.2.2) are said to form a *fundamental set of n linearly independent solutions* of (IV.2.1) on the interval $\mathcal{I}$.

Observation IV-2-3. Let $\Phi(t)$ be a solution of (IV.2.2) on $\mathcal{I}$. Also, let $\Psi(t)$ be a solution of the *adjoint equation* of (IV.2.2):

$$\text{(IV.2.3)} \qquad \frac{dZ}{dt} = -ZA(t)$$

on the interval $\mathcal{I}$, where Z is an $n \times n$ unknown matrix. Then,

$$\frac{d}{dt}[\Psi(t)\Phi(t)] = -\Psi(t)A(t)\Phi(t) + \Psi(t)A(t)\Phi(t) = O.$$

This implies that the matrix $\Psi(t)\Phi(t)$ is independent of t. Therefore, $\Psi(t)\Phi(t) = \Psi(\tau)\Phi(\tau)$ for any fixed point $\tau \in \mathcal{I}$ and for all $t \in \mathcal{I}$. Note that the initial values $\Phi(\tau)$ and $\Psi(\tau)$ at $t = \tau$ can be prescribed arbitrarily. In particular, in the case when $\Phi(\tau) \in \mathrm{GL}(n, \mathbb{C})$, by choosing $\Psi(\tau) = \Phi(\tau)^{-1}$, we obtain $\Psi(t)\Phi(t) = I_n$ for all $t \in \mathcal{I}$. Thus, we proved the following lemma.

Lemma IV-2-4. *Let an $n \times n$ matrix $\Phi(t)$ be a solution of (IV.2.2) on the interval $\mathcal{I}$. Then, $\Phi(t)$ is invertible for all $t \in \mathcal{I}$ (i.e., a fundamental matrix solution of (IV.2.1)) if $\Phi(\tau)$ is invertible for some $\tau \in \mathcal{I}$. Furthermore, $\Phi(t)^{-1}$ is the unique solution of (IV.2.3) on $\mathcal{I}$ satisfying the initial condition $Z(\tau) = \Phi(\tau)^{-1}$.*

Observation IV-2-5. Denote by $\Phi(t;\tau)$ the unique solution of the initial-value problem

$$\text{(IV.2.4)} \qquad \frac{dY}{dt} = A(t)Y, \qquad Y(\tau) = I_n,$$

where $\tau \in \mathcal{I}$. Then, $\Phi(t;\tau) \in \mathrm{GL}(n, \mathbb{C})$ for all $t \in \mathcal{I}$. The general structure of solutions of (IV.2.1) and (IV.2.2) are given by the following theorem, which can be easily verified.

Theorem IV-2-6. *The $\mathbb{C}^n$-valued function $\vec{y}(t) = \Phi(t;\tau)\vec{\eta}$ is the unique solution of the initial-value problem*

$$\frac{d\vec{y}}{dt} = A(t)\vec{y}, \qquad \vec{y}(\tau) = \vec{\eta},$$

where $\vec{\eta} \in \mathbb{C}^n$. Also, the $n \times n$ matrix $Y = \Phi(t;\tau)\Gamma$ is the unique solution of the initial-value problem

$$\frac{dY}{dt} = A(t)Y, \qquad Y(\tau) = \Gamma,$$

where $\Gamma \in \mathcal{M}_n(\mathbb{C})$.

Theorem IV-2-1 is a corollary of Theorem IV-2-6. $\square$

Remark IV-2-7.
   (1) The general form of a fundamental matrix solution of (IV.2.1) is given by $Y(t) = \Phi(t;\tau)\Gamma$, where $\Gamma \in \mathrm{GL}(n, \mathbb{C})$.
   (2) If a fundamental matrix solution is given by $Y(t) = \Phi(t;\tau)\Gamma$, then $Y(\tau) = \Gamma$. Hence,

$$\text{(IV.2.5)} \qquad \Phi(t;\tau) = Y(t)Y(\tau)^{-1} \qquad (t,\ \tau \in \mathcal{I})$$

for any fundamental matrix solution $Y(t)$. In particular,

$$(\text{IV.2.6}) \qquad \Phi(t;\tau) \; = \; \Phi(t;\tau_1)\Phi(\tau;\tau_1)^{-1} \qquad \text{for} \quad t, \, \tau, \, \tau_1 \in \mathcal{I}.$$

(3) In the case when $A(t)$ is a scalar (i.e., $n=1$), we obtain easily

$$(\text{IV.2.7}) \qquad \Phi(t;\tau) = \exp\left[\int_\tau^t A(s)ds)\right].$$

In the general case, we define $\exp\left[\int_\tau^t A(s)ds\right]$ by

$$\exp[B(t)] \; = \; I_n \; + \; \sum_{m=1}^{+\infty} \frac{1}{m!}B(t)^m, \qquad \text{where} \qquad B(t) \; = \; \int_\tau^t A(s)ds.$$

However, generally speaking, (IV.2.7) holds only in the case when $B(t)$ and $B'(t) = A(t)$ commute. In particular, $\Phi(t;\tau) = \exp[(t-\tau)A]$ if $A = A(t)$ is independent of t. In §IV-3, we shall explain how to calculate $\exp[(t-\tau)A]$, using the S-N decomposition of A. Also, (IV.2.7) holds in the case when $A(t)$ is diagonal on the interval $\mathcal{I}$. A less trivial case is given in Exercise IV-9. It is easy to see that $B(t)$ and $A(t)$ commute if $A(t)$ is a 2×2 upper-triangular matrix with an eigenvalue of multiplicity 2. For example, the matrix $A(t) = \begin{bmatrix} \cos t & 1 \\ 0 & \cos t \end{bmatrix}$ satisfies the requirement. In this case, $\Phi(t;\tau) \; = \; \exp\left[\int_\tau^t A(s)ds\right] \; = \; \exp\left(\begin{bmatrix} \sin t - \sin\tau & t-\tau \\ 0 & \sin t - \sin\tau \end{bmatrix}\right) =$ $\exp(\sin t - \sin\tau)\begin{bmatrix} 1 & t-\tau \\ 0 & 1 \end{bmatrix}.$

(4) $\det Y(t) = \det Y(\tau)\exp\left\{\int_\tau^t \operatorname{tr}A(s)ds\right\}$ if $Y(t)$ satisfies (IV.2.2), where $\det A$ and $\operatorname{tr}A$ are the determinant and trace of the matrix A. This formula is known as *Abel's formula* (cf. [CL, p. 28]).

Proof.

Regarding $\det Y(t)$ as a function of n column vectors $\{\vec{y}_1(t),\dots,\vec{y}_n(t)\}$ of $Y(t)$, set $\det Y(t) = \mathcal{F}(\vec{y}_1(t),\dots,\vec{y}_n(t))$. Then,

$$(\text{IV.2.8}) \qquad \frac{d\det Y(t)}{dt} \; = \; \sum_{m=1}^{n} \mathcal{F}(\dots, A(t)\vec{y}_m(t),\dots).$$

Denote the right-hand side of (IV.2.8) by $\mathcal{G}(\vec{y}_1(t),\dots,\vec{y}_n(t))$. Then, $\mathcal{G}$ is multilinear and alternating in $\vec{y}_1(t),\dots,\vec{y}_n(t)$. Furthermore, $\mathcal{G} = \operatorname{tr}A(t)$ if $Y(t) = I_n$. Therefore, $\mathcal{G} = \operatorname{tr}A(t)\det Y(t)$. Solving the differential equation $\frac{d\det Y(t)}{dt} = \operatorname{tr}A(t)\det Y(t)$, we obtain Abel's formula. $\square$

IV-3. Homogeneous systems with constant coefficients

For an $n \times n$ matrix A, we define $\exp[A]$ by

$$(IV.3.1) \qquad \exp[A] \;=\; I_n \;+\; \sum_{h=1}^{\infty} \frac{1}{h!}\, A^h.$$

It is easy to show that the matrix $\exp[A]$ satisfies the condition $\exp[A + B] = \exp[A]\exp[B]$ if A and B commute. This implies that $\exp[A]$ is invertible and $(\exp[A])^{-1} = \exp[-A]$. Thus, we obtain a fundamental matrix solution $Y = \exp[tA]$ of the system $\dfrac{d\vec{y}}{dt} \;=\; A\vec{y}$ with a constant matrix A by solving the initial-value problem

$$(IV.3.2) \qquad \frac{dY}{dt} \;=\; AY, \qquad Y(0) \;=\; I_n.$$

This, in turn, implies that the unique solution of the initial-value problem

$$(IV.3.3) \qquad \frac{d\vec{y}}{dt} \;=\; A\vec{y}, \qquad \vec{y}(\tau) \;=\; \vec{p}$$

is given by $\vec{y} = \exp[(t - \tau)A]\vec{p}$, where $\vec{p} \in \mathbb{C}^n$ is a constant vector.

In this section, we explain how to calculate $\exp[tA]$ for a given constant matrix A, using the S-N decomposition of A. Assume that an $n \times n$ matrix A has k distinct eigenvalues $\lambda_1, \lambda_2, \ldots, \lambda_k$. Let $A = S + N$ be the S-N decomposition of A. Also, let $P_j(A)$ $(j = 1, 2, \ldots, k)$ be the projections defined in §IV-1 (cf. (IV.1.4)). Then,

$$(IV.3.4) \qquad I_n \;=\; \sum_{j=1}^{k} P_j(A), \qquad S \;=\; \sum_{j=1}^{k} \lambda_j P_j(A), \qquad N \;=\; A - S,$$

and

$$(IV.3.5) \qquad P_j(A)P_h(A) \;=\; \begin{cases} P_j(A) & \text{if} \quad h = j \\ O & \text{if} \quad h \neq j \end{cases} \qquad (j, h = 1, 2, \ldots, k).$$

The two matrices S and N commute.

Denote by $\mathcal{V}_j$ the image of the mapping $P_j(A) : \mathbb{C}^n \to \mathbb{C}^n$ (cf. Lemma IV-1-10). It is known that $S\vec{p} = \lambda_j \vec{p}$ for $\vec{p}$ in $\mathcal{V}_j$. Hence, $S^{\ell}\vec{p} = \lambda_j^{\ell}\vec{p}$ and

$$\exp[tS]\vec{p} \;=\; \left\{ 1 + \sum_{h=1}^{+\infty} \frac{(\lambda_j t)^h}{h!} \right\}\vec{p} \;=\; e^{\lambda_j t}\, \vec{p}.$$

On the other hand, $\exp[tN] = I_n + \displaystyle\sum_{h=1}^{n-1} \frac{t^h}{h!} N^h$ since N is nilpotent. Therefore,

$$\exp[tA]\vec{p} = \exp[t(S + N)]\vec{p} \;=\; \exp[tN]\exp[tS]\vec{p} = e^{\lambda_j t}\exp[tN]\vec{p}$$

$$(IV.3.6) \qquad = e^{\lambda_j t}\left[I_n + \sum_{h=1}^{n-1} \frac{t^h}{h!} N^h \right]\vec{p} \qquad \text{for} \quad \vec{p} \in \mathcal{V}_j.$$

Applying (IV.3.4) and (IV.3.6) to a general $\vec{p} \in \mathbb{C}^n$, we derive

$$(IV.3.7) \qquad \exp[tA]\vec{p} \;=\; \sum_{j=1}^{k} e^{\lambda_j t}\left[I_n \;+\; \sum_{h=1}^{n-1}\frac{t^h}{h!}N^h \right] P_j(A)\vec{p} \qquad \text{for} \quad \vec{p} \in \mathbb{C}^n.$$

Thus, we proved the following theorem.

Theorem IV-3-1. *The matrix* $\exp[tA]$ *is calculated by formula*

$$(IV.3.8) \qquad \exp[tA] \;=\; \sum_{j=1}^{k} e^{\lambda_j t}\left[I_n \;+\; \sum_{h=1}^{n-1}\frac{t^h}{h!}N^h \right] P_j(A).$$

Since the general solution of the differential equation

$$(IV.3.9) \qquad \qquad \frac{d\vec{y}}{dt} \;=\; A\vec{y}$$

is given by (IV.3.7), the following important result is obtained.

Theorem IV-3-2.
(i) *If* $\Re(\lambda_j) < 0$ *for* $j = 1, 2, \ldots, k$, *then every solution of (IV.3.9) tends to* $\vec{0}$ *as* $t \to +\infty$,
(ii) *if* $\Re(\lambda_j) > 0$ *for some* j, *some solutions of (IV.3.9) tend to* ∞ *as* $t \to +\infty$,
(iii) *every solution of (IV.3.9) is bounded for* $t \geq 0$ *if and only if* $\Re(\lambda_j) \leq 0$ *for* $j = 1, 2, \ldots, k$ *and* $NP_j(A) = 0$ *if* $\Re(\lambda_j) = 0$.

Now, we illustrate calculation of $\exp[tA]$ in two examples. Note that in the case when A has nonreal eigenvalues, we must use complex numbers in our calculation. Nevertheless, if A is a real matrix, then $\exp[tA]$ is also real. Hence, at the end of our calculation, we obtain real-valued solutions of (IV.3.9) if A is real.

Example IV-3-3. Consider the matrix $A = \begin{bmatrix} -2 & 1 & 0 \\ 0 & -2 & 0 \\ 3 & 2 & 1 \end{bmatrix}$. The characteristic polynomial of A is $p_A(\lambda) = (\lambda-1)(\lambda+2)^2$. By using the partial fraction decomposition of $\dfrac{1}{p_A(\lambda)}$, we derive $1 = \dfrac{(\lambda+2)^2 \;-\; (\lambda+5)(\lambda-1)}{9}$. Setting $P_1(\lambda) = \dfrac{(\lambda+2)^2}{9}$ and $P_2(\lambda) = -\dfrac{(\lambda+5)(\lambda-1)}{9}$, we obtain

$$P_1(A) \;=\; \begin{bmatrix} 0 & 0 & 0 \\ 0 & 0 & 0 \\ 1 & 1 & 1 \end{bmatrix}, \quad P_2(A) \;=\; \begin{bmatrix} 1 & 0 & 0 \\ 0 & 1 & 0 \\ -1 & -1 & 0 \end{bmatrix}.$$

Set

$$S \;=\; P_1(A) \;-\; 2P_2(A) \;=\; \begin{bmatrix} -2 & 0 & 0 \\ 0 & -2 & 0 \\ 3 & 3 & 1 \end{bmatrix}, \quad N \;=\; A - S \;=\; \begin{bmatrix} 0 & 1 & 0 \\ 0 & 0 & 0 \\ 0 & -1 & 0 \end{bmatrix}.$$

Note that $N^2 = 0$. Hence,

$$\exp[tA] = e^t [\, I_3 + tN \,] P_1(A) + e^{-2t} [\, I_3 + tN \,] P_2(A)$$

$$= \begin{bmatrix} e^{-2t} & te^{-2t} & 0 \\ 0 & e^{-2t} & 0 \\ e^t - e^{-2t} & e^t - (1+t)e^{-2t} & e^t \end{bmatrix}.$$

The solution of the initial-value problem $\dfrac{d\vec{y}}{dt} = A\vec{y}$, $\vec{y}(0) = \vec{\eta}$ is $\vec{y}(t) = \exp[tA]\vec{\eta}$. To find a solution satisfying the condition $\lim\limits_{t \to +\infty} \vec{y}(0) = \vec{0}$, we must choose $\vec{\eta}$ so that $P_1(A)\vec{\eta} = \vec{0}$. Such an $\vec{\eta}$ is given by $\vec{\eta} = P_2(A)\vec{c}$, where $\vec{c}$ is an arbitrary constant vector in $\mathbb{C}^3$.

Example IV-3-4. Next, consider the matrix $A = \begin{bmatrix} 0 & -1 & 1 \\ 1 & 0 & -1 \\ -1 & 1 & 0 \end{bmatrix}$. The characteristic polynomial of A is $p_A(\lambda) = \lambda(\lambda^2 + 3) = \lambda(\lambda - i\sqrt{3})(\lambda + i\sqrt{3})$. Using the partial fraction decomposition of $\dfrac{1}{p_A(\lambda)}$, we obtain

$$1 = \frac{1}{3}(\lambda - i\sqrt{3})(\lambda + i\sqrt{3}) - \frac{1}{6}\lambda(\lambda + i\sqrt{3}) - \frac{1}{6}\lambda(\lambda - i\sqrt{3}).$$

Setting

$$P_1(\lambda) = \frac{1}{3}(\lambda^2 + 3), \quad P_2(\lambda) = -\frac{1}{6}\lambda(\lambda + i\sqrt{3}), \quad P_3(\lambda) = -\frac{1}{6}\lambda(\lambda - i\sqrt{3}),$$

we obtain

$$P_1(A) = \frac{1}{3}\begin{bmatrix} 1 & 1 & 1 \\ 1 & 1 & 1 \\ 1 & 1 & 1 \end{bmatrix}, \quad P_2(A) = -\frac{1}{6}\begin{bmatrix} -2 & 1 - i\sqrt{3} & 1 + i\sqrt{3} \\ 1 + i\sqrt{3} & -2 & 1 - i\sqrt{3} \\ 1 - i\sqrt{3} & 1 + i\sqrt{3} & -2 \end{bmatrix},$$

and $P_3(A)$ is the complex conjugate of $P_2(A)$. If we set

$$S = (i\sqrt{3})P_2(A) - (i\sqrt{3})P_3(A),$$

then $S = A$. This implies that $N = 0$. Thus, we obtain

$$\exp[tA] = P_1(A) + e^{i\sqrt{3}t}P_2(A) + e^{-i\sqrt{3}t}P_3(A)$$

$$= P_1(A) + 2\Re\left(e^{i\sqrt{3}t}P_2(A)\right).$$

Using

$$\begin{cases} \Re\left(e^{i\sqrt{3}t}(1 + i\sqrt{3})\right) = \cos(\sqrt{3}t) - \sqrt{3}\sin(\sqrt{3}t), \\ \Re\left(e^{i\sqrt{3}t}(1 - i\sqrt{3})\right) = \cos(\sqrt{3}t) + \sqrt{3}\sin(\sqrt{3}t), \end{cases}$$

we find

$$\exp[tA] \;=\; \frac{1}{3} \begin{bmatrix} a(t) & b(t) & c(t) \\ c(t) & a(t) & b(t) \\ b(t) & c(t) & a(t) \end{bmatrix},$$

where

$$\begin{cases} a(t) \;=\; 1 + 2\cos(\sqrt{3}\,t), \\[2mm] b(t) \;=\; 1 - \left\{ \cos(\sqrt{3}\,t) + \sqrt{3}\sin(\sqrt{3}\,t) \right\}, \\[2mm] c(t) \;=\; 1 - \left\{ \cos(\sqrt{3}\,t) - \sqrt{3}\sin(\sqrt{3}\,t) \right\}. \end{cases}$$

Remark IV-3-5. *Functions of a matrix.* In this remark, we explain how to define functions of a matrix A.

I. *A particular case*: Let λ_0, I_n, and N be a number, the $n \times n$ identity matrix, and an $n \times n$ nilpotent matrix, respectively. Also, consider a function $f(\lambda)$ in a neighborhood of λ_0. Assume that $f(\lambda)$ has the Taylor series expansion (i.e., f is analytic at λ_0)

$$f(\lambda) \;=\; f(\lambda_0) \;+\; \sum_{h=1}^{\infty} \frac{f^{(h)}(\lambda_0)}{h!}(\lambda - \lambda_0)^h.$$

In this case, define $f(\lambda_0 I_n + N)$ by

$$f(\lambda_0 I_n + N) \;=\; f(\lambda_0)I_n \;+\; \sum_{h=1}^{\infty} \frac{f^{(h)}(\lambda_0)}{h!} N^h \;=\; f(\lambda_0)I_n \;+\; \sum_{h=1}^{n-1} \frac{f^{(h)}(\lambda_0)}{h!} N^h.$$

Since N is nilpotent, the matrix $\displaystyle\sum_{h=1}^{n-1} \frac{f^{(h)}(\lambda_0)}{h!} N^h$ is also nilpotent. Therefore, the characteristic polynomial $p_{f(\lambda_0 I + N)}(\lambda)$ of $f(\lambda_0 I + N)$ is

$$p_{f(\lambda_0 I + N)}(\lambda) \;=\; (\lambda \,-\, f(\lambda_0))^n.$$

II. *The general case*: Assume that the characteristic polynomial $p_A(\lambda)$ of an $n \times n$ matrix A is

$$p_A(\lambda) \;=\; (\lambda - \lambda_1)^{m_1}(\lambda - \lambda_2)^{m_2}\cdots(\lambda - \lambda_k)^{m_k},$$

where $\lambda_1, \ldots, \lambda_k$ are distinct eigenvalues of A. Construct $P_j(A)\,(j = 1, \ldots, k)$, S, and N as above. Then,

$$A \;=\; (\lambda_1 I_n + N)P_1(A) \;+\; (\lambda_2 I_n + N)P_2(A) \;+\; \cdots \;+\; (\lambda_k I_n + N)P_k(A).$$

Therefore,

$$A^\ell \;=\; (\lambda_1 I_n + N)^\ell P_1(A) \;+\; (\lambda_2 I_n + N)^\ell P_2(A) \;+\; \cdots \;+\; (\lambda_k I_n + N)^\ell P_k(A)$$

for every integer ℓ.

Assuming that a function $f(\lambda)$ has the Taylor series expansion

$$f(\lambda) = f(\lambda_j) + \sum_{h=1}^{\infty} \frac{f^{(h)}(\lambda_j)}{h!} (\lambda - \lambda_j)^h$$

at $\lambda = \lambda_j$ for every $j = 1, \ldots, k$, we define $f(A)$ by

$$
\text{(IV.3.10)} \qquad
\begin{aligned}
f(A) = {}& f(\lambda_1 I_n + N)P_1(A) + f(\lambda_2 I_n + N)P_2(A) \\
& + \cdots + f(\lambda_k I_n + N)P_k(A).
\end{aligned}
$$

Since $P_j(S) = P_j(A)$ (cf. Observation IV-1-15), this definition applied to S yields

$$f(S) = f(\lambda_1 I_n)P_1(A) + f(\lambda_2 I_n)P_2(A) + \cdots + f(\lambda_k I_n)P_k(A)$$

and $f(A) - f(S)$ has a form $N \times$ (a polynomial in S and N). Therefore, $f(A) - f(S)$ is nilpotent. Furthermore, $f(S)$ and $f(A)$ commute. This implies that

$$f(A) = f(S) + (f(A) - f(S))$$

is the S-N decomposition of $f(A)$. Thus,

$$p_{f(A)}(\lambda) = p_{f(S)}(\lambda) = (\lambda - f(\lambda_1))^{m_1} \cdots (\lambda - f(\lambda_k))^{m_k}.$$

Example IV-3-6. In the case when $f(\lambda) = \log(\lambda)$, define $\log(A)$ by

$$\log(A) = \log(\lambda_1 I_n + N)P_1(A) + \log(\lambda_2 I_n + N)P_2(A) + \cdots + \log(\lambda_k I_n + N)P_k(A),$$

where we must assume that A is invertible so that $\lambda_j \neq 0$ for all eigenvalues of A. Let us look at $\log(\lambda_0 I_n + N)$ more closely, assuming that $\lambda_0 \neq 0$. Since

$$\log(\lambda_0 + \mu) = \log(\lambda_0) + \log\left(1 + \frac{\mu}{\lambda_0}\right) = \log(\lambda_0) + \sum_{m=1}^{+\infty} \frac{(-1)^{m+1}}{m} \left(\frac{\mu}{\lambda_0}\right)^m,$$

we obtain

$$\log(\lambda_0 I_n + N) = \log(\lambda_0)I_n + \sum_{m=1}^{n-1} \frac{(-1)^{m+1}}{m} \left(\frac{N}{\lambda_0}\right)^m.$$

It is not difficult to show that $\exp[\log(A)] = A$. In fact, since

$$
\begin{aligned}
(\log(A))^m = {}& (\log(\lambda_1 I_n + N))^m P_1(A) + (\log(\lambda_2 I_n + N))^m P_2(A) \\
& + \cdots + (\log(\lambda_k I_n + N))^m P_k(A),
\end{aligned}
$$

it is sufficient to show that $\exp[\log(\lambda_0 I_n + N)] = \lambda_0 I_n + N$. This can be proved by using $\exp[\log(\lambda_0 + \mu)] = \lambda_0 + \mu$.

Observation IV-3-7. In the definition of $\log(A)$ in Example IV-3-6, we used $\log(\lambda_j)$. The function $\log(\lambda)$ is not single-valued. Therefore, the definition of $\log(A)$ is not unique.

Observation IV-3-8. Let $A = S + N$ be the S-N decomposition of A. If A is invertible, S is also invertible. Therefore, we can write A as $A = S(I_n + M)$, where $M = S^{-1}N = NS^{-1}$. Since S and N commute, two matrices S and M commute. Furthermore, M is nilpotent. Using this form, we can define $\log(A)$ by

$$\log(A) \; = \; \log(S) \; + \; \log(I_n \; + \; M),$$

where

$$\log(S) \; = \; \log(\lambda_1)P_1(A) \; + \; \log(\lambda_2)P_2(A) \; + \; \cdots \; + \; \log(\lambda_k)P_k(A)$$

and

$$\log(I_n \; + \; M) \; = \; \sum_{m=1}^{n-1} \frac{(-1)^{m+1}}{m} M^m.$$

This definition and the previous definition give the same function $\log(A)$ if the same definition of $\log(\lambda_j)$ is used.

Example IV-3-9. Let us calculate $\sin(A)$ for $A = \begin{bmatrix} 3 & 4 & 3 \\ 2 & 7 & 4 \\ -4 & 8 & 3 \end{bmatrix}$ (cf. Example IV-1-19). The matrix A has two eigenvalues 11 and 1. The corresponding projections are

$$P_1(A) \; = \; \begin{bmatrix} 0 & \dfrac{14}{25} & \dfrac{7}{25} \\[2mm] 0 & \dfrac{19}{25} & \dfrac{19}{50} \\[2mm] 0 & \dfrac{12}{25} & \dfrac{6}{25} \end{bmatrix}, \qquad P_2(A) \; = \; \begin{bmatrix} 1 & \dfrac{-14}{25} & \dfrac{-7}{25} \\[2mm] 0 & \dfrac{6}{25} & \dfrac{-19}{50} \\[2mm] 0 & \dfrac{-12}{25} & \dfrac{19}{25} \end{bmatrix}.$$

Define $S = 11P_1(A) + P_2(A)$ and $N = A - S$. Then $N^2 = O$. Also,

$$\begin{cases} \sin(11 + x) = & -0.99999 + 0.0044257x + 0.499995x^2 + O(x^3), \\ \sin(1 + x) = & 0.841471 + 0.540302x - 0.420735x^2 + O(x^3). \end{cases}$$

Therefore,

$$\sin(A) \; = \; (-0.99999I_3 + 0.0044257N)P_1(A) + (0.841471I_3 + 0.540302N)P_2(A)$$

$$= \; \begin{bmatrix} 1.92207 & -1.8957 & -0.407549 \\ 1.0806 & -1.42252 & -0.591695 \\ -2.1612 & 0.845065 & 0.1834 \end{bmatrix}.$$

It is known that $\sin x$ has the series expansion

$$\sin x \; = \; \sum_{h=0}^{+\infty} \frac{(-1)^h}{(2h+1)!} x^{2h+1}.$$

Therefore, we can also define $\sin(A)$ by

$$\sin(A) \;=\; \sum_{h=0}^{+\infty} \frac{(-1)^h}{(2h+1)!} A^{2h+1}.$$

However, this approximation is not quite satisfactory if we notice that

$$\sin(11) \;=\; -0.99999 \qquad \text{and} \qquad \sum_{h=0}^{9} \frac{(-1)^h}{(2h+1)!} 11^{2h+1} \;=\; -117.147.$$

IV-4. Systems with periodic coefficients

In this section, we explain how to construct a fundamental matrix solution of a system

$$(IV.4.1) \qquad\qquad \frac{d\vec{y}}{dt} \;=\; A(t)\vec{y}$$

in the case when the $n \times n$ matrix $A(t)$ satisfies the following conditions:
  (1) entries of $A(t)$ are continuous on the entire real line $\mathbb{R}$,
  (2) entries of $A(t)$ are periodic in t of a (positive) period ω, i.e.,

$$(IV.4.2) \qquad\qquad A(t+\omega) \;=\; A(t) \qquad \text{for} \quad t \in \mathbb{R}.$$

Look at the unique $n \times n$ fundamental matrix solution $\Phi(t)$ defined by the initial-value problem

$$(IV.4.3) \qquad\qquad \frac{dY}{dt} \;=\; A(t)Y, \qquad Y(0) \;=\; I_n.$$

Since $\Phi'(t+\omega) = A(t+\omega)\Phi(t+\omega) = A(t)\Phi(t+\omega)$ and $\Phi(0+\omega) = \Phi(\omega)$, the matrix $\Phi(t+\omega)$ is also a fundamental matrix solution of (IV.4.3). As mentioned in (1) of Remark IV-2-7, there exists a constant matrix Γ such that $\Phi(t+\omega) = \Phi(t)\Gamma$ and, consequently, $\Gamma = \Phi(\omega)$. Thus,

$$(IV.4.4) \qquad\qquad \Phi(t+\omega) \;=\; \Phi(t)\Phi(\omega) \qquad \text{for} \quad t \in \mathbb{R}.$$

Setting $B = \omega^{-1}\log[\Phi(\omega)]$ (cf. Example IV-3-6), define an $n \times n$ matrix $P(t)$ by

$$(IV.4.5) \qquad\qquad P(t) \;=\; \Phi(t)\exp[-tB].$$

Then, $P(t+\omega) = \Phi(t+\omega)\exp[-(t+\omega)B] = \Phi(t)\Phi(\omega)\exp[-\omega B]\exp[-tB] = \Phi(t)\exp[-tB] = P(t)$. This shows that $P(t)$ is periodic in t of period ω. Thus, we proved the following theorem.

Theorem IV-4-1 (G. Floquet [Fl]). *Under assumptions (1) and (2), the fundamental matrix solution $\Phi(t)$ of (IV.4.1) defined by the initial-value problem (IV.4.3) has the form*

$$\text{(IV.4.6)} \qquad \Phi(t) = P(t)\exp[tB],$$

where $P(t)$ and B are $n \times n$ matrices such that
(α) $P(t)$ is invertible, continuous, and periodic of period ω in t,
(β) B is a constant matrix such that $\Phi(\omega) = \exp[\omega B]$.

Observation IV-4-2. As was explained in Observation IV-3-8, letting $\Phi(\omega) = S + N = S(I_n + M)$ be the S-N decomposition of $\Phi(\omega)$, we define $\log(\Phi(\omega))$ by $\log(\Phi(\omega)) = \log(S) + \log(I_n + M)$, where

$$\log(S) = \log(\lambda_1)P_1(\Phi(\omega)) + \log(\lambda_2)P_2(\Phi(\omega)) + \cdots + \log(\lambda_k)P_k(\Phi(\omega))$$

and

$$\log(I_n + M) = \sum_{m=1}^{n-1} \frac{(-1)^{m+1}}{m} M^m.$$

In the case when $A(t)$ is a real matrix, the unique solution $\Phi(t)$ of problem (IV.4.3) is also a real matrix. Therefore, the entries of $\Phi(\omega)$ are real. Since S and N are real matrices, the matrix $M = S^{-1}N$ is real. Therefore, $\log(I_n + M)$ is also real. Let us look at $\log(S)$ more closely. If λ_j is a complex eigenvalue of $\Phi(\omega)$, its complex conjugate $\overline{\lambda_j}$ is also an eigenvalue of $\Phi(\omega)$. In this case, set $\lambda_{j+1} = \overline{\lambda_j}$. It is easy to see that the projection $P_{j+1}(\Phi(\omega))$ is also the complex conjugate of $P_j(\Phi(\omega))$. However, if some eigenvalues of $\Phi(\omega)$ are negative, $\log[S]$ is not real. To see this more clearly, rewrite $\log[S]$ in the form

$$\log[S] = \sum_{\lambda_j < 0} \log[\lambda_j]P_j(\Phi(\omega)) + \sum_{\text{other } j} \log[\lambda_j]P_j(\Phi(\omega)).$$

The matrix $\displaystyle\sum_{\text{other } j} \log[\lambda_j]P_j(\Phi(\omega))$ is a real matrix, while the matrix $\displaystyle\sum_{\lambda_j < 0} \log[\lambda_j]P_j(\Phi(\omega))$ is not real. Therefore, $\log[S]$ is not real. To rectify this situation, let us look at S^2. By virtue of the relations given in Lemma IV-1-9, we obtain

$$S^2 = \sum_{\lambda_j < 0} (\lambda_j)^2 P_j(\Phi(\omega)) + \sum_{\text{other } j} (\lambda_j)^2 P_j(\Phi(\omega)).$$

Notice that

$$\log[S^2] = \sum_{\lambda_j < 0} \log[(\lambda_j)^2]P_j(\Phi(\omega)) + 2\sum_{\text{other } j} \log[\lambda_j]P_j(\Phi(\omega))$$

is a real matrix. Therefore, we can find a real matrix

$$\log[\Phi(\omega)^2] = \log[S^2] + 2\log[I_n + M].$$

Now, observe that $\Phi(\omega)^2 = \Phi(2\omega)$ and $\Phi(t + 2\omega) = \Phi(t)\Phi(2\omega)$. Thus, setting

$$\text{(IV.4.7)} \qquad B = \frac{1}{2\omega}\log[\Phi(\omega)^2] \qquad \text{and} \qquad P(t) = \Phi(t)\exp[-tB],$$

we obtain the following theorem.

Theorem IV-4-3. *In the case when the matrix $A(t)$ is real, the matrix $\Phi(t)$ has the form*
$$\Phi(t) = P(t)\exp[tB],$$
where $P(t)$ and B are $n \times n$ real matrices such that
(α) $P(t)$ is invertible, continuous, and periodic of period 2ω in t,
(β) B is a constant matrix such that $\Phi(\omega)^2 = \exp[2\omega B]$.

Remark IV-4-4. In the case when $\Phi(t) = P(t)\exp[tB]$, we have $\dfrac{dP(t)}{dt} = A(t)P(t) - P(t)B$. Therefore, (IV.4.1) is changed to

(IV.4.8)
$$\frac{d\vec{z}}{dt} = B\vec{z}$$

by the transformation $\vec{y} = P(t)\vec{z}$. As noted above, $P(t)$ is periodic in t of period ω (or, respectively, 2ω when $A(t)$ is real), the behavior of solutions of (IV.4.1) is derived from the behavior of solutions of (IV.4.8).

Definition IV-4-5. *The eigenvalues $\mu_1, \ldots, \mu_n$ of the matrix B are called the* characteristic exponents *of equation (IV.4.1). The eigenvalues $\rho_1 = \exp[\omega\mu_1], \ldots, \rho_n = \exp[\omega\mu_n]$ of the matrix $\Phi(\omega)$ (or, respectively, when $A(t)$ is real, the eigenvalues $\rho_1 = \exp[2\omega\mu_1], \ldots, \rho_n = \exp[2\omega\mu_n]$ of $\Phi(\omega)^2$) are called the* multipliers *of equation (IV.4.1).*

The following basic result is a corollary of Theorem IV-3-1.

Corollary IV-4-6. *All solutions of equation (IV.4.1) tend to $\vec{0}$ as $t \to +\infty$ if and only if $|\rho_j| < 1$ for all j (or $\Re(\mu_j) < 0$ for all j).*

Observation IV-4-7. Let $\vec{p}$ be an eigenvector of $\Phi(\omega)$ associated with a multiplier ρ. Then, the solution $\vec{\phi}(t) = \Phi(t)\vec{p}$ of (IV.4.1) satisfies the condition $\vec{\phi}(t + \omega) = \rho\vec{\phi}(t)$. The terminology *multiplier* came from this fact. (In the case when $A(t)$ is real, choosing an eigenvector $\vec{p}$ of $\Phi(\omega)^2$ associated with a multiplier ρ, we obtain $\vec{\phi}(t + 2\omega) = \rho\vec{\phi}(t)$ if $\vec{\phi}(t) = \Phi(t)\vec{p}$.)

Example IV-4-8. Consider the initial-value problem

(IV.4.9)
$$\frac{dY}{dt} = A(t)Y, \qquad Y(0) = I_2,$$

where
$$A(t) = \begin{bmatrix} \cos t & 1 \\ 0 & \cos t \end{bmatrix}.$$

The solution of (IV.4.9) is given by

$$\Phi(t) = \exp\left[\int_0^t A(s)ds\right] = \exp(\sin t)\begin{bmatrix} 1 & t \\ 0 & 1 \end{bmatrix}$$

(cf. (3) of Remark IV-2-7). Therefore,

$$\Phi(2\pi) = \begin{bmatrix} 1 & 2\pi \\ 0 & 1 \end{bmatrix}, \qquad B = \frac{1}{2\pi}\log[\Phi(2\pi)].$$

Since $p_\Phi(\lambda) = (\lambda - 1)^2$, $P_1(\Phi) = I_2$, we obtain

$$S = I_2, \quad M = \begin{bmatrix} 0 & 2\pi \\ 0 & 0 \end{bmatrix}, \quad \text{and} \quad \log[\Phi(2\pi)] = \log[I_2 + M] = M = \begin{bmatrix} 0 & 2\pi \\ 0 & 0 \end{bmatrix}.$$

This gives $B = \dfrac{1}{2\pi} M = \begin{bmatrix} 0 & 1 \\ 0 & 0 \end{bmatrix}$. Therefore,

$$P(t) = \Phi(t) \exp \begin{bmatrix} 0 & -t \\ 0 & 0 \end{bmatrix} = \Phi(t) \begin{bmatrix} 1 & -t \\ 0 & 1 \end{bmatrix} = \exp(\sin t) I_2$$

and $\Phi(t) = P(t) \exp[tB] = \exp(\sin t) \exp \begin{bmatrix} 0 & t \\ 0 & 0 \end{bmatrix}$. Moreover, the characteristic exponents are 0, 0, and the multipliers of (IV.4.9) are 1, 1. Note that in this case, as both of two multipliers are positive, it is not necessary to take $\Phi(2\pi)^2$ and period 4π to find B.

IV-5. Linear Hamiltonian systems with periodic coefficients

This section is divided into two parts. In Part 1, we explain the structure of solutions of a linear Hamiltonian system. The basic facts are found in the theory of real symplectic groups and their Lie algebra. If we denote by J the real $(2n) \times (2n)$ matrix $\begin{bmatrix} O & I_n \\ -I_n & O \end{bmatrix}$, where I_n is the $n \times n$ identity matrix, the *Lie algebra* of the real symplectic group $Sp(2n, \mathbb{R})$ of order $2n$ is the set of all real $2n \times 2n$ matrices of the form JH, where H is a real $(2n) \times (2n)$ symmetric matrix. A linear system $\dfrac{d\vec{y}}{dt} = A(t)\vec{y}$ on $\vec{y} \in \mathbb{R}^{2n}$ is a linear *Hamiltonian system*, if there exists a real $(2n) \times (2n)$ symmetric matrix $H(t)$ such that $A(t) = JH(t)$. Consequently, the fundamental matrix $\Phi(t)$ of this system belongs to $Sp(2n, \mathbb{R})$ if $\Phi(0) = I_{2n}$. We explain all these elementary facts of Hamiltonian systems in Part 1. We also prove that if $G \in Sp(2n, \mathbb{R})$ and $G = S + N = S(I_{2n} + M)$ is the S-N decomposition of G, then S and $I_{2n} + M$ belong to $Sp(2n, \mathbb{R})$. In Part 2, we refine the Floquet theorem (cf. Theorem IV-4-1) for the linear periodic Hamiltonian systems in such a way that the periodic transformation $\vec{y} = P(t)\vec{z}$ given in Remark IV-4-4 is canonical. This means that $P(t)$ belongs to $Sp(2n, \mathbb{R})$. The important consequence is that the linear system $\dfrac{d\vec{z}}{dt} = B\vec{z}$ of Remark IV-4-4 is a Hamiltonian system.

Part 1

Consider a linear Hamiltonian system $\dfrac{d\vec{p}}{dt} = \dfrac{\partial \mathcal{H}}{\partial \vec{q}}(t, \vec{y})$, $\dfrac{d\vec{q}}{dt} = -\dfrac{\partial \mathcal{H}}{\partial \vec{p}}(t, \vec{y})$, where $\vec{p}$ and $\vec{q}$ are unknown vectors in $\mathbb{R}^n$, $\vec{y} = \begin{bmatrix} \vec{p} \\ \vec{q} \end{bmatrix}$, $\mathcal{H}(t, \vec{y}) = \dfrac{1}{2} \vec{y}^T H(t) \vec{y}$ with a real $(2n) \times (2n)$ symmetric matrix $H(t)$, and $\dfrac{\partial \mathcal{H}}{\partial \vec{p}} = \begin{bmatrix} \dfrac{\partial \mathcal{H}}{\partial p_1} \\ \vdots \\ \dfrac{\partial \mathcal{H}}{\partial p_n} \end{bmatrix}$, $\dfrac{\partial \mathcal{H}}{\partial \vec{q}} = \begin{bmatrix} \dfrac{\partial \mathcal{H}}{\partial q_1} \\ \vdots \\ \dfrac{\partial \mathcal{H}}{\partial q_n} \end{bmatrix}$. Here,

$\vec{y}^T$ is the transpose of the column vector $\vec{y}$. Since

$$\frac{\partial \mathcal{H}}{\partial \vec{q}}(t, \vec{y}) \;=\; [O, \; I_n]\, H(t)\vec{y}, \qquad \frac{\partial \mathcal{H}}{\partial \vec{p}}(t, \vec{y}) = [I_n, \; O]\, H(t)\vec{y},$$

the Hamiltonian system can be written in the form

(IV.5.1)
$$\frac{d\vec{y}}{dt} \;=\; JH(t)\vec{y},$$

where $\vec{y}$ is a unknown vector in $\mathbb{R}^{2n}$, $H(t)$ is a real $(2n) \times (2n)$ symmetric matrix, and

(IV.5.2)
$$J = \begin{bmatrix} O & I_n \\ -I_n & O \end{bmatrix}.$$

The matrix J has the following important properties:

(IV.5.3)
$$J^T \;=\; -J, \qquad J^{-1} \;=\; -J,$$

where J^T is the transpose of J.

Let $\Phi(t)$ be the unique real $(2n) \times (2n)$ matrix such that

$$\frac{d\Phi(t)}{dt} \;=\; JH(t)\Phi(t), \qquad \Phi(0) \;=\; I_{2n}.$$

Lemma IV-5-1. *The matrix $\Phi(t)$ satisfies the condition $\Phi(t)^T J\Phi(t) = J$, where $\Phi(t)^T$ is the transpose of $\Phi(t)$.*

Proof.

Differentiate $\Phi(t)^T J\Phi(t)$ to derive

$$\frac{d}{dt}\left[\Phi(t)^T J\Phi(t)\right] \;=\; \Phi(t)^T H(t)(-J)J\Phi(t) \;+\; \Phi(t)^T JJH(t)\Phi(t)$$
$$=\; \Phi(t)^T H(t)\Phi(t) \;-\; \Phi(t)^T H(t)\Phi(t) = O.$$

Then, $\Phi(t)^T J\Phi(t) = \Phi(0)^T J\Phi(0) = J.$ $\quad\square$

The converse of Lemma IV-5-1 is shown in the following lemma.

Lemma IV-5-2. *Assume that a real $(2n) \times (2n)$ matrix $G(t)$ satisfies the following conditions:*

(i) the entries of $G(t)$ are continuously differentiable in t on an interval $\mathcal{I}$,
(ii) $G(t)^T JG(t) = J$ for $t \in \mathcal{I}$.

Then, $G(t)$ is invertible and $H(t) = -J\dfrac{dG(t)}{dt}G(t)^{-1}$ is symmetric for $t \in \mathcal{I}$.

Proof.

Computing the determinant of both sides of condition (ii), we obtain $[\det G(t)]^2 = 1$. Therefore, $G(t)$ is invertible. To show that $H(t)$ is symmetric, differentiating both sides of $G(t)^T JG(t) = J$, we obtain

$$\left[\frac{dG(t)}{dt}\right]^T JG(t) + G(t)^T J\frac{dG(t)}{dt} \;=\; O.$$

Since $\left[G(t)^{-1}\right]^T = \left[G(t)^T\right]^{-1}$, we further obtain

$$\left[G(t)^{-1}\right]^T \left[\frac{dG(t)}{dt}\right]^T (-J) \;=\; J\frac{dG(t)}{dt}G(t)^{-1}.$$

Observing that the left-hand side of this relation is the transpose of the right-hand side, we conclude that $H(t)$ is symmetric. $\quad\square$

Observation 1. If $G^T JG = J$, then $(G^{-1})^T JG^{-1} = J$ and $GJG^T = J$. In fact, $G^T JG = J$ implies that $(G^{-1})^T JG^{-1} = J$. Then, taking the inverse of both sides of the first relation, we obtain the second relation (cf. (IV.5.3)).

Definition IV-5-3. *A $(2n) \times (2n)$ real invertible matrix G such that $G^T JG = J$ is called a* real symplectic matrix of order $2n$. *The set of all real symplectic matrices of order $2n$ is denoted by $Sp(2n, \mathbb{R})$.*

It is easy to show that $Sp(2n, \mathbb{R})$ is a subgroup of $\mathrm{GL}(2n, \mathbb{R})$ (cf. Observation 1). Hence, $Sp(2n, \mathbb{R})$ is called the real symplectic group of order $2n$.

Observation 2. If we change Hamiltonian system (IV.5.1) by a linear transformation

$$(\text{IV.5.4}) \qquad\qquad \vec{y} \;=\; P(t)\vec{z},$$

(IV.5.1) becomes

$$\frac{d\vec{z}}{dt} \;=\; \left[P(t)^{-1}JH(t)P(t) \;-\; P(t)^{-1}\frac{dP(t)}{dt}\right]\vec{z}.$$

Furthermore, if $P(t) \in Sp(2n, \mathbb{R})$ for $t \in \mathcal{I}$, this system becomes

$$(\text{IV.5.5}) \qquad \frac{d\vec{z}}{dt} \;=\; J\left[P(t)^T H(t)P(t) \;-\; J\frac{dP(t)^{-1}}{dt}P(t)\right]\vec{z}.$$

Since $P^{-1}(t)P(t) = I_{2n}$ and $P(t)^T \in Sp(2n, \mathbb{R})$, we obtain

$$\begin{cases} P(t)^{-1}\dfrac{dP(t)}{dt} + \dfrac{dP(t)^{-1}}{dt}P(t) \;=\; \dfrac{dI_{2n}}{dt} \;=\; O, \\[2mm] P(t)^{-1}J \;=\; JP(t)^T. \end{cases}$$

Observe that $-J\dfrac{dP(t)^{-1}}{dt}P(t)$ is symmetric since $P(t)^{-1} \in Sp(2n, \mathbb{R})$ (cf. Lemma IV-5-2). This implies that (IV.5.5) is a Hamiltonian system.

Definition IV-5-4. *Transformation (IV.5.4) is called a* canonical transformation *if $P(t) \in Sp(2n, \mathbb{R})$.*

Observation 3. For a given constant matrix G in $Sp(2n, \mathbb{R})$, let us construct the S-N decomposition $G = S + N$, where S is diagonalizable, N is nilpotent, and $SN = NS$. Since G is invertible, S is invertible (cf. Remark IV-1-13). If we set $M = S^{-1}N$, then M is also nilpotent, and we derive the unique decomposition

$$(\text{IV.5.6}) \qquad\qquad G \;=\; S(I_{2n} + M), \qquad SM \;=\; MS.$$

The identity $G^T J G = J$ implies $J G J^{-1} = (G^T)^{-1}$. Using (IV.5.6), we can write this last relation in the form

$$\left(J S J^{-1}\right)\left(J(I_{2n} + M)J^{-1}\right) = (S^T)^{-1}\left(I_{2n} + M^T\right)^{-1}.$$

Set $S_1 = J S J^{-1}$, $M_1 = J(I_{2n} + M)J^{-1} - I_{2n}$, $S_2 = (S^T)^{-1}$, and $M_2 = \left(I_{2n} + M^T\right)^{-1} - I_{2n}$. It is not difficult to show that S_j $(j = 1, 2)$ are diagonalizable and that M_j $(j = 1, 2)$ are nilpotent. Furthermore, $S_j M_j = M_j S_j$ $(j = 1, 2)$. Hence, the uniqueness of S-N decomposition implies that $S_1 = S_2$ and $I_{2n} + M_1 = I_{2n} + M_2$. Therefore,

$$(IV.5.7) \qquad J S J^{-1} = (S^T)^{-1}, \qquad J(I_{2n} + M)J^{-1} = \left(I_{2n} + M^T\right)^{-1}.$$

Thus, we proved the following lemma.

Lemma IV-5-5. *If we write a symplectic matrix G of order $2n$ in form (IV.5.6) by using the unique S-N decomposition of G (where $M = S^{-1}N$), the two matrices S and $I_{2n} + M$ are also symplectic matrices of order $2n$.*

Remark IV-5-6. Let J be the $2n \times 2n$ matrix defined by (III.5.2) and let H be a $2n \times 2n$ symmetric matrix. Then, it is important to know that if λ is an eigenvalue of JH with multiplicity m, then $-\lambda$ is also an eigenvalue of JH with multiplicity m. In fact, $(JH)^T = -HJ = HJ^{-1}$. Hence, $J^{-1}(JH)^T J = -JH$, where A^T denotes the transpose of A.

Remark IV-5-7. Let G be a $2n \times 2n$ symplectic matrix. Then, it is important to know also that if λ is an eigenvalue of G with multiplicity m, then $\dfrac{1}{\lambda}$ is also an eigenvalue of G with multiplicity m. In fact, $J G J^{-1} = (G^T)^{-1}$.

Remark IV-5-8. Assuming that the set of $2n \times 2n$ symplectic matrices with $2n$ distinct eigenvalues is dense in the symplectic group $Sp(2n, \mathbb{R})$, it can be shown that if 1 (and/or -1) is an eigenvalue of a symplectic matrix G, its multiplicity is even. Also, $\det G = 1$. In fact, $\lim\limits_{\lambda \to \pm 1} \dfrac{1}{\lambda} = \pm 1$. It is also known that $\det A$ is equal to the product of all eigenvalues of A.

Remark IV-5-9. Assuming that $Sp(2n, \mathbb{R})$ is a connected set, we can show that $\det G = 1$ for all $G \in Sp(2n, \mathbb{R})$. In fact, $(\det G)^2 = 1$ and the $2n \times 2n$ identity matrix is in $Sp(2n, \mathbb{R})$.

Part 2

Now, assume that the real symmetric matrix $H(t)$ of system (IV.5.1) is periodic in t of a positive period ω, i.e., $H(t + \omega) = H(t)$. Then, applying Theorem IV-4-3 and Remark IV-4-4 to system (IV.5.1), we conclude that the unique $(2n) \times (2n)$ matrix solution $\Phi(t)$ of the initial-value problem $\dfrac{d\Phi(t)}{dt} = JH(t)\Phi(t)$, $\Phi(0) = I_{2n}$ can be written in the form $\Phi(t) = P(t)\exp[tB]$, where $P(t)$ and B are real $(2n) \times (2n)$ matrices such that $P(t)$ is invertible, $P(t+2\omega) = P(t)$ for all $t \in \mathbb{R}$, and B is a constant matrix such that $\Phi(\omega)^2 = \exp[2\omega B]$. Furthermore, system (IV.5.1) is changed to

$$(IV.5.8) \qquad\qquad \frac{d\vec{z}}{dt} = B\vec{z}$$

by the transformation

$$(IV.5.9) \qquad \vec{y} = P(t)\vec{z}.$$

The main concern of Part 2 is to show that a matrix $P(t)$ can be chosen so that the transformation (IV.5.9) is canonical. Note that $Y = \exp[tB]$ is the unique solution of the initial-value problem $\dfrac{dY}{dt} = BY, \ \ Y(0) = I_{2n}$. Therefore, if JB is symmetric, the matrix $\exp[tB]$ is symplectic. Hence, if JB is symmetric, the matrix $P(t) = \Phi(t)\exp[-tB]$ is symplectic, since $\Phi(t)$ is symplectic. This implies that in order to show that the transformation (IV.5.9) is canonical, it suffices to show that we can choose $P(t)$ so that JB is symmetric.

Hereafter we use notations and results given in §IV-4. For examples,

$$(IV.5.10) \qquad \begin{cases} I_{2n} = \displaystyle\sum_{j=1}^{k} P_j(\Phi(\omega)), \quad P_j(\Phi(\omega))P_\ell(\Phi(\omega)) = O \quad (j \neq \ell), \\ P_j(\Phi(\omega))^2 = P_j(\Phi(\omega)). \end{cases}$$

Taking the transpose of (IV.5.10), we obtain

$$(IV.5.11) \qquad \begin{cases} I_{2n} = \displaystyle\sum_{j=1}^{k} P_j(\Phi(\omega))^T, \quad P_j(\Phi(\omega))^T P_\ell(\Phi(\omega))^T = O \quad (j \neq \ell), \\ \left(P_j(\Phi(\omega))^T\right)^2 = P_j(\Phi(\omega))^T. \end{cases}$$

Note that $p_{A^T}(\lambda) = p_A(\lambda)$ for any square matrix A. Also,

$$\begin{cases} \log[\Phi(\omega)^2] = \log[S^2] + 2\log[I_{2n} + M], \\ \log[S^2] = \displaystyle\sum_{\lambda_j<0} \log[(\lambda_j)^2]P_j(\Phi(\omega)) + 2\sum_{\text{other } j} \log[\lambda_j]P_j(\Phi(\omega)), \\ \log[I_{2n} + M] = \displaystyle\sum_{m=1}^{n-1} \frac{(-1)^{m+1}}{m}M^m, \end{cases}$$

and

$$B = \frac{1}{2\omega}\log[\Phi(\omega)^2].$$

In order to prove that JB is symmetric, it suffices to prove that $J\log[S^2]$ and $J\log[I_{2n} + M]$ are symmetric. To do this, let us first prove the following lemma.

Lemma IV-5-10. *For the symplectic matrix $\Phi(\omega)$, we have*

$$(IV.5.12) \qquad P_j(\Phi(\omega)^T)JP_\ell(\Phi(\omega)) = O \qquad if \ \ \lambda_j\lambda_\ell \neq 1,$$

where $\lambda_j \ (j = 1, 2, \dots, k)$ are distinct eigenvalues of $\Phi(\omega)$.

Proof.

Upon applying (IV.5.6) to $G = \Phi(\omega)$, write $\Phi(\omega)$ in the form $\Phi(\omega) = S(I_{2n}+M)$. Then, S is symplectic, i.e., $S^T JS = J = I_{2n}JI_{2n}$ (cf. Lemma IV-5-5). Using the formulas

$$(\text{IV.5.13}) \qquad \begin{cases} I_{2n} = \sum_{\ell=1}^{k} P_\ell(\Phi(\omega)), & I_{2n} = \sum_{j=1}^{k} P_j(\Phi(\omega))^T, \\[2mm] S = \sum_{\ell=1}^{k} \lambda_\ell P_\ell(\Phi(\omega)), & S^T = \sum_{j=1}^{k} \lambda_j P_j(\Phi(\omega))^T, \end{cases}$$

rewrite the identity $S^T JS = J = I_{2n}JI_{2n}$ in the following form:

$$(\text{IV.5.14}) \qquad \sum_{j,\ell=1}^{k} \lambda_j \lambda_\ell P_j(\Phi(\omega))^T JP_\ell(\Phi(\omega)) = \sum_{j,\ell=1}^{k} P_j(\Phi(\omega))^T JP_\ell(\Phi(\omega)).$$

Fixing a pair of indices j and ℓ and multiplying both sides of (IV.5.14) by $P_j(\Phi(\omega))^T$ from the left and by $P_\ell(\Phi(\omega))$ from the right, we derive

$$\lambda_j \lambda_\ell P_j(\Phi(\omega))^T JP_\ell(\Phi(\omega)) = P_j(\Phi(\omega))^T JP_\ell(\Phi(\omega)).$$

Hence, (IV.5.12) follows. $\square$

Observation 4. Let us prove that $J\log[S^2]$ is symmetric. To do this, by using (IV.5.12) and (IV.5.13), we write $J\log[S^2]$ and $\log[(S^2)^T](-J)$ in the following form:

$$\begin{cases} I_{2n}J\ \log[S^2] = \sum_{\lambda_j \lambda_\ell=1} \log[(\lambda_\ell)^2]P_j(\Phi(\omega))^T JP_\ell(\Phi(\omega)), \\[2mm] \log[(S^2)^T](-J)I_{2n} = \sum_{\lambda_j \lambda_\ell=1} \log[(\lambda_j)^2]P_j(\Phi(\omega))^T (-J)P_\ell(\Phi(\omega)). \end{cases}$$

Now, choose $\log[(\lambda_j)^2]$ for $j = 1, 2, \ldots, k$ in such a way that

$$\log[(\lambda_\ell)^2] = -\log[(\lambda_j)^2] \quad \text{if} \quad \lambda_j \lambda_\ell = 1.$$

Then,

$$\log[(S^2)^T](-J) = \sum_{\lambda_j \lambda_\ell=1} \log[(\lambda_\ell)^2]P_j(\Phi(\omega))^T JP_\ell(\Phi(\omega)) = J\log[S^2].$$

Note that for any square matrix A, we have $[\log(A)]^T = \log(A^T)$. Therefore, $[J\log(S^2)]^T = -\log[(S^2)^T]J$. Hence, $J\log[S^2]$ is symmetric.

Observation 5. Let us prove that $J\log[I_{2n} + M]$ is symmetric. To do this, we write first $[I_{2n} + M]^T J[I_{2n} + M] = J$ in the form $J[I_{2n} + M] = [I_{2n} + M^T]^{-1}J$.

Then, write $[I_{2n} + M^T]^{-1}$ in the form $[I_{2n} + M^T]^{-1} = I_{2n} + \hat{M}$. Using (IV.5.7), it is not difficult to show that $\hat{M}$ is nilpotent and $JM = \hat{M}J$. Hence,

$$JM^m = (\hat{M})^m J \qquad \text{for} \quad m = 1, 2, \ldots .$$

Therefore,

$$(\text{IV.5.15}) \qquad J \sum_{m=1}^{\infty} \frac{(-1)^{m+1}}{m} M^m = \left(\sum_{m=1}^{\infty} \frac{(-1)^{m+1}}{m} \hat{M}^m \right) J$$

and

$$\begin{aligned} J \log[I_{2n} + M] &= \left(\log[I_{2n} + \hat{M}] \right) J \\ &= \left(\log \left[(I_{2n} + M^T)^{-1} \right] \right) J = - \left(\log \left[I_{2n} + M^T \right] \right) J. \end{aligned}$$

Thus,

$$[J \log(I_{2n} + M)]^T = \log(I_{2n} + M^T)(-J) = J \log(I_{2n} + M).$$

This proves that $J \log[I_{2n} + M]$ is symmetric. Here, use was made of the fact that if $\dfrac{1}{1+x} = 1 + y(x)$, then

$$\log \left[\frac{1}{1+x} \right] = \log[1 + y(x)] = - \log[1 + x]$$

as a power series in x.

Thus, we arrived at the following conclusion.

Theorem IV-5-11. *In the case when the matrix $H(t)$ of system (IV.5.1) is real, symmetric, and periodic in t of a positive period ω, there exist $(2n) \times (2n)$ real matrices $P(t)$ and B such that*
  (a) $P(t) \in Sp(2n, \mathbb{R})$ for all $t \in \mathbb{R}$,
  (b) $P(t + 2\omega) = P(t)$ for all $t \in \mathbb{R}$,
  (c) B is a constant matrix such that JB is symmetric,
  (d) the canonical transformation $\vec{y} = P(t)\vec{z}$ changes system (IV.5.1) to $\dfrac{d\vec{z}}{dt} = B\vec{z}$.

For more information of exponentials in algebraic matrix groups, see [Mar] and [Si4].

IV-6. Nonhomogeneous equations

In this section, we explain how to solve an initial-value problem

$$(\text{IV.6.1}) \qquad \frac{d\vec{y}}{dt} = A(t)\vec{y} + \vec{b}(t), \qquad \vec{y}(\tau) = \vec{\eta},$$

where the entries of the $n \times n$ matrix $A(t)$ and the $\mathbb{C}^n$-valued function $\vec{b}(t)$ are continuous on an interval $\mathcal{I} = \{t : a \le t \le b\}$ and $\tau \in \mathcal{I}$, $\vec{\eta} \in \mathbb{C}^n$. As it was explained in §IV-2, let the $n \times n$ matrix $\Phi(t;\tau)$ be the unique fundamental matrix solution of the homogeneous system

$$\text{(IV.6.2)} \qquad \frac{d\vec{y}}{dt} = A(t)\vec{y}$$

determined by the initial-value problem

$$\text{(IV.6.3)} \qquad \frac{dY}{dt} = A(t)Y, \qquad Y(\tau) = I_n,$$

where $\tau \in \mathcal{I}$. System (IV.6.2) is called the *associated homogeneous equation* of (IV.6.1). We treat the solution of (IV.6.1) by using the knowledge of the fundamental matrix solution $\Phi(t;\tau)$ of (IV.6.2) (cf. §§IV-2 and IV-3).

Consider the transformation

$$\text{(IV.6.4)} \qquad \vec{y} = \Phi(t;\tau)\vec{z}.$$

Then, the problem (IV.6.1) is changed to

$$\text{(IV.6.5)} \qquad \frac{d\vec{z}}{dt} = \Phi(t;\tau)^{-1}\vec{b}(t), \qquad \vec{z}(\tau) = \vec{\eta}.$$

In fact, differentiating both sides of (IV.6.4), we obtain

$$A(t)\Phi(t;\tau)\vec{z} + \vec{b}(t) = A(t)\Phi(t;\tau)\vec{z} + \Phi(t;\tau)\frac{d\vec{z}}{dt}.$$

Since the unique solution of problem (IV.6.5) is given by

$$\vec{z} = \vec{\eta} + \int_\tau^t \Phi(s;\tau)^{-1}\vec{b}(s)ds,$$

the unique solution of problem (IV.6.1) is given by

$$\text{(IV.6.6)} \qquad \begin{aligned} \vec{y} &= \Phi(t;\tau)\left[\vec{\eta} + \int_\tau^t \Phi(s;\tau)^{-1}\vec{b}(s)ds\right] \\ &= \Phi(t;\tau)\vec{\eta} + \int_\tau^t \Phi(t;s)\vec{b}(s)ds. \end{aligned}$$

Here, use was made of the fact that $\Phi(t;s) = \Phi(t;\tau)\Phi(s;\tau)^{-1}$ (cf. (2) of Remark IV-2-7; in particular (IV.2.6)).

In a similar way, using (IV.6.6), we can change a nonlinear initial-value problem

$$\frac{d\vec{y}}{dt} = A(t)\vec{y} + \vec{g}(t,\vec{y}), \qquad \vec{y}(\tau) = \vec{\eta}$$

to an integral equation

$$\text{(IV.6.7)} \qquad \vec{y} = \Phi(t;\tau)\vec{\eta} + \int_\tau^t \Phi(t;s)\vec{g}(s,\vec{y}(s))ds.$$

The function $\vec{g}(t, \vec{y})$ is the nonlinear term which satisfies some suitable condition(s).

In the case when the matrix A is independent of t, formulas (IV.6.6) and (IV.6.7) become

$$(IV.6.8) \qquad \vec{y} = \exp[(t - \tau)A]\vec{\eta} + \int_{\tau}^{t} \exp[(t - s)A]\vec{b}(s)ds$$

and

$$(IV.6.9) \qquad \vec{y} = \exp[(t - \tau)A]\vec{\eta} + \int_{\tau}^{t} \exp[(t - s)A]\vec{g}(s, \vec{y}(s))ds,$$

respectively.

Example IV-6-1. Let us solve the initial-value problem

$$(IV.6.10) \qquad \frac{d\vec{y}}{dt} = A\vec{y} + \vec{b}(t), \qquad \vec{y}(0) = \vec{\eta}_0,$$

where

$$A = \begin{bmatrix} -2 & 1 & 0 \\ 0 & -2 & 0 \\ 3 & 2 & 1 \end{bmatrix}, \quad \vec{b}(t) = \begin{bmatrix} 2 \\ 0 \\ t \end{bmatrix}, \quad \vec{\eta}_0 = \begin{bmatrix} 1 \\ 1 \\ 0 \end{bmatrix}.$$

The matrix A was given in Example IV-3-3. As computed in §IV-3, a fundamental matrix solution of the associated homogeneous equation is

$$\Phi(t; 0) = \exp(tA) = \begin{bmatrix} e^{-2t} & te^{-2t} & 0 \\ 0 & e^{-2t} & 0 \\ e^{t} - e^{-2t} & e^{t} - (1+t)e^{-2t} & e^{t} \end{bmatrix}.$$

Therefore, the solution of (IV.6.10) is

$$\exp[tA]\left\{ \vec{\eta}_0 + \int_{0}^{t} \exp[-sA]\vec{b}(s)ds \right\} = \begin{bmatrix} 1 + te^{-2t} \\ e^{-2t} \\ -4 - t + 5e^{t} - (1+t)e^{-2t} \end{bmatrix}.$$

IV-7. Higher-order scalar equations

In this section, we explain how to solve the initial-value problem of an n-th order linear ordinary differential equation

$$(IV.7.1) \qquad a_0(t)u^{(n)} + a_1(t)u^{(n-1)} + \cdots + a_{n-1}(t)u' + a_n(t)u = b(t),$$

$$u(\tau) = \eta_1, \ u'(\tau) = \eta_2, \ \ldots, \ u^{(n-1)}(\tau) = \eta_n,$$

where the coefficients $a_0(t), a_1(t), \ldots, a_n(t)$ and the nonhomogeneous term $b(t)$ are continuous on an interval $\mathcal{I} = \{t : a \leq t \leq b\}$. In order to reduce (IV.7.1) to

a system, setting $y_1 = u$ and $y_2 = u', \ldots, y_n = u^{(n-1)}$, we introduce a vector
$\vec{y} = \begin{bmatrix} y_1 \\ \vdots \\ y_n \end{bmatrix}$. Then, problem (IV.7.1) is equivalent to

$$(IV.7.2) \qquad \frac{d\vec{y}}{dt} = A(t)\vec{y} + \vec{b}(t), \qquad \vec{y}(\tau) = \vec{\eta},$$

where

$$A(t) = \begin{bmatrix} 0 & 1 & 0 & 0 & \cdots & 0 \\ 0 & 0 & 1 & 0 & \cdots & 0 \\ \vdots & \vdots & \vdots & \vdots & \cdots & \vdots \\ 0 & 0 & 0 & 0 & \cdots & 1 \\ -\dfrac{a_n(t)}{a_0(t)} & -\dfrac{a_{n-1}(t)}{a_0(t)} & -\dfrac{a_{n-3}(t)}{a_0(t)} & -\dfrac{a_{n-2}(t)}{a_0(t)} & \cdots & -\dfrac{a_1(t)}{a_0(t)} \end{bmatrix},$$

$$\vec{b}(t) = \begin{bmatrix} 0 \\ 0 \\ \vdots \\ 0 \\ \dfrac{b(t)}{a_0(t)} \end{bmatrix}, \qquad \vec{\eta} = \begin{bmatrix} \eta_1 \\ \eta_2 \\ \vdots \\ \eta_{n-1} \\ \eta_n \end{bmatrix}.$$

Assuming $a_0(t) \neq 0$ on $\mathcal{I}$, let $\Phi(t) = [\vec{\phi}_1(t)\ \vec{\phi}_2(t)\ \ldots\ \vec{\phi}_n(t)]$ be a fundamental matrix solution of the associated homogeneous equation

$$(IV.7.3) \qquad \frac{d\vec{y}}{dt} = A(t)\vec{y}$$

of (IV.7.2). Using $\Phi(t)$, we can solve problem (IV.7.2) (cf. §IV-6). The first component of the solution of (IV.7.2) is the solution of problem (IV.7.1). Also, the first components of n column vectors of the matrix $\Phi(t)$ give the n linearly independent solutions of the associated homogeneous equation

$$(IV.7.4) \qquad a_0(t)u^{(n)} + a_1(t)u^{(n-1)} + \cdots + a_{n-1}(t)u' + a_n(t)u = 0$$

of (IV.7.1).

Example IV-7-1. Let us solve the initial-value problem

$$(IV.7.5) \qquad u''' - 2u'' - 5u' + 6u = 3t, \qquad u(0) = 1,\ u'(0) = 2,\ u''(0) = 0.$$

To start with, set

$$y_1 = u,\ y_2 = u',\ y_2 = u'', \quad \text{and} \quad \vec{y} = \begin{bmatrix} y_1 \\ y_2 \\ y_3 \end{bmatrix}.$$

Then, problem (IV.7.5) is equivalent to

(IV.7.6)
$$\frac{d\vec{y}}{dt} = A\vec{y} + \vec{b}(t), \qquad \vec{y}(0) = \vec{\eta},$$

where

(IV.7.7)
$$A = \begin{bmatrix} 0 & 1 & 0 \\ 0 & 0 & 1 \\ -6 & 5 & 2 \end{bmatrix}, \qquad \vec{b}(t) = \begin{bmatrix} 0 \\ 0 \\ 3t \end{bmatrix}, \qquad \text{and} \qquad \vec{\eta} = \begin{bmatrix} 1 \\ 2 \\ 0 \end{bmatrix}.$$

Three eigenvalues of the matrix A are 1, -2, and 3. The corresponding projections are

$$P_1(A) = -\frac{1}{6} \begin{bmatrix} -6 & -1 & 1 \\ -6 & -1 & 1 \\ -6 & -1 & 1 \end{bmatrix}, \quad P_2(A) = \frac{1}{15} \begin{bmatrix} 3 & -4 & 1 \\ -6 & 8 & -2 \\ 12 & -16 & 4 \end{bmatrix},$$

$$P_3(A) = \frac{1}{10} \begin{bmatrix} -2 & 1 & 1 \\ -6 & 3 & 3 \\ -18 & 9 & 9 \end{bmatrix},$$

and $N = O$. Therefore,

$$\Phi(t; 0) = \exp[tA] = e^t P_1(A) + e^{-2t} P_2(A) + e^{3t} P_3(A)$$

$$= -\frac{e^t}{6} \begin{bmatrix} -6 & -1 & 1 \\ -6 & -1 & 1 \\ -6 & -1 & 1 \end{bmatrix} + \frac{e^{-2t}}{15} \begin{bmatrix} 3 & -4 & 1 \\ -6 & 8 & -2 \\ 12 & -16 & 4 \end{bmatrix} + \frac{e^{3t}}{10} \begin{bmatrix} -2 & 1 & 1 \\ -6 & 3 & 3 \\ -18 & 9 & 9 \end{bmatrix}.$$

Thus,

$$\int_0^t \Phi(s; 0)^{-1} \vec{b}(s)\, ds$$

$$= -\frac{1}{2} \begin{bmatrix} 1 - (t+1)e^{-t} \\ 1 - (t+1)e^{-t} \\ 1 - (t+1)e^{-t} \end{bmatrix} + \frac{1}{20} \begin{bmatrix} 1 + (2t-1)e^{2t} \\ -2[1 + (2t-1)e^{2t}] \\ 4[1 + (2t-1)e^{2t}] \end{bmatrix} + \frac{1}{30} \begin{bmatrix} 1 - (3t+1)e^{-3t} \\ 3[1 - (3t+1)e^{-3t}] \\ 9[1 - (3t+1)e^{-3t}] \end{bmatrix},$$

and, consequently,

$$\vec{y}(t) = \exp[tA] \left\{ \begin{bmatrix} 1 \\ 2 \\ 0 \end{bmatrix} - \frac{1}{2} \begin{bmatrix} 1 - (t+1)e^{-t} \\ 1 - (t+1)e^{-t} \\ 1 - (t+1)e^{-t} \end{bmatrix} + \frac{1}{20} \begin{bmatrix} 1 + (2t-1)e^{2t} \\ -2[1 + (2t-1)e^{2t}] \\ 4[1 + (2t-1)e^{2t}] \end{bmatrix} \right.$$

$$\left. + \frac{1}{30} \begin{bmatrix} 1 - (3t+1)e^{-3t} \\ 3[1 - (3t+1)e^{-3t}] \\ 9[1 - (3t+1)e^{-3t}] \end{bmatrix} \right\} = \frac{1}{60} \begin{bmatrix} 30t + 25 - 17e^{-2t} + 50e^t + 2e^{3t} \\ 2(15 + 17e^{-2t} + 25e^t + 3e^{3t}) \\ 2(-34e^{-2t} + 25e^t + 9e^{3t}) \end{bmatrix}.$$

The first component of $\vec{y}(t)$ gives the solution of (IV.7.5).

Remark IV-7-2. In the case of a second-order linear differential equation

(IV.7.8)
$$a_0(t)u'' + a_1(t)u' + a_2(t)u = b(t),$$

a fundamental matrix solution of (IV.7.3) has the form $\Phi(t) = \begin{bmatrix} \phi_1(t) & \phi_2(t) \\ \phi_1'(t) & \phi_2'(t) \end{bmatrix}$, where $\phi_1(t)$ and $\phi_2(t)$ are two linearly independent solutions of the associated homogeneous equation (IV.7.4). Since $\Phi(t)^{-1} = \dfrac{1}{W(t)} \begin{bmatrix} \phi_2'(t) & -\phi_2(t) \\ -\phi_1'(t) & \phi_1(t) \end{bmatrix}$, where $W(t) = \det(\Phi(t))$, the first component of the formula

$$\vec{y}(t) = \Phi(t)\vec{\eta} + \Phi(t) \int^t \Phi(s)^{-1}\vec{b}(s)ds$$

gives the general solution of (IV.7.8) in the form

(IV.7.9)
$$\begin{aligned} u(t) = {} & \eta_1\phi_1(t) + \eta_2\phi_2(t) \\ & - \phi_1(t) \int^t \frac{\phi_2(s)}{a_0(s)W(s)}b(s)ds + \phi_2(t) \int^t \frac{\phi_1(s)}{a_0(s)W(s)}b(s)ds, \end{aligned}$$

where η_1 and η_2 are two arbitrary constants. This is known as the *formula of variation of parameters* (see, for example, [Rab, pp. 241-246]). Moreover,

(IV.7.10)
$$W(t) = W(\tau)\exp\left[- \int_\tau^t \frac{a_1(s)}{a_0(s)}ds \right],$$

as in (4) of Remark IV-2-7.

Remark IV-7-3. Write (IV.7.10) in the form

(IV.7.11)
$$\phi_1(t)\phi_2'(t) - \phi_1'(t)\phi_2(t) = c_2 \exp\left[- \int^t \frac{a_1(s)}{a_0(s)}ds \right],$$

where c_2 is a constant. Regarding (IV.7.11) as a differential equation for $\phi_2(t)$, we obtain

$$\phi_2(t) = c_1\phi_1(t) + c_2\phi_1(t) \int^t \frac{1}{\phi_1(\tau)^2} \exp\left[- \int^t \frac{a_1(s)}{a_0(s)}ds \right]d\tau,$$

where c_1 is another constant. Therefore, $\phi_1(t)$ and $\phi_1(t) \int^t \dfrac{1}{\phi_1(\tau)^2}$ $\times \exp\left[- \int^t \frac{a_1(s)}{a_0(s)}ds \right]d\tau$ form a fundamental set of solutions of $a_0(t)u'' + a_1(t)u' + a_2(t)u = 0$.

Remark IV-7-4. Let $\phi_1(t)$ and $\phi_2(t)$ be linearly independent solutions of a third-order linear homogeneous differential equation

(IV.7.12)
$$a_0(t)u''' + a_1(t)u'' + a_2(t)u' + a_3(t)u = 0.$$

Also, let $\phi(t)$ be any solution of (IV.7.12). Then, (4) of Remark IV-2-7 implies that

$$(\text{IV.7.13}) \qquad \begin{vmatrix} \phi & \phi_1 & \phi_2 \\ \phi' & \phi_1' & \phi_2' \\ \phi'' & \phi_1'' & \phi_2'' \end{vmatrix} = c_3 \exp\left[-\int^t \frac{a_1(s)}{a_0(s)} ds\right],$$

where c_3 is a constant. Write (IV.7.13) in the form

$$(\text{IV.7.14}) \qquad A_0(t)\phi'' + A_1(t)\phi' + A_2(t)\phi = c_3 \exp\left[-\int^t \frac{a_1(s)}{a_0(s)} ds\right],$$

where

$$A_1(t) = \begin{vmatrix} \phi_1 & \phi_2 \\ \phi_1' & \phi_2' \end{vmatrix}, \quad A_2(t) = -\begin{vmatrix} \phi_1 & \phi_2 \\ \phi_1'' & \phi_2'' \end{vmatrix}, \quad A_3(t) = \begin{vmatrix} \phi_1' & \phi_2' \\ \phi_1'' & \phi_2'' \end{vmatrix}.$$

A fundamental set of solutions of the homogeneous part of (IV.7.14) is given by $\{\phi_1, \phi_2\}$. Therefore, using (IV.7.9), we obtain

$$\phi(t) = c_1\phi_1(t) + c_2\phi_2(t) + c_3\left\{-\phi_1(t)\int^t \frac{\phi_2(\tau)}{A_0(\tau)^2} \exp\left[-\int^\tau \frac{a_1(s)}{a_0(s)} ds\right] d\tau \right.$$
$$\left. + \phi_2(t)\int^t \frac{\phi_1(\tau)}{A_0(\tau)^2} \exp\left[-\int^\tau \frac{a_1(s)}{a_0(s)} ds\right] d\tau\right\},$$

where c_1 and c_2 are constants. This implies that $\phi_1(t)$, $\phi_2(t)$, and

$$\phi_1(t)\int^t \frac{\phi_2(\tau)}{A_0(\tau)^2} \exp\left[-\int^\tau \frac{a_1(s)}{a_0(s)} ds\right] d\tau - \phi_2(t)\int^t \frac{\phi_1(\tau)}{A_0(\tau)^2} \exp\left[-\int^\tau \frac{a_1(s)}{a_0(s)} ds\right] d\tau$$

form a fundamental set of solutions of (IV.7.12).

EXERCISES IV

IV-1. Let A be an $n \times n$ matrix, and let $\lambda_1, \lambda_2, \ldots, \lambda_k$ be all the distinct eigenvalues of A. On the complex λ-plane, consider k small closed disks $\Delta_j = \{\lambda : |\lambda - \lambda_j| \leq r_j\}$ $(j = 1, \ldots, k)$. Assume that $\Delta_j \cap \Delta_\ell = \emptyset$ for $j \neq \ell$. Set

$$\begin{cases} P_j = \dfrac{1}{2\pi i} \oint_{|\lambda - \lambda_j| = r_j} (\lambda I_n - A)^{-1} d\lambda & (j = 1, \cdots, k), \\ S = \lambda_1 P_1 + \cdots + \lambda_k P_k \quad \text{and} \quad N = A - S, \end{cases}$$

where the integrals are taken in the counterclockwise orientation. Show that
  (i) $P_1 + \cdots + P_k = I_n$,
  (ii) $P_j P_\ell = O$ if $(j \neq \ell)$,
  (iii) $A = S + N$ is the S-N decomposition of A.

Hint. Note that

$$\frac{1}{\tau - \sigma}\{(\sigma I_n - A)^{-1} - (\tau I_n - A)^{-1}\} = (\sigma I_n - A)^{-1}(\tau I_n - A)^{-1}.$$

Also, if $p(\lambda)$ is a polynomial in λ with constant coefficients, then

$$p(A) = \sum_{j=1}^{k} \frac{1}{2\pi i} \oint_{|\lambda - \lambda_j| = r_j} p(\lambda)(\lambda I_n - A)^{-1} d\lambda.$$

For general information, see [Ka, pp. 38–43].

IV-2. Under the same assumptions as in Exercise IV-1, show that if $f(\lambda)$ is analytic in a neighborhood of each eigenvalue λ_j, and an $n \times n$ matrix B is in a sufficiently small neighborhood of A, then $f(B)$ is well defined and $\lim_{B \to A} f(B) = f(A)$.

Hint. If $f(\lambda)$ is analytic in a neighborhood of each eigenvalue λ_j, then

$$f(A) = \sum_{j=1}^{k} \frac{1}{2\pi i} \oint_{|\lambda - \lambda_j| = r_j} f(\lambda)(\lambda I_n - A)^{-1} d\lambda.$$

IV-3. Let $A = S + N$ be the S-N decomposition of an $n \times n$ matrix A. Show that, as functions of A, the matrices S and N are not continuous.

Hint. Use Lemma IV-1-4.

IV-4. Let A be an $n \times n$ matrix with n eigenvalues $\lambda_1, \ldots, \lambda_n$. Fix a real number τ satisfying the condition $\tau > \Re[\lambda_j]$ $(j = 1, \ldots, n)$. Show that

$$(\text{F}) \qquad \exp[tA] = \frac{e^{\tau t}}{2\pi} \int_{-\infty}^{+\infty} e^{i\sigma t}((\tau + i\sigma)I_n - A)^{-1} d\sigma$$

for $t \in \mathbb{R}$.

Remark. This result means that $\exp[tA]$ is the inverse Laplace transform of $(sI_n - A)^{-1}$. For example, if $A = \begin{bmatrix} 0 & 1 \\ -1 & 0 \end{bmatrix}$, then $(sI_2 - A)^{-1} = \dfrac{1}{s^2 + 1}\begin{bmatrix} s & 1 \\ -1 & s \end{bmatrix}$. Hence, taking the inverse Laplace transform, we obtain $\exp[tA] = \begin{bmatrix} \cos t & \sin t \\ -\sin t & \cos t \end{bmatrix}$. Formula (F) is useful in the study of $\exp[tA]$ of an unbounded operator A defined on a Banach space (cf. [HilP, Chapter XII, pp. 356-386]).

IV-5. Show that $\displaystyle\int_0^t (A^2 + \tau^2 I_n)^{-1} d\tau = A^{-1}\arctan(tA^{-1})$ if every eigenvalue of an $n \times n$ matrix A is a nonzero real number.

IV-6. For an $n \times n$ matrix A, show that $\displaystyle\lim_{m \to +\infty}\left(I_n + \frac{A}{m}\right)^m = e^A$.

IV-7. Find the solution of the following initial-value problem

$$\frac{d^2\vec{y}}{dt^2} - 2A\frac{d\vec{y}}{dt} + A^2\vec{y} = e^{tA}\vec{c}, \qquad \vec{y}(0) = \vec{\eta}, \qquad \vec{y}'(0) = \vec{\zeta},$$

where A is a constant $n \times n$ matrix, $\{\vec{c}, \vec{\eta}, \vec{\zeta}\}$ are three constant vectors in $\mathbb{C}^n$, and $\vec{y} \in \mathbb{C}^n$ is the unknown quantity.

Hint. If we set $\vec{y} = e^{tA}\vec{u}$, the given problem is changed to

$$\frac{d^2\vec{u}}{dt^2} = \vec{c}, \qquad \vec{u}(0) = \vec{\eta}, \qquad \vec{u}'(0) = \vec{\zeta} - A\vec{\eta}.$$

IV-8. Find the solution of the following initial-value problem

$$\frac{d^2\vec{y}}{dx^2} + A^2\vec{y} = \vec{0}, \qquad \vec{y}(0) = \vec{\eta}_0, \quad \frac{d\vec{y}}{dx}(0) = \vec{\eta}_1,$$

where $\vec{y} \in \mathbb{C}^n$ is an unknown quantity, A is a constant $n \times n$ matrix such that $\det A = 0$, and $\{\vec{\eta}_0, \vec{\eta}_1\}$ are two constant vectors in $\mathbb{C}^n$.

Hint. Note that $\cos(xA)$ and $\sin(xA)$ are not linearly independent when $\det A = 0$. If we define $F(u) = \dfrac{\sin u}{u}$, then the general solution of the given differential equation is

$$\vec{y}(x) = \cos(xA)\vec{c}_1 + xF(xA)\vec{c}_2.$$

IV-9. Find explicitly a fundamental matrix solution of the system $\dfrac{d\vec{y}}{dt} = A(t)\vec{y}$ if there exists a constant matrix $P \in GL(n, \mathbb{C})$ such that $P^{-1}A(t)P$ is in Jordan canonical form, i.e.,

$$P^{-1}A(t)P = \text{diag}[\lambda_j(t)I_1 + N_1, \lambda_2(t)I_2 + N_2, \dots, \lambda_k(t)I_k + N_k],$$

where for each j, $\lambda_j(t)$ is a $\mathbb{C}$-valued continuous function on the interval $a \leq t \leq b$, I_j is the $n_j \times n_j$ identity matrix, and N_j is an $n_j \times n_j$ matrix whose entries are 1 on the superdiagonal and zero everywhere else.

Hint. See [GH].

IV-10. Find $\log\left(\begin{bmatrix} 1 & 1 \\ -1 & 3 \end{bmatrix}\right)$, $\cos\left(\begin{bmatrix} 3 & 0 & -1 \\ 1 & 2 & -1 \\ 0 & 1 & 1 \end{bmatrix}\right)$, and

$$\arctan\left(\begin{bmatrix} 252 & 498 & 4134 & 698 \\ -234 & -465 & -3885 & -656 \\ 15 & 30 & 252 & 42 \\ -10 & -20 & -166 & -25 \end{bmatrix}\right).$$

IV-11. In the case when two invertible matrices A and B commute, find

$$\log(AB) - \log(A) - \log B$$

if these three logarithms are defined as in Example IV-3-6.

IV-12. Given that

$$
A = \begin{bmatrix} 0 & 0 & 1 & 0 \\ 0 & 0 & 0 & 1 \\ 2 & 3 & 0 & 0 \\ 4 & -2 & 0 & 0 \end{bmatrix}, \qquad
\vec{y} = \begin{bmatrix} y_1 \\ y_2 \\ y_3 \\ y_4 \end{bmatrix},
$$

find a nonzero constant vector $\vec{\eta} \in \mathbb{C}^4$ in such a way that the solution $\vec{y}(t)$ of the initial-value problem

$$
\frac{d\vec{y}}{dt} = A\vec{y}, \qquad \vec{y}(0) = \vec{\eta}
$$

satisfies the condition $\lim\limits_{t \to +\infty} \vec{y}(t) = \vec{0}$.

IV-13. Assume that $A(t), B(t)$, and $F(t)$ are $n \times n$, $m \times m$, and $n \times m$ matrices, respectively, whose entries are continuous on the interval $a < t < b$. Let $\Phi(t)$ be an $n \times n$ fundamental matrix of $\dfrac{d\Phi}{dt} = A(t)\Phi$ and $\Psi(t)$ be an $m \times m$ fundamental matrix of $\dfrac{d\Psi}{dt} = B(t)\Psi$. Show that the general solution of the differential equation on an $n \times m$ matrix Y

(E) $$\frac{dY}{dt} = A(t)Y - YB(t) + F(t),$$

is given by

$$
Y(t) = \Phi(t)\, C\, \Psi(t)^{-1} + \int^{t} \Phi(t)\Phi(s)^{-1}F(s)\Psi(s)\Psi(t)^{-1}ds,
$$

where C is an arbitrary constant $n \times m$ matrix.

IV-14. Given the $n \times n$ linear system

$$
\frac{dY}{dt} = A(t)Y - YB(t),
$$

where $A(t)$ and $B(t)$ are $n \times n$ matrices continuous in the interval $a < t < b$,

(1) show that, if $Y(t_0)^{-1}$ exists at some point t_0 in the interval $a < t < b$, then $Y(t)^{-1}$ exists for all points of the interval $a < t < b$,

(2) show that $Z = Y^{-1}$ satisfies the differential equation

$$
\frac{dZ}{dt} = B(t)Z - ZA(t).
$$

IV-15. Find the multipliers of the periodic system $\dfrac{d\vec{y}}{dt} = A(t)\vec{y}$, where $A(t) = \begin{bmatrix} \cos t & \sin t \\ -\sin t & \cos t \end{bmatrix}$.

Hint. $A(t) = \exp\left[t\begin{bmatrix} 0 & 1 \\ -1 & 0 \end{bmatrix}\right]$.

IV-16. Let $A(t)$ and $B(t)$ be $n \times n$ and $n \times m$ matrices whose entries are real-valued and continuous on the interval $0 \leq t < +\infty$. Denote by $\mathcal{U}$ the set of all $\mathbb{R}^m$-valued measurable functions $\vec{u}(t)$ such that $|\vec{u}(t)| \leq 1$ for $0 \leq t < +\infty$. Fix a vector $\vec{\xi} \in \mathbb{R}^n$. Denote by $\vec{\phi}(t,u)$ the unique $\mathbb{R}^n$-valued function which satisfies the initial-value problem $\dfrac{d\vec{x}}{dt} = A(t)\vec{x} + B(t)\vec{u}(t)$, $\vec{x}(0) = \vec{\xi}$, where $\vec{u} \in \mathcal{U}$. Also, set $\mathcal{R} = \{(t, \vec{\phi}(t,u)) : 0 \leq t < +\infty, \vec{u} \in \mathcal{R}\}$. Show that $\mathcal{R}$ is closed in $\mathbb{R}^{n+1}$.

Hint. See [LM2, Theorem 1 of Chapter 2 on pp. 69-72 and Lemma 3A of Appendix of Chapter 2 on pp. 161-163].

IV-17. Let $u(t)$ be a real-valued, continuous, and periodic of period $\omega > 0$ in t on $\mathbb{R}$. Also, for every real τ, let $\phi(t,\tau)$ and $\psi(t,\tau)$ be two solutions of the differential equation

(Eq)
$$\frac{d^2y}{dt^2} - u(t)\, y = 0$$

such that

$$\phi(\tau,\tau) = 1, \quad \phi'(\tau,\tau) = 0, \qquad \text{and} \qquad \psi(\tau,\tau) = 0, \quad \psi'(\tau,\tau) = 1,$$

where the prime denotes $\dfrac{d}{dt}$. Set

$$\begin{cases} C(\tau) = \begin{bmatrix} \phi(0,\tau) & \psi(0,\tau) \\ \phi'(0,\tau) & \psi'(0,\tau) \end{bmatrix}, \\[2mm] M(\tau) = \begin{bmatrix} \phi(\tau+\omega,\tau) & \psi(\tau+\omega,\tau) \\ \phi'(\tau+\omega,\tau) & \psi'(\tau+\omega,\tau) \end{bmatrix}. \end{cases}$$

(I) Show that

$$M(\tau) = C(\tau)^{-1}M(0)C(\tau).$$

(II) Let Λ_+ and Λ_- be two eigenvalues of the matrix $M(\tau)$. Let $\begin{bmatrix} 1 \\ K_+(\tau) \end{bmatrix}$ and $\begin{bmatrix} 1 \\ K_-(\tau) \end{bmatrix}$ be eigenvectors of $M(\tau)$ corresponding to Λ_+ and Λ_-, respectively. Show that

(i) $K_\pm(\tau)$ are two periodic solutions of period ω of the differential equation

$$\frac{dK}{d\tau} + K^2 + u(\tau) = 0,$$

(ii) if we set

$$\beta_\pm(t,\tau) = \exp\left[\int_\tau^t K_\pm(\zeta)\, d\zeta\right],$$

these two functions satisfy the differential equation (Eq) and the conditions

$$\beta_\pm(t+\omega,\tau) = \Lambda_\pm\beta_\pm(t,\tau).$$

Hint for (I). Set $\Phi(t,\tau) = \begin{bmatrix} \phi(t,\tau) & \psi(t,\tau) \\ \phi'(t,\tau) & \psi'(t,\tau) \end{bmatrix}$. Then, $\Phi(t,\tau)$ is a fundamental matrix solution of the system

$$\frac{dy_1}{dt} = y_2, \qquad \frac{dy_2}{dt} = u(t)y_1,$$

and $\vec{y} = \Phi(t,\tau)\vec{c}$ is the solution satisfying the initial condition $\vec{y}(\tau) = \vec{c}$. Note that $C(\tau) = \Phi(0,\tau)$ and $M(\tau) = \Phi(\tau+\omega,\tau)$. Therefore, $\Phi(t,\tau) = \Phi(t,0)C(\tau)$ and $\Phi(t+\omega,\tau) = \Phi(t,\tau)M(\tau)$. Now, it is not difficult to see $\Phi(\omega,\tau) = \Phi(\omega,0)C(\tau) = \Phi(0,\tau)M(\tau) = C(\tau)M(\tau)$ and $\Phi(\omega,0) = M(0)$.

Hint for (II). Since eigenvalues of $M(\tau)$ and $M(0)$ are the same, the eigenvalues of $M(\tau)$ is independent of τ. Solutions $\eta_+(t)$ of (Eq) satisfying the condition $\eta_+(t+\omega) = \Lambda_+\eta_+(t)$ are linearly dependent on each other. Hence, $\dfrac{\eta'_+(t)}{\eta_+(t)} = K_+(t)$ is independent of any particular choice of such solutions $\eta_+(t)$. In particular $K_+(t+\omega) = \dfrac{\eta'_+(t+\omega)}{\eta_+(t+\omega)} = K_+(t)$. Problem (II) claims that the quantity $K_+(\tau)$ can be found by calculating eigenvectors of $M(\tau)$ corresponding to Λ_+. Furthermore, $\Lambda_+ = \exp\left[\displaystyle\int_0^\omega K_+(\tau)d\tau\right]$. The same remark applies to Λ_-. The solutions $\beta_\pm(x,\tau)$ of (Eq) are called the *Block solutions*.

IV-18. Show that
  (a) A real 2×2 matrix A is symplectic (i.e., $A \in Sp(2,\mathbb{R})$) if and only if $\det A = 1$;
  (b) the matrix $\begin{bmatrix} 0 & 1 \\ -1 & 0 \end{bmatrix} N$ is symmetric for any 2×2 real nilpotent matrix N.

Hint for (b). Note that $\det[e^{tN}] = 1$ for any real 2×2 nilpotent matrix N.

IV-19. Let G, H, and J are real $(2n) \times (2n)$ matrices such that G is symplectic, H is symmetric, and $J = \begin{bmatrix} O & I_n \\ -I_n & O \end{bmatrix}$. Show that $G^{-1}JHGJ$ is symmetric.

IV-20. Suppose that the $(2n) \times (2n)$ matrix $\Phi(t)$ is the unique solution of the initial-value problem $\dfrac{d\Phi}{dt} = JH(t)\Phi$, $\Phi(0) = I_{2n}$, where $J = \begin{bmatrix} O & I_n \\ -I_n & O \end{bmatrix}$ and $H(t)$ is symmetric. Set $L(t) = \Phi(t)^{-1}JH(t)\Phi(t)J$. Show that $\dfrac{d\Phi(t)^T}{dt} = JL(t)\Phi(t)^T$, where $\Phi(t)^T$ is the transpose of $\Phi(t)$.

CHAPTER V

SINGULARITIES OF THE FIRST KIND

In this chapter, we consider a system of differential equations

$$\text{(E)} \qquad x\frac{d\vec{y}}{dx} = \vec{f}(x,\vec{y}),$$

assuming that the entries of the $\mathbb{C}^n$-valued function $\vec{f}$ are convergent power series in complex variables $(x,\vec{y}) \in \mathbb{C}^{n+1}$ with coefficients in $\mathbb{C}$, where x is a complex independent variable and $\vec{y} \in \mathbb{C}^n$ is an unknown quantity. The main tool is calculation with power series in x. In §I-4, using successive approximations, we constructed power series solutions. However, generally speaking, in order to construct a power series solution $\vec{y}(x) = \sum_{m=1}^{\infty} x^m \vec{a}_m$, this expression is inserted into the given differential equation to find relationships among the coefficients $\vec{a}_m$, and the coefficients $\vec{a}_m$ are calculated by using these relationships. In this stage of the calculation, we do not pay any attention to the convergence of the series. This process leads us to the concept of formal power series solutions (cf. §V-1). Having found a formal power series solution, we estimate $|\vec{a}_m|$ to test its convergence. As the function $x^{-1}\vec{f}(x,\vec{y})$ is not analytic at $x = 0$, Theorem I-4-1 does not apply to system (E). Furthermore, the existence of formal power series solutions of (E) is not always guaranteed. Nevertheless, it is known that if a formal power series solution of (E) exists, then the series is always convergent. This basic result is explained in §V-2 (cf. [CL, Theorem 3.1, pp. 117-119] and [Was1, Theorem 5.3, pp. 22-25] for the case of linear differential equations]). In §V-3, we define the S-N decomposition for a lower block-triangular matrix of infinite order. Using such a matrix, we can represent a linear differential operator

$$\text{(LDO)} \qquad \mathcal{L}[\vec{y}] = x\frac{d\vec{y}}{dx} + \Omega(x)\vec{y},$$

where $\Omega(x)$ is an $n \times n$ matrix whose entries are formal power series in x with coefficients in $\mathbb{C}^n$. In this way, we derive the S-N decomposition of $\mathcal{L}$ in §V-4 and a normal form of $\mathcal{L}$ in §V-5 (cf. [HKS]). The S-N decomposition of $\mathcal{L}$ was originally defined in [GérL]. In §V-6, we calculate the normal form of a given operator $\mathcal{L}$ by using a method due to M. Hukuhara (cf. [Si17, §3.9, pp. 85-89]). We explain the classification of singularities of homogeneous linear differential equations in §V-7. Some basic results concerning linear differential equations given in this chapter are also found in [CL, Chapter 4].

108

V-1. Formal solutions of an algebraic differential equation

We denote by $\mathbb{C}[[x]]$ the set of all formal power series in x with coefficients in $\mathbb{C}$. For two formal power series $f = \sum_{m=0}^{\infty} a_m x^m$ and $g = \sum_{m=0}^{\infty} b_m x^m$, we define $f = g$ by the condition $a_m = b_m$ for all $m \geq 0$. Also, the sum $f + g$ and the product fg are defined by $f + g = \sum_{m=0}^{\infty} (a_m + b_m)x^m$ and $fg = \sum_{m=0}^{\infty} (\sum_{n=0}^{m} a_{m-n} b_n)x^m$, respectively. Furthermore, for $c \in \mathbb{C}$ and $f = \sum_{m=0}^{\infty} a_m x^m \in \mathbb{C}[[x]]$, we define cf by $cf = \sum_{m=0}^{\infty} ca_m x^m$. With these three operations, $\mathbb{C}[[x]]$ is a *commutative algebra* over $\mathbb{C}$ with the identity element given by $\sum_{m=0}^{\infty} \delta_m x^m$, where $\delta_0 = 1$ and $\delta_m = 0$ if $m \geq 1$. (For commutative algebra, see, for example, [AM].) Also, we define the derivative $\dfrac{df}{dx}$ of f with respect to x by $\dfrac{df}{dx} = \sum_{m=0}^{\infty} (m+1)a_{m+1} x^m$ and the integral $\int_0^x f(x)dx$ by $\int_0^x f(x)dx = \sum_{m=1}^{\infty} \dfrac{a_{m-1}}{m} x^m$ for $f = \sum_{m=0}^{\infty} a_m x^m \in \mathbb{C}[[x]]$. Then, $\mathbb{C}[[x]]$ is a *commutative differential algebra* over $\mathbb{C}$ with the identity element. There are some subalgebras of $\mathbb{C}[[x]]$ that are useful in applications. For example, denote by $\mathbb{C}\{x\}$ the set of all power series in $\mathbb{C}[[x]]$ that have nonzero radii of convergence. Also, denote by $\mathbb{C}[x]$ the set of all polynomials in x with coefficients in $\mathbb{C}$. Then, $\mathbb{C}\{x\}$ is a subalgebra of $\mathbb{C}[[x]]$ and $\mathbb{C}[x]$ is a subalgebra of $\mathbb{C}\{x\}$. Consequently, $\mathbb{C}[x]$ is also a subalgebra of $\mathbb{C}[[x]]$.

Let $F(x, y_0, y_1, \ldots, y_n)$ be a polynomial in $y_0, y_1, \ldots, y_n$ with coefficients in $\mathbb{C}[[x]]$. Then, a differential equation

$$(\text{V.1.1}) \qquad F\left(x, y, \frac{dy}{dx}, \ldots, \frac{d^n y}{dx^n}\right) = 0$$

is called an *algebraic differential equation*, where y is the unknown quantity and x is the independent variable. If $F\left(x, f, \ldots, \dfrac{d^n f}{dx^n}\right) = 0$ for some f in $\mathbb{C}[[x]]$, then such an f is called a *formal solution* of equation (V.1.1). In this definition, it is not necessary to assume that the coefficients of F are in $\mathbb{C}\{x\}$.

Example V-1-1. To find a formal solution of

$$(\text{V.1.2}) \qquad x\frac{dy}{dx} + y - x = 0,$$

set $y = \sum_{m \geq 0} a_m x^m$. Then, it follows from (V.1.2) that $a_0 = 0$, $2a_1 = 1$, and $(m+1)a_m = 0$ ($m \geq 2$). Hence, $y = \dfrac{1}{2}x$ is a formal solution of equation (V.1.2).

In general, if $f \in \mathbb{C}\{x\}$ is a formal solution of (V.1.1), then the sum of f as a convergent series is an actual solution of (V.1.1) if all coefficients of the polynomial F are in $\mathbb{C}\{x\}$.

Denote by $x^\sigma\mathbb{C}[[x]]$ the set of formal series $x^\sigma f(x)$, where $f(x) \in \mathbb{C}[[x]]$, σ is a complex number, and $x^\sigma = \exp[\sigma \log x]$. For $\phi = x^\sigma f \in x^\sigma\mathbb{C}[[x]]$, $\psi = x^\sigma g \in x^\sigma\mathbb{C}[[x]]$, and $h \in \mathbb{C}[[x]]$, define $\phi + \psi = x^\sigma(f + g)$, $h\phi = x^\sigma hf$, and $x\dfrac{d\phi}{dx} = \sigma x^\sigma f + x^\sigma x\dfrac{df}{dx}$. Then, $x^\sigma\mathbb{C}[[x]]$ is a *commutative differential module* over the algebra $\mathbb{C}[[x]]$. Similarly, let $x^\sigma\mathbb{C}\{x\}$ denote the set of convergent series $x^\sigma f(x)$, where $f(x) \in \mathbb{C}\{x\}$. The set $x^\sigma\mathbb{C}\{x\}$ is a *commutative differential module* over the algebra $\mathbb{C}\{x\}$. Furthermore, if σ is a non-negative integer, then $x^\sigma\mathbb{C}[[x]] \subset \mathbb{C}[[x]]$. If $F(x, y_0, \dots, y_n)$ is a formal power series in $(x, y_0, \dots, y_n)$ and if $f_0(x) \in x\mathbb{C}[[x]]$, $\dots$, $f_n(x) \in x\mathbb{C}[[x]]$, then $F(x, f_0(x), \dots, f_n(x)) \in \mathbb{C}[[x]]$. Also, if $F(x, y_0, \dots, y_n)$ is a convergent power series in $(x, y_0, \dots, y_n)$ and if $f_0(x) \in x\mathbb{C}\{x\}$, $\dots, f_n(x) \in x\mathbb{C}\{x\}$, then $F(x, f_0(x), \dots, f_n(x)) \in \mathbb{C}\{x\}$. Therefore, using the notation $x^\sigma\mathbb{C}[[x]]^n$ to denote the set of all vectors with n entries in $x^\sigma\mathbb{C}[[x]]$, we can define a formal solution $\vec{\phi}(x) \in x\mathbb{C}[[x]]^n$ of system (E) by the condition $x\dfrac{d\vec{\phi}}{dx} = \vec{f}(x, \vec{\phi})$ in $\mathbb{C}[[x]]^n$. (Similarly, we define $x^\sigma\mathbb{C}\{x\}^n$.) Also, in the case of a homogeneous system of linear differential equations $x\dfrac{d\vec{y}}{dx} = A(x)\vec{y}$, a series $\vec{\phi}(x) \in x^\sigma\mathbb{C}[[x]]^n$ is said to be a formal solution if $x\dfrac{d\vec{\phi}}{dx} = A(x)\vec{\phi}$ in $x^\sigma\mathbb{C}[[x]]^n$. Now, let us prove the following theorem.

Theorem V-1-2. *Suppose that $A(x)$ is an $n \times n$ matrix whose entries are formal power series in x and that λ is an eigenvalue of $A(0)$. Assume also that $\lambda + k$ are not eigenvalues of $A(0)$ for all positive integers k. Then, the differential equation*

$$(V.1.3) \qquad\qquad x\frac{d\vec{y}}{dx} = A(x)\vec{y}$$

has a nontrivial formal solution $\vec{\phi}(x) = x^\lambda \vec{f}(x) \in x^\lambda\mathbb{C}[[x]]^n$.

Proof.

Insert $\vec{y} = x^\lambda \displaystyle\sum_{m=0}^{\infty} x^m \vec{a}_m$ into (V.1.3). Setting $A(x) = \displaystyle\sum_{m=0}^{\infty} x^m A_m$, where $A_m \in \mathcal{M}_n(\mathbb{C})$ and $A_0 = A(0)$, we obtain

$$x^\lambda \sum_{m=0}^{\infty} (\lambda + m) x^m \vec{a}_m = x^\lambda \sum_{m=0}^{\infty} x^m \left[\sum_{h=0}^{m} A_{m-h} \vec{a}_h \right].$$

Therefore, in order to construct a formal solution, the coefficients $\vec{a}_m$ must be determined by the equations

$$\lambda\vec{a}_0 = A_0\vec{a}_0 \quad \text{and} \quad (\lambda + m)\vec{a}_m = A_0\vec{a}_m + \sum_{h=0}^{m-h} A_{m-h}\vec{a}_h \qquad (m \geq 1).$$

Hence, $\vec{a}_0$ must be an eigenvector of A_0 associated with the eigenvalue λ, whereas

$$\vec{a}_m = ((\lambda + m)I_n - A_0)^{-1}\left[\sum_{h=0}^{m-h} A_{m-h}\vec{a}_h\right] \text{ for } m \geq 1. \quad \Box$$

For an eigenvalue λ_0 of $A(0)$, let h be the maximum integer such that $\lambda_0 + h$ is also an eigenvalue of $A(0)$. Then, Theorem V-1-2 applies to $\lambda = \lambda_0 + h$. The convergence of the formal solution $\vec{\phi}(x)$ of Theorem V-1-2 will be proved in §V-2, assuming that the entries of the matrix $A(x)$ are in $\mathbb{C}\{x\}$ (cf. Remark V-2-9).

Let $P = \displaystyle\sum_{h=0}^{n} a_h(x)\delta^h$ $\left(\delta = x\dfrac{d}{dx}\right)$ be a differential operator with the coefficients $a_h(x)$ in $\mathbb{C}[[x]]$. Assume that $n \geq 1$ and $a_n(x) \neq 0$. Set

$$P[x^s] = x^s \sum_{m=n_0}^{\infty} f_m(s)\, x^m,$$

where the coefficients $f_m(s)$ are polynomials in s and n_0 is a non-negative integer such that $f_{n_0} \neq 0$. Then, we can prove the following theorem.

Theorem V-1-3.
   (i) *The degree of f_{n_0} in s is not greater than n.*
   (ii) *If zeros of f_{n_0} do not differ by integers, then, for each zero r of f_{n_0}, there exists a formal series $x^r\phi(x) \in x^{r+1}\mathbb{C}[[x]]$ such that $P[x^r(1 + \phi(x))] = 0$.*

Proof.

(i) Since $\delta^h[x^s] = s^h x^s$, it is evident that the degree of f_{n_0} in s is not greater than n.

(ii) For a formal power series $\phi(x) = \displaystyle\sum_{k=1}^{+\infty} c_k x^k$, we have

$$P[x^s(1 + \phi(x))] = P[x^s] + \sum_{k=1}^{+\infty} c_k P[x^{s+k}]$$

$$= x^s\left\{\sum_{m=n_0}^{+\infty} f_m(s)x^m + \sum_{k=1}^{+\infty} c_k x^k\left(\sum_{m=n_0}^{+\infty} f_m(s+k)x^m\right)\right\}$$

$$= x^s\left\{f_{n_0}(s)x^{n_0} + \sum_{m=n_0+1}^{+\infty}\left(f_m(s) + \sum_{k=1}^{m-n_0} c_k f_{m-k}(s+k)\right)x^m\right\}$$

$$= x^{s+n_0}\left\{f_{n_0}(s) + \sum_{m=1}^{+\infty}\left(f_{n_0+m}(s) + \sum_{k=1}^{m} c_k f_{n_0+m-k}(s+k)\right)x^m\right\}.$$

In order that $y = x^r(1 + \phi(x))$ be a formal solution of $P[y] = 0$, it is necessary and sufficient that the coefficients c_m and r satisfy the equations

$$f_{n_0}(r) = 0, \qquad f_{n_0+m}(r) + \sum_{k=1}^{m} c_k f_{n_0+m-k}(r+k) = 0 \qquad (m \geq 1).$$

It is assumed that $f_{n_0}(r + m) \neq 0$ for nonzero integer m if r is a zero of $f_{n_0}(s)$. Therefore, r and c_m are determined by

$$f_{n_0}(r) = 0 \quad \text{and} \quad c_m = -\frac{1}{f_{n_0}(r+m)} \sum_{k=1}^{m-1} c_k f_{n_0+m-k}(r+k) \quad (m \geq 1). \quad \square$$

Convergence of the formal solution $x^r(1 + \phi(x))$ of Theorem V-1-3 will be proved at the end of §V-7, assuming that the degree of $f_{n_0}(s)$ in s is n and the coefficients $a_h(x)$ of the operator P belong to $\mathbb{C}\{x\}$. The polynomial $f_{n_0}(s)$ is called the *indicial polynomial* of the operator P.

Remark V-1-4. Formal solutions of algebraic differential equation (V.1.1) as defined above are not necessarily convergent, even if all coefficients of the polynomial F are in $\mathbb{C}\{x\}$. For example, $y = \sum_{m=0}^{\infty} (-1)^m (m!) x^{m+1}$ is a formal solution of the differential equation $x^2 \dfrac{d^2 y}{dx^2} + y - x = 0$. Also, the formal solution $x^r(1 + \phi(x))$ of Theorem V-1-3 is not necessarily convergent if the degree of $f_{n_0}(s)$ in s is less than n. In order that $f = \sum_{m=0}^{\infty} a_m x^m$ be convergent, it is necessary and sufficient that $|a_m| \leq K A^m$ for all non-negative integers m, where K and A are non-negative numbers. Also, it is known that some power series such as $\sum_{m=0}^{\infty} (m!)^m x^m$ do not satisfy any algebraic differential equation. The following result gives a reasonable necessary condition that a power series be a formal solution of an algebraic differential equation.

Theorem V-1-5 (E. Maillet [Mai]). *Let $F(x, y_0, y_1, \ldots, y_n)$ be a nonzero polynomial in $y_0, y_1, \ldots, y_n$ with coefficients in $\mathbb{C}\{x\}$, and let $f = \sum_{m=0}^{\infty} a_m x^m \in \mathbb{C}[[x]]$ be a formal solution of the differential equation $F\left(x, y, \dfrac{dy}{dx}, \ldots, \dfrac{d^n y}{dx^n}\right) = 0$. Then, there exist non-negative numbers K, p, and A such that*

$$(V.1.4) \qquad\qquad |a_m| \leq K(m!)^p A^m \quad (m \geq 0).$$

We shall return to this result later in §XIII-8.

Remark V-1-6. In various applications, including some problems in analytic number theory, sharp estimates of lower and upper bounds of coefficients a_m of a formal solution $f = \sum_{m=0}^{\infty} a_m x^m$ of an algebraic differential equations are very important. For those results, details are found, for example, in [Mah], [Pop], [SS1], [SS2], and [SS3]. The book [GérT] contains many informations concerning upper estimates of coefficients $|a_m|$.

V-2. Convergence of formal solutions of a system of the first kind

In this section, we prove convergence of formal solutions of a system of differential equations

$$(\text{V.2.1}) \qquad x\frac{d\vec{y}}{dx} \;=\; \vec{f}(x,\vec{y}),$$

where $\vec{y} \in \mathbb{C}^n$ is an unknown quantity and the entries of the $\mathbb{C}^n$-valued function $\vec{f}$ are convergent power series in $(x,\vec{y})$ with coefficients in $\mathbb{C}$. A formal power series

$$(\text{V.2.2}) \qquad \vec{\phi}(x) \;=\; \sum_{m=1}^{\infty} x^m \vec{c}_m \;\in\; x\mathbb{C}[[x]]^n \qquad (\, \vec{c}_m \in \mathbb{C}^n \,)$$

is a formal solution of system (V.2.1) if

$$(\text{V.2.3}) \qquad x\frac{d\vec{\phi}}{dx} \;=\; \vec{f}(x,\vec{\phi})$$

as a formal power series.

To achieve our main goal, we need some preparations.

Observation V-2-1. In order that a formal power series (V.2.2) satisfy condition (V.2.3), it is necessary that $\vec{f}(0,\vec{0}) = \vec{0}$. Therefore, write $\vec{f}$ in the form

$$\vec{f}(x,\vec{y}) \;=\; \vec{f}_0(x) \;+\; A(x)\vec{y} \;+\; \sum_{|\wp|\geq 2} \vec{y}^{\wp} \vec{f}_{\wp}(x),$$

where

(1) $\wp = (p_1,\dots,p_n)$ and the p_j are non-negative integers,

(2) $|\wp| = p_1 + \cdots + p_n$ and $\vec{y}^{\wp} = y_1^{p_1} \cdots y_n^{p_n}$, where $y_1,\dots,y_n$ are the entries of $\vec{y}$,

(3) $\vec{f}_0 \in x\mathbb{C}\{x\}^n$ and $\vec{f}_{\wp} \in \mathbb{C}\{x\}^n$,

(4) $A(x)$ is an $n \times n$ matrix with the entries in $\mathbb{C}\{x\}$.

Note that

$$\vec{f}_0(x) \;=\; \vec{f}(x,\vec{0}) \qquad \text{and} \qquad A(x) \;=\; \frac{\partial \vec{f}}{\partial \vec{y}}(x,\vec{0}).$$

Setting

$$A(x) \;=\; \sum_{m=0}^{\infty} x^m A_m \qquad \left(A_0 \;=\; \frac{\partial \vec{f}}{\partial \vec{y}}(0,\vec{0}) \right),$$

where the coefficients A_m are in $\mathcal{M}_n(\mathbb{C})$, write condition (V.2.3) in the form

$$x\frac{d\vec{\phi}}{dx} \;=\; A_0\vec{\phi} \;+\; \vec{f}(x,\vec{\phi}) \;-\; A_0\vec{\phi}.$$

Then,

$$(\text{V.2.4}) \qquad m\vec{c}_m \;=\; A_0\vec{c}_m \;+\; \vec{\gamma}_m \qquad \text{for} \quad m = 1,2,\dots,$$

where

$$\vec{f}(x,\vec{\phi}) \, - \, A_0\vec{\phi} \, = \, \vec{f}_0(x) \, + \, [A(x) - A_0]\,\vec{\phi}(x) \, + \, \sum_{|\rho|\geq 2} \left(\vec{\phi}(x)\right)^\rho \vec{f}_\rho(x)$$

$$= \, \sum_{m=1}^\infty x^m \vec{\gamma}_m$$

and $\vec{\gamma}_m \in \mathbb{C}^n$. Note that $\vec{\gamma}_m$ is determined when $\vec{c}_1,\ldots,\vec{c}_{m-1}$ are determined and that the matrices $mI_n - A_0$ are invertible if positive integers m are sufficiently large. This implies that there exists a positive integers m_0 such that if $\vec{c}_1,\ldots,\vec{c}_{m_0}$ are determined, then $\vec{c}_m$ is uniquely determined for all integers m greater than m_0. Therefore, the system of a finite number of equations

$$(\text{V.2.5}) \qquad\qquad m\vec{c}_m \, = \, A_0\vec{c}_m \, + \, \vec{\gamma}_m \qquad (m = 1, 2, \ldots, m_0)$$

decides whether a formal solution $\vec{\phi}(x)$ exists. If system (V.2.5) has a solution $\{\vec{c}_1,\ldots,\vec{c}_{m_0}\}$, those m_0 constants vectors determine a formal solution $\vec{\phi}(x)$ uniquely.

Observation V-2-2. Supposing that formal power series (V.2.2) is a formal solution of (V.2.1), set $\vec{\phi}_N(x) = \sum_{m=1}^N x^m \vec{c}_m$. Since

$$\vec{f}(x,\vec{\phi}(x)) - \vec{f}(x,\vec{\phi}_N(x)) = A(x)(\vec{\phi}(x) - \vec{\phi}_N(x)) + \sum_{|\rho|\geq 2}\left[\vec{\phi}(x)^\rho - \vec{\phi}_N(x)^\rho\right]\vec{f}_\rho(x),$$

it follows that $\vec{f}(x,\vec{\phi}(x)) - \vec{f}(x,\vec{\phi}_N(x)) \in x^{N+1}\mathbb{C}[[x]]^n$. Also,

$$\vec{f}(x,\vec{\phi}(x)) \, - \, x\frac{d\vec{\phi}_N(x)}{dx} \, = \, x\frac{d\vec{\phi}(x)}{dx} \, - \, x\frac{d\vec{\phi}_N(x)}{dx} \, \in \, x^{N+1}\mathbb{C}[[x]]^n.$$

Hence, $\vec{f}(x,\vec{\phi}_N(x)) \, - \, x\dfrac{d\vec{\phi}_N(x)}{dx} \, \in \, x^{N+1}\mathbb{C}\{x\}^n$. Set

$$(\text{V.2.6}) \qquad\qquad \vec{g}_{N,0}(x) \, = \, \vec{f}(x,\vec{\phi}_N(x)) \, - \, x\frac{d\vec{\phi}_N(x)}{dx}.$$

Now, by means of the transformation $\vec{y} = \vec{z} + \vec{\phi}_N(x)$, change system (V.2.1) to the system

$$(\text{V.2.7}) \qquad\qquad x\frac{d\vec{z}}{dx} \, = \, \vec{g}_N(x,\vec{z})$$

on $\vec{z} \in \mathbb{C}^n$, where

$$\vec{g}_N(x,\vec{z}) \, = \, \vec{f}(x,\vec{z} + \vec{\phi}_N(x)) \, - \, x\frac{d\vec{\phi}_N(x)}{dx}$$

$$= \, \vec{g}_{N,0}(x) \, + \, \vec{f}(x,\vec{z} + \vec{\phi}_N(x)) \, - \, \vec{f}(x,\vec{\phi}_N(x))$$

$$= \, \vec{g}_{N,0}(x) \, + \, A(x)\vec{z} \, + \, \sum_{|\rho|\geq 2}\left[\left(\vec{z} + \vec{\phi}_N(x)\right)^\rho - \vec{\phi}_N(x)^\rho\right]\vec{f}_\rho(x).$$

As in Observation V-2-1, write $\vec{g}_N$ in the form

$$\vec{g}_N(x, \vec{z}) \;=\; \vec{g}_{N,0}(x) \;+\; B_N(x)\vec{z} \;+\; \sum_{|\wp|\geq 2} \vec{z}^{\,\wp}\, \vec{g}_{N,\wp}(x),$$

where
  (1) $\vec{g}_{N,0} \in x^{N+1}\mathbb{C}\{x\}^n$ and $\vec{g}_{N,\wp} \in \mathbb{C}\{x\}^n$,
  (2) $B_N(x)$ is an $n \times n$ matrix with the entries in $\mathbb{C}\{x\}$,
  (3) the entries of the matrix $B_N(x) - A_0$ are contained in $x\mathbb{C}\{x\}$.

Observation V-2-3. System (V.2.7) has a formal solution

$$(V.2.8) \qquad \vec{\psi}_N(x) \;=\; \vec{\phi}(x) \;-\; \vec{\phi}_N(x) \;=\; \sum_{m=N+1}^{\infty} x^m \vec{c}_m \in x^{N+1}\mathbb{C}[[x]]^n.$$

The coefficients $\vec{c}_m$ are determined recursively by

$$m\vec{c}_m \;=\; A_0\vec{c}_m \;+\; \vec{\gamma}_m \qquad \text{for} \quad m = N+1, N+2, \dots\,,$$

where

$$\begin{aligned}
\vec{g}_N(x, \vec{\psi}_N(x)) - A_0\vec{\psi}_N(x) &\;=\; \vec{f}(x, \vec{\phi}(x)) - A_0\vec{\psi}_N(x) - x\frac{d\vec{\phi}_N(x)}{dx} \\
&\;=\; \vec{f}(x, \vec{\phi}(x)) - A_0\vec{\phi}(x) + A_0\vec{\phi}_N(x) - x\frac{d\vec{\phi}_N(x)}{dx} \\
&\;=\; \vec{g}_{N,0}(x) + [B_N(x) - A_0]\,\vec{\psi}_N(x) + \sum_{|\wp|\geq 2} \left(\vec{\psi}_N(x)\right)^{\wp} \vec{g}_{N,\wp}(x) \\
&\;=\; \sum_{m=N+1}^{\infty} x^m \vec{\gamma}_m.
\end{aligned}$$

Note that the matrices $mI_n - A_0$ are invertible for $m = N+1, N+2, \dots$, if N is sufficiently large.

Observation V-2-4. Suppose that system (V.2.1) has an actual solution $\vec{\eta}(x)$ such that the entries of $\vec{\eta}(x)$ are analytic at $x = 0$ and that $\vec{\eta}(0) = \vec{0}$. Then, the Taylor expansion $\vec{\phi}(x) = \sum_{m=1}^{\infty} \frac{x^m}{m!}\frac{d^m\vec{\eta}}{dx^m}(0)$ of $\vec{\eta}(x)$ at $x = 0$ is a formal solution of (V.2.1). Furthermore, $\vec{\phi}$ is convergent and $\vec{\phi} \in x\mathbb{C}\{x\}^n$.

Keeping these observations in mind, let us prove the following theorem.

Theorem V-2-5. *Suppose that $\vec{f}_0(x) = \vec{f}(x, \vec{0}) \in x^{N+1}\mathbb{C}\{x\}^n$ and that the matrices $mI_n - A_0\,(m \geq N+1)$ are invertible, where $A_0 = \dfrac{\partial \vec{f}}{\partial \vec{y}}(0, \vec{0})$. Then, system (V.2.1) has a unique formal solution*

$$(V.2.9) \qquad \vec{\phi}(x) \;=\; \sum_{m=N+1}^{\infty} x^m \vec{c}_m \in x^{N+1}\mathbb{C}[[x]]^n.$$

Furthermore, $\vec{\phi} \in x^{N+1}\mathbb{C}\{x\}^n$.

Remark V-2-6. Under the assumptions of Theorem V-2-5, system (V.2.1) possibly has many formal solutions in $x\mathbb{C}[[x]]^n$. However, Theorem V-2-5 states that there is only one formal solution in $x^{N+1}\mathbb{C}[[x]]^n$.

Proof of Theorem V-2-5.

We prove this theorem in six steps.

Step 1. Using the argument of Observation V-2-1, we can prove the existence and uniqueness of formal solution (V.2.9). In fact,

$$\vec{f}(x,\vec{\phi}) \ - \ A_0\vec{\phi} \ = \ \vec{f}_0(x) \ + \ [A(x) \ - \ A_0]\vec{\phi}(x) \ + \ \sum_{|\wp|\geq 2} \left(\vec{\phi}(x)\right)^\wp \vec{f}_\wp(x)$$

$$= \ \sum_{m=N+1}^{\infty} x^m \vec{\gamma}_m.$$

This implies that $\vec{\gamma}_m = 0$ for $m = 1, 2, \ldots, N$. Hence, $\vec{c}_m$ $(m \geq N+1)$ are uniquely determined by

$$m\vec{c}_m \ = \ A_0\vec{c}_m \ + \ \vec{\gamma}_m \qquad \text{for} \quad m = N+1, N+2, \ldots .$$

Step 2. Suppose that system (V.2.1) has an actual solution $\vec{\eta}(x)$ satisfying the following conditions:
  (i) the entries of $\vec{\eta}(x)$ are analytic at $x = 0$,
  (ii) there exist two positive numbers K and δ such that

$$|\vec{\eta}(x)| \ \leq \ K|x|^{N+1} \qquad \text{for} \qquad |x| \ \leq \ \delta.$$

Then, the Taylor expansion $\displaystyle\sum_{m=N+1}^{\infty} \frac{x^m}{m!} \frac{d^m\vec{\eta}}{dx^m}(0)$ of $\vec{\eta}(x)$ at $x = 0$ is a formal solution of (V.2.1) (cf. Observation V-2-4). Since such a formal solution is unique, it follows that $\vec{\phi}(x) = \displaystyle\sum_{m=N+1}^{\infty} \frac{x^m}{m!} \frac{d^m\vec{\eta}}{dx^m}(0)$. Because the Taylor expansion of $\vec{\eta}(x)$ at $x = 0$ is convergent, the formal solution $\vec{\phi}$ is convergent and $\vec{\phi} \in x^{N+1}\mathbb{C}\{x\}^n$.

Step 3. Hereafter, we shall construct an actual solution $\vec{\eta}(x)$ of (V.2.1) that satisfies conditions (i) and (ii) of Step 2. To do this, first notice that there exist three positive numbers H, δ, and ρ such that

(I) $$\qquad\qquad |\vec{f}(x,\vec{0})| \ \leq \ H|x|^{N+1} \qquad \text{for} \quad |x| \ \leq \ \delta$$

and

(II) $$\quad |\vec{f}(x,\vec{y}_1) \ - \ \vec{f}(x,\vec{y}_2)| \ \leq \ (|A_0| \ + \ 1) \, |\vec{y}_1 \ - \ \vec{y}_2|$$
$$\text{for} \quad |x| \ \leq \ \delta \quad \text{and} \quad |\vec{y}_j| \ \leq \ \rho \quad (j = 1, 2).$$

Hence,

(III) $$\quad |\vec{f}(x,\vec{y})| \ \leq \ H|x|^{N+1} + (|A_0|+1)|\vec{y}| \quad \text{for} \ |x| \ \leq \ \delta \ \text{and} \ |\vec{y}| \ \leq \ \rho.$$

Using the transformation of Observation V-2-2, N can be made as large as we want without changing the matrix A_0. Hence, assume without loss of any generality that

$$(\text{V.2.10}) \qquad \frac{|A_0|+1}{N+1} < \frac{1}{2}.$$

Also, fix two positive numbers K and δ so that

$$(\text{V.2.11}) \qquad K > \frac{H+[|A_0|+1]K}{N+1} \quad \text{and} \quad K\delta^{N+1} \leq \rho.$$

Step 4. Change system (V.2.1) to an integral equation

$$(\text{V.2.12}) \qquad \vec{\eta}(x) = \int_0^x \frac{1}{\xi}\vec{f}(\xi,\vec{\eta}(\xi))d\xi.$$

Define successive approximations

$$\vec{\eta}_0(x) = 0 \quad \text{and} \quad \vec{\eta}_{k+1}(x) = \int_0^x \frac{1}{\xi}\vec{f}(\xi,\vec{\eta}_k(\xi))d\xi \quad \text{for} \quad k=0,1,2,\dots.$$

Now, we shall show that

$$|\vec{\eta}_k(x)| \leq K|x|^{N+1} \qquad \text{for} \quad |x| \leq \delta \quad \text{and} \quad k=0,1,2\dots.$$

Since this is true for $k=0$, we show this recursively with respect to k as follows. First if this is true for k, then

$$|\vec{\eta}_k(x)| \leq K|x|^{N+1} \leq K\delta^{N+1} \leq \rho \qquad \text{for} \quad |x| \leq \delta.$$

Hence,

$$\left|\frac{1}{\xi}\vec{f}(\xi,\vec{\eta}_k(\xi))\right| \leq \{H+[|A_0|+1]K\}|\xi|^N \qquad \text{for} \quad |\xi| \leq \delta.$$

Therefore,

$$|\vec{\eta}_{k+1}(x)| \leq \frac{H+[|A_0|+1]K}{N+1}|x|^{N+1} \leq K|x|^{N+1} \qquad \text{for} \quad |x| \leq \delta.$$

Step 5. Set

$$\|\vec{\eta}_{k+1} - \vec{\eta}_k\| = \max\left\{\frac{|\vec{\eta}_{k+1}(x)-\vec{\eta}_k(x)|}{|x|^{N+1}} : |x| \leq \delta\right\}.$$

Then, since

$$|\vec{\eta}_{k+1}(x) - \vec{\eta}_k(x)| = \left|\int_0^x \frac{1}{\xi}\left(\vec{f}(\xi,\vec{\eta}_k(\xi)) - \vec{f}(\xi,\vec{\eta}_{k-1}(\xi))\right)d\xi\right|,$$

we obtain

$$\|\vec{\eta}_{k+1} - \vec{\eta}_k\| \leq \frac{|A_0|+1}{N+1}\|\vec{\eta}_k - \vec{\eta}_{k-1}\| \leq \frac{1}{2}\|\vec{\eta}_k - \vec{\eta}_{k-1}\|.$$

This implies that

$$\lim_{k\to+\infty} \frac{\vec{\eta}_k(x)}{x^{N+1}} = \sum_{\ell=0}^{\infty} \frac{\vec{\eta}_{\ell+1}(x) - \vec{\eta}_\ell(x)}{x^{N+1}}$$

exists uniformly for $|x| \leq \delta$.

Step 6. Setting

$$\vec{\eta}(x) \;=\; x^{N+1}\left(\lim_{k\to+\infty}\frac{\vec{\eta}_k(x)}{x^{N+1}}\right) \;=\; \lim_{k\to+\infty}\vec{\eta}_k(x),$$

it is easy to show that $\vec{\eta}(x)$ satisfies integral equation (V.2.12). It is also evident that $\vec{\eta}(x)$ is analytic for $|x| < \delta$. Thus, the proof of Theorem V-2-5 is completed. $\quad\Box$

Now, finally, by using the argument given in Observations V-2-2 and V-2-3, we obtain the following theorem.

Theorem V-2-7. *Every formal solution $\vec{\phi} \in x\mathbb{C}[[x]]^n$ of system (V.2.1) is convergent, i.e., $\vec{\phi} \in x\mathbb{C}\{x\}^n$.*

Remark V-2-8. In general, (V.2.1) may not have any formal solutions. However, Theorem V-2-7 states that if (V.2.1) has formal solutions, then every formal solution is convergent.

Remark V-2-9. If $\vec{\phi}(x) = x^{\lambda}\vec{f}(x) \in x^{\lambda}\mathbb{C}[[x]]^n$ is a formal solution of a linear system $x\dfrac{d\vec{y}}{dx} = A(x)\vec{y}$, the formal power series $\vec{f}$ is a formal solution of $x\dfrac{d\vec{z}}{dx} = [A(x) - \lambda I_n]\vec{z}$. Therefore, $\vec{f} \in \mathbb{C}\{x\}^n$ by virtue of Theorem V-2-7. This proves convergence of the formal solution constructed in Theorem V-1-2.

V-3. The S-N decomposition of a matrix of infinite order

In §V-4, we shall define the S-N decomposition of a linear differential operator. As a linear differential operator will be represented by a lower block-triangular matrix of infinite order, we derive, in this section, the S-N decomposition of such a matrix

$$A \;=\; \begin{bmatrix} A_{11} & O & O & \cdots & \cdots & \cdots \\ A_{21} & A_{22} & O & \cdots & \cdots & \cdots \\ \vdots & \vdots & \vdots & \vdots & \vdots & \vdots \\ A_{m1} & A_{m2} & \cdots & A_{mm} & O & \cdots \\ \vdots & \vdots & \vdots & \vdots & \vdots & \vdots \end{bmatrix},$$

where for each (j,k), the quantity A_{jk} is an $n_j \times n_k$ constant matrix. Set

$$\begin{cases} A_1 \;=\; A_{11}, \\ A_m \;=\; \begin{bmatrix} A_{11} & O & O & \cdots & O \\ A_{21} & A_{22} & O & \cdots & O \\ \vdots & \vdots & \vdots & \vdots & \vdots \\ A_{m1} & A_{m2} & \cdots & \cdots & A_{mm} \end{bmatrix} & (m \geq 2). \end{cases}$$

For $m \geq 2$, we write $A_m = \begin{bmatrix} A_{m-1} & O \\ B_m & A_{mm} \end{bmatrix}$. Set $N_m = \displaystyle\sum_{\ell=1}^{m} n_\ell$. Then, A_m is an $N_m \times N_m$ matrix, while B_m is an $n_m \times N_{m-1}$ matrix. Also, let $A_{mm} = S_{mm} + \mathcal{N}_{mm}$

and $A_m = S_m + N_m$ be the $S\text{-}N$ decompositions of A_{mm} and A_m, respectively, where S_{mm} is an $n_m \times n_m$ diagonalizable matrix, S_m is an $N_m \times N_m$ diagonalizable matrix, N_{mm} is an $n_m \times n_m$ nilpotent matrix, N_m is an $N_m \times N_m$ nilpotent matrix, $S_{mm}N_{mm} = N_{mm}S_{mm}$, and $S_m N_m = N_m S_m$. The following lemma shows how the two matrices S_m and N_m look.

Lemma V-3-1. *The matrices S_m and N_m have the following forms:*

$$
\begin{cases}
S_1 = S_{11}, & S_m = \begin{bmatrix} S_{m-1} & O \\ C_m & S_{mm} \end{bmatrix} & (m \geq 2), \\[2ex]
N_1 = N_{11}, & N_m = \begin{bmatrix} N_{m-1} & O \\ \mathcal{F}_m & N_{mm} \end{bmatrix} & (m \geq 2),
\end{cases}
$$

where C_m and $\mathcal{F}_m$ are $n_m \times N_{m-1}$ matrices.

Proof.

Consider the case $m \geq 2$. Since the matrices S_m and N_m are polynomials in A_m with constant coefficients, it follows that

$$
S_m = \begin{bmatrix} \mathcal{B}_{m-1} & O \\ C_m & \mu_m \end{bmatrix} \quad \text{and} \quad N_m = \begin{bmatrix} \mathcal{D}_{m-1} & O \\ \mathcal{F}_m & \nu_m \end{bmatrix},
$$

where $\mathcal{B}_{m-1}$ and $\mathcal{D}_{m-1}$ are $N_{m-1} \times N_{m-1}$ matrices, C_m and $\mathcal{F}_m$ are $n_m \times N_{m-1}$ matrices, and μ_m and ν_m are $n_m \times n_m$ matrices. Furthermore, $\mathcal{D}_{m-1}$ and ν_m are nilpotent. Also, $\mathcal{B}_{m-1}\mathcal{D}_{m-1} = \mathcal{D}_{m-1}\mathcal{B}_{m-1}$ and $\mu_m \nu_m = \nu_m \mu_m$. Hence, it suffices to show that $\mathcal{B}_{m-1}$ and μ_m are diagonalizable.

Note that since S_m is diagonalizable, S_m has N_m linearly independent eigenvectors. An eigenvector of S_m has one of two forms $\begin{bmatrix} \vec{p} \\ \vec{r} \end{bmatrix}$ and $\begin{bmatrix} \vec{0} \\ \vec{q} \end{bmatrix}$, where $\vec{p}$ is an eigenvector of $\mathcal{B}_{m-1}$, whereas $\vec{q}$ is an eigenvector of μ_m. Therefore, if we count those independent eigenvectors, it can be shown that $\mathcal{B}_{m-1}$ has N_{m-1} linearly independent eigenvectors, while μ_m has n_m linearly independent eigenvectors. Note that $N_m = N_{m-1} + n_m$. This completes the proof of Lemma V-3-1. $\square$

Lemma V-3-1 implies that the matrices S_m and N_m have the following forms:

$$
\begin{cases}
S_1 = S_{11}, \\[1ex]
S_m = \begin{bmatrix}
S_{11} & O & O & \cdots & O \\
C_{21} & S_{22} & O & \cdots & O \\
\vdots & \vdots & \vdots & \vdots & \vdots \\
C_{m1} & C_{m2} & \cdots & \cdots & S_{mm}
\end{bmatrix} & (m \geq 2)
\end{cases}
$$

and

$$
\begin{cases}
N_1 = N_{11}, \\[1ex]
N_m = \begin{bmatrix}
N_{11} & O & O & \cdots & O \\
\mathcal{F}_{21} & N_{22} & O & \cdots & O \\
\vdots & \vdots & \vdots & \vdots & \vdots \\
\mathcal{F}_{m1} & \mathcal{F}_{m2} & \cdots & \cdots & N_{mm}
\end{bmatrix} & (m \geq 2),
\end{cases}
$$

where C_{jk} and $\mathcal{F}_{jk}$ are $n_j \times n_k$ matrices.

Set

$$
\left\{
\begin{aligned}
\mathcal{S} &=
\begin{bmatrix}
\mathcal{S}_{11} & O & O & \cdots & \cdots & \cdots \\
C_{21} & \mathcal{S}_{22} & O & \cdots & \cdots & \cdots \\
\vdots & \vdots & \vdots & \vdots & \vdots & \vdots \\
C_{m1} & C_{m2} & \cdots & \mathcal{S}_{mm} & O & \cdots \\
\vdots & \vdots & \vdots & \vdots & \vdots & \vdots
\end{bmatrix}, \\[2ex]
\mathcal{N} &=
\begin{bmatrix}
\mathcal{N}_{11} & O & O & \cdots & \cdots & \cdots \\
\mathcal{F}_{21} & \mathcal{N}_{22} & O & \cdots & \cdots & \cdots \\
\vdots & \vdots & \vdots & \vdots & \vdots & \vdots \\
\mathcal{F}_{m1} & \mathcal{F}_{m2} & \cdots & \mathcal{N}_{mm} & O & \cdots \\
\vdots & \vdots & \vdots & \vdots & \vdots & \vdots
\end{bmatrix}.
\end{aligned}
\right.
$$

Then,

$$
\text{(V.3.1)} \qquad A = \mathcal{S} + \mathcal{N} \quad \text{and} \quad \mathcal{S}\mathcal{N} = \mathcal{N}\mathcal{S}.
$$

We call (V.3.1) the *S-N decomposition* of the matrix A.

V-4. The S-N decomposition of a differential operator

Consider a differential operator

$$
\text{(V.4.1)} \qquad \mathcal{L}[\vec{y}] = x\frac{d\vec{y}}{dx} + \Omega(x)\vec{y},
$$

where $\Omega(x) = \displaystyle\sum_{\ell=0}^{\infty} x^{\ell}\Omega_{\ell}$ and $\Omega_{\ell} \in \mathcal{M}_n(\mathbb{C})$. Let us identify a formal power series

$\vec{p} = \displaystyle\sum_{\ell\geq 0} x^{\ell}\vec{a}_{\ell}$ with the vector $\wp = \begin{bmatrix} \vec{a}_0 \\ \vec{a}_1 \\ \vec{a}_2 \\ \vdots \end{bmatrix} \in \mathbb{C}^{\infty}$. Then, the operator $\mathcal{L}$ is represented

by the matrix

$$
A =
\begin{bmatrix}
\Omega_0 & O & O & O & \cdots & \cdots & \cdots & \cdots \\
\Omega_1 & I_n + \Omega_0 & O & O & \cdots & \cdots & \cdots & \cdots \\
\Omega_2 & \Omega_1 & 2I_n + \Omega_0 & O & \cdots & \cdots & \cdots & \cdots \\
\vdots & \vdots & \vdots & \vdots & \vdots & \vdots & \vdots & \vdots \\
\Omega_m & \Omega_{m-1} & \Omega_{m-2} & \cdots & \Omega_1 & mI_n + \Omega_0 & O & \cdots \\
\vdots & \vdots & \vdots & \vdots & \vdots & \vdots & \vdots & \vdots
\end{bmatrix},
$$

where I_n is the $n \times n$ identity matrix. Let $A = S + N$ be the S-N decomposition of A. Since $A = \begin{bmatrix} \Omega_0 & O \\ \tilde{\Omega} & I_\infty + A \end{bmatrix}$, where $\tilde{\Omega} = \begin{bmatrix} \Omega_1 \\ \Omega_2 \\ \vdots \end{bmatrix}$ and I_∞ is the $\infty \times \infty$ identity matrix, the matrices S and N have the forms

$$
(\text{V.4.2}) \quad S = \begin{bmatrix}
S_0 & O & O & O & \cdots & \cdots & \cdots & \cdots \\
\sigma_1 & I_n + S_0 & O & O & \cdots & \cdots & \cdots & \cdots \\
\sigma_2 & \sigma_1 & 2I_n + S_0 & O & \cdots & \cdots & \cdots & \cdots \\
\vdots & \vdots & \vdots & \vdots & \vdots & \vdots & \vdots & \vdots \\
\sigma_m & \sigma_{m-1} & \sigma_{m-2} & \cdots & \sigma_1 & mI_n + S_0 & O & \cdots \\
\vdots & \vdots & \vdots & \vdots & \vdots & \vdots & \vdots & \vdots
\end{bmatrix},
$$

and

$$
(\text{V.4.3}) \quad N = \begin{bmatrix}
N_0 & O & O & O & \cdots & \cdots & \cdots & \cdots \\
\nu_1 & N_0 & O & O & \cdots & \cdots & \cdots & \cdots \\
\nu_2 & \nu_1 & N_0 & O & \cdots & \cdots & \cdots & \cdots \\
\vdots & \vdots & \vdots & \vdots & \vdots & \vdots & \vdots & \vdots \\
\nu_m & \nu_{m-1} & \nu_{m-2} & \cdots & \nu_1 & N_0 & O & \cdots \\
\vdots & \vdots & \vdots & \vdots & \vdots & \vdots & \vdots & \vdots
\end{bmatrix},
$$

respectively, where the σ_j and ν_j are $n \times n$ matrices and $\Omega_0 = S_0 + N_0$ is the S-N decomposition of the matrix Ω_0.

Set $\sigma(x) = S_0 + \sum_{\ell=1}^\infty x^\ell \sigma_\ell$ and $\nu(x) = N_0 + \sum_{\ell=1}^\infty x^\ell \nu_\ell$. Then, S represents a differential operator $\mathcal{L}_0[\vec{y}] = x\dfrac{d\vec{y}}{dx} + \sigma(x)\vec{y}$, while N represents multiplication by $\nu(x)$ (i.e., the operator: $\vec{y} \to \nu(x)\vec{y}$). Since $A = S + N$ and $SN = NS$, it follows that

$$
(\text{V.4.4}) \qquad \mathcal{L}[\vec{y}] = \mathcal{L}_0[\vec{y}] + \nu(x)\vec{y} \qquad \text{and} \qquad \mathcal{L}_0[\nu(x)\vec{y}] = \nu(x)\mathcal{L}_0[\vec{y}].
$$

We call (V.4.4) the S-N *decomposition* of operator (V.4.1). We shall show in the next section that $\mathcal{L}_0[\vec{y}]$ is diagonalizable and $\nu(x)^n = O$.

V-5. A normal form of a differential operator

Let us again consider the differential operator

$$
(\text{V.5.1}) \qquad\qquad \mathcal{L}[\vec{y}] = x\frac{d\vec{y}}{dx} + \Omega(x)\vec{y},
$$

where $\Omega(x) = \sum_{\ell=0}^\infty x^\ell \Omega_\ell$ and $\Omega_\ell \in \mathcal{M}_n(\mathbb{C})$. In §V-4, we derived the S-N decomposition

$$
(\text{V.5.2}) \qquad \mathcal{L}[\vec{y}] = \mathcal{L}_0[\vec{y}] + \nu(x)\vec{y} \qquad \text{and} \qquad \mathcal{L}_0[\nu(x)\vec{y}] = \nu(x)\mathcal{L}_0[\vec{y}],
$$

where

$$(V.5.3) \qquad \mathcal{L}_0[\vec{y}] \ = \ x\frac{d\vec{y}}{dx} \ + \ \sigma(x)\vec{y}$$

and

$$\sigma(x) \ = \ \mathcal{S}_0 \ + \ \sum_{\ell=1}^{\infty} x^\ell \sigma_\ell \qquad \text{and} \qquad \nu(x) \ = \ \mathcal{N}_0 \ + \ \sum_{\ell=1}^{\infty} x^\ell \nu_\ell$$

(cf. (V.4.4)). Notice that $\Omega_0 = \mathcal{S}_0 + \mathcal{N}_0$ is the S-N decomposition of Ω_0 and that the operator $\mathcal{L}_0[\vec{u}]$ and multiplication by $\nu(x)$ are represented respectively by matrices (V.4.2) and (V.4.3).

Set

$$\mathcal{S}_m \ = \ \begin{bmatrix} \mathcal{S}_0 & O & O & O & \cdots & \cdots \\ \sigma_1 & I_n + \mathcal{S}_0 & O & O & \cdots & \cdots \\ \sigma_2 & \sigma_1 & 2I_n + \mathcal{S}_0 & O & \cdots & \cdots \\ \vdots & \vdots & \vdots & \vdots & \vdots & \vdots \\ \sigma_m & \sigma_{m-1} & \sigma_{m-2} & \cdots & \sigma_1 & mI_n + \mathcal{S}_0 \end{bmatrix}.$$

Since $\mathcal{S}_0$ is diagonalizable, there exist an invertible $n \times n$ matrix P_0 and a diagonal matrix Λ_0 such that

$$\mathcal{S}_0 P_0 \ = \ P_0 \Lambda_0, \qquad \Lambda_0 \ = \ \text{diag}[\lambda_1, \lambda_2, \ldots, \lambda_n].$$

Note that λ_j $(j = 1, 2, \ldots, n)$ are eigenvalues of $\mathcal{S}_0$. Hence, λ_j $(j = 1, 2, \ldots, n)$ are also eigenvalues of Ω_0. For every positive integer m, $\mathcal{S}_m$ is diagonalizable. Hence, there exists an $(m+1)n \times n$ matrix $\mathcal{P}_m = \begin{bmatrix} P_0 \\ P_1 \\ \vdots \\ P_m \end{bmatrix}$ such that $\mathcal{S}_m \mathcal{P}_m = \mathcal{P}_m \Lambda_0$. If m is sufficiently large, we can further determine matrices P_ℓ for $\ell \geq m+1$ by the equations

$$(V.5.4) \qquad (\ell I_n + \mathcal{S}_0)P_\ell \ + \ \sum_{h=1}^{\ell} \sigma_h P_{\ell-h} \ = \ P_\ell \Lambda_0.$$

Equation (V.5.4) can be solved with respect to P_ℓ, since the linear operator $P_\ell \rightarrow (\ell I + \mathcal{S}_0)P_\ell - P_\ell \Lambda_0$ is invertible if m is sufficiently large. In this way, we can find an $\infty \times n$ matrix $\mathcal{P} = \begin{bmatrix} P_0 \\ P_1 \\ \vdots \\ P_m \\ \vdots \end{bmatrix}$ such that $\mathcal{S}\mathcal{P} = \mathcal{P}\Lambda_0$. Set $P(x) = \sum_{\ell=0}^{\infty} x^\ell P_\ell$. Then, entries of $P(x)^{-1}$ are also formal power series in x and

$$(V.5.5) \qquad \mathcal{L}_0[P(x)] \ = \ P(x)\Lambda_0.$$

Define two differential operators and an $n \times n$ matrix by

$$\begin{cases} \mathcal{K}[\vec{u}] & = P(x)^{-1}\mathcal{L}[P(x)\vec{u}], \quad \mathcal{K}_0[\vec{u}] = P(x)^{-1}\mathcal{L}_0[P(x)\vec{u}], \\ \nu_0(x) & = P(x)^{-1}\nu(x)P(x), \end{cases}$$

respectively. Then $\mathcal{K}[\vec{u}] = \mathcal{K}_0[\vec{u}] + \nu_0(x)\vec{u}$. Observe that

$$\begin{aligned} \text{(V.5.6)} \quad \mathcal{K}_0[\vec{u}] & = P(x)^{-1}\mathcal{L}_0[P(x)\vec{u}] = P(x)^{-1}\left[P(x)\left(x\frac{d\vec{u}}{dx} \right) + \mathcal{L}_0[P(x)]\vec{u} \right] \\ & = x\frac{d\vec{u}}{dx} + \Lambda_0\vec{u}. \end{aligned}$$

This shows that the operator $\mathcal{L}_0[\vec{y}]$ is diagonalizable. Furthermore,

$$\begin{aligned} \text{(V.5.7)} \quad x\frac{d\nu_0(x)}{dx} + \Lambda_0\nu_0(x) & = P(x)^{-1}\mathcal{L}_0[P(x)\nu_0(x)] = P(x)^{-1}\mathcal{L}_0[\nu(x)P(x)] \\ & = P(x)^{-1}\nu(x)\mathcal{L}_0[P(x)] = \nu_0(x)P(x)^{-1}\mathcal{L}_0[P(x)] = \nu_0(x)\Lambda_0. \end{aligned}$$

This shows that the entry $\nu_{jk}(x)$ on the j-th row and k-th column of the matrix $\nu_0(x)$ must have the form $\nu_{jk}(x) = \gamma_{jk}x^{\lambda_k - \lambda_j}$, where γ_{jk} is a constant. Since $\nu_0(x)$ is a formal power series in x, it follows that

$$\text{(V.5.8)} \qquad \nu_{jk}(x) = 0 \quad \text{if} \quad \lambda_k - \lambda_j \text{ is not a non-negative integer.}$$

Observe further that the matrix $\nu_0(x)$ can be written in the form

$$\nu_0(x) = x^{-\Lambda_0}\Gamma x^{\Lambda_0}, \quad \text{where} \quad \Gamma = \begin{bmatrix} \gamma_{11} & \cdots & \gamma_{1n} \\ \vdots & \vdots & \vdots \\ \gamma_{n1} & \cdots & \gamma_{nn} \end{bmatrix}.$$

Hence, for any non-negative integer p, we have $\nu_0(x)^p = x^{-\Lambda_0}\Gamma^p x^{\Lambda_0}$, where

$$z^{\Lambda_0} = \exp\left[(\log x)\Lambda_0 \right] = \mathrm{diag}[z^{\lambda_1}, z^{\lambda_2}, \ldots, z^{\lambda_n}].$$

On the other hand, since $\mathcal{N}_0$ is nilpotent, $\nu_0(x)^p$ can be written in a form $\nu_0(x)^p = x^{m_p}Q_p(x)$, where m_p is a non-negative integer such that $\lim_{p \to +\infty} m_p = +\infty$ and the entries of the matrix $Q_p(x)$ are power series in x with constant coefficients. Therefore, $\nu_0(x)^p = O$ if p is sufficiently large. This implies that the matrix Γ is nilpotent. Since $\nu(x) = P(x)\nu_0(x)P(x)^{-1}$, we obtain $\nu(x)^n = O$.

Thus, we arrive at the following conclusion.

Theorem V-5-1. *For a given differential operator (V.5.1), let $\Omega_0 = \mathcal{S}_0 + \mathcal{N}_0$ be the S-N decomposition of the matrix Ω_0. Then, there exists an $n \times n$ matrix $P(x)$ such that*

(1) the entries of $P(x)$ are formal power series in x with constant coefficients,

(2) $P(0)$ is invertible and $\mathcal{S}_0 P(0) = P(0)\Lambda_0$, where Λ_0 is a diagonal matrix whose diagonal entries are eigenvalues of Ω_0,

(3) the transformation

(V.5.9)
$$\vec{y} = P(x)\vec{u}$$

changes the differential operator (V.5.1) to another differential operator

(V.5.10)
$$P(x)^{-1}\mathcal{L}[P(x)\vec{u}] = \mathcal{K}_0[\vec{u}] + \nu_0(x)\vec{u},$$

where

$$\mathcal{K}_0[\vec{u}] = x\frac{d\vec{u}}{dx} + \Lambda_0\vec{u}, \qquad \nu_0(x) = x^{-\Lambda_0}\Gamma x^{\Lambda_0},$$

and

$$\Gamma = \begin{bmatrix} \gamma_{11} & \cdots & \gamma_{1n} \\ \vdots & \vdots & \vdots \\ \gamma_{n1} & \cdots & \gamma_{nn} \end{bmatrix}$$

with constants γ_{jk} such that

(V.5.11) $\gamma_{jk} = 0$ *if* $\lambda_k - \lambda_j$ *is not a non-negative integer.*

Furthermore, the matrix Γ is nilpotent.

Remark V-5-2. It is easily verified that the matrix $P(x)$ is a formal solution of the system
$$x\frac{dP(x)}{dx} = P(x)(\Lambda_0 + \nu_0(x)) - \Omega(x)P(x).$$

Since the entries of $\nu_0(x)$ are polynomials in x, the power series $P(x)$ is convergent if $\Omega(x)$ is convergent (cf. Theorem V-2-7). Therefore, in such a case, $\nu(x)$ is convergent and, hence, $\sigma(x)$ is convergent.

Observation V-5-3. Choose integers $\ell_1, \ldots, \ell_n$ so that $\lambda_j + \ell_j = \lambda_k + \ell_k$ if $\lambda_j - \lambda_k$ is an integer. Then, $\nu_0(x) = x^L \Gamma x^{-L}$, where $L = \text{diag}[\ell_1, \ell_2, \ldots, \ell_n]$. If we set $\mathcal{H}[\vec{v}] = x^{-L}\mathcal{K}[x^L \vec{v}]$ and $\mathcal{H}_0[\vec{v}] = x^{-L}\mathcal{K}_0[x^L \vec{v}]$, it follows that

$$\mathcal{H}_0[\vec{v}] = x^{-L}\left\{x^L x\frac{d\vec{v}}{dx} + x^L L\vec{v} + \Lambda_0 x^L \vec{v}\right\} = x\frac{d\vec{v}}{dx} + (\Lambda_0 + L)\vec{v}.$$

Hence,

(V.5.12)
$$\mathcal{H}[\vec{v}] = x\frac{d\vec{v}}{dx} + (\Lambda_0 + L + \Gamma)\vec{v}.$$

Note that
$$\Lambda_0 + L = \text{diag}[\lambda_1 + \ell_1, \lambda_2 + \ell_2, \ldots, \lambda_n + \ell_n]$$

and

(V.5.13)
$$(\Lambda_0 + L)\Gamma = \Gamma(\Lambda_0 + L).$$

It was already showen in §V-4 that Γ is nilpotent. Thus, the following theorem is proved.

Theorem V-5-4. *The transformation $\vec{y} = P(x)x^L\vec{v}$ changes the system*

$$(V.5.14) \qquad \mathcal{L}[\vec{y}] = x\frac{d\vec{y}}{dx} + \Omega(x)\vec{y} = \vec{0}$$

to

$$(V.5.15) \qquad \mathcal{H}[\vec{v}] = x\frac{d\vec{v}}{dx} + (\Lambda_0 + L + \Gamma)\vec{v} = \vec{0}.$$

Observation V-5-5. The matrix

$$\Phi_0(x) = x^{-\Lambda_0 - L - \Gamma} = x^{-\Lambda_0 - L}\left[I + \sum_{h=1}^{n-1}(-1)^h\frac{[\log x]^h}{h!}\Gamma^h\right]$$

is a fundamental matrix solution of system (V.5.15). Note that if $(\lambda_j + \ell_j) - (\lambda_k + \ell_k)$ is an integer, then $\lambda_j + \ell_j = \lambda_k + \ell_k$ in $\Lambda_0 + L$.

Remark V-5-6. A fundamental matrix solution $\Phi(x)$ of system (V.5.14) is given by $\Phi(x) = P(x)x^L\Phi_0(x)$, which can be written in the form

$$\Phi(x) = P(x)x^L x^{-\Lambda_0 - L}\left[I_n + \sum_{h=1}^{n-1}(-1)^h\frac{[\log x]^h}{h!}\Gamma^h\right].$$

Since $L - \Lambda_0 - L = -\Lambda_0$, the matrix $\Phi(x)$ can be also written in the form

$$\Phi(x) = P(x)x^{-\Lambda_0}\left[I_n + \sum_{h=1}^{n-1}(-1)^h\frac{[\log x]^h}{h!}\Gamma^h\right].$$

However, $x^{-\Lambda_0}$ and $\left[I_n + \sum_{h=1}^{n-1}(-1)^h\frac{[\log x]^h}{h!}\Gamma^h\right]$ do not commute, whereas $x^{-\Lambda_0 - L}$

and $\left[I_n + \sum_{h=1}^{n-1}(-1)^h\frac{[\log x]^h}{h!}\Gamma^h\right]$ commute.

Remark V-5-7. The methods used in §§V-3, V-4, and V-5 are based on the original idea given in [GérL].

Example V-5-8.

In order to illustrate the results of this section, consider the differential equation of the Bessel functions

$$(V.5.16) \qquad z\frac{d}{dz}\left(z\frac{dy}{dz}\right) + (z^2 - a^2)y = 0,$$

where a is a non-negative integer. If we change the independent variable z by $x = z^2$, (V.5.16) becomes

$$x\frac{d}{dx}\left(x\frac{dy}{dx}\right) + \frac{x - a^2}{4}y = 0.$$

This equation is equivalent to the system

$$(\text{V.5.17}) \qquad x\frac{d\vec{y}}{dx} + \Omega(x)\vec{y} = \vec{0}, \qquad \vec{y} = \begin{bmatrix} y \\ x\dfrac{dy}{dx} \end{bmatrix},$$

where $\Omega(x) = \begin{bmatrix} 0 & -1 \\ \dfrac{x-a^2}{4} & 0 \end{bmatrix} = \Omega_0 + x\Omega_1$, $\Omega_0 = \begin{bmatrix} 0 & -1 \\ -\dfrac{a^2}{4} & 0 \end{bmatrix}$ and $\Omega_1 = \begin{bmatrix} 0 & 0 \\ \dfrac{1}{4} & 0 \end{bmatrix}$.

To begin with, let us remark that the S-N decomposition of the matrix Ω_0 is given by $\Omega_0 = S_0 + \mathcal{N}_0$, where

$$S_0 = \begin{cases} O & \text{if} \quad a = 0, \\ \Omega_0 & \text{if} \quad a > 0, \end{cases} \qquad \text{and} \qquad \mathcal{N}_0 = \begin{cases} \Omega_0 & \text{if} \quad a = 0, \\ O & \text{if} \quad a > 0. \end{cases}$$

Also, set $\Lambda_0 = \begin{bmatrix} -\dfrac{a}{2} & 0 \\ 0 & \dfrac{a}{2} \end{bmatrix}$. Note that two eigenvalues of Ω_0 are $\pm\dfrac{a}{2}$.

Now, we calculate in the following steps:

Step 1. Fix a non-negative integer m so that $m \geq \dfrac{a}{2} - \left(-\dfrac{a}{2}\right) = a$.

Step 2. Find three $2(m+1) \times 2(m+1)$ matrices A_m, S_m, and $\mathcal{N}_m$ (cf. §§V-4 and V-5).

Step 3. Find a $2(m+1) \times 2$ matrix $\mathcal{P}_m = \begin{bmatrix} P_0 \\ P_1 \\ \vdots \\ P_m \end{bmatrix}$, where the P_ℓ are 2×2 matrices such that $S_m\mathcal{P}_m = \mathcal{P}_m\Lambda_0$.

Step 4. Find two 2×2 matrices $M_m(x) = \mathcal{N}_0 + \displaystyle\sum_{\ell=1}^{m} x^\ell \nu_\ell$ and $Q_m(x) = \displaystyle\sum_{\ell=0}^{m} x^\ell P_\ell$ (cf. §§V-4 and V-5).

Step 5. We must obtain $Q_m(x)^{-1}M_m(x)Q_m(x) = \nu_0(x) + O(x^{m+1})$, where

$$\nu_0(x) = \begin{cases} P_0^{-1}\Omega_0 P_0 & \text{if} \quad a = 0, \\[2mm] \begin{bmatrix} 0 & \alpha(a)x^a \\ 0 & 0 \end{bmatrix} & \text{if} \quad a > 0, \end{cases}$$

where $\alpha(a)$ is a real constant depending on a (cf. Theorem V-5-1).

After these calculations have been completed, we come to the following conclusion.

Conclusion V-5-9. *There exists a unique 2×2 matrix $P(x) = \sum_{\ell=0}^{+\infty} x^\ell P_\ell$ such that*

(1)   the matrices P_ℓ for $\ell = 0, 1, \ldots, m$ are given in Step 3,
(2)   the power series $P(x)$ converges for every x,
(3)   the transformation $\vec{y} = P(x)\vec{u}$ changes (V.5.17) to

$$(V.5.18) \qquad x\frac{d\vec{u}}{dx} + (\Lambda_0 + \nu_0(x))\,\vec{u} = \vec{0}.$$

We illustrate the scheme given above in the case when $a = 0$. First we fix $m = 2$. Then,

$$A_2 = \begin{bmatrix} 0 & -1 & 0 & 0 & 0 & 0 \\ 0 & 0 & 0 & 0 & 0 & 0 \\ 0 & 0 & 1 & -1 & 0 & 0 \\ \dfrac{1}{4} & 0 & 0 & 1 & 0 & 0 \\ 0 & 0 & 0 & 0 & 2 & -1 \\ 0 & 0 & \dfrac{1}{4} & 0 & 0 & 2 \end{bmatrix}, \quad S_2 = \begin{bmatrix} 0 & 0 & 0 & 0 & 0 & 0 \\ 0 & 0 & 0 & 0 & 0 & 0 \\ \dfrac{1}{4} & -\dfrac{1}{2} & 1 & 0 & 0 & 0 \\ \dfrac{1}{4} & -\dfrac{1}{4} & 0 & 1 & 0 & 0 \\ -\dfrac{3}{32} & \dfrac{3}{32} & \dfrac{1}{4} & -\dfrac{1}{2} & 2 & 0 \\ -\dfrac{1}{16} & \dfrac{3}{32} & \dfrac{1}{4} & -\dfrac{1}{4} & 0 & 2 \end{bmatrix},$$

and

$$\mathcal{N}_2 = \begin{bmatrix} 0 & -1 & 0 & 0 & 0 & 0 \\ 0 & 0 & 0 & 0 & 0 & 0 \\ -\dfrac{1}{4} & \dfrac{1}{2} & 0 & -1 & 0 & 0 \\ 0 & \dfrac{1}{4} & 0 & 0 & 0 & 0 \\ \dfrac{3}{32} & -\dfrac{3}{32} & -\dfrac{1}{4} & \dfrac{1}{2} & 0 & -1 \\ \dfrac{1}{16} & -\dfrac{3}{32} & 0 & \dfrac{1}{4} & 0 & 0 \end{bmatrix}.$$

Calculating eigenvectors of $\mathcal{S}_2$, we find a 6×2 matrix $\mathcal{P}_2 = \begin{bmatrix} 192 & -320 \\ 64 & -128 \\ -16 & 16 \\ -32 & 48 \\ 0 & 1 \\ 1 & 0 \end{bmatrix}$ such

that $\mathcal{S}_2\mathcal{P}_2 = O$. Note that, in this case, $\Lambda_0 = O$.

Set

$$\begin{cases} \nu_1 = \begin{bmatrix} -\dfrac{1}{4} & \dfrac{1}{2} \\[2mm] 0 & \dfrac{1}{4} \end{bmatrix}, & \nu_2 = \begin{bmatrix} \dfrac{3}{32} & -\dfrac{3}{32} \\[2mm] \dfrac{1}{16} & -\dfrac{3}{32} \end{bmatrix}, \\[6mm] M_2(x) = \Omega_0 + x\nu_1 + x^2\nu_2 \end{cases}$$

and

$$(\text{V.5.19}) \quad \begin{cases} P_0 = \begin{bmatrix} 192 & -320 \\ 64 & -128 \end{bmatrix}, & P_1 = \begin{bmatrix} -16 & 16 \\ -32 & 48 \end{bmatrix}, & P_2 = \begin{bmatrix} 0 & 1 \\ 1 & 0 \end{bmatrix}, \\[3mm] Q_2(x) = P_0 + xP_1 + x^2P_2. \end{cases}$$

Then, $Q_2^{-1}M_2(x)Q_2(x) = \begin{bmatrix} -2 & 4 \\ -1 & 2 \end{bmatrix} + O(x^3)$. Note that

$$P_0^{-1}\Omega_0 P_0 = \begin{bmatrix} \dfrac{1}{32} & -\dfrac{5}{64} \\[2mm] \dfrac{1}{64} & -\dfrac{3}{64} \end{bmatrix} \begin{bmatrix} 0 & -1 \\ 0 & 0 \end{bmatrix} \begin{bmatrix} 192 & -320 \\ 64 & -128 \end{bmatrix} = \begin{bmatrix} -2 & 4 \\ -1 & 2 \end{bmatrix}.$$

Thus, we arrived at the following conclusion.

Conclusion V-5-10. *There exists a unique 2×2 matrix $P(x) = \sum_{\ell=0}^{\infty} x^\ell P_\ell$ such that*

(1)   the matrices P_ℓ for $\ell = 0, 1, 2$ are given by (V.5.19),
(2)   the power series $P(x)$ converges for every x,
(3)   the transformation $\vec{y} = P(x)\vec{u}$ changes the system

$$(\text{V.5.20}) \qquad x\frac{d\vec{y}}{dx} + \begin{bmatrix} 0 & -1 \\ \dfrac{x}{4} & 0 \end{bmatrix} \vec{y} = \vec{0}$$

to

$$(\text{V.5.21}) \qquad x\frac{d\vec{u}}{dx} + \begin{bmatrix} -2 & 4 \\ -1 & 2 \end{bmatrix} \vec{u} = \vec{0}.$$

Furthermore, the transformation $\vec{y} = P(x)P_0^{-1}\vec{v}$ changes (V.5.20) to

$$(\text{V.5.22}) \qquad x\frac{d\vec{v}}{dx} + \begin{bmatrix} 0 & -1 \\ 0 & 0 \end{bmatrix} \vec{v} = \vec{0}.$$

In the case when $a = 1$, we fix $m = 2$. Then

$$A_2 = \begin{bmatrix} 0 & -1 & 0 & 0 & 0 & 0 \\ -\dfrac{1}{4} & 0 & 0 & 0 & 0 & 0 \\ 0 & 0 & 1 & -1 & 0 & 0 \\ \dfrac{1}{4} & 0 & -\dfrac{1}{4} & 1 & 0 & 0 \\ 0 & 0 & 0 & 0 & 2 & -1 \\ 0 & 0 & \dfrac{1}{4} & 0 & -\dfrac{1}{4} & 2 \end{bmatrix}, \quad S_2 = \begin{bmatrix} 0 & -1 & 0 & 0 & 0 & 0 \\ -\dfrac{1}{4} & 0 & 0 & 0 & 0 & 0 \\ -\dfrac{1}{8} & \dfrac{1}{4} & 1 & -1 & 0 & 0 \\ \dfrac{3}{16} & \dfrac{1}{8} & -\dfrac{1}{4} & 1 & 0 & 0 \\ \dfrac{1}{16} & -\dfrac{1}{16} & -\dfrac{1}{8} & \dfrac{1}{4} & 2 & -1 \\ \dfrac{3}{64} & -\dfrac{1}{16} & \dfrac{3}{16} & \dfrac{1}{8} & -\dfrac{1}{4} & 2 \end{bmatrix},$$

and

$$\mathcal{N}_2 = \begin{bmatrix} 0 & 0 & 0 & 0 & 0 & 0 \\ 0 & 0 & 0 & 0 & 0 & 0 \\ \dfrac{1}{8} & -\dfrac{1}{4} & 0 & 0 & 0 & 0 \\ \dfrac{1}{16} & -\dfrac{1}{8} & 0 & 0 & 0 & 0 \\ -\dfrac{1}{16} & \dfrac{1}{16} & \dfrac{1}{8} & -\dfrac{1}{4} & 0 & 0 \\ -\dfrac{3}{64} & \dfrac{1}{16} & \dfrac{1}{16} & -\dfrac{1}{8} & 0 & 0 \end{bmatrix}.$$

Calculating eigenvectors of S_2, we find a 6×2 matrix $\mathcal{P}_2 = \begin{bmatrix} 384 & 32 \\ 192 & -16 \\ -48 & -12 \\ -72 & -14 \\ 2 & 0 \\ 5 & 1 \end{bmatrix}$ such

that $S_2 \mathcal{P}_2 = \mathcal{P}_2 \Lambda_0$. Note that, in this case, $\Lambda_0 = \begin{bmatrix} -\dfrac{1}{2} & 0 \\ 0 & \dfrac{1}{2} \end{bmatrix}$. Set

$$\begin{cases} \nu_1 = \begin{bmatrix} \dfrac{1}{8} & -\dfrac{1}{4} \\ \dfrac{1}{16} & -\dfrac{1}{8} \end{bmatrix}, \quad \nu_2 = \begin{bmatrix} -\dfrac{1}{16} & \dfrac{1}{16} \\ -\dfrac{3}{64} & \dfrac{1}{16} \end{bmatrix}, \\ M_2(x) = x\nu_1 + x^2\nu_2, \end{cases}$$

and

$$(V.5.23) \quad \begin{cases} P_0 = \begin{bmatrix} 384 & 32 \\ 194 & -16 \end{bmatrix}, \quad P_1 = \begin{bmatrix} -48 & -12 \\ -72 & -14 \end{bmatrix}, \quad P_2 = \begin{bmatrix} 2 & 0 \\ 5 & 1 \end{bmatrix}, \\ Q_2(x) = P_0 + xP_1 + x^2P_2. \end{cases}$$

Then, $Q_2^{-1}M_2(x)Q_2(x) = \begin{bmatrix} 0 & \dfrac{x}{48} \\ 0 & 0 \end{bmatrix} + O(x^3)$. Thus, we arrived at the following conclusion.

Conclusion V-5-11. *There exists a unique 2×2 matrix $P(x) = \displaystyle\sum_{\ell=0}^{+\infty} x^\ell P_\ell$ such that*

    (1)   the matrices P_ℓ for $\ell = 0, 1, 2$ are given by (V.5.23),
    (2)   the power series $P(x)$ converges for every x,
    (3)   the transformation $\vec{y} = P(x)\vec{u}$ changes the system

$$(\text{V.5.24}) \qquad x\frac{d\vec{y}}{dx} + \begin{bmatrix} 0 & -1 \\ \dfrac{x-1}{4} & 0 \end{bmatrix} \vec{y} = \vec{0}$$

to

$$(\text{V.5.25}) \qquad x\frac{d\vec{u}}{dx} + \begin{bmatrix} -\dfrac{1}{2} & \dfrac{x}{48} \\ 0 & \dfrac{1}{2} \end{bmatrix} \vec{u} = \vec{0}.$$

For a further discussion, see [HKS]. A computer might help the reader to calculate $\mathcal{S}_2$, $\mathcal{N}_2$, and $\mathcal{P}_2$ in the cases when $a = 0$ and $a = 1$. Such a calculation is not difficult in these cases since eigenvalues of $\mathcal{A}_2$ are found easily (cf. §IV-1).

V-6. Calculation of the normal form of a differential operator

In this section, we present another proof of Theorem V-5-1. The main idea is to construct a power series $P(x)$ as a formal solution of the system

$$x\frac{dP(x)}{dx} = P(x)(\Lambda_0 + \nu_0(x)) - \Omega(x)P(x)$$

(cf. Remark V-5-2).

Another proof of Theorem V-5-1.

To simplify the presentation, we assume that $\mathcal{S}_0 = \Lambda_0 = \text{diag}[\mu_1 I_1, \mu_2 I_2, \ldots, \mu_k I_k]$, where $\mu_1, \mu_2, \ldots, \mu_k$ are distinct eigenvalues of Ω_0 with multiplicities m_j, respectively, and the matrix I_j is $m_j \times m_j$ identity matrix. Since $\Lambda_0 \mathcal{N}_0 = \mathcal{N}_0 \Lambda_0$, the matrix $\mathcal{N}_0$ must have the form $\mathcal{N}_0 = \text{diag}[\mathcal{N}_{01}, \mathcal{N}_{02}, \ldots, \mathcal{N}_{0k}]$, where $\mathcal{N}_{0j}$ is an $m_j \times m_j$ nilpotent matrix. Let us determine two matrices $P(x) = I_n + \displaystyle\sum_{m=1}^{\infty} x^m P_m$

and $B(x) = \Omega_0 + \displaystyle\sum_{m=1}^{\infty} x^m B_m$ by the equation

$$x\frac{d}{dx}\left(\sum_{m=1}^{\infty} x^m P_m \right) = \left(I_n + \sum_{m=1}^{\infty} x^m P_m \right)\left(\Omega_0 + \sum_{m=1}^{\infty} x^m B_m \right)$$
$$- \left(\Omega_0 + \sum_{m=1}^{\infty} x^m \Omega_m \right)\left(I_n + \sum_{m=1}^{\infty} x^m P_m \right).$$

This equation is equivalent to

$$mP_m = P_m\Omega_0 - \Omega_0 P_m + B_m - \Omega_m + \sum_{\ell=1}^{m-1}(P_h B_{m-h} - \Omega_{m-h}P_h) \qquad (m \geq 1).$$

Therefore, it suffices to solve the equation

(V.6.1) $\qquad mX + (\Lambda_0 + \mathcal{N}_0)X - X(\Lambda_0 + \mathcal{N}_0) - Y = H,$

where X and Y are $n \times n$ unknown matrices, whereas the matrix H is given. If we write X, Y, and H in the block-form

$$X = \begin{bmatrix} X_{11} & \cdots & X_{1k} \\ \vdots & \vdots & \vdots \\ X_{k1} & \cdots & X_{kk} \end{bmatrix}, \quad Y = \begin{bmatrix} Y_{11} & \cdots & X_{1k} \\ \vdots & \vdots & \vdots \\ Y_{k1} & \cdots & Y_{kk} \end{bmatrix}, \quad H = \begin{bmatrix} H_{11} & \cdots & X_{1k} \\ \vdots & \vdots & \vdots \\ H_{k1} & \cdots & H_{kk} \end{bmatrix},$$

where X_{jh}, Y_{jh}, and H_{jh} are $m_j \times m_h$ matrices, equation (V.6.1) becomes

$$(m + \mu_j - \mu_h)X_{jh} + \mathcal{N}_{0j}X_{jh} - X_{jh}\mathcal{N}_{0h} - Y_{jh} = H_{jh},$$

where $j, h = 1, \ldots, k$. We can determine X_{jh} and Y_{jh} by setting $Y_{jh} = O$ if $m + \mu_j - \mu_h \neq 0$, and $X_{jh} = O$ if $m + \mu_j - \mu_h = 0$. More precisely speaking, if $m + \mu_j - \mu_h \neq 0$, we determine X_{jh} uniquely by solving

$$(m + \mu_j - \mu_h)X_{jh} + \mathcal{N}_{0j}X_{jh} - X_{jh}\mathcal{N}_{0h} = H_{jh}.$$

If $m + \mu_j - \mu_h = 0$, we set $Y_{jh} = H_{jh}$. In this way, we can determine $P(x)$ and $B(x)$. In particular, $\mathcal{N}_0 + B(x)$ has the form $x^{-\Lambda_0}\Gamma x^{\Lambda_0}$, where Γ is a constant $n \times n$ nilpotent matrix. Furthermore, the operator $x\dfrac{d}{dx} + \Lambda_0$ and the multiplication operator by $\mathcal{N}_0 + B(x)$ commute. $\square$

The idea of this proof is due to M. Hukuhara (cf. [Si17, §3.9, pp. 85–89]).

Remark V-6-1. In the case of a second-order linear homogeneous differential equation at a regular singular point $x = a$, there exists a solution of the form

$$\phi_1(x) = (x - a)^\alpha \sum_{n=0}^{+\infty} c_n(x - a)^n, \text{ where the coefficients } c_n \text{ are constants}, c_0 \neq 0,$$

and the power series is convergent. If there is no other linearly independent solution of this form, a second solution can be constructed by using the idea explained in Remark IV-7-3. This second solution contains a logarithmic term. Similarly, a third-order linear homogeneous differential equation has a solution of the form

$$\phi_1(x) = (x - a)^\alpha \sum_{n=0}^{+\infty} c_n(x - a)^n$$ at a regular singular point $x = a$. Using this solution, the given equation can be reduced to a second-order equation. In particular, if there exists another solution $\phi_2(x)$ of this kind such that ϕ_1 and ϕ_2 are linearly independent, then the idea given in Remark IV-7-4 can be used to find a fundamental set of solutions.

In general, a fundamental matrix solution of system (V.5.14) can be constructed if the transformation $\vec{y} = P(x)x^L\vec{v}$ of Theorem V-5-4 is found. In fact, if the definition of $\Phi_0(x)$ of Observation V-5-5 is used, $P(x)x^Lx^{-\Lambda_0-L-\Gamma}$ is a fundamental matrix solution of (V.5.14). The matrix $P(x)$ can be calculated by using the method of Hukuhara, which was explained earlier.

V-7. Classification of singularities of homogeneous linear systems

In this chapter, so far we have studied a system

$$(V.7.1) \qquad\qquad x\,\frac{d\vec{y}}{dx} \;=\; A(x)\vec{y}, \qquad \vec{y} \in \mathbb{C}^n,$$

where the entries of the $n \times n$ matrix $A(x)$ are convergent power series in x with complex coefficients. In this case, the singularity at $x = 0$ is said to be of the *first kind*. If a system has the form

$$(V.7.2) \qquad\qquad x^{k+1}\,\frac{d\vec{y}}{dx} \;=\; A(x)\vec{y}, \qquad \vec{y} \in \mathbb{C}^n,$$

where k is a positive integer and the entries of the $n \times n$ matrix $A(x)$ are convergent power series in x with complex coefficients, then the singularity at $x = 0$ is said to be of the *second kind*.

In §V-5, we proved the following theorem (cf. Theorem V-5-4).

Theorem V-7-1. *For system (V.7.1), there exist a constant $n \times n$ matrix A_0 and an $n \times n$ invertible matrix $P(x)$ such that*

(i) the entries of $P(x)$ and $P(x)^{-1}$ are analytic and single-valued in a domain $0 < |x| < r$ and have, at worst, a pole at $x = 0$, where r is a positive number,

(ii) the transformation

$$(V.7.3) \qquad\qquad\qquad \vec{y} \;=\; P(x)\vec{u}$$

changes (V.7.1) to a system

$$(V.7.4) \qquad\qquad\qquad x\,\frac{d\vec{u}}{dx} \;=\; A_0\vec{u}.$$

Theorem V-7-1 can be generalized to system (V.7.2) as follows.

Theorem V-7-2. *For system (V.7.2), there exist a constant $n \times n$ matrix A_0 and an $n \times n$ invertible matrix $P(x)$ such that*
  (i) the entries of $P(x)$ and $P(x)^{-1}$ are analytic and single-valued in a domain
      $\mathcal{D} = \{x : 0 < |x| < r\}$, *where r is a positive number,*
  (ii) the transformation

(V.7.5)
$$\vec{y} = P(x)\vec{u}$$

  changes (V.7.2) to (V.7.4).

Proof.

Let $\Phi(x)$ be a fundamental matrix of (V.7.2) in $\mathcal{D}$. Since $A(x)$ is analytic and single-valued in $\mathcal{D}$, $\hat{\Phi}(x) = \Phi(xe^{2\pi i})$ is also a fundamental matrix of (V.7.2). Therefore, there exists an invertible constant matrix C such that $\hat{\Phi}(x) = \Phi(x)C$ (cf. (1) of Remark IV-2-7). Choose a constant matrix A_0 so that $C = \exp[2\pi i A_0]$ (cf. Example IV-3-6) and let $P(x) = \Phi(x)\exp[-(\log x)A_0]$. Then, $P(x)$ and $P(x)^{-1}$ are analytic and single-valued in $\mathcal{D}$. Furthermore,

$$\begin{aligned}
\frac{dP(x)}{dx} &= \frac{d\Phi(x)}{dx}\exp[-(\log x)A_0] - P(x)(x^{-1}A_0) \\
&= x^{-(k+1)}A(x)P(x) - P(x)(x^{-1}A_0).
\end{aligned}$$

This completes the proof of the theorem. $\square$
An important difference between Theorems V-7-1 and V-7-2 is the fact that the matrix $P(x)$ in Theorem V-7-2 possibly has an essential singularity at $x = 0$.

The proof of Theorem V-7-2 immediately suggests that Theorem V-7-2 can be extended to a system

(V.7.6)
$$\frac{d\vec{y}}{dx} = F(x)\vec{y},$$

where every entry of the $n \times n$ matrix $F(x)$ is analytic and single-valued on the domain $\mathcal{D}$ even if such an entry of $F(x)$ possibly has an essential singularity at $x = 0$. More precisely speaking, for system (V.7.6), there exist a constant $n \times n$ matrix A_0 and an $n \times n$ invertible matrix $P(x)$ satisfying conditions (i) and (ii) of Theorem V-7-2 such that transformation (V.7.5) changes (V.7.6) to (V.7.4).

Now, we state a definition of regular singularity of (V.7.6) at $x = 0$ as follows.

Definition V-7-3. *Let $P(x)$ be a matrix satisfying conditions (i) and (ii) such that transformation (V.7.5) changes (V.7.6) to (V.7.4). Then, the singularity of (V.7.6) at $x = 0$ is said to be* regular *if every entry of $P(x)$ has, at worst, a pole at $x = 0$.*

Remark V-7-4. Theorem V-7-1 implies that a singularity of the first kind is a regular singularity. The converse is not true. However, it can be proved easily that a regular singularity is, at worst, a singularity of the second kind. Furthermore, if (V.7.2) has a regular singularity at $x = 0$, then the matrix $A(0)$ is nilpotent. This is a consequence of the following theorem.

Theorem V-7-5. *Let $A(x)$ and $B(x)$ be two $n \times n$ matrices whose entries are formal power series in x with constant coefficients. Also, let r and s be two positive integers. Suppose that there exists an $n \times n$ matrix $P(x)$ such that*

   (a) the entries of $P(x)$ are formal power series in x with constant coefficients,

   (b) $\det[P(x)] \neq 0$ as a formal power series in x,

   (c) the transformation $\vec{y} = P(x)\vec{u}$ changes the system $x^r \dfrac{d\vec{y}}{dx} = A(x)\vec{y}$ to $x^s \dfrac{d\vec{u}}{dx} = B(x)\vec{u}$.

Suppose also that $s > r$. Then, the matrix $B(0)$ must be nilpotent.

Proof.

Step 1. Applying to the matrix $P(x)$ suitable elementary row and column operations successively, we can prove the following lemma.

Lemma V-7-6. *There exist two $n \times n$ matrices*

$$T(x) \; = \; \sum_{m=0}^{+\infty} x^m T_m \qquad and \qquad S(x) \; = \; \sum_{m=0}^{+\infty} x^m S_m,$$

and n integers $\lambda_1, \lambda_2, \ldots, \lambda_n$ such that

   (i) the entries of $n \times n$ matrices T_m and S_m are constants,

   (ii) $\det T_0 \neq 0$ and $\det S_0 \neq 0$,

   (iii)

$$T(x)P(x)S(x) \; = \; \Lambda(x) \; = \; \mathrm{diag}[x^{\lambda_1}, x^{\lambda_2}, \ldots, x^{\lambda_n}],$$

   (iv) $\lambda_1 \leq \lambda_2 \leq \cdots \leq \lambda_n$.

The proof of this lemma is left to the reader as an exercise.

Step 2. Change the two systems $x^r \dfrac{d\vec{y}}{dx} = A(x)\vec{y}$ and $x^s \dfrac{d\vec{u}}{dx} = B(x)\vec{u}$, respectively, to $x^r \dfrac{d\vec{z}}{dx} = C(x)\vec{z}$ and $x^s \dfrac{d\vec{v}}{dx} = D(x)\vec{v}$ by the transformations $\vec{z} = T(x)\vec{y}$ and $\vec{u} = S(x)\vec{v}$. Then, $\vec{z} = \Lambda(x)\vec{v}$. Furthermore, if the matrix $D(0)$ is nilpotent, the matrix $B(0)$ is also nilpotent.

Step 3. Look at $D(x) = x^{s-r}\Lambda(x)^{-1}C(x)\Lambda(x) - x^s\Lambda(x)^{-1}\dfrac{d\Lambda(x)}{dx}$. This shows clearly that if $s > r$, the matrix $D(0)$ is nilpotent. $\square$

The following corollary of Theorem V-7-5 is important.

Corollary V-7-7. *Assume that conditions (a), (b), and (c) of Theorem V-7-5 are satisfied. Also, assume that $A(0)$ and $B(0)$ are not nilpotent. Then, $r = s$.*

Definition V-7-8. *If a singularity of the second kind is not a regular singularity, this singularity is said to be* irregular. *In particular, if the matrix $A(0)$ of (V.7.2) is not nilpotent, the singularity of (V.7.2) at $x = 0$ is said to be* irregular of order k. *Also, a regular singularity is said to be of order zero.*

In order to define the order of singularity at $x = 0$ for all systems (V.7.2), the independent variable x must be replaced by $x^{1/p}$ with a suitable positive integer p.

For example, for a differential equation

$$(V.7.7) \qquad \sum_{h=0}^{n} a_h(x)\delta^h y = 0,$$

assume that the coefficients a_h are convergent power series in x and that $a_n(x) \neq 0$. Set $y_j = \delta^{j-1} y$ $(j = 0, \dots, n-1)$. Then, (V.7.7) becomes the system

$$(V.7.8) \qquad \delta \vec{y} = \begin{bmatrix} 0 & 1 & 0 & \cdots & 0 \\ \vdots & \vdots & \vdots & \vdots & \vdots \\ 0 & 0 & 0 & \cdots & 0 \\ 0 & 0 & 0 & \cdots & 1 \\ -\dfrac{a_0}{a_n} & \cdots & \cdots & \cdots & -\dfrac{a_{n-1}}{a_n} \end{bmatrix} \vec{y}, \qquad \vec{y} = \begin{bmatrix} y_1 \\ \vdots \\ y_n \end{bmatrix}.$$

If we change (V.7.8) by the transformation $\vec{y} = \mathrm{diag}[1, x^{-\sigma}, x^{-2\sigma}, \dots, x^{-(n-1)\sigma}]\vec{u}$, then

$$(V.7.9)$$
$$x^\sigma \delta \vec{u}$$

$$= \left\{ \begin{bmatrix} 0 & 1 & 0 & \cdots & 0 \\ \vdots & \vdots & \vdots & \vdots & \vdots \\ 0 & 0 & 0 & \cdots & 0 \\ 0 & 0 & 0 & \cdots & 1 \\ -x^{n\sigma}\dfrac{a_0}{a_n} & \cdots & \cdots & \cdots & -x^\sigma \dfrac{a_{n-1}}{a_n} \end{bmatrix} + x^\sigma \mathrm{diag}[1, \sigma, 2\sigma, \dots, (n-1)\sigma] \right\} \vec{u}.$$

Set $b_h(x) = \dfrac{a_h(x)x^{(n-h)\sigma}}{a_n(x)}$. Choose a non-negative rational number σ so that $b_h(x)$ $(h = 0, \dots, n-1)$ are bounded in a neighborhood of $x = 0$. Also, if σ must be positive, choose σ so that $b_h(0) \neq 0$ for some h. Then, the matrix on the right-hand side of (V.7.9) is not nilpotent at $x = 0$ if $\sigma > 0$. Hence, we define σ as the order of the singularity of (V.7.7) at $x = 0$. System (V.7.2) can be reduced to an equation (V.7.7) (cf. §XIII-5).

The following theorem due to J. Moser [Mo] concerns the order of a given singularity.

Theorem V-7-9. *Let* $A(x) = x^{-p} \displaystyle\sum_{\nu=0}^{\infty} x^\nu A_\nu$ *be an* $n \times n$ *matrix, where p is an integer, A_ν are $n \times n$ constant matrices, $A_0 \neq O$, and the power series is convergent. Set*

$$\begin{cases} m(A) & = \ \max\left(p - 1 + \dfrac{r}{n}, 0\right), \\ \mu(A) & = \ \min_{T}\left[m\left(T^{-1}AT - T^{-1}\dfrac{dT}{dx}\right)\right]. \end{cases}$$

Here, $r = \operatorname{rank}(A_0)$ and min is taken over all $n \times n$ invertible matrices T of the form $T(x) = x^{-q} \sum_{\nu=0}^{\infty} x^{\nu} T_{\nu}$, where q is an integer, T_{ν} are constant $n \times n$ matrices, and the power series is convergent. Note that $\det T(x) \neq 0$, but $\det T_0$ may be 0. Assume that $m(A) > 1$. Then, we have $m(A) > \mu(A)$ if and only if the polynomial

$$\mathcal{P}(\lambda) \;=\; x^r \det(\lambda I_n + x^{p-1} A(x))|_{x=0}$$

vanishes identically in λ, where I_n is the $n \times n$ identity matrix.

The main idea is to find out $\mu(A)$ in a finite number of steps. Using the quantity $\mu(A)$, we can calculate the order of the singularity at $x = 0$. The criterion for $m(A) > \mu(A)$ is given in terms of a finite number of conditions on the coefficients of $A(x)$. There is a computer program to work with those conditions. In particular, in this way, we can decide, in a finite number of steps, whether a given singularity at $x = 0$ is regular.

Notice that $|P(x)|$ can be estimated at $x = 0$ for the matrix $P(x)$ of Theorem V-7-2, using similar estimates for fundamental matrix solutions of (V.7.2) and (V.7.4). Therefore, an analytic criterion that a singularity of second kind at $x = 0$ be a regular singularity is that in any sectorial domain $\mathcal{D}$ with the vertex at $x = 0$, every solution $\vec{\phi}(x)$ satisfies an estimate

$$|\vec{\phi}(x)| \;\leq\; K|x|^m \qquad (x \in \mathcal{D})$$

for some positive number K and a real number m. These two numbers may depend on ϕ and $\mathcal{D}$. In fact, Theorem V-7-2 and its remark imply that a fundamental matrix solution of (V.7.6) is $P(x)x^{A_0}$. The matrix $P(x)$ has, at worst, a pole at $x = 0$ if and only if $|P(x)| \leq K|x|^p$ in a neighborhood of $x = 0$ for some positive number K and a real number p (cf. [CL, §2 of Chapter 4, pp. 111-141]). Another criterion which depends only on a finite number of coefficients of power series expansion of the matrix $A(x)$ of (V.7.2) was given in a very concrete form by W. Jurkat and D. A. Lutz [JL] (see also [Si17, Chapter V, pp. 115-141]).

Now, let us look into the problem of convergence of the formal solution which was constructed in Theorem V-1-3. Let

$$P = \sum_{h=0}^{n} a_h(x)\delta^h \qquad \left(\delta = x\frac{d}{dx}\right)$$

be a differential operator with coefficients $a_h(x)$ in $\mathbb{C}\{x\}$. Assume that $n \geq 1$ and $a_n(x) \neq 0$. Defining the indicial polynomial $f_{n_0}(s)$ of the operator P as in §V-1, we prove the following theorem.

Theorem V-7-10.

    (i) *In the case when the degree of f_{n_0} in s is equal to n, if we change the equation $P[y] = 0$ to a system by setting $y_1 = y$ and $y_j = \delta^{j-1}[y]$ $(j = 2, \ldots, n)$, the system has a singularity of the first kind at $x = 0$.*

(ii) In the case when the degree of f_{n_0} in s is less than n, if we change the equation $P[y] = 0$ by setting $y_1 = y$ and $y_j = \delta^{j-1}[y]$ $(j = 2, \ldots, n)$, the system has an irregular singularity at $x = 0$.

Proof.

Look at $P[x^s] = \sum_{h=0}^{n} a_h(x)\delta^h[x^s] = x^s\left[\sum_{h=0}^{n} a_h(x)s^h\right] = x^s \sum_{m=n_0}^{\infty} f_m(s)x^m$ and set $a_h(x) = x^{n_0}b_h(x)$. Then, the functions $b_h(x)$ are analytic at $x = 0$. Furthermore, if the degree of f_{n_0} is n, we must have $b_n(0) \neq 0$. Therefore, in this case we can write the equation $P[y] = 0$ in the form $\delta^n[y] = -\dfrac{1}{b_n(x)}\sum_{h=0}^{n-1} b_h(x)\delta^h[y]$. Claim (i) follows immediately from this form of the equation.

If the degree of f_{n_0} is less than n, we must have $b_n(0) = 0$ and $b_j(0) \neq 0$ for some j such that $0 \leq j < n$. To show (ii), change y_h further by $z_h = x^{(h-1)\sigma}y_h$ with a suitable positive rational number σ so that the system for $(z_1, \ldots, z_n)$ has a singularity of the second kind of a positive order (cf. the arguments given right after Definition V-7-8, and also §XIII-7). $\square$

From Theorem V-7-10, we conclude that the formal solution $x^r(1 + \phi(x))$ of Theorem V-1-3 is convergent if the degree of $f_{n_0}(s)$ in s is n. The following corollary of Theorem V-7-10 is a basic result due to L. Fuchs [Fu].

Corollary V-7-11. *The differential equation $P[y] = 0$ has, at worst, a regular singularity at $x = 0$ if and only if the functions $\dfrac{a_h(x)}{a_n(x)}$ $(h = 0, \ldots, n-1)$ are analytic at $x = 0$.*

Some of the results of this section are also found in [CL, Chapter 4, pp. 108-137].

EXERCISES V

V-1. Show that if λ is a nonzero constant, H is a constant $n \times m$ matrix, and N_1 and N_2 are $n \times n$ and $m \times m$ nilpotent matrices, respectively, then there exists one and only one $n \times m$ matrix X satisfying the equation $\lambda X + N_1X - XN_2 = H$.

V-2. Show that the convergent power series

$$y = F(\alpha, \beta, \gamma, x) = 1 + \sum_{m=1}^{\infty} \frac{\alpha(\alpha + 1)\cdots(\alpha + m - 1)\beta(\beta + 1)\cdots(\beta + m - 1)}{m!\gamma(\gamma + 1)\cdots(\gamma + m - 1)}x^m$$

satisfies the differential equation

$$x(1 - x)\frac{d^2y}{dx^2} + [\gamma - (\alpha + \beta + 1)]\frac{dy}{dx} - \alpha\beta y = 0,$$

where α, β, and γ are complex constants.

Comment. The series $F(\alpha, \beta, \gamma, x)$ is called the *hypergeometric series* (see, for example, [CL; p. 135], [Ol; p. 159], and [IKSY]).

V-3. For each of the following differential equations, find all formal solutions of the form $x^r \left[1 + \sum_{m=1}^{\infty} c_m x^m \right]$. Examine also if they are convergent.

(i)
$$x\delta^2 y + \alpha \delta y + \beta y = 0,$$

(ii)
$$\delta^2 y + \alpha \delta y + \beta y = \gamma x^m y,$$

where $\delta = x\dfrac{d}{dx}$, the quantities α, β, and γ are nonzero complex constants and m is a positive integer.

V-4. Given the system

(E) $\qquad \dfrac{d\vec{y}}{dt} = (N + R(t))\vec{y}, \qquad N = \begin{bmatrix} 0 & 1 \\ 0 & 0 \end{bmatrix}, \quad$ and $\quad R(t) = t^{-3}\begin{bmatrix} 0 & 0 \\ 1 & 0 \end{bmatrix},$

show that
(i) (E) has two linearly independent solutions

$$\vec{\eta}_1(t) = \begin{bmatrix} \phi_1(t) \\ \phi_1'(t) \end{bmatrix}, \qquad \text{where} \qquad \phi_1(t) = \sum_{m=0}^{+\infty} \frac{t^{-m}}{m!(m+1)!},$$

and

$$\vec{\eta}_2(t) = \begin{bmatrix} \phi_2(t) \\ \phi_2'(t) \end{bmatrix}, \qquad \text{where} \qquad \phi_2(t) = t + \sum_{m=1}^{+\infty} b_m t^{-m} - \phi_1(t) \log t,$$

with the constants b_m determined by $b_{-1} = 1, b_0 = 0$, and
$$m(m+1)b_m - b_{m-1} = -\frac{2m+1}{m!(m+1)!} \qquad (m = 1, 2, \dots),$$

(ii) $\lim\limits_{t \to +\infty} t^{-1}(Y(t) - e^{tN}) = O$ for the fundamental matrix solution $Y(t) = [\vec{\eta}_1(t) \ \vec{\eta}_2(t)]$,

(iii) the limit of $e^{-tN}Y(t)$ as $t \to +\infty$ does not exists.

V-5. Let $\vec{y}$ be a column vector with n entries and let $\vec{f}(x, \vec{y})$ be a vector with n entries which are formal power series in $n + 1$ variables $(x, \vec{y})$ with coefficients in $\mathbb{C}$. Also, let $\vec{u}$ be a vector with n entries and let $\vec{P}(x, \vec{u})$ be a vector with n entries which are formal power series in $n + 1$ variables $(x, \vec{u})$ with coefficients in $\mathbb{C}$. Find the most general $P(x, \vec{u})$ such that the transformation $\vec{y} = \vec{u} + xP(x, \vec{u})$ changes the differential equation $\dfrac{d\vec{y}}{dx} = \vec{f}(x, \vec{y})$ to $\dfrac{d\vec{u}}{dx} = \vec{0}$.

Hint. Expand $xP(x, \vec{u})$ and $\vec{f}(x, \vec{u} + xP(x, \vec{u}))$ as power series in $\vec{u}$. Identify coefficients of $\dfrac{d[xP(x, \vec{u})]}{dx}$ with those of $\vec{f}(x, \vec{u} + xP(x, \vec{u}))$ to derive differential equations which are satisfied by coefficients of $xP(x, \vec{u})$.

V-6. Suppose that three $n \times n$ matrices A, B, and $P(x)$ satisfy the following conditions:

(a) the entries of A and B are constants,

(b) the entries of $P(x)$ are analytic and single-valued in $0 < |x| < r$ for some positive number r,

(c) the transformation $\vec{y} = P(x)\vec{u}$ changes the system $x\dfrac{d\vec{y}}{dx} = A\vec{y}$ to $x\dfrac{d\vec{u}}{dx} = B\vec{u}$.

Show that there exists an integer p such that the entries of $x^p P(x)$ are polynomials in x.

Hint. The three matrices $P(x)$, A, and B satisfy the equation $x\dfrac{dP(x)}{dx} = AP(x) - P(x)B$. Setting $P(x) = \displaystyle\sum_{m=-\infty}^{+\infty} x^m P_m$, we must have $mP_m = AP_m - P_m B$ for all integers m. Hence, there exists a large positive integer p such that $P_m = O$ for $|m| \geq p$.

V-7. Let $\vec{y}$ be a column vector with n entries $\{y_1, \dots, y_n\}$ and let $A(\vec{y})$ be an $n \times n$ matrix whose entries are convergent power series in $\{y_1, \dots, y_n\}$ with coefficients in $\mathbb{C}$. Assume that $A(\vec{0})$ has an eigenvalue λ such that $m\lambda$ is not an eigenvalue of $A(\vec{0})$ for any positive integer m. Show that there exists a nontrivial vector $\vec{\phi}(x)$ with n entries in $\mathbb{C}\{x\}$ such that $\vec{y} = \vec{\phi}(\exp[\lambda t])$ satisfies the system $\dfrac{d\vec{y}}{dt} = A(\vec{y})\vec{y}$.

Hint. Calculate the derivative of $\vec{\phi}(\exp[\lambda t])$ to derive the system $\lambda x\dfrac{d\vec{\phi}}{dx} = A(\vec{\phi})\vec{\phi}$ for $\vec{\phi}$.

V-8. Consider a nonzero differential operator $P = \displaystyle\sum_{k=0}^{N} a_k(x)D^k$ with coefficients $a_k(x) \in \mathbb{C}[[x]]$, where $D = \dfrac{d}{dx}$. Regarding $\mathbb{C}[[x]]$ as a vector space over $\mathbb{C}$, define a homomorphism $P : \mathbb{C}[[x]] \to \mathbb{C}[[x]]$. Show that $\mathbb{C}[[x]]/P[\mathbb{C}[[x]]]$ is a finite-dimensional vector space over $\mathbb{C}$.

Hint. Show that the equation $P[y] = x^N \phi(x)$ has a solution in $\mathbb{C}[[x]]$ for any $\phi(x) \in \mathbb{C}[[x]]$ if a positive integer N is sufficiently large. To do this, use the indicial polynomial of P.

V-9. Consider a nonzero differential operator $P = \displaystyle\sum_{k=0}^{N} a_k(x)\delta^k$ with coefficients $a_k(x) \in \mathbb{C}[[x]]$, where $\delta = x\dfrac{d}{dx}$, $n \geq 1$, and $a_n(x) \neq 0$. Define the indicial polynomial $f_{n_0}(s)$ as in Theorem V-1-3. Show that if the degree of $f_{n_0}(s)$ is n and if n zeros $\{\lambda_1, \dots, \lambda_n\}$ of $f_{n_0}(s)$ do not differ by integers, we can factor P in the following form:

$$P = a_n(x)(\delta - \phi_1(x))(\delta - \phi_2(x)) \cdots (\delta - \phi_n(x)),$$

where all the functions $\phi_j(x)$ $(j = 1, 2, \ldots, n)$ are convergent power series in x and $\phi_j(0) = \lambda_j$.

Hint. Without any loss of generality, we can assume that $a_n(x) = 1$. Then, $\lambda^n + \sum_{k=0}^{n-1} a_k(0)\lambda^k = (\lambda - \lambda_1) \cdots (\lambda - \lambda_n)$. Define constants $\{\gamma_0, \ldots, \gamma_{n-2}\}$ by $\lambda^{n-1} + \sum_{k=0}^{n-2} \gamma_k \lambda^k = (\lambda - \lambda_2) \cdots (\lambda - \lambda_n)$. For $\psi(x) \in x\mathbb{C}[[x]]$ and $c_h(x) \in x\mathbb{C}[[x]]$ $(h = 0, \ldots, n-2)$, solve the equation

$$\text{(C)} \qquad P = (\delta - \lambda_1 - \psi(x))\left(\delta^{n-1} + \sum_{h=0}^{n-2}(\gamma_h + c_h(x))\delta^h\right).$$

If we eliminate $\psi(x)$ by $\lambda_1 + \psi(x) = \gamma_{n-2} + c_{n-2}(x) - a_{n-1}(x)$, condition (C) becomes the differential equations

$$\begin{cases} \delta[c_0(x)] = a_0(x) + (\gamma_{n-2} + c_{n-2}(x) - a_{n-1}(x))(\gamma_0 + c_0(x)), \\ \delta[c_h(x)] = a_h(x) + (\gamma_{n-2} + c_{n-2}(x) - a_{n-1}(x))(\gamma_h + c_h(x)) - (\gamma_{h-1} + c_{h-1}(x)), \end{cases}$$

where $h = 1, \ldots, n-2$.

V-10. Consider a linear differential operator $\mathcal{P} = \sum_{\ell=0}^{n} a_\ell \delta^\ell$, where $a_0, \ldots, a_n$ are complex numbers and $\delta = x\dfrac{d}{dx}$. The differential equation

$$\text{(P)} \qquad\qquad\qquad \mathcal{P}[y] = 0$$

is called the *Cauchy-Euler* differential equation. Find a fundamental set of solutions of equation (P).

Hint. If we set $t = \log x$, then $\delta = \dfrac{d}{dt}$.

V-11. Find a fundamental set of solutions of the differential equation

$$(\delta - \alpha)(\delta - \beta)(\delta - \alpha - \beta)[y] = x^m y,$$

where $\delta = x\dfrac{d}{dx}$ and m is a positive integer, whereas α and β are complex numbers such that they are not integers and $\Re[\alpha] < 0 < \Re[\beta]$.

V-12. Find the fundamental set of solutions of the differential equation

$$\text{(LGE)} \qquad\qquad \frac{d}{dx}\left[(1 - x^2)\frac{dy}{dx}\right] + a(a + 1)y = 0$$

at $x = 0$, where a is a complex parameter.

Remark. Differential equation (LGE) is called *the Legendre equation* (cf. [AS, pp. 331–338] or [Ol, pp. 161–189]).

V-13. Show that for a non-negative integer n, the polynomial $P_n(x) = \dfrac{1}{2^n n!}$ $\times \dfrac{d^n}{dx^n}[(x^2 - 1)^n]$ satisfies the Legendre equation

$$\frac{d}{dx}\left[(1 - x^2)\frac{dP_n}{dx}\right] + n(n+1)P_n = 0.$$

Show also that these polynomials satisfy the following conditions:

(1) $P_n(-x) = (-1)^n P_n(x)$,   (2) $P_n(1) = 1$,   (3) $|P_n(t)| \geq 1$ for $|x| \geq 1$,

(4) $\dfrac{1}{\sqrt{1 - 2xt + t^2}} = \displaystyle\sum_{n=0}^{+\infty} P_n(x)t^n$,   and   (5) $|P_n(x)| < 1$ for $|x| < 1$.

Hint. Set $g(x) = (1 - x^2)^n$. Then, $(1 - x^2)g'(x) + 2nxg(x) = 0$. Differentiate this relation $(n + 1)$ times with respect to x to obtain

$$(1 - x^2)g^{(n+2)}(x) - 2(n+1)xg^{(n+1)}(x) - n(n+1)g^{(n)}(x)$$
$$+ 2nxg^{(n+1)}(x) + 2n(n+1)g^{(n)}(x) = 0$$

or

$$(1 - x^2)g^{(n+2)}(x) - 2ng^{(n+1)}(x) + n(n+1)g^{(n)}(x) = 0.$$

Statements (1), (2), (3), and (4) can be proved with straight forward calculations. Statement (5) also can be proved similarly by using (4) (cf. [WhW, Chapter XV, Example 2 on p. 303]). However, the following proof is shorter. To begin with, set

$$F(x) = P_n(x)^2 + \frac{1}{n(n+1)(1 - x^2)}((1 - x^2)P_n'(x))^2.$$

Then, $F(\pm 1) = P_n(\pm 1)^2 = 1$, $F'(x) = \dfrac{2x(P_n'(x))^2}{n(n+1)}$, and $F(x) = P_n(x)^2$ if $P_n'(x) = 0$. Therefore, for $0 \leq x \leq 1$, local maximal values of $P_n(x)^2$ are less than $F(1) = 1$, whereas, for $-1 \leq x \leq 0$, local maximal values of $P_n(x)^2$ are less than $F(-1) = 1$. Hence, $P_n(x)| < 1$ for $|x| < 1$ (cf. [Sz, §§7.2–7.3, pp. 161-172; in particular, Theorems 7.2 and 7.3, pp. 161-162]. See also [NU, pp. 45-46].)

The polynomials $P_n(x)$ are called *the Legendre polynomials*.

V-14. Find a fundamental set of solutions of (LGE) at $x = \infty$. In particular, show what happends when a is a non-negative integer.

V-15. Show that the differential equation $\delta^n y = xy$ has a fundamental set of solutions consisting of n solutions of the following form:

$$\sum_{h=0}^{k-1} \frac{(\log x)^{k-1-h}}{(k - 1 - h)!}\phi_h(x) \qquad (k = 1, \ldots, n),$$

where the functions $\phi_h(x)$ $(h = 0, \ldots, n-1)$ are entire in x, $\phi_0(0) = 1$, and $\phi_h(0) = 0$ $(h = 1, \ldots, n-1)$. Also, find $\phi_0(x)$.

V-16. Find the order of singularity at $x = 0$ of each of the following three equations.

(i) $x^5\{\delta^6 y - 3\delta^2 y + 4y\} = y$,

(ii) $x^5\{\delta^6 y - 3\delta^2 y + 4y\} + x^7\{\delta^3 y - 5\delta y\} = y$,

(iii) $x^5 y''' + 5x^2 y'' + \delta y + 20y = 0$.

V-17. Find a fundamental matrix solution of the system

$$\begin{cases} x^2 \dfrac{dy_1}{dx} = xy_1 + y_2, \\[2mm] x^2 \dfrac{dy_2}{dx} = 2xy_2 + 2y_3, \\[2mm] x^2 \dfrac{dy_3}{dx} = x^3 y_1 + 3y_3. \end{cases}$$

V-18. Let $A(x)$ be an $n \times n$ matrix whose entries are holomorphic at $x = 0$. Also, let Λ be an $n \times n$ diagonal matrix whose entries are non-negative integers. Show that for every sufficiently large positive integer N and every $\mathbb{C}^n$-valued function $\vec{\phi}(x)$ whose entries are polynomials in x such that those of $x^\Lambda \vec{\phi}(x)$ are of degree N, there exists a $\mathbb{C}^n$-valued function $\vec{f}(x; A, \vec{\phi}, N)$ such that

(a) the entries of $\vec{f}$ are polynomials in x of degree $N-1$ with coefficients depending on A, N, and $\vec{\phi}$,

(b) $\vec{f}$ is linear and homogeneous in $\vec{\phi}$,

(c) the linear system

$$x^\Lambda \frac{d\vec{y}}{dz} = A(x)\vec{y} + \vec{f}(x; A, \vec{\phi}, N)$$

has a solution $\vec{\eta}(x)$ whose entries are holomorphic at $x = 0$ and $\vec{\eta}(x)$ is linear and homogeneous in $\vec{\phi}$,

(d) $x^\Lambda(\vec{\eta}(x) - \vec{\phi}(x)) = O(x^{N+1})$ as $x \to 0$.

Hint. See [HSW].

V-19. Let $A(x)$ and Λ be the same as in Exercise V-18. Assuming that $n > \mathrm{trace}(\Lambda)$, show that the system $x^\Lambda \dfrac{d\vec{y}}{dz} = A(x)\vec{y}$ has at least $n - \mathrm{trace}(\Lambda)$ linearly independent solutions holomorphic at $x = 0$.

Hint. This result is due to F. Lettenmeyer [Let]. To solve Exercise V-19, calculate $\vec{f}(x; A, \vec{\phi}, N)$ of Exercise V-18 and solve $\vec{f}(x; A, \vec{\phi}, N) = \vec{0}$ to determine a suitable function $\vec{\phi}$.

V-20. Suppose that for a formal power series $\phi(x) = \displaystyle\sum_{m=0}^{\infty} c_m x^m \in \mathbb{C}[[x]]$, there exist two nonzero differential operators $P = \displaystyle\sum_{k=0}^{N} a_k(x) \left(\frac{d}{dx}\right)^k$ and $Q = \displaystyle\sum_{k=0}^{M} b_k(x) \left(\frac{d}{dx}\right)^k$ with coefficients $a_k(x)$ and $b_k(x)$ in $\mathbb{C}\{x\}$ such that $P[\phi] = 0$ and $Q\left[\dfrac{1}{\phi}\right] = 0$. Show that ϕ is convergent.

Hint. The main ideas are
(a) derive two algebraic (nonlinear) ordinary differential equations

$$(E) \qquad F(x, v, v', \dots, v^{(n)}) = 0 \qquad \text{and} \qquad G(x, v, v', \dots, v^{(n)}) = 0$$

for $v = \dfrac{\phi'}{\phi}$ from the given equations $P[\phi] = 0$ and $Q\left[\dfrac{1}{\phi}\right] = 0$.

(b) eliminate all derivatives of v from (E) to derive a nontrivial purely algebraic equation $H(x, v) = 0$ on v. See [HaS1] and [HaS2] for details.

CHAPTER VI

BOUNDARY-VALUE PROBLEMS OF LINEAR
DIFFERENTIAL EQUATIONS OF THE SECOND-ORDER

In this chapter, we explain (1) oscillation of solutions of a homogeneous second-order linear differential equation (§VI-1), (2) the Sturm-Liouville problems (§§VI-2–VI-4, topics including Green's functions, self-adjointness, distribution of eigenvalues, and eigenfunction expansion), (3) scattering problems (§§VI-5–VI-9, mostly focusing on reflectionless potentials), and (4) periodic potentials (§VI-10). The materials concerning these topics are also found in [CL, Chapters 7, 8, and 11], [Har2 Chapter XI], [Cop1, Chapter 1], [Bel2, Chapter 6], and [TD]. Singular self-adjoint boundary-value problems (in particular continuous spectrum, limit-point and limit-circle cases) are not explained in this book. For these topics, see [CL, Chapter 9].

VI-1. Zeros of solutions

It is known that real-valued solutions of the differential equation $\dfrac{d^2y}{dx^2} + y = 0$ are linear combinations of $\sin x$ and $\cos x$. These solutions have infinitely many zeros on the real line $\mathbb{R}$. It is also known that real-valued solutions of the differential equation $\dfrac{d^2y}{dx^2} - y = 0$ are linear combination of e^x and e^{-x}. Therefore, nontrivial solutions has at most one zero on the real line $\mathbb{R}$. Furthermore, solutions of the differential equation $\dfrac{d^2y}{dx^2} + 2y = 0$ have more zeros than solutions of $\dfrac{d^2y}{dx^2} + y = 0$. In this section, keeping these examples in mind, we explain the basic results concerning zeros of solutions of the second-order homogeneous linear differential equations. In §§VI-1–VI-4, every quantity is supposed to take real values only.

We start with the most well-known comparison theorem concerning a homogeneous second-order linear differential equation

$$(\text{VI.1.1}) \qquad\qquad \frac{d^2y}{dx^2} + g(x)y = 0.$$

Theorem VI-1-1. *Suppose that*

(i) $g_1(x)$ and $g_2(x)$ are continuous and $g_2(x) > g_1(x)$ on an interval $a < x < b$,

(ii) $\dfrac{d^2\phi_1(x)}{dx^2} + g_1(x)\phi_1(x) = 0$ and $\dfrac{d^2\phi_2(x)}{dx^2} + g_2(x)\phi_2(x) = 0$ on $a < x < b$,

(iii) ξ_1 and ξ_2 are successive zeros of $\phi_1(x)$ on $a < x < b$.

Then, $\phi_2(x)$ must vanish at some point ξ_3 between ξ_1 and ξ_2.

Proof.

Assume without any loss of generality that $\xi_1 < \xi_2$ and $\phi_1(x) > 0$ on $\xi_1 < x < \xi_2$. Notice that $\phi_1(\xi_1) = 0$, $\dfrac{d\phi_1}{dx}(\xi_1) > 0$, $\phi_1(\xi_2) = 0$, and $\dfrac{d\phi_1}{dx}(\xi_2) < 0$. A contradiction will be derived from the assumption that $\phi_2(x) > 0$ on $\xi_1 < x < \xi_2$. In fact, assumption (ii) implies

$$(VI.1.2) \qquad \phi_2(x)\frac{d^2\phi_1(x)}{dx^2} - \phi_1(x)\frac{d^2\phi_2(x)}{dx^2} = [g_2(x) - g_1(x)]\phi_1(x)\phi_2(x)$$

and, hence,

$$(VI.1.3) \qquad \phi_2(\xi_2)\frac{d\phi_1}{dx}(\xi_2) - \phi_2(\xi_1)\frac{d\phi_1}{dx}(\xi_1) = \int_{\xi_1}^{\xi_2} [g_2(x) - g_1(x)]\phi_1(x)\phi_2(x)dx.$$

The left-hand side of (VI.1.3) is nonpositive, but the right-hand side of (VI.1.3) is positive. This is a contradiction. $\square$

A similar argument yields the following theorem.

Theorem VI-1-2. *Suppose that*
(i) $g(x)$ is continuous on an interval $a < x < b$,
(ii) $\phi_1(x)$ and $\phi_2(x)$ are two linearly independent solutions of (VI.1.1),
(iii) ξ_1 and ξ_2 are successive zeros of $\phi_1(x)$ on $a < x < b$.
Then, $\phi_2(x)$ must vanish at some point ξ_3 between ξ_1 and ξ_2.

Proof.

Assume without any loss of generality that $\xi_1 < \xi_2$ and $\phi_1(x) > 0$ on $\xi_1 < x < \xi_2$. Notice that $\phi_1(\xi_1) = 0$, $\dfrac{d\phi_1}{dx}(\xi_1) > 0$, $\phi_1(\xi_2) = 0$, and $\dfrac{d\phi_1}{dx}(\xi_2) < 0$. Note also that if $\phi_2(\xi_1) = 0$ or $\phi_2(\xi_2) = 0$, then ϕ_1 and ϕ_2 are linearly dependent. Now, a contradiction will be derived from the assumption that $\phi_2(x) > 0$ on $\xi_1 \leq x \leq \xi_2$. In fact, assumption (ii) implies $\phi_2(x)\dfrac{d^2\phi_1(x)}{dx^2} - \phi_1(x)\dfrac{d^2\phi_2(x)}{dx^2} = 0$ and, hence,

$$(VI.1.4) \qquad \phi_2(\xi_2)\frac{d\phi_1}{dx}(\xi_2) - \phi_2(\xi_1)\frac{d\phi_1}{dx}(\xi_1) = 0.$$

Since the left-hand side of (VI.1.4) is positive, this is a contradiction. $\square$

The following result is a simple consequence of Theorem VI.1.2.

Corollary VI-1-3. *Let $g(x)$ be a real-valued and continuous function on the interval $\mathcal{I}_0 = \{x : 0 \leq x < +\infty\}$. Then,*
(a) if a nontrivial solution of the differential equation (VI.1.1) has infinitely many zeros on $\mathcal{I}_0$, then every solution of (VI.1.1) has an infinitely many zeros on $\mathcal{I}_0$,
(b) if a solution of (VI.1.1) has m zeros on an open subinterval $\mathcal{I} = \{x : \alpha < x < \beta\}$ of $\mathcal{I}_0$, then every nontrivial solution of (VI.1.1) has at most $m + 1$ zeros on $\mathcal{I}$.

Denote by $W(x) = W(x; \phi_1, \phi_2) = \begin{vmatrix} \phi_1(x) & \phi_2(x) \\ \phi_1'(x) & \phi_2'(x) \end{vmatrix}$ the *Wronskian* of the set of functions $\{\phi_1(x), \phi_2(x)\}$. For further discussion, we need the following lemma.

Lemma VI-1-4. *Let $g(x)$ be a real-valued and continuous function on the interval $\mathcal{I}_0 = \{x : 0 \leq x < +\infty\}$, and let $\eta_1(x)$ and $\eta_2(x)$ be two solutions of differential equation (VI.1.1). Then,*

(a) $W(x; \eta_1, \eta_2)$, the Wronskian of $\{\eta_1(x), \eta_2(x)\}$, is independent of x,

(b) if we set $\xi(x) = \dfrac{\eta_1(x)}{\eta_2(x)}$, then $\dfrac{d\xi(x)}{dx} = \dfrac{c}{\eta_2(x)^2}$, where c is a constant.

Also, if $\eta(x)$ is a nontrivial solution of (VI.1.1) and if we set $w(x) = \dfrac{\eta'(x)}{\eta(x)}$, then we obtain

$$(\text{VI.1.5}) \qquad\qquad \frac{dw(x)}{dx} + w(x)^2 + g(x) = 0.$$

Proof.

(a) It can be easily shown that

$$\frac{d}{dx} \begin{vmatrix} \eta_1(x) & \eta_2(x) \\ \eta_1'(x) & \eta_2'(x) \end{vmatrix} = \begin{vmatrix} \eta_1(x) & \eta_2(x) \\ \eta_1''(x) & \eta_2''(x) \end{vmatrix} = -g(x) \begin{vmatrix} \eta_1(x) & \eta_2(x) \\ \eta_1(x) & \eta_2(x) \end{vmatrix} = 0.$$

(b) Note that $\dfrac{d\xi(x)}{dx} = -\dfrac{1}{\eta_2(x)^2} \begin{vmatrix} \eta_1(x) & \eta_2(x) \\ \eta_1'(x) & \eta_2'(x) \end{vmatrix}$ and that $\dfrac{dw(x)}{dx} = \dfrac{\eta''(x)}{\eta(x)} - \left(\dfrac{\eta'(x)}{\eta(x)}\right)^2 = -g(x) - w(x)^2.$ $\square$

The following theorem shows the structure of solutions in the case when every nontrivial solution of differential equation (VI.1.1) has only a finite number of zeros on $\mathcal{I}_0 = \{x : 0 \leq x < +\infty\}$.

Theorem VI-1-5. *Let $g(x)$ be a real-valued and continuous function on the interval $\mathcal{I}_0$. Assume that every nontrivial solution of differential equation (VI.1.1) has only a finite number of zeros on $\mathcal{I}_0$. Then, (VI.1.1) has two linearly independent solutions $\eta_1(x)$ and $\eta_2(x)$ such that*

(a) $\displaystyle\lim_{x \to +\infty} \frac{\eta_1(x)}{\eta_2(x)} = 0$,

(b)

$$(\text{VI.1.6}) \qquad \int_{x_0}^{+\infty} \frac{dx}{\eta_1(x)^2} = +\infty \qquad \text{and} \qquad \int_{x_0}^{+\infty} \frac{dx}{\eta_2(x)^2} < +\infty,$$

where x_0 is a sufficiently large positive number,

(c) the solution $\eta_2(x)$ is unbounded on $\mathcal{I}_0$.

Proof.

(a) Let $\zeta_1(x)$ and $\zeta_2(x)$ be two linearly independent solutions of (VI.1.1) such that $\zeta_j(x) > 0 \; (j = 1, 2)$ for $x \geq x_0 > 0$. Using (b) of Lemma VI-1-4, we can derive the following three possibilities (i) $\displaystyle\lim_{x\to+\infty}\frac{\zeta_1(x)}{\zeta_2(x)} = 0$, (ii) $\displaystyle\lim_{x\to+\infty}\frac{\zeta_1(x)}{\zeta_2(x)} = +\infty$, and (iii)

$\displaystyle\lim_{x\to+\infty}\frac{\zeta_1(x)}{\zeta_2(x)} = \gamma > 0$. Set

$$\eta_1(x) = \begin{cases} \zeta_1(x) & \text{in case (i),} \\ \zeta_2(x) & \text{in case (ii),} \\ \zeta_1(x) - \gamma\zeta_2(x) & \text{in case (iii),} \end{cases} \qquad \eta_2(x) = \begin{cases} \zeta_2(x) & \text{in case (i),} \\ \zeta_1(x) & \text{in case (ii),} \\ \zeta_2(x) & \text{in case (iii).} \end{cases}$$

Then, (a) follows immediately.

(b) Conclusion (b) of Lemma VI-1-4 implies that

$$\frac{1}{\eta_1(x)^2} = \frac{1}{c}\frac{d}{dx}\left(\frac{\eta_2(x)}{\eta_1(x)}\right) \qquad \text{and} \qquad \frac{1}{\eta_2(x)^2} = \frac{-1}{c}\frac{d}{dx}\left(\frac{\eta_1(x)}{\eta_2(x)}\right),$$

where c is the Wronskian of η_1 and η_2. Hence, (VI.1.6) follows.

(c) If η_2 is bounded, we must have $\displaystyle\int_{x_0}^{+\infty}\frac{dx}{\eta_2(x)^2} = +\infty$. $\square$

The following theorem gives a simple sufficient condition that solutions of (VI.1.1) have infinitely many zeros on the interval $\mathcal{I}_0$.

Theorem VI-1-6. *Let $g(x)$ be a real-valued and continuous function on the interval $\mathcal{I}_0$. If $\displaystyle\int_0^{+\infty} g(x)dx = +\infty$, then every solution of the differential equation (VI.1.1) has infinitely many zeros on $\mathcal{I}_0$.*

Proof.

Suppose that a solution $\eta(x)$ satisfies the condition that $\eta(x) > 0$ for $x \geq x_0 > 0$. Set $w(x) = \dfrac{\eta'(x)}{\eta(x)}$. Then, $w(x)$ satisfies differential equation (VI.1.5). Hence, $\displaystyle\lim_{x\to+\infty} w(x) = -\infty$. This implies that $\displaystyle\lim_{x\to+\infty} \eta(x) = 0$ since $\eta(x) = \eta(x_0)$

$\times \exp\left[\displaystyle\int_{x_0}^{x} w(\xi)d\xi\right]$. This contradicts (c) of Theorem VI-1-5. $\square$

The converse of Theorem VI-1-6 is not true, as shown by the following example.

Example VI-1-7. Solutions of the differential equation

(VI.1.7) $$\frac{d^2y}{dx^2} + \frac{Ay}{1+x^2} = 0,$$

where A is a constant, have infinitely many zeros on the interval $-\infty < x < +\infty$ if and only if $A > \dfrac{1}{4}$.

Proof.

Look at (VI.1.7) near $x = \infty$. To do this, set $t = \dfrac{1}{x^2}$ to change (VI.1.7) to

(VI.1.8) $$\left[4\delta^2 + 2\delta + \frac{A}{1+t}\right]y = 0,$$

where $\delta = t\dfrac{d}{dt}$. Note that $t = 0$ is a regular singular point of (VI.1.8) and the indicial equation is $4s^2 + 2s + A = 0$ whose roots are $s = -\dfrac{1}{2} \pm \sqrt{\dfrac{1}{4} - A}$. These roots are real if and only if $A \le \dfrac{1}{4}$. This verifies the claim. $\square$

Note that $\displaystyle\int_0^{+\infty} \dfrac{1}{1 + x^2}\, dx = \dfrac{\pi}{2}$.

The following theorem shows some structure of solutions in the case when

$$\int_0^{+\infty} |g(x)|dx < +\infty.$$

Theorem VI-1-8. *Let $g(x)$ be a real-valued and continuous function on the interval $\mathcal{I}_0$. If $\displaystyle\int_0^{+\infty} |g(x)|dx < +\infty$, then there exist unbounded solutions of differential equation (VI.1.1) on $\mathcal{I}_0$.*

Proof.

The assumption of this theorem and (VI.1.1) imply that $\displaystyle\lim_{x\to+\infty} \dfrac{d\eta(x)}{dx} = \gamma$ exists for any bounded solution $\eta(x)$ of (VI.1.1). If $\gamma \ne 0$, then $\eta(x)$ is unbounded. Hence, $\displaystyle\lim_{x\to+\infty} \dfrac{d\eta(x)}{dx} = 0$. Therefore, calculating the Wronskian of two bounded solutions of (VI.1.1), we find that those bounded solutions are linearly dependent on each other. This implies that there must be unbounded solutions. $\square$

Remark VI-1-9. Theorems VI-1-1 and VI-1-2 are also explained in [CL, §1 of Chapter 8, pp. 208-211] and [Har2, Chapter XI, pp. 322-403]. For details concerning other results in this section, see also [Cop2, Chapter 1, pp. 4-33] and [Bel2, Chapter 6, pp. 107-142].

VI-2. Sturm-Liouville problems

A Sturm-Liouville problem is a boundary-value problem

(BP)
$$\begin{cases} \dfrac{d}{dx}\left(p(x)\dfrac{dy}{dx}\right) + u(x)y = f(x), \\ y(a)\cos\alpha - p(a)y'(a)\sin\alpha = 0, \\ y(b)\cos\beta - p(b)y'(b)\sin\beta = 0, \end{cases}$$

under the assumptions:
  (i) the quantities a, b, α, and β are real numbers such that $a < b$,
 (ii) two functions $u(x)$ and $f(x)$ are real-valued and continuous on the interval $\mathcal{I}(a,b) = \{x : a \le x \le b\}$,
(iii) the function $p(x)$ is real-valued and continuously differentiable, and $p(x) > 0$ on $\mathcal{I}(a,b)$.
In this section, we explain some basic results concerning problem (BP).

Let $\phi(x)$ and $\psi(x)$ be two solutions of the homogeneous linear differential equation

$$\text{(VI.2.1)} \qquad \frac{d}{dx}\left(p(x)\frac{dy}{dx}\right) + u(x)y = 0$$

such that

$$\text{(VI.2.2)} \quad \phi(a) = \sin\alpha, \quad p(a)\phi'(a) = \cos\alpha, \quad \psi(b) = \sin\beta, \quad p(b)\psi'(b) = \cos\beta,$$

respectively. Then, these two solutions satisfy the boundary conditions

$$\text{(VI.2.3)} \qquad \phi(a)\cos\alpha - p(a)\phi'(a)\sin\alpha = 0, \quad \psi(b)\cos\beta - p(b)\psi'(b)\sin\beta = 0.$$

The two solutions $\phi(x)$ and $\psi(x)$ are linearly independent if and only if

$$\text{(VI.2.4)} \qquad \phi(b)\cos\beta - p(b)\phi'(b)\sin\beta \neq 0 \quad \text{or} \quad \psi(a)\cos\alpha - p(a)\psi'(a)\sin\alpha \neq 0.$$

The first basic result of this section is the following theorem, which concerns the existence and uniqueness of solution of (BP).

Theorem VI-2-1. *If the two solutions $\phi(x)$ and $\psi(x)$ of (VI.2.1) are linearly independent, then problem (BP) has one and only one solution on the interval $\mathcal{I}(a,b)$.*

Proof.

Using the method of variation of parameters (cf. Remark IV-7-2), write the general solution $y(x)$ of the differential equation of (BP) and its derivative $y'(x)$ respectively in the following form:
$$\text{(VI.2.5)}$$
$$\begin{cases} y(x) = c_1\phi(x) + c_2\psi(x) + \phi(x)\int_x^b \frac{\psi(\xi)f(\xi)}{p(\xi)W(\xi)}d\xi + \psi(x)\int_a^x \frac{\phi(\xi)f(\xi)}{p(\xi)W(\xi)}d\xi, \\ y'(x) = c_1\phi'(x) + c_2\psi'(x) + \phi'(x)\int_x^b \frac{\psi(\xi)f(\xi)}{p(\xi)W(\xi)}d\xi + \psi'(x)\int_a^x \frac{\phi(\xi)f(\xi)}{p(\xi)W(\xi)}d\xi, \end{cases}$$

where c_1 and c_2 are arbitrary constants and $W(x)$ denotes the Wronskian of $\{\phi(x), \psi(x)\}$. Now, using (VI.2.3) and (VI.2.4), it can be shown that solution (VI.2.5) satisfies the boundary conditions of (BP) if and only if $c_1 = 0$ and $c_2 = 0$. $\square$

Observation VI-2-2. It can be shown easily that $p(x)W(x)$ is independent of x.

Observation VI-2-3. Under the assumption that the two solutions $\phi(x)$ and $\psi(x)$ of (VI.2.1) are linearly independent, the unique solution of (BP) is given by

$$\text{(VI.2.6)} \qquad \begin{cases} y(x) = \phi(x)\int_x^b \frac{\psi(\xi)f(\xi)}{p(\xi)W(\xi)}d\xi + \psi(x)\int_a^x \frac{\phi(\xi)f(\xi)}{p(\xi)W(\xi)}\xi, \\ y'(x) = \phi'(x)\int_x^b \frac{\psi(\xi)f(\xi)}{p(\xi)W(\xi)}d\xi + \psi'(x)\int_a^x \frac{\phi(\xi)f(\xi)}{p(\xi)W(\xi)}d\xi. \end{cases}$$

Setting

$$(VI.2.7) \qquad G(x,\xi) = \begin{cases} \dfrac{\phi(x)\psi(\xi)}{p(\xi)W(\xi)} & \text{if} \quad x \le \xi \le b, \\[2ex] \dfrac{\psi(x)\phi(\xi)}{p(\xi)W(\xi)} & \text{if} \quad a \le \xi \le x, \end{cases}$$

we can write (VI.2.6) in the form

$$y(x) = \int_a^b G(x,\xi)f(\xi)d\xi, \qquad y'(x) = \int_a^b \frac{\partial G}{\partial x}(x,\xi)f(\xi)d\xi.$$

The function $G(x,\xi)$ is called *Green's function* of problem (BP). The following theorem gives the characterization of Green's function.

Theorem VI-2-4. *The function $G(x,\xi)$ given by (VI.2.7) satisfies the following conditions:*

  (i) $G(x,\xi)$ is continuous with respect to (x,ξ) on the region $\mathcal{D} = \{(x,\xi) : a \le x \le b,\ a \le \xi \le b\}$,

  (ii) $\dfrac{\partial G}{\partial x}(x,\xi)$ is continuous with respect to (x,ξ) on the region $\mathcal{D} - \{(\xi,\xi) : a \le \xi \le b\}$,

  (iii) at (ξ,ξ), we have

$$\frac{\partial G}{\partial x}(\xi+0,\xi) \;-\; \frac{\partial G}{\partial x}(\xi-0,\xi) = \frac{1}{p(\xi)} \qquad for \qquad a \le \xi \le b,$$

  (iv) as a function of x, $G(x,\xi)$ satisfies homogeneous linear differential equation (VI.2.1) if $x \neq \xi$,

  (v) as a function of x, $G(x,\xi)$ satisfies the boundary conditions of (BP), i.e.,

$$G(a,\xi)\cos\alpha - p(a)\frac{\partial G}{\partial x}(a,\xi)\sin\alpha = 0, \quad G(b,\xi)\cos\beta - p(b)\frac{\partial G}{\partial x}(b,\xi)\sin\beta = 0$$

  for $a \le \xi \le b$.

Furthermore, the function $G(x,\xi)$ is uniquely determined by these five conditions.

Proof.

It is evident that the function $G(x,\xi)$ satisfies these five conditions. Suppose $H(x,\xi)$ satisfies the five conditions. Conditions (iv) and (v) imply that

$$H(x,\xi) = \begin{cases} C_1(\xi)\phi(x) & \text{for} \quad a \le x < \xi, \\ C_2(\xi)\psi(x) & \text{for} \quad \xi < x \le b \end{cases}$$

for $(x,\xi) \in \mathcal{D}$, where $C_1(\xi)$ and $C_2(\xi)$ must be determined by conditions (i), (ii), and (iii). This means that $C_1(\xi)$ and $C_2(\xi)$ must be determined by

$$C_1(\xi)\phi(\xi) - C_2(\xi)\psi(\xi) = 0 \quad \text{and} \quad C_1(\xi)\phi'(\xi) - C_2(\xi)\psi'(\xi) = -\frac{1}{p(\xi)}.$$

This yields $C_1(\xi) = \dfrac{\psi(\xi)}{p(\xi)W(\xi)}$ and $C_2(\xi) = \dfrac{\phi(\xi)}{p(\xi)W(\xi)}$. Since $\{(x,\xi) : a \leq x \leq \xi\}$ $= \{(x,\xi) : x \leq \xi \leq b\}$ and $\{(x,\xi) : \xi \leq x \leq b\} = \{(x,\xi) : a \leq \xi \leq x\}$ for $(x,\xi) \in \mathcal{D}$, we obtain $H(x,\xi) = G(x,\xi)$. $\square$

The second basic result concerns self-adjointness of problem (BP). Define a differential operator $\mathcal{L}[y] = \dfrac{d}{dx}\left(p(x)\dfrac{dy}{dx}\right) + u(x)y$ and a vector space $\mathcal{V}(a,b)$ over the real number field $\mathbb{R}$ by

$$\mathcal{V}(a,b) = \{f \in \mathcal{C}^2(a,b) : f(a)\cos\alpha - p(a)f'(a)\sin\alpha = 0,$$
$$f(b)\cos\beta - p(b)f'(b)\sin\beta = 0\},$$

where $\mathcal{C}^2(a,b)$ denotes the set of all real-valued functions which are twice continuously differentiable on the interval $\mathcal{I}(a,b) = \{x : a \leq x \leq b\}$. Define also an *inner product* (f,g) for two real-valued continuous functions f and g on $\mathcal{I}(a,b)$ by $(f,g) = \displaystyle\int_a^b f(\xi)g(\xi)d\xi$. Since

$$(f,\mathcal{L}[g]) = \int_a^b f(\xi)\left[\frac{d}{d\xi}\left(p(\xi)\frac{dg}{d\xi}\right) + u(\xi)g(\xi)\right]d\xi$$

$$= p(b)f(b)g'(b) - p(a)f(a)g'(a) + \int_a^b [-p(\xi)f'(\xi)g'(\xi) + u(\xi)f(\xi)g(\xi)]\,d\xi$$

for $f \in \mathcal{V}(a,b)$ and $g \in \mathcal{V}(a,b)$, we obtain

$$(f,\mathcal{L}[g]) - (\mathcal{L}[f],g)$$
$$= \{p(b)f(b)g'(b) - p(b)g(b)f'(b)\} - \{p(a)f(a)g'(a) - p(a)g(a)f'(a)\}.$$

Also, since

$$\left\{\begin{array}{ll} f(a)\cos\alpha - p(a)f'(a)\sin\alpha = 0, & f(b)\cos\beta - p(b)f'(b)\sin\beta = 0, \\ g(a)\cos\alpha - p(a)g'(a)\sin\alpha = 0, & g(b)\cos\beta - p(b)g'(b)\sin\beta = 0 \end{array}\right.$$

for $f \in \mathcal{V}(a,b)$ and $g \in \mathcal{V}(a,b)$, we obtain

$$p(a)f(a)g'(a) - p(a)g(a)f'(a) = 0, \qquad p(b)f(b)g'(b) - p(b)g(b)f'(b) = 0.$$

Thus, we proved the following theorem.

Theorem VI-2-5 (self-adjointness). *The operator $\mathcal{L}$ has the following property:*

$$(f,\mathcal{L}[g]) = (\mathcal{L}[f],g) \qquad for \qquad f \in \mathcal{V}(a,b) \quad and \quad g \in \mathcal{V}(a,b).$$

Observation VI-2-6. In Theorem VI.2.1, it was assumed that the two solutions $\phi(x)$ and $\psi(x)$ are linearly independent. Consider the case when this assumption is not satisfied. So, assume that $\phi(x)$ and $\psi(x)$ are linearly dependent. This means that

$$\phi(a)\cos\alpha - p(a)\phi'(a)\sin\alpha = 0 \quad \text{and} \quad \phi(b)\cos\beta - p(b)\phi'(b)\sin\beta = 0.$$

In other words,

(VI.2.8) $$\mathcal{L}[\phi] = 0 \quad \text{and} \quad \phi \in \mathcal{V}(a,b).$$

The boundary-value problem (BP) can be written in the form

(VI.2.9) $$\mathcal{L}[y] = f \quad \text{and} \quad y \in \mathcal{V}(a,b).$$

Using (VI.2.8) and (VI.2.9), we obtain

$$(f,\phi) = (\mathcal{L}[y],\phi) = (y,\mathcal{L}[\phi]) = 0,$$

if there exists a solution y of problem (VI.2.9). The converse is also true, as shown in the following theorem.

Theorem VI-2-7. *Assume that $\phi(x)$ satisfies condition (VI.2.8). Then, if a real-valued continuous function $f(x)$ on $\mathcal{I}(a,b)$ satisfies the condition*

(VI.2.10) $$(f,\phi) = 0,$$

problem (BP) has solutions depending on an arbitrary constant.

Proof.

Using the method of variation of parameters, write the general solution $y(x)$ of the differential equation of (BP) and its derivative $y'(x)$ respectively in the following form:

(VI.2.11)
$$\begin{cases} y(x) = c_1\phi(x) + c_2\mu(x) + \phi(x)\int_x^b \dfrac{\mu(\xi)f(\xi)}{p(\xi)\tilde{W}(\xi)}d\xi + \mu(x)\int_a^x \dfrac{\phi(\xi)f(\xi)}{p(\xi)\tilde{W}(\xi)}d\xi, \\[2mm] y'(x) = c_1\phi'(x) + c_2\mu'(x) + \phi'(x)\int_x^b \dfrac{\mu(\xi)f(\xi)}{p(\xi)\tilde{W}(\xi)}d\xi + \mu'(x)\int_a^x \dfrac{\phi(\xi)f(\xi)}{p(\xi)\tilde{W}(\xi)}d\xi, \end{cases}$$

where $\mu(x)$ is a solution of the linear homogeneous differential equation (VI.2.1) such that ϕ and μ are linearly independent, two quantities c_1 and c_2 are arbitrary constants, and $\tilde{W}(x)$ denotes the Wronskian of $\{\phi(x),\mu(x)\}$. Note that $p(x)\tilde{W}(x)$ is independent of x and that

(VI.2.12) $$\mu(a)\cos\alpha - p(a)\mu'(a)\sin\alpha \neq 0, \quad \mu(b)\cos\beta - p(b)\mu'(b)\sin\beta \neq 0.$$

From (VI.2.10) and (VI.2.11), we derive

$$\begin{cases} y(a) = c_1\phi(a) + c_2\mu(a) + \phi(a)\int_a^b \dfrac{\mu(\xi)f(\xi)}{p(\xi)\tilde{W}(\xi)}d\xi, \\[2mm] y'(a) = c_1\phi'(a) + c_2\mu'(a) + \phi'(a)\int_a^b \dfrac{\mu(\xi)f(\xi)}{p(\xi)\tilde{W}(\xi)}d\xi, \\[2mm] y(b) = c_1\phi(b) + c_2\mu(b), \quad y'(b) = c_1\phi'(b) + c_2\mu'(b). \end{cases}$$

The condition that $y \in \mathcal{V}(a,b)$ and (VI.2.12) imply that c_1 is arbitrary and $c_2 = 0$. $\square$

Remark VI-2-8. The materials of this section are also found in [CL, Chapters 7 and 11] and [Har2, Chapter XI].

VI-3. Eigenvalue problems

In this section, we consider the eigenvalue problem

$$
\text{(EP)} \qquad
\begin{cases}
\dfrac{d}{dx}\left(p(x)\dfrac{dy}{dx}\right) + u(x)y = \lambda y, \\[2mm]
y(a)\cos\alpha - p(a)y'(a)\sin\alpha = 0, \\[2mm]
y(b)\cos\beta - p(b)y'(b)\sin\beta = 0,
\end{cases}
$$

under the assumptions:
- (i) the quantities a, b, α, and β are real numbers such that $a < b$,
- (ii) $u(x)$ is a real-valued and continuous function on the interval $\mathcal{I}(a,b) = \{x : a \leq x \leq b\}$,
- (iii) the function $p(x)$ is real-valued and continuously differentiable, and $p(x) > 0$ on $\mathcal{I}(a,b)$.

The quantity λ is the eigenvalue parameter.

Let $\phi(x,\lambda)$ and $\psi(x,\lambda)$ be two solutions of the homogeneous linear differential equation

$$
\text{(VI.3.1)} \qquad \frac{d}{dx}\left(p(x)\frac{dy}{dx}\right) + u(x)y = \lambda y
$$

such that

$$
\text{(VI.3.2)} \qquad
\begin{cases}
\phi(a,\lambda) = \sin\alpha, & p(a)\phi'(a,\lambda) = \cos\alpha, \\[2mm]
\psi(b,\lambda) = \sin\beta, & p(b)\psi'(b,\lambda) = \cos\beta,
\end{cases}
$$

respectively. Then,

$$
\text{(VI.3.3)} \qquad
\begin{cases}
\phi(a,\lambda)\cos\alpha - p(a)\phi'(a,\lambda)\sin\alpha = 0, \\[2mm]
\psi(b,\lambda)\cos\beta - p(b)\psi'(b,\lambda)\sin\beta = 0.
\end{cases}
$$

These two solutions $\phi(x,\lambda)$ and $\psi(x,\lambda)$ are analytic in λ everywhere in $\mathbb{C}$. Also, they are linearly independent if and only if

$$
\text{(VI.3.4)} \qquad
\begin{cases}
\phi(b,\lambda)\cos\beta - p(b)\phi'(b,\lambda)\sin\beta \neq 0 \quad \text{or} \\[2mm]
\psi(a,\lambda)\cos\alpha - p(a)\psi'(a,\lambda)\sin\alpha \neq 0.
\end{cases}
$$

Therefore, we obtain the following result.

Theorem VI-3-1. *In order that λ be an eigenvalue of (EP), it is necessary and sufficient that λ satisfies the equation*

$$
\text{(VI.3.5)} \qquad \phi(b,\lambda)\cos\beta - p(b)\phi'(b,\lambda)\sin\beta = 0
$$

or, equivalently,

$$
\psi(a,\lambda)\cos\alpha - p(a)\psi'(a,\lambda)\sin\alpha = 0.
$$

Observation VI-3-2. All roots of (VI.3.5) are simple, since, if λ is a root of (VI.3.5), we can prove

$$\frac{\partial \phi(b, \lambda)}{\partial \lambda} \cos \beta \, - \, p(b) \frac{\partial \phi'(b, \lambda)}{\partial \lambda} \sin \beta \, \neq \, 0$$

in the following way. Let λ be a root of (VI.3.5). Then, notice that $z = \dfrac{\partial \phi(x, \lambda)}{\partial \lambda}$ is a solution of the differential equation

$$\frac{d}{dx}\left(p(x)\frac{dz}{dx}\right) \, + \, u(x)z \, = \, \lambda z \, + \, \phi(x, \lambda).$$

This implies that
(VI.3.6)
$$\begin{cases} z(x) = c_1 \phi(x, \lambda) + c_2 \mu(x) - \phi(x, \lambda) \displaystyle\int_a^x \frac{\mu(\xi)\phi(\xi, \lambda)}{p(\xi)\tilde{W}(\xi)}d\xi + \mu(x) \int_a^x \frac{\phi(\xi, \lambda)^2}{p(\xi)\tilde{W}(\xi)}d\xi, \\[2mm] z'(x) = c_1 \phi'(x, \lambda) + c_2 \mu'(x) - \phi'(x, \lambda) \displaystyle\int_a^x \frac{\mu(\xi)\phi(\xi, \lambda)}{p(\xi)\tilde{W}(\xi)}d\xi + \mu'(x) \int_a^x \frac{\phi(\xi, \lambda)^2}{p(\xi)\tilde{W}(\xi)}d\xi, \end{cases}$$

where $\mu(x)$ is a solution of homogeneous linear differential equation (VI.3.1) such that $\phi(x, \lambda)$ and $\mu(x)$ are linearly independent, two quantities c_1 and c_2 are constants, and $\tilde{W}(x)$ denotes the Wronskian of $\{\phi(x, \lambda), \mu(x)\}$. Note that $p(x)\tilde{W}(x)$ is independent of x and that

(VI.3.7) $\mu(a) \cos \alpha - p(a)\mu'(a) \sin \alpha \neq 0$, $\mu(b) \cos \beta - p(b)\mu'(b) \sin \beta \neq 0$.

Formulas (VI.3.6) imply that
(VI.3.8)
$$\begin{cases} z(a) = c_1 \phi(a, \lambda) + c_2 \mu(a), \quad z'(a) = c_1 \phi'(a, \lambda) + c_2 \mu'(a), \\[2mm] z(b) = c_1 \phi(b, \lambda) + c_2 \mu(b) - \phi(b, \lambda) \displaystyle\int_a^b \frac{\mu(\xi)\phi(\xi, \lambda)}{p(\xi)\tilde{W}(\xi)}d\xi + \mu(b) \int_a^b \frac{\phi(\xi, \lambda)^2}{p(\xi)\tilde{W}(\xi)}d\xi, \\[2mm] z'(b) = c_1 \phi'(b, \lambda) + c_2 \mu'(b) - \phi'(b, \lambda) \displaystyle\int_a^b \frac{\mu(\xi)\phi(\xi, \lambda)}{p(\xi)\tilde{W}(\xi)}d\xi + \mu'(b) \int_a^b \frac{\phi(\xi, \lambda)^2}{p(\xi)\tilde{W}(\xi)}d\xi. \end{cases}$$

Since (VI.3.3) is true for all values of λ, it follows that $z(a) \cos \alpha - p(a)z'(a) \sin \alpha = 0$. Hence, from (VI.3.7) and (VI.3.8), we conclude that $c_2 = 0$. Therefore,

$$z(b) \cos \beta - p(b)z'(b) \sin \beta = (\mu(b) \cos \beta - p(b)\mu'(b) \sin \beta) \int_a^b \frac{\phi(\xi, \lambda)^2}{p(\xi)\tilde{W}(\xi)}d\xi \neq 0. \quad \square$$

Remark VI-3-3. If $\lambda = 0$ is not an eigenvalue of (EP), we can define Green's function $G(x, \xi)$ of problem (BP) of §V-2 so that (EP) is changed equivalently to the integral equation $y(x) = \lambda \displaystyle\int_a^b G(x, \xi)y(\xi)d\xi$.

To prove that all eigenvalues of (EP) are real, it is convenient to extend Theorem VI-2-5 (self-adjointness) to complex-valued functions f and g. Define a differential operator $\mathcal{L}[y] = \dfrac{d}{dx}\left(p(x)\dfrac{dy}{dx}\right) + u(x)y$ and a vector space $\mathcal{U}(a,b)$ over the complex number field $\mathbb{C}$ by

$$\mathcal{U}(a,b) = \{f \in C^2(a,b;\mathbb{C}) : f(a)\cos\alpha - p(a)f'(a)\sin\alpha = 0,$$
$$f(b)\cos\beta - p(b)f'(b)\sin\beta = 0\},$$

where $C^2(a,b;\mathbb{C})$ denotes the set of all complex-valued functions which are twice continuously differentiable on the interval $\mathcal{I}(a,b) = \{x : a \leq x \leq b\}$. Also, define an *inner product* (f,g) for two complex-valued continuous functions f and g on $\mathcal{I}(a,b)$ by

$$(f,g) \;=\; \int_a^b f(\xi)\overline{g(\xi)}d\xi,$$

where $\overline{g(\xi)}$ denotes the complex conjugate of $g(\xi)$. Since

$$(f,\mathcal{L}[g]) \;=\; \int_a^b f(\xi)\left[\frac{d}{d\xi}\left(p(\xi)\frac{d\overline{g}}{d\xi}\right) \;+\; u(\xi)\overline{g(\xi)}\right]d\xi$$
$$=\; p(b)f(b)\overline{g'(b)} \;-\; p(a)f(a)\overline{g'(a)} \;+\; \int_a^b\left[-p(\xi)f'(\xi)\overline{g'(\xi)} \;+\; u(\xi)f(\xi)\overline{g(\xi)}\right]d\xi,$$

for $f \in \mathcal{U}(a,b)$ and $g \in \mathcal{U}(a,b)$, we obtain

$$(f,\mathcal{L}[g]) \;-\; (\mathcal{L}[f],g)$$
$$=\; \{p(b)f(b)\overline{g'(b)} \;-\; p(b)\overline{g(b)}f'(b)\} \;-\; \{p(a)f(a)\overline{g'(a)} \;-\; p(a)\overline{g(a)}f'(a)\}.$$

Also, since

$$\left\{\begin{array}{ll} f(a)\cos\alpha \;-\; p(a)f'(a)\sin\alpha \;=\; 0, & f(b)\cos\beta \;-\; p(b)f'(b)\sin\beta \;=\; 0, \\[4pt] \overline{g(a)}\cos\alpha \;-\; p(a)\overline{g'(a)}\sin\alpha \;=\; 0, & \overline{g(b)}\cos\beta \;-\; p(b)\overline{g'(b)}\sin\beta \;=\; 0 \end{array}\right.$$

for $f \in \mathcal{U}(a,b)$ and $g \in \mathcal{U}(a,b)$, we obtain

$$p(a)f(a)\overline{g'(a)} \;-\; p(a)\overline{g(a)}f'(a) \;=\; 0, \qquad p(b)f(b)\overline{g'(b)} \;-\; p(b)\overline{g(b)}f'(b) \;=\; 0.$$

Thus, we extended Theorem VI-2-5 as follows.

Theorem VI-3-4 (self-adjointness). *The operator $\mathcal{L}$ has the following property:*

$$(f,\mathcal{L}[g]) = (\mathcal{L}[f],g) \quad for \quad f \in \mathcal{U}(a,b) \quad and \quad g \in \mathcal{U}(a,b).$$

Theorem VI-3-4 implies the following conclusion.

Theorem VI-3-5. *All eigenvalues of (EP) are real.*

Proof.

Let λ be an eigenvalue of (EP) and let $y(x)$ be an eigenfunction corresponding to λ. We may assume that $(y, y) = 1$. Then,

$$(y, \mathcal{L}[y]) \; = \; (y, \lambda y) \; = \; \overline{\lambda}(y, y) \; = \; \overline{\lambda},$$

$$(\mathcal{L}[y], y) \; = \; (\lambda y, y) \; = \; \lambda(y, y) \; = \; \lambda,$$

and Theorem VI-3-4 imply that $\overline{\lambda} = \lambda$. $\quad\square$

In order to explain distribution of eigenvalues on the real line $\mathbb{R}$, look at the basic comparison theorem (Theorem VI-1-1) more closely. Let us rewrite the differential equation

$$(\text{VI.3.9}) \qquad\qquad \frac{d}{dx}\left(p(x)\frac{dy}{dx}\right) \; + \; g(x)y \; = \; 0$$

in the form

$$p(x)\frac{dr}{dx} \; = \; (1 \; - \; p(x)g(x))r\sin(\theta)\cos(\theta)$$

and

$$(\text{VI.3.10}) \qquad\qquad p(x)\frac{d\theta}{dx} \; = \; 1 \; + \; (p(x)g(x) \; - \; 1)\sin^2(\theta)$$

by setting

$$y \; = \; r\sin(\theta), \qquad p(x)\frac{dy}{dx} \; = \; r\cos(\theta).$$

The right-hand side of (VI.3.10) is independent of r. Therefore, (VI.3.10) is a first-order nonlinear differential equation on θ. Our main concern is the behavior of solutions of (VI.3.10).

Remark VI-3-6. Set $w = \dfrac{1}{y}\left(p(x)\dfrac{dy}{dx}\right)$. Then, differential equation (VI.3.9) becomes $p(x)\dfrac{dw}{dx} + w^2 + p(x)g(x) = 0$. This equation is further changed to (VI.3.10) by the transformation $w = \cot(\theta)$.

Remark VI-3-7. If the function $g(x)$ is continuous on the interval $\mathcal{I}(a, b)$, then every solution $\theta(x)$ of (VI.3.10) exists on $\mathcal{I}(a, b)$.

Now, we prove the following basic lemma.

Lemma VI-3-8. *Assume that*
(1) four functions $g_1(x)$, $g_2(x)$, $p_1(x)$, and $p_2(x)$ are continuous on the interval $\mathcal{I}(a, b)$,
(2) $g_2(x) \geq g_1(x)$ and $p_1(x) \geq p_2(x) > 0$ on $\mathcal{I}(a, b)$,
(3) $\theta_j(x)$ $(j = 1, 2)$ are solutions of the differential equations

$$p_j(x)\frac{d\theta_j}{dx} \; = \; 1 \; + \; [p_j(x)g_j(x) \; - \; 1]\sin^2(\theta_j) \qquad (j = 1, 2),$$

respectively.

Then,

$$\theta_2(x) \geq \theta_1(x) \qquad on \quad \mathcal{I}(a,b) \qquad if \quad \theta_2(a) \geq \theta_1(a).$$

Proof.

Set

$$\begin{cases} A(x,\theta_1,\theta_2) = \left(\dfrac{p_1(x)g_1(x) - 1}{p_1(x)}\right)\left(\dfrac{\sin^2(\theta_2) - \sin^2(\theta_1)}{\theta_2 - \theta_1}\right), \\[2mm] h(x,\theta_2) = (g_2(x) - g_1(x))\sin^2(\theta_2) + \left(\dfrac{1}{p_2(x)} - \dfrac{1}{p_1(x)}\right)\cos^2(\theta_2). \end{cases}$$

Then, the function $\omega(x) = \theta_2(x) - \theta_1(x)$ satisfies the differential equation

$$\frac{d\omega}{dx} - A(x,\theta_1(x),\theta_2(x))\,\omega = h(x,\theta_2(x)) \geq 0.$$

Therefore,

$$\omega(x)\exp\left[-\int_a^x A(\xi,\theta_1(\xi),\theta_2(\xi))d\xi\right]$$
$$\geq \omega(a) + \int_a^x h(s,\theta_2(s))\exp\left[-\int_a^s A(\xi,\theta_1(\xi),\theta_2(\xi))d\xi\right]ds \geq 0$$

on $\mathcal{I}(a,b)$. $\quad\square$

Remark VI-3-9. The proof given above also showed that if $p_1(x) \geq p_2(x) > 0$ on $\mathcal{I}(a,b)$, then we obtain

(α) $\theta_2(x) > \theta_1(x)$ on $\mathcal{I}(a,b)$ if $\theta_2(a) > \theta_1(a)$,

(β) $\theta_2(x) > \theta_1(x)$ for $a < x \leq b$ if $\theta_2(a) \geq \theta_1(a)$ and $g_2(x) > g_1(x)$ on $a < x < b$,

(γ) in the case when $g_2(x) > g_1(x)$ on $a < x < b$, if $\theta_1(a) = 0$ and $\theta_1(b) = \pi$ and if $0 \leq \theta_2(a) < \pi$, then $\theta_2(b) > \pi$.

Also, $y = r\sin(\theta) > 0$ if and only if $0 < \theta < \pi$ (mod 2π) and $p(x)\dfrac{d\theta}{dx} = 1 > 0$ at $\theta = 0$ (mod π). Theorem VI-1-1 follows from (γ).

Setting $g(x,\lambda) = u(x) - \lambda$, apply Lemma VI-3-8 to problem (EP).

Lemma VI-3-10. *Let $\theta(x,\lambda)$ be the unique solution to the initial-value problem*

$$p(x)\frac{d\theta}{dx} = 1 + (p(x)(u(x) - \lambda) - 1)\sin^2(\theta), \qquad \theta(a,\lambda) = \alpha$$

on the interval $\mathcal{I}(a,b)$, where $u(x)$ is continuous on $\mathcal{I}(a,b)$, $p(x)$ is continuous and positive on $\mathcal{I}(a,b)$, λ is a real parameter, and α is a fixed real number such that $0 \leq \alpha < \pi$. Then, for every $c \in \mathcal{I}(a,b)$ such that $c > a$,

(i) $\theta(c,\lambda)$ *is a continuous and strictly decreasing function of λ for $-\infty < \lambda < +\infty$,*

(ii) $\lim\limits_{\lambda \to -\infty} \theta(c,\lambda) = +\infty$,

(iii) $\lim\limits_{\lambda \to +\infty} \theta(c,\lambda) = 0$.

Proof.

We prove this lemma in three steps.

Step 1. To prove (i), apply Lemma VI-3-8 to $p_1(x) = p_2(x) = p(x)$, $g_1(x) = g(x, \lambda_1)$, and $g_2(x) = g(x, \lambda_2)$. If $\lambda_2 < \lambda_1$, then $g_2(x) > g_1(x)$ on $\mathcal{I}(a, b)$. Hence, $\theta(x, \lambda_2) > \theta(x, \lambda_1)$ on $a < x \leq b$, since $\theta(a, \lambda_2) = \theta(a, \lambda_1) = \alpha$.

Step 2. To prove (ii), choosing a real number m and a positive number P so that $u(x) \geq m$ and $p(x) \leq P$ on $\mathcal{I}(a, b)$, determine $\theta_0(x, \lambda)$ by the initial-value problem

$$P\frac{d\theta_0}{dx} = 1 + (P(m - \lambda) - 1)\sin^2(\theta_0), \qquad \theta_0(a, \lambda) = 0.$$

Then, since $g(x, \lambda) \geq m - \lambda$ and $P \geq p(x) > 0$ for $x \in \mathcal{I}(a, b)$, $-\infty < \lambda < +\infty$ and $\alpha \geq 0$, it follows from Lemma VI-3-8 that

$$\theta(c, \lambda) \geq \theta_0(c, \lambda) \qquad \text{for} \quad a \leq c \leq b \quad \text{and} \quad -\infty < \lambda < +\infty.$$

Observe that $v = P\tan(\theta_0)$ satisfies the differential equation $\dfrac{dv}{dx} = 1 + \left(\dfrac{m - \lambda}{P}\right)v^2$.

Hence, $\tan(\theta_0(x, \lambda)) = \dfrac{1}{\sqrt{P(m - \lambda)}}\tan\left(\sqrt{\dfrac{m - \lambda}{P}}\,(x - a)\right)$ for $\lambda < m$. Note that

$\tan(\theta_0(x, \lambda)) = 0$ if and only if $\tan\left(\sqrt{\dfrac{m - \lambda}{P}}\,(x - a)\right) = 0$, i.e.,

$$\tan(\theta_0(x, \lambda)) = 0 \qquad \text{at} \qquad \sqrt{\dfrac{m - \lambda}{P}}\,(x - a) = n\pi \quad \text{for} \quad n = 1, 2, \ldots$$

and that $\theta_0(x, \lambda)$ is strictly decreasing with respect to λ if $x > a$ (cf. Step 1). Furthermore, since $\tan(\theta_0(x, m)) = \dfrac{x - a}{P}$, we must have $0 < \theta_0(x, m) < \dfrac{\pi}{2}$ for $a < x$. Therefore,

$$\theta_0\left(x,\ m - P\left(\frac{n\pi}{x - a}\right)^2\right) = n\pi \qquad \text{for} \quad n = 1, 2, \ldots\,.$$

Hence,

$$\theta\left(x,\ m - P\left(\frac{n\pi}{x - a}\right)^2\right) \geq n\pi \qquad \text{for} \quad n = 1, 2, \ldots\,.$$

Thus, we conclude that $\lim\limits_{\lambda \to -\infty} \theta(c, \lambda) = +\infty$ if $c > a$.

Step 3. To prove (iii), first notice that $\theta(c, \lambda) > 0$ for $-\infty < \lambda < +\infty$, since $\theta(a, \lambda) = \alpha \geq 0$ and $\dfrac{d\theta}{dx} = \dfrac{1}{p(x)} > 0$ if $\theta = 0$. Choose a positive number M so that

$$\left|u(x)\sin^2(\theta) + \frac{1 - \sin^2(\theta)}{p(x)}\right| \leq M \qquad \text{on} \quad \mathcal{I}(a, b).$$

Then,

$$\frac{d\theta}{dx} \leq M - \lambda\sin^2(\theta) \qquad \text{on} \quad \mathcal{I}(a, b).$$

For any positive number δ such that $0 \le \alpha < \pi - \delta$, fix another positive number $\Lambda(\delta)$ so that

$$M \,-\, \lambda \sin^2(\theta) \;<\; -\,\frac{10}{c-a} \qquad \text{for} \qquad \delta \le \theta \le \pi - \delta \quad \text{and} \quad \lambda \ge \Lambda(\delta).$$

This implies that $\dfrac{d\theta}{dx} < 0$ if $\theta = \delta$. Hence, $\theta(x,\lambda) > \delta$ for $a \le x \le c$ if $\theta(c,\lambda) > \delta$, whereas $\theta(c,\lambda) < \alpha - 10 < \pi - 10 < 0$ if $\theta(x,\lambda) > \delta$ for $a \le x \le c$. This is a contradiction. Therefore, $0 < \theta(c,\lambda) \le \delta$ for $\lambda \ge \Lambda(\delta)$. Thus, we conclude that $\lim\limits_{\lambda \to +\infty} \theta(c,\lambda) = 0$. $\square$

Now, using Lemma VI-3-10, we prove the following theorem concerning problem (EP).

Theorem VI-3-11. *Assume that $u(x)$ is continuous on the interval $\mathcal{I}(a.b) = \{x : a \le x \le b\}$ and that $p(x)$ is continuously differentiable and positive on $\mathcal{I}(a,b)$. Then,*

(1) problem (EP) has infinitely many eigenvalues:

$$\lambda_1 > \lambda_2 > \cdots > \lambda_n > \cdots,$$

(2) $\lim\limits_{n \to +\infty} \lambda_n = -\infty,$

(3) every eigenfunction $\phi_n(x)$ corresponding to the eigenvalue λ_n has exactly $n-1$ zeros on the open interval $a < x < b$.

Proof.

Assume without any loss of generality that $0 \le \alpha < \pi$ and $0 < \beta \le \pi$. Define $\theta(x,\lambda)$ by the initial-value problem

$$(\text{VI.3.11}) \qquad p(x)\frac{d\theta}{dx} \;=\; 1 \,+\, (p(x)(u(x) - \lambda) \,-\, 1)\sin^2(\theta), \qquad \theta(a,\lambda) \;=\; \alpha,$$

and, then, define $r(x,\lambda)$ by

$$(\text{VI.3.12}) \qquad p(x)\frac{dr}{dx} \;=\; (1 \,-\, p(x)(u(x) - \lambda))r\sin(\theta)\cos(\theta), \qquad r(a,\lambda) = 1.$$

Then,

$$(\text{VI.3.13}) \qquad\qquad y \;=\; \phi(x,\lambda) \;=\; r(x,\lambda)\sin(\theta(x,\lambda))$$

is a nontrivial solution of the differential equation

$$(\text{VI.3.14}) \qquad\qquad \frac{d}{dx}\left(p(x)\frac{dy}{dx}\right) \,+\, u(x)y \;=\; \lambda y$$

that satisfies the condition $y(a)\cos(\alpha) - p(a)\dfrac{dy}{dx}(a)\sin(\alpha) = 0$. Note that from (VI.3.11), (VI.3.12), and (VI.3.13), it follows that $p(x)\dfrac{d\phi(x,\lambda)}{dx} = r(x,\lambda)\cos(\theta)(x,\lambda)$.

The eigenvalues of (EP) are determined by the condition $\theta(b, \lambda) = \beta$ (mod π). Since $\theta(b, \lambda)$ is strictly decreasing as $\lambda \to +\infty$ and since $\theta(b, \lambda)$ takes all positive values (cf. Lemma VI-3-10), the eigenvalues λ_n are determined by

$$\theta(b, \lambda_n) \;=\; \beta \,+\, (n-1)\pi, \qquad n = 1, 2, \ldots .$$

Observe that

(I) $\phi(x, \lambda) = 0$ if and only if $\theta(x, \lambda) = 0$ (mod π),

(II) $\dfrac{d\theta}{dx} = \dfrac{1}{p(x)} > 0$ if $\theta = 0$ (mod π).

This implies that $\phi(x, \lambda)$ has exactly k zeros on the open interval $a < x < b$ if and only if

$$k\pi \;<\; \theta(b, \lambda) \;\leq\; (k+1)\pi.$$

Observe also that

$$(n-1)\pi \;<\; \beta \,+\, (n-1)\pi \;\leq\; n\pi.$$

Hence, the eigenfunction $\phi_n(x) = \phi(x, \lambda_n)$ has $n - 1$ zeros on the open interval $a < x < b$. $\square$

Observation VI-3-12. As shown earlier, if $u(x) \geq m$ and $p(x) \leq P$ on $\mathcal{I}(a.b)$, then

$$\theta\left(b, \; m \,-\, P\left(\frac{n\pi}{b-a}\right)^2\right) \;\geq\; n\pi \qquad \text{for} \quad n = 1, 2, \ldots .$$

This implies that

$$(VI.3.15) \qquad \lambda_n \;\geq\; m \,-\, P\left(\frac{n\pi}{b-a}\right)^2, \qquad n \;=\; 1,\, 2, \ldots .$$

Observation VI-3-13. If $u(x) \leq \mu$ and $p(x) \geq \rho > 0$ on $\mathcal{I}(a.b)$, determine $\theta_1(x, \lambda)$ by the initial-value problem

$$\rho\frac{d\theta_1}{dx} \;=\; 1 \,+\, (\rho(\mu - \lambda) \,-\, 1)\sin^2(\theta_1), \qquad \theta_1(a, \lambda) \;=\; \pi.$$

Then, since $u(x) - \lambda \leq \mu - \lambda$ and $p(x) \geq \rho > 0$ for $x \in \mathcal{I}(a, b)$, $-\infty < \lambda < +\infty$, and $\alpha < \pi$, it follows from Lemma VI-3-8 that

$$\theta(c, \lambda) \;<\; \theta_1(c, \lambda) \qquad \text{for} \quad a \,<\, c \,\leq\, b \quad \text{and} \quad -\infty \,<\, \lambda \,<\, +\infty.$$

Observe that $v = \rho\tan(\theta_1)$ satisfies the differential equation $\dfrac{dv}{dx} = 1 + \left(\dfrac{\mu - \lambda}{\rho}\right)v^2$. Hence,

$$\tan(\theta_1(x, \lambda)) \;=\; \frac{1}{\sqrt{\rho(\mu - \lambda)}} \, \tan\left(\sqrt{\frac{\mu - \lambda}{\rho}}\,(x - a)\right) \qquad \text{for} \quad \lambda < \mu.$$

Note that $\tan(\theta_1(x, \lambda)) = 0$ at $\sqrt{\dfrac{\mu - \lambda}{\rho}}\,(x - a) = n\pi$ for $n = 1, 2, \dots$ and that $\theta_1(x, \lambda)$ is strictly decreasing with respect to λ if $x > a$. Furthermore, $\pi < \theta_1(x, \mu) < \dfrac{3\pi}{2}$ for $a < x$ since $\tan(\theta_1(x, \mu)) = \dfrac{x - a}{\rho}$. Therefore,

$$\theta_1\left(x,\ \mu\ -\ \rho\left(\frac{n\pi}{x - a}\right)^2\right)\ =\ (n+1)\pi \qquad \text{for}\quad n = 1, 2, \dots .$$

Hence,

$$\theta\left(x,\ \mu\ -\ \rho\left(\frac{n\pi}{x - a}\right)^2\right)\ \le\ (n+1)\pi \qquad \text{for}\quad n = 1, 2, \dots .$$

This implies that

$$(\text{VI.3.16}) \qquad\qquad \lambda_{n+2}\ <\ \mu\ -\ \rho\left(\frac{n\pi}{b - a}\right)^2 \qquad \text{for} \qquad n = 1, 2, \dots .$$

If Λ_1 and Λ_2 are determined by

$$(\text{VI.3.17}) \qquad\qquad \theta_1(b, \Lambda_1)\ =\ \beta \qquad \text{and} \qquad \theta_1(b, \Lambda_2)\ =\ \beta\ +\ \pi,$$

then $\lambda_1 \le \Lambda_1$ and $\lambda_2 \le \Lambda_2$. From (VI.3.15) and (VI.3.16) we derive the following result concerning distribution of eigenvalues of (EP).

Theorem VI-3-14. *For $n \ge 3$, eigenvalues λ_n satisfy the following estimates:*

$$(i) \qquad \left|\frac{\lambda_n}{n^2} + P\left(\frac{\pi}{b - a}\right)^2\right|\ \le\ \frac{|m|}{n^2} \qquad \text{if} \qquad \frac{\lambda_n}{n^2}\ \le\ -P\left(\frac{\pi}{b - a}\right)^2,$$

and

$$(ii) \qquad \left|\frac{\lambda_n}{n^2} + \rho\left(\frac{\pi}{b - a}\right)^2\right|\ \le\ \frac{|\mu|}{n^2}\ +\ 4\rho\left(\frac{\pi}{b - a}\right)^2\left(1 - \frac{1}{n}\right)\frac{1}{n}$$

$$\text{if} \qquad \frac{\lambda_n}{n^2}\ \ge\ -\rho\left(\frac{\pi}{b - a}\right)^2,$$

where m, μ, P and ρ are real numbers such that

$$P\ \ge\ p(x)\ \ge\ \rho\ >\ 0 \qquad \text{and} \qquad \mu\ \ge\ u(x)\ \ge\ m \qquad \text{for} \qquad x\ \in\ \mathcal{I}(a, b).$$

Remark VI-3-15. Lemma VI-3-8, Lemma VI-10, and Theorem VI-3-11 are also found in [CL, §§1 and 2 of Chapter 8] and [Har2, Chapter XI].

VI-4. Eigenfunction expansions

Let us define a differential operator $\mathcal{L}[y] = \dfrac{d}{dx}\left(p(x)\dfrac{dy}{dx}\right) + u(x)y$ and the vector space $\mathcal{V}(a,b)$ over the real number field $\mathbb{R}$ in the same way as in §VI-2. Also, as in §VI-2, define the inner product (f,g) for two real-valued continuous functions f and g on $\mathcal{I}(a,b) = \{x : a \leq x \leq b\}$ by $(f,g) = \displaystyle\int_a^b f(\xi)g(\xi)d\xi$. It is known that the operator $\mathcal{L}$ has the following property:

$$(f, \mathcal{L}[g]) = (\mathcal{L}[f], g) \qquad \text{for} \qquad f \in \mathcal{V}(a,b) \quad \text{and} \quad g \in \mathcal{V}(a,b)$$

(cf. Theorem VI-2-5). Define a norm of a continuous function on $\mathcal{I}(a,b)$ by $\|f\| = \sqrt{(f,f)}$. Then, $|(f,g)| \leq \|f\|\|g\|$. The following theorem is a basic result in the theory of self-adjoint boundary-value problems.

Theorem VI-4-1. *Let λ_1 and λ_2 be two distinct eigenvalues of the problem*

$$\text{(EP)} \qquad\qquad \mathcal{L}[y] = \lambda y, \qquad y \in \mathcal{V}(a,b).$$

Let $\eta_1(x)$ and $\eta_2(x)$ be eigenfunctions corresponding to λ_1 and λ_2, respectively. Then, $(\eta_1, \eta_2) = 0$.

Proof.

Note that $(\eta_1, \mathcal{L}[\eta_2]) = \lambda_2(\eta_1, \eta_2)$ and $(\mathcal{L}[\eta_1], \eta_2) = \lambda_1(\eta_1, \eta_2)$. This implies that $\lambda_2(\eta_1, \eta_2) = \lambda_1(\eta_1, \eta_2)$. Therefore, $(\eta_1, \eta_2) = 0$ since $\lambda_1 \neq \lambda_2$. $\square$

It is known that problem (EP) has real eigenvalues

$$\lambda_1 > \lambda_2 > \cdots > \lambda_n > \cdots \qquad \left(\lim_{n \to +\infty} \lambda_n = -\infty\right).$$

Let $\eta_1(x), \eta_2(x), \ldots, \eta_n(x), \ldots$ be the eigenfunctions corresponding to $\lambda_1, \lambda_2, \ldots$ such that $\|\eta_n\| = 1$ $(n = 1, 2, \ldots)$. Theorem VI-4-1 implies that $(\eta_h, \eta_k) = 0$ if $h \neq k$. From these properties of the eigenfunctions η_n, we obtain

$$\left\|f - \sum_{h=1}^{k}(f, \eta_h)\eta_h\right\|^2 = \|f\|^2 - \sum_{h=1}^{k}(f, \eta_h)^2 \geq 0$$

and hence

$$\sum_{h=1}^{+\infty}(f, \eta_h)^2 \leq \|f\|^2 \qquad\qquad \text{(the \textit{Bessel inequality})}$$

for any continuous function f on $\mathcal{I}(a,b)$.

In this section, we explain the generalized Fourier expansion of a function $f(x)$ in terms of the orthonormal sequence $\{\eta_n(x) : n = 1, 2, \ldots\}$. As a preparation, we prove the following theorem.

Theorem VI-4-2. *Let μ_0 not be an eigenvalue of (EP) and let $f(x)$ be a continuous function on $\mathcal{I}(a, b)$. Then, the series*

$$\sum_{h=1}^{+\infty} \frac{(f, \eta_h)\eta_h(x)}{\lambda_h - \mu_0}$$

is uniformly convergent on the interval $\mathcal{I}(a, b)$.

Proof.

Define a linear differential operator $\mathcal{L}_0$ by $\mathcal{L}_0[y] = \mathcal{L}[y] - \mu_0 y$. Then, there exists Green's function of the boundary-value problem

$$\text{(BP)} \qquad\qquad \mathcal{L}_0[y] \;=\; f(x), \qquad y \;\in\; \mathcal{V}(a, b)$$

such that the unique solution of problem (BP) is given by

$$y(x) \;=\; \int_a^b G(x, \xi) f(\xi) d\xi \;=\; (G(x, \cdot), f)$$

(cf. §VI-2; in particular, Observation VI-2-3). Define the operator $\mathcal{G}[f]$ by

$$\mathcal{G}[f] \;=\; (G(x, \cdot), f) \;=\; \int_a^b G(x, \xi) f(\xi) d\xi$$

for any continuous function $f(x)$ on $\mathcal{I}(a, b)$. Since $\mathcal{L}_0[\mathcal{G}[f]] = f$, we obtain

$$(f, \mathcal{G}[g]) = (\mathcal{L}_0[\mathcal{G}[f]], \mathcal{G}[g]) = (\mathcal{G}[f], \mathcal{L}_0[\mathcal{G}[g]]) = (\mathcal{G}[f], g)$$

for any continuous functions $f(x)$ and $g(x)$ on $\mathcal{I}(a, b)$. Also, since $\mathcal{L}_0[\eta_h] = \mathcal{L}[\eta_h] - \mu_0\eta_h = (\lambda_h - \mu_0)\eta_h$, we obtain $\mathcal{G}[\eta_h] = \dfrac{\eta_h}{\lambda_h - \mu_0}$. Hence,

$$\left| \sum_{h=\ell_1}^{\ell_2} \frac{(f, \eta_h)\eta_h(x)}{\lambda_h - \mu_0} \right| = \left| \mathcal{G}\left[\sum_{h=\ell_1}^{\ell_2} (f, \eta_h)\eta_h \right] \right|$$

$$\leq K \left\| \sum_{h=\ell_1}^{\ell_2} (f, \eta_h)\eta_h \right\| = K \sqrt{\sum_{h=\ell_1}^{\ell_2} (f, \eta_h)^2},$$

where K is a positive constant such that $\|G(x, \cdot)\| \leq K$ on $\mathcal{I}(a, b)$. Finally,

the Bessel inequality implies that $\displaystyle\lim_{\ell_1, \ell_2 \to +\infty} \left| \sum_{h=\ell_1}^{\ell_2} \frac{(f, \eta_h)\eta_h(x)}{\lambda_h - \mu_0} \right| = 0$ uniformly on

$\mathcal{I}(a, b)$. $\square$

Now, we claim that the limit of the uniformly convergent series of Theorem VI-4-2 is equal to $\mathcal{G}[f]$.

Theorem VI-4-3. *For any continuous function $f(x)$ on $\mathcal{I}(a,b)$, the sum $\displaystyle\sum_{h=1}^{+\infty} \frac{(f,\eta_h)\eta_h}{\lambda_h - \mu_0}$ is equal to $\mathcal{G}[f]$.*

Proof.

Set $\displaystyle \mathcal{G}_\ell[f] = \mathcal{G}[f] - \sum_{h=1}^{\ell} \frac{(f,\eta_h)\eta_h}{\lambda_h - \mu_0}$. Then, it suffices to show

$$(\mathrm{VI.4.1}) \qquad \lim_{\ell \to +\infty} \|\mathcal{G}_\ell[f]\| \;=\; 0$$

for every continuous function $f(x)$ on $\mathcal{I}(a,b)$. We prove (VI.4.1) in four steps. Without any loss of generality, we assume that $\lambda_h - \mu_0 < 0$ for $h > \ell$. Also, note that

$$(\mathrm{VI.4.2}) \qquad (f, \mathcal{G}_\ell[g]) = (\mathcal{G}_\ell[f], g)$$

for any real-valued continuous functions f and g on $\mathcal{I}(a,b)$.

Step 1. It can be shown easily that $\mathcal{G}_\ell[\eta_h] = c_h \eta_h$, where

$$c_h \;=\; \begin{cases} 0 & \text{for} \quad h = 1, 2, \ldots, \ell, \\[2mm] \dfrac{1}{\lambda_h - \mu_0} & \text{for} \quad h > \ell. \end{cases}$$

Let γ be an eigenvalue of $\mathcal{G}_\ell$ and let $\phi(x)$ be an eigenfunction of $\mathcal{G}_\ell$ associated with γ. Then, $\phi \in \mathcal{V}(a,b)$, $\|\phi\| \neq 0$, and $\displaystyle \gamma\phi = \mathcal{G}[\phi] - \sum_{h=1}^{\ell} \frac{(\phi,\eta_h)\eta_h}{\lambda_h - \mu_0}$. Also, (VI.4.2) implies that $(\phi, \mathcal{G}_\ell[\eta_h]) = (\mathcal{G}_\ell[\phi], \eta_h)$. Hence, $c_h(\phi, \eta_h) = \gamma(\phi, \eta_h)$. Consequently, $(\phi, \eta_h) = 0$ if $\gamma \neq c_h$. Therefore, $(\phi, \eta_h) = 0$ for $h = 1, \ldots, \ell$ and $\gamma\phi = \mathcal{G}[\phi]$ if $\gamma \neq 0$. Hence, either $\gamma = 0$ or $\gamma = c_h$ $(h > \ell)$.

Step 2. In this step, we prove that

$$(\mathrm{VI.4.3}) \qquad \sup_{\|f\|=1} \|\mathcal{G}_\ell[f]\| \;=\; \sup_{\|f\|=1} |(\mathcal{G}_\ell[f], f)|.$$

In fact, the inequality $|(\mathcal{G}_\ell[f], f)| \leq \|\mathcal{G}_\ell[f]\|\|f\|$ implies $\displaystyle \sup_{\|f\|=1} \|\mathcal{G}_\ell[f]\| \geq \sup_{\|f\|=1} |(\mathcal{G}_\ell[f], f)|$. Also, since $(\mathcal{G}_\ell[f \pm g], f \pm g) = (\mathcal{G}_\ell[f], f) + (\mathcal{G}_\ell[g], g) \pm 2(\mathcal{G}_\ell[f], g)$, we obtain

$$\begin{aligned} 4(\mathcal{G}_\ell[f], g) \;&=\; (\mathcal{G}_\ell[f+g], f+g) - (\mathcal{G}_\ell[f-g], f-g) \\[2mm] &\leq\; \left(\sup_{\|w\|=1} |(\mathcal{G}_\ell[w], w)| \right) \left(\|f+g\|^2 + \|f-g\|^2 \right) \\[2mm] &=\; 2 \left(\sup_{\|w\|=1} |(\mathcal{G}_\ell[w], w)| \right) \left(\|f\|^2 + \|g\|^2 \right). \end{aligned}$$

Suppose that $\mathcal{G}_\ell[f] \neq 0$ and $\|f\|\| = 1$, and set $g = \dfrac{\mathcal{G}_\ell[f]}{\|\mathcal{G}_\ell[f]\|}$. Then, $\|\mathcal{G}_\ell[f]\| \leq$

$\sup\limits_{\|w\|=1} |(\mathcal{G}_\ell[w], w)|$. This shows that $\sup\limits_{\|f\|=1} \|\mathcal{G}_\ell[f]\| \leq \sup\limits_{\|f\|=1} |(\mathcal{G}_\ell[f], f)|$. Thus, the

proof of (VI.4.3) is completed.

Step 3. In this step, we prove that $-\sup\limits_{\|f\|=1} |(\mathcal{G}_\ell[f], f)|$ is an eigenvalue of $\mathcal{G}_\ell$. To

do this, set $c = \sup\limits_{\|f\|=1} |(\mathcal{G}_\ell[f], f)|$. Note first that $c \geq |(\mathcal{G}_\ell[\eta_{\ell+1}], \eta_{\ell+1})| = |c_{\ell+1}| >$

0. Also, note that $c = \sup\limits_{\|f\|=1} (\mathcal{G}_\ell[f], f)$ or $-c = \inf\limits_{\|f\|=1} (\mathcal{G}_\ell[f], f)$. Suppose that

$c = \sup\limits_{\|f\|=1} (\mathcal{G}_\ell[f], f)$. Then, there exists a sequence f_j $(j = 1, 2, \dots)$ of continuous

functions on $\mathcal{I}(a, b)$ such that $\lim\limits_{j \to +\infty} (\mathcal{G}_\ell[f_j], f_j) = c$ and $\|f_j\| = 1$ $(j = 1, 2, \dots)$.

Since $\{\mathcal{G}_\ell[f_j] : j = 1, 2, \dots\}$ is bounded and equicontinuous on $\mathcal{I}(a, b)$, assume

without any loss of generality that $\lim\limits_{j \to +\infty} \mathcal{G}_\ell[f_j] = g \in \mathcal{G}(a, b)$ uniformly on $\mathcal{I}(a, b)$.

Let us look at

$$(\text{VI.4.4}) \quad \|\mathcal{G}_\ell[f_j] - cf_j\|^2 = \|\mathcal{G}_\ell[f_j]\|^2 + c^2 - 2c(\mathcal{G}_\ell[f_j], f_j) \leq 2c^2 - 2c(\mathcal{G}_\ell[f_j], f_j).$$

This implies that $\lim\limits_{j \to +\infty} \|\mathcal{G}_\ell[f_j] - cf_j\| = 0$ and, hence, $\lim\limits_{j \to +\infty} \|g - cf_j\| = 0$. Thus,

$\mathcal{G}_\ell[g] = cg$. Also, from (VI.4.4), it follows that $\|g\|^2 = c^2 > 0$. Therefore, c must be

an eigenvalue of $\mathcal{G}_\ell$. However, since all eigenvalues of $\mathcal{G}_\ell$ are nonpositive, this is a

contradiction. Therefore, $-c = \inf\limits_{\|f\|=1} (\mathcal{G}_\ell[f], f)$. We can now prove in a similar way

that

$$-c = c_{\ell+1} = \frac{1}{\lambda_{\ell+1} - \mu_0}.$$

Step 4. Since $\sup\limits_{\|f\|=1} \|\mathcal{G}_\ell[f]\| = \dfrac{1}{\mu_0 - \lambda_{\ell+1}}$, we conclude that $\lim\limits_{\ell \to +\infty} \sup\limits_{\|f\|=1} \|\mathcal{G}_\ell[f]\| = 0$

and the proof of (VI.4.1) is completed. $\square$

The following theorem is the basic result concerning eigenfunction expansion.

Theorem VI-4-4. *For every $f \in \mathcal{V}(a, b)$, the series $f = \sum\limits_{h=1}^{+\infty} (f, \eta_h)\eta_h$ converges to*

f uniformly on $\mathcal{I}(a, b)$.

Proof.

Set $g = \mathcal{L}_0[f]$, then $f = \mathcal{G}[g]$. Therefore,

$$f = \sum_{h=1}^{+\infty} \frac{(g, \eta_h)\eta_h}{\lambda_h - \mu_0} = \sum_{h=1}^{+\infty} \frac{(\mathcal{L}_0[f], \eta_h)\eta_h}{\lambda_h - \mu_0}$$

$$= \sum_{h=1}^{+\infty} \frac{(f, \mathcal{L}_0[\eta_h])\eta_h}{\lambda_h - \mu_0} = \sum_{h=1}^{+\infty} (f, \eta_h)\eta_h. \qquad \square$$

For every continuous function f on $\mathcal{I}(a,b)$ and a positive number ϵ, there exists an $f_\epsilon \in \mathcal{V}(a,b)$ such that $\|f - f_\epsilon\| \leq \epsilon$. Since

$$f - \sum_{h=1}^{\ell}(f,\eta_h)\eta_h = f - f_\epsilon + f_\epsilon - \sum_{h=1}^{\ell}(f_\epsilon,\eta_h)\eta_h + \sum_{h=1}^{\ell}(f_\epsilon - f,\eta_h)\eta_h,$$

we obtain $\displaystyle\lim_{\ell\to+\infty}\left\|f - \sum_{h=1}^{\ell}(f,\eta_h)\eta_h\right\| = 0$. Therefore, $f = 0$ if $(f,\eta_h) = 0$ for $h = 1,2,\ldots$. This proves the following theorem.

Theorem VI-4-5. *If a continuous function f on $\mathcal{I}(a,b)$ satisfies the condition $(f,\eta_h) = 0$ for $h = 1,2,\ldots$, then f is identically equal to zero on the interval $\mathcal{I}(a,b)$.*

Remark VI-4-6. Also, we have

$$\|f\|^2 = \sum_{h=1}^{+\infty}(f,\eta_h)^2 \qquad \text{(the } Parseval \text{ equality).}$$

Furthermore, this, in turn, implies that if $f(x)$ and $g(x)$ are continuous on $\mathcal{I}(a,b)$, then

$$(f,g) = \sum_{h=1}^{+\infty}(f,\eta_h)(g,\eta_h).$$

Example VI-4-7. Using the eigenvalue problem $\dfrac{d^2y}{dx^2} = \lambda y$, $y'(0) = 0$, $y'(\pi) = 0$, we construct an orthogonal sequence $\{\cos(nx) : n = 0,1,2,\ldots\}$ of eigenfunctions. This yields the Fourier cosine series $\dfrac{a_0}{2} + \displaystyle\sum_{n=1}^{\infty}a_n\cos(nx)$ of a function $f(x)$, where $a_n = \dfrac{2}{\pi}\displaystyle\int_0^{\pi} f(x)\cos(nx)ds$. By virtue of Theorem VI-4-4, this series converges uniformly to $f(x)$ on $\mathcal{I}(0,\pi)$ if $f(x)$ is twice continuously differentiable on $\mathcal{I}(0,\pi)$ and $f'(0) = 0$ and $f'(\pi) = 0$.

Example VI-4-8. Using the eigenvalue problem $\dfrac{d^2y}{dx^2} = \lambda y$, $y(0) = 0$, $y(\pi) = 0$, we construct an orthogonal sequence $\{\sin(nx) : n = 1,2,\ldots\}$ of eigenfunctions. This yields the Fourier sine series $\displaystyle\sum_{n=1}^{\infty}b_n\sin(nx)$ of a function $f(x)$, where $b_n = \dfrac{2}{\pi}\displaystyle\int_0^{\pi} f(x)\sin(nx)ds$. By virtue of Theorem VI-4-4, this series converges uniformly to $f(x)$ on $\mathcal{I}(0,\pi)$ if $f(x)$ is twice continuously differentiable on $\mathcal{I}(0,\pi)$ and $f(0) = 0$ and $f(\pi) = 0$.

Example VI-4-9. If $f(x)$ is twice continuously differentiable on $\mathcal{I}(-\pi,\pi)$, $f(-\pi) = f(\pi)$, and $f'(\pi) = f'(-\pi)$, set $f_e(x) = \dfrac{1}{2}[f(x)+f(-x)]$ and $f_o(x) = \dfrac{1}{2}[f(x)-f(-x)]$.

Then, f_e satisfies the conditions of Example VI-4-7, while $f_o(x)$ satisfies the conditions of Example VI-4-8. Thus, we obtain the Fourier series

$$\frac{a_0}{2} + \sum_{n=1}^{\infty} [a_n \cos(nx) + b_n \sin(nx)],$$

where $a_n = \dfrac{1}{\pi} \displaystyle\int_{-\pi}^{\pi} f(x) \cos(nx) ds$ and $b_n = \dfrac{1}{\pi} \displaystyle\int_{-\pi}^{\pi} f(x) \sin(nx) ds$. The Fourier series converges uniformly to $f(x)$ on $\mathcal{I}(-\pi, \pi)$.

Remark VI-4-10. The Fourier series of Example VI-4-9 can be constructed also from the eigenvalue problem $\dfrac{d^2 y}{dx^2} = \lambda y$, $y(-\pi) = y(\pi)$, $y'(-\pi) = y'(\pi)$. Including this eigenvalue problem, more general cases are systematically explained in [CL, Chapters 7 and 11]. For the uniform convergence of the Fourier series, it is not necessary to assume that $f(x)$ be twice continuously differentiable. For example, it suffices to assume that $f(x)$ is continuous and $f'(x)$ is piecewise continuous on $\mathcal{I}(-\pi, \pi)$, and $f(\pi) = f(-\pi)$. For those informations, see [Z].

Remark VI-4-11. The results in §§VI-2, VI-3, and VI-4 can be extended to the case when $p(x)$ is continuous on $\mathcal{I}(a, b)$, $p(x) > 0$ on $a < x < b$, and $p(a)p(b) = 0$ under some suitable assumptions and with some suitable boundary conditions. Although we do not go into these cases in this book, the following example illustrates such a case.

Example VI-4-12. Let us consider the boundary-value problem

$$(\text{VI.4.5}) \qquad \begin{cases} \dfrac{d}{dx}\left(x \dfrac{dy}{dx}\right) + u(x)y = f(x), \\[2mm] y(x) \text{ is bounded in a neighborhood of } x = 0, \quad \text{and} \quad y(1) = 0, \end{cases}$$

where $u(x)$ and $f(x)$ are real-valued and continuous on the interval $\mathcal{I}(0, 1)$.

Step 1. Consider the linear homogeneous equation

$$(\text{VI.4.6}) \qquad \frac{d}{dx}\left(x \frac{dv}{dx}\right) + u(x)v = 0$$

on the interval $\mathcal{I}(0, 1)$. Since 1 and $\log x$ form a fundamental set of solutions of the differential equation $\dfrac{d}{dx}\left(x \dfrac{dy}{dx}\right) = 0$, a general solution of (VI.4.6) can be constructed by solving the integral equation

$$v(x) = c_1 + c_2 \log x + \int_0^x (\log \xi) u(\xi) v(\xi) d\xi - (\log x) \int_0^x u(\xi) v(\xi) d\xi,$$

where c_1 and c_2 are arbitrary constants. In this way, we find two solutions $\phi(x)$ and $\psi(x)$ of (VI.4.6) such that $\phi(x)$ and $\phi'(x)$ are continuous on $\mathcal{I}(0, 1)$, $\lim\limits_{x \to 0^+} \phi(x) = 1$, and $\lim\limits_{x \to 0^+} x\phi'(x) = 0$, whereas $\psi(x)$ and $\psi'(x)$ are continuous for $0 < x \le 1$, $\lim\limits_{x \to 0^+} (\psi(x) - \log x) = 0$, and $\lim\limits_{x \to 0^+} (x\psi'(x) - 1) = 0$. In general, this step of analysis is very important. The behavior of solutions at $x = 0$ determines the nature of eigenvalues of the given problem. Denote also by $\rho(x)$ the unique soltion of (VI.4.6) such that $\rho(1) = 0$ and $\rho'(1) = 1$.

Step 2. Assuming that $\phi(1) \neq 0$, set

$$G(x,\xi) = \begin{cases} \dfrac{\phi(x)\rho(\xi)}{\xi W(\xi)} & \text{for} \quad 0 \leq x \leq \xi \leq 1, \\[3mm] \dfrac{\phi(\xi)\rho(x)}{\xi W(\xi)} & \text{for} \quad 0 \leq \xi \leq x \leq 1, \end{cases}$$

where $W(x)$ is the Wronskian of $\{\phi, \rho\}$ and $xW(x) = \phi(1)$. Then, $G(x,\xi)$ is Green's function of problem (VI.4.5). The unique solution of (VI.4.5) is given by $y(x) = \displaystyle\int_0^1 G(x,\xi)f(\xi)d\xi$. Furthermore,

$$(VI.4.7) \qquad\qquad \int_0^1 \int_0^1 G(x,\xi)^2 dx d\xi < +\infty.$$

Step 3. It is easy to prove the self-adjointness of problem (VI.4.5). Hence, all eigenvalues are real, and the orthogonality of eigenfunctions follows. By virtue of (VI.4.7), eigenfunction expansions can be derived in the exactly same way as in §VI-4.

Step 4. We can also derive Theorem VI-3-11 in the exactly same way as in §VI-3. To do this, consider the differential equation

$$(VI.4.8) \qquad\qquad \frac{d}{dx}\left(x\frac{dy}{dx}\right) + u(x)y = \lambda y.$$

For every value of λ, there exists a unique solution $\phi(x,\lambda)$ such that $\lim\limits_{x\to 0^+} \phi(x,\lambda) = 1$ and $\lim\limits_{x\to 0^+} x\phi'(x,\lambda) = 0$ (cf. Step 1). It is easy to see that $\phi(x,\lambda)$ and $x\phi'(x,\lambda)$ are continuous in (x,λ) for $0 \leq x \leq 1$ and $-\infty < \lambda < +\infty$. Define $\theta(x,\lambda)$ by $\cot(\theta(x,\lambda)) = \dfrac{x\phi'(x,\lambda)}{\phi(x,\lambda)}$. We fix $\theta(0,\lambda) = \dfrac{\pi}{2}$. Using the same method as in §VI-3, we can verify that $\phi(1,\lambda)$ is strictly decreasing and takes all values between 0 and $-\infty$. Eigenvalues of problem (VI.4.8) are determined by the equations $\theta(1,\lambda) = m\pi \, (m = 1, 2, \dots)$.

Theorem VI-3-14 cannot be extended to the present case, since there is no positive lower bound of x on $\mathcal{I}(0,1)$.

VI-5. Jost solutions

So far, we have studied boundary-value problems on a bounded interval on the real line $\mathbb{R}$. Hereafter, we consider the scattering problem, which is a problem on the entire real line. To explain the essential part of the problem, let us consider a homogeneous linear differential equation $\dfrac{d^2 y}{dx^2} + (\zeta^2 - u(x))y = 0$, where ζ is a complex parameter and $u(x)$ is a real-valued continuous function of x such that $u(x) = 0$ for all sufficiently large values of $|x|$. It is evident that $e^{-i\zeta x}$ and $e^{i\zeta x}$ are two linearly independent solutions of this equation if $|x|$ is large. Therefore, if a positive number M is sufficiently large, the solution $y = e^{-i\zeta x}$ for $x < -M$ becomes a linear combination $a(\zeta)e^{-i\zeta x} + b(\zeta)e^{i\zeta x}$ of two solutions $e^{-i\zeta x}$ and $e^{i\zeta x}$ for $x > M$. The main problems are (i) the properties of $a(\zeta)$ and $b(\zeta)$ as functions of ζ and (ii) construction of $u(x)$ for given data $\{a(\zeta), b(\zeta)\}$. Keeping this introduction in mind, let us consider a differential operator

$$\text{(VI.5.1)} \qquad \mathcal{L} = -D^2 + u(x),$$

where $D = \dfrac{d}{dx}$ and $u(x)$ is real-valued and belongs to $C^\infty(-\infty, +\infty)$ such that

$$\text{(VI.5.2)} \qquad \int_{-\infty}^{+\infty} (1 + |x|)|u(x)|dx < +\infty.$$

We study the differential equation

$$\text{(VI.5.3)} \qquad \mathcal{L}y = \zeta^2 y,$$

where ζ is a complex parameter. First, we construct two basic solutions which are called the *Jost solutions* of (VI.5.3).

Theorem VI-5-1. *There exist two solutions $f_+(x, \zeta)$ and $f_-(x, \zeta)$ of (VI.5.3) such that*

(i) $f_\pm$ are continuous for

$$\text{(VI.5.4)} \qquad -\infty < x < +\infty, \quad \Im(\zeta) \geq 0,$$

(ii) $f_\pm$ are analytic in ζ for $\Im(\zeta) > 0$,
(iii)

$$\text{(VI.5.5)} \qquad \begin{cases} |f_+(x, \zeta) - e^{i\zeta x}| \leq C(x)\rho(x)\dfrac{e^{-\eta x}}{1 + |\zeta|}, \\ |f_-(x, \zeta) - e^{-i\zeta x}| \leq C(-x)\tilde{\rho}(x)\dfrac{e^{\eta x}}{1 + |\zeta|} \end{cases}$$

for (VI.5.4), where $\eta = \Im(\zeta)$, $C(x)$ is non-negative and nonincreasing,

$$\rho(x) = \int_x^{+\infty} (1 + |\tau|)|u(\tau)|d\tau, \quad \text{and} \quad \tilde{\rho}(x) = \int_{-\infty}^x (1 + |\tau|)|u(\tau)|d\tau.$$

Proof.

Step 1. *Construction of f_+:* Solve the integral equation

$$y(x, \zeta) = e^{i\zeta x} - \int_x^{+\infty} \frac{\sin(\zeta(x - \tau))}{\zeta} u(\tau)y(\tau, \zeta)d\tau$$

by setting

$$y_0(x, \zeta) = e^{i\zeta x},$$

$$y_{m+1}(x, \zeta) = -\int_x^{+\infty} \frac{\sin(\zeta(x - \tau))}{\zeta} u(\tau)y_m(\tau, \zeta)d\tau \quad (m \geq 0),$$

and

$$f_+(x, \zeta) = \sum_{m=0}^\infty y_m(x, \zeta).$$

Step 2. *Proof of (i) and (ii):* To prove (i) and (ii), it suffices to prove that if

$$\tilde{C}(x) = \begin{cases} 2|x| + 2 & \text{for } x \leq 0, \\ 2 & \text{for } x \geq 0, \end{cases}$$

then

(VI.5.6) $$|y_m(x,\zeta)| \leq \frac{1}{m!}(\tilde{C}(x)\rho(x))^m e^{-\eta x}, \quad m = 0, 1, 2, \ldots,$$

for (VI.5.4).

Inequality (VI.5.6) is true for $m = 0$. Assume that (VI.5.6) is true for an m; then,

$$|y_{m+1}(x,\zeta)| \leq \frac{1}{m!}\int_x^{+\infty}\left|\frac{\sin(\zeta(x-\tau))}{\zeta}\right||u(\tau)|(\tilde{C}(\tau)\rho(\tau))^m e^{-\eta x}d\tau.$$

Note that

$$\frac{\sin(\zeta(x-\tau))}{\zeta} = \frac{e^{i\zeta(x-\tau)}}{2i\zeta}(1 - e^{-2i(x-\tau)\zeta}) = e^{i\zeta(x-\tau)}\int_0^{x-\tau}e^{-2i\zeta z}dz.$$

Hence,

$$\left|\frac{\sin(\zeta(x-\tau))}{\zeta}\right| \leq e^{-\eta(x-\tau)}(\tau - x) \quad \text{for } \eta \geq 0.$$

Therefore,

$$|y_{m+1}(x,\zeta)| \leq \frac{1}{m!}\tilde{C}(x)^m e^{-\eta x}\int_x^{+\infty}(\tau - x)|u(\tau)|\rho(\tau)^m d\tau.$$

Since

$$\int_x^{+\infty}(\tau - x)|u(\tau)|\rho(\tau)^m d\tau \leq \tilde{C}(x)\int_x^{+\infty}(1 + |\tau|)|u(\tau)|\rho(\tau)^m d\tau$$

$$= \frac{1}{m+1}\tilde{C}(x)\rho(x)^{m+1},$$

we obtain

$$|y_{m+1}(x,\zeta)| \leq \frac{1}{(m+1)!}(\tilde{C}(x)\rho(x))^{m+1}e^{-\eta x}.$$

Remark VI-5-2. At the last estimate, the following argument was used:
(a) $\tau - x \leq \tau \leq 1 + \tau$ for $x \geq 0$,
(b)

$$\int_x^{+\infty} = \int_x^{-x} + \int_{-x}^{+\infty} \leq 2|x|\int_x^{-x}|u(\tau)|\rho(\tau)^m d\tau + 2\int_{-x}^{+\infty}\tau|u(\tau)|\rho(\tau)^m d\tau$$

for $x \leq 0$.

Step 3. *Proof of (iii):* First, the estimate

$$(VI.5.7) \qquad \begin{aligned} |f_+(x,\zeta) - e^{i\zeta x}| &\leq \sum_{m=1}^{\infty} \frac{1}{m!}(\tilde{C}(x)\rho(x))^m e^{-\eta x} \\ &\leq \tilde{C}(x)\rho(x)\left[\sum_{m=1}^{\infty}\frac{1}{m!}(\tilde{C}(x)\rho(x))^{m-1}\right]e^{-\eta x} \end{aligned}$$

follows from (VI.5.6). However, this estimate is not enough to prove (iii). So, let us derive another estimate for large $|\zeta|$. To do this, we shall prove that

$$(VI.5.8) \qquad |y_m(x,\zeta)| \leq \frac{1}{m!}\left(\frac{\sigma(x)}{|\zeta|}\right)^m e^{-\eta x}, \quad m = 0,1,2,\ldots,$$

if $\Im(\zeta) \geq 0$ and $\zeta \neq 0$, where

$$\sigma(x) = \int_x^{+\infty} |u(\tau)|d\tau.$$

Inequality (VI.5.8) is true for $m = 0$. Assume that (VI.5.8) is true for an m. Then,

$$|y_{m+1}(x,\zeta)| \leq \frac{1}{m!}\frac{1}{|\zeta|^{m+1}}\int_x^{+\infty} |\sin(\zeta(x-\tau))||u(\tau)|e^{-\eta\tau}\sigma(\tau)^m d\tau.$$

Note that

$$|\sin(\zeta(x-\tau))e^{-\eta x}| = \frac{e^{-\eta x}}{2}|1 - e^{-2i\zeta(x-\tau)}| \leq e^{-\eta x} \qquad \text{if} \qquad \Im(\zeta) \geq 0.$$

Hence,

$$|y_{m+1}(x,\zeta)| \leq \frac{1}{m!}\frac{e^{-\eta x}}{|\zeta|^{m+1}}\int_x^{+\infty} |u(\tau)|\sigma(\tau)^m d\tau = \frac{1}{(m+1)!}\left(\frac{\sigma(x)}{|\zeta|}\right)^{m+1}e^{-\eta x}.$$

This establishes (VI.5.8). Therefore,

$$(VI.5.9) \qquad |f_+(x,\zeta) - e^{i\zeta x}| \leq \frac{\sigma(x)}{|\zeta|}\left[\sum_{m=1}^{\infty}\frac{1}{m!}\left(\frac{\sigma(x)}{|\zeta|}\right)^{m-1}\right]e^{-\eta x}.$$

The proof of Theorem VI-5-1 for $f_+(x,\zeta)$ can be completed by using (VI.5.7) and (VI.5.9).

For $f_-(x,\zeta)$, change x by $-x$ to derive

$$(VI.5.10) \qquad -\frac{d^2 y}{dx^2} + u(-x)y = \zeta^2 y.$$

The solution f_+ for (VI.5.10) is the solution f_- for (VI.5.3). $\square$

Remark VI-5-3. The solutions $f_\pm$ are uniquely determined by the conditions given in Theorem VI-5-1. Also,

$$(VI.5.11) \qquad \frac{df_+}{dx}(x,\zeta) = i\zeta e^{i\zeta x} - \int_x^{+\infty} \cos(\zeta(x-\tau))u(\tau)f_+(\tau,\zeta)d\tau$$

and

$$(VI.5.12) \qquad \left| \frac{df_+}{dx}(x,\zeta) - i\zeta e^{i\zeta x} + \int_x^{+\infty} \cos(\zeta(x-\tau))u(\tau)e^{i\zeta x}d\tau \right|$$
$$\leq C(x)\rho(x)\sigma(x)\frac{e^{-\eta x}}{1+|\zeta|}$$

for (VI.5.4).

Remark VI-5-4. If $u(x) = 0$ identically on the real line, then $f_\pm(x,\zeta) = e^{\pm i\zeta x}$. If $u(x) = 0$ for $|x| \geq M$ for some positive constant M, then $f_+(x,\zeta) = e^{i\zeta x}$ for $x \geq M$, while $f_-(x,\zeta) = e^{-i\zeta x}$ for $x \leq -M$.

VI-6. Scattering data

For real ξ, $\bar{f}_+(x,\xi) = f_+(x,-\xi)$, where $\bar{f}$ denotes the conjugate complex of f. Both f_+ and $\bar{f}_+$ are solutions of $\mathcal{L}y = \xi^2 y$ and

$$\begin{cases} |\bar{f}_+(x,\xi) - e^{-i\xi x}| \leq \dfrac{C(x)\rho(x)}{1+|\xi|}, \\[2mm] \left| \dfrac{d\bar{f}_+}{dx}(x,\xi) + i\xi e^{-i\xi x} + \displaystyle\int_x^{+\infty} \cos(\xi(x-\tau))u(\tau)e^{i\xi\tau}d\tau \right| \leq \dfrac{C(x)\rho(x)\sigma(x)}{1+|\xi|} \end{cases}$$

for $-\infty < x < +\infty$ and $-\infty < \xi < +\infty$. Let $W[f,g]$ denotes the Wronskian of $\{f,g\}$. Then,

$$W[f_+,\bar{f}_+] = \begin{vmatrix} f_+(x,\xi) & \bar{f}_+(x,\xi) \\ f'_+(x,\xi) & \bar{f}'_+(x,\xi) \end{vmatrix} = -2i\xi \qquad (-\infty < \xi < +\infty).$$

This implies that $\{f_+,\bar{f}_+\}$ is linearly independent if $\xi \neq 0$. This, in turn, implies that the solution f_- is a linear combination of $\{f_+,\bar{f}_+\}$. Set

$$(VI.6.1) \qquad f_-(x,\xi) = a(\xi)\bar{f}_+(x,\xi) + b(\xi)f_+(x,\xi).$$

It is easy to see that

$$(VI.6.2) \qquad a(\zeta) = \frac{i}{2\zeta}W[f_+,f_-] = \frac{i}{2\zeta}\begin{vmatrix} f_+(x,\zeta) & f_-(x,\zeta) \\ f'_+(x,\zeta) & f'_-(x,\zeta) \end{vmatrix}$$

and

$$(VI.6.3) \qquad b(\xi) = W[\bar{f}_+,f_-] = \frac{-i}{2\xi}\begin{vmatrix} \bar{f}_+(x,\xi) & f_-(x,\xi) \\ \bar{f}'_+(x,\xi) & f'_-(x,\xi) \end{vmatrix}.$$

The function $a(\zeta)$ is analytic for $\Im(\zeta) > 0$ and

$$(VI.6.4) \qquad a(\zeta) = 1 + O(\zeta^{-1}) \qquad \text{as} \quad \zeta \to \infty \quad \text{on} \quad \Im(\zeta) > 0,$$

whereas the function $b(\xi)$ is defined and continuous on $-\infty < \xi < +\infty$. Furthermore, $b(\xi) = O(|\xi|^{-1})$ as $|\xi| \to +\infty$. To simplify the situation, we introduce the following assumption.

Assumption VI-6-1. *We assume that*

$$|u(x)| \leq Ae^{-k|x|} \qquad (-\infty < x < +\infty)$$

for some positive numbers A and k.

The boundary-value problem

$$(VI.6.5) \qquad \mathcal{L}y = \lambda y, \qquad y \in L^2(-\infty, +\infty)$$

is self-adjoint, where $L^2(-\infty, +\infty)$ denotes the set of all complex-valued functions $f(x)$ satisfying the condition $\displaystyle\int_{-\infty}^{+\infty} |f(x)|^2 dx < +\infty$. The self-adjointness of problem (VI.6.5) can be proved by using an inner product in the vector space $L^2(-\infty, +\infty)$ in the same way as the proof of Theorem VI-3-4. Therefore, eigenvalues of problem (VI.6.5) are real. Furthermore, all eigenvalues are negative. In fact, if $\xi \neq 0$ is real, then $\lambda = \xi^2 > 0$ and the general solution $c_1 f_+ + c_2 \bar{f}_+$ is asymptotically equal to $c_1 e^{i\xi x} + c_2 e^{-i\xi x}$ as $x \to +\infty$. If $\xi = 0$, then $f_+(x, 0)$ is asymptotically equal to 1 as $x \to +\infty$. Moreover, another solution $\displaystyle f_+ \int^x \frac{d\tau}{f_+^2}$ is asymptotically equal to x as $x \to +\infty$. Therefore, all eigenvalues are $\lambda = (i\eta)^2 < 0$. Furthermore, all eigenvalues of (VI.6.5) are determined by

$$(VI.6.6) \qquad a(i\eta) = 0.$$

This implies that all zeros of $a(\zeta)$ for $\Im(\zeta) > 0$ are purely imaginary.

Under Assumption VI-6-1, $f_\pm(x, \zeta)$ are analytic for $\Im(\zeta) > -\dfrac{k}{2}$. Hence, $a(\zeta)$ has only a finite number of zeros for $\Im(\zeta) > 0$ (cf. (VI.6.4)). Let $\zeta = i\eta_j$ ($j = 1, 2, \ldots, N$) be the zeros of $a(\zeta)$ for $\Im(\zeta) > 0$. Then, $f_+(x, i\eta_j)$ are real-valued and $f_+(x, i\eta_j) \in L^2(-\infty, +\infty)$. Set

$$(VI.6.7) \qquad \begin{cases} c_j = \dfrac{1}{\displaystyle\int_{-\infty}^{+\infty} f_+(x, i\eta_j)^2 dx} & (j = 1, 2, \ldots, N), \\[4ex] r(\xi) = \dfrac{b(\xi)}{a(\xi)} & (-\infty < \xi < +\infty). \end{cases}$$

Observation VI-6-2. Every eigenvalue of (VI.6.5) is simple, i.e.,

$$\frac{da(\zeta)}{d\zeta} \neq 0 \qquad \text{if} \qquad a(\zeta) = 0.$$

Proof.

From $a(\zeta) = 0$, it follows that

$$\frac{da(\zeta)}{d\zeta} = \frac{i}{2\zeta}\frac{d}{d\zeta}W[f_+, f_-] - \frac{i}{2\zeta^2}W[f_+, f_-] = \frac{i}{2\zeta}(W[f_{+\zeta}, f_-] + W[f_+, f_{-\zeta}]),$$

where f_ζ denotes $\dfrac{df}{d\zeta}$. Also, two relations

$$-f''_{+\zeta} + uf_{+\zeta} = \zeta^2 f_{+\zeta} + 2\zeta f_+ \qquad \text{and} \qquad -f''_{-\zeta} + uf_{-\zeta} = \zeta^2 f_{-\zeta} + 2\zeta f_-$$

imply that

$$\frac{d}{dx} W[f_{+\zeta}, f_-] = 2\zeta f_+ f_- \qquad \text{and} \qquad \frac{d}{d\zeta} W[f_+, f_{-\zeta}] = -2\zeta f_+ f_-.$$

Since $a(\zeta) = 0$, there exists a constant $d(\zeta)$ such that $f_-(x, \zeta) = d(\zeta) f_+(x, \zeta)$. Therefore,

$$W[f_{+\zeta}, f_-] = -2\zeta \int_x^{+\infty} f_+ f_- d\tau \qquad \text{and} \qquad W[f_+, f_{-\zeta}] = -2\zeta \int_{-\infty}^x f_+ f_- d\tau.$$

Thus, we obtain

$$\frac{da(\zeta)}{d\zeta} = -i \int_{-\infty}^{+\infty} f_+ f_- d\tau = -id(\zeta) \int_{-\infty}^{+\infty} f_+^2 d\tau = -\frac{id(\zeta)}{c_j} \neq 0. \qquad \square$$

Observation VI-6-3. The quantities $a(\xi)$ and $b(\xi)$ satisfy the following relation:

$$(\text{VI.6.8}) \qquad |a(\xi)|^2 - |b(\xi)|^2 = 1 \qquad \text{for} \qquad -\infty < \xi < +\infty.$$

Proof.

From $\bar{f}_-(x, \xi) = f_-(x, -\xi)$, it follows that

$$\bar{a}(\xi) = a(-\xi), \qquad \bar{b}(\xi) = b(-\xi), \qquad \text{and} \qquad W[f_-, \bar{f}_-] = 2i\xi.$$

Therefore, (VI.6.8) follows from

$$\begin{aligned}
f_- &= a\bar{f}_+ + b(a\bar{f}_- - \bar{b}f_-) = a(\bar{f}_+ + b\bar{f}_-) - |b|^2 f_- \\
&= a\bar{a}f_- - |b|^2 f_- = \{|a|^2 - |b|^2\} f_-. \qquad \square
\end{aligned}$$

Observation VI-6-4. The formula

$$(\text{VI.6.9}) \qquad a(\zeta) = \left[\prod_{j=1}^N \frac{\zeta - i\eta_j}{\zeta + i\eta_j} \right] \exp\left[\frac{1}{2\pi i} \int_{-\infty}^{+\infty} \frac{\log(1 - |r(\xi)|^2)}{\zeta - \xi} d\xi \right]$$

for $\Im(\zeta) > 0$ shows that the quantities η_j and $r(\xi)$ determine $a(\zeta)$.

Proof.

Set $f(\zeta) = a(\zeta) \displaystyle\prod_{j=1}^N \frac{\zeta + i\eta_j}{\zeta - i\eta_j}$. Then,

(i) $f(\zeta)$ is analytic for $\Im(\zeta) > -\dfrac{k}{2}$ and $\zeta \neq 0$,

(ii) $\zeta(f(\zeta) - 1)$ is bounded for $\Im(\zeta) > -\dfrac{k}{2}$,

(iii) $f(\zeta) \neq 0$ for $\Im(\zeta) \geq 0$ and $\zeta \neq 0$.

Set also $F(\zeta) = \log(f(\zeta))$. Then, $F(\xi) = O(\log(\xi))$ near $\xi = 0$ and $F(\zeta) = O(\zeta^{-1})$ as $\zeta \to \infty$ on $\Im(\zeta) \geq 0$. Observe that $f(\zeta) - 1 = O(\zeta^{-1})$. From this observation, it follows without any complication that

$$F(\zeta) = \frac{1}{2\pi i} \int_{-\infty}^{+\infty} \frac{2\log|f(\xi)|}{\xi - \zeta} d\xi \qquad \text{for} \qquad \Im(\zeta) > 0$$

(cf. Exercise VI-18). Now, (VI.6.9) follows from $|f(\xi)| = |a(\xi)|$ and

$$\log|a(\xi)|^2 = -\log\left(\frac{1}{|a(\xi)|^2}\right) = -\log\left(\frac{|a(\xi)|^2 - |b(\xi)|^2}{|a(\xi)|^2}\right) = -\log(1 - |r(\xi)|^2).$$

Definition VI-6-5. *The set* $\{r(\xi),\ (\eta_1, \eta_2, \ldots, \eta_N),\ (c_1, c_2, \ldots, c_N)\}$ *is called the scattering data* associated with the *potential* $u(x)$.

VI-7. Reflectionless potentials

The function $r(\xi)$ is called the *reflection coefficient*. If this coefficient is zero, the potential $u(x)$ becomes a function of simple form. Let us look into this situation.

Observation VI-7-1. If $r(\xi) = 0$, then $b(\xi) = 0$. This means that $f_-(x,\xi) = a(\xi)\bar{f}_+(x,\xi) = a(\xi)f_+(x,-\xi)$ (cf. (§VI-6)). Therefore, using the relation

$$f_+(x,\zeta) = \frac{f_-(x,-\zeta)}{a(-\zeta)} \qquad \text{for} \qquad \Im(\zeta) \leq 0,$$

we can extend f_+ for all ζ as a meromorphic function in ζ. Furthermore, if $i\eta_j$ ($j = 1, 2, \ldots, N$) are zeros of $a(\zeta)$ in $\Im(\zeta) > 0$, then $-i\eta_j$ ($j = 1, 2, \ldots, N$) are simple poles of $f_+(x,\zeta)$ in ζ and

$$\text{Residue of } f_+ \text{ at } -i\eta_j = -\frac{f_-(x, i\eta_j)}{a_\zeta(i\eta_j)} = \frac{c_j f_-(x, i\eta_j)}{id(i\eta_j)} = -ic_j f_+(x, i\eta_j).$$

Observation VI-7-2. Set $g_j(x) = e^{-\eta_j x} c_j f_+(x, i\eta_j)$. Then, from the fact that $e^{-i\zeta x} f_+(x,\zeta) \to 1$ as $|\zeta| \to +\infty$, it follows that

$$(\text{VI.7.1}) \qquad f_+(x,\zeta) = e^{i\zeta x}\left[1 - i\sum_{j=1}^{N} \frac{g_j(x)}{\zeta + i\eta_j}\right].$$

Setting $\zeta = i\eta_\ell$ in (VI.7.1), we obtain

$$(\text{VI.7.2}) \qquad \frac{e^{2\eta_\ell x}}{c_\ell} g_\ell(x) + \sum_{j=1}^{N} \frac{1}{\eta_\ell + \eta_j} g_j(x) = 1 \qquad (\ell = 1, 2, \ldots, N).$$

Observation VI-7-3. If we set $F(x, \zeta) = e^{-i\zeta x} f_+(x, \zeta)$, then $-F'' - 2i\zeta F' + uF = 0$. Since $F(x, \zeta) = 1 - i\displaystyle\sum_{j=1}^{N} \frac{g_j(x)}{\zeta + i\eta_j}$, it follows that

$$(VI.7.3) \qquad\qquad u(x) \;=\; 2\sum_{j=1}^{N} g_j'(x).$$

Observation VI-7-4. Let us solve (VI.7.2) for the $g_j(x)$. First, set

$$h_j \;=\; g_j e^{\eta_j x} \qquad (j = 1, 2, \ldots, N).$$

Then, (VI.7.2) becomes

$$(VI.7.4) \qquad h_\ell \;+\; \sum_{j=1}^{N} \frac{c_\ell e^{-(\eta_\ell + \eta_j)x}}{\eta_\ell + \eta_j} h_j \;=\; c_\ell e^{-\eta_\ell x} \qquad (\ell = 1, 2, \ldots, N).$$

Write the coefficient matrix of (VI.7.4) in the form $I_N + C(x)$, where I_N is the $N \times N$ identity matrix. Since

$$\sum_{j,\ell=1}^{N} \frac{\gamma_j \gamma_\ell}{\eta_j + \eta_\ell} \;=\; \int_0^{+\infty} \left[\sum_{p=1}^{N} \gamma_p e^{-\eta_p x}\right]^2 dx$$

and η_j $(j = 1, 2, \ldots, N)$ are distinct, the matrix $I_N + C(x)$ is invertible for $-\infty < x < +\infty$. Set $\triangle(x) = \det(I_N + C(x))$. Then, manipulating with Cramer's rule, we can write g_j $(j = 1, 2, \ldots, N)$ and $\dfrac{d\triangle}{dx}$ in the following forms:

$$(VI.7.5) \qquad\qquad g_j(x) \;=\; \frac{\triangle_j(x)}{\triangle(x)} \qquad (j = 1, 2, \ldots, N)$$

and

$$(VI.7.6) \qquad\qquad \frac{d\triangle(x)}{dx} \;=\; -\sum_{j=1}^{N} \triangle_j(x).$$

Thus finally, from (VI.7.3), it follows that

$$u(x) \;=\; 2\sum_{j=1}^{N} g_j'(x) \;=\; -2\left(\frac{\triangle'(x)}{\triangle(x)}\right)'$$

or

$$(VI.7.7) \qquad\qquad u(x) \;=\; -2\frac{d^2}{dx^2}(\log(\triangle(x))).$$

This implies that $u(x)$ is a rational function of exponential functions and satisfies Assumption VI-6-1 of §VI-6. To see this, the following remarks are also useful:

(a) we have the identity

$$\sum_{j=1}^{N} g_j'(x) = -2\sum_{j=1}^{N} \frac{\eta_j}{c_j} e^{2\eta_j x} g_j(x)^2,$$

(b) $g_j(-\infty) = \lim_{x \to -\infty} g_j(x)$ $(j = 1, 2, \dots, N)$ exist and

$$\sum_{j=1}^{N} \frac{1}{\eta_\ell + \eta_j} g_j(-\infty) = 1 \qquad (\ell = 1, 2, \dots, N).$$

To show (a), derive

$$\frac{e^{2\eta_\ell x}}{c_\ell} g_\ell'(x) + \sum_{j=1}^{N} \frac{1}{\eta_\ell + \eta_j} g_j'(x) = -\frac{2\eta_\ell}{c_\ell} e^{2\eta_\ell x} g_\ell(x) \qquad (\ell = 1, 2, \dots, N)$$

from (VI.7.2). Multiplying both sides by g_ℓ, adding them up over ℓ, and interchanging the orders of summation, we obtain (a). To show (b), calculate the inverse of the coefficient matrix of (VI.7.2). Using Cramer's rule on (VI.7.4), it can be shown that $h_j(x) \to 0$ exponentially as $x \to \infty$. Also, (b) implies that $e^{2\eta_j x} g_j(x)^2$ is exponentially small as $x \to -\infty$. Therefore, (a) implies that $u(x)$ satisfies Assumption VI-6-1.

Example VI-7-5. In the case $N = 1$, if we determine $g(x)$ by $\left(\dfrac{e^{2\eta x}}{c} + \dfrac{1}{2\eta}\right) g(x) = 1$, then $u(x) = 2g'(x)$. Since $g(x) = \dfrac{2\eta c}{2\eta e^{2\eta x} + c}$, we obtain $u(x) = -\dfrac{4\eta}{c} e^{2\eta x} g(x)^2$.

Also, $\dfrac{e^{\eta x}}{\sqrt{2\eta c}}\left(\sqrt{\dfrac{2\eta}{c}} e^{\eta x} + \sqrt{\dfrac{c}{2\eta}} e^{-\eta x}\right) g(x) = 1$ implies that $g(x) = \sqrt{\dfrac{\eta c}{2}} e^{-\eta x}$

$\times \mathrm{sech}(\eta(x + \rho))$, where $\rho = \dfrac{\ln\left(\dfrac{2\eta}{c}\right)}{2\eta}$, and, hence,

$$(VI.7.8) \qquad u(x) = -2\eta^2 \,\mathrm{sech}^2(\eta(x + \rho)).$$

In particular, formula (VI.7.8) yields $u(x) = -8\,\mathrm{sech}\left(2\left(x + \dfrac{\ln\left(\frac{4}{5}\right)}{4}\right)\right)$ in the case when $N = 1$, $\eta_1 = 2$, and $c_1 = 5$. On the other hand, a straightforward calculation using formula (VI.7.7) yields $u(x) = -\dfrac{640 e^{4x}}{(5 + 4e^{4x})^2}$. Also, $u(x) = -2\,\mathrm{sech}(x)$ is the *reflectionless potential* corresponding to the data $\eta = 1$ and $c = 2$.

Example VI-7-6. If $N = 2$, $\eta_1 = 2$, $\eta_2 = 3$, $c_1 = 5$, and $c_2 = 2$, we obtain

$$u(x) = -\frac{40 e^{4x}(16 + 135 e^{2x} + 600 e^{6x} + 2160 e^{10x} + 3600 e^{12x})}{(1 + 20 e^{4x} + 75 e^{6x} + 60 e^{10x})^2}$$

by calculating (VI.7.7).

Remark VI-7-7. If the reflection coefficient $r(\xi) = 0$ and if there is no eigenvalue for the problem

$$-\frac{d^2y}{dx^2} + u(x)y = \lambda y, \qquad y \in L^2(-\infty, +\infty),$$

then $u(x) = 0$ identically for $-\infty < x < +\infty$. In fact, $f_+(x, \zeta)$ is analytic in ζ everywhere in the ζ-plane. Furthermore, $|e^{-i\zeta x}f_+(x, \zeta) - 1| \to 0$ as $\zeta \to \infty$. Hence, $f_+(x, \zeta) = e^{i\zeta x}$. This shows that $u(x) = 0$.

Remark VI-7-8. If $u(x) \neq 0$ but $u(x) = 0$ for $|x| \geq M$, where M is a positive number, then the reflection coefficient $r(\xi) \neq 0$. In fact, if $r(\xi) = 0$, then $u(x)$ is analytic for $-\infty < x < +\infty$. Hence, $u(x) = 0$. This is a contradiction. This remark reveals that the problem posed at the beginning of §VI-5 is not as simple as it looks.

VI-8. Construction of a potential for given data

If scattering data $\{0, (\eta_1, \dots, \eta_N), (c_1, \dots, c_N)\}$ are given, formula (VI.7.3) gives the corresponding reflectionless potential $u(x)$ as it is shown in Examples VI-7-5 and VI-7-6. However, in these two examples, we assumed implicitly that such a potential exist. Since the existence of $u(x)$ has not been shown yet, it must be proved. To do this, for given data

$$\eta_1 > 0, \ \eta_2 > 0, \ \dots, \ \eta_N > 0 \quad \text{and} \quad c_1 > 0, \ c_2 > 0, \ \dots, \ c_N > 0,$$

define g_j $(j = 1, 2, \dots, N)$ by (VI.7.2) and define f_+ and u by (VI.7.1) and (VI.7.3). Hereafter, the first thing that we have to do is to show that $y = f_+$ is one of the Jost solutions of $\mathcal{L}y = \zeta^2 y$.

Observation VI-8-1. Let g_ℓ $(\ell = 1, 2, \dots, N)$ be determined by (VI.7.2) and $u(x)$ be defined by (VI.7.3). Then,

(VI.8.1) $$g_\ell'' + 2\eta_\ell g_\ell' - ug_\ell = 0 \qquad (\ell = 1, 2, \dots, N).$$

Proof.

Write (VI.7.2) in the form

(VI.8.2) $$\sum_{j=1}^{N} a_{\ell,j}(x)g_j(x) = 1 \qquad (\ell = 1, 2, \dots, N)$$

and differentiate both sides with reaspect to x. Then,

(VI.8.3) $$\sum_{j=1}^{N} a_{\ell,j}(x)g_j'(x) + 2\eta_\ell \frac{e^{2\eta_\ell x}}{c_\ell} g_\ell(x) = 0 \qquad (\ell = 1, 2, \dots, N)$$

and, hence,

$$\sum_{j=1}^{N} a_{\ell,j}(x)g_j'(x) + 2\eta_\ell \left[1 - \sum_{j=1}^{N} \frac{1}{\eta_\ell + \eta_j} g_j(x)\right] = 0 \qquad (\ell = 1, 2, \dots, N).$$

From (VI.8.2), it follows that

$$\sum_{j=1}^{N} a_{\ell,j}(x)\{g_j'(x) + 2\eta_\ell g_j(x)\} - 2\eta_\ell \sum_{j=1}^{N} \frac{1}{\eta_\ell + \eta_j} g_j(x) = 0$$

$$(\ell = 1, 2, \ldots, N)$$

or

$$\sum_{j=1}^{N} a_{\ell,j}(x)\{g_j'(x) + 2\eta_\ell g_j(x)\} - 2\sum_{j=1}^{N} g_j(x) + 2\sum_{j=1}^{N} \frac{\eta_j}{\eta_\ell + \eta_j} g_j(x) = 0$$

$$(\ell = 1, 2, \ldots, N).$$

Differentiating again, we obtain

$$\sum_{j=1}^{N} a_{\ell,j}(x)\{g_j''(x) + 2\eta_\ell g_j'(x)\} + 2\eta_\ell \frac{e^{2\eta_\ell x}}{c_\ell}(g_\ell'(x) + 2\eta_\ell g_\ell(x)) - u(x)$$

$$+ 2\sum_{j=1}^{N} \frac{\eta_j}{\eta_\ell + \eta_j} g_j'(x) = 0 \qquad (\ell = 1, 2, \ldots, N)$$

or

$$\sum_{j=1}^{N} a_{\ell,j}(x)\{g_j''(x) + 2\eta_\ell g_j'(x)\} - \sum_{j=1}^{N} a_{\ell,j}(x)u(x)g_j(x)$$

$$+ \sum_{j=1}^{N} a_{\ell,j}(x)2\eta_j g_j'(x) + 2\eta_\ell \left(2\eta_\ell \frac{e^{2\eta_\ell x}}{c_\ell} g_\ell(x) \right) = 0 \qquad (\ell = 1, 2, \ldots, N).$$

Thus, we derive

$$\sum_{j=1}^{N} a_{\ell,j}(x)\{g_j''(x) + 2\eta_j g_j'(x) - u(x)g_j(x)\}$$

$$+ 2\eta_\ell \left[\sum_{j=1}^{N} a_{\ell,j}(x)g_j'(x) + 2\eta_\ell \frac{e^{2\eta_\ell x}}{c_\ell} g_\ell(x) \right] = 0 \qquad (\ell = 1, 2, \ldots, N),$$

and (VI.8.3) implies that

$$\sum_{j=1}^{N} a_{\ell,j}(x)\{g_j''(x) + 2\eta_j g_j'(x) - u(x)g_j(x)\} = 0 \qquad (\ell = 1, 2, \ldots, N).$$

Therefore, (VI.8.1) follows. $\square$

Observation VI-8-2. If we further define f_+ by (VI.7.1), then $\mathcal{L}f_+ = \zeta^2 f_+$.

Proof.

$$
\begin{aligned}
(e^{-i\zeta x} f_+)'' &+ 2i\zeta(e^{-i\zeta x} f_+)' - u(e^{-i\zeta x} f_+) \\
&= -i\sum_{j=1}^{N} \frac{g_j'' + 2i\zeta g_j' - u g_j}{\zeta + i\eta_j} - 2\sum_{j=1}^{N} g_j' \\
&= -i\sum_{j=1}^{N} \frac{g_j'' + 2\eta_j g_j' - u g_j}{\zeta + i\eta_j} = 0. \quad \square
\end{aligned}
$$

Observation VI-8-3. Since

$$
f_+(x, i\eta_\ell) = e^{-\eta_\ell x}\left[1 - \sum_{j=1}^{N} \frac{g_j(x)}{\eta_\ell + \eta_j}\right] = \frac{e^{\eta_\ell x}}{c_\ell} g_\ell(x) \in L^2(-\infty, +\infty),
$$

N numbers $-\eta_j^2$ $(j = 1, 2, \ldots, N)$ are eigenvalues.

Observation VI-8-4. Set $a(\zeta) = \displaystyle\prod_{j=1}^{N} \frac{\zeta - i\eta_j}{\zeta + i\eta_j}$ (cf. (VI.6.9) with $r(\xi) = 0$) and

$$
f_-(x, \zeta) = a(\zeta) f_+(x, -\zeta) = a(\zeta) e^{-\zeta x}\left[1 + i\sum_{j=1}^{N} \frac{g_j(x)}{\zeta - i\eta_j}\right]
$$

(cf. Observation VI-7-1). Then, $\mathcal{L}f_- = \zeta^2 f_-$. Furthermore, since

$$
\sum_{j=1}^{N} \frac{g_j(-\infty)}{\eta_\ell + \eta_j} = 1 \qquad (\ell = 1, 2, \ldots, N)
$$

(cf. (VI.7.2)), we obtain

$$
\text{(VI.8.4)} \qquad\qquad a(\zeta) = 1 - i\sum_{j=1}^{N} \frac{g_j(-\infty)}{\zeta + \eta_j}.
$$

In fact, this follows from the fact that both sides of (VI.8.4) are rational functions in ζ with the same zeros, the same poles, and the same limits as $\zeta \to \infty$.

Observation VI-8-5. The functions $f_\pm(x, \zeta)$ are the Jost solutions of $\mathcal{L}y = \zeta^2 y$.

Proof.

Note first that $\displaystyle\lim_{x \to +\infty} f_+(x, \zeta) e^{-i\zeta x} = 1$ (cf. (VI.7.1)). Also, we have

$$
\lim_{x \to -\infty} f_+(x, \zeta) e^{-i\zeta x} = 1 - i\sum_{j=1}^{N} \frac{g_j(-\infty)}{\zeta + i\eta_j} = a(\zeta) \qquad \text{(cf. (VI.7.1) and (VI.8.4))}.
$$

This implies that

$$
\lim_{x \to -\infty} f_-(x, \zeta) e^{i\zeta x} = a(\zeta) \lim_{x \to -\infty} f_+(x, -\zeta) e^{i\zeta x} = a(\zeta) a(-\zeta) = 1. \quad \square
$$

Observation VI-8-6. Note that $f_-(x,\xi) = a(\xi)f_+(x,-\xi) = a(\xi)\bar{f}_+(x,\xi)$ for $\zeta = \xi$ real. This means that $b(\xi) = 0$. Hence, $r(\xi) = 0$. Thus, we conclude that $u(x)$ is reflectionless.

Remark VI-8-7. For the general $r(\xi)$, the potential $u(x)$ can be constructed by solving the integral equation of Gel'fand-Levitan:

$$K(x,\tau) \; + \; F(x+\tau) \; + \; \int_0^{+\infty} F(x+\tau+s)K(x,s)ds \; = \; 0,$$

where

$$F(x) \; = \; \frac{1}{\pi}\int_{-\infty}^{+\infty} r(\xi)e^{2i\xi x}d\xi \; + \; 2\sum_{j=1}^{N} c_j e^{-2\eta_j x}.$$

Find $K(x,\tau)$ by this equation. Then, the potential is given by

$$u(x) \; = \; -\frac{\partial K}{\partial x}(x,0).$$

If we set $r(\xi) = 0$ in this integral equation, we can derive (VI.7.2) and (VI.7.3). Details are left to the reader as exercises (cf. [GelL]).

VI-9. Differential equations satisfied by reflectionless potentials

Suppose that $u(x)$ is a reflectionless potential whose associated scattering data are given by $\{r(\xi) = 0, \ (\eta_1, \eta_2, \ldots, \eta_N), \ (c_1, c_2, \ldots, c_N)\}$, where $0 < \eta_1 < \eta_2 < \cdots < \eta_N$. It was proven that one of the Jost solutions is given by

$$f_+(x,\zeta) \; = \; e^{i\zeta x}\left[1 \; - \; i\sum_{j=1}^{N}\frac{g_j(x)}{\zeta + i\eta_j}\right].$$

Furthermore,

$$f_+(x,-\zeta) \; = \; e^{-i\zeta x}\left[1 \; - \; i\sum_{j=1}^{N}\frac{g_j(x)}{-\zeta + i\eta_j}\right]$$

is also a solution of

$$\frac{d^2y}{dx^2} \; - \; (u(x) \; - \; \lambda)y \; = \; 0, \quad \text{where} \quad \lambda \; = \; \zeta^2.$$

Therefore,

(VI.9.1) $$\mathcal{P}(x,\lambda) = \prod_{j=1}^{N}\left(1 + \frac{\eta_j^2}{\lambda}\right)f_+(x,\zeta)f_+(x,-\zeta)$$

satisfies the differential equation

(E) $$-\frac{1}{2}\frac{d^3P}{dx^3} \; + \; 2(u(x) \; - \; \lambda)\frac{dP}{dx} \; + \; \frac{du(x)}{dx}P \; = \; 0.$$

Let

$$(S) \qquad P = \sum_{n=0}^{+\infty} \frac{p_n}{\lambda^n} \qquad (p_0 \neq 0)$$

be a formal solution in powers of λ^{-1} of differential equation (E). Then,

$$0 = P\left[-\frac{P'''}{2} + 2(u(x) - \lambda)P' + u'(x)P\right]$$

$$= \left[-\frac{PP''}{2} + \frac{(P')^2}{4} + (u(x) - \lambda)P^2\right]'$$

and, hence,

$$\lambda(P^2)' = \left[-\frac{PP''}{2} + \frac{(P')^2}{4} + u(x)P^2\right]'.$$

Therefore,

(i) the coefficients p_n can be determined successively by

$$\begin{cases} p_0 = c_0, \qquad \text{where } c_0 \text{ is a constant,} \\ 2c_0 p'_{n+1} = \left[-\frac{1}{2}\sum_{\ell=0}^{n} p_\ell p''_{n-\ell} + \frac{1}{4}\sum_{\ell=0}^{n} p'_\ell p'_{n-\ell} + u\sum_{\ell=0}^{n} p_\ell p_{n-\ell} - \sum_{\ell=1}^{n} p_\ell p'_{n+1-\ell}\right]' \\ \qquad\qquad\qquad\qquad\qquad\qquad\qquad\qquad (n = 0, 1, 2, \ldots), \end{cases}$$

(ii) the coefficients p_n are polynomials in u and its derivatives with constant coefficients,

(iii) the formal power series (S) is uniquely determined by the condition $P = 1$ for $u = 0$,

(iv) if we denote the unique P of (iii) by $G(u, \lambda) = \sum_{n=0}^{+\infty} \frac{G_n(u)}{\lambda^n}$ $(G_0 = 1)$, then the general formal solution of (S) is given by $P(u, \lambda) = P(0, \lambda)G(u, \lambda)$.

Example VI-9-1. A straight forward calculation yields

$$G_1 = \frac{u}{2}, \quad G_2 = \frac{3u^2 - u''}{8}, \quad G_3 = \frac{10u^3 - 10uu'' - 5(u')^2 + u^{(4)}}{32}.$$

Observe that $\mathcal{P}(x, \lambda)$ of (VI.9.1) is a polynomial in $\frac{1}{\lambda}$ of degree N. Therefore, the potential $u(x)$ satisfies a differential equation

$$(\text{VI.9.2}) \qquad G_{N+1}(u) + \alpha_1 G_N(u) + \alpha_2 G_{N-1}(u) + \cdots + \alpha_N G_1(u) = 0$$

for some suitable constants $\alpha_1, \alpha_2, \ldots, \alpha_N$. More precisely speaking, it is shown above that $\mathcal{P}(u, \lambda) = \mathcal{P}(0, \lambda)G(u, \lambda)$. Compute the coefficients of $\lambda^{-(N+1)}$ on both sides of this identity. In fact,

$$u(x) = 2\sum_{j=1}^{N} g'_j(x) = -4\sum_{j=1}^{N} \frac{\eta_j}{c_j} e^{2\eta_j x} g_j(x)^2$$

implies that $u = 0$ if and only if $g_\ell = 0 \, (\ell = 1, 2, \cdots, N)$. This, in turn, implies that

$$\mathcal{P}(0, \lambda) = \prod_{j=1}^{N} \left(1 + \frac{\eta_j^2}{\lambda} \right)$$

and

$$G(u, \lambda) \;=\; f_+(x, \zeta) f_+(x, -\zeta) \,, \qquad \text{where} \qquad \lambda = \zeta^2 \,.$$

Hence, $\alpha_0 = 1$ and

$$1 + \sum_{j=1}^{N} \frac{\alpha_j}{\lambda^j} \;=\; \prod_{j=1}^{N} \left(1 + \frac{\eta_j^2}{\lambda} \right) .$$

Example VI-9-2. The function $u(x) = -2\eta^2 \operatorname{sech}^2(\eta(x + \rho))$ is a reflectionless potential corresponding to the data $\{\eta, c\}$, where $\rho = \dfrac{1}{2\eta} \ln \left(\dfrac{2\eta}{c} \right)$ (cf. Example VI-7-5). In this case, $G_2(u) + \eta^2 G_1(u) = 0$. Also, $G_3(u) + (\eta_1^2 + \eta_2^2) G_2(u) + \eta_1^2 \eta_2^2 G_1(u) = 0$ in the case when $N = 2$ and $\{r(\xi) = 0, (\eta_1, \eta_2), (c_1, c_2)\}$ are the scattering data. In particular, $G_3(u) + 13 G_2(u) + 36 G_1(u) = 0$ if $\eta_1 = 2$ and $\eta_2 = 3$.

Remark VI-9-3. The materials of §§VI-5–VI-9 are also found in [TD].

VI-10. Periodic potentials

In this section, we consider the differential equation

$$(\text{VI.10.1}) \qquad \frac{d^2 y}{dx^2} + (\lambda - u(x))\, y = 0$$

under the assumption that
   (I) $u(x)$ is continuous for $-\infty < x < +\infty$,
   (II) $u(x)$ is periodic of period ℓ, i.e., $u(x + \ell) = u(x)$ for $-\infty < x < +\infty$.
The period ℓ is a positive number and λ is a real parameter. Denote by $\phi_1(x, \lambda)$ and $\phi_2(x, \lambda)$ the two linearly independent solutions of (VI.10.1) such that

$$(\text{VI.10.2}) \quad \phi_1(0, \lambda) = 1, \quad \frac{d\phi_1}{dx}(0, \lambda) = 0, \quad \phi_2(0, \lambda) = 0, \quad \frac{d\phi_2}{dx}(0, \lambda) = 1.$$

Set

$$\Phi(\lambda) \;=\; \begin{bmatrix} \phi_1(\ell, \lambda) & \phi_2(\ell, \lambda) \\[4pt] \dfrac{d\phi_1}{dx}(\ell, \lambda) & \dfrac{d\phi_2}{dx}(\ell, \lambda) \end{bmatrix} .$$

The two eigenvalues $Z_1(\lambda)$ and $Z_2(\lambda)$ of the matrix $\Phi(\lambda)$ are the multipliers of the periodic system

$$(\text{VI.10.3}) \qquad \frac{d}{dx} \begin{bmatrix} y_1 \\ y_2 \end{bmatrix} = \begin{bmatrix} 0 & 1 \\ u(x) - \lambda & 0 \end{bmatrix} \begin{bmatrix} y_1 \\ y_2 \end{bmatrix}$$

(cf. Definition IV-4-5). It is easy to show that $\det[\Phi(\lambda)] = 1$ for $-\infty < \lambda < +\infty$. Hence, the two multipliers $Z_1(\lambda)$ and $Z_2(\lambda)$ are determined by the equation $Z^2 - f(\lambda)Z + 1 = 0$, where

$$(VI.10.4) \qquad f(\lambda) = \phi_1(\ell, \lambda) + \frac{d\phi_2}{dx}(\ell, \lambda).$$

Note that $f(\lambda)$ is continuous for $-\infty < \lambda < \infty$ (cf. Theorem II-1-2). We derive first the following conclusion.

Lemma VI-10-1. *The two multipliers $Z_1(\lambda)$ and $Z_2(\lambda)$ of system (VI.10.3) are*

$$\begin{cases} Z_1(\lambda) = \dfrac{1}{2}\left[\, f(\lambda) + \sqrt{f(\lambda)^2 - 4}\,\right], \\[2mm] Z_2(\lambda) = \dfrac{1}{2}\left[\, f(\lambda) - \sqrt{f(\lambda)^2 - 4}\,\right], \end{cases}$$

where $f(\lambda)$ is given by (VI.10.4). Therefore,
(i) if $|\, f(\lambda)\,| > 2$, two multipliers are real and distinct, i.e.,

$$\begin{cases} \quad Z_1(\lambda) > 1, \qquad 0 < Z_2(\lambda) < 1 \qquad \text{if } f(\lambda) > 2, \\ -1 < Z_1(\lambda) < 0, \qquad Z_2(\lambda) < -1 \qquad \text{if } f(\lambda) < -2, \end{cases}$$

(ii) if $|\, f(\lambda)\,| < 2$, two multipliers are complex and distinct, and

$$|\, Z_1(\lambda)\,| = 1 \qquad and \qquad |\, Z_2(\lambda)\,| = 1,$$

(iii) if $f(\lambda) = 2$, then $Z_1(\lambda) = Z_2(\lambda) = 1$,
(iv) if $f(\lambda) = -2$, then $Z_1(\lambda) = Z_2(\lambda) = -1$.

Observation VI-10-2. If $f(\lambda) = 2$, let $\begin{bmatrix} c_1 \\ c_2 \end{bmatrix}$ be an eigenvector of $\Phi(\lambda)$ associated with the eigenvalue 1. This means that c_1 and c_2 are two real numbers not both zero and that

$$c_1\phi_1(\ell, \lambda) + c_2\phi_2(\ell, \lambda) = c_1, \qquad c_1\frac{d\phi_1}{dx}(\ell, \lambda) + c_2\frac{d\phi_2}{dx}(\ell, \lambda) = c_2.$$

Set $\phi(x, \lambda) = c_1\phi_1(x, \lambda) + c_2\phi_2(x, \lambda)$. Then, $\phi(x, \lambda)$ is a nontrivial solution of (VI.10.1) such that

$$\phi(\ell, \lambda) = \phi(0, \lambda) \qquad and \qquad \frac{d\phi}{dx}(\ell, \lambda) = \frac{d\phi}{dx}(0, \lambda).$$

This implies that $\phi(x, \lambda)$ is a nontrivial periodic solution of (VI.10.1) of period ℓ.

Observation VI-10-3. If $f(\lambda) = -2$, it can be shown that equation (VI.10.1) has a nontrivial solution $\phi(x, \lambda)$ such that

$$\phi(\ell, \lambda) = -\phi(0, \lambda) \qquad and \qquad \frac{d\phi}{dx}(\ell, \lambda) = -\frac{d\phi}{dx}(0, \lambda).$$

This implies that $\phi(x, \lambda)$ satisfies the condition
$$\phi(x + \ell, \lambda) = -\phi(x, \lambda) \qquad \text{for} -\infty < x < +\infty.$$

Observation VI-10-4. Cconsider the eigenvalue problem
$$(VI.10.5) \qquad \frac{d^2 y}{dx^2} + (\lambda - u(x))\, y = 0, \qquad y(0) = 0, \qquad y(\ell) = 0.$$

Applying Theorem VI-3-11 (with $-\lambda$ instead of λ) to problem (VI.10.5), it can be shown that (VI.10.5) has infinitely many eigenvalues
$$(VI.10.6) \qquad \mu_1 < \mu_2 < \cdots < \mu_n < \cdots$$

which are determined by the equation
$$(VI.10.7) \qquad \phi_2(\ell, \lambda) = 0.$$

The eigenfunction $\phi_2(x, \mu_n)$ has exactly $n - 1$ zeros on the open interval $0 < x < \ell$.

Since $\dfrac{d\phi_2}{dx}(0, \lambda) = 1 > 0$, it follows that
$$\frac{d\phi_2}{dx}(\ell, \mu_n) \begin{cases} < 0 & \text{for } n \text{ is odd,} \\ > 0 & \text{for } n \text{ is even.} \end{cases}$$

Furthermore, condition (VI.10.7) implies that $\det[\Phi(\lambda)] = \phi_1(\ell, \lambda)\dfrac{d\phi_2}{dx}(\ell, \lambda) = 1$.

Thus, $f(\lambda) = \phi_1(\ell, \lambda) + \dfrac{1}{\phi_1(\ell, \lambda)}$ follows from (VI.10.4). Therefore,
$$f(\lambda) \geq 2 \quad \text{if} \quad \frac{d\phi_2}{dx}(\ell, \lambda) > 0 \quad \text{and} \quad f(\lambda) \leq -2 \quad \text{if} \quad \frac{d\phi_2}{dx}(\ell, \lambda) < 0,$$

i.e.,
$$f(\mu_n) \begin{cases} \geq 2 & \text{if } n \text{ is even,} \\ \leq -2 & \text{if } n \text{ is odd} \end{cases}$$

(cf. (VI.10.7)). Here, use was made of the fact that $|\alpha| + \dfrac{1}{|\alpha|} = \left(\sqrt{|\alpha|} - \dfrac{1}{\sqrt{|\alpha|}}\right)^2 + 2$

if $\alpha \neq 0$.

Observation VI-10-5. If
$$(VI.10.8) \qquad \lambda < \min\{u(x) : -\infty < x < +\infty\},$$

then $f(\lambda) > 2$. In fact, $\dfrac{d^2\phi_1}{dx^2}(x, \lambda) = (u(x) - \lambda)\phi_1(x, \lambda) > 0$ as long as $\phi_1(x, \lambda) > 0$ and, hence, $\dfrac{d\phi_1}{dx}(x, \lambda) > 0$. Thus, we obtain $\phi_1(\ell, \lambda) > 1$. Similarly, $\dfrac{d\phi_2}{dx}(\ell, \lambda) > 1$ if (VI.10.8) is satisfied. In this way, a rough picture of the graph of the function $f(\lambda)$ is obtained (cf. Figure 1).

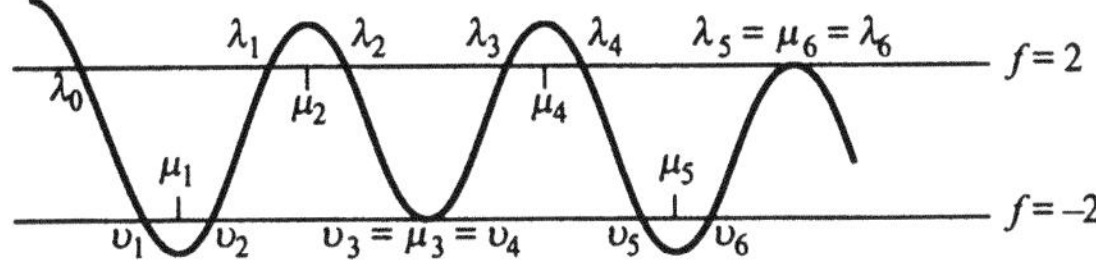

FIGURE 1.

Actually, we can prove the following theorem.

Theorem VI-10-6. *Assuming that $u(x)$ is continuous and periodic of period $\ell > 0$ on the entire real line $\mathbb{R}$, consider three boundary-value problems:*

(A) $$\frac{d^2y}{dx^2} + (\lambda - u(x))\, y = 0, \qquad y(0) = 0, \quad y(\ell) = 0,$$

(B) $$\frac{d^2y}{dx^2} + (\lambda - u(x))\, y = 0, \qquad y(\ell) = y(0), \quad \frac{dy}{dx}(\ell) = \frac{dy}{dx}(0),$$

(C) $$\frac{d^2y}{dx^2} + (\lambda - u(x))\, y = 0, \qquad y(\ell) = -y(0), \quad \frac{dy}{dx}(\ell) = -\frac{dy}{dx}(0).$$

Then, each of these three problems has infinitely many eigenvalues:

(VI.10.9)
$$
\begin{cases}
\mu_1 < \mu_2 < \mu_3 \cdots < \mu_n < \cdots, \\
\lambda_0 < \lambda_1 \leq \lambda_2 < \lambda_3 \leq \lambda_4 < \cdots < \lambda_{2n-1} \leq \lambda_{2n} < \cdots, \\
\nu_1 \leq \nu_2 < \nu_3 \leq \nu_4 < \cdots < \nu_{2n-1} \leq \nu_{2n} < \cdots,
\end{cases}
$$

respectively. Furthermore, if we denote by $\alpha_n(x)$, $\beta_n(x)$, and $\gamma_n(x)$ the eigenfunctions associated with the eigenvalues μ_n, λ_n, and ν_n, respectively, then

(1) $\lambda_0 < \nu_1$,

(2) $\nu_{2n-1} \leq \mu_{2n-1} \leq \nu_{2n} < \lambda_{2n-1} \leq \mu_{2n} \leq \lambda_{2n} < \nu_{2n+1} \leq \mu_{2n+1} \leq \nu_{2(n+1)}$ for $n = 1, 2, \ldots$,

(3) $\beta_{2n-1}(x)$ and $\beta_{2n}(x)$ are linearly independent if $\lambda_{2n-1} = \lambda_{2n}$,

(4) $\gamma_{2n-1}(x)$ and $\gamma_{2n}(x)$ are linearly independent if $\nu_{2n-1} = \nu_{2n}$,

(5) $\beta_0(x)$ does not have any zero on the interval $0 \leq x < \ell$,

(6) $\beta_{2n-1}(x)$ and $\beta_{2n}(x)$ have exactly $2n$ zeros on the interval $0 \leq x < \ell$,

(7) $\gamma_{2n-1}(x)$ and $\gamma_{2n}(x)$ have exactly $2n - 1$ zeros on the interval $0 \leq x < \ell$.

Proof.

We prove this theorem in five steps. Note first that (VI.10.9) is obtained from Figure 1.

Step 1. If $\phi(x, \lambda)$ is a solution of differential equation (VI.10.1), then $\dfrac{d\phi}{d\lambda}(x, \lambda)$ is a solution of the initial-value problem
(VI.10.10)

$$\frac{d^2w}{dx^2} + (\lambda - u(x))\, w + \phi(x, \lambda) = 0, \qquad
\begin{cases}
w(0, \lambda) = \dfrac{d\phi}{d\lambda}(0, \lambda), \\[2mm]
\dfrac{dw}{dx}(0, \lambda) = \dfrac{d}{d\lambda}\left(\dfrac{d\phi}{dx}(0, \lambda)\right)
\end{cases}$$

(cf. §II-2). Therefore, using the variation of parameters method, we obtain
(VI.10.11)

$$
\begin{cases}
\dfrac{d\phi_1}{d\lambda}(x, \lambda) = \phi_1(x, \lambda) \displaystyle\int_0^x \phi_1(t, \lambda)\phi_2(t, \lambda)\,dt - \phi_2(x, \lambda) \displaystyle\int_0^x \phi_1(t, \lambda)^2\,dt, \\[4mm]
\dfrac{d\phi_2}{d\lambda}(x, \lambda) = \phi_1(x, \lambda) \displaystyle\int_0^x \phi_2(t, \lambda)^2\,dt - \phi_2(x, \lambda) \displaystyle\int_0^x \phi_1(t, \lambda)\,\phi_2(t, \lambda)\,dt
\end{cases}
$$

and, hence,

$$\frac{d}{d\lambda}\left(\frac{d\phi_2}{dx}(x,\lambda)\right) = \frac{d\phi_1}{dx}(x,\lambda)\int_0^x \phi_2(t,\lambda)^2 dt - \frac{d\phi_2}{dx}(x,\lambda)\int_0^x \phi_1(t,\lambda)\phi_2(t,\lambda)dt.$$

Therefore, $\dfrac{df(\lambda)}{d\lambda} = \displaystyle\int_0^\ell Q(\phi_1(t,\lambda),\phi_2(t,\lambda))dt$ follows from (VI.10.4), where

$$Q(Y_1,Y_2) = -\phi_2(\ell,\lambda)Y_1^2 + \frac{d\phi_1}{dx}(\ell,\lambda)Y_2^2 + \left[\phi_1(\ell,\lambda) - \frac{d\phi_2}{dx}(\ell,\lambda)\right]Y_1Y_2.$$

Step 2. The discriminant of the quadratic form Q is $f(\lambda)^2 - 4$. In fact,

$$\left[\phi_1(\ell,\lambda) - \frac{d\phi_2}{dx}(\ell,\lambda)\right]^2 + 4\frac{d\phi_1}{dx}(\ell,\lambda)\phi_2(\ell,\lambda)$$

$$= \left[\phi_1(\ell,\lambda) - \frac{d\phi_2}{dx}(\ell,\lambda)\right]^2 + 4\phi_1(\ell,\lambda)\frac{d\phi_2}{dx}(\ell,\lambda) - 4 = f(\lambda)^2 - 4.$$

Note that

$$\phi_1(\ell,\lambda)\frac{d\phi_2}{dx}(\ell,\lambda) - \frac{d\phi_1}{dx}(\ell,\lambda)\phi_2(\ell,\lambda) = \det[\Phi(\lambda)] = 1.$$

Thus, $Q(\phi_1(x,\lambda),\phi_2(x,\lambda))$ does not change sign for $0 \le x \le \ell$ if $|f(\lambda)| < 2$. It follows that

(I) $$\frac{df(\lambda)}{d\lambda} \neq 0 \quad \text{if} \quad |f(\lambda)| < 2.$$

Step 3. Also, it can be proven that

(II) $$\begin{cases} \dfrac{df(\lambda)}{d\lambda} < 0 & \text{if} \quad \lambda < \mu_1 \text{ and } f(\lambda)^2 = 4, \\[2mm] (-1)^j\dfrac{df(\lambda)}{d\lambda} < 0 & \text{if} \quad \mu_j < \lambda < \mu_{j+1} \text{ and } f(\lambda)^2 = 4. \end{cases}$$

To prove this, notice that Q is a perfect square if $f(\lambda)^2 = 4$ and that $\phi_2(\ell,\lambda) \neq 0$ if $\lambda \neq \mu_j$ for every j. Hence, $\dfrac{df(\lambda)}{d\lambda}\phi_2(\ell,\lambda) < 0$. Also, using Lemma VI-3-11 (with $-\lambda$ instead of λ), we obtain

$$\begin{cases} \phi_2(\ell,\lambda) > 0 & \text{if} \quad \lambda < \mu_1, \\ (-1)^j\phi_2(\ell,\lambda) > 0 & \text{if} \quad \mu_j < \lambda < \mu_{j+1}. \end{cases}$$

Thus, (1) is verified.

Step 4. Let us prove that

(III)
$$
\frac{d^2 f(\lambda)}{d\lambda^2} \begin{cases} < 0 & \text{if} \quad f(\lambda) = 2 \quad \text{and} \quad \dfrac{df(\lambda)}{d\lambda} = 0, \\[2mm] > 0 & \text{if} \quad f(\lambda) = -2 \quad \text{and} \quad \dfrac{df(\lambda)}{d\lambda} = 0. \end{cases}
$$

First observe that Q is a perfect square if $|f(\lambda)| = 2$. Hence,

$$
\frac{df}{d\lambda}(\lambda) = \pm \int_0^\ell [c_1\phi_1(t,\lambda) + c_2\phi_2(t,\lambda)]^2 \, dt = 0,
$$

where c_1 and c_2 are some real numbers. Therefore, $c_1 = 0$ and $c_2 = 0$, since ϕ_1 and ϕ_2 are linearly independent. This means that

(IV)
$$
\phi_2(\ell,\lambda) = 0, \quad \frac{d\phi_1}{dx}(\ell,\lambda) = 0, \quad \phi_1(\ell,\lambda) = \frac{d\phi_2}{dx}(\ell,\lambda).
$$

Since $\det[\Phi(\lambda)] = 1$, we conclude that

(V)
$$
\Phi(\lambda) = \begin{cases} I_2 & \text{if} \quad f(\lambda) = 2 \quad \text{and} \quad \dfrac{df}{d\lambda}(\lambda) = 0, \\[2mm] - I_2 & \text{if} \quad f(\lambda) = -2 \quad \text{and} \quad \dfrac{df}{d\lambda}(\lambda) = 0, \end{cases}
$$

where I_2 is the 2×2 identity matrix. Furthermore,

$$
\frac{d^2 f}{d\lambda^2}(\lambda) = \int_0^\ell \left[-\frac{d\phi_2}{d\lambda}(\ell,\lambda)\phi_1(t,\lambda)^2 + \frac{d}{d\lambda}\left(\frac{d\phi_1}{dx}\right)(\ell,\lambda)\phi_2(t,\lambda)^2 \right.
$$
$$
\left. + \left\{ \frac{d\phi_1}{d\lambda}(\ell,\lambda) - \frac{d}{d\lambda}\left(\frac{d\phi_2}{dx}\right)(\ell,\lambda) \right\} \phi_1(t,\lambda)\phi_2(t,\lambda) \right] dt.
$$

Also, if $f(\lambda) = 2$ and $\dfrac{df}{d\lambda}(\lambda) = 0$, it follows from (VI.10.11) and (IV) that

$$
\begin{cases} \dfrac{d\phi_1}{d\lambda}(\ell,\lambda) = \displaystyle\int_0^\ell \phi_1(s,\lambda)\phi_2(s,\lambda)ds, \\[3mm] \dfrac{d\phi_2}{d\lambda}(\ell,\lambda) = \displaystyle\int_0^\ell \phi_2(s,\lambda)^2 ds, \\[3mm] \dfrac{d}{d\lambda}\left(\dfrac{d\phi_1}{dx}\right)(\ell,\lambda) = -\displaystyle\int_0^\ell \phi_1(s,\lambda)^2 ds, \\[3mm] \dfrac{d}{d\lambda}\left(\dfrac{d\phi_2}{dx}\right)(\ell,\lambda) = -\displaystyle\int_0^\ell \phi_1(s,\lambda)\phi_2(s,\lambda)ds. \end{cases}
$$

Hence,

$$
\frac{d^2 f}{d\lambda^2}(\lambda) = -\int_0^\ell \int_0^\ell [\phi_1(t,\lambda)\phi_2(s,\lambda) - \phi_1(s,\lambda)\phi_2(t,\lambda)]^2 dt ds < 0
$$

if $f(\lambda) = 2$ and $\dfrac{df}{d\lambda}(\lambda) = 0$. Similarly,

$$\frac{d^2 f}{d\lambda^2}(\lambda) = \int_0^\ell \int_0^\ell [\phi_1(t,\lambda)\phi_2(s,\lambda) - \phi_1(s,\lambda)\phi_2(t,\lambda)]^2 \, dt\, ds \; > \; 0$$

if $f(\lambda) = -2$ and $\dfrac{df}{d\lambda}(\lambda) = 0$. Note that, if $\lambda_{2n-1} = \lambda_{2n}$, it follows from (VI.10.2), (IV), and (V) that $\phi_1(x,\lambda)$ and $\phi_2(x,\lambda)$ are two linearly independent solutions for problem (B). Similarly, if $\nu_{2n-1} = \nu_{2n}$, it follows from (VI.10.2), (IV), and (V) that $\phi_1(x,\lambda)$ and $\phi_2(x,\lambda)$ are two linearly independent solutions for problem (C). Thus, (3) and (4) are verified.

Step 5. The functions $\beta_n(x)$ (respectively $\gamma_n(x)$) have an even (respectively odd) number of zeros on the interval $0 \le x < \ell$, since $\beta_n(0) = \beta_n(\ell)$ and $\gamma_n(0) = -\gamma_n(\ell)$. As can be seen in Figure 1,

(VI) $$\mu_{2n-1} \; < \; \lambda_{2n-1} \le \lambda_{2n} \; < \; \mu_{2n+1}.$$

Therefore, by virtue of Theorems VI-3-11 and VI-1-1, we conclude that β_{2n-1} and β_{2n} have more than $2n - 1$, and less than $2n + 2$, zeros on the interval $0 \le x < \ell$. Hence, they have exactly $2n$ zeros there. Similarly, since

(VII) $$\mu_{2n-2} \; < \; \nu_{2n-1} \le \nu_{2n} \; < \; \mu_{2n},$$

γ_{2n-1} and γ_{2n} have exactly $2n - 1$ zeros on the interval $0 \le x < \ell$. The function β_0 does not have any zero on the interval $0 \le x < \ell$ since $\lambda_0 < \mu_1$. Thus, (5), (6), and (7) are verified. Finally, (2) follows from (VI.10.9), (VI), and (VII). $\square$

Definition VI-10-7.
  (I) *The set* $\{\lambda : f(\lambda)^2 < 4\}$ *is called the* stability region *of the differential equation (VI.10.1).*
  (II) *The set* $\{\lambda : f(\lambda)^2 > 4\}$ *is called the* instability region *of the differential equation (VI.10.1).*

Example VI-10-8. A periodic function $u(x)$ is called a *finite-zone potential* if the function $f(\lambda)^2 - 4$ of the differential equation $\dfrac{d^2 y}{dx^2} + (\lambda - u(x))y = 0$ has a finite number of simple zeros (i.e., all other zeros are double). For example, consider the case when $u(x) = a$, where a is a constant. The differential equation becomes $\dfrac{d^2 y}{dx^2} + (\lambda - a)y = 0$ and, hence,

$$\phi_1(x,\lambda) \; = \; \begin{cases} \cos(\sqrt{\lambda - a}\, x) & \text{if} \quad \lambda > a, \\ 1 & \text{if} \quad \lambda = a, \\ \cosh(\sqrt{a - \lambda}\, x) & \text{if} \quad \lambda < a \end{cases}$$

and

$$\phi_2(x,\lambda) \;=\; \begin{cases} \dfrac{\sin(\sqrt{\lambda-a}\,x)}{\sqrt{\lambda-a}} & \text{if } \lambda > a, \\[2ex] x & \text{if } \lambda = a, \\[2ex] \dfrac{\sinh(\sqrt{a-\lambda}\,x)}{\sqrt{a-\lambda}} & \text{if } \lambda < a. \end{cases}$$

Therefore,

$$f(\lambda) \;=\; \phi_1(\ell,\lambda) \;+\; \frac{d\phi_2}{dx}(\ell,\lambda) \;=\; \begin{cases} 2\cos(\sqrt{\lambda-a}\,\ell) & \text{if } \lambda > a, \\[1ex] 2 & \text{if } \lambda = a, \\[1ex] 2\cosh(\sqrt{a-\lambda}\,\ell) & \text{if } \lambda < a. \end{cases}$$

Thus, we conclude that in this case, $f(\lambda)^2 - 4$ has only one simple zero a (cf. Figure 2).

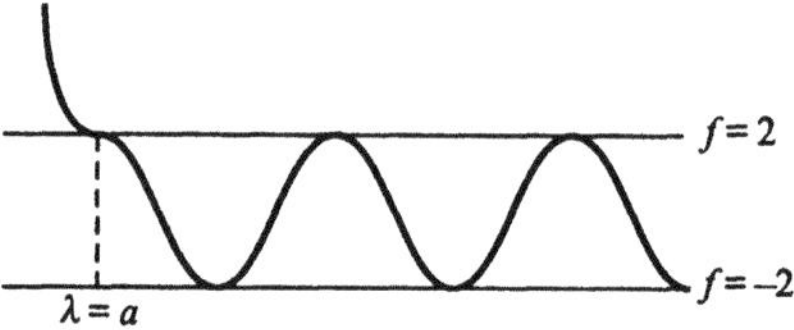

FIGURE 2.

The materials in this section are also found in [CL, Chapter 8].

EXERCISES VI

VI-1. Assume that $u(x)$ is a real-valued continuous function on the interval $\mathcal{I}_0 = \{x : 0 \le x < +\infty\}$ such that $u(x) \ge m_0$ for $x \ge x_0$ for some positive numbers m_0 and x_0. Show that

(1) every nontrivial solution of the differential equation

$$\text{(E)} \qquad\qquad \frac{d^2y}{dx^2} \;-\; u(x)y \;=\; 0$$

has at most a finite number of zeros on $\mathcal{I}_0$,

(2) the differential equation (E) has a nontrivial solution $\eta(x)$ such that
$\displaystyle \lim_{x \to +\infty} \eta(x) = 0$.

Hint.

(1) Note that if $y(x_0) > 0$, then $y''(x_0) > 0$. Hence, $y(x) > 0$ for $x \ge x_0$ if $y'(x_0) > 0$.

(2) It is sufficient to find a solution $\phi(x)$ such that $\phi(x) > 0$ and $\phi'(x) < 0$ for $x \ge x_0$.

VI-2. For the eigenvalue-problem

$$\text{(EP)} \qquad \frac{d^2y}{dx^2} + u(x)y = \lambda y, \qquad y(0) = y(1), \quad y'(0) = y'(1),$$

where $u(x)$ is real-valued and continuous on the interval $0 \le x \le 1$,
  (1) construct Green's function,
  (2) show that (EP) is self-adjoint,
  (3) show that (EP) has infinitely many eigenvalues.

VI-3. Let $\lambda_1 > \lambda_2 > \cdots > \lambda_n > \cdots$ be eigenvalues of the boundary-value problem

$$\frac{d^2y}{dx^2} + u(x)y = \lambda y, \qquad y(a) = 0, \quad y'(b) = 0,$$

where $u(x)$ is real-valued and continuous on the interval $a \le x \le b$. Show that there exists a positive number K such that

$$\left| \frac{\lambda_n}{n^2} + \left(\frac{\pi}{b-a} \right)^2 \right| \le \frac{K}{n} \qquad \text{for} \quad n = 1, 2, 3, \ldots .$$

VI-4. Assuming that $u(x)$ is real-valued and continuous on the interval $0 \le x < +\infty$ and that $\lim\limits_{x \to +\infty} u(x) = +\infty$, consider the eigenvalue-problem

$$\frac{d^2y}{dx^2} - u(x)y = \lambda y, \qquad y(0)\cos\alpha - y'(0)\sin\alpha = 0, \qquad \lim\limits_{x \to +\infty} y(x) = 0.$$

where α is a non-negative constant. Show that
  (a) there exist infinitely many real eigenvalues $\lambda_1 > \lambda_2 > \cdots$ such that $\lim\limits_{n \to +\infty} \lambda_n = -\infty$,
  (b) eigenfunctions corresponding to the eigenvalue λ_n have exactly $n-1$ zeros on the interval $0 < x < +\infty$.

Hint. Let $\lambda_1(b) > \lambda_2(b) > \cdots$ be the eigenvalues of

$$\frac{d^2y}{dx^2} - u(x)y = \lambda y, \qquad y(0)\cos\alpha - y'(0)\sin\alpha = 0, \qquad y(b) = 0,$$

where $b > 0$. Define λ_m by $\lim\limits_{b \to +\infty} \lambda_m(b)$. See [CL, Problem 1 on p. 254].

VI-5. Show that if a function $\phi(x)$ is real-valued, twice continuously differentiable, and $\phi''(x) + e^{-x}\phi(x) = 0$ on the interval $\mathcal{I}_0 = \{x : 0 \le x < +\infty\}$ and if $\int_0^{+\infty} \phi(x)^2 dx < +\infty$, then $\phi(x)$ is identically equal to zero on $\mathcal{I}_0$.

VI-6. Using the notations and definitions of §VI-4, show that

$$(f, \mathcal{L}[f]) = \sum_{n=1}^{+\infty} \lambda_n (f, \eta_n)^2$$

$$= \int_a^b \{u(x)f(x)^2 - p(x)f'(x)^2\}dx + p(b)f(b)f'(b) - p(a)f(a)f'(a),$$

if $f \in \mathcal{V}(a, b)$.

VI-7. Assume that $u(x)$ is real-valued and continuous and $u(x) \leq 0$ on the interval $\mathcal{I}(a, b)$, where $a < b$. Denote by $\phi(x, \lambda)$ the unique solution of the initial-value problem $\dfrac{d^2 y}{dx^2} + u(x)y = \lambda y$, $y(a) = 0$, $y'(a) = 1$, where λ is a comlex parameter. Show that

   (i) $\phi(b, \lambda)$ is an entire function of λ,

   (ii) $\phi(b, \lambda) \neq 0$ if λ is a positive real number,

  (iii) $\phi(b, \lambda)$ has infinitely many zeros λ_n such that $0 \geq \lambda_0 > \lambda_1 > \lambda_2 > \cdots$ and

$$\lim_{n \to +\infty} \frac{\lambda_n}{n^2} = -\left(\frac{\pi}{b-a}\right)^2.$$

VI-8. Find the unique solution $\phi(x)$ of the differential equation $\dfrac{d}{dx}\left(x\dfrac{dy}{dx}\right) = y$ such that $\phi(x)$ is analytic at $x = 0$ and $\phi(0) = 1$. Also, show that

   (i) $\phi(x)$ is an entire function of x,

   (ii) $\phi(x) \neq 0$ if x is a positive real number,

  (iii) $\phi(x)$ has infinitely many zeros λ_n such that $0 \geq \lambda_0 > \lambda_1 > \lambda_2 > \cdots$ and $\lim\limits_{n \to +\infty} \lambda_n = -\infty$,

  (iv) $\displaystyle\int_0^1 \phi(\lambda_n x)\phi(\lambda_m x)dx = 0$ if $n \neq m$.

VI-9. Show that the Legendre polynomials

$$P_n(x) = \frac{1}{2^n n!}\frac{d^n}{dx^n}[(x^2 - 1)^n] \qquad (n = 0, 1, 2, \ldots)$$

satisfy the following conditions:

   (i) $\deg P_n(x) = n$ $(n = 0, 1, 2, \ldots)$,

   (ii) $\displaystyle\int_{-1}^1 P_n(x)P_m(x)dx = 0$ if $n \neq m$,

  (iii) $\displaystyle\int_{-1}^1 P_n(x)^2 dx = \frac{1}{n + \frac{1}{2}}$ $(n = 0, 1, 2, \ldots)$,

  (iv) $\displaystyle\int_{-1}^1 x^k P_n(x)dx = 0$ for $k = 0, \ldots, n-1$,

   (v) $P_n(x)$ $(n \geq 1)$ has n simple zeros in the interval $|x| < 1$,

(vi) if $f(x)$ is real-valued and continuous on the interval $|x| \leq 1$, then

$$\lim_{N \to +\infty} \int_{-1}^{1} \left(f(x) - \sum_{n=0}^{N} \left(n + \frac{1}{2} \right) (f, P_n) P_n(x) \right)^2 dx = 0,$$

where $(f, P_n) = \displaystyle\int_{-1}^{1} f(x) P_n(x) dx$,

(vii) the series $\displaystyle\sum_{n=0}^{+\infty} \left(n + \frac{1}{2} \right) (f, P_n) P_n(x)$ converges to f uniformly on the interval

$|x| \leq 1$ if f, f', and f'' are continuous on the interval $|x| \leq 1$.

Hint. See Exercise V-13. Also, note that if $f(x)$ is continuous on the interval $|x| \leq 1$, then $f(x)$ can be approximated on this interval uniformly by a polynomial in x. To prove (vii), construct the Green function $G(x, \xi)$ for the boundary-value problem

$$\begin{cases} \dfrac{d}{dx} \left((1 - x^2) \dfrac{dy}{dx} \right) + \alpha_0 y = f(x), \\ y(x) \text{ is bounded in the neighborhood of } x = \pm 1, \end{cases}$$

where α_0 is not a non-negative integer. Show that $\displaystyle\int_{-1}^{1} \int_{-1}^{1} G(x, \xi)^2 dx d\xi < +\infty$. Then, we can use a method similar to that of §VI-4.

VI-10. Assume that (1) $p(x)$ and $p'(x)$ are continuous on an interval $\mathcal{I}_0(a, b) = \{x : a \leq x \leq b\}$, (2) $p(x) > 0$ on $\mathcal{I}_0$, and (3) $u(x, \lambda)$ is a real-valued and continuous function of (x, λ) on the region $\mathcal{I}_0 \times \mathbb{R} = \{(x, \lambda) : x \in \mathcal{I}_0, \lambda \in \mathbb{R}\}$ such that $\lim_{\lambda \to \pm\infty} u(x, \lambda) = \mp\infty$ uniformly for $x \in \mathcal{I}_0$. Assume also that $u(x, \lambda)$ is strictly decreasing in $\lambda \in \mathbb{R}$ for each fixed x on $\mathcal{I}_0$. Denote by $\phi(x, \lambda)$ the unique solution of the intial-value problem

$$\frac{d}{dx} \left(p(x) \frac{dy}{dx} \right) + u(x, \lambda) y = 0 \quad y(a) = 0, \quad y'(a) = 1.$$

Show that there exists a sequence $\{\mu_n : n = 0, 1, 2, \dots\}$ of real numbers such that

(i) $\mu_n < \mu_{n-1}$ $(n = 1, 2, \dots)$, and $\displaystyle\lim_{n \to +\infty} \mu_n = -\infty$,

(ii) $\phi(b, \mu_n) = 0$ $(n = 0, 1, 2, \dots)$,

(iii) $\phi(x, \lambda) \neq 0$ on $a < x \leq b$ for $\lambda > \mu_0$, and $\phi(x, \lambda)$ $(n \geq 1)$ has n simple zeros on $a < x \leq b$ for $\mu_n < \lambda \leq \mu_{n-1}$,

(iv) $\dfrac{p(b) \phi'(b, \lambda)}{\phi(b, \lambda)}$ strictly decreases from $+\infty$ to $-\infty$ as λ decreases from μ_{n-1} to μ_n.

VI-11. Assume that $p(x)$ and $u(x)$ are real-valued and continuous on the interval $\mathcal{I}(0, 1) = \{x : 0 \leq x \leq 1\}$ and that $p'(x)$ is also continuous on $\mathcal{I}(0, 1)$, $p(x) > 0$ for $0 < x \leq 1$, $p(0) = 0$, and $p'(0) < 0$. Show that the differential equation

$\dfrac{d}{dx}\left(p(x)\dfrac{dy}{dx}\right)+u(x)y=0$ has a fundamental set $\{\phi,\psi\}$ of solutions on the interval $0<x\le 1$ such that

(i) $\displaystyle\lim_{x\to 0+}\phi(x)=1$ and $\displaystyle\lim_{x\to 0+}p(x)\phi'(x)=0$,

(ii) $\displaystyle\lim_{x\to 0+}\left(\psi(x)-\int_1^x\dfrac{d\xi}{p(\xi)}\right)=0$ and $\displaystyle\lim_{x\to 0+}p(x)\psi'(x)=1$.

VI-12. Set

$$u(x)=\begin{cases}1 & \text{for} & 0\le x\le\dfrac{1}{2},\\[2ex] -1 & \text{for} & \dfrac{1}{2}\le x\le 1.\end{cases}$$

Denote by $\phi_1(x,\lambda)$ and $\phi_2(x,\lambda)$ the two unique solutions of the differential equation $\dfrac{d^2y}{dx^2}+(\lambda-u(x))y=0$ satisfying the initial conditions $\phi_1(0,\lambda)=1$, $\dfrac{d\phi_1}{dx}(0,\lambda)=0$, $\phi_2(0,\lambda)=0$, and $\dfrac{d\phi_2}{dx}(0,\lambda)=1$, where λ is a real parameter. Sketch the graph of the function $f(\lambda)=\phi(1,\lambda)+\dfrac{d\phi}{dx}(1,\lambda)$.

VI-13. For the scattering data $\{r(\xi)=0,(1,2,3),(1,1,1)\}$, find the potential $u(x)$ and the Jost solutions $f_\pm(x,\zeta)$.

VI-14. Calculate the scattering data for each of $u(x+1)$ and $u(-x)$, assuming that $\{r(\xi),(\eta_1,\dots,\eta_N),(c_1,\dots,c_N)\}$ are the scattering data for $u(x)$ and that $u(x)$ satisfies a condition $|u(x)|\le Ae^{-k|x|}$ for some positive numbers A and k.

Hint. Let $f_\pm(x,\zeta)$ be the Jost solutions for $u(x)$. Two quantities $a(\zeta)$ and $b(\xi)$ are given by

$$a(\zeta)=\frac{i}{2\zeta}\begin{vmatrix}f_+(x,\zeta) & f_-(x,\zeta)\\ f'_+(x,\zeta) & f'_-(x,\zeta)\end{vmatrix}\quad\text{and}\quad b(\xi)=\frac{-i}{2\xi}\begin{vmatrix}\bar f_+(x,\xi) & f_-(x,\xi)\\ \bar f'_+(x,\xi) & f'_-(x,\xi)\end{vmatrix},$$

respectively.

The Jost solutions for $u(x+1)$ are $e^{\mp i\zeta}f_\pm(x+1,\zeta)$. Therefore, the scattering data for $u(x+1)$ are

$$\{e^{2i\xi}r(\xi),\ (\eta_1,\dots,\eta_N),\ (e^{-2\eta_1}c_1,\dots,e^{-2\eta_N}c_N)\}.$$

The Jost solutions for $u(-x)$ are $f_\mp(-x,\zeta)$. Hence, the scattering data for $u(-x)$ are

$$\left\{-\frac{\bar b(\xi)}{a(\xi)},\ (\eta_1,\dots,\eta_N),\ \left(-\frac{1}{c_1a_\zeta(i\eta_1)^2},\dots,-\frac{1}{c_Na_\zeta(i\eta_N)^2}\right)\right\}.$$

Note that

$$f_-(x,i\eta_j)=ic_ja_\zeta(i\eta_j)f_+(x,i\eta_j).$$

VI-15. Let $u(x)$ be real-valued, continuous, and periodic of period $\ell > 0$. Also, for every real ξ, let $\psi(x, \xi, \lambda)$ be the solution of the differential equation

$$\text{(Eq)} \qquad \frac{d^2y}{dx^2} + (\lambda - u(x))y = 0$$

satisfying the initial conditions $\psi(\xi, \xi, \lambda) = 0$ and $\psi'(\xi, \xi, \lambda) = 1$. Let

$$\lambda = \mu_1(\xi) < \mu_2(\xi) < \mu_3(\xi) < \cdots$$

be all roots of $\psi(\xi + \ell, \xi, \lambda) = 0$ with respect to λ. Show that

$$\psi(\xi + \ell, \xi, \lambda) = \ell \prod_{m=1}^{+\infty} \left[\frac{\mu_m(\xi) - \lambda}{\left(\dfrac{m\pi}{\ell}\right)^2} \right].$$

Hint.
Step 1. Let us construct $\psi(x, \xi, \lambda)$ for negative λ. To do this, change (Eq) to the integral equation

$$\psi(x, \xi, \lambda) = \frac{\sinh(\mu(x - \xi))}{\mu} + \frac{1}{\mu} \int_{\xi}^{x} \sinh(\mu(x - s))u(s)\psi(s, \xi, \lambda)ds,$$

where $\mu = \sqrt{-\lambda} > 0$. Since $e^{-\mu(x-\xi)} \sinh(\mu(x - \xi))$ is bounded as $\mu \to +\infty$, it can be shown that

$$e^{-\mu(x-\xi)}\psi(x, \xi, \lambda) = e^{-\mu(x-\xi)} \frac{\sinh(\mu(x - \xi))}{\mu} + O\left(-\frac{1}{\lambda}\right)$$

on the interval $0 \leq x - \xi \leq \ell$ as $\lambda \to -\infty$. Thus, we derive

$$\lim_{\lambda \to -\infty} \frac{\psi(\xi + \ell, \xi, \lambda)}{\left(\dfrac{\sinh(\mu\ell)}{\mu}\right)} = 1.$$

Step 2. There exists an entire function $R(\lambda)$ of λ such that

$$R(\lambda) = \begin{cases} \dfrac{\sin(\ell\sqrt{\lambda})}{\sqrt{\lambda}} & \text{for} \quad \lambda > 0, \\[3mm] \dfrac{\sinh(\ell\sqrt{-\lambda})}{\sqrt{-\lambda}} & \text{for} \quad \lambda < 0. \end{cases}$$

This function $R(\lambda)$ has the following factorization:

$$R(\lambda) = \ell \prod_{m=1}^{+\infty} \left[\frac{c_m^2 - \lambda}{c_m^2} \right],$$

where $c_m = \dfrac{m\pi}{\ell}$.

Step 3. If the function $\psi(\xi+\ell,\ell,\lambda)$ is entire in λ and $\psi(\xi+\ell,\ell,\lambda) = O(\exp(\ell\sqrt{|\lambda|}))$ as $\lambda \to \infty$, we can write $\psi(\xi+\ell,\ell,\lambda)$ in the following form:

$$\psi(\xi+\ell,\xi,\lambda) = c \prod_{m=1}^{+\infty} \left[\frac{\mu_m(\xi) - \lambda}{c_m^2}\right],$$

where c is independent of λ. Here, we used the fact that $|\mu_m(\xi) - c_m^2|$ is bounded as $m \to +\infty$ (cf. Theorem VI-3-14).

Step 4. Note that

(1)
$$\frac{\psi(\xi+\ell,\ell,\lambda)}{R(\lambda)} = \left(\frac{c}{\ell}\right) \prod_{m=1}^{+\infty} \left[\frac{\mu_m(\xi) - \lambda}{c_m^2 - \lambda}\right].$$

Note also that

$$\frac{\mu_m(\xi) - \lambda}{c_m^2 - \lambda} = 1 - \frac{c_m^2 - \mu_m(\xi)}{c_m^2 - \lambda}.$$

This implies the uniform convergence of the infinite product on the right-hand side of (1) for $-\infty < \lambda < 0$. Thus, $\displaystyle\lim_{\lambda \to -\infty} \frac{\psi(\xi+\ell,\ell,\lambda)}{R(\lambda)} = 1 = \frac{c}{\ell}$.

Remark. For expressions of analytic functions in infinite products, see, for example, [Pa, pp. 490-504].

VI-16. The functions $G_n(u)$ are defined in §VI-9. Calculate $G_4(u)$ and $G_5(u)$.

VI-17. Use the same notations as in Theorem VI-10-6. Show that if $\displaystyle\int_0^\ell u(t)dt = 0$, then $\lambda_0 \leq 0$. Also show that if $\displaystyle\int_0^\ell u(t)dt = 0$ and $\lambda_0 = 0$, then $u(x) = 0$ identically on $-\infty < x < +\infty$.

Hint. Since $\beta_0(x) \neq 0$ on $-\infty < x < +\infty$, set $w(x) = \dfrac{\beta_0'(x)}{\beta_0(x)}$. Then, $w'(x) + w(x)^2 + \lambda - u(x) = 0$ (cf. [MW, Theorem 4.4, p. 62]).

VI-18. Set $F(\zeta) = \log f(\zeta)$ for $\Im\zeta \geq 0, \zeta \neq 0$, where
   (i) $f(\zeta)$ is continuous for $\Im\zeta \geq 0, \zeta \neq 0$,
   (ii) $f(\zeta)$ is analytic for $\Im\zeta > 0$,
   (iii) $\zeta(f(\zeta) - 1)$ is bounded for $\Im\zeta \geq 0$,
   (iv) $f(\zeta) \neq 0$ for $\Im\zeta \geq 0, \zeta \neq 0$.
Denote by $\mathcal{S}_R$ the semicircle $\{\zeta : |\zeta| = R, 0 \leq \arg\zeta \leq \pi\}$ which is oriented counterclockwise, where R is a positive number. Show that

(a)
$$\frac{1}{2\pi i}\int_{-R}^R \frac{F(\xi)}{\xi - \zeta}d\xi + \frac{1}{2\pi i}\int_{\mathcal{S}_R} \frac{F(z)}{z - \zeta}dz = \begin{cases} F(\zeta) & (\Im\zeta > 0, |\zeta| < R), \\ 0 & (\Im\zeta < 0, |\zeta| < R), \end{cases}$$

(b)
$$F(\zeta) = \frac{1}{2\pi i}\int_{-\infty}^\infty \frac{F(\xi)}{\xi - \zeta}d\xi + \frac{1}{2\pi i}\int_{-\infty}^\infty \frac{\overline{F(\xi)}}{\xi - \zeta}d\xi \quad \text{for} \quad \Im\zeta > 0,$$

where $\overline{F(\xi)}$ is the complex conjugate of $F(\xi)$.

CHAPTER VII

ASYMPTOTIC BEHAVIOR OF
SOLUTIONS OF LINEAR SYSTEMS

In this chapter, we explain the behavior of solutions of a homogeneous linear system $\dfrac{d\vec{y}}{dt} = A(t)\vec{y}$ as $t \to +\infty$ in the case when the coefficient matrix $A(t)$ has a limit A_0 as $t \to +\infty$. The purpose is to show how much information we can glean from the limit matrix A_0. We are interested in the exponential growth of solutions and the asymptotic behavior of solutions. In order to measure the exponential growth of a function, we use Liapounoff's type numbers which was originally introduced by A. Liapounoff in [Lia]. Liapounoff's type numbers are explained in §VII-1. Also, we explain Liapounoff's type numbers of solutions of a homogeneous linear system in §VII-2. (Liapounoff's type numbers are also found in L. Cesari [Ce, pp. 50-55].) In §VII-3, assuming that $\lim_{t \to +\infty} A(t) = A_0$, we calculate Liapounoff's type numbers of solutions in terms of the eigenvalues of A_0 (cf. [Huk3], [Har2, Chapter X], and [Si9]). In §VII-4, we explain how to derive the asymptotic behavior of solutions by diagonalizing the given system. A theorem of M. Hukuhara and N. Nagumo gives the original motivation (cf. Theorem VII-4-1). The main result is Theorem of N. Levinson (cf. Theorem VII-4-2). (Generalizations and refinements of Theorem of Levinson are found, for example, in [Bel1], [Bel2], [CK], [Cop1], [Dev], [DK], [E1], [E2], [Gi], [GHS], [HarL1], [HarL2], [HarL3], [HW1], [HW2], [HX1], [HX2], and [HX3].) In §VII-5, the Theorem of Levinson is applied to a system whose matrix has a limit as $t \to +\infty$ and its derivative is absolutely integrable on the interval $0 \le t < \infty$. The topics of §§VII-4 and VII-5 are also found in [CL, §8 of Chapter 3, pp. 91-97]. In §VII-6, we explain how we can reduce some problems such as the differential equation $\dfrac{d^2\eta}{dt^2} + \{\lambda + h(t)\sin(\alpha t)\}\eta = 0$ to the Theorem of Levinson, even if the derivative of $h(t)\sin(\alpha t)$ is small but not absolutely integrable on the interval $0 \le t < \infty$. The main idea is to apply the Floquet theorem (Theorem IV-4-1) to $\dfrac{d^2\eta}{dt^2} + \{1 + \epsilon\sin(\alpha t)\}\eta = 0$ to eliminate the periodic parts of coefficients so that we can use the Theorem of Levinson (cf. [HaS3]; see also [HarL1], [HarL2], [HarL3]).

VII-1. Liapounoff's type numbers

In order to measure the exponential growth of a function, let us introduce *Liapounoff's type numbers*.

Definition VII-1-1. *Let $\vec{f}(t)$ be a $\mathbb{C}^n$-valued function whose entries are continuous on an interval $\mathcal{I} = \{t : t_0 \le t < +\infty\}$. Let us denote by $\mathcal{A}$ the set of all real* *numbers α such that* $\exp[-\alpha t]\,\vec{f}(t)$ *is bounded on the interval $\mathcal{I}$. Set*

$$(\text{VII.1.1}) \qquad \lambda\left(\vec{f}\right) = \begin{cases} +\infty & \text{if } \mathcal{A} = \emptyset, \\ \inf\{\alpha : \alpha \in \mathcal{A}\} & \text{if } \mathcal{A} \neq \mathbb{R} \ \text{ and } \ \mathcal{A} \neq \emptyset, \\ -\infty & \text{if } \mathcal{A} = \mathbb{R}. \end{cases}$$

The quantity $\lambda\left(\vec{f}\right)$ is called Liapounoff's type number *of $\vec{f}$ at $t = +\infty$.*

Note that if $\alpha \in \mathcal{A}$ and $\alpha < \beta$, then $\beta \in \mathcal{A}$.

Example VII-1-2.

$$\begin{cases} \lambda\,(0) = -\infty, \quad \lambda\left(\exp[-t^2]\right) = -\infty, \quad \lambda\,(t^m) = 0 \ \text{ (for all constants } m), \\ \lambda\,(\exp[\alpha t]) = \alpha, \quad \lambda\left(\exp[t^2]\right) = +\infty. \end{cases}$$

Here, it was assumed implicitly that $t_0 > 0$ if $m < 0$.
The following lemma can be proved easily.

Lemma VII-1-3. *Let $\lambda\left(\vec{f}\right)$ be Liapounoff's type number of a $\mathbb{C}^n$-valued function $\vec{f}(t)$ whose entries are continuous on an interval $\mathcal{I} = \{t : t_0 \leq t < +\infty\}$. Then,*

(i) $\exp\left[-\left(\lambda\left(\vec{f}\right)+\epsilon\right)t\right]\vec{f}(t)$ *is bounded on $\mathcal{I}$ if $\epsilon > 0$, and unbounded if $\epsilon < 0$, whenever $\lambda\left(\vec{f}\right) \neq -\infty$,*

(ii) $\lambda\left(\displaystyle\sum_{1 \leq j \leq m} \vec{f}_j\right) \leq \max\left\{\lambda\left(\vec{f}_j\right) : j = 1, 2, \ldots, m\right\},$

(iii) $\lambda\left(\displaystyle\sum_{1 \leq j \leq m} \vec{f}_j\right) = \lambda\left(\vec{f}_1\right), \ \ \text{if } \lambda\left(\vec{f}_1\right) > \lambda\left(\vec{f}_j\right) \text{ for } j = 2, \ldots, m,$

(iv) $\vec{f}_1, \vec{f}_2, \ldots, \vec{f}_m$ *are linearly independent on the interval $\mathcal{I}$ if $\lambda\left(\vec{f}_1\right), \ldots, \lambda\left(\vec{f}_m\right)$ are mutually distinct,*

(v) $\lambda\,(f_1 f_2 \cdots f_m) \leq \displaystyle\sum_{j=1}^{m} \lambda\,(f_j)$ *in the case when $f_1, \ldots, f_m$ are $\mathbb{C}$-valued functions,*

(vi) $\lambda\left(P(t)\vec{f}\right) = \lambda\left(\vec{f}\right)$, *if the entries of an $n \times n$ matrix $P(t)$ and the entries of its inverse $P(t)^{-1}$ are bounded on the interval $\mathcal{I}$.*

The following lemma characterizes Liapounoff's type number.

Lemma VII-1-4. *If $\lambda(\vec{f})$ is Liapounoff's type number of a function $\vec{f}(t)$ at $t = +\infty$, then*

$$\lambda(\vec{f}) = \lim_{t \to +\infty}\left\{\sup_{t \leq s < +\infty} \frac{\log |\vec{f}(s)|}{s}\right\}.$$

Proof.

Since there exists a positive constant K such that $|\vec{f}(s)| \leq Ke^{(\lambda(\vec{f})+\epsilon)s}$ for $\epsilon > 0$ and large values of s, it follows that

$$\frac{\log|\vec{f}(s)|}{s} \leq \lambda(\vec{f}) + \epsilon + \frac{\log K}{s} \leq \lambda(\vec{f}) + \epsilon + \frac{\log K}{t} \qquad \text{for} \quad t \leq s.$$

Hence,

$$\lim_{t \to +\infty} \left\{ \sup_{t \leq s < +\infty} \frac{\log|\vec{f}(s)|}{s} \right\} \leq \lambda(\vec{f}).$$

Also, for a fixed positive number ϵ and any positive integer m, there exists a large value of s_m such that $me^{(\lambda(\vec{f})-\epsilon)s_m} \leq |\vec{f}(s_m)|$ and $\lim\limits_{m \to +\infty} s_m = +\infty$. Hence, $\dfrac{\log m}{s_m} +$

$\lambda(\vec{f}) - \epsilon \leq \dfrac{\log|\vec{f}(s_m)|}{s_m}$. Therefore, we obtain

$$\lim_{t \to +\infty} \left\{ \sup_{t \leq s < +\infty} \frac{\log|\vec{f}(s)|}{s} \right\} \geq \lambda(\vec{f}).$$

Thus, Lemma VII-1-4 is proved. $\square$

VII-2. Liapounoff's type numbers of a homogeneous linear system

In this section, we explain Liapounoff's type numbers of solutions of a homogeneous linear system

$$\text{(VII.2.1)} \qquad\qquad \frac{d\vec{y}}{dt} = A(t)\vec{y}$$

under the assumption that the entries of the $n \times n$ matrix $A(t)$ are continuous and bounded on an interval $\mathcal{I} = \{t : t_0 \leq t < +\infty\}$. Let us start with the following fundamental result.

Theorem VII-2-1. *If $\vec{y} = \vec{\phi}(t)$ is a nontrivial solution of system (VII.2.1) and if $|A(t)| \leq K$ on the interval $\mathcal{I} = \{t : t_0 \leq t < +\infty\}$ for some non-negative number K, then*

$$\text{(VII.2.2)} \qquad\qquad \left|\lambda\left(\vec{\phi}\right)\right| \leq K.$$

Proof.

Let the $n \times n$ invertible matrix $\Phi(t)$ be the unique solution of the initial-value problem $\dfrac{dY}{dt} = A(t)Y$, $Y(t_0) = I_n$, where I_n is the $n \times n$ identity matrix. Then,

$\Phi(t)^{-1}$ is the unique solution of the initial-value problem $\dfrac{dZ}{dt} = -ZA(t), \quad Z(t_0) = I_n$ (cf. Lemma IV-2-4). Since $|I_n| = 1$, we obtain

$$\begin{cases} |\Phi(t)| \leq 1 + K \displaystyle\int_{t_0}^{t} |\Phi(t)|\, dt, \\[2mm] |\Phi(t)^{-1}| \leq 1 + K \displaystyle\int_{t_0}^{t} |\Phi(t)^{-1}|\, dt \end{cases} \qquad \text{for} \quad t \in \mathcal{I}.$$

Therefore,

$$|\Phi(t)| \leq \exp[K(t - t_0)] \quad \text{and} \quad |\Phi(t)^{-1}| \leq \exp[K(t - t_0)] \quad \text{for} \quad t \in \mathcal{I}$$

(cf. Lemma I-1-5). Observe that $\vec{\phi}(t) = \Phi(t)\vec{\phi}(t_0)$ and $\vec{\phi}(t_0) = \Phi(t)^{-1}\vec{\phi}(t)$. From the definition of the norm of a matrix, it follows that

$$\left|\vec{\phi}(t_0)\right| \exp[-K(t - t_0)] \leq \left|\vec{\phi}(t)\right| \leq \left|\vec{\phi}(t_0)\right| \exp[K(t - t_0)] \quad \text{for} \quad t \in \mathcal{I}.$$

Therefore, (VII.2.2) follows, since $\left|\vec{\phi}(t_0)\right| \neq 0$. $\square$

Set $\Lambda = \left\{ \lambda\left(\vec{\phi}\right) : \vec{\phi} \text{ is a nontrivial solution of (VII.2.1)} \right\}$. Then, (iv) of Lemma VII-1-3 implies that Λ is a nonempty subset of $\mathbb{R}$ which contains at most n numbers. Let $\lambda_1 > \lambda_2 > \cdots > \lambda_m$ $(1 \leq m \leq n)$ be all of the distinct numbers in Λ. Set

$$(VII.2.3) \quad \mathcal{V}_j = \left\{ \vec{\phi} : \vec{\phi} \text{ is a solution of (VII.2.1) such that } \lambda\left(\vec{\phi}\right) \leq \lambda_j \right\}$$

for $j = 1, 2, \ldots, m$. Then, (ii) and (vi) of Lemma VII-1-3 imply that $\mathcal{V}_j$ is a vector space over $\mathbb{C}$. Set

$$(VII.2.4) \qquad \gamma_j = \dim_{\mathbb{C}} \mathcal{V}_j \qquad (j = 1, 2, \ldots, m).$$

The following lemma states that there exists a particular basis for each space $\mathcal{V}_j$ which consists of γ_j solutions whose type numbers are equal to λ_j.

Lemma VII-2-2. *For system (VII.2.1) and γ_j given by (VII.2.4), it holds that*
 (i) $\gamma_1 = n$,
 (ii) $\gamma_m < \gamma_{m-1} < \cdots < \gamma_1$,
 (iii) for each j, there exists a basis for $\mathcal{V}_j$ that consists of γ_j linearly independent
   solutions $\vec{\phi}_{j,\nu}$ $(\nu = 1, 2, \ldots, \gamma_j)$ of (VII.2.1) such that $\lambda\left(\vec{\phi}_{j,\nu}\right) = \lambda_j$.

Proof.

It is easy to derive (i) and (ii) from the definition of $\mathcal{V}_j$ and the definition of the numbers $\lambda_1, \ldots, \lambda_m$. To prove (iii), let $\vec{\psi}_{j,\nu}$ $(\nu = 1, 2, \ldots, \gamma_j)$ be a basis for $\mathcal{V}_j$. Assume that

$$\lambda\left(\vec{\psi}_{j,\nu}\right) \begin{cases} = \lambda_j & \text{for} \quad \nu = 1, \ldots \ell, \\[2mm] < \lambda_j & \text{for} \quad \nu = \ell + 1, \ldots, \gamma_j. \end{cases}$$

It follows from (ii) of Lemma VII-1-3 and definitions of $\mathcal{V}_j$ and λ_j that $\ell \geq 1$. Set

$$\vec{\phi}_{j,\nu} = \begin{cases} \vec{\psi}_{j,\nu} & \text{for} \quad \nu = 1, \ldots, \ell, \\[2mm] \vec{\psi}_{j,1} + \vec{\psi}_{j,\nu} & \text{for} \quad \nu = \ell + 1, \ldots, \gamma_j. \end{cases}$$

Then, $\vec{\phi}_{j,\nu}$ $(\nu = 1, \ldots, \gamma_j)$ satisfy all the requirements of (iii). $\square$

Observation VII-2-3. The maximum number of linearly independent solutions of (VII.2.1) having Liapounoff's type number λ_j is γ_j.

Definition VII-2-4. *The numbers $\lambda_1, \lambda_2, \ldots, \lambda_m$ are called Liapounoff's type numbers of system (VII.2.1) at $t = +\infty$. For every $j = 1, 2, \ldots, m$, the multiplicity of Liapounoff's type number λ_j is defined by*

$$(VII.2.5) \qquad h_j = \begin{cases} \gamma_j - \gamma_{j+1} & \text{for} \quad j = 1, 2, \ldots, m-1, \\ \gamma_m & \text{for} \quad j = m. \end{cases}$$

The structure of solutions of (VII.2.1) according with their type numbers is given in the following result.

Theorem VII-2-5. *Let $\{\vec{\phi}_1, \vec{\phi}_2, \ldots, \vec{\phi}_n\}$ be a fundamental set of n linearly independent solutions of system (VII.2.1). Then, the following four conditions are mutually equivalent:*

(1) for every j, the total number of those $\vec{\phi}_\ell$ such that $\lambda\left(\vec{\phi}_\ell\right) = \lambda_j$ is h_j (cf. (VII.2.5)),

(2) for every j, the subset $\left\{\vec{\phi}_\ell : \lambda\left(\vec{\phi}_\ell\right) \leq \lambda_j\right\}$ is a basis for $\mathcal{V}_j$,

(3) $\lambda\left(\sum\limits_{\lambda(\vec{\phi}_\ell)>\lambda_j} c_\ell \vec{\phi}_\ell\right) > \lambda_j$ if the constants c_ℓ are not all zero,

(4) $\lambda\left(\sum\limits_{\ell=1}^{n} c_\ell \vec{\phi}_\ell\right) = \max\left\{\lambda\left(\vec{\phi}_\ell\right) : c_\ell \neq 0\right\}$ for every nontrivial linear combination $\sum\limits_{\ell=1}^{n} c_j \vec{\phi}_\ell$ of $\{\vec{\phi}_1, \vec{\phi}_2, \ldots, \vec{\phi}_n\}$.

Proof.

Assume that (1) is satisfied. Then, the total number of those $\vec{\phi}_\ell$ such that $\lambda\left(\vec{\phi}_\ell\right) \leq \lambda_j$ is $\sum\limits_{\ell=j}^{m} h_\ell = \gamma_j = \dim_{\mathbb{C}} \mathcal{V}_j$. Hence, (2) is also satisfied. Conversely, assume that (2) is satisfied. Then, the total number of those $\vec{\phi}_\ell$ such that $\lambda\left(\vec{\phi}_\ell\right) \leq \lambda_j$ is equal to $\dim_{\mathbb{C}} \mathcal{V}_j = \gamma_j$. Hence, (1) is also satisfied (cf. (VII.2.5)).

Assume that (2) is satisfied. Then, if $\lambda\left(\sum\limits_{\lambda(\vec{\phi}_\ell)>\lambda_j} c_\ell \vec{\phi}_\ell\right) \leq \lambda_j$ for some c_ℓ, it follows that $\sum\limits_{\lambda(\vec{\phi}_\ell)>\lambda_j} c_\ell \vec{\phi}_\ell \in \mathcal{V}_j$, and hence

$$\sum_{\lambda(\vec{\phi}_\ell)>\lambda_j} c_\ell \vec{\phi}_\ell = \sum_{\lambda(\vec{\phi}_\ell)\leq\lambda_j} \hat{c}_\ell \vec{\phi}_\ell$$

for some $\hat{c}_\ell$. Since $\{\vec{\phi}_\ell : \ell = 1, 2, \ldots, n\}$ is linearly independent, all constants c_ℓ must be zero. Thus, (3) is satisfied.

Assume that (3) is satisfied. Write a linear combination $\sum_{\ell=1}^{n} c_\ell \vec{\phi}_\ell$ in the form

$$\sum_{j=1}^{m} \sum_{\lambda(\vec{\phi}_\ell)=\lambda_j} c_\ell \vec{\phi}_\ell. \quad \text{Then, } \lambda \left(\sum_{\lambda(\vec{\phi}_\ell)=\lambda_j} c_\ell \vec{\phi}_\ell \right) = \lambda_j, \text{ if } \sum_{\lambda(\vec{\phi}_\ell)=\lambda_j} c_\ell \vec{\phi}_\ell \neq 0. \text{ Hence, (4)}$$

follows from (iii) of Lemma VII-1-3.

Finally, assume that (4) is satisfied. Then, every solution $\vec{\phi}$ of (VII.2.1) with $\lambda\left(\vec{\phi}\right) \leq \lambda_j$ must be a linear combination of the subset $\left\{\vec{\phi}_\ell : \lambda\left(\vec{\phi}_\ell\right) \leq \lambda_j\right\}$. Hence, (2) is satisfied. $\square$

Definition VII-2-6. *A fundamental set $\{\phi_1, \phi_2, \ldots, \phi_n\}$ of n linearly independent solutions of system (VII.2.1) is said to be* normal *if one of four conditions (1) - (4) of Theorem VII-2-5 is satisfied.*

Since $\mathcal{V}_m \subset \mathcal{V}_{m-1} \subset \cdots \subset \mathcal{V}_2 \subset \mathcal{V}_1$, it is easy to construct a fundamental set of (VII.2.1) that satisfies condition (1) of Theorem VII-2-5. Thus, we obtain the following theorem.

Theorem VII-2-7. *If the entries of the matrix $A(t)$ are continuous and bounded on an interval $\mathcal{I} = \{t : t_0 \leq t < +\infty\}$, system (VII.2.1) has a normal fundamental set of n linearly independent solutions on the interval $\mathcal{I}$.*

Example VII-2-8. For a system $\dfrac{d\vec{y}}{dt} = A\vec{y}$ with a constant matrix A, Liapounoff's type numbers $\lambda_1, \lambda_2, \ldots, \lambda_m$ at $t = +\infty$ and their respective multiplicities $h_1, h_2, \ldots, h_m$ are determined in the following way.

Let $\mu_1, \mu_2, \ldots, \mu_k$ be the distinct eigenvalues of A and $m_1, m_2, \ldots, m_k$ be their respective multiplicities. Set $\nu_j = \Re(\mu_j)$ $(j = 1, 2, \ldots, k)$. Let $\lambda_1 > \lambda_2 > \cdots > \lambda_m$ be the distinct real numbers in the set $\{\nu_1, \nu_2, \ldots, \nu_k\}$. Set $h_j = \displaystyle\sum_{\nu_\ell=\lambda_j} m_\ell$ for

$j = 1, 2, \ldots, m$. Then, $\lambda_1, \lambda_2, \ldots, \lambda_m$ are Liapounoff's type numbers of $\dfrac{d\vec{y}}{dt} = A\vec{y}$ at $t = +\infty$ and $h_1, h_2, \ldots, h_m$ are their respective multiplicities. The proof of this result is left to the reader as an exercise.

Example VII-2-9. For a system

$$\text{(VII.2.6)} \qquad\qquad \frac{d\vec{y}}{dt} = A(t)\vec{y}$$

with a matrix $A(t)$ whose entries are continuous and periodic of a positive period ω on the entire real line $\mathbb{R}$, Liapounoff's type numbers $\lambda_1, \lambda_2, \ldots, \lambda_m$ at $t = +\infty$ and their respective multiplicities $h_1, h_2, \ldots, h_m$ are determined in the following way:

There exists an $n \times n$ matrix $P(t)$ such that

(i) the entries of $P(t)$ are continuous and periodic of period ω,

(ii) $P(t)$ is invertible for all $t \in \mathbb{R}$,

(iii) the transformation $\vec{y} = P(t)\vec{z}$ changes system (VII.2.6) to

$$\text{(VII.2.7)} \qquad\qquad \frac{d\vec{z}}{dt} = B\vec{z}$$

with a constant matrix B.

Therefore, (vi) of Lemma VII-1-3 implies that systems (VII.2.6) and (VII.2.7) have the same Liapounoff's type numbers at $t = +\infty$ with the same respective multiplicities. Liapounoff's type numbers of system (VII.2.7) can be determined by using Example VII-2-8. Note that if $\rho_1, \rho_2, \ldots, \rho_n$ are the multipliers of system (VII.2.6), then $\mu_j = \dfrac{1}{\omega}\log[\rho_j]$ $(j = 1, 2, \ldots, n)$ are the characteristic exponents of system (VII.2.6), i.e., the eigenvalues of B, if we choose $\log[\rho_j]$ in a suitable way. Hence,

$$\Re(\mu_j) \;=\; \frac{1}{\omega}\log[|\rho_j|] \qquad (j = 1, 2, \ldots, n).$$

Those numbers are independent of the choice of branches of $\log[\rho_j]$.

Example VII-2-10. For the system $\dfrac{d\vec{y}}{dt} = \begin{bmatrix} 1 & 0 \\ 0 & 2 \end{bmatrix}\vec{y}$, the fundamental set $\left\{ e^t \begin{bmatrix} 1 \\ 0 \end{bmatrix}, \right.$ $\left. e^{2t}\begin{bmatrix} 0 \\ 1 \end{bmatrix} \right\}$ is normal, but the fundamental set $\left\{ \begin{bmatrix} e^t \\ e^{2t} \end{bmatrix}, \begin{bmatrix} e^t \\ -e^{2t} \end{bmatrix} \right\}$ is not normal.

VII-3. Calculation of Liapunoff's type numbers of solutions

The main concern of this section is to show that Liapounoff's type numbers of a system $\dfrac{d\vec{y}}{dt} = B(t)\vec{y}$ at $t = +\infty$ and their respective multiplicities are exactly the same as those of the system $\dfrac{d\vec{y}}{dt} = A\vec{y}$ with a constant matrix A if $\lim\limits_{t\to+\infty} B(t) = A$. It is known that any constant matrix A is similar to a block-diagonal form

$$\text{diag}[\mu_1 I_{m_1} + M_1, \mu_2 I_{m_2} + M_2, \ldots, \mu_k I_{m_k} + M_k],$$

where $\mu_1, \ldots, \mu_k$ are distinct eigenvalues of A whose respective multiplicities are $m_1, \ldots, m_k$, I_{m_j} is the $m_j \times m_j$ identity matrix, and M_j is an $m_j \times m_j$ nilpotent matrix (cf. (IV.1.10)).

Consider a system of the form

$$\text{(VII.3.1)} \qquad \frac{d\vec{y}_j}{dt} \;=\; A_j\vec{y}_j + \sum_{\ell=1}^{m} B_{j\ell}(t)\vec{y}_\ell \qquad (j = 1, 2, \ldots, m),$$

where $\vec{y}_j \in \mathbb{C}^{n_j}$, A_j is an $n_j \times n_j$ constant matrix, and $B_{j\ell}(t)$ is an $n_j \times n_\ell$ matrix whose entries are continuous on the interval $\mathcal{I}_0 = \{t : 0 \leq t < +\infty\}$, under the following assumption.

Assumption 1.

(i) *For each j, the matrix A_j has the form*

$$\text{(VII.3.2)} \qquad A_j = \lambda_j I_{n_j} + E_j + N_j \qquad (j = 1, 2, \ldots, m),$$

where λ_j is a real number, I_{n_j} is the $n_j \times n_j$ identity matrix, E_j is an $n_j \times n_j$ constant diagonal matrix whose entries on the main diagonal are all purely imaginary, and N_j is an $n_j \times n_j$ nilpotent matrix,

(ii) $\lambda_m < \lambda_{m-1} < \cdots < \lambda_2 < \lambda_1$,

(iii) $N_j E_j = E_j N_j \quad (j = 1, 2, \ldots, m)$,

(iv) $\displaystyle\lim_{t \to +\infty} B_{j\ell}(t) = O \quad (j, \ell = 1, 2, \ldots, m)$.

The following result is a basic block-diagonalization theorem.

Theorem VII-3-1. *Under Assumption 1, there exist a non-negative number t_0 and a linear transformation*

$$\text{(VII.3.3)} \qquad \vec{y}_j = \vec{z}_j + \sum_{\ell \neq j} T_{j\ell}(t)\vec{z}_\ell \qquad (j = 1, 2, \ldots, m)$$

with $n_j \times n_\ell$ matrices $T_{j\ell}(t)$ such that

(1) *for every pair (j, ℓ) such that $j \neq \ell$, the derivative $\dfrac{dT_{j\ell}}{dt}(t)$ exists and the entries of $T_{j\ell}$ and $\dfrac{dT_{j\ell}}{dt}$ are continuous on the interval $\mathcal{I} = \{t : t_0 \leq t < +\infty\}$,*

(2) $\displaystyle\lim_{t \to +\infty} T_{j\ell}(t) = O \quad (j \neq \ell)$,

(3) *transformation (VII.3.3) changes system (VII.3.1) to*

$$\text{(VII.3.4)} \qquad \frac{d\vec{z}_j}{dt} = \left[A_j + B_{jj}(t) + \sum_{h \neq j} B_{jh}(t)T_{hj}(t) \right] \vec{z}_j \qquad (j = 1, 2, \ldots, m).$$

Proof.

We prove this theorem in eight steps.

Step 1. Differentiating both sides of (VII.3.3), we obtain

$$\frac{d\vec{z}_j}{dt} + \sum_{\ell \neq j} T_{j\ell}\frac{d\vec{z}_\ell}{dt} + \sum_{\ell \neq j} \frac{dT_{j\ell}}{dt}\vec{z}_\ell$$

$$\text{(VII.3.5)} \qquad = A_j \left[\vec{z}_j + \sum_{\ell \neq j} T_{j\ell}\vec{z}_\ell \right] + \sum_{\ell = 1}^{m} B_{j\ell} \left[\vec{z}_\ell + \sum_{\nu \neq \ell} T_{\ell\nu}\vec{z}_\nu \right]$$

$$= \left[A_j + B_{jj} + \sum_{h \neq j} B_{jh}T_{hj} \right] \vec{z}_j + \sum_{\ell \neq j} \left[A_j T_{j\ell} + B_{j\ell} + \sum_{h \neq \ell} B_{jh}T_{h\ell} \right] \vec{z}_\ell,$$

where $j = 1, 2, \ldots, m$. Note that

$$\sum_{\ell=1}^{m} B_{j\ell} \sum_{\nu \neq \ell} T_{\ell\nu} \vec{z}_\nu \ = \ \sum_{\nu=1}^{m} \left[\sum_{h \neq \nu} B_{jh} T_{h\nu} \right] \vec{z}_\nu.$$

Rewrite (VII.3.5) in the form

$$\left\{ \frac{d\vec{z}_j}{dt} - [A_j + F_j]\,\vec{z}_j \right\} + \sum_{\ell \neq j} T_{j\ell} \left\{ \frac{d\vec{z}_\ell}{dt} - [A_\ell + F_\ell]\,\vec{z}_\ell \right\}$$

$$= \sum_{\ell \neq j} \left[A_j T_{j\ell} + B_{j\ell} + \sum_{h \neq \ell} B_{jh} T_{h\ell} - T_{j\ell}\,(A_\ell + F_\ell) - \frac{dT_{j\ell}}{dt} \right] \vec{z}_\ell,$$

where $j = 1, 2, \ldots, m$, and

$$F_j \ = \ B_{jj} \ + \ \sum_{h \neq j} B_{jh} T_{hj} \qquad (j = 1, 2, \ldots, m).$$

Define the $T_{j\ell}$ by the following system of differential equations:

$$(\text{VII.3.6}) \qquad \frac{dT_{j\ell}}{dt} = A_j T_{j\ell} - T_{j\ell} A_\ell + B_{j\ell} + \sum_{h \neq \ell} B_{jh} T_{h\ell} - T_{j\ell} F_\ell \qquad (j \neq \ell).$$

Then,

$$\left\{ \frac{d\vec{z}_j}{dt} - [A_j + F_j]\,\vec{z}_j \right\} + \sum_{\ell \neq j} T_{j\ell} \left\{ \frac{d\vec{z}_\ell}{dt} - [A_\ell + F_\ell]\,\vec{z}_\ell \right\} = 0 \quad (j = 1, 2, \ldots, m).$$

This implies that we can derive (VII.3.4) on the interval $\mathcal{I}$ if the $T_{j\ell}$ satisfy (VII.3.6) and condition (2) of Theorem VII-3-1, and t_0 is sufficiently large.

Step 2. Let us find a solution T of system (VII.3.6) that satisfies condition (2) of Theorem VII-3-1. To do this, change (VII.3.6) to a system of nonlinear integral equations

$$(\text{VII.3.7}) \qquad T_{j\ell}(t) \ = \ \int_{\tau_{j\ell}}^{t} \exp[(t - s)A_j] U_{j\ell}(s, T(s)) \exp[-(t - s)A_\ell]\,ds,$$

for $j \neq \ell$, where the initial points $\tau_{j\ell}$ are to be specified and

$$U_{j\ell}(t, T) \ = \ B_{j\ell}(t) \ + \ \sum_{h \neq \ell} B_{jh}(t) T_{h\ell} - T_{j\ell} \left\{ B_{\ell\ell}(t) + \sum_{h \neq \ell} B_{\ell h}(t) T_{h\ell} \right\}.$$

Using conditions given in Assumption 1, rewrite (VII.3.7) in the form

$$(\text{VII.3.7}') \qquad T_{j\ell}(t) \ = \ \int_{\tau_{j\ell}}^{t} \exp\left[\frac{1}{2}(\lambda_j - \lambda_\ell)(t - s) \right] W_{j\ell}(t - s, s, T(s))\,ds,$$

where $j \neq \ell$ and

$$W_{j\ell}(\tau, s, T)$$
$$(\text{VII.3.8}) \qquad = \exp\left[\frac{1}{2}(\lambda_j - \lambda_\ell)\tau\right] \exp[\tau E_j] \exp[\tau N_j] U_{j\ell}(s, T) \exp[-\tau E_\ell] \exp[-\tau N_\ell].$$

On the right-hand side of (VII.3.7$'$), choose the initial points $\tau_{j\ell}$ in the following way

$$(\text{VII.3.9}) \qquad \tau_{j\ell} \ = \ \begin{cases} +\infty & \text{if} \quad \lambda_j > \lambda_\ell \quad (\text{i.e., } j < \ell), \\ t_0 & \text{if} \quad \lambda_j < \lambda_\ell \quad (\text{i.e., } j > \ell). \end{cases}$$

Step 3. Let us prove the following lemma.

Lemma VII-3-2. *Let α be a positive number and let $\vec{f}(t, s)$ be continuous in (t, s) on the region $\mathcal{I} \times \mathcal{I}$, where $\mathcal{I} = \{t : t_0 \leq t < +\infty\}$. Then,*

$$(\text{VII.3.10}) \qquad \left| \int_{t_0}^{t} \exp[-\alpha(t - s)]\vec{f}(t, s)ds \right| \leq \frac{1}{\alpha} \max\left\{ \left|\vec{f}(t, s)\right| : t_0 \leq s \leq t \right\}$$

for each fixed $t \in \mathcal{I}$. Also, if $\left|\vec{f}(t, s)\right|$ is bounded for $t_0 \leq t \leq s < +\infty$, then

$$(\text{VII.3.11}) \qquad \left| \int_{+\infty}^{t} \exp[\alpha(t - s)]\vec{f}(t, s)ds \right| \leq \frac{1}{\alpha} \sup\left\{ \left|\vec{f}(t, s)\right| : t_0 \leq t \leq s < +\infty \right\}$$

for each fixed $t \in \mathcal{I}$.

Proof.

In fact, estimates (VII.3.10) and (VII.3.11) follow respectively from

$$\begin{cases} \displaystyle\int_{t_0}^{t} \exp[-\alpha(t - s)]ds = \frac{1}{\alpha}\{1 - \exp[-\alpha(t - t_0)]\} \leq \frac{1}{\alpha}, \\ \displaystyle\int_{+\infty}^{t} \exp[\alpha(t - s)]ds = -\frac{1}{\alpha}. \qquad\qquad \square \end{cases}$$

Step 4. Let us estimate $W_{j\ell}(\tau, s, T)$ for $t \leq s < +\infty$ if $\lambda_j > \lambda_\ell$, and for $t_0 \leq s \leq t$ if $\lambda_j < \lambda_\ell$. If $\lambda_j > \lambda_\ell$, $\tau \leq 0$, and $s \geq t$, it follows that

$$|W_{j\ell}(\tau, s, T)| \leq \mathcal{K}_0\left(1 + |\tau|^{m_j + m_\ell - 2}\right) \exp\left[\frac{1}{2}(\lambda_j - \lambda_\ell)\tau\right] |U_{j\ell}(s, T)|$$

for some positive number $\mathcal{K}_0$. It can be shown easily that there exist a positive number $\mathcal{K}_1$ and a non-negative valued function $\beta(t)$ such that

$$\begin{cases} \left(1 + |\tau|^{m_j + m_\ell - 2}\right) \exp\left[\frac{1}{2}(\lambda_j - \lambda_\ell)\tau\right] \leq \mathcal{K}_1 \quad \text{for} \quad \tau \leq 0, \\ |B_{pq}(s)| \leq \beta(t) \quad \text{for} \quad t \leq s < +\infty \text{ and all pairs } (p, q), \\ \lim_{t \to +\infty} \beta(t) = 0. \end{cases}$$

Hence, if $\lambda_j > \lambda_\ell$, $\tau \leq 0$, and $s \geq t$, we obtain $|W_{j\ell}(\tau, s, T)| \leq \mathcal{K}_2\beta(t)\left\{1 + |T|^2\right\}$ for some positive number $\mathcal{K}_2$, where $|T| = \max_{\ell \neq j}|T_{j\ell}|$. Similarly, the estimate $|W_{j\ell}(\tau, s, T)| \leq \mathcal{K}_2\beta(t_0)\left\{1 + |T|^2\right\}$ is obtained by choosing a positive number $\mathcal{K}_2$ sufficiently large, if $\lambda_j < \lambda_\ell$, $\tau \geq 0$, and $t_0 \leq s \leq t$.

Step 5. In a manner similar to Step 4, we can derive the following estimates:

$$\left| W_{j\ell}(\tau, s, T) - W_{j\ell}(\tau, s, \hat{T}) \right|$$

$$\leq \begin{cases} \mathcal{K}_3 \beta(t) \left\{ 1 + |T| + |\hat{T}| \right\} |T - \hat{T}| & \text{if } \lambda_j > \lambda_\ell, \ \tau \leq 0, \ s \geq t, \\[2mm] \mathcal{K}_3 \beta(t_0) \left\{ 1 + |T| + |\hat{T}| \right\} |T - \hat{T}| & \text{if } \lambda_j < \lambda_\ell, \ \tau \geq 0, \ t_0 \leq s \leq t \end{cases}$$

for some positive number $\mathcal{K}_3$.

Step 6. Define successive approximations on the interval $\mathcal{I}$ as follows:

$$\begin{cases} T_{0;j\ell}(t) = O, \\[2mm] T_{p+1;j\ell}(t) = \displaystyle\int_{\tau_{j\ell}}^{t} \exp\left[\frac{1}{2}(\lambda_j - \lambda_\ell)(t - s) \right] W_{j\ell}(t - s, s, T_p(s)) ds, \end{cases}$$

where $j \neq \ell$ and $p = 1, 2, \dots$. Suppose that

$$(\text{VII.3.12}) \qquad | T_{p;j\ell}(t) | \leq C \qquad \text{on the interval } \mathcal{I} = \{ t : t_0 \leq t < +\infty \}$$

for some positive number C. Then, Lemma VII-3-2 and Step 4 imply that

$$| T_{p+1;j\ell}(t) | \leq \begin{cases} \dfrac{2}{\lambda_j - \lambda_\ell} \mathcal{K}_2 \beta(t) \{ 1 + C^2 \} & \text{if } \lambda_j > \lambda_\ell, \\[4mm] \dfrac{2}{\lambda_\ell - \lambda_j} \mathcal{K}_2 \beta(t_0) \{ 1 + C^2 \} & \text{if } \lambda_j < \lambda_\ell \end{cases}$$

on the interval $\mathcal{I}$. Hence, choosing t_0 so large that

$$\frac{2}{|\lambda_j - \lambda_\ell|} \mathcal{K}_2 \beta(t) \{ 1 + C^2 \} \leq C \qquad \text{on } \mathcal{I},$$

we obtain

$$|T_{p+1;j\ell}(t)| \leq C \qquad \text{on the interval } \mathcal{I}$$

from (VII.3.12).

Step 7. Suppose that (VII.3.12) holds for $p = 0, 1, 2, \dots$. Then,

$$|T_{p+1;j\ell}(t) - T_{p;j\ell}(t)| \leq \frac{1}{2} \sup_{s \in \mathcal{I}} |T_p(s) - T_{p-1}(s)| \leq \left(\frac{1}{2} \right)^p C \quad \text{on } \mathcal{I},$$

where $|T_p| = \max_{\ell \neq j} |T_{p;j\ell}|$ if t_0 is so large that $\mathcal{K}_3 \beta(t) \{ 1 + 2C \} \leq \dfrac{1}{2}$ on $\mathcal{I}$. Since

$T_{p;j\ell}(t) = \displaystyle\sum_{q=1}^{p} \{ T_{q;j\ell}(t) - T_{q-1;j\ell}(t) \}$, it can be shown easily that $\displaystyle\lim_{p \to +\infty} T_{p;j\ell}(t) =$ $T_{j\ell}(t)$ exist for all (j, ℓ) such that $j \neq \ell$ uniformly on the interval $\mathcal{I}$. The limit $T_{j\ell}(t)$ satisfies integral equation (VII.3.7).

Step 8. In this final step, we prove that the bounded solution $T_{j\ell}$ of (VII.3.7) satisfies condition (2) of Theorem VII-3-1. It easily follows from Steps 6 and 4 that

$$\lim_{t \to +\infty} T_{j\ell}(t) = \lim_{t \to +\infty} \int_{+\infty}^{t} \exp\left[\frac{1}{2}(\lambda_j - \lambda_\ell)(t - s)\right] W_{j\ell}(t - s, s, T(s))ds = O$$

if $\lambda_j > \lambda_\ell$.

For (j, ℓ) such that $\lambda_j < \lambda_\ell$, write the right-hand side of (VII.3.7′) in the following form:

$$\int_{t_0}^{t} \exp\left[\frac{1}{2}(\lambda_j - \lambda_\ell)(t - s)\right] W_{j\ell}(t - s, s, T(s))ds$$

$$= \int_{t_0}^{\sigma} \exp\left[\frac{1}{2}(\lambda_j - \lambda_\ell)(t - s)\right] W_{j\ell}(t - s, s, T(s))ds$$

$$+ \int_{\sigma}^{t} \exp\left[\frac{1}{2}(\lambda_j - \lambda_\ell)(t - s)\right] W_{j\ell}(t - s, s, T(s))ds$$

for any σ such that $t_0 \leq \sigma \leq t$. Observe that

$$\left| \int_{t_0}^{\sigma} \exp\left[\frac{1}{2}(\lambda_j - \lambda_\ell)(t - s)\right] W_{j\ell}(t - s, s, T(s))ds \right|$$

$$\leq \exp\left[\frac{1}{2}(\lambda_j - \lambda_\ell)(t - \sigma)\right] \left| \int_{t_0}^{\sigma} \exp\left[\frac{1}{2}(\lambda_j - \lambda_\ell)(\sigma - s)\right] W_{j\ell}(t - s, s, T(s))ds \right|$$

$$\leq \frac{2\mathcal{K}_2\beta(t_0)}{\lambda_\ell - \lambda_j}\{1 + C^2\} \exp\left[\frac{1}{2}(\lambda_j - \lambda_\ell)(t - \sigma)\right],$$

if $|T(t)| \leq C$ on $\mathcal{I}$ (cf. (VII.3.10)). Observe also that

$$\left| \int_{\sigma}^{t} \exp\left[\frac{1}{2}(\lambda_j - \lambda_\ell)(t - s)\right] W_{j\ell}(t - s, s, T(s))ds \right| \leq \frac{2\mathcal{K}_2\beta(\sigma)}{\lambda_\ell - \lambda_j}\{1 + C^2\}$$

if $|T(t)| \leq C$ on $\mathcal{I}$ (cf. (VII.3.10)). Hence,

$$\left| \int_{t_0}^{t} \exp\left[\frac{1}{2}(\lambda_j - \lambda_\ell)(t - s)\right] W_{j\ell}(t - s, s, T(s))ds \right|$$

$$\leq \frac{2\mathcal{K}_2\{1 + C^2\}}{\lambda_\ell - \lambda_j}\left\{ \beta(t_0) \exp\left[\frac{1}{2}(\lambda_j - \lambda_\ell)(t - \sigma)\right] + \beta(\sigma) \right\}.$$

if $|T(t)| \leq C$ on $\mathcal{I}$. Therefore, letting σ and t tend to $+\infty$, we obtain

$$\lim_{t \to +\infty} \int_{t_0}^{t} \exp\left[\frac{1}{2}(\lambda_j - \lambda_\ell)(t - s)\right] W_{j\ell}(t - s, s, T(s))ds = O.$$

This completes the proof of Theorem VII-3-1. $\square$

In order to find Liapounoff's type numbers of system (VII.3.4), it is necessary to establish the following lemma.

Lemma VII-3-3. *If N is a nilpotent matrix, then for any positive number ϵ such that $0 < \epsilon \leq 1$, there exists an invertible matrix $P(\epsilon)$ such that*

$$\left| P(\epsilon)^{-1} N P(\epsilon) \right| \leq \mathcal{K}\epsilon$$

for some positive number $\mathcal{K}$ independent of ϵ.

Proof.

Assume without any loss of generality that N is in an upper-triangular form with zeros on the main diagonal (cf. Lemma IV-1-8). Set $\Lambda(\epsilon) = \text{diag}[1, \epsilon, \dots, \epsilon^{n-1}]$. Then, $\Lambda(\epsilon)^{-1} N \Lambda(\epsilon) = \left[\, \epsilon^{k-j}\nu_{jk} \,\right]$ (a *shearing transformation*). Hence, as $0 < \epsilon \leq 1$ and $j < k$, we obtain $\left| \Lambda(\epsilon)^{-1} N \Lambda(\epsilon) \right| \leq \epsilon |N|$. $\square$

Let us find Liapounoff's type numbers of system (VII.3.4), i.e.,

$$\frac{d\vec{z}_j}{dt} = \left[A_j + B_{jj}(t) + \sum_{h \neq j} B_{jh}(t) T_{hj}(t) \right] \vec{z}_j \qquad (j = 1, 2, \dots, m).$$

Theorem VII-3-4. *A system of the form*

$$\frac{d\vec{z}}{dt} = \left[\lambda I_n + E + N + B(t) \right] \vec{z}$$

has only one Liapounoff's type number λ at $t = +\infty$ if
  (i) λ is a real number,
  (ii) I_n is the $n \times n$ identity matrix, E is an $n \times n$ constant diagonal matrix whose entries on the main diagonal are all purely imaginary, N is an $n \times n$ constant nilpotent matrix, and $EN = NE$,
  (iii) the entries of the $n \times n$ matrix $B(t)$ are continuous on the interval $\mathcal{I} = \{t : t_0 \leq t < +\infty\}$,
  (iv) $\lim\limits_{t \to +\infty} B(t) = O$.

Proof.

Set $\vec{z} = e^{\lambda t} e^{tE} \vec{u}$. Then,

$$(VII.3.13) \qquad \frac{d\vec{u}}{dt} = [N + C(t)]\,\vec{u}, \qquad C(t) = e^{-tE} B(t) e^{tE}.$$

Let μ be any Liapounoff's type number of (VII.3.13) at $t = +\infty$. Then, $|\mu| \leq \sup\limits_{t \in \mathcal{I}} |N + C(t)|$ (cf. Theorem VII-2-1). This is true even if $t_0 \to +\infty$. Since $\lim\limits_{t \to +\infty} C(t) = O$, we conclude that $|\mu| \leq |N|$. Using Lemma VII-3-3, it can be shown that any Liapounoff's type number μ of (VII.3.13) at $t = +\infty$ is zero. This, in turn, completes the proof of Theorem VII-3-4. $\square$

Remark VII-3-5. For a nilpotent matrix N, the two matrices N and ϵN are similar to each other for any nonzero number ϵ. To prove this, use the Jordan canonical form of N.

From the argument given above, we conclude that Liapounoff's type numbers of system (VII.3.4) are $\lambda_1, \dots, \lambda_m$ and their respective multiplicities are $n_1, \dots, n_m$. Thus, the following theorem was proved.

Theorem VII-3-6. *Liapounoff's type numbers of a system $\dfrac{d\vec{y}}{dt} = B(t)\vec{y}$ at $t = +\infty$ and their respective multiplicities are exactly the same as those of the system $\dfrac{d\vec{y}}{dt} = A\vec{y}$ with a constant matrix A if the entries of the matrix $B(t)$ are continuous on the interval $\mathcal{I}_0 = \{t : 0 \leq t < +\infty\}$ and $\lim\limits_{t \to +\infty} B(t) = A$.*

Applying Lemma VII-1-4 to the solutions of the system $\dfrac{d\vec{y}}{dt} = B(t)\vec{y}$, we obtain the following corollary of Theorem VII-3-6.

Corollary VII-3-7. *If an $n \times n$ matrix $B(t)$ satisfies the conditions*
(i) the entries of $B(t)$ are continuous on the interval $\mathcal{I}_0 = \{t : 0 \leq t < +\infty\}$,
(ii) $\lim\limits_{t \to +\infty} B(t) = A$ exists,

then, for every nontrivial solution $\vec{\phi}(t)$ of the system $\dfrac{d\vec{y}}{dt} = B(t)\vec{y}$, $\lim\limits_{t \to +\infty} \dfrac{\log |\vec{\phi}(t)|}{t} = \mu$ exists and μ is the real part of an eigenvalue of the matrix A.

Remark VII-3-8. The conclusion of Theorem VII-3-1 still holds even if condition (iv) of Assumption 1 of this section is replaced by $|B_{jk}(t)| \leq f(t)$ $(j, k = 1, 2, \dots, m)$ on the interval $\mathcal{I}_0$, where $f(t)$ satisfies

$$(\text{VII.3.14}) \qquad \sup_{p \geq t}(1 + p - t)^{-1} \int_t^p f(s)\,ds \to 0 \quad \text{as} \quad t \to +\infty.$$

To see this, let us assume that a positive-valued function $f(t)$ satisfies condition (VII.3.14). Set $h(t) = \sup\limits_{p \geq t} \left((1 + p - t)^{-1} \int_t^p f(s)\,ds \right)$ for a fixed positive number t. Also, set $E(t) = \sup\limits_{\tau \geq t} h(\tau)$. Then, $\lim\limits_{t \to +\infty} E(t) = 0$, and $E(t_2) \leq E(t_1)$ if $t_1 < t_2$. Now, it is sufficient to prove the following lemma.

Lemma VII-3-9. *For any positive numbers t, t_0, and c, it holds that*

(a)
$$\int_{t_0}^t e^{-c(t-s)} f(s)\,ds \ \leq \ \left(1 + \frac{1}{c}\right) E(t_0),$$

(b)
$$\int_t^{+\infty} e^{c(t-s)} f(s)\,ds \ \leq \ \left(1 + \frac{1}{c}\right) E(t).$$

Furthermore,

(c)
$$\lim_{t \to +\infty} \int_{t_0}^t e^{-c(t-s)} f(s)\,ds \ = \ 0.$$

Proof of (a).

Set $\phi(t) = \int_{t_0}^{t} e^{-c(t-s)} f(s)ds$. Then, $\phi'(t) = -c\phi(t) + f(t)$ and, hence,

$$\phi(\tau) - \phi(\sigma) = -c \int_{\sigma}^{\tau} \phi(s)ds + \int_{\sigma}^{\tau} f(s)ds$$

for $t_0 < \sigma < \tau$. Suppose that there exists a positive number δ such that

$$\phi(\tau) = \left(1 + \frac{1}{c} + \delta\right) E(t_0), \qquad \phi(\sigma) = \left(\frac{1}{c} + \delta\right) E(t_0),$$

and

$$\phi(s) > \left(\frac{1}{c} + \delta\right) E(t_0) \qquad \text{for} \qquad \sigma < s \leq \tau.$$

Then,

$$E(t_0) + c\left(\frac{1}{c} + \delta\right) E(t_0)(\tau - \sigma) < E(\sigma)(1 + (\tau - \sigma)).$$

This is a contradiction.

Proof of (b).

Set $\psi(t) = \int_{t}^{T} e^{c(t-s)} f(s)ds$ for $0 \leq t \leq T$ for a fixed $T > 0$. Then, $\psi'(t) = c\psi(t) - f(t)$ and, hence, $\psi(\tau) - \psi(t) = c \int_{t}^{T} \psi(s)ds - \int_{t}^{T} f(s)ds$ for $t \leq \tau \leq T$. Suppose that there exists a positive number δ such that

$$\psi(\tau) = \left(\frac{1}{c} + \delta\right) E(t), \qquad \psi(t) = \left(1 + \frac{1}{c} + \delta\right) E(t),$$

and

$$\psi(s) > \left(\frac{1}{c} + \delta\right) E(t) \qquad \text{for} \qquad t \leq s \leq \tau.$$

Then,

$$E(t) + c\left(\frac{1}{c} + \delta\right) E(t)(\tau - t) < E(t)(1 + (\tau - t)).$$

This is a contradiction. This, in turn, proves that

$$\int_{t}^{T} e^{c(t-s)} f(s)ds \leq \left(1 + \frac{1}{c}\right) E(t) \qquad \text{for} \qquad T \geq t.$$

Since $\lim_{t \to +\infty} E(t) = 0$, the integral $\int_{t}^{+\infty} e^{c(t-s)} f(s)ds$ exists and

$$\int_{t}^{+\infty} e^{c(t-s)} f(s)ds \leq \left(1 + \frac{1}{c}\right) E(t).$$

The proof of (c) is left to the reader as an exercise. $\square$
For the argument given above, see [Har1].

VII-4. A diagonalization theorem

Liapounoff's type number of a solution is useful information since it provides some idea about the behavior of the solution as $t \to +\infty$. However, it is not quite enough when we look for a more specific information. For example, Liapounoff's type number of the second-order differential equation

$$\text{(VII.4.1)} \qquad \frac{d^2\eta}{dt^2} + \{1 + R(t)\}\eta = 0$$

is 0 and its multiplicity is 2 if the function $R(t)$ is continuous on the interval $0 \leq t < +\infty$ and $\lim_{t \to +\infty} R(t) = 0$ (cf. Theorem VII-3-6). However, this does not imply the boundedness of solutions as $t \to \infty$. O. Perron [Per3] constructed a function $R(t)$ satisfying the two conditions given above in such a way that solutions of (VII.4.1) are not bounded as $t \to +\infty$. Looking for better information concerning equation (VII.4.1), M. Hukuhara and M. Nagumo [HN1] proved the following theorem.

Theorem VII-4-1. *Every solution of differential equation (VII.4.1) is bounded as* $t \to +\infty$, *if* $\displaystyle\int_0^{+\infty} |R(t)|dt < +\infty$.

Proof.

First, fix $t_0 \geq 0$ in such a way that $\alpha = \displaystyle\int_{t_0}^{+\infty} |R(t)|dt < 1$. Write a solution $\phi(t)$ of (VII.4.1) in the form

$$\phi(t) = \phi(t_0)\cos(t - t_0) + \phi'(t_0)\sin(t - t_0) - \int_{t_0}^{t} R(s)\phi(s)\sin(t - s)ds.$$

Choose a positive number K so that $|\phi(t_0)| + |\phi'(t_0)| \leq K$ and choose another positive number M so that $M > \dfrac{K}{1 - \alpha} > K$. Then,

$$|\phi(t)| \leq K + M\int_{t_0}^{t} |R(s)|ds < M \qquad \text{for} \quad t_0 \leq t \leq t_1$$

if $|\phi(t)| \leq M$ for $t_0 \leq t \leq t_1$. Hence, $|\phi(t)| < M$ for $t_0 \leq t < +\infty$. $\square$

In this section, we explain the behavior of solutions of a system of linear differential equations under a condition similar to the Hukuhara-Nagumo condition. Precisely speaking, we consider a system of the form

$$\text{(VII.4.2)} \qquad \frac{d\vec{y}}{dt} = [\Lambda(t) + R(t)]\vec{y}$$

under the following assumptions.

Assumption 2. *Assume that $\Lambda(t)$ is an $n \times n$ diagonal matrix*

(VII.4.3) $\qquad\qquad \Lambda(t) \ = \ \mathrm{diag}[\lambda_1(t), \lambda_2(t), \dots, \lambda_n(t)],$

$R(t)$ is an $n \times n$ matrix whose entries are continuous on the interval $\mathcal{I}_0 = \{t : 0 \leq t < +\infty\}$, and

(VII.4.4) $\qquad\qquad\qquad \displaystyle\int_0^{+\infty} |R(t)| dt \ < \ +\infty.$

Set

(VII.4.5) $\quad \lambda_{jk}(t) = \lambda_j(t) - \lambda_k(t) \ \ \text{and} \ \ D_{jk}(t) = \Re\,(\lambda_{jk}(t)) \quad (j, k = 1, 2, \dots, n).$

Concerning the functions $\lambda_j(t)\,(j = 1, 2, \dots, n)$, the following is the main assumption.

Assumption 3. *The functions $\lambda_1(t), \lambda_2(t), \dots, \lambda_n(t)$ are continuous on the interval $\mathcal{I}_0$. Furthermore, for each fixed j, the set of all positive integers not greater than n is the union of two disjoint subsets $\mathcal{P}_{j1}$ and $\mathcal{P}_{j2}$, where*

(i) $k \in \mathcal{P}_{j1}$ if

$$\lim_{t \to +\infty} \int_0^t D_{jk}(\tau)d\tau = -\infty \quad \text{and} \quad \int_s^t D_{jk}(\tau)d\tau < K \qquad \text{for} \ \ 0 \leq s \leq t$$

for some positive number K,

(ii) $k \in \mathcal{P}_{j2}$ if

$$\int_s^t D_{jk}(\tau)d\tau \ < \ K \qquad \text{for} \ \ s \geq t \geq 0$$

for some positive number K.

Remark VII-4-2. Assume that the functions $\lambda_1(t), \dots, \lambda_n(t)$ are continuous on the interval $\mathcal{I}_0$ and that $\lim_{t \to +\infty} \lambda_j(t) = \mu_j\,(j = 1, 2, \dots, n)$ exist. Then, the functions $\lambda_1(t), \dots, \lambda_n(t)$ satisfy Assumption 3 if the real parts of $\mu_1, \dots, \mu_n$ are mutually distinct. The proof of this fact is left to the reader as an exercise.

The main concern in this section is to prove the following theorem due to N. Levinson.

Theorem VII-4-3 ([Levi1]). *Under Assumptions 2 and 3, there exists an $n \times n$ matrix $Q(t)$ such that*

(1) the derivative $\dfrac{dQ(t)}{dt}$ exists and the entries of Q and $\dfrac{dQ}{dt}$ are continuous on the interval $\mathcal{I}_0$,

(2) $\lim\limits_{t \to +\infty} Q(t) = O$,

(3) the transformation

(VII.4.6) $\qquad\qquad\qquad \vec{y} \ = \ [I_n \ + \ Q(t)]\,\vec{z}$

changes system (VII.4.2) to

(VII.4.7) $\qquad\qquad\qquad\qquad \dfrac{d\vec{z}}{dt} \ = \ \Lambda(t)\vec{z}$

on the interval $\mathcal{I}_0$, where I_n is the $n \times n$ identity matrix.

Proof.

We prove this theorem in six steps.

Step 1. Differentiating both sides of (VII.4.6), derive

$$\frac{dQ}{dt}\vec{z} + [I_n + Q]\frac{d\vec{z}}{dt} = [\Lambda(t) + R(t)][I_n + Q]\vec{z}.$$

Then, it follows from (VII.4.7) that Q should satisfy the linear differential equation

$$(VII.4.8) \qquad \frac{dQ}{dt} = [\Lambda(t) + R(t)][I_n + Q] - [I_n + Q]\Lambda(t)$$

or, equivalently,

$$(VII.4.9) \qquad \frac{dQ}{dt} = \Lambda(t)Q - Q\,\Lambda(t) + R(t)[I_n + Q].$$

The general solution $Q(t)$ of (VII.4.9) can be written in the form

$$(VII.4.9') \qquad Q(t) = \Phi(t)C\Psi(t)^{-1} + \int^t \Phi(t)\Phi(s)^{-1}R(s)\Psi(s)\Psi(t)^{-1}ds,$$

where C is an arbitrary constant matrix, $\Phi(t)$ is an $n \times n$ fundamental matrix of $\dfrac{d\Phi}{dt} = [\Lambda(t) + R(t)]\Phi$, and $\Psi(t)$ is an $n \times n$ fundamental matrix of $\dfrac{d\Psi}{dt} = \Lambda(t)\Psi$ (cf. Exercise IV-13). Thus, the general solution $Q(t)$ of (VII.4.9) exists and satisfies condition (1) of Theorem VII-4-3 on $\mathcal{I}_0$. Therefore, the proof of Theorem VII-4-3 will be completed if we prove the existence of a solution of (VII.4.9) which satisfies condition (2) of Theorem VII-4-3 on an interval $\mathcal{I} = \{t : t_0 \leq t < +\infty\}$ for a large t_0.

Step 2. Now, construct $Q(t)$ by using equation (VII.4.9) and condition (2) of Theorem VII-4-3. To do this, let $\Phi(t,s)$ be the unique solution of the initial-value problem

$$\frac{dY}{dt} = \Lambda(t)Y, \qquad Y(s) = I_n.$$

Then, (VII.4.9) is equivalent to the following linear integral equation:

$$(VII.4.10) \qquad Q(t) = \int^t \Phi(t,s)R(s)[I_n + Q(s)]\Phi(t,s)^{-1}ds,$$

where

$$\begin{cases} \Phi(t,s) = \mathrm{diag}[F_1(t,s), F_2(t,s), \dots, F_n(t,s)], \\[2mm] F_j(t,s) = \exp\left[\int_s^t \lambda_j(\tau)d\tau\right] \qquad (j = 1, 2, \dots, n). \end{cases}$$

Step 3. Letting $q_{jk}(t)$ and $r_{jk}(t)$ be the entries on the j-th row and the k-th column of $Q(t)$ and $R(t)$, respectively, write the integral equation (VII.4.10) in the form

$$(VII.4.10') \quad q_{jk}(t) = \int_{\tau_{jk}}^t \exp\left[\int_s^t \lambda_{jk}(\tau)d\tau\right]\left[r_{jk}(s) + \sum_{h=1}^n r_{jh}(s)q_{hk}(s)\right]ds,$$

where $j, k = 1, 2, \ldots, n$ and the $\lambda_{jk}(t)$ are defined by (VII.4.5). Note that

$$\left| \exp\left[\int_s^t \lambda_{jk}(\tau)d\tau \right] \right| = \exp\left[\int_s^t D_{jk}(\tau)d\tau \right] \leq e^K$$

if

$$0 \leq s \leq t \quad \text{for} \quad k \in \mathcal{P}_{j1} \qquad \text{and} \qquad t \leq s < +\infty \quad \text{for} \quad k \in \mathcal{P}_{j2},$$

where $j, k = 1, 2, \ldots, n$ and the $D_{jk}(t)$ are defined by (VII.4.5). The initial points τ_{jk} are chosen as follows:

$$\tau_{jk} = \begin{cases} t_0 & \text{if} \quad k \in \mathcal{P}_{j1}, \\ +\infty & \text{if} \quad k \in \mathcal{P}_{j2} \end{cases}$$

for some $t_0 \geq 0$.

Step 4. Define successive approximations by

$$\begin{cases} q_{0;jk}(t) = 0, \\ q_{p;jk}(t) = \displaystyle\int_{\tau_{jk}}^t \exp\left[\int_s^t \lambda_{jk}(\tau)d\tau \right]\left[r_{jk}(s) + \sum_{h=1}^n r_{jh}(s)q_{p-1;hk}(s) \right] ds, \end{cases}$$

where $p = 1, 2, \ldots$. Then, we obtain $| q_{p;jk}(t) | \leq e^K[1 + nC]\displaystyle\int_{t_0}^{+\infty} r(s)ds$ on the interval $\mathcal{I} = \{t : t_0 \leq t < +\infty\}$ if

(VII.4.11) $|q_{p-1;jk}(t)| \leq C$ on the interval $\mathcal{I}$,

where $r(t) = \max_{(j,k)} |r_{jk}(t)|$. Using assumption (VII.4.4), choose t_0 so large that

$$e^K[1 + nC] \int_{t_0}^{+\infty} r(s)ds \leq C.$$

Then, from (VII.4.11), it follows that $|q_{p;jk}(t)| \leq C$ on the interval $\mathcal{I}$.

Step 5. Similarly, if (VII.4.11) holds for $p = 1, 2, \ldots$, we obtain

$$|q_{p+1;jk}(t) - q_{p;jk}(t)| \leq \left(\frac{1}{2}\right)^p C$$

on the interval $\mathcal{I} = \{t : t_0 \leq t < +\infty\}$ if t_0 is chosen sufficiently large. Hence, $\lim_{p \to +\infty} q_{p;jk}(t) = q_{jk}(t)$ exists uniformly on the interval $\mathcal{I}$, and the limit satisfies integral equation (VII.4.10′).

Step 6. Let us prove that the bounded solution $q_{jk}(t)$ satisfies condition (2) of Theorem VII-4-3. If $k \in \mathcal{P}_{j2}$, then $\tau_{jk} = +\infty$. Hence,

$$\lim_{t \to +\infty} q_{jk}(t)$$

$$= \lim_{t \to +\infty} \int_{+\infty}^{t} \exp\left[\int_{s}^{t} \lambda_{jk}(\tau)d\tau\right]\left[r_{jk}(s) + \sum_{h=1}^{n} r_{jh}(s)q_{hk}(s)\right] ds = 0.$$

In the case when $k \in \mathcal{P}_{j1}$, note that

$$q_{jk}(t) = \int_{t_0}^{t} \exp\left[\int_{s}^{t} \lambda_{jk}(\tau)d\tau\right]\left[r_{jk}(s) + \sum_{h=1}^{n} r_{jh}(s)q_{hk}(s)\right] ds$$

and

$$\int_{t_0}^{t} \exp\left[\int_{s}^{t} \lambda_{jk}(\tau)d\tau\right]\left[r_{jk}(s) + \sum_{h=1}^{n} r_{jh}(s)q_{hk}(s)\right] ds$$

$$= \int_{t_0}^{\sigma} \exp\left[\int_{s}^{t} \lambda_{jk}(\tau)d\tau\right]\left[r_{jk}(s) + \sum_{h=1}^{n} r_{jh}(s)q_{hk}(s)\right] ds$$

$$+ \int_{\sigma}^{t} \exp\left[\int_{s}^{t} \lambda_{jk}(\tau)d\tau\right]\left[r_{jk}(s) + \sum_{h=1}^{n} r_{jh}(s)q_{hk}(s)\right] ds$$

for any σ such that $t_0 \leq \sigma \leq t$. Observe that

$$\left|\int_{t_0}^{\sigma} \exp\left[\int_{s}^{t} \lambda_{jk}(\tau)d\tau\right]\left[r_{jk}(s) + \sum_{h=1}^{n} r_{jh}(s)q_{hk}(s)\right] ds\right|$$

$$= \exp\left[\int_{\sigma}^{t} D_{jk}(\tau)d\tau\right]\left|\int_{t_0}^{\sigma} \exp\left[\int_{s}^{\sigma} \lambda_{jk}(\tau)d\tau\right]\left[r_{jk}(s) + \sum_{h=1}^{n} r_{jh}(s)q_{hk}(s)\right] ds\right|$$

$$\leq \exp\left[\int_{\sigma}^{t} D_{jk}(\tau)d\tau\right]e^{K}[1 + nC]\int_{t_0}^{+\infty} r(s)ds$$

if $|q_{jk}(t)| \leq C$ on $\mathcal{I}$. Observe also that

$$\left|\int_{\sigma}^{t} \exp\left[\int_{s}^{t} \lambda_{jk}(\tau)d\tau\right]\left[r_{jk}(s) + \sum_{h=1}^{n} r_{jh}(s)q_{hk}(s)\right] ds\right|$$

$$\leq e^{K}[1 + nC]\int_{\sigma}^{+\infty} r(s)ds$$

if $|q_{jk}(t)| \leq C$ on $\mathcal{I}$. For a given positive number ϵ, fix a positive number σ so large that $e^{K}[1 + nC]\int_{\sigma}^{+\infty} r(s)ds \leq \frac{\epsilon}{2}$. Since

$$\lim_{t \to +\infty} \int_{\sigma}^{t} D_{jk}(\tau)d\tau = -\infty \qquad \text{if} \quad k \in \mathcal{P}_{j1}$$

for a fixed σ, we obtain

$$\exp\left[\int_{\sigma}^{t} D_{jk}(\tau)d\tau\right]e^{K}[1 + nC]\int_{t_0}^{+\infty} r(s)ds \leq \frac{\epsilon}{2}$$

for large t. Therfore, for any positive number ϵ, there exists $t(\epsilon)$ such that $|q_{jk}(t)| \leq \epsilon$ for $t \geq t(\epsilon)$. This complete the proof of Theorem IV-4-3. $\square$

Remark VII-4-4. The $n \times n$ matrix $\Psi(t,s) = [I_n + Q(t)]\Phi(t,s)$ is the unique solution of the initial-value problem

$$\frac{dY}{dt} = [\Lambda(t) + R(t)]Y, \qquad Y(s) = I_n + Q(s),$$

where $\Phi(t,s)$ is the diagonal matrix defined in Step 2 of the proof of Theorem VII-4-3 (cf. (VII.4.10)). Since $\det[\Psi(t,s)] \neq 0$ for large t, the matrix $\Psi(t,s)$ is invertible for all (t,s) in $\mathcal{I}_0 \times \mathcal{I}_0$ (cf. Exercises IV-8). This, in turn, implies that the matrix $I_n + Q(t)$ is invertible for all t in $\mathcal{I}_0$.

Remark VII-4-5. Theorem VII-4-3 has been shown to be the basis for many results concerning asymptotic integration (cf. [E1, E2] and [HarL1, HarL2, HarL3]).

Remark VII-4-6. Using the results of globally analytic simplifications of matrix functions in [GH], H. Gingold, et al. [GHS] shows some results similar to Theorem VII-4-3 with $Q(t)$ analytic on the entire interval $\mathcal{I}_0$ under suitable conditions.

Remark VII-4-7. Instead of (VII.4.4), [HX1] and [HX2] assume only the integrability at $t = \infty$ for above (or below) the diagonal entries of $R(t)$ and obtain the results similar to Theorem VII-4-3. More results were obtained in [HX3] applying a result of [Si9].

The following example illustrates applications of Theorem VII-4-3.

Example VII-4-8. Let us look at a second-order linear differential equation

$$(VII.4.12) \qquad \frac{d^2\eta}{dt^2} + p(t)\eta = 0.$$

If we set

$$\vec{y} = \begin{bmatrix} \eta \\ \dfrac{d\eta}{dt} \end{bmatrix}, \qquad A(t) = \begin{bmatrix} 0 & 1 \\ -p(t) & 0 \end{bmatrix},$$

equation (VII.4.12) becomes the system

$$(VII.4.13) \qquad \frac{d\vec{y}}{dt} = A(t)\vec{y}.$$

The two eigenvalues and corresponding eigenvectors of the matrix $A(t)$ are

$$\begin{cases} \lambda_1(t) = i\,p(t)^{1/2}, & \vec{p}_1(t) = \begin{bmatrix} 1 \\ i\,p(t)^{1/2} \end{bmatrix}, \\[2ex] \lambda_2(t) = -i\,p(t)^{1/2}, & \vec{p}_2(t) = \begin{bmatrix} 1 \\ -i\,p(t)^{1/2} \end{bmatrix}. \end{cases}$$

Set $P_0(t) = \begin{bmatrix} 1 & 1 \\ i\,p(t)^{1/2} & -ip(t)^{1/2} \end{bmatrix}$. Then,

$$\frac{dP_0(t)}{dt} = \frac{ip'(t)}{2p(t)^{1/2}} \begin{bmatrix} 0 & 0 \\ 1 & -1 \end{bmatrix}, \qquad P_0(t)^{-1} = \frac{-i}{2p(t)^{1/2}} \begin{bmatrix} ip(t)^{1/2} & 1 \\ ip(t)^{1/2} & -1 \end{bmatrix},$$

and

$$P_0(t)^{-1}A(t)P_0(t) \;=\; \begin{bmatrix} i\,p(t)^{1/2} & 0 \\ 0 & -i\,p(t)^{1/2} \end{bmatrix}.$$

The transformation $\vec{y} = P_0(t)\vec{z}$ changes system (VII.4.13) to

$$\frac{d\vec{z}}{dt} \;=\; P_0(t)^{-1}\left[A(t)P_0(t) \;-\; \frac{dP_0(t)}{dt}\right]\vec{z}.$$

Using the computations given above, we can write this system in the form

$$(VII.4.14) \qquad \frac{d\vec{z}}{dt} \;=\; \left\{ i\,p(t)^{1/2}\begin{bmatrix} 1 & 0 \\ 0 & -1 \end{bmatrix} - \frac{p'(t)}{4p(t)}\begin{bmatrix} 1 & -1 \\ -1 & 1 \end{bmatrix}\right\}\vec{z}.$$

Suppose that
(1)   a function $p(t)$ is continuous on the interval $\mathcal{I}_0 = \{t:\ 0 \le t < +\infty\}$,
(2)   there exists a positive number c such that $p(t) \ge c > 0$ on the interval $\mathcal{I}_0$,
(3)   the derivative $p'(t)$ of $p(t)$ is absolutely integrable on $\mathcal{I}_0$,
Then, Theorem VII-4-3 applies to system (VII.4.14) and yields the following theorem (cf. [HN2]).

Theorem VII-4-9. *If a function $p(t)$ satisfies the conditions (1), (2), and (3) given above, every solution of equation (IV.4.12) and its derivative are bounded on the interval $\mathcal{I}_0$.*

The proof of this theorem is left to the reader as an exercise. Note that condition (3) implies the boundedness of $p(t)$ on the interval $\mathcal{I}_0$.

If $p'(t)$ is not absolutely integrable, set

$$(VII.4.15) \qquad\qquad \vec{z} \;=\; [I_2 + q(t)E]\vec{u},$$

where I_2 is the 2×2 identity matrix, $q(t)$ is a unknown complex-valued function, and E is a constant 2×2 unknown matrix. Then, the transformation (VII.4.15) changes system (VII.4.14) to
(VII.4.16)

$$\frac{d\vec{u}}{dt} = [I_2 + q(t)E]^{-1}\left\{\left[ip(t)^{1/2}A_0 - \frac{p'(t)}{4p(t)}A_1\right][I_2 + q(t)E] - \frac{dq(t)}{dt}E\right\}\vec{u},$$

where

$$A_0 \;=\; \begin{bmatrix} 1 & 0 \\ 0 & -1 \end{bmatrix}, \qquad A_1 \;=\; \begin{bmatrix} 1 & -1 \\ -1 & 1 \end{bmatrix}.$$

Anticipating that
(i) $|q(t)|$ and $|p'(t)|$ are of the same size,
(ii) $|q'(t)|$ and $|p''(t)|$ are of the same size,
choose $q(t)$ and E so that two off-diagonal entries of the matrix on the right-hand side of (VII.4.16) become as small as $|p'(t)|^2 + |p''(t)|$. In fact, choosing

$$q(t) \;=\; \frac{-i\,p'(t)}{8\,p(t)^{3/2}}, \qquad E \;=\; \begin{bmatrix} 0 & -1 \\ 1 & 0 \end{bmatrix},$$

we obtain

$$[I_2 + q(t)E]^{-1} = \frac{1}{1 + q(t)^2}[I_2 - q(t)E]$$

and

$$[I_2 + q(t)E]^{-1}\left\{\left[ip(t)^{1/2}A_0 - \frac{p'(t)}{4p(t)}A_1\right][I_2 + q(t)E] - \frac{dq(t)}{dt}E\right\}$$

$$= \frac{1}{1 + q(t)^2}\left\{ip(t)^{1/2}A_0 - \frac{p'(t)}{8p(t)}EA_0 - \frac{p'(t)}{4p(t)}A_1\right.$$

$$\left. + \frac{p'(t)}{8p(t)}A_0E + \mathcal{E}(p, p', p'') - q(t)\frac{dq(t)}{dt}I_2\right\}$$

$$= \frac{1}{1 + q(t)^2}\left\{ip(t)^{1/2}A_0 - \frac{p'(t)}{4p(t)}I_2 + \mathcal{E}(p, p', p'') - q(t)\frac{dq(t)}{dt}I_2\right\},$$

where

$$E^2 = -I_2, \qquad EA_0 = -A_0E = \begin{bmatrix} 0 & 1 \\ 1 & 0 \end{bmatrix},$$

and $\mathcal{E}(p, p', p'')$ is the sum of a finite number of terms of the form

$$\frac{\alpha\,[\,p'(t)\,]^2 + \beta\,p''(t)}{p(t)^{h/2}}$$

with some rational numbers α and β and some positive integers h. Applying Theorem VII-4-3 to system (VII.4.16), we can prove the following theorem.

Theorem VII-4-10. *Suppose that*
(1)   a function $p(t)$ is continuous on the interval $\mathcal{I}_0 = \{t : 0 \leq t < +\infty\}$,
(2)   there exists a positive number c such that $p(t) \geq c > 0$ on the interval $\mathcal{I}_0$,
(3)

$$\int_0^{+\infty}\{|p'(t)|^2 + |p''(t)|\}\,dt < +\infty,$$

(4) $\lim\limits_{t \to +\infty} p'(t) = 0$.
Then, every solution of equation (VII.4.12) is bounded on the interval $\mathcal{I}_0$.

The proof of this theorem is left to the reader. Condition (3) of Theorem VII-4-10 does not imply the boundedness of $p(t)$ on the interval I_0. For example, Theorem VII-4-10 applies to equation (VII.4.12) with $p(t) = \log[2 + t]$.

VII-5. Systems with asymptotically constant coefficients

In this section, we apply Theorem VII-4-3 to a system of the form

$$(\text{VII.5.1}) \qquad\qquad \frac{d\vec{y}}{dt} = [A + V(t)]\vec{y},$$

where A is a constant $n \times n$ matrix and $V(t)$ is an $n \times n$ matrix whose entries and their derivatives are continuous in t on the interval $\mathcal{I}_0 = \{t : 0 \leq t < +\infty\}$ under the following assumption.

Assumption 4. *The matrix A has n mutually distinct eigenvalues $\mu_1, \mu_2, \ldots, \mu_n$, and the matrix $V(t)$ satisfies the conditions*

$$\lim_{t \to +\infty} V(t) \; = \; 0 \qquad and \qquad \int_0^{+\infty} |V'(t)| dt \; < \; +\infty.$$

Let $\lambda_1(t), \lambda_2(t), \ldots, \lambda_n(t)$ be the eigenvalues of the matrix $A + V(t)$. Then, these are continuous on the interval $\mathcal{I}_0$. Furthermore, it can be assumed that

$$\lim_{t \to +\infty} \lambda_j(t) \; = \; \mu_j \qquad (j = 1, 2, \ldots, n).$$

Choose $t_0 \geq 0$ so that $\lambda_1(t), \lambda_2(t), \ldots, \lambda_n(t)$ are mutually distinct on the interval $\mathcal{I} = \{t : t_0 \leq t < +\infty\}$. Set

$$F(t, \lambda) \; = \; \det[\lambda I_n - A - V(t)].$$

Then, $F(t, \lambda_j(t)) \; = \; 0$ on $\mathcal{I}_0$ for $j = 1, 2, \ldots, n$, and

$$\frac{\partial F}{\partial t}(t, \lambda_j(t)) \; + \; \frac{\partial F}{\partial \lambda}(t, \lambda_j(t)) \, \lambda_j'(t) \; = \; 0 \quad (j = 1, 2, \ldots, n) \quad \text{on } \mathcal{I}_0.$$

Also,

$$\lim_{t \to +\infty} \frac{\partial F}{\partial \lambda}(t, \lambda_j(t)) \; \neq \; 0 \qquad (j = 1, 2, \ldots, n),$$

since $\mu_1, \mu_2, \ldots, \mu_n$ are mutually distinct. Observe that $\lambda_j'(t) = -\dfrac{\dfrac{\partial F}{\partial t}(t, \lambda_j(t))}{\dfrac{\partial F}{\partial \lambda}(t, \lambda_j(t))}$ is linear homogeneous in the entries of the matrix $V'(t)$. In this way, we obtain the following lemma.

Lemma VII-5-1. *Under Assumption 4, the derivatives of the eigenvalues of the matrix $A + V(t)$ are absolutely integrable over the interval $\mathcal{I} = \{t : t_0 \leq t < +\infty\}$, i.e.,*

$$\int_{t_0}^{+\infty} |\lambda_j'(t) dt| \; < \; +\infty \qquad (j = 1, 2, \ldots, n).$$

An eigenvector $\vec{p}_j(t)$ of the matrix $A + V(t)$ associated with the eigenvalue $\lambda_j(t)$ can be constructed in the following manner. Observe that the characteristic polynomial of $A + V(t)$ can be factored as

$$F(t, \lambda) \; = \; (\lambda - \lambda_1(t))(\lambda - \lambda_2(t)) \cdots (\lambda - \lambda_n(t)).$$

Hence, by virtue of Theorem IV-1-5 (Cayley-Hamilton), we obtain

$$(VII.5.2) \quad (A + V(t) - \lambda_1(t) I_n)(A + V(t) - \lambda_2(t) I_n) \cdots (A + V(t) - \lambda_n(t) I_n) \; = \; 0$$

on $\mathcal{I}_0$. Furthermore, if $t \geq t_0$,

$$\begin{cases} \displaystyle\prod_{h \neq j}(A + V(t) - \lambda_h(t)I_n) \neq 0, \\[2mm] \displaystyle\lim_{t \to +\infty}\prod_{h \neq j}(A + V(t) - \lambda_h(t)I_n) = \prod_{h \neq j}(A - \mu_h I_n) \neq 0, \end{cases}$$

for $j = 1, 2, \ldots, n$. From (VII.5.2), it follows that

$$[A + V(t)]\prod_{h \neq j}(A + V(t) - \lambda_h(t)I_n) = \lambda_j(t)\prod_{h \neq j}(A + V(t) - \lambda_h(t)I_n).$$

Hence, choosing a suitable column vector $\vec{p}_j(t)$ of $\displaystyle\prod_{h \neq j}(A + V(t) - \lambda_h(t)I_n)$, we obtain

$$\vec{p}_j(t) \neq 0 \quad \text{and} \quad [A + V(t)]\vec{p}_j(t) = \lambda_j(t)\vec{p}_j(t)$$

on the interval $\mathcal{I} = \{t : t_0 \leq t < +\infty\}$ if $t_0 \geq 0$ is sufficiently large. Furthermore,

$$\lim_{t \to +\infty}\vec{p}_j(t) = \vec{q}_j \qquad (j = 1, 2, \ldots, n)$$

are eigenvectors of the matrix A associated with the eigenvalues μ_j $(j = 1, 2, \ldots, n)$, respectively. Observe that the entries of the vectors $\vec{p}_j(t)$ $(j = 1, 2, \ldots, n)$ are polynomials in the entries of $V(t)$ and $\lambda_1(t), \ldots, \lambda_n(t)$ with constant coefficients. Hence,

$$\int_{t_0}^{+\infty} |\vec{p}_j{}'(t)|\, dt < +\infty \qquad (j = 1, 2, \ldots, n).$$

Thus, we proved the following lemma.

Lemma VII-5-2. *Under Assumption 4, there exists a non-negative number t_0 such that*
(1) the matrix $A + V(t)$ has n mutually distinct eigenvalues $\lambda_1(t), \ldots, \lambda_n(t)$ on the interval $\mathcal{I} = \{t : t_0 \leq t < +\infty\}$,
(2) the eigenvalues $\lambda_1(t), \ldots, \lambda_n(t)$ are continuously differentiable on $\mathcal{I}$ and $\displaystyle\lim_{t \to +\infty}\lambda_j(t) = \mu_j\,(j = 1, 2, \ldots, n)$,
(3) the matrix $A + V(t)$ has n eigenvectors $\vec{p}_1(t), \ldots, \vec{p}_n(t)$ associated with the eigenvalues $\lambda_1(t), \ldots, \lambda_n(t)$, respectively, such that $\displaystyle\lim_{t \to +\infty}\vec{p}_j(t) = \vec{q}_j\,(j = 1, 2, \ldots, n)$ are eigenvectors of the matrix A associated with the eigenvalues $\mu_j\,(j = 1, 2, \ldots, n)$,
(4) the derivatives of the eigenvalues $\lambda_j(t)\,(j = 1, 2, \ldots, n)$ and the derivatives of the eigenvectors $\vec{p}_j(t)\,(j = 1, 2, \ldots, n)$ with respect to t are absolutely integrable on the interval $\mathcal{I}$.

Set

$$P_0(t) = [\,\vec{p}_1(t)\ \vec{p}_2(t)\ \ldots\ \vec{p}_n(t)\,], \qquad Q = [\,\vec{q}_1\ \vec{q}_2\ \ldots\ \vec{q}_n\,],$$
$$\Lambda(t) = \mathrm{diag}[\lambda_1(t), \lambda_2(t), \ldots, \lambda_n(t)], \qquad M = \mathrm{diag}[\mu_1, \mu_2, \ldots, \mu_n].$$

Then,

$$\begin{cases} \lim_{t\to+\infty} P_0(t) = Q, & \lim_{t\to+\infty} \Lambda(t) = M, \\[2mm] P_0(t)^{-1}[A + V(t)]P_0(t) = \Lambda(t), & Q^{-1}AQ = M, \\[2mm] \displaystyle\int_{t_0}^{+\infty} \left| P_0(t)^{-1}\frac{dP_0(t)}{dt} \right| dt < +\infty. \end{cases}$$

Observe that the transformation $\vec{y} = P_0(t)\vec{z}$ changes system (VII.5.1) to

$$\frac{d\vec{z}}{dt} = P_0(t)^{-1}\left[[A + V(t)]P_0(t) - \frac{dP_0(t)}{dt}\right]\vec{z} = \left[\Lambda(t) - P_0(t)^{-1}\frac{dP_0(t)}{dt}\right]\vec{z}.$$

Applying Theorem VII-4-3 to this system, we obtain the following theorem.

Theorem VII-5-3. *Under Assumption 4, if the eigenvalues $\lambda_1(t),\dots,\lambda_n(t)$ of the matrix $A + V(t)$ satisfy all requirements given in Assumption 3 (cf. §VII-4) on the interval $\mathcal{I} = \{t : t_0 \leq t < +\infty\}$, a fundamental matrix solution of (VII.5.1) will be given by*

$$\Phi(t) = P_0(t)[I_n + H(t)] \exp\left[\int^t \Lambda(s)ds\right],$$

where $H(t)$ is an $n \times n$ matrix whose entries are continuously differentiable on the interval $\mathcal{I}$ and $\lim_{t\to+\infty} H(t) = O$.

Remark VII-5-4.

(a) As $t \to +\infty$, the matrix $\exp\left[-\int^t \Lambda(s)ds\right]\Phi(t)$ approaches the matrix Q,

(b) we can prove a result similar to Theorem VII-5-3 even if system (VII.5.1) is replaced by

$$\frac{d\vec{y}}{dt} = [A + V(t) + R(t)]\vec{y},$$

where the matrix $A + V(t)$ satisfies all the requirements of Assumption 4 and the entries of the matrix $R(t)$ are absolutely integrable for $t \geq 0$, i.e.,

$$\int_0^{+\infty} |R(t)|dt < +\infty.$$

Observation VII-5-5. Let us look into the case when the matrix A has multiple eigenvalues. To do this, consider system (VII.3.1) under Assumption 1 given in §VII-3. By virtue of Theorem VII-3-1, system (VII.3.1) is changed to system (VII.3.4) by transformation (VII.3.3). Furthermore, (VII.3.4) is changed to

(VII.5.3) $$\qquad \frac{d\vec{u}_j}{dt} = [N_j + R_j(t)]\vec{u}_j \qquad (j = 1, 2, \dots, m),$$

where

$$R_j(t) = e^{-tE_j}\left[B_{jj}(t) + \sum_{h\neq j} B_{jh}(t)T_{hj}(t)\right]e^{tE_j} \qquad (j = 1, 2, \dots, m)$$

by the transformation

$$\vec{z}_j = \exp[t(\lambda_j I_{n_j} + E_j)]\vec{u}_j \qquad (j = 1, 2, \dots, m).$$

Therefore, we look into the following two cases.

Case VII-5-5-1. If N is an $n \times n$ nilpotent matrix such that $N^r = O$ and $R(t)$ is an $n \times n$ matrix whose entries are continuous on the interval $0 \leq t < +\infty$ and satisfy the condition

$$(\text{VII.5.4}) \qquad \int_0^{+\infty} t^{2(r-1)}|R(t)|dt \; < \; +\infty,$$

we can construct a fundamental matrix solution $\Phi(t)$ of the system

$$(\text{VII.5.5}) \qquad \frac{d\vec{y}}{dt} \; = \; (N + R(t))\vec{y}$$

such that $\lim\limits_{t \to +\infty} e^{-tN}\Phi(t) \; = \; I_n$, where I_n is the $n \times n$ identity matrix. In fact, the transformation $\vec{y} = e^{tN}\vec{u}$ changes system (VII.5.5) to $\dfrac{d\vec{u}}{dt} \; = \; e^{-tN}R(t)e^{tN}\vec{u}$. Since $|e^{\pm tN}| \leq K|t|^{r-1}$ for some positive number K, the integral equation $X(t) = I_n - \int_t^{+\infty} e^{-\tau N}R(\tau)e^{\tau N}X(\tau)d\tau$ can be solved easily.

Case VII-5-5-2. Let N be an $n \times n$ nilpotent matrix such that $N^r = O$ and let $R(t)$ be an $n \times n$ matrix whose entries are continuous on the interval $0 \leq t < +\infty$ and satisfy the condition

$$(\text{VII.5.6}) \qquad \int_0^{+\infty} t^{r-1}|R(t)|dt \; < \; +\infty.$$

Exercise V-4 shows that this case is different from Case VII-5-5-1. As a matter of fact, in this case, we can construct a fundamental matrix solution $\Psi(t)$ of system (VII.5.5) such that

$$(\text{VII.5.7}) \qquad \lim\limits_{t \to +\infty} t^{-(k-1)}\left(\Psi(t) - e^{tN}\right)\vec{c} \; = \; \vec{0}, \qquad \text{whenever} \quad N^k\vec{c} \; = \; \vec{0},$$

where $k \leq r$. We prove this result in three steps.

Step 1. First of all, if an $n \times n$ matrix $\Psi(t)$ satisfies condition (VII.5.7), then $\vec{c} = \vec{0}$ if $\Psi(t)\vec{c} = \vec{0}$ for large t. In fact, noice first that $\lim\limits_{t \to +\infty} \dfrac{e^{tN}\vec{c}}{t^\ell} \; = \; \dfrac{N^\ell\vec{c}}{\ell!}$ if $N^p\vec{c} = \vec{0}$ for $p \; = \; \ell + 1, \dots, r - 1$. Also, note that $\lim\limits_{t \to +\infty} \dfrac{e^{tN}\vec{c}}{t^{k-1}} = \vec{0}$ if $\Psi(t)\vec{c} = \vec{0}$ and $N^k\vec{c} = \vec{0}$. Therefore, $N^{r-1}\vec{c} = \vec{0}$ if $k = r$. Hence, $N^{r-2}\vec{c} = \vec{0}$. In this way, we obtain $\vec{c} = \vec{0}$. Thus, we showed that if the matrix $\Psi(t)$ satisfies the differential equation $\dfrac{d\Psi}{dt} = (N + R(t))\Psi$ and condition (VII.5.7), then $\Psi(t)$ is a fundamental matrix solution of (VII.5.5).

Step 2. To prove the existence of such a matrix $\Psi(t)$, we estimate integrals of $s^k e^{(t-s)N}R(s)$ with respect to s, where $0 \leq k \leq r - 1$. Observe that

$$s^k e^{(t-s)N} \; = \; \sum_{h=0}^{r-1} \frac{s^k(t-s)^h}{h!}N^h.$$

Let us look at the quantity $s^k(t - s)^h$. Note that $0 \leq k \leq r - 1$ and $0 \leq h \leq r - 1$.

Case 1. $k + h \leq r - 1$. In this case, since $s^k(t-s)^h = s^{k+h}\left(\dfrac{t}{s} - 1\right)$, define

$$\int_\eta^t \frac{s^k(t-s)^h}{h!} N^h R(s)ds = \int_{+\infty}^t \frac{s^k(t-s)^h}{h!} N^h R(s)ds.$$

Case 2. $k + h \geq r$. In this case, look at $s^k(t-s)^h = \sum_{p=0}^h (-1)^p \binom{h}{p} s^{k+p} t^{h-p}$.

Subcase 2(i). $k + p \leq r - 1$. In this case, $s^{k+p} t^{h-p} = s^{r-1}\left(\dfrac{t}{s}\right)^\mu t^\nu$, where $r - 1 - \mu = k + p, \mu + \nu = h - p$. Since

$$\nu = h - p - \mu = (h-p) + (k+p) - (r-1) = (k+h) - (r-1) = k - (r-1-h),$$

we obtain $0 < \nu \leq k$. Now, in this case, define

$$\int_\eta^t \frac{\binom{h}{p}}{h!} s^{k+p} t^{h-p} N^h R(s)ds = \int_{+\infty}^t \frac{\binom{h}{p}}{h!} s^{k+p} t^{h-p} N^h R(s)ds.$$

Subcase 2(ii). $k + p \geq r$. In this case,

$$s^{k+p} t^{h-p} = s^{r-1}\left(\dfrac{s}{t}\right)^\mu t^\nu, \qquad \text{where} \qquad r - 1 + \mu = k + p, \quad \nu - \mu = h - p.$$

Since

$$\nu = h - p + \mu = (h-p) + (k+p) - (r-1) = (k+h) - (r-1) = k - (r-1-h),$$

we obtain $0 < \nu \leq k$. Also, note that $h - p \leq (r - 1) + (k - r) = k - 1$. Now, in this case, define

$$\int_\eta^t \frac{\binom{h}{p}}{h!} s^{k+p} t^{h-p} N^h R(s)ds = \int_{t_0}^t \frac{\binom{h}{p}}{h!} s^{k+p} t^{h-p} N^h R(s)ds, \qquad \text{where} \qquad t \geq t_0.$$

From the definition given above, it follows that

$$\begin{cases} \displaystyle\int_\eta^t s^k e^{(t-s)N} R(s)ds = e^{tN} \int_\eta^t s^k e^{-sN} R(s)ds, \\[2ex] \displaystyle\lim_{t\to+\infty} \frac{1}{t^k} \int_\eta^t s^k e^{(t-s)N} R(s)ds = O \quad (k = 0, \ldots, r - 1), \end{cases}$$

Setting

$$U(t) = \int_\eta^t s^k e^{(t-s)N} R(s)ds,$$

we obtain

$$\frac{dU(t)}{dt} = NU(t) + t^k R(t).$$

Step 3. Let us construct $n \times n$ matrices $U_k(t)$ $(k = 0, 1, 2, \ldots, r-1)$ by the integral equations

$$U_k(t) = N^k + \frac{1}{t^k} \int_\eta^t s^k e^{(t-s)N} R(s) U_k(s) ds.$$

Then,

$$\lim_{t \to +\infty} U_k(t) = N^k \qquad \text{and} \qquad U_k(t)\vec{c} = O \quad \text{if} \quad N^k \vec{c} = \vec{0}.$$

Observe that

$$\frac{t^k U_k(t)}{k!} = \frac{t^k N^k}{k!} + \int_\eta^t e^{(t-s)N} R(s) \frac{s^k U_k(s)}{k!} ds.$$

Hence, setting

$$\Psi(t) = \sum_{k=0}^{r-1} \frac{t^k U_k(t)}{k!},$$

we obtain

$$\Psi(t) = e^{tN} + \sum_{k=0}^{r-1} \int_\eta^t e^{(t-s)N} R(s) \frac{s^k U_k(s)}{k!} ds.$$

This implies that

$$\frac{d\Psi(t)}{dt} = (N + R(t))\Psi(t).$$

It can be easily shown that

$$\lim_{t \to +\infty} \frac{\Psi(t)\vec{c}}{t^k} = \lim_{t \to +\infty} \frac{e^{tN}\vec{c}}{t^k} = \frac{N^k \vec{c}}{k!}$$

if $N^{k+1}\vec{c} = \vec{0}$. Thus, the construction of $\Psi(t)$ is completed.

Case VII-5-5-2 was given as Exercise 35 in [CL, p. 106].

VII-6. An application of the Floquet theorem

The method discussed in §VII-5 does not apply directly to the scalar differential equation

$$\frac{d^2\eta}{dt^2} + \{1 + h(t)\sin(\alpha t)\}\eta = 0$$

when $h(t)$ is a small function such as $\dfrac{1}{t}$, $\dfrac{1}{t^{1/2}}$, $\dfrac{1}{\log t}$, and $\dfrac{1}{\log t}\sin(t^{1/2})$ since the derivative of $h(t)\sin(\alpha t)$ is not absolutely integrable over the interval $0 \leq t < +\infty$. In this section, using the Floquet theorem (cf. Theorem IV-4-1), we eliminate periodic parts of coefficients so that Theorem VII-4-3 applies. Keeping this scheme in mind, consider a system of the form

$$(\text{VII.6.1}) \qquad \frac{d\vec{y}}{dt} = A(t, \vec{h}(t))\vec{y}$$

under the following assumption.

Assumption VII-6-1.

   (1) The entries of an $n \times n$ matrix $A(t, \vec{\epsilon})$ are continuous in $(t, \vec{\epsilon}) \in \mathbb{R} \times \triangle(r)$ and analytic in $\vec{\epsilon} \in \triangle(r)$ for each fixed $t \in \mathbb{R}$, where $\vec{\epsilon} \in \mathbb{C}^m$ with the entries $\epsilon_1, \ldots, \epsilon_m$, $\triangle(r) = \{\vec{\epsilon} : |\vec{\epsilon}| < r\}$, and r is a positive number.

   (2) The entries of $A(t, \vec{\epsilon})$ are periodic in t of a positive period ω.

   (3) The entries $h_1, \ldots, h_m$ of a $\mathbb{C}^m$-valued function $\vec{h}(t)$ are continuous on the interval $\mathcal{I}(t_0) = \{t : t_0 \leq t < +\infty\}$ for some non-negative number t_0 and $\lim\limits_{t \to +\infty} \vec{h}(t) = \vec{0}$.

Let us consider the following two cases.

Case 1. The function $\vec{h}(t)$ is supposed to be continuously differentiable on $\mathcal{I}(t_0)$ and satisfy the condition

$$(\text{VII.6.2}) \qquad \int_{t_0}^{+\infty} |\vec{h}(t)| dt = +\infty \qquad \text{and} \qquad \int_{t_0}^{+\infty} |\vec{h}'(t)| dt < +\infty.$$

Case 2. The function $\vec{h}(t)$ is supposed to be twice continuously differentiable on $\mathcal{I}(t_0)$ and satisfy the condition

$$(\text{VII.6.3}) \qquad \begin{cases} \lim\limits_{t \to +\infty} \vec{h}'(t) = \vec{0}, \\[2mm] \displaystyle\int_0^{+\infty} |\vec{h}(t)| dt = +\infty, \qquad \int_0^{+\infty} |\vec{h}'(t)| dt = +\infty, \\[2mm] \displaystyle\int_0^{+\infty} |\vec{h}''(t)| dt < +\infty, \qquad \int_0^{+\infty} |\vec{h}'(t)|^2 dt < +\infty. \end{cases}$$

For example, the function $h(t) = \dfrac{1}{t}$ satisfies (VII.6.2), whereas $h(t) = \dfrac{\sin(\sqrt{t})}{\sqrt{t}}$ satisfies (VII.6.3).

Observation in Case 1. In Case 1, we use the following lemma.

Lemma VII-6-2. *If a matrix $A(x, \vec{\epsilon})$ satisfies conditions (1) and (2) of Assumption VII-6-1, there exist $n \times n$ matrices $P(t, \vec{\epsilon})$ and $H(\vec{\epsilon})$ such that*

   (i) the entries of $P(t, \vec{\epsilon})$ are continuous in $(t, \vec{\epsilon}) \in \mathbb{R} \times \triangle(\hat{r})$ and analytic in $\vec{\epsilon} \in \triangle(\hat{r})$ for each fixed $t \in \mathbb{R}$, where $\hat{r}$ is a suitable positive number,

   (ii) $P(t + \omega, \vec{\epsilon}) = P(t, \vec{\epsilon})$ for $(t, \vec{\epsilon}) \in \mathbb{R} \times \triangle(\hat{r})$,

   (iii) $P(t, \vec{\epsilon})$ is invertible for every $(t, \vec{\epsilon}) \in \mathbb{R} \times \triangle(\hat{r})$,

   (iv) the entries of $H(\vec{\epsilon})$ are analytic in $\vec{\epsilon} \in \triangle(\hat{r})$,

   (v) any two distinct eigenvalues of $H(0)$ do not differ by integral multiples of $\dfrac{2\pi i}{\omega}$,

   (vi) $\dfrac{\partial}{\partial t} P(t, \vec{\epsilon})$ exists for $(t, \vec{\epsilon}) \in \mathbb{R} \times \triangle(\hat{r})$ and given by

$$(\text{VII.6.4}) \qquad \frac{\partial}{\partial t} P(t, \vec{\epsilon}) = A(t, \vec{\epsilon}) P(t, \vec{\epsilon}) - P(t, \vec{\epsilon}) H(\vec{\epsilon})$$

   for $(t, \vec{\epsilon}) \in \mathbb{R} \times \triangle(\hat{r})$.

Proof.

In order to prove this lemma, construct a fundamental matrix solution $\Phi(t,\vec{\epsilon})$ of the differential equation $\dfrac{d\vec{y}}{dt} = A(t,\vec{\epsilon})\vec{y}$ by solving the initial-value problem $\dfrac{dX}{dt} = A(t,\vec{\epsilon})X$, $X(0) = I_n$, where I_n is the $n \times n$ identity matrix. The entries of $\Phi(\omega,\vec{\epsilon})$ are analytic in $\triangle(r)$. Define $H(\vec{\epsilon})$ by $H(\vec{\epsilon}) = \dfrac{\log(\Phi(\omega,\vec{\epsilon}))}{\omega}$ and $P(t,\vec{\epsilon}) = \Phi(t,\vec{\epsilon})\exp[-tH(\vec{\epsilon})]$. Then, (VII.6.4) follows. $\square$

The most delicate part of this proof is the definition of $H(\vec{\epsilon})$. Details are left for the reader as an exercise (cf. [Si1]).

Changing system (VII.6.1) by the transformation

$$(\text{VII.6.5}) \qquad\qquad \vec{y} = P(t,\vec{h}(t))\vec{u}$$

we obtain the following theorem.

Theorem VII-6-3. *Transformation (VII.6.5) changes system (VII.6.1) to*

$$(\text{VII.6.6}) \qquad \frac{d\vec{u}}{dt} = \left\{ H(\vec{h}(t)) - P(t,\vec{h}(t))^{-1} \sum_{1 \le j \le m} h_j'(t)\frac{\partial P}{\partial \epsilon_j}(t,\vec{h}(t)) \right\}\vec{u}.$$

Proof.

In fact,

$$
\begin{aligned}
\frac{d\vec{u}}{dt} &= P(t,\vec{h}(t))^{-1}\left\{ A(t,\vec{h}(t))P(t,\vec{h}(t)) - \frac{d}{dt}[P(t,\vec{h}(t))] \right\}\vec{u}\\[2mm]
&= P(t,\vec{h}(t))^{-1}\left\{ A(t,\vec{h}(t))P(t,\vec{h}(t)) - \frac{\partial P}{\partial t}(t,\vec{h}(t))\right.\\[2mm]
&\qquad\qquad \left. - \sum_{1 \le j \le m} h_j'(t)\frac{\partial P}{\partial \epsilon_j}(t,\vec{h}(t)) \right\}\vec{u}.
\end{aligned}
$$

(VII.6.7)

Since $H(\vec{\epsilon})$ is given by (VII.6.4), equation (VII.6.6) follows from (VII.6.7). $\square$

Observe that

$$(\text{IV.6.8}) \qquad \int_0^{+\infty} \left| P(t,\vec{h}(t))^{-1} \sum_{1 \le j \le m} h_j'(t)\frac{\partial P}{\partial \epsilon_j}(t,\vec{h}(t)) \right| dt < +\infty$$

under assumption (VII.6.2). Also, observe that $H(\vec{h}(t))$ does not contain any periodic quantities and $\dfrac{dH(\vec{h}(t))}{dt}$ is absolutely integrable. Therefore, if eigenvalues of $H(\vec{0})$ satisfy suitable conditions, the argument given in §VII-5 applies to system (VII.6.6).

Observation in Case 2. Set

$$(VII.6.9) \qquad \begin{cases} B_0(\vec{\epsilon}) \;=\; H(\vec{\epsilon}), \\[2mm] B_1(t,\vec{\epsilon},\vec{\mu}) \;=\; -P(t,\vec{\epsilon})^{-1} \displaystyle\sum_{1 \le j \le m} \mu_j \frac{\partial P}{\partial \epsilon_j}(t,\vec{\epsilon}), \end{cases}$$

where $\vec{\mu} \in \mathbb{C}^m$ and the entries of $\vec{\mu}$ are $\mu_1, \mu_2, \dots, \mu_m$. Then, differential equation (VII.6.6) can be written in the form

$$(VII.6.10) \qquad \frac{d\vec{u}}{dt} = \{B_0(\vec{h}(t)) \;+\; B_1(t,\vec{h}(t),\vec{h}'(t))\}\vec{u}.$$

Notice that the entries of the matrix $B_1(t,\vec{\epsilon},\vec{\mu})$ are periodic in t of period ω and that $B_1(t,\vec{\epsilon},\vec{0}) = O$. Furthermore, any two distinct eigenvalues of $B_0(\vec{0})$ do not differ by integral multiples of $\dfrac{2\pi i}{\omega}$. Set $\mathcal{D}(r) = \{(\vec{\epsilon},\vec{\mu}) : |\vec{\epsilon}| + |\vec{\mu}| < r\}$.

In Case 2, the following lemma is used.

Lemma VII-6-4. *There exist $n \times n$ matrices $P_1(t,\vec{\epsilon},\vec{\mu})$ and $H_1(\vec{\epsilon},\vec{\mu})$ such that*
  (i) the entries of $P_1(t,\vec{\epsilon},\vec{\mu})$ are continuous in $(t,\vec{\epsilon},\vec{\mu}) \in \mathbb{R} \times \mathcal{D}(\tilde{r})$ and analytic in $(\vec{\epsilon},\vec{\mu}) \in \mathcal{D}(\tilde{r})$ for each fixed $t \in \mathbb{R}$, where $\tilde{r}$ is a suitable positive number,
  (ii) $P_1(t + \omega,\vec{\epsilon},\vec{\mu}) = P_1(t,\vec{\epsilon},\vec{\mu})$ for $(t,\vec{\epsilon},\vec{\mu}) \in \mathbb{R} \times \mathcal{D}(\tilde{r})$,
  (iii) $P_1(t,\vec{\epsilon},\vec{0}) = O$ for $(t,\vec{\epsilon}) \in \mathbb{R} \times \triangle(\tilde{r})$,
  (iv) the entries of $H_1(\vec{\epsilon},\vec{\mu})$ are analytic in $(\vec{\epsilon},\vec{\mu}) \in \mathcal{D}(\tilde{r})$ and $H_1(\vec{\epsilon},\vec{0}) = O$ for $\vec{\epsilon} \in \triangle(\tilde{r})$,
  (v) $\dfrac{\partial}{\partial t} P_1(t,\vec{\epsilon},\vec{\mu})$ exists for $(t,\vec{\epsilon},\vec{\mu}) \in \mathbb{R} \times \mathcal{D}(\tilde{r})$ and is given by

$$(VII.6.11) \qquad \begin{aligned} \frac{\partial}{\partial t} P_1(t,\vec{\epsilon},\vec{\mu}) =& \{B_0(\vec{\epsilon}) + B_1(t,\vec{\epsilon},\vec{\mu})\}\{I_n + P_1(t,\vec{\epsilon},\vec{\mu})\} \\ & - \{I_n + P_1(t,\vec{\epsilon},\vec{\mu})\}\{B_0(\vec{\epsilon}) + H_1(\vec{\epsilon},\vec{\mu})\} \end{aligned}$$

for $(t,\vec{\epsilon},\vec{\mu}) \in \mathbb{R} \times \mathcal{D}(\tilde{r})$, where I_n is the $n \times n$ identity matrix.

Remark VII-6-5. Equation (VII.6.11) can be simplified as

$$(VII.6.12) \qquad \begin{aligned} \frac{\partial}{\partial t} P_1(t,\vec{\epsilon},\vec{\mu}) =& \; B_0(\vec{\epsilon}) P_1(t,\vec{\epsilon},\vec{\mu}) - P_1(t,\vec{\epsilon},\vec{\mu}) B_0(\vec{\epsilon}) \\ & + \{B_1(t,\vec{\epsilon},\vec{\mu}) P_1(t,\vec{\epsilon},\vec{\mu}) - P_1(t,\vec{\epsilon},\vec{\mu}) H_1(\vec{\epsilon},\vec{\mu}) + B_1(t,\vec{\epsilon},\vec{\mu})\} - H_1(\vec{\epsilon},\vec{\mu}). \end{aligned}$$

Proof of Lemma VII-6-4.

Given $\wp = (p_1, \dots, p_m)$, where the p_j are non-negative integers, denote $\displaystyle\sum_{j=1}^{m} |p_j|$ and $\mu_1^{p_1} \cdots \mu_m^{p_m}$ by $|\wp|$ and $\vec{\mu}^{\wp}$, respectively. Set

$$\begin{cases} P_1(t,\vec{\epsilon},\vec{\mu}) = \displaystyle\sum_{|\wp| \ge 1} \vec{\mu}^{\wp} P_{1,\wp}(t,\vec{\epsilon}), \qquad B_1(t,\vec{\epsilon},\vec{\mu}) = \displaystyle\sum_{|\wp| \ge 1} \vec{\mu}^{\wp} B_{1,\wp}(t,\vec{\epsilon}), \\[3mm] H_1(\vec{\epsilon},\vec{\mu}) = \displaystyle\sum_{|\wp| \ge 1} \vec{\mu}^{\wp} H_{1,\wp}(\vec{\epsilon}). \end{cases}$$

Then, equation (VII.6.12) is equivalent to

$$(\text{VII.6.13}) \qquad \frac{\partial P_{1,\wp}}{\partial t} = B_0(\vec{\epsilon})P_{1,\wp} - P_{1,\wp}B_0(\vec{\epsilon}) + \mathcal{Q}_{1,\wp}(t,\vec{\epsilon}) - H_{1,\wp}(\vec{\epsilon}),$$

where

$$(\text{VII.6.14}) \quad \mathcal{Q}_{1,\wp}(t,\vec{\epsilon}) = \sum_{\wp_1+\wp_2=\wp,\ (|\wp_1|,|\wp_2|\geq 1)} (B_{1,\wp_1}P_{1,\wp_2} - P_{1,\wp_2}H_{1,\wp_1}) + B_{1,\wp}.$$

Hence,

$$(\text{VII.6.15}) \qquad
\begin{aligned}
P_{1,\wp}(t,\vec{\epsilon}) = \exp[tB_0(\vec{\epsilon})]&\left\{ C(\vec{\epsilon}) + \int_0^t \exp[-sB_0(\vec{\epsilon})] \right.\\
&\left. \times \left(\mathcal{Q}_{1,\wp}(s,\vec{\epsilon}) - H_{1,\wp}(\vec{\epsilon}) \right) \exp[sB_0(\vec{\epsilon})]ds \right\} \exp[-t\,B_0(\vec{\epsilon})],
\end{aligned}$$

where $C(\vec{\epsilon})$ and $H_{1,\wp}(\vec{\epsilon})$ are $n \times n$ matrices to be determined by the condition that $P_{1,\wp}(t,\vec{\epsilon})$ is periodic in t of period ω, i.e.,

$$(\text{VII.6.16}) \qquad
\begin{aligned}
&\exp[\omega B_0(\vec{\epsilon})]C(\vec{\epsilon}) - C(\vec{\epsilon})\exp[\omega B_0(\vec{\epsilon})]\\
&\quad - \exp[\omega B_0(\vec{\epsilon})]\int_0^\omega \exp[-sB_0(\vec{\epsilon})]H_{1,\wp}(\vec{\epsilon})\exp[sB_0(\vec{\epsilon})]ds\\
&= -\exp[\omega B_0(\vec{\epsilon})]\int_0^\omega \exp[-sB_0(\vec{\epsilon})]\mathcal{Q}_{1,\wp}(s,\vec{\epsilon})\exp[sB_0(\vec{\epsilon})]ds.
\end{aligned}$$

It is not difficult to see that condition (VII.6.16) determines the matrices $C(\vec{\epsilon})$ and $H_{1,\wp}(\vec{\epsilon})$. Then, the matrix $P_{1,\wp}(t,\vec{\epsilon})$ is determined by (VII.6.15). The convergence of power series P_1 and H_1 can be shown by using suitable majorant series. $\quad\square$

In the same way as the proof of Theorem VII-6-3, the following theorem can be proven.

Theorem VII-6-6. *The transformation*

$$(\text{VII.6.17}) \qquad \vec{u} = \{I_n + P_1(t,\vec{h}(t),\vec{h}'(t))\}\vec{v}$$

changes differential equation (VII.6.10) to

$$(\text{VII.6.18}) \qquad
\begin{aligned}
\frac{d\vec{v}}{dt} = &\left\{ H(\vec{h}(t)) + H_1(\vec{h}(t),\vec{h}'(t)) \right.\\
&- \left[I_n + P_1(t,\vec{h}(t),\vec{h}'(t))\right]^{-1} \sum_{1\leq j\leq m} \left[h'_j(t)\frac{\partial P_1}{\partial \epsilon_j}(t,\vec{h}(t),\vec{h}'(t)) \right.\\
&\left.\left. + h''_j(t)\frac{\partial P_1}{\partial \mu_j}(t,\vec{h}(t),\vec{h}'(t)) \right] \right\} \vec{v},
\end{aligned}$$

where $P_1(t, \vec{\epsilon}, \vec{\mu})$ and $H_1(\vec{\epsilon}, \vec{\mu})$ are those two matrices given in Lemma VII-6-4.

Observe that under assumption (VII.6.3), we obtain

$$\int_0^{+\infty} \left| \left[I_n + P_1(t, \vec{h}(t), \vec{h}'(t)) \right]^{-1} \sum_{1 \leq j \leq m} \left[h_j'(t) \frac{\partial P_1}{\partial \epsilon_j}(t, \vec{h}(t), \vec{h}'(t)) \right. \right.$$

(VII.6.19)

$$\left. \left. + h_j''(t) \frac{\partial P_1}{\partial \mu_j}(t, \vec{h}(t), \vec{h}'(t)) \right] \right| \, dt \; < \; +\infty.$$

Observe also that the matrix $H(\vec{h}(t)) + H_1(\vec{h}(t), \vec{h}'(t))$ does not contain any periodic terms. Furthermore, $H(\vec{0}) + H_1(\vec{0}, \vec{0}) = H(\vec{0})$. It is clear that the derivative of $H(\vec{h}(t))$ is not necessarily absolutely integrable over $t_0 \leq t < +\infty$. Therefore, in order to apply the argument of §VII-5, the matrix $H(\vec{h}(t))$ must be examined more closely in each application. Details are left to the reader for further observation.

EXERCISES VII

VII-1. Find the Liapounoff's type number of each of the following four functions $f(t)$ at $t = +\infty$:

(1) $\exp\left[t^2 \sin\left(\frac{1}{1+t} \right) \right]$,   (2) $\exp\left[\int_0^t |\sin t| dt \right]$,   (3) $\min(\exp[3t], \exp[5t] \sin[6\pi t])$,

(4) the solution of the initial-value problem $y'' - y' - 6y = e^{3t}$, $y(0) = 1$, $y'(0) = 4$.

VII-2. Find a normal fundamental set of four linearly independent solutions of the system $\dfrac{d\vec{y}}{dt} = A\vec{y}$ on the interval $0 \leq t < +\infty$, where

$$A = \begin{bmatrix} 252 & 498 & 4134 & 698 \\ -234 & -465 & -3885 & -656 \\ 15 & 30 & 252 & 42 \\ -10 & -20 & -166 & -25 \end{bmatrix}.$$

Hint. See Example IV-1-18.

VII-3. Let $\Phi(x)$ be a fundamental matrix solution of the system

$$\frac{d\vec{y}}{dx} = \begin{bmatrix} \log(1+x) & e^x & \exp x^2 \\ x^3 & 1 + \dfrac{1}{1+x} & \sin x \\ \exp(e^x) & \cos x & \arctan x \end{bmatrix} \vec{y}$$

on the interval $0 \leq x < +\infty$. Find Liapunoff's type number of $\det \Phi(x)$ at $x = +\infty$.

VII-4. Assuming that the entries of an $n \times n$ matrix $A(t)$ are convergent power series in t^{-1}, calculate $\lim\limits_{t \to +\infty} \dfrac{\log |\vec{\phi}(t)|}{t}$ for a nontrivial solution $\vec{\phi}(t)$ of the differential equation $t\dfrac{d\vec{y}}{dt} = A(t)\vec{y}$ at $t = +\infty$.

Hint. Set $t = e^s$.

VII-5. Let $\dfrac{d\vec{y}}{dx} = x^{p-1}A(x)\vec{y}$ be a system of linear differential equations such that $\vec{y} \in \mathbb{C}^n$ is an unknown quantity, p is a positive integer, the entries of an $n \times n$ matrix $A(x)$ are convergent power series in x^{-1}, and $\lim\limits_{x \to \infty} A(x)$ is not nilpotent. Show that there exists a solution $\vec{y}(x)$ of this system such that $\lim\limits_{r \to +\infty} \dfrac{\ln(|\vec{y}(re^{i\theta})|)}{r^p}$ is a positive number for some real number θ.

Hint. Set $x = re^{i\theta}$ for a fixed θ and $t = r^p$. Then, the given system becomes $\dfrac{d\vec{y}}{dt} = \dfrac{e^{ip\theta}}{p}A(re^{i\theta})\vec{y}$. Note that $\lim\limits_{r \to +\infty} \dfrac{e^{ip\theta}}{p}A(re^{i\theta}) = \dfrac{e^{ip\theta}}{p}A(\infty)$. The eigenvalues of $\dfrac{e^{ip\theta}}{p}A(\infty)$ are $\dfrac{e^{ip\theta}\lambda_j}{p}$, where the λ_j are the eigenvalues of $A(\infty)$. Now, apply Corollary VII-3-7.

VII-6. For each of two matrices $A(t)$ given below, find a unitary matrix $U(t)$ analytic on $(-\infty, \infty)$ for each matrix such that $U^{-1}(t)A(t)U(t)$ is diagonal or upper-triangular.

(a) $A(t) = \begin{bmatrix} 0 & it \\ -it & 0 \end{bmatrix}$, (b) $A(t) = \begin{bmatrix} 1 & t & 0 \\ t & 1+2t & 0 \\ 0 & 0 & 1+t^2 \end{bmatrix}$.

Hint. See [GH].

VII-7. Show that if a function $p(t)$ is continuous on the interval $0 \le t < +\infty$ and $\lim\limits_{t \to +\infty} t^{-p}p(t) = 1$ for some positive integer p, the differential equation $\dfrac{d^2 y}{dt^2} + p(t)y = 0$ has two linearly independent solutions $\eta_\pm(t)$ such that

$$\begin{cases} \eta_\pm(t) = p(t)^{-1/4}(1 + o(1))\exp\left[\pm i \int_{t_0}^t \sqrt{p(s)}\, ds\right], \\[2mm] \eta'_\pm(t) = p(t)^{1/4}(\pm 1 + o(1))\exp\left[\pm i \int_{t_0}^t \sqrt{p(s)}\, ds\right], \end{cases}$$

as $t \to +\infty$, where t_0 is a sufficiently large positive number and $o(1)$ denotes a quantity which tends to zero as $t \to +\infty$.

Hint. Use an argument similar to Example VII-4-8.

VII-8. Let $Q(x) = x^m + \sum\limits_{h=0}^{m-1} a_h x^h$ and $P(x, \epsilon) = \epsilon x^{m+1} + Q(x)$, where m is a positive integer, x is an independent variable, and $(\epsilon, a_1, \ldots, a_{m-1})$ are complex parameters. Set $A(x, \epsilon) = \begin{bmatrix} 0 & 1 \\ P(x, \epsilon) & 0 \end{bmatrix}$. Also, let R_0, ρ_0, and α_0 be arbitrary but fixed positive numbers. Suppose that $0 < \rho_0 < \dfrac{\pi}{2}$. Show that the system

$$\frac{d\vec{y}}{dx} = A(x, \epsilon)\vec{y}, \qquad \vec{y} = \begin{bmatrix} \eta_1 \\ \eta_2 \end{bmatrix}$$

has two solutions $\vec{y}_j(x, \epsilon)$ $(j = 1, 2)$ satisfying the following conditions:
 (a) the entries of $\vec{y}_j(x, \epsilon)$ are holomorphic with respect to $(x, \epsilon, a_0, \dots, a_{m-1})$ in the domain

$$\mathcal{D} = \{(x, \epsilon, a_0, \dots, a_{m-1}) : \ x \in \mathbb{C}, \ 0 < |\epsilon| < \alpha_0, \ |\arg \epsilon| < \rho_0,$$
$$|a_0| + \cdots + |a_{m-1}| < R_0\},$$

 (b) $\vec{y}_j(x, \epsilon)$ $(j = 1, 2)$ possess the asymptotic representations

$$\begin{cases} \vec{y}_1(x, \epsilon) = \exp\left[\int_{x_0}^{x} P(t, \epsilon)^{1/2} dt\right] \begin{bmatrix} [1 + o(1)]P(x, \epsilon)^{-1/4} \\ [1 + o(1)]P(x, \epsilon)^{1/4} \end{bmatrix}, \\[2em] \vec{y}_2(x, \epsilon) = \exp\left[-\int_{x_0}^{x} P(t, \epsilon)^{1/2} dt\right] \begin{bmatrix} [1 + o(1)]P(x, \epsilon)^{-1/4} \\ [-1 + o(1)]P(x, \epsilon)^{1/4} \end{bmatrix}, \end{cases}$$

respectively as $x \to +\infty$ on the positive real line in the x-plane, where x_0 is a positive number depending on (R_0, ρ_0, α_0) and $o(1)$ denotes a quantity which tends to 0 as $x \to +\infty$ on the positive real line uniformly with respect to $(\epsilon, a_0, \dots, a_{m-1})$ for $|\epsilon| < \alpha_0$, $|\arg \epsilon| < \rho_0$, and $|a_0| + \cdots + |a_{m-1}| < R_0$.

Hint. Use a method similar to that for Exercise VII-7. See, also, [Mu] and [Si13, Chapter 3].

VII-9. Let $\lambda_1(t), \dots, \lambda_n(t)$ be n continuous functions of t on the interval $\mathcal{I}_0 = \{t \in \mathbb{R} : 0 \leq t < +\infty\}$ such that $Re[\lambda_j(t) - \lambda_{j+1}(t)] > 1$ $(j = 1, \dots, n - 1)$ on $\mathcal{I}_0$. Also, let $A(t)$ be an $n \times n$ matrix such that the entries are continuous in t on $\mathcal{I}_0$ and that

$$\sup_{p \geq t}(1 + p - t)^{-1} \int_1^p |A(s)| ds \to 0 \quad \text{as} \quad t \to +\infty.$$

Show that there exist a non-negative number t_0 and an $n \times n$ matrix $T(t)$ such that
 (1) the derivative $\dfrac{dT}{dt}(t)$ exists and the entries of $T(x)$ and $\dfrac{dT}{dt}(x)$ are continuous in t on the interval $\mathcal{I} = \{t : t_0 \leq t < +\infty\}$,
 (2) $\displaystyle\lim_{t \to +\infty} T(t) = O$,
 (3) transformation $\vec{y} = (I_n + T(x))\vec{z}$ changes the system

$$\frac{d\vec{y}}{dt} = (\text{diag}[\lambda_1(t), \lambda_2(t), \dots, \lambda_n(t)] + T(t))\,\vec{y}$$

to

$$\frac{d\vec{z}}{dt} = \text{diag}[\lambda_1(t) + b_1(t), \lambda_2(t) + b_2(t), \dots, \lambda_n(t) + b_n(t)]\vec{z},$$

where $\vec{y} \in \mathbb{C}$, $\vec{z} \in \mathbb{C}$, n functions $b_1(t), \dots, b_n(t)$ are complex-valued and continuous in t on $\mathcal{I}$, and $\displaystyle\lim_{t \to +\infty} b_j(t) = 0$ $(j = 1, \dots, n)$.

VII-10. Show that if $R(t)$ is a real-valued and continuous function on $\mathcal{I}_0 = \{t \in \mathbb{R} :$ $0 \leq t < +\infty\}$ satisfying the condition $\displaystyle\int_0^{+\infty} |R(t)| < +\infty$, the differential equation $\dfrac{d^2y}{dt^2} + (1 + R(t))y = 0$ has a solution $\phi(t)$ such that $\displaystyle\lim_{t \to +\infty} (\phi(t) - \sin t) = 0$. Also, show that $\phi(t)$ has infinitely many positive zeros λ_n such that $\displaystyle\lim_{n \to +\infty} \dfrac{\lambda_n}{n} = \pi$.

VII-11. Show that every solution of a differetial equation $\dfrac{d^2y}{dt^2} + R(t)y = 0$ has at most a finite number of zeros on the interval $\mathcal{I}_0 = \{t \in \mathbb{R} : 0 \leq t < +\infty\}$ if $R(t)$ is a real-valued and continuous function on $\mathcal{I}_0$ satisfying the condition $\displaystyle\int_0^{+\infty} t|R(t)| < +\infty$.

Remark. See [CL, Problem 28 on p. 103]. For the case when $\displaystyle\int_0^{+\infty} t|R(t)| = +\infty$, but $\displaystyle\int_0^{+\infty} |R(t)| < +\infty$, see Example VI-1-7.

VII-12. Suppose that $u(t)$ is a real-valued, continuous, and bounded function of t on the interval $0 \leq t < +\infty$. Also, assume that $\displaystyle\int_0^{+\infty} |u(t)|dt < +\infty$. Show that if λ is an eigenvalue of the eigenvalue problem

$$\frac{d^2y}{dt^2} + u(t)y = \lambda y, \quad y(0) = 0, \quad \int_0^{+\infty} y(t)^2 dt < +\infty,$$

then $0 \leq \lambda \leq \displaystyle\sup_{0 \leq t < +\infty} |u(t)|$.

VII-13. Let $A(t)$ be an $n \times n$ matrix whose entries are real-valued, continuous, and periodic of period 1 on $\mathbb{R}$. Show that there exist two $n \times n$ matrices $P(t, \epsilon)$ and $B(\epsilon)$ such that
   (a) the entries of $P(t, \epsilon)$ and $B(\epsilon)$ are power series in ϵ which are uniformly convergent for $-\infty < t < +\infty$ and small $|\epsilon|$,
   (b) $P(t, 0) = I_n$ $(-\infty < t < +\infty)$, where I_n is the $n \times n$ identity matrix,
   (c) the entries of $P(t, \epsilon)$ is periodic in t of period 1,
   (d) $P(t, \epsilon)$ and $B(\epsilon)$ satisfy the equation

$$\frac{\partial P}{\partial t} = \epsilon[A(t)P(t, \epsilon) - P(t, \epsilon)B(\epsilon)].$$

VII-14. Apply Lemma VII-6-2 to the following system:

$$\frac{d\vec{y}}{dt} = [nL + (\alpha + \beta\cos(2t))(K - L)]\vec{y}, \quad \vec{y} = \begin{bmatrix} y_1 \\ y_2 \end{bmatrix},$$

where n is a positive integer, the two quantities α and β are real parameters, and

$$K = \begin{bmatrix} 0 & 1 \\ 1 & 0 \end{bmatrix} \quad \text{and} \quad L = \begin{bmatrix} 0 & 1 \\ -1 & 0 \end{bmatrix}.$$

VII-15. Using Theorem VII-6-3, find the asymptotic behavior of solutions of the differential equation

$$\frac{d^2\eta}{dt^2} + \{1 + h(t)\sin(\alpha t)\}\,\eta = 0, \qquad h(t) = \frac{1}{\ln(2+t)}$$

as $t \to +\infty$, where α is a real parameter.

CHAPTER VIII

STABILITY

In the previous chapter, we explained the behavior of solutions of linear systems as $t \to +\infty$. In this chapter, we look into similar problems for nonlinear systems. To start with, in §VIII-1, we introduce the concepts of stability and asymptotic stability of a given particular solution as $t \to +\infty$. We illustrate those concepts with simple examples. Reducing the given solution to the trivial solution by a simple transformation, we concentrate our explanation on the stability property of the trivial solution. It is well known that the trivial solution is asymptotically stable as $t \to +\infty$ if real parts of eigenvalues of the leading matrix of the given system are all negative. This basic result is given as Theorem VIII-2-1 in §VIII-2. The case when some of those real parts are not negative is treated in §VIII-3. In particular, we discuss the stable and unstable manifolds. In §VIII-4, we look into the structure of stable manifolds more closely for analytic differential equations. First we change a given system by an analytic transformation to a simple standard form. By virtue of such a simplification, we can construct the stable manifold in a simple analytic form. This idea is applied to analytic systems in $\mathbb{R}^2$ in §VIII-6. In §§VIII-7–VIII-10, using the polar coordinates, we explain continuous perturbations of linear systems in $\mathbb{R}^2$. In §VIII-5, we summarize some known facts concerning linear systems with constant coefficients in $\mathbb{R}^2$. The topics discussed in this chapter are also found in [CL, pp. 371-388], [Har2, pp. 160-161, 220-227], and [SC, pp. 49-96]. The materials in §§VIII-4 and VIII-6 are also found in [Du], [Huk5], and [Si2].

VIII-1. Basic definitions

Let us consider a system of differential equations

$$\text{(VIII.1.1)} \qquad \frac{d\vec{y}}{dt} = \vec{f}(t, \vec{y}),$$

where $\vec{y} \in \mathbb{R}^n$ and the $\mathbb{R}^n$-valued function $\vec{f}(t, \vec{y})$ is continuous on a region

$$\triangle(r_0) = \mathcal{I}_0 \times \mathcal{D}(r_0) = \{(t, \vec{y}) \in \mathbb{R}^{n+1} : 0 \leq t < +\infty, \ |\vec{y}| < r_0\}.$$

Also, assume that a solution $\vec{\phi}_0(t)$ of (VIII.1.1) is defined on the entire interval $\mathcal{I}_0$ and that $(t, \vec{\phi}_0(t)) \in \triangle(r_0)$ on $\mathcal{I}_0$. The main topic in this chapter is the behavior of solutions of the initial-value problem

$$\text{(VIII.1.2)} \qquad \frac{d\vec{y}}{dt} = \vec{f}(t, \vec{y}), \qquad \vec{y}(0) = \vec{\eta}$$

as $t \to +\infty$. To start, we introduce the concept of stability.

235

Definition VIII-1-1. *The solution $\vec{\phi}_0(t)$ is said to be* stable *as $t \to +\infty$ if, for any given positive number ϵ, there exists another positive number $\delta(\epsilon)$ such that whenever $|\vec{\phi}_0(0) - \vec{\eta}| \leq \delta(\epsilon)$, every solution $\vec{\phi}(t)$ of initial-value problem (VIII.1.2) exists on the entire interval $\mathcal{I}_0$ and satisfies the condition*

(VIII.1.3) $$|\vec{\phi}_0(t) - \vec{\phi}(t)| \leq \epsilon \qquad on \quad \mathcal{I}_0.$$

Remark VIII-1-2. Suppose that the initial-value problem

(VIII.1.2.τ) $$\frac{d\vec{y}}{dt} = \vec{f}(t, \vec{y}), \qquad \vec{y}(\tau) = \vec{\eta}$$

has the unique solution $\vec{y} = \vec{\phi}(t, \tau, \vec{\eta})$ if $(\tau, \vec{\eta}) \in \triangle(r_0)$. Then, $\vec{\phi}(t, \tau, \vec{\eta})$ is continuous with respect to $(t, \tau, \vec{\eta})$. Therefore, for any $\tau \in \mathcal{I}_0$ and any given positive numbers T and ϵ, there exists a positive number $\rho(\tau, T, \epsilon)$ such that whenever $|\vec{\phi}_0(\tau) - \vec{\eta}| \leq \rho(\tau, T, \epsilon)$, the solution $\vec{\phi}(t, \tau, \vec{\eta})$ exists on the interval $0 \leq t \leq T$ and $|\vec{\phi}_0(t) - \vec{\phi}(t, \tau, \vec{\eta})| \leq \epsilon$ on the interval $0 \leq t \leq T$ (cf. §II-1). This implies that if the solution $\vec{\phi}_0(t)$ is stable as $t \to +\infty$, then for any $\tau \in \mathcal{I}_0$ and any positive number ϵ, there exists another positive number $\delta(\tau, \epsilon)$ such that whenever $|\vec{\phi}_0(\tau) - \vec{\eta}| \leq \delta(\tau, \epsilon)$, the solution $\vec{\phi}(t, \tau, \vec{\eta})$ exists on the entire interval $\mathcal{I}_0$ and $\vec{\phi}(t, \tau, \vec{\eta})$ satisfies the condition $|\vec{\phi}_0(t) - \vec{\phi}(t, \tau, \vec{\eta})| \leq \epsilon$ on $\mathcal{I}_0$.

We also introduce the concept of asymptotic stability.

Definition VIII-1-3. *The solution $\vec{\phi}_0(t)$ is said to be* asymptotically stable *as $t \to +\infty$ if*

  (i) the solution $\vec{\phi}_0(t)$ is stable as $t \to +\infty$,

  (ii) there exists a positive number r such that whenever $|\vec{\phi}_0(0) - \vec{\eta}| \leq r$, every solution $\vec{\phi}(t)$ of initial-value problem (VIII.1.2) satisfies the condition

(VIII.1.4) $$\lim_{t \to +\infty} |\vec{\phi}_0(t) - \vec{\phi}(t)| = 0.$$

Remark VIII-1-4. Set $\vec{y} = \vec{z} + \vec{\phi}_0(t)$. Then, system (VIII.1.1) is changed to

(VIII.1.5) $$\frac{d\vec{z}}{dt} = \vec{f}(t, \vec{z} + \vec{\phi}_0(t)) - \vec{f}(t, \vec{\phi}_0(t)).$$

Hence, the study of the solution $\vec{\phi}_0(t)$ of (VIII.1.1) is reduced to that of the trivial solution $\vec{z}(t) = \vec{0}$ of (VIII.1.5). Thus, the solution $\vec{\phi}_0(t)$ of (VIII.1.1) is stable (respectively asymptotically stable) as $t \to +\infty$ if and only if the trivial solution of (VIII.1.5) is stable (respectively asymptotically stable) as $t \to +\infty$. In the following sections, we shall study stability and asymptotic stability of the trivial solution.

The following three examples illustrate stability and asymptotic stability.

Example VIII-1-5. The first example is the system given by

(VIII.1.6) $$\frac{dy_1}{dt} = -y_1, \qquad \frac{dy_2}{dt} = (y_1 - y_2^2)y_2.$$

To find the general solution of (VIII.1.6), solve the first-order equation

(VIII.1.7)
$$\frac{dy_2}{dy_1} = -y_2 + \frac{y_2^3}{y_1}.$$

The transformation $u = \dfrac{1}{y_2^2}$ changes this equation to a first-order linear differential

equation $\dfrac{du}{dy_1} = 2u - \dfrac{2}{y_1}$. Thus, we obtain $u(y_1) = e^{2y_1}\left(c - 2\displaystyle\int_\gamma^{y_1} \frac{e^{-2\eta}}{\eta}\,d\eta\right)$. Since

$u = y_2^{-2} > 0$, the quantity c must be positive. It is evident that $y_1(t) = \gamma\,e^{-t}$
satisfies the first equation of system (VIII.1.6) with the initial value $y_1(0) = \gamma$.
Therefore,

(VIII.1.8)
$$y_1(t) = \gamma\,e^{-t}, \qquad y_2(t) = \pm\frac{1}{\sqrt{u(y_1(t))}}$$

is a solution of (VIII.1.6). Observe that

$$\gamma \le y_1(t) < 0 \quad\text{if}\quad \gamma < 0 \quad\text{and}\quad 0 < y_1(t) \le \gamma \quad\text{if}\quad \gamma > 0$$

for $t \ge 0$. Therefore, $\displaystyle\int_\gamma^{y_1(t)} \frac{e^{-2\eta}}{\eta}\,d\eta \le 0$ for $t \ge 0$. Thus, $y_2(t)^2 \le \dfrac{e^{-2y_1(t)}}{c} \le$

$\dfrac{e^{2|\gamma|}}{c} = e^{2|\gamma|}e^{2\gamma}y_2(0)^2$ for $t \ge 0$. This proves that the trivial solution of system

(VIII.1.6) is stable as $t \to +\infty$. Furthermore, since

$$\lim_{t\to+\infty} \int_\gamma^{y_1(t)} \frac{e^{-2\eta}}{\eta}\,d\eta = \lim_{y_1\to 0} \int_\gamma^{y_1} \frac{e^{-2\eta}}{\eta}\,d\eta = -\infty,$$

the trivial solution of system (VIII.1.6) is asymptotically stable as $t \to +\infty$. This
result is shown also by Figure 1.

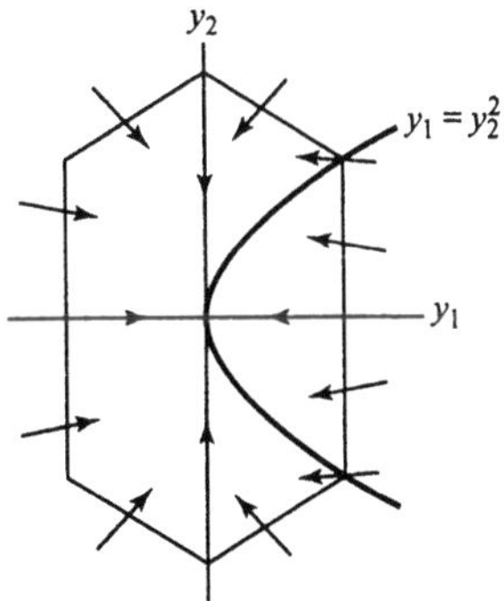

FIGURE 1.

Example VIII-1-6. The second example is a second-order differential equation

(VIII.1.9)
$$\frac{d^2\eta}{dt^2} + g(\eta) = 0,$$

where $g(\eta)$ is a real-valued and continuously differentiable function of η on the real line $\mathbb{R}$. Equation (VIII.1.9) is equivalent to the system

$$\text{(VIII.1.10)} \qquad \frac{dy_1}{dt} = y_2, \qquad \frac{dy_2}{dt} = -g(y_1).$$

If $g(\alpha) = 0$, then system (VIII.1.10) has a constant solution $y_1 = \alpha$, $y_2 = 0$. Set $y_1 = z_1 + \alpha$ and $y_2 = z_2$. Then, system (VIII.1.10) becomes

$$\text{(VIII.1.11)} \qquad \frac{d\vec{z}}{dt} = A\vec{z} + \vec{g}(\vec{z}),$$

where

$$\vec{z} = \begin{bmatrix} z_1 \\ z_2 \end{bmatrix}, \qquad A = \begin{bmatrix} 0 & 1 \\ -g'(\alpha) & 0 \end{bmatrix}, \qquad \vec{g}(\vec{z}) = \begin{bmatrix} 0 \\ -[g(z_1 + \alpha) - g'(\alpha)z_1] \end{bmatrix}.$$

The eigenvalues of A are $\pm(-g'(\alpha))^{1/2}$. Note also that

$$\lim_{z_1 \to 0} \frac{g(z_1 + \alpha) - g'(\alpha)z_1}{z_1} = 0.$$

Set

$$\text{(VIII.1.12)} \qquad G(\eta) = \int_0^\eta g(s)ds.$$

Then, $G(\eta)$ takes a local minimum (respectively a local maximum) at $\eta = \alpha$ if $g(\alpha) = 0$ and $g'(\alpha) > 0$ (respectively $g'(\alpha) < 0$). Also, set

$$\text{(VIII.1.13)} \qquad H(y_1, y_2) = G(y_1) + \frac{y_2^2}{2}.$$

Then,

$$\frac{dH(y_1(t), y_2(t))}{dt} = g(y_1(t))\frac{dy_1(t)}{dt} + y_2(t)\frac{dy_2(t)}{dt}.$$

Therefore,

$$\frac{dH(y_1(t), y_2(t))}{dt} = g(y_1(t))y_2(t) - y_2(t)g(y_1(t)) = 0$$

if $(y_1(t), y_2(t))$ is a solution of (VIII.1.10). This means that

$$\text{(VIII.1.14)} \quad G(z_1(t) + \alpha) + \frac{z_2(t)^2}{2} = G(z_1(0) + \alpha) + \frac{z_2(0)^2}{2} \qquad \text{for all } t$$

if $\vec{z}(t)$ is a solution of (VIII.1.11).

Case when $g'(\alpha) < 0$: If $g'(\alpha) < 0$, there exists a positive number μ_0 such that $G(\eta) < G(\alpha)$ for $0 < |\eta - \alpha| \leq \mu_0$. Let $\vec{\zeta}(t)$ be the unique solution of (VIII.1.11) satisfying the initial condition $z_1(0) = 0$, $z_2(0) = \zeta_0 > 0$. Since $\zeta_2(t) = \zeta_1'(t)$, we obtain $\zeta_1'(t) = \sqrt{2(G(\alpha) - G(\zeta_1(t) + \alpha)) + \zeta_0^2} \geq \zeta_0$ as long as $0 \leq \zeta_1(t) \leq \mu_0$. Hence, there must be a positive number t_0 depending on ζ_0 such that $\zeta_1(t_0) = \mu_0$ no matter how small ζ_0 may be. This implies that the trivial solution of (VIII.1.11) is not stable.

Case when $g'(\alpha) > 0$: For a given positive number ρ, let $\mathcal{D}(\rho)$ be the connected component containing 0 of the set $\{\zeta : G(\zeta + \alpha) \leq G(\alpha) + \rho\}$. Then, $\mathcal{D}(\rho_1) \subset \mathcal{D}(\rho_2)$ if $\rho_1 \leq \rho_2$. Furthermore, if $g(\alpha) = 0$, $g'(\alpha) > 0$, and if a positive number ρ is sufficiently small, there exist two positive numbers $\epsilon_1(\rho)$ and $\epsilon_2(\rho)$ such that

(1) $\epsilon_2(\rho) < \epsilon_1(\rho)$,
(2) $\lim_{\rho \to 0} \epsilon_1(\rho) = 0$,
(3) $\{\zeta : |\zeta| < \epsilon_2(\rho)\} \subset \mathcal{D}(\rho) \subset \{\zeta : |\zeta| < \epsilon_1(\rho)\}$ (cf. Figure 2).

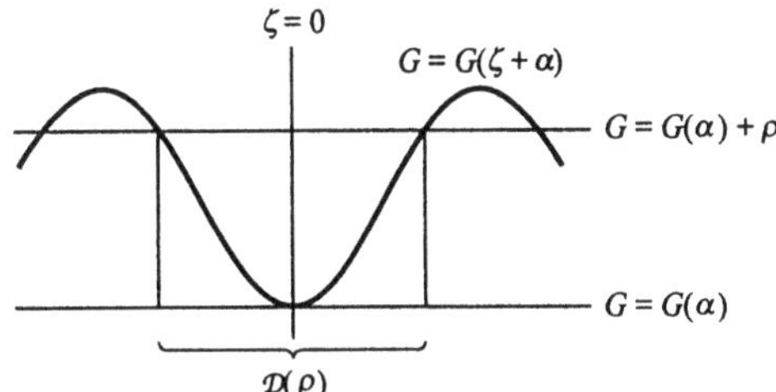

FIGURE 2.

Observe also that $G(\alpha) \leq G(\zeta + \alpha) \leq G(\alpha) + \rho$ for $\zeta \in \mathcal{D}(\rho)$ if ρ is a sufficiently small positive number.

For a given positive number ϵ, choose another positive number ρ so that $\epsilon_1(\rho) \leq \epsilon$ and $\rho \leq \epsilon$. Choose also the initial value $\vec{z}(0)$ so that $|z_1(0)| \leq \epsilon_2\left(\dfrac{\rho}{2}\right)$ and $z_2(0)^2 \leq \rho$. Then, $G(z_1(0) + \alpha) \leq G(\alpha) + \dfrac{\rho}{2}$ since $z_1(0) \in \mathcal{D}\left(\dfrac{\rho}{2}\right)$. Hence, (VIII.1.14) implies that $G(z_1(t) + \alpha) \leq G(\alpha) + \rho$ for all t. Observe that the set $\{z_1(t) : \text{ all } t\}$ is connected and $z_1(0) \in \mathcal{D}\left(\dfrac{\rho}{2}\right) \subset \mathcal{D}(\rho)$. Therefore, $z_1(t) \in \mathcal{D}(\rho)$ for all t. Hence, $|z_1(t)| \leq \epsilon_1(\rho) \leq \epsilon$. On the other hand, $G(\alpha) \leq G(z_1(t) + \alpha) \leq G(\alpha) + \rho$ since $z_1(t) \in \mathcal{D}(\rho)$ for all t. Hence, (VIII.1.14) implies that $\dfrac{z_2(t)^2}{2} \leq G(z_1(0) + \alpha) - G(\alpha) + \dfrac{\rho}{2} \leq \rho \leq \epsilon$. This proves that the trivial solution of system (VIII.1.11) is stable as $t \to +\infty$.

The analysis given in Example VIII-1-6 is an example of an application of the *Liapounoff functions* to which we will return in Chapters IX and X.

The third example is the following result.

Theorem VIII-1-7. *If $f(x, y)$ and $g(x, y)$ are real-valued continuously differentiable functions such that*
(i) $f(0, 0) = 0$ and $g(0, 0) = 0$,
(ii) $(f(x, y), g(x, y)) \neq (0, 0)$ if $(x, y) \neq (0, 0)$,
(iii) $\dfrac{\partial f}{\partial x}(x, y) + \dfrac{\partial g}{\partial y}(x, y) = 0$,
then, the trivial solution $(x, y) = (0, 0)$ of the system

$$(S) \qquad \frac{dx}{dt} = f(x, y), \qquad \frac{dy}{dt} = g(x, y)$$

is not asymptotically stable as $t \to +\infty$.

Proof.

A contradiction will be derived from the assumption that the trivial solution is asymptotically stable. Let $\vec{\phi}(t, c_1, c_2) = \begin{bmatrix} x(t, c_1, c_2) \\ y(t, c_1, c_2) \end{bmatrix}$ be the unique solution of system (S) satisfying the initial condition $\vec{\phi}(0, c_1, c_2) = \begin{bmatrix} c_1 \\ c_2 \end{bmatrix}$. If the trivial solution is asymptotically stable as $t \to +\infty$, the trivial solution is also stable as $t \to +\infty$. Therefore, for every positive number ϵ, there exists another positive number $\delta(\epsilon)$ such that $|\vec{\phi}(t, c_1, c_2)| \leq \epsilon$ for $0 \leq t < +\infty$ whenever $\max\{|c_1|, |c_2|\} \leq \delta(\epsilon)$. Since f and g are independent of t, $\vec{\phi}(t - \tau, c_1, c_2)$ is also a solution of (S) and satisfies the initial condition $(x(\tau), y(\tau)) = (c_1, c_2)$. This implies that $|\vec{\phi}(t, c_1, c_2)| \leq \epsilon$ for $\tau \leq t < +\infty$ if $|\vec{\phi}(\tau, c_1, c_2)| \leq \delta(\epsilon)$. Denote by $\Delta(r)$ the disk $\{(c_1, c_2) \in \mathbb{R}^2 : |c_1|^2 + |c_2|^2 \leq r\}$. Also, for a fixed value τ of t, let us denote by $\mathcal{D}(r, \tau)$ the set $\{\vec{\phi}(\tau, c_1, c_2) : (c_1, c_2) \in \Delta(r)\}$. Then, the mapping $(c_1, c_2) \to \vec{\phi}(\tau, c_1, c_2)$ is a homeomorphism of $\Delta(r)$ onto $\mathcal{D}(r, \tau)$ (cf. Exercise II-4). Fix a sufficiently small positive number r_0. Since $(0, 0)$ is asymptotically stable as $t \to +\infty$ and the disk $\Delta(r_0)$ is compact, there exists a positive number τ_0 such that $\mathcal{D}(r_0, \tau_0) \subset \Delta\left(\frac{r_0}{2}\right)$. This implies that the area of $\mathcal{D}(r_0, \tau_0)$ is definitely smaller than the area of $\Delta(r_0)$. Observe that

(VIII.1.15)
$$\text{area of } \mathcal{D}(r_0, \tau_0) \; = \; \int_{\Delta(r_0)} \left| \det \begin{bmatrix} \dfrac{\partial \vec{\phi}(\tau_0, c_1, c_2)}{\partial c_1} & \dfrac{\partial \vec{\phi}(\tau_0, c_1, c_2)}{\partial c_2} \end{bmatrix} \right| dc_1 dc_2.$$

It is known that the matrix $\Psi(t, c_1, c_2) = \begin{bmatrix} \dfrac{\partial \vec{\phi}(\tau_0, c_1, c_2)}{\partial c_1} & \dfrac{\partial \vec{\phi}(\tau_0, c_1, c_2)}{\partial c_2} \end{bmatrix}$ is the unique solution of the initial-value problem

$$\frac{dX}{dt} = \begin{bmatrix} \dfrac{\partial f}{\partial x} & \dfrac{\partial f}{\partial y} \\ \dfrac{\partial g}{\partial x} & \dfrac{\partial g}{\partial y} \end{bmatrix}\Bigg|_{(x,y)=\vec{\phi}(t,c_1,c_2)} X, \qquad X(0) = I_2,$$

where I_2 is the 2×2 identity matrix (cf. Theorem II-2-1). Therefore,

$$\det \Psi(t, c_1, c_2) \; = \; \exp\left[\int_0^t \left\{ \frac{\partial f}{\partial x}(x, y) + \frac{\partial g}{\partial y}(x, y) \right\}\Bigg|_{(x,y)=\vec{\phi}(t,c_1,c_2)} dt \right] \; = \; 1$$

(cf. (4) of Remark IV-2-7). Now, it follows from (VIII.1.15) that the area of $\mathcal{D}(r_0, \tau_0)$ is equal to the area of $\Delta(r_0)$. This is a contradiction. $\qquad \square$

VIII-2. A sufficient condition for asymptotic stability

In this section, we prove a basic sufficient condition for asymptotic stability. Let us consider a system of differential equations

(VIII.2.1) $$\frac{d\vec{y}}{dt} = A\vec{y} + \vec{g}(t, \vec{y})$$

under the assumptions that
  (i) A is a constant $n \times n$ matrix,
  (ii) the entries of the $\mathbb{R}^n$-valued function $\vec{g}(t, \vec{y})$ are continuous and satisfy the estimate

(VIII.2.2) $\qquad |\vec{g}(t, \vec{y})| \leq k(t)|\vec{y}| + c_0|\vec{y}|^{1+\nu_0} \qquad$ for $\quad (t, \vec{y}) \in \triangle(r_0)$,

where, $\nu_0 > 0$, $c_0 \geq 0$, $r_0 > 0$, $k(t) \geq 0$ and bounded for $t \geq 0$, $\displaystyle\lim_{t \to +\infty} k(t) = 0$, and $\triangle(r_0) = \{(t, \vec{y}) : 0 \leq t < +\infty, |\vec{y}| \leq r_0\}$.

The following theorem is the main result in this section.

Theorem VIII-2-1. *If the real part of every eigenvalue of A is negative, the trivial solution $\vec{y} = \vec{0}$ of (VIII.2.1) is asymptotically stable as $t \to +\infty$.*

Proof.

We prove this theorem in four steps.

Step 1. Let λ_j $(j = 1, 2, \ldots, n)$ be the eigenvalues of A. Then, $\Re[\lambda_j]$ $(j = 1, 2, \ldots, n)$ are Liapounoff's type numbers of the system $\dfrac{d\vec{y}}{dt} = A\vec{y}$ (cf. Example VII-2-8). Therefore, choosing a positive number μ so that $0 < \mu < -\Re[\lambda_j]$ for $j = 1, 2, \ldots, n$, we obtain

$$|\exp[At]| \leq Ke^{-\mu t} \qquad \text{on the interval } \mathcal{I}_0 = \{t : 0 \leq t < +\infty\}$$

for some positive constant K.

Step 2. Fixing a non-negative number T, change system (VIII.2.1) to the integral equation

$$\vec{y}(t) = e^{(t-T)A}\vec{y}(T) + \int_T^t e^{(t-s)A}\vec{g}(s, \vec{y}(s))ds.$$

If $|\vec{y}(t)| \leq \delta$ for some positive number δ on an interval $T \leq t \leq T_1$, then

$$|\vec{y}(t)| \leq Ke^{-\mu(t-T)}|\vec{y}(T)| + K\int_T^t e^{-\mu(t-s)} \left\{k(s) + c_0\delta^{\nu_0}\right\} |\vec{y}(s)|ds$$

on the same interval $T \leq t \leq T_1$. Setting $\rho(T) = \sup_{s \geq T} k(s)$, change this inequality to the form

$$e^{\mu(t-T)}|\vec{y}(t)| \leq K|\vec{y}(T)| + K\left\{\rho(T) + c_0\delta^{\nu_0}\right\} \int_T^t e^{\mu(s-T)}|\vec{y}(s)|ds$$

for $T \leq t \leq T_1$. Then,

$$e^{\mu(t-T)}|\vec{y}(t)| \leq K|\vec{y}(T)|\exp\left[K(\rho(T) + c_0\delta^{\nu_0})(t - T)\right]$$

and, hence,

(VIII.2.3) $\qquad |\vec{y}(t)| \leq K|\vec{y}(T)|\exp\left[-(\mu - K(\rho(T) + c_0\delta^{\nu_0}))(t - T)\right]$

for $T \leq t \leq T_1$ (cf. Lemma I-1-5).

Step 3. Fix a non-negative number T and two positive numbers δ and δ_1 in such a way that

$$\mu - K\left(\rho(T) + c_0\delta^{\nu_0}\right) > 0, \qquad 0 < \delta_1 < \delta, \qquad K\delta_1 < \delta.$$

Assume that $|\vec{y}(T)| \leq \delta_1$. Then, inequality (VIII.2.3) holds for $T \leq t \leq T_1$ as long as $|\vec{y}(t)| \leq \delta$ on the interval $T \leq t \leq T_1$. This, in turn, implies that

$$|\vec{y}(t)| \leq K\delta_1 < \delta \qquad \text{for} \quad T \leq t \leq T_1.$$

This is true for all T_1 not less than T. Hence, inequality (VIII.2.3) holds for $t \geq T$.

Step 4. If $|\vec{y}(t)| \leq \delta < 1$ for $0 \leq t \leq T$, there exists a positive number κ such that $\left|\dfrac{d\vec{y}(t)}{dt}\right| \leq \kappa|\vec{y}(t)|$, for $0 \leq t \leq T$. This implies that $|\vec{y}(t)| \leq |\vec{y}(0)|e^{\kappa t}$ as long as $|\vec{y}(t)| \leq \delta < 1$ for $0 \leq t \leq T$. Therefore, $|\vec{y}(T)|$ is small if $|\vec{y}(0)|$ is small. Thus, it was proven that (VIII.2.3) holds for $t \geq T$ if $|\vec{y}(0)|$ is small. This completes the proof of Theorem VIII-2-1. $\square$

Example VIII-2-2. For the following system of differential equations $\dfrac{d\vec{y}}{dt} = A\vec{y} + \vec{g}(\vec{y})$, where

$$\vec{y} = \begin{bmatrix} y_1 \\ y_2 \end{bmatrix}, \qquad A = \begin{bmatrix} -0.4 & -2 \\ 1 & -0.2 \end{bmatrix}, \qquad \vec{g}(\vec{y}) = (y_1^2 + y_2^2)\begin{bmatrix} 1 \\ -1 \end{bmatrix},$$

the trivial solution $\vec{y} = \vec{0}$ is asymptotically stable as $t \to +\infty$. In fact, the characteristic polynomial of the matrix A is $p_A(\lambda) = (\lambda + 0.3)^2 + 1.99$. Therefore, the real part of two eigenvalues are negative.

Remark VIII-2-3. The same conclusion as Theorem VIII-2-1 can be proven, even if (VIII.2.2) is replaced by

(VIII.2.4) $\qquad |\vec{g}(t,\vec{y})| \leq (k(t) + h(t)|\vec{y}|^{\mu})|\vec{y}| \qquad \text{for} \quad (t,\vec{y}) \in \Delta(r_0),$

where μ is a positive number, and two functions $h(t)$ and $k(t)$ are continuous for $t \geq 0$ such that $k(t) > 0$ for $t \geq 0$, $\displaystyle\lim_{t \to +\infty} k(t) = 0$, $h(t) > 0$ for $t \geq 0$, and Liapounoff's type number of $h(t)$ at $t = +\infty$ is not positive (cf. [CL, Theorem 1.3 on pp. 318-319]).

Remark VIII-2-4. The converse of Theorem VIII-2-1 is not true, as clearly shown in Example VIII-1-5. In that example, the eigenvalues of the matrix A are -1 and 0, but the trivial solution is asymptotically stable as $t \to +\infty$. Also, in that example, solutions starting in a neighborhood of $\vec{0}$ do not tend to $\vec{0}$ exponentially as $t \to +\infty$.

Remark VIII-2-5. Even if the matrix A is diagonalizable and its eigenvalues are all purely imaginary, the trivial solution is not necessarily stable as $t \to +\infty$. In such a case, we must frequently go through tedious analysis to decide if the trivial solution is stable as $t \to +\infty$ (cf. the case when $g'(\alpha) > 0$ in Example VIII-1-6). We shall return to such cases in $\mathbb{R}^2$ later in §§VIII-6 and VIII-10 .

Remark VIII-2-6. If the real part of an eigenvalue of A is positive, then the trivial solution is not stable as it is claimed in the following theorem.

Theorem VIII-2-7. *Assume that*
(i) A is a constant $n \times n$ matrix,
(ii) the entries of $\vec{g}(t, \vec{y})$ are continuous and satisfy the estimate

$$|\vec{g}(t, \vec{y})| \leq \epsilon(t, \vec{y})|\vec{y}| \quad for \quad (t, \vec{y}) \in \Delta(r_0) = \{(t, \vec{y}) : 0 \leq t < +\infty, |\vec{y}| \leq r_0\},$$

where $r_0 > 0$, $\epsilon(t, \vec{y}) \geq 0$ for $(t, \vec{y}) \in \Delta(r_0)$, and $\lim_{t^{-1}+|\vec{y}| \to 0} \epsilon(t, \vec{y}) = 0$,

(iii) the real part of an eigenvalue of the matrix A is positive.

Then, the trivial solution of the system $\dfrac{d\vec{y}}{dt} = A\vec{y} + \vec{g}(t, \vec{y})$ is not stable as $t \to +\infty$.

A proof of this theorem is given in [CL, Theorem 1.2, pp. 317-318]. We shall prove this theorem for a particular case in the next section (cf. (vi) of Remark VIII-3-2). An example of instability covered by this theorem is the case when $g'(\alpha) < 0$ of Example VIII-1-6. The converse of Theorem VIII-2-7 is not true. In fact, Figure 3 shows that the trivial solution of the system

$$\frac{dy_1}{dt} = -y_1, \qquad \frac{dy_2}{dt} = (y_1 + y_2^2)y_2$$

is not stable as $t \to +\infty$. Note that, in this case, eigenvalues of the matrix A are -1 and 0.

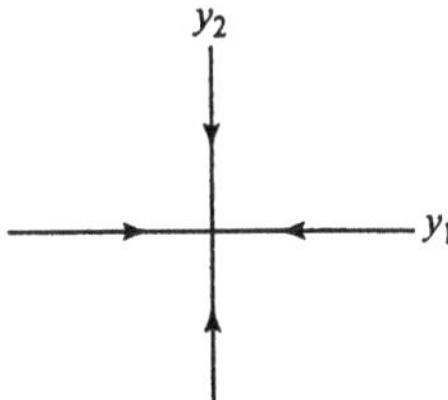

FIGURE 3.

VIII-3. Stable manifolds

A *stable manifold* of the trivial solution is a set of points such that solutions starting from them approach the trivial solution as $t \to +\infty$. In order to illustrate such a manifold, consider a system of the form

$$(\text{VIII.3.1}) \qquad \frac{d\vec{x}}{dt} = A_1\vec{x} + \vec{g}_1(\vec{x}, \vec{y}), \qquad \frac{d\vec{y}}{dt} = A_2\vec{y} + \vec{g}_2(\vec{x}, \vec{y}),$$

where $\vec{x} \in \mathbb{R}^n$, $\vec{y} \in \mathbb{R}^m$, entries of $\mathbb{R}^n$-valued function $\vec{g}_1$ and $\mathbb{R}^m$-valued function $\vec{g}_2$ are continuous in $(\vec{x}, \vec{y})$ for $\max(|\vec{x}|, |\vec{y}|) \leq \rho_0$ and satisfy the Lipschitz condition

$$|\vec{g}_j(\vec{x}, \vec{y}) - \vec{g}_j(\vec{\xi}, \vec{\eta})| \leq L(\rho) \max(|\vec{x} - \vec{\xi}|, |\vec{y} - \vec{\eta}|) \qquad (j = 1, 2)$$

for $\max(|\vec{x}|, |\vec{y}|) \leq \rho_0$ and $\max(|\vec{\xi}|, |\vec{\eta}|) \leq \rho_0$, where ρ_0 is a positive number and $\lim\limits_{\rho \to 0} L(\rho) = 0$. Furthermore, assume that $\vec{g}_j(\vec{0}, \vec{0}) = \vec{0}$ $(j = 1, 2)$. Two matrices A_1 and A_2 are respectively constant $n \times n$ and $m \times m$ matrices satisfying the following condition:

$$(\text{VIII.3.2}) \qquad \begin{cases} |e^{(t-s)A_1}| \leq K_1 e^{-\sigma_1(t-s)} & \text{for} \quad t \geq s, \\ |e^{(t-s)A_2}| \leq K_2 e^{-\sigma_2(t-s)} & \text{for} \quad t \leq s, \end{cases}$$

where K_j and σ_j $(j = 1, 2)$ are positive constants. Condition (VIII.3.2) implies that the real parts of eigenvalues of A_1 are not greater than $-\sigma_1$, whereas the real parts of eigenvalues of A_2 are not less than $-\sigma_2$. Assume that

$$(\text{VIII.3.3}) \qquad \sigma_1 > \sigma_2.$$

Let us change (VIII.3.1) to the following system of integral equations:

$$(\text{VIII.3.4}) \qquad \begin{cases} \vec{x}(t, \vec{c}) = e^{tA_1}\vec{c} + \displaystyle\int_0^t e^{(t-s)A_1} \vec{g}_1(\vec{x}(s, \vec{c}), \vec{y}(s, \vec{c}))ds, \\[2mm] \vec{y}(t, \vec{c}) = \displaystyle\int_{+\infty}^t e^{(t-s)A_2} \vec{g}_2(\vec{x}(s, \vec{c}), \vec{y}(s, \vec{c}))ds. \end{cases}$$

The main result in this section is the following theorem.

Theorem VIII-3-1. *Fix a positive number ϵ so that $\sigma_1 > \sigma_2 + \epsilon$. Then, there exists another positive number $\rho(\epsilon)$ such that if an arbitrary constant vector $\vec{c}$ in $\mathbb{R}^n$ satisfies the condition $K_1|\vec{c}| \leq \dfrac{\rho(\epsilon)}{2}$, a solution $(\vec{x}(t, \vec{c}), \vec{y}(t, \vec{c}))$ of (VIII.3.1) can be constructed so that*

$$(\text{VIII.3.5}) \qquad \vec{x}(0, \vec{c}) = \vec{c} \quad and \quad \max(|\vec{x}(t, \vec{c})|, |\vec{y}(t, \vec{c})|) \leq \rho(\epsilon)e^{-(\sigma_1 - \epsilon)t}$$

for $0 \leq t < +\infty$. Furthermore, this solution $(\vec{x}(t, \vec{c}), \vec{y}(t, \vec{c}))$ is uniquely determined by condition (VIII.3.5).

Proof.

Observe that if

$$\max(|\vec{x}(t, \vec{c})|, |\vec{y}(t, \vec{c})|) \leq \rho e^{-(\sigma_1 - \epsilon)t} \qquad \text{for} \qquad 0 \leq t < +\infty,$$

then

$$\left| \int_0^t e^{(t-s)A_1} \vec{g}_1(\vec{x}(s, \vec{c}), \vec{y}(s, \vec{c}))ds \right| \leq K_1 L(\rho)\rho \int_0^t e^{-\sigma_1(t-s)} e^{-(\sigma_1 - \epsilon)s}ds$$

$$= \frac{K_1 L(\rho)\rho}{\epsilon} e^{-\sigma_1 t} \left(e^{\epsilon t} - 1 \right) \leq \frac{K_1 L(\rho)\rho}{\epsilon} e^{-(\sigma_1 - \epsilon)t}$$

and

$$\left| \int_{+\infty}^t e^{(t-s)A_2} \vec{g}_2(\vec{x}(s, \vec{c}), \vec{y}(s, \vec{c}))ds \right| \leq K_2 L(\rho)\rho \int_t^{+\infty} e^{-\sigma_2(t-s)} e^{-(\sigma_1 - \epsilon)s}ds$$

$$= \frac{K_2 L(\rho)\rho}{\sigma_1 - \sigma_2 - \epsilon} e^{-(\sigma_1 - \epsilon)t}.$$

This implies that if a positive number ρ is chosen so small that $\dfrac{K_2 L(\rho)}{\sigma_1 - \sigma_2 - \epsilon} \leq \dfrac{1}{2}$ and $\dfrac{K_1 L(\rho)}{\epsilon} \leq \dfrac{1}{2}$ and if the arbitrary vector $\vec{c}$ in $\mathbb{R}^n$ satisfies the condition $K_1|\vec{c}| \leq \dfrac{\rho}{2}$, then, using successive approximations, a solution $(\vec{x}(t,\vec{c}), \vec{y}(t,\vec{c}))$ of (VIII.3.4) can be constructed so that

$$\vec{x}(0,\vec{c}) = \vec{c} \quad \text{and} \quad \max(|\vec{x}(t,\vec{c})|, |\vec{y}(t,\vec{c})|) \leq \rho e^{-(\sigma_1 - \epsilon)t} \qquad \text{for} \quad 0 \leq t < +\infty.$$

Details are left to the reader as an exercise. $\quad\square$

Remark VIII-3-2.
(i) The positive number ϵ is given to start with and the choice of ρ depends on ϵ. However, since this solution approaches the trivial solution, the constant ϵ may be eventually replaced by any smaller number, since the right-hand side of (VIII.3.1) is independent of t. This implies that the curve $(\vec{x}(t,\vec{c}), \vec{y}(t,\vec{c}))$ is independent of ϵ as $t \to +\infty$. More precisely, if a solution $(\vec{x}(t), \vec{y}(t))$ of (VIII.3.1) satisfies a condition

$$\max(|\vec{x}(t)|, |\vec{y}(t)|) \leq K e^{-(\sigma_1 - \epsilon_0)t} \qquad \text{for} \qquad 0 \leq t < +\infty$$

with some positive constants K and ϵ_0 such that $\sigma_1 - \sigma_2 - \epsilon_0 > 0$, then for every positive ϵ smaller than ϵ_0, there exists $t_0 \geq 0$ such that $(\vec{x}(t_0 + t), \vec{y}(t_0 + t)) = (\vec{x}(t,\vec{c}), \vec{y}(t,\vec{c}))$, where $\vec{c} = \vec{x}(t_0)$.
(ii) The initial value of $\vec{y}(t,\vec{c})$ is given by

$$\vec{y}(0,\vec{c}) = \int_{+\infty}^{0} e^{-sA_2} \vec{g}_2(\vec{x}(s,\vec{c}), \vec{y}(s,\vec{c}))ds.$$

(iii) If $\dfrac{\partial \vec{g}_j}{\partial \vec{x}}$ and $\dfrac{\partial \vec{g}_j}{\partial \vec{y}}$ exist and are continuous in a neighborhood of $(\vec{0}, \vec{0})$ and if $\dfrac{\partial \vec{g}_j}{\partial \vec{x}}(\vec{0}, \vec{0}) = \vec{0}$ and $\dfrac{\partial \vec{g}_j}{\partial \vec{y}}(\vec{0}, \vec{0}) = \vec{0}$, then $\vec{x}(t,\vec{c})$ and $\vec{y}(t,\vec{c})$ are continuously differentiable with respect to $\vec{c}$.
(iv) If the real parts of all eigenvalues of the matrix A_1 of (VIII.3.1) are negative and the real parts of all eigenvalues of the matrix A_2 are positive, then the stable manifold of the trivial solution of system (VIII.3.1) is given by $\mathcal{S} = \{(\vec{c}, \vec{y}(0,\vec{c})) : |\vec{c}| \leq \rho\}$, where ρ is a sufficiently small positive number.
(v) Consider a system

$$\text{(VIII.3.6)} \qquad \qquad \frac{d\vec{y}}{dt} = A\vec{y} + \vec{g}(\vec{y})$$

in the following situation:
(a) A is a constant $n \times n$ matrix,
(b) the entries of $\vec{g}(\vec{y})$ are continuous for $|\vec{y}| \leq \rho_0$ and satisfy the Lipschitz condition $|\vec{g}(\vec{y}) - \vec{g}(\vec{\eta})| \leq L(\rho)|\vec{y} - \vec{\eta}|$ for $|\vec{y}| \leq \rho_0$ and $|\vec{\eta}| \leq \rho_0$, where ρ_0 is a positive number and $\lim_{\rho \to 0} L(\rho) = 0$,

(c) $\vec{g}(\vec{0}, \vec{0}) = \vec{0}$.

Suppose further that A has an eigenvalue with positive real part. Then, applying Theorem VIII-3-1 to the system $\dfrac{d\vec{y}}{dt} = -A\vec{y} - \vec{g}(\vec{y})$, we can construct the stable manifold $\mathcal{U}$ of the trivial solution. This means that if a solution $\vec{\phi}(t)$ of (VIII.3.6) starts from a point on $\mathcal{U}$, then $\lim\limits_{t \to -\infty} \vec{\phi}(t) = \vec{0}$. This shows that the trivial solution of (VIII.3.6) is not stable as $t \to +\infty$, and Theorem VIII-2-7 is proved for (VIII.3.6). The set $\mathcal{U}$ is called the *unstable manifold* of the trivial solution of (VIII.3.6).

The materials in this section are also found in [CL, §§4 and 5 of Chapter 8] and [Har2, Chapter IX; in particular Theorem 6.1 on p. 242].

VIII-4. Analytic structure of stable manifolds

In order to look closely into the structure of the stable manifold of the trivial solution of a system of analytic differential equations

$$\text{(VIII.4.1)} \qquad \frac{d\vec{y}}{dt} = A\vec{y} + \vec{f}(\vec{y}),$$

let us construct a formal simplification of system (VIII.4.1). To do this, consider system (VIII.4.1) under the following assumption.

Assumption VIII-4-1. *The unknown quantity $\vec{y}$ is a vector in $\mathbb{C}^n$ with entries $\{y_1, \ldots, y_n\}$, A is a constant $n \times n$ matrix, and*

$$\text{(VIII.4.2)} \qquad \vec{f}(\vec{y}) = \sum_{|\wp| \geq 2} \vec{y}^{\wp} \vec{f}_{\wp}$$

is a formal power series with coefficients $\vec{f}_{\wp} \in \mathbb{C}^n$, where $\wp = (p_1, \ldots, p_n)$ with non-negative integers $p_1, \ldots, p_n$, $|\wp| = \sum\limits_{h=1}^{n} p_h$, and $\vec{y}^{\wp} = y_1^{p_1} \cdots y_n^{p_n}$.

The following theorem is a basic result concerning formal simplifications of system (VIII.4.1).

Theorem VIII-4-2. *Under Assumption VIII-4-1, there exists a formal power series*

$$\text{(VIII.4.3)} \qquad \vec{P}(\vec{u}) = P_0 \vec{u} + \sum_{|\wp| \geq 2} \vec{u}^{\wp} \vec{P}_{\wp}$$

in a vector $\vec{u} \in \mathbb{C}^n$ with entries $\{u_1, \ldots, u_n\}$ such that
  (i) P_0 is an invertible constant $n \times n$ matrix and $\vec{P}_{\wp} \in \mathbb{C}^n$,
  (ii) the formal transformation $\vec{y} = \vec{P}(\vec{u})$ reduces system (VIII.4.1) to

$$\text{(VIII.4.4)} \qquad \frac{d\vec{u}}{dt} = B_0 \vec{u} + \vec{g}(\vec{u})$$

with a constant $n \times n$ matrix B_0 and the formal power series $\vec{g}(\vec{u}) = \sum_{|\wp| \geq 2} u^\wp \vec{g}_\wp$

with coefficients $\vec{g}_\wp$ in $\mathbb{C}^n$ such that

(iia) the matrix B_0 is lower triangular with the diagonal entries $\lambda_1, \ldots, \lambda_n$, and the entry $b_0(j,k)$ on the j-th row and k-th column of B_0 is zero whenever $\lambda_j \neq \lambda_k$,

(iib) for $\wp$ with $|\wp| \geq 2$, the j-th entry $g_{\wp,j}$ of the vector $\vec{g}_\wp$ is zero whenever

$$\text{(VIII.4.5)} \qquad \lambda_j \neq \sum_{h=1}^{n} p_h \lambda_h.$$

Proof.

Observe that if $\vec{y} = \vec{P}(\vec{u})$, then

$$\frac{d\vec{y}}{dt} = \left\{ P_0 + \sum_{|\wp| \geq 2} \left[\frac{p_1 u^\wp}{u_1} \vec{P}_\wp \quad \frac{p_2 u^\wp}{u_2} \vec{P}_\wp \quad \cdots \quad \frac{p_n u^\wp}{u_n} \vec{P}_\wp \right] \right\} \left\{ B_0 \vec{u} + \sum_{|\wp| \geq 2} u^\wp \vec{g}_\wp \right\}$$

and

$$A\vec{y} + \vec{f}(\vec{y}) = A \left\{ P_0 \vec{u} + \sum_{|\wp| \geq 2} \vec{u}^\wp \vec{P}_\wp \right\} + \sum_{|\wp| \geq 2} \vec{P}(\vec{u})^\wp \vec{f}_\wp.$$

Furthermore,

$$\left\{ P_0 + \sum_{|\wp| \geq 2} \left[\frac{p_1 u^\wp}{u_1} \vec{P}_\wp \quad \frac{p_2 u^\wp}{u_2} \vec{P}_\wp \quad \cdots \quad \frac{p_n u^\wp}{u_n} \vec{P}_\wp \right] \right\} B_0 \vec{u}$$

$$= P_0 B_0 \vec{u} + \sum_{|\wp| \geq 2} \left\{ \left(\sum_{h=1}^{n} p_h \lambda_h \right) \vec{u}^\wp \vec{P}_\wp + \left[\sum_{1 \leq h < j \leq n} p_j \beta_{j,h} \frac{u_h}{u_j} \right] \vec{u}^\wp \vec{P}_\wp \right\},$$

where $\beta_{j,k}$ is the entry of B_0 on the j-th row and k-th column.

Let us introduce a linear order $\wp_1 \prec \wp_2$ for $\wp_j = (p_{j1}, \ldots, p_{jn})$ $(j = 1, 2)$ by the relation

$$p_{1h} = p_{2h} \quad \text{for} \quad (h < h_0) \quad \text{and} \quad p_{1h_0} < p_{2h_0}.$$

Now, calculating the coefficients of $\vec{u}^\wp$ $(|\wp| \geq 1)$ on both sides of (VIII.4.1), we obtain

$$\text{(VIII.4.6)} \qquad P_0 B_0 = A P_0$$

and

$$\text{(VIII.4.7)} \qquad \left(\sum_{h=1}^{n} p_h \lambda_h \right) \vec{P}_\wp + P_0 \vec{g}_\wp - A\vec{P}_\wp$$

$$= \vec{\mathcal{F}}_\wp(\vec{P}_{\wp'} : \wp' \prec \wp) + \vec{\mathcal{G}}_\wp(\vec{P}_{\wp'}, \vec{g}_{\wp'} : |\wp'| < |\wp|)$$

for $|\wp| \geq 2$. From (VIII.4.6), it is concluded that the diagonal entries $\lambda_1, \ldots, \lambda_n$ of B_0 are eigenvalues of A and that this allows us to set $A = B_0$ and $P_0 = I_n$, where I_n is the $n \times n$ identity matrix. Then, (VIII.4.7) becomes

$$\text{(VIII.4.8)} \qquad \left(\sum_{h=1}^{n} p_h \lambda_h \right) \vec{P}_\wp + \vec{g}_\wp - B_0 \vec{P}_\wp$$
$$= \vec{\mathcal{F}}_\wp(\vec{P}_{\wp'} : \wp' \prec \wp) + \vec{\mathcal{G}}_\wp(\vec{P}_{\wp'}, \vec{g}_{\wp'} : |\wp'| < |\wp|).$$

Solve (VIII.4.8) by solving equations of the form

$$\text{(VIII.4.9)} \qquad \left(\sum_{h=1}^{n} p_h \lambda_h - \lambda_j \right) P_{\wp,j} + g_{\wp,j} = F_{\wp,j}$$

successively, where $P_{\wp,j}$ and $g_{\wp,j}$ are the j-th entries of the vector $\vec{P}_\wp$ and $\vec{g}_\wp$, respectively, and $F_{\wp,j}$ are known quantities. If $\sum_{h=1}^{n} p_h \lambda_h - \lambda_j \neq 0$, set $g_{\wp,j} = 0$ and solve (VIII.4.9). If $\sum_{h=1}^{n} p_h \lambda_h - \lambda_j = 0$, then set $g_{\wp,j} = F_{\wp,j}$ and choose $P_{\wp,j}$ in any way. This completes the proof of Theorem VIII.4.2. $\square$

Observation VIII-4-3. Assume that $\Re(\lambda_j) < 0$ for $j = 1, \ldots, r$ and $\Re(\lambda_j) \geq 0$ for $j \neq 1, \ldots, r$. In this case, if $u_h = 0$ for $h \neq 1, \ldots, r$, the ℓ-th entry of the vector $B_0 \vec{u} + \vec{g}(\vec{u})$ is zero for $\ell \neq 1, \ldots, r$. In fact, look at $\vec{g}(\vec{u})$. Then, the ℓ-th entry of the coefficient $\vec{g}_\wp$ is zero if $\lambda_\ell \neq \sum_{h=1}^{n} p_h \lambda_h$. Note that, if $\ell > r$ and $\lambda_\ell = \sum_{h=1}^{n} p_h \lambda_h$, then $(p_{r+1}, \ldots, p_n) \neq (0, \ldots, 0)$. Hence, in such a case, $u^\wp = 0$ if $(u_{r+1}, \ldots, u_n) = (0, \ldots, 0)$. Therefore, the ℓ-th entry of the vector $B_0 \vec{u} + \vec{g}(\vec{u})$ is zero for $\ell > r$ if $(u_{r+1}, \ldots, u_n) = (0, \ldots, 0)$.

Observation VIII-4-4. Under the same assumption on the λ_j as in Observation VIII-4-3, set $(u_{r+1}, \ldots, u_n) = (0, \ldots, 0)$. Then, the system of differential equations on $(u_1, \ldots, u_r)$ has the form

$$\text{(VIII.4.10)} \qquad \begin{aligned} \frac{du_j}{dt} &= \lambda_j u_j + \sum_{\lambda_h = \lambda_j, h < j} \beta_{j,h} u_h \\ &+ \sum_{\lambda_j = p_1 \lambda_1 + \cdots + p_r \lambda_r, |\wp| \geq 2} g_{(p_1, \ldots, p_r), j} u_1^{p_1} \cdots u_r^{p_r} \quad (j = 1, \ldots, r). \end{aligned}$$

Observe that since $\lambda_j - \sum_{h=1}^{r} p_h \lambda_h \neq 0$ if $\sum_{h=1}^{r} p_h$ is sufficiently large, the right-hand members of (VIII.4.10) are polynomials in $(u_1, \ldots, u_r)$.

Observation VIII-4-5. Assume that $\Re[\lambda_{h+1}] \leq \Re[\lambda_h]$. Then, (VIII.4.10) can be written in the form

$$\frac{du_1}{dt} = \lambda_1 u_1 \quad \text{and} \quad \frac{du_j}{dt} = \lambda_j u_j + \mathcal{F}_j(u_1, \dots, u_{j-1}) \quad \text{for} \quad j = 2, \dots, r.$$

Hence, system (VIII.4.10) can be solved with an elementary method. To see the structure of solutions of (VIII.4.10) more clearly, change $(u_1, \dots, u_r)$ by $u_j = e^{\lambda_j t} v_j \, (j = 1, \dots, r)$. This transformation changes (VIII.4.10) to

$$\frac{dv_j}{dt} = \sum_{\lambda_h = \lambda_j, h < j} \beta_{j,h} v_h$$

$$+ \sum_{\lambda_j = p_1 \lambda_1 + \cdots + p_r \lambda_r, |\wp| \geq 2} g_{(p_1, \dots, p_r), j} v_1^{p_1} \cdots v_r^{p_r} \qquad (j = 1, \dots, r).$$

This, in turn, shows that the general solution of (VIII.4.10) has the form $u_j = e^{\lambda_j t} \psi_j(t, c_1, \dots, c_r) \, (j = 1, \dots, r)$, where $(c_1, \dots, c_r)$ are arbitrary constants and $\psi_j(t, c_1, \dots, c_r)$ are polynomials in $(t, c_1, \dots, c_r)$.

An analytic justification of the formal series $\vec{P}(\vec{u})$ is given by the following theorem.

Theorem VIII-4-6. *In the case when the entries of the $\mathbb{C}^n$-valued function $\vec{f}(\vec{y})$ on the right-hand side of (VIII.4.1) are analytic in a neighborhood of $\vec{0}$, under the same assumption on the λ_j as in Observation VIII-4-3, the power series $\vec{P}(\vec{u})$ is convergent if $(u_{r+1}, \dots, u_n) = (0, \dots, 0)$.*

The proof of this theorem is straight forward but lengthy (cf. [Si2]). The key fact is the inequality

$$\left| \lambda_j - \sum_{h=1}^{r} p_h \lambda_h \right| \geq \sigma \sum_{h=1}^{r} p_h$$

for some positive number σ if $\displaystyle\sum_{h=1}^{r} p_h$ is large. In this case, a majorant series for $\vec{P}$ can be constructed.

The construction of such a majorant series is illustrated for a simple case of a system

$$\frac{d\vec{y}}{dt} = \lambda \vec{y} + \vec{f}(\vec{y}),$$

where λ is a nonzero complex number and $\vec{f}(\vec{y})$ is a convergent power series in $\vec{y}$ given by (VIII.4.2). According to Theorem VIII-4-2, in this case there exists a formal transfomation $\vec{y} = \vec{u} + \vec{Q}(\vec{u})$ such that $\dfrac{d\vec{u}}{dt} = \lambda \vec{u}$, where $\vec{Q}(\vec{u}) = \displaystyle\sum_{|\wp| \geq 2} \vec{u}^\wp \vec{Q}_\wp$.

This implies that $\displaystyle\sum_{|\wp| \geq 2} \lambda |\wp| \vec{u}^\wp \vec{Q}_\wp = \lambda \vec{Q}(\vec{u}) + \vec{f}(\vec{u} + \vec{Q}(\vec{u}))$. Set $\vec{f}(\vec{u} + \vec{Q}(\vec{u})) = $

$$\sum_{|\wp|\geq 2} \vec{u}^{\wp}\vec{A}_{\wp}. \quad \text{Then,}$$

$$\vec{Q}_{\wp} = \frac{\vec{A}\wp}{\lambda(|\wp|-1)} \quad \text{for all} \ \wp \ (|\wp|\geq 2).$$

Set

$$\vec{F}(\vec{y}) = \left[\sum_{|\wp|\geq 2}|\vec{f}_{\wp}|\vec{y}^{\wp}\right]\begin{bmatrix}1\\ \vdots \\ 1\end{bmatrix}.$$

Then, $\vec{F}(\vec{y})$ is a majorant series for $\vec{f}(\vec{y})$. Determine a power series $\vec{v}(\vec{u}) = \sum_{|\wp|\geq 2}\vec{u}^{\wp}\vec{v}_{\wp}$

by the equation $\vec{v} = \dfrac{1}{|\lambda|}\vec{F}(\vec{u}+\vec{v})$. Set $\vec{F}(\vec{u}+\vec{v}(\vec{u})) = \sum_{|\wp|\geq 2}\vec{u}^{\wp}\vec{B}_{\wp}$. Then,

$$\vec{v}_{\wp} = \frac{\vec{B}_{\wp}}{|\lambda|} \quad \text{for all} \ \wp \ (|\wp|\geq 2).$$

It can be shown easily that $\vec{v}(\vec{u})$ is a convergent majorant series of $\vec{Q}(\vec{u})$. This proves the convergence of $\vec{Q}(\vec{u})$.

Putting $\vec{P}(\vec{u})$ and the general solution of (VIII.4.10) together, we obtain a particular solution $\vec{y} = \vec{P}(\vec{\phi}(t,\vec{c}))$ of (VIII.4.1), where

$$\vec{\phi}(t,\vec{c}) = (e^{\lambda_1 t}\psi_1(t,c_1,\dots,c_r),\dots,e^{\lambda_r t}\psi_r(t,c_1,\dots,c_r),0,\dots,0).$$

This particular solution is depending on r arbitrary constants $\vec{c} = (c_1,\dots,c_r)$. Furthermore, this solution represents the stable manifold of the trivial solution of (VIII.4.1) if $\Re(\lambda_j) > 0$ for $j \neq 1,\dots,r$.

Remark VIII-4-7. In the case when $\vec{y}$, A, and $\vec{f}(\vec{y})$ are real, but A has some eigenvalues which are not real, then $\vec{P}(\vec{y})$ must be constructed carefully so that the particular solution $\vec{P}(\vec{\phi}(t,\vec{c}))$ is also real-valued. For example, if $A = \begin{bmatrix} a & -b \\ b & a \end{bmatrix}$, the eigenvalues of A are $a \pm ib$. If $\vec{y} = \begin{bmatrix} y_1 \\ y_2 \end{bmatrix}$ is changed by $u_1 = y_1 + iy_2$, and $u_2 = y_1 - iy_2$, system (VIII.4.1) becomes

(VIII.4.11)
$$\begin{cases} \dfrac{du_1}{dt} = (a+ib)u_1 + \displaystyle\sum_{p_1+p_2\geq 2} g_{p_1,p_2}u_1^{p_1}u_2^{p_2}, \\[3ex] \dfrac{du_2}{dt} = (a-ib)u_2 + \displaystyle\sum_{p_1+p_2\geq 2} \overline{g_{p_1,p_2}}u_2^{p_1}u_1^{p_2}, \end{cases}$$

where $\overline{g_{p_1,p_2}}$ is the complex conjugate of g_{p_1,p_2}. If $a \neq 0$, using Theorem VIII-4-2, simplify (VIII.4.11) to

(VIII.4.12)
$$\frac{dv_1}{dt} = (a+ib)v_1, \qquad \frac{dv_2}{dt} = (a-ib)v_2$$

by the transformation

$$\text{(VIII.4.13)} \qquad \begin{cases} u_1 = v_1 + \displaystyle\sum_{p_1+p_2\geq 2} P_{p_1,p_2} v_1^{p_1} v_2^{p_2}, \\[4mm] u_2 = v_2 + \displaystyle\sum_{p_1+p_2\geq 2} \overline{P_{p_1,p_2}} v_2^{p_1} v_1^{p_2}. \end{cases}$$

Now, system (VIII.4.12), in turn, is changed back to $\dfrac{d\vec{w}}{dt} = A\vec{w}$ by $w_1 = \dfrac{v_1 + v_2}{2}$ and $w_2 = \dfrac{v_1 - v_2}{2i}$, where (w_1, w_2) are the entries of the vector $\vec{w}$. Observe that

$$\begin{cases} y_1 = \dfrac{u_1 + u_2}{2} = w_1 + \displaystyle\sum_{p_1+p_2\geq 2} q_{1,p_1,p_2} w_1^{p_1} w_2^{p_2}, \\[4mm] y_2 = \dfrac{u_1 - u_2}{2i} = w_2 + \displaystyle\sum_{p_1+p_2\geq 2} q_{2,p_1,p_2} w_1^{p_1} w_2^{p_2}, \end{cases}$$

where q_{1,p_1,p_2} and q_{2,p_1,p_2} are real numbers. Similar arguments can be used in general cases to construct real-valued solutions. (For complexification, see, for example, [HirS, pp. 64–65].)

For classical works related with the materials in this section as well as more general problems, see, for example, [Du].

VIII-5. Two-dimensional linear systems with constant coefficients

Throughout the rest of this chapter, we shall study the behavior of solutions of nonlinear systems in $\mathbb{R}^2$. The $\mathbb{R}^2$-plane is called the *phase plane* and a solution curve projected to the phase plane is called an *orbit* of the system of equations. A diagram that shows the orbits in the phase plane is called a *phase portrait of the orbits* of the system of equations. As a preparation, in this section, we summarize the basic facts concerning linear systems with constant coefficients in $\mathbb{R}^2$.

Consider a linear system

$$\text{(VIII.5.1)} \qquad\qquad \frac{d\vec{y}}{dt} = A\vec{y},$$

where $\vec{y} \in \mathbb{R}^2$ and A is a real, constant, and invertible 2×2 matrix. Set $p = \text{trace}(A)$ and $q = \det(A)$, where $q \neq 0$. Then, the characteristic equation of the matrix A is $\lambda^2 - p\lambda + q = 0$. Hence, two eigenvalues of A are given by

$$\lambda_1 = \frac{p}{2} + \sqrt{\frac{p^2}{4} - q} \qquad \text{and} \qquad \lambda_2 = \frac{p}{2} - \sqrt{\frac{p^2}{4} - q}.$$

It is known that

$$p = \lambda_1 + \lambda_2, \quad q = \lambda_1\lambda_2, \quad \text{and} \quad \lambda_1 - \lambda_2 = 2\sqrt{\frac{p^2}{4} - q}.$$

Also, let $\vec{\xi}$ and $\vec{\eta}$ be two eigenvectors of A associated with the eigenvalues λ_1 and λ_2, respectively, i.e., $A\vec{\xi} = \lambda_1\vec{\xi}$, $\vec{\xi} \neq \vec{0}$, and $A\vec{\eta} = \lambda_2\vec{\eta}$, $\vec{\eta} \neq \vec{0}$. Observe that, if $\vec{y}(t)$ is a solution of (VIII.5.1), then $c\vec{y}(t+\tau)$ is also a solution of (VIII.5.1) for any constants c and τ. This fact is useful in order to find orbits of equation (VIII.5.1) in the phase plane.

Case 1. Assume that two eigenvalues λ_1 and λ_2 are real and distinct. In this case, two eigenvectors $\vec{\xi}$ and $\vec{\eta}$ are linearly independent and the general solution of differential equation (VIII.5.1) is given by

$$\vec{y}(t) = c_1 e^{\lambda_1 t}\vec{\xi} + c_2 e^{\lambda_2 t}\vec{\eta} = e^{\lambda_1 t}[c_1\vec{\xi} + c_2 e^{(\lambda_2-\lambda_1)t}\vec{\eta}] = e^{\lambda_2 t}[c_1 e^{(\lambda_1-\lambda_2)t}\vec{\xi} + c_2\vec{\eta}],$$

where c_1 and c_2 are arbitrary constants and $-\infty < t < +\infty$.

1a: In the case when $\lambda_1 > \lambda_2 > 0$ (i.e., $p > 0$, $q > 0$ and $\dfrac{p^2}{4} > q$), the phase portrait of orbits of (VIII.5.1) is shown by Figure 4. The arrow indicates the direction in which t increases. The trivial solution $\vec{0}$ is unstable as $t \to +\infty$. Note that as $t \to -\infty$, the solutions $\vec{y}(t)$ tends to $\vec{0}$ in one of the four directions of $\vec{\xi}$, $-\vec{\xi}$, $\vec{\eta}$, and $-\vec{\eta}$. The point $(0,0)$ is called an *unstable improper node*.

1b: In the case when $0 > \lambda_1 > \lambda_2$ (i.e., $p < 0$, $q > 0$ and $\dfrac{p^2}{4} > q$), the phase portrait of orbits of (VIII.5.1) is shown by Figure 5. The trivial solution $\vec{0}$ is stable as $t \to +\infty$. The point $(0,0)$ is called a *stable improper node*.

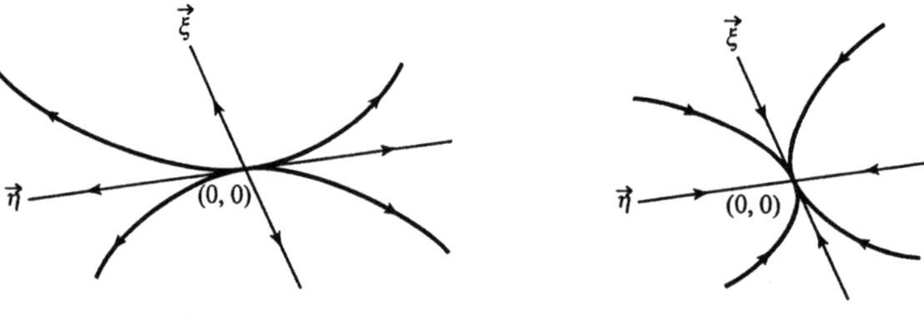

FIGURE 4.   FIGURE 5.

1c: In the case when $\lambda_1 > 0 > \lambda_2$ (i.e., $q < 0$), the phase portrait of orbits of (VIII.5.1) is shown by Figure 6. The trivial solution $\vec{0}$ is unstable as $|t| \to +\infty$. Note that as $t \to -\infty$ (or $+\infty$), only two orbits of (VIII.5.1) tend to $\vec{0}$. The point $(0,0)$ is called a *saddle point*.

Case 2. Assume that two eigenvalues λ_1 and λ_2 are equal. Then, $q = \dfrac{p^2}{4}$ and $\lambda_1 = \lambda_2 = \dfrac{p}{2} \neq 0$.

2a: Assume furhter that A is diagonalizable; i.e., $A = \dfrac{p}{2}I_2$, where I_2 is the 2×2 identity matrix. Then, every nonzero vector $\vec{c}$ is an eigenvector of A, and the general solution of (VIII.5.1) is given by $\vec{y}(t) = \exp\left[\dfrac{p}{2}t\right]\vec{c}$. In this case, the phase portrait of orbits of (VIII.5.1) is shown by Figures 7-1 and 7-2. As $t \to +\infty$, the trivial solution $\vec{0}$ is unstable (respectively stable) if $p > 0$ (respectively $p < 0$). Note that, for every direction $\vec{n}$, there exists an orbit which tends to $\vec{0}$ in the direction $\vec{n}$ as t tends to $-\infty$ (respectively $+\infty$). The point $(0,0)$ is called an *unstable* (respectively *stable*) *proper node* if $p > 0$ (respectively $p < 0$).

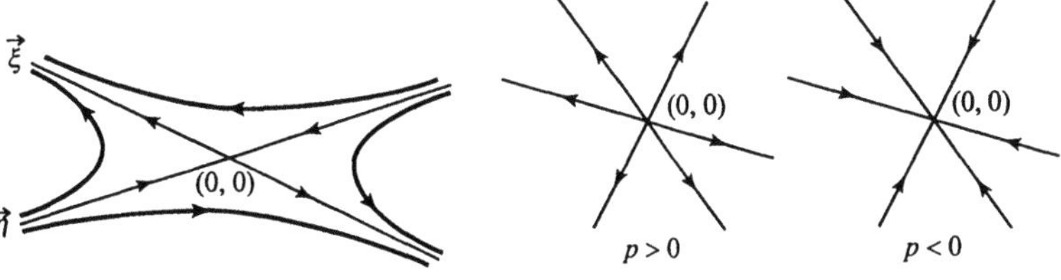

FIGURE 6.   FIGURE 7-1.   FIGURE 7-2.

2b: Assume that A is not diagonalizable; i.e., $A = \dfrac{p}{2}I_2 + N$, where I_2 is the 2×2 identity matrix and N is a 2×2 nilpotent matrix. Note that $N \neq O$ and $N^2 = O$. Hence, $\exp[tA] = \exp\left[\dfrac{p}{2}t\right]\{I_2 + tN\}$. Observe also that a nonzero vector $\vec{c}$ is an eigenvector of A if and only if $N\vec{c} = \vec{0}$. Since $N(N\vec{c}) = \vec{0}$, the vector $N\vec{c}$ is either $\vec{0}$ or an eigenvector of A. Hence, $N\vec{c} = \alpha(\vec{c})\vec{\xi}$, where $\vec{\xi}$ is the eigenvector of A which was given at the beginning of this section and $\alpha(\vec{c})$ is a real-valued linear homogeneous function of $\vec{c}$. Observe also that $\alpha(\vec{c}) = 0$ if and only if $\vec{c}$ is a constant multiple of the eigenvector $\vec{\xi}$. The general solution of (VIII.5.1) is given by $\vec{y}(t) = \exp\left[\dfrac{p}{2}t\right]\{\vec{c} + t\alpha(\vec{c})\vec{\xi}\}$, where $\vec{c}$ is an arbitrary constant vector. In this case, the phase portrait of orbits of (VIII.5.1) is shown by Figures 8-1 and 8-2. The trivial solution $\vec{0}$ is unstable (respectively stable) as $t \to +\infty$ if $p > 0$ (respectively $p < 0$). The point $(0,0)$ is called an *unstable* (respectively *stable*) *improper node* if $p > 0$ (respectively $p < 0$).

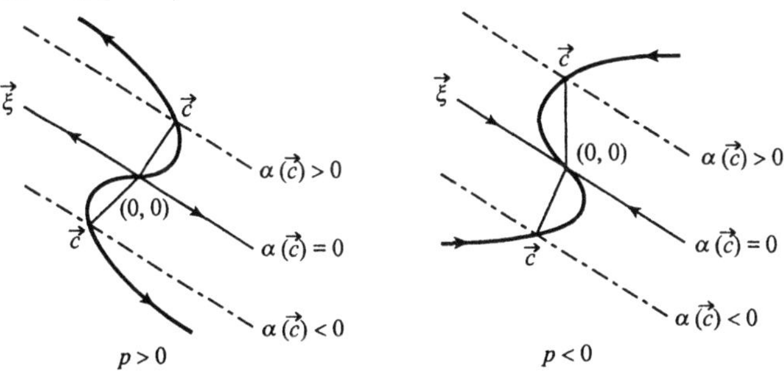

FIGURE 8-1.   FIGURE 8-2.

Case 3. If two eigenvalues λ_1 and λ_2 are not real, then $q > \dfrac{p^2}{4}$ and

$$\lambda_1 = a + ib, \qquad \lambda_2 = a - ib, \qquad a = \frac{p}{2}, \qquad b = \sqrt{q - \frac{p^2}{4}}.$$

Note that $b > 0$. Set $A = aI_2 + B$. Then, $B^2 = -b^2 I_2$, since two eigenvalues of B

are ib and $-ib$. Therefore, $\exp[tB] = (\cos(bt))I_2 + \dfrac{\sin(bt)}{b}B$.

3a: Assume that $a = 0$ (i.e., $p = 0$). Then, the general solution of (VIII.5.1) is given by $\vec{y}(t) = \exp[tB]\vec{c} = (\cos(bt))\vec{c} + \dfrac{\sin(bt)}{b}B\vec{c}$, which is periodic of period $\dfrac{2\pi}{b}$ in t. The phase portrait of orbits of (VIII.5.1) is shown by Figures 9-1 and 9-2. The trivial solution $\vec{0}$ is stable as $|t| \to +\infty$. The point $(0,0)$ is called a *center*. It is important to notice that every orbit $\vec{y}(t)$ is invariant by the operator $\dfrac{B}{b}$. In fact,

$$\frac{B}{b}\vec{y}(t) = \frac{\cos(bt)}{b}B\vec{c} - (\sin(bt))\vec{c} = \left(\cos\left(bt + \frac{\pi}{2}\right)\right)\vec{c} + \frac{\sin\left(bt + \frac{\pi}{2}\right)}{b}B\vec{c} = \vec{y}\left(t + \frac{\pi}{2b}\right).$$

In other words, $\dfrac{d\vec{y}(t)}{dt} = b\vec{y}\left(t + \dfrac{\pi}{2b}\right)$. Note that $\dfrac{\pi}{2b}$ is $\dfrac{1}{4}$ of the period $\dfrac{2\pi}{b}$.

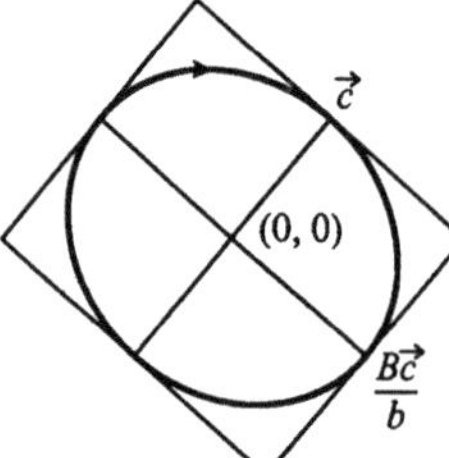

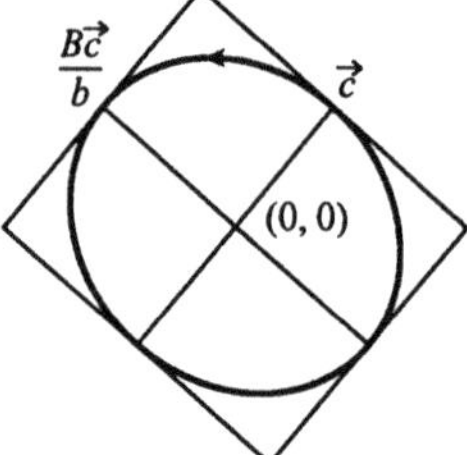

FIGURE 9-1.   FIGURE 9-2.

3b: Assume that $a \neq 0$ (i.e., $p \neq 0$). Then, the general solution of (VIII.5.1) is given by $\vec{y}(t) = \exp[at]\vec{u}(t)$, where $\vec{u}(t)$ is the general solution of $\dfrac{d\vec{u}}{dt} = B\vec{u}$. The solution $\vec{0}$ is unstable (respectively stable) as $t \to +\infty$ if $p > 0$ (respectively $p < 0$). The orbits, as shown by Figures 10-1 and 10-2, go around the point $(0,0)$ infinitely many times as $\vec{y}(t) \to \vec{0}$. The point $(0,0)$ is called an *unstable* (respectively *stable*) *spiral point* if $p > 0$ (respectively $p < 0$).

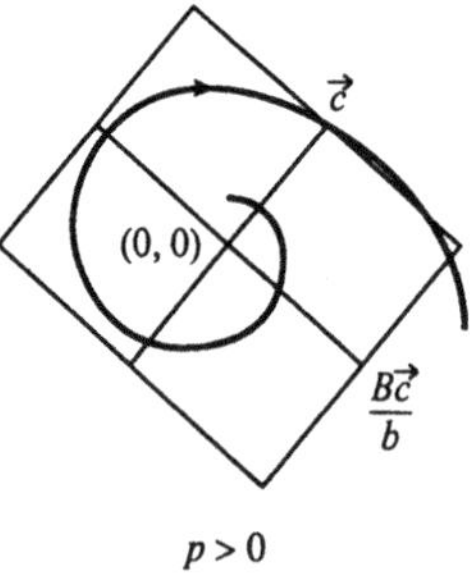

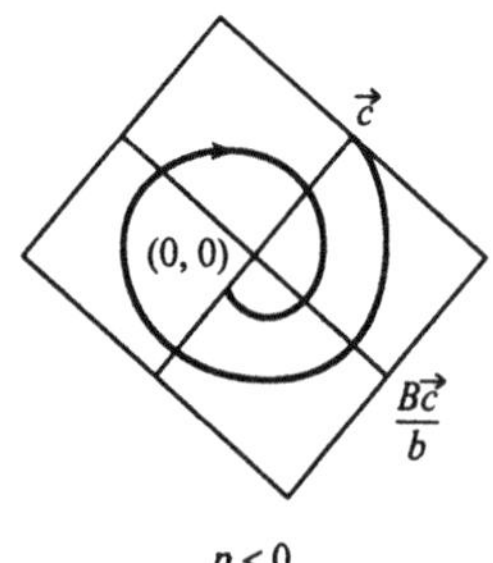

FIGURE 10-1.   FIGURE 10-2.

Let us summarize the results given above by using Figure 11:
(1) $(0,0)$ is an improper node.
(2) $(0,0)$ is a saddle point.

(3) $(0,0)$ is a proper or improper node.
(4) $(0,0)$ is a center.
(5) $(0,0)$ is a spiral point.

Stability of the trivial solution $\vec{0}$ is summarized as follows:

(I) If $q < 0$, the trivial solution $\vec{0}$ is unstable as $|t| \to +\infty$.
(II) If $q > 0$ and $p > 0$, the trivial solution $\vec{0}$ is unstable as $t \to +\infty$.
(III) If $q > 0$ and $p < 0$, the trivial solution $\vec{0}$ is stable as $t \to +\infty$.
(IV) If $p = 0$ and $q > 0$, the trivial solution $\vec{0}$ is stable as $|t| \to +\infty$.

(Cf. Figure 12).

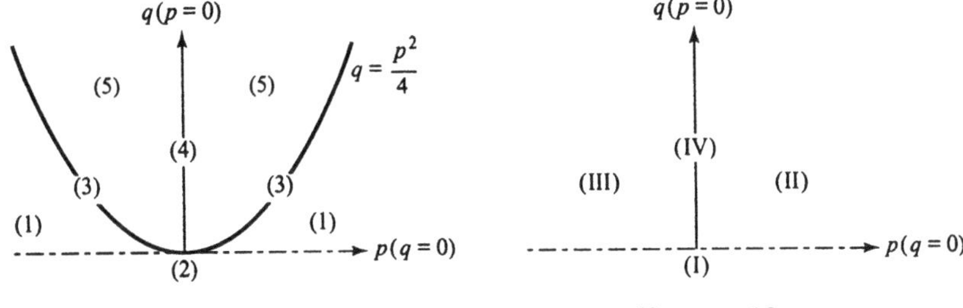

FIGURE 11.   FIGURE 12.

VIII-6. Analytic systems in $\mathbb{R}^2$

In this section, we apply Theorems VIII-4-2 and VIII-4-6 to an analytic system in $\mathbb{R}^2$. Consider a system

$$\text{(VIII.6.1)} \qquad \frac{d\vec{y}}{dt} = A\vec{y} + \sum_{|\wp| \geq 2} \vec{y}^{\wp} \vec{f}_{\wp},$$

where $\vec{y} \in \mathbb{R}^2$ with the entries y_1 and y_2, A is a real, invertible, and constant 2×2 matrix, $\wp = (p_1, p_2)$ with two non-negative integers p_1 and p_2, $|\wp| = p_1 + p_2$, $\vec{y}^{\wp} = y_1^{p_1} y_2^{p_2}$, the entries of vectors $\vec{f}_{\wp} \in \mathbb{R}^2$ are real constants independent of t, and the series on the right-hand side of (VIII.6.1) is uniformly convergent in a domain $\Delta(\rho_0) = \{\vec{y} \in \mathbb{R}^2 : |\vec{y}| < \rho_0\}$ for some positive number ρ_0. Let us look into the structure of solutions of (VIII.6.1) in the following five cases.

Case 1. If the point $(0,0)$ is a *stable proper node* of the linear system $\frac{d\vec{y}}{dt} = A\vec{y}$, then $A = \lambda I_2$, where λ is a negative number and I_2 is the 2×2 identity matrix. To apply Theorem VIII-4-2 to this case, look at the equation $\lambda = p_1 \lambda + p_2 \lambda$ on non-negative integers p_1 and p_2 such that $p_1 + p_2 \geq 2$. Since no such (p_1, p_2) exists, there exists an $\mathbb{R}^2$-valued function $\vec{P}(\vec{u})$ whose entries are convergent power series in a vector $\vec{u} \in \mathbb{R}^2$ with real coefficients such that $\frac{\partial \vec{P}}{\partial \vec{u}}(\vec{0}) = I_2$ and that the transformation $\vec{y} = \vec{P}(\vec{u})$ reduces system (VIII.6.1) to $\frac{d\vec{u}}{dt} = \lambda \vec{u}$. This, in turn, implies that the point $(0,0)$ is also a stable proper node of (VIII.6.1) and that the general solution of (VIII.6.1) is $\vec{y} = \vec{P}(e^{\lambda t}\vec{c})$, where $\vec{c}$ is an arbitrary constant vector in $\mathbb{R}^2$.

Case 2. If the point $(0,0)$ is a *stable improper node* of the linear system $\dfrac{d\vec{y}}{dt} = A\vec{y}$, then we may assume that either (1) $A = \lambda I_2 + N$, where λ is a negative number and $N \neq O$ is a 2×2 nilpotent matrix, or (2) $A = \begin{bmatrix} \lambda_1 & 0 \\ 0 & \lambda_2 \end{bmatrix}$, where λ_1 and λ_2 are real negative numbers such that $\lambda_1 > \lambda_1$. In case (1), the same conclusion is obtained concerning the existence of an $\mathbb{R}^2$-valued function $\vec{P}(\vec{u})$. Therefore, the point $(0,0)$ is a stable improper node of (VIII.6.1), and the general solution of (VIII.6.1) is $\vec{y} = \vec{P}(e^{\lambda t}(I_2 + tN)\vec{c})$, where $\vec{c}$ is an arbitrary constant vector in $\mathbb{R}^2$. In case (2), looking at the equations $\lambda_1 = p_1\lambda_1 + p_2\lambda_2$ and $\lambda_2 = p_1\lambda_1 + p_2\lambda_2$ on non-negative integers p_1 and p_2 such that $p_1 + p_2 \geq 2$, it is concluded that there exists an $\mathbb{R}^2$-valued function $\vec{P}(\vec{u})$ whose entries are convergent power series in a vector $\vec{u} \in \mathbb{R}^2$ with real coefficients such that $\dfrac{\partial \vec{P}}{\partial \vec{u}}(\vec{0}) = I_2$ and that the transformation $\vec{y} = \vec{P}(\vec{u})$ reduces system (VIII.6.1) to $\dfrac{d\vec{u}}{dt} = \begin{bmatrix} \lambda_1 u_1 \\ \gamma u_1^M + \lambda_2 u_2 \end{bmatrix}$, where (u_1, u_2) are the entries of $\vec{u}$, M is a positive integer such that $\lambda_2 = M\lambda_1$, and γ is a real constant which must be 0 if $\lambda_2 \neq M\lambda_1$ for any positive integer M. Therefore, in case (2), the general solution of (VIII.1.6) is $\vec{y} = \vec{P}\left(\begin{bmatrix} c_1 e^{\lambda_1 t} \\ (c_1^M \gamma t + c_2)e^{\lambda_2 t} \end{bmatrix}\right)$, where c_1 and c_2 are arbitrary constants. This, in turn, implies that the point $(0,0)$ is a stable improper node of (VIII.6.1).

Case 3. If the point $(0,0)$ is a *stable spiral point* of the linear system $\dfrac{d\vec{y}}{dt} = A\vec{y}$, then we may assume that $A = \begin{bmatrix} a & -b \\ b & a \end{bmatrix}$, where a and b are real numbers such that $a < 0$ and $b \neq 0$. The eigenvalues of A are $\lambda_{\pm} = a \pm ib$. This implies that there are no non-negative integers p_1 and p_2 satisfying the condition $\lambda_{\pm} = p_1\lambda_+ + p_2\lambda_-$ and $p_1 + p_2 \geq 2$. Therefore, there exists an $\mathbb{R}^2$-valued function $\vec{P}(\vec{u})$ whose entries are convergent power series in a vector $\vec{u} \in \mathbb{R}^2$ with real coefficients such that $\dfrac{\partial \vec{P}}{\partial \vec{u}}(\vec{0}) = I_2$, and the transformation $\vec{y} = \vec{P}(\vec{u})$ reduces system (VIII.6.1) to $\dfrac{d\vec{u}}{dt} = A\vec{u}$ (cf. Remark VIII.4.7). This, in turn, implies that the point $(0,0)$ is also a stable spiral point of (VIII.6.1) and that the general solution of (VIII.6.1) is $\vec{y} = \vec{P}(e^{tA}\vec{c})$, where $\vec{c}$ is an arbitrary constant vector in $\mathbb{R}^2$.

Case 4. If the point $(0,0)$ is a *saddle point* of the linear system $\dfrac{d\vec{y}}{dt} = A\vec{y}$, the eigenvalues λ_1 and λ_2 of A are real and $\lambda_1 < 0 < \lambda_2$. Construct two nontrivial and real-valued convergent power series $\vec{\phi}(x)$ and $\vec{\psi}(x)$ in a variable x so that $\vec{\phi}(e^{\lambda_1 t}c_1)\,(t \geq 0)$ and $\vec{\psi}(e^{\lambda_2 t}c_2)\,(t \leq 0)$ are solutions of (VIII.6.1), where c_1 and c_2 are arbitrary constants (cf. Exercise V-7). The solution $\vec{y} = \vec{\phi}(e^{\lambda_1 t}c_1)$ represents the stable manifold of the trivial solution of (VIII.6.1), while the solution $\vec{y} = \vec{\psi}(e^{\lambda_2 t}c_2)$ represents the unstable manifold of the trivial solution of (VIII.6.1). The point $(0,0)$

is a saddle point of (VIII.6.1). In the next section, we shall explain the behavior of solutions in a neighborhood of a saddle point in a more general case.

Case 5. If the point $(0,0)$ is a *center* of the linear system $\dfrac{d\vec{y}}{dt} = A\vec{y}$, both eigenvalues of A are purely imaginary. Assume that they are $\pm i$ and $A = \begin{bmatrix} 0 & -1 \\ 1 & 0 \end{bmatrix}$. Set $\vec{y} = \dfrac{1}{2} \begin{bmatrix} 1 & 1 \\ -i & i \end{bmatrix} \vec{v}$, where $\vec{v} = \begin{bmatrix} v_1 \\ v_2 \end{bmatrix}$. Then, the given system (VIII.6.1) is changed to

$$(\text{VIII.6.2}) \qquad \frac{d\vec{v}}{dt} = \vec{g}(\vec{v}) = \begin{bmatrix} g(v_1, v_2) \\ \bar{g}(v_2, v_1) \end{bmatrix},$$

where

$$g(v_1, v_2) = iv_1 + \sum_{p+q=2}^{+\infty} g_{p,q} v_1^p v_2^q, \qquad \bar{g}(v_2, v_1) = -iv_2 + \sum_{p+q=2}^{+\infty} \overline{g_{p,q}} v_2^p v_1^q.$$

Here, $\bar{a}$ denotes the complex conjugate of a complex number a. Let us apply Theorem VIII-4-2 to system (VIII.6.2).

Observation VIII-6-1. First, setting $\lambda_1 = i$ and $\lambda_2 = -i$, look at $\lambda_1 = p_1\lambda_1 + p_2\lambda_2$ and $\lambda_2 = q_1\lambda_1 + q_2\lambda_2$, where p_1, p_2, q_1, and q_2 are non-negative integers such that $p_1 + p_2 \geq 2$ and $q_1 + q_2 \geq 2$. Then, $p_1 - 1 = p_2$ and $q_2 - 1 = q_1$. This implies that system (VIII.6.2) can be changed to

$$(\text{VIII.6.3}) \qquad \frac{du_1}{dt} = i\omega(u_1 u_2)u_1, \qquad \frac{du_2}{dt} = -i\bar{\omega}(u_1 u_2)u_2,$$

where $\omega(z) = 1 + \displaystyle\sum_{m=1}^{+\infty} \omega_m z^m$, by a formal transformation

$$(\text{T}) \qquad \vec{v} = \vec{f}(\vec{u}) = \begin{bmatrix} u_1 + h(u_1, u_2) \\ u_2 + \bar{h}(u_2, u_1) \end{bmatrix}.$$

Here,

$$h(u_1, u_2) = \sum_{p+q=2}^{+\infty} h_{p,q} u_1^p u_2^q, \qquad \bar{h}(u_2, u_1) = \sum_{p+q=2}^{+\infty} \overline{h_{p,q}} u_1^q u_2^p.$$

In particular, $h(u_1, u_2)$ can be construced so that

$$(\text{VIII.6.4}) \qquad \text{the quantities } h_{p+1,p} \text{ are real for all positive integers } p.$$

Observation VIII-6-2. We can show that if one of the ω_m is not real, then $\vec{y} = \vec{0}$ is a spiral point (cf. Exercises VIII-14). Hence, let us look into the case when the ω_m are all real. Furthermore, if a formal power series $\alpha(\zeta) = 1 + \displaystyle\sum_{m=1}^{+\infty} \alpha_m \zeta^m$ with real coefficients α_m is chosen in a suitable way, the transformation

$$(\text{VIII.6.5}) \qquad u_1 = \alpha(\beta_1\beta_2)\beta_1, \qquad u_2 = \alpha(\beta_1\beta_2)\beta_2$$

changes (VIII.6.3) to another system

$$\frac{d\beta_1}{dt} = i\Omega(\beta_1\beta_2)\beta_1, \qquad \frac{du_2}{dt} = -i\Omega(\beta_1\beta_2)\beta_2,$$

where $\Omega(\zeta)$ is a polynomial in ζ with real coefficients which has one of the following two forms:

$$\Omega(\zeta) = \begin{cases} 1, & \text{Case I} \\ 1 + c_0\zeta^{m_0}, & \text{Case II} \end{cases}$$

where c_0 is a nonzero real number and m_0 is a positive integer.

Observation VIII-6-3. Let us look into Case I. Assume that system (VIII.6.3) is

$$\text{(VIII.6.3.1)} \qquad \frac{du_1}{dt} = iu_1, \qquad \frac{du_2}{dt} = -iu_2.$$

Note that transformation (VIII.6.5) does not change (VIII.6.3.1). Using a transformation of form (VIII.6.5), change $h(u_1, u_2)$ so that

$$\text{(VIII.6.6)} \qquad h_{p+1,p} = 0 \qquad \text{for all positive integers } p.$$

Since $(u_1, u_2) = (ce^{it}, \bar{c}e^{-it})$ is a solution of system (VIII.6.3.1) with a complex arbitrary constant c, the formal series

$$\text{(FS-1)} \qquad \vec{v} = \vec{f}(\vec{u}) = \begin{bmatrix} ce^{it} + h(ce^{it}, \bar{c}e^{-it}) \\ \bar{c}e^{-it} + \bar{h}(\bar{c}e^{-it}, ce^{it}) \end{bmatrix}$$

is a formal solution of system (VIII.6.2) which depends on two real arbitrary constants.

Let

$$\text{(S)} \qquad \vec{v}(t, \xi, \bar{\xi}) = \begin{bmatrix} \xi e^{it} + H(t, \xi, \bar{\xi}) \\ \bar{\xi}e^{-it} + \bar{H}(t, \bar{\xi}, \xi) \end{bmatrix}$$

be the solution of system (VIII.6.2) satisfying the initial condition

$$\text{(C)} \qquad \vec{v}(0, \xi, \bar{\xi}) = \begin{bmatrix} \xi \\ \bar{\xi} \end{bmatrix}.$$

Note that

$$H(t, \xi, \bar{\xi}) = \sum_{p+q=2}^{+\infty} H_{p,q}(t)\xi^p\bar{\xi}^q, \qquad \bar{H}(t, \bar{\xi}, \xi) = \sum_{p+q=2}^{+\infty} \overline{H_{p,q}(t)}\xi^q\bar{\xi}^p$$

are power series in ξ and $\bar{\xi}$ which are convergent uniformly on any fixed bounded interval on the real t line.

Using (VIII.6.6), we obtain

$$\int_0^{2\pi} h(ce^{it}, \bar{c}e^{-it})e^{-it}dt = 0, \qquad \int_0^{2\pi} \bar{h}(\bar{c}e^{-it}, ce^{it})e^{it}dt = 0.$$

Fix c and $\bar{c}$ by the equations

$$c = \xi + \frac{1}{2\pi}\int_0^{2\pi} H(s,\xi,\bar{\xi})e^{-is}ds \quad \text{and} \quad \bar{c} = \bar{\xi} + \frac{1}{2\pi}\int_0^{2\pi} \bar{H}(s,\bar{\xi},\xi)e^{is}ds.$$

Then,

$$\xi = c + \Xi(c,\bar{c}) \quad \text{and} \quad \bar{\xi} = \bar{c} + \bar{\Xi}(\bar{c},c),$$

where $\Xi(c,\bar{c})$ and $\bar{\Xi}(\bar{c},c)$ are convergent power series in $(c,\bar{c})$.

Now, we can prove that two formal solutions

$$\vec{v}(t, c + \Xi(c,\bar{c}), \bar{c} + \bar{\Xi}(c,\bar{c})) \quad \text{and} \quad \begin{bmatrix} ce^{it} + h(ce^{it}, \bar{c}e^{-it}) \\ \bar{c}e^{-it} + \bar{h}(\bar{c}e^{-it}, ce^{it}) \end{bmatrix}$$

of system (VIII.6.2) are identical. The first of these two is convergent; hence, the second is also convergent. This finishes Case I.

Observation VIII-6-4. Let us look into Case II. Assume that system (VIII.6.3) has the form

$$(\text{VIII.6.3.2}) \quad \frac{du_1}{dt} = iu_1(1 + c_0(u_1u_2)^{m_0}), \qquad \frac{du_2}{dt} = -iu_2(1 + c_0(u_1u_2)^{m_0}).$$

Note that we have (VIII.6.4).

Since $(u_1, u_2) = (ce^{it(1+c_0(c\bar{c})^{m_0})}, \bar{c}e^{-it(1+c_0(c\bar{c})^{m_0})})$ is a solution of system (VIII.6.3.2) with a complex arbitrary constant c, the formal series

$$(\text{FS-2}) \quad \vec{v} = \vec{f}(\vec{u}) = \begin{bmatrix} ce^{it(1+c_0(c\bar{c})^{m_0})} + h(ce^{it(1+c_0(c\bar{c})^{m_0})}, \bar{c}e^{-it(1+c_0(c\bar{c})^{m_0})}) \\ \bar{c}e^{-it(1+c_0(c\bar{c})^{m_0})} + \bar{h}(\bar{c}e^{-it(1+c_0(c\bar{c})^{m_0})}, ce^{it(1+c_0(c\bar{c})^{m_0})}) \end{bmatrix}$$

is a formal solution of system (VIII.6.2) which depends on two real arbitrary constants.

Again, as in Observation VIII-6-3, let (S) be the solution of system (VIII.6.2) satisfying the initial condition (C). Set

$$(\text{VIII.6.7}) \quad \xi = \Xi[c,\bar{c}] = c + h(c,\bar{c}) \quad \text{and} \quad \bar{\xi} = \bar{\Xi}[\bar{c},c] = \bar{c} + \bar{h}(\bar{c},c).$$

Then,

$$\vec{v}(t, \Xi[c,\bar{c}], \bar{\Xi}[\bar{c},c]) = \begin{bmatrix} ce^{it(1+c_0(c\bar{c})^{m_0})} + h(ce^{it(1+c_0(c\bar{c})^{m_0})}, \bar{c}e^{-it(1+c_0(c\bar{c})^{m_0})}) \\ \bar{c}e^{-it(1+c_0(c\bar{c})^{m_0})} + \bar{h}(\bar{c}e^{-it(1+c_0(c\bar{c})^{m_0})}, ce^{it(1+c_0(c\bar{c})^{m_0})}) \end{bmatrix}$$

as formal power series in $(c, \bar{c})$. This series has a formal period
$$T(c\bar{c}) = \frac{2\pi}{1 + c_0(c\bar{c})^{m_0}} \text{ with respect to } t, \text{ i.e.,}$$

$$\text{(VIII.6.8)} \qquad \vec{v}(T(c\bar{c}), \Xi[c, \bar{c}], \bar{\Xi}[\bar{c}, c]) = \begin{bmatrix} \Xi(c, \bar{c}) \\ \bar{\Xi}(\bar{c}, c) \end{bmatrix}.$$

Solving (VIII.6.7) with respect to $(c, \bar{c})$, we obtain two power series in $(\xi, \bar{\xi})$:

$$c = c(\xi, \bar{\xi}) \qquad \text{and} \qquad \bar{c} = \bar{c}(\bar{\xi}, \xi).$$

Set $\mathcal{P}(\xi, \bar{\xi}) = T(c(\xi, \bar{\xi})\bar{c}(\bar{\xi}, \xi))$. Then, (VIII.6.8) becomes

$$\text{(VIII.6.9)} \qquad \vec{v}(\mathcal{P}(\xi, \bar{\xi}), \xi, \bar{\xi}) = \begin{bmatrix} \xi \\ \bar{\xi} \end{bmatrix}.$$

Using (VIII.6.9) as equations for $\mathcal{P}(\xi, \bar{\xi})$, it can be shown that the formal power series $\mathcal{P}(\xi, \bar{\xi})$ is convergent. This implies that

$$c(\xi, \bar{\xi})\bar{c}(\bar{\xi}, \xi) = \left[\left(\frac{1}{c_0} \right) \left(\frac{2\pi}{\mathcal{P}(\xi, \bar{\xi})} - 1 \right) \right]^{1/m_0}$$

is convergent. (Here, some details are left to the reader as an exercise.)

On the other hand,

$$\int_0^{T(c\bar{c})} \left\{ \xi e^{is} + H(s, \xi, \bar{\xi}) \right\} e^{-is(1 + c_0(c\bar{c})^{m_0})} ds$$

$$= \int_0^{T(c\bar{c})} \left\{ c e^{is(1 + c_0(c\bar{c})^{m_0})} + h(c e^{is(1 + c_0(c\bar{c})^{m_0})}, \bar{c} e^{-is(1 + c_0(c\bar{c})^{m_0})}) \right\}$$

$$\times e^{-is(1 + c_0(c\bar{c})^{m_0})} ds = c \left(1 + \sum_{p=1}^{+\infty} h_{p+1,p}(c\bar{c})^p \right)$$

and

$$\int_0^{T(c\bar{c})} \left\{ \bar{\xi} e^{-is} + \bar{H}(s, \bar{\xi}, \xi) \right\} e^{is(1 + c_0(c\bar{c})^{m_0})} ds = \bar{c} \left(1 + \sum_{p=1}^{+\infty} h_{p+1,p}(c\bar{c})^p \right)$$

since the quantity $h_{p+1,p}$ are real (cf. (VIII.6.4)). This proves that $\dfrac{c(\xi, \bar{\xi})}{\bar{c}(\bar{\xi}, \xi)}$ is convergent. Hence, $c(\xi, \bar{\xi})$ and $\bar{c}(\bar{\xi}, \xi)$ are convergent. Thus, the proof of the convergence of $\vec{f}(\vec{u})$ in Case II is completed.

Thus, it was proved that if all of the coefficients ω_m on the right-hand side of system (VIII.6.3) are real, the point $(0, 0)$ is a center of system (VIII.6.1). The general solution of (VIII.6.1) can be constructed by using various transformations of (VII.6.1) which bring the system to either (VIII.6.3.1) or (VIII.6.3.2). Periods of solutions in t are independent of each solution in Case I, but depend on each solution in Case II.

Remark VIII-6-5. The argument given above does not apply to the case when the right-hand side of (VIII.6.1) is of $C^{(\infty)}$ but not analytic. A counterexample is given by

$$\frac{d\vec{y}}{dt} = \begin{bmatrix} h(r^2) & -1 \\ 1 & h(r^2) \end{bmatrix} \vec{y}, \qquad r^2 = y_1^2 + y_2^2,$$

where $h(r^2) = e^{-1/r^2} \sin\left(\dfrac{1}{r^2}\right)$. Using the polar coordinates (r, θ), this system is changed to

$$\frac{dr}{dt} = rh(r^2) \qquad \text{and} \qquad \frac{d\theta}{dt} = 1$$

and periodic solutions are given by $r^2 = \dfrac{1}{m\pi}$ with any integers m. This is an example of centers in the sense of Bendixson (cf. [Ben2, p. 26] and §VIII-10).

For more information concerning analytic systems in $\mathbb{R}^2$, see [Huk5].

VIII-7. Perturbations of an improper node and a saddle point

In the previous section, the structure of solutions of an analytic system in $\mathbb{R}^2$ was showen by means of the method with power series. Hereafter, in this chapter, we consider a system

$$(VIII.7.1) \qquad \frac{d\vec{y}}{dt} = A\vec{y} + \vec{g}(\vec{y}),$$

under the assumption that

(i) A is a real constant 2×2 matrix,
(ii) the entries of the $\mathbb{R}^2$-valued function $\vec{g}(\vec{y})$ are continuous in $\vec{y} \in \mathbb{R}^2$ near $\vec{0}$ and satisfies the condition

$$(VIII.7.2) \qquad \lim_{\vec{y} \to \vec{0}} \frac{\vec{g}(\vec{y})}{|\vec{y}|} = \vec{0}.$$

In this section, it is also assumed that the initial-value problem

$$(VIII.7.3) \qquad \frac{d\vec{y}}{dt} = A\vec{y} + \vec{g}(\vec{y}), \qquad \vec{g}(0) = \vec{\eta}$$

has one and only one solution as long as $\vec{\eta}$ belongs to a sufficiently small neighborhood of $\vec{0}$.

Remark VIII-7-1. In the case, when the entries of an $\mathbb{R}^2$-valued function $\vec{f}(\vec{z})$ are continuously differentiable with respect to the entries z_1 and z_2 of $\vec{z}$ in a neighborhood of $\vec{z} = \vec{0}$ and $\vec{f}(\vec{0}) = 0$, write the differential equation $\dfrac{d\vec{z}}{dt} = \vec{f}(\vec{z})$ in form (VIII.7.1) by setting

$$A = \frac{\partial \vec{f}}{\partial \vec{z}}(\vec{0}) = \begin{bmatrix} \dfrac{\partial f_1}{\partial z_1}(\vec{0}) & \dfrac{\partial f_1}{\partial z_2}(\vec{0}) \\[2mm] \dfrac{\partial f_2}{\partial z_1}(\vec{0}) & \dfrac{\partial f_2}{\partial z_2}(\vec{0}) \end{bmatrix} \qquad \text{and} \qquad \vec{g}(\vec{y}) = \vec{f}(\vec{y}) - A\vec{y}.$$

Using the polar coordinates (r, θ) in the $\vec{y}$-plane, write the vector $\vec{y}$ in the form $\vec{y} = r \begin{bmatrix} \cos\theta \\ \sin\theta \end{bmatrix}$. Then, system (VIII.7.1) is written in terms of (r, θ) as follows:

(VIII.7.4)
$$
\begin{cases}
\dfrac{dr}{dt} = r[a_{11}\cos^2(\theta) + a_{22}\sin^2(\theta) + (a_{12} + a_{21})\sin(\theta)\cos(\theta)] \\
\qquad\quad + g_1(\vec{y})\cos(\theta) + g_2(\vec{y})\sin(\theta), \\
r\dfrac{d\theta}{dt} = r[-a_{12}\sin^2(\theta) + a_{21}\cos^2(\theta) + (a_{22} - a_{11})\sin(\theta)\cos(\theta)] \\
\qquad\quad - g_1(\vec{y})\sin(\theta) + g_2(\vec{y})\cos(\theta),
\end{cases}
$$

where $A = \begin{bmatrix} a_{11} & a_{12} \\ a_{21} & a_{22} \end{bmatrix}$ and $\vec{g}(\vec{y}) = \begin{bmatrix} g_1(\vec{y}) \\ g_2(\vec{y}) \end{bmatrix}$. Here, use was made of the formula

$$
\frac{dr}{dt} = \frac{dy_1}{dt}\cos(\theta) + \frac{dy_2}{dt}\sin(\theta), \qquad r\frac{d\theta}{dt} = -\frac{dy_1}{dt}\sin(\theta) + \frac{dy_2}{dt}\cos(\theta).
$$

In this section, we consider the case when the two eigenvalues λ_1 and λ_2 of the matrix A are real, distinct, and at least one of them is negative, i.e., $\lambda_2 > \lambda_1$ and $\lambda_1 < 0$. Throughout this section, assume that $a_{11} = \lambda_1$, $a_{22} = \lambda_2$, and $a_{12} = a_{21} = 0$. Then, in terms of the polar coordinates (r, θ), system (VIII.7.1) can be writen in the form

(VIII.7.5)
$$
\begin{cases}
\dfrac{dr}{dt} = r\left[\lambda_1(\cos(\theta))^2 + \lambda_2(\sin(\theta))^2 + \dfrac{1}{r}(g_1(\vec{y})\cos(\theta) + g_2(\vec{y})\sin(\theta))\right], \\
\dfrac{d\theta}{dt} = (\lambda_2 - \lambda_1)\sin(\theta)\cos(\theta) + \dfrac{1}{r}(-g_1(\vec{y})\sin(\theta) + g_2(\vec{y})\cos(\theta)).
\end{cases}
$$

Observation VIII-7-2. The point $(0,0)$ is not a proper node of (VIII.7.1). In fact, if $(r(t), \theta(t))$ is a solution of (VIII.7.5) such that $r(t) \to 0$ and $\theta(t) \to \omega$ as $t \to +\infty$ or $-\infty$, then $\dfrac{d\theta}{dt} \to (\lambda_2 - \lambda_1)\sin\omega\cos\omega$ as $t \to +\infty$ or $-\infty$. Hence, $\sin\omega\cos\omega = 0$. This implies that $\omega = -\pi, -\dfrac{\pi}{2}, 0,$ or $\dfrac{\pi}{2}$.

Observation VIII-7-3. For any given positive number $\epsilon > 0$, there exists another positive number $\rho(\epsilon) > 0$ such that

$$
\frac{d\theta}{dt}
\begin{cases}
> 0 & \text{for} \quad -\pi + \epsilon \leq \theta \leq -\dfrac{\pi}{2} - \epsilon, \\
< 0 & \text{for} \quad -\dfrac{\pi}{2} + \epsilon \leq \theta \leq -\epsilon, \\
> 0 & \text{for} \quad \epsilon \leq \theta \leq \dfrac{\pi}{2} - \epsilon, \\
< 0 & \text{for} \quad \dfrac{\pi}{2} + \epsilon \leq \theta \leq \pi - \epsilon,
\end{cases}
$$

if $0 < r \leq \rho(\epsilon)$ (cf. Figure 13).

Observation VIII-7-4. If $0 > \lambda_2 > \lambda_1$, there exists a positive number $r_0 > 0$ such that $\dfrac{dr}{dt} < \dfrac{\lambda_2 r}{2}$ for $0 < r \leq r_0$.

Observation VIII-7-5. In the case when $\lambda_2 > 0 > \lambda_1$, find a real number ω_0 such that $0 < \omega_0 < \dfrac{\pi}{2}$ and $\tan\omega_0 = \sqrt{-\dfrac{\lambda_1}{\lambda_2}}$. Then, for any given positive number $\epsilon > 0$, there exists another positive number $\rho(\epsilon) > 0$ such that

$$\frac{dr}{dt} \begin{cases} > 0 & \text{for} \quad (\omega_0 - \pi) + \epsilon \leq \theta \leq -\omega_0 - \epsilon, \\ < 0 & \text{for} \quad -\omega_0 + \epsilon \leq \theta \leq \omega_0 - \epsilon, \\ > 0 & \text{for} \quad \omega_0 + \epsilon \leq \theta \leq (\pi - \omega_0) - \epsilon, \\ < 0 & \text{for} \quad (\pi - \omega_0) + \epsilon \leq \theta \leq (\pi + \omega_0)\, \epsilon, \end{cases}$$

if $0 < r \leq \rho(\epsilon)$ (cf. Figure 14).

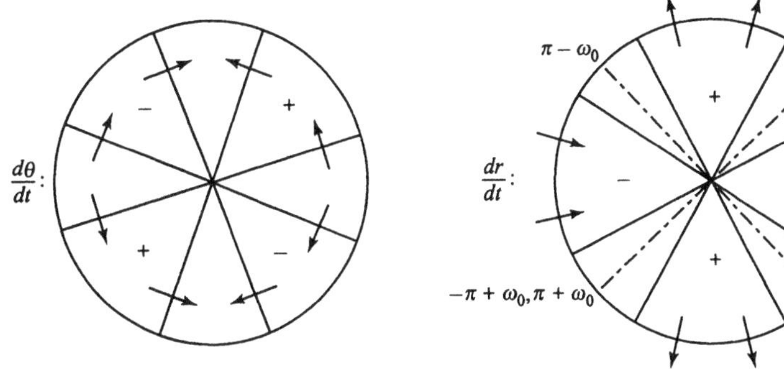

FIGURE 13.   FIGURE 14.

Observation VIII-7-6. If two positive numbers ϵ and ρ are sufficiently small, Observations VIII-7-2 and VIII-7-3 show that the behavior of orbits of (VIII.7.1) in the sectorial region $\mathcal{S} = \{(r\cos\theta, r\sin\theta) : |\theta| \leq \epsilon, 0 < r \leq \rho\}$ looks like Figure 15.

Denote by Ω the set of all real numbers ω such that

(i) $|\omega| \leq \epsilon$,

(ii) the orbit $(r(t), \theta(t))$ of (VIII.7.1) such that $r(0) = \rho$ and $\theta(0) = \omega$ leaves the sectorial region $\mathcal{S}$ through the boundary $\theta = -\epsilon$.

Set $\alpha = \sup\{\omega : \omega \in \Omega\}$. Then, $|\alpha| < \epsilon$. Furthermore, the orbit $(r(t), \theta(t))$ of (VIII.7.1) such that $r(0) = \rho$ and $\theta(0) = \alpha$ satisfies the condition $\lim\limits_{t \to +\infty} r(t) = 0$ and $\lim\limits_{t \to +\infty} \theta(t) = 0$. This can be shown by using the continuity of orbits with respect to initial data (cf. Figures 16-1 and 16-2). Here, use was made of the uniqueness of solutions of initial-value problem (VIII.7.3).

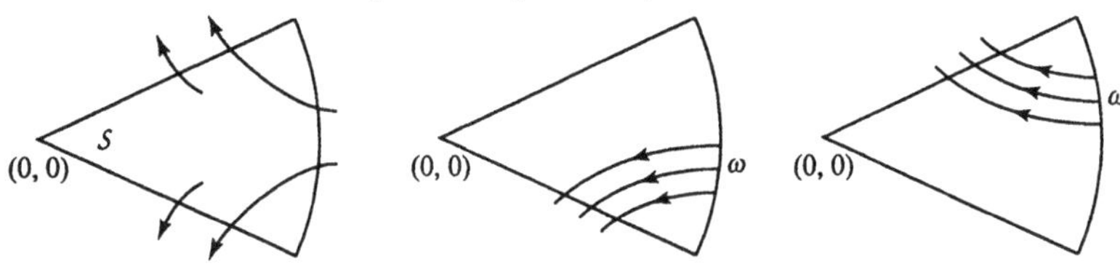

FIGURE 15.   FIGURE 16-1.   FIGURE 16-2.

Observation VIII-7-7. In the case when $0 > \lambda_2 > \lambda_1$, the point $(0,0)$ is a stable improper node of (VIII.7.1) as $t \to +\infty$ (cf. Figure 17).

Observation VIII-7-8. In the case when $\lambda_2 > 0 > \lambda_1$, the point $(0,0)$ is a saddle point of (VIII.7.1) (cf. Figure 18).

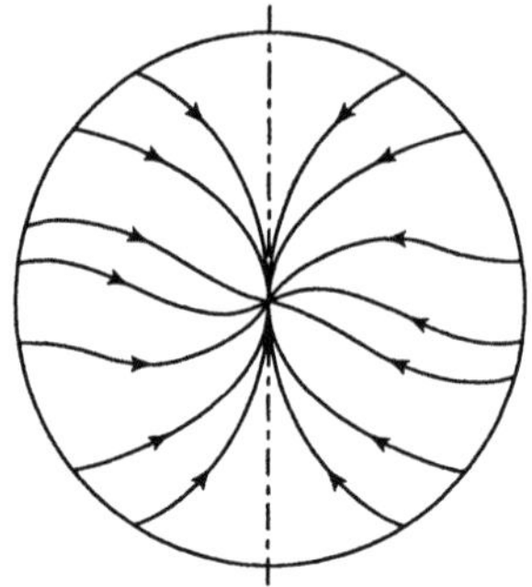

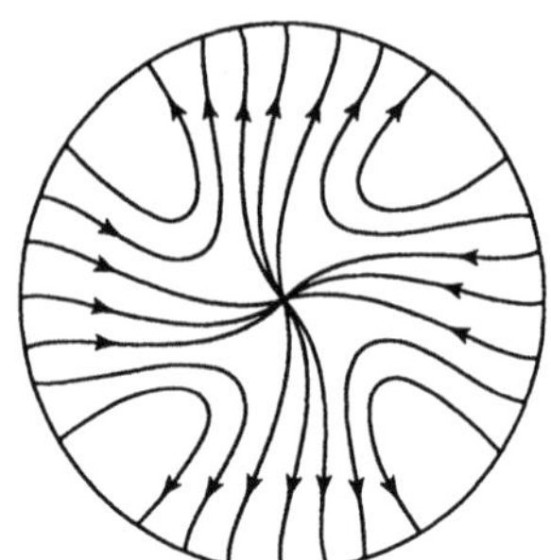

FIGURE 17.FIGURE 18.

Observation VIII-7-9. Figures 17 and 18 indicate that there are many orbits tend to the point $(0,0)$ in the direction $\theta = 0$ as $t \to +\infty$. This is the general situation, as the following example shows.

Fixing two positive numbers ϵ and ν, define two regions $\mathcal{D}_0$ and $\mathcal{D}_1$ in the $\vec{y}$-plane as follows (cf. Figure 19):

$$\begin{cases} \mathcal{D}_0 = \{\vec{y}: |y_2| < y_1^{1+\nu}, \ 0 < y_1\}, \\ \mathcal{D}_1 = \mathbb{R}^2 - \{\vec{y}: |y_2| \leq (1+\epsilon)y_1^{1+\nu}, \ 0 \leq y_1\}. \end{cases}$$

Choose three real numbers λ_1, λ_2, and λ_3 so that $\lambda_1 < 0$, $\lambda_2 > \lambda_1$, and

$$\text{(VIII.7.6)} \qquad\qquad \frac{\lambda_3}{\lambda_1} > 1 + \nu.$$

Let $\mu(\vec{y})$ be a real-valued function such that (a) μ is bounded on $\mathbb{R}^2$, (b) μ is continuously differentiable in $\vec{y}$ on $\mathbb{R}^2 - \{(0,0)\}$, (c) $\mu(\vec{y}) = \lambda_3$ on $\mathcal{D}_0$, and (d) $\mu(\vec{y}) = \lambda_2$ on $\mathcal{D}_1$. Set $\vec{g}(\vec{y}) = \begin{bmatrix} 0 \\ (\mu(\vec{y}) - \lambda_2)\, y_2 \end{bmatrix}$. Then, $\vec{g}(\vec{y}) = \vec{0}$ if $\vec{y} \in \mathcal{D}_1$. Furthermore, $0 \leq y_1$, $|y_2| \leq (1+\epsilon)y_1^{1+\nu}$ if $\vec{y} \notin \mathcal{D}_1$, and

$$|\vec{g}(\vec{y})| = |[\mu(\vec{y}) - \lambda_2]y_2| \leq K|y_2| \leq K(1+\epsilon)y_1^{1+\nu} \leq K(1+\epsilon)|\vec{y}|^{1+\nu},$$

where K is a suitable positive constant.

Consider the system

$$\text{(VIII.7.7)} \qquad \frac{dy_1}{dt} = \lambda_1 y_1, \qquad \frac{dy_2}{dt} = \lambda_2 y_2 + [\mu(\vec{y}) - \lambda_2]y_2.$$

If $\vec{y}(0) \in \mathcal{D}_0$, then $\dfrac{dy_2}{dy_1} = \left(\dfrac{\lambda_3}{\lambda_1}\right)\dfrac{y_2}{y_1}$. Hence, $y_2 = y_1^{\lambda_3/\lambda_1}$. Now, condition (VIII.7.6) implies that $\vec{y}(t) \in \mathcal{D}_0$ for $t \geq 0$, $\lim\limits_{t \to +\infty} r(t) = 0$, and $\lim\limits_{t \to +\infty} \theta(t) = 0$. (cf. Figure 20).

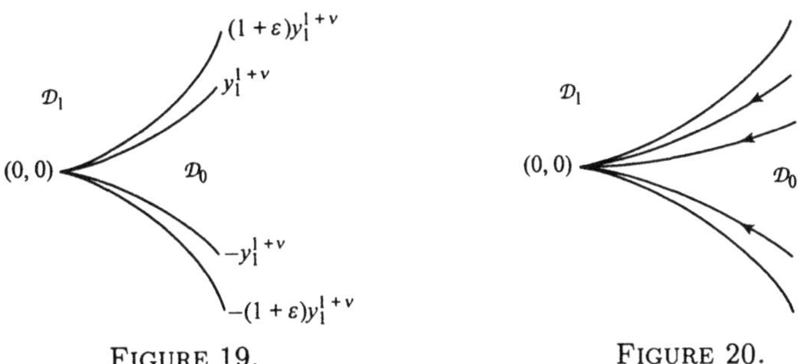

$$\text{FIGURE 19.} \qquad\qquad \text{FIGURE 20.}$$

Observation VIII-7-10. If we assume some condition on smoothness of $\vec{g}(\vec{y})$, there is only one orbit of (VIII.7.1) that tends to the point $(0,0)$ in the direction $\theta = 0$ as $t \to +\infty$.

Theorem VIII-7-11. *Assume that $a_{11} = \lambda_1$, $a_{22} = \lambda_2$, and $a_{12} = a_{21} = 0$, and that*

(1) $\lambda_2 > \lambda_1$ and $\lambda_1 < 0$,

(2) $\vec{g}(\vec{y})$ is continuous in a neighborhood of $\vec{0}$,

(3) $\lim\limits_{\vec{y} \to \vec{0}} \dfrac{\vec{g}(\vec{y})}{|\vec{y}|} = \vec{0}$,

(4) $\dfrac{\partial \vec{g}(\vec{y})}{\partial y_2}$ exists and is continuous in a neighborhood of $\vec{0}$.

Then, there exists a unique orbit $\vec{y}(t) = \begin{bmatrix} \phi_1(t) \\ \phi_2(t) \end{bmatrix}$ such that

$$\begin{cases} \phi_1(t) > 0 \quad \text{for} \quad t \geq 0, \\[2mm] \lim\limits_{t \to +\infty} \phi_1(t) = 0, \quad \lim\limits_{t \to +\infty} \phi_2(t) = 0, \quad \lim\limits_{t \to +\infty} \dfrac{\phi_2(t)}{\phi_1(t)} = 0. \end{cases}$$

Proof.

The existence of such orbit is seen in previous observations. To show the uniqueness of such orbit, rewrite (VIII.7.1) in the form

$$\text{(VIII.7.8)} \qquad \frac{dy_2}{dy_1} = \frac{\lambda_2 y_2 + g_2(\vec{y})}{\lambda_1 y_1 + g_1(\vec{y})} = \left(\frac{\lambda_2}{\lambda_1}\right)\frac{y_2}{y_1} + \frac{\dfrac{1}{\lambda_1 y_1}g_2(\vec{y}) - \dfrac{\lambda_2 y_2}{(\lambda_1 y_1)^2}g_1(\vec{y})}{1 + \dfrac{1}{\lambda_1 y_1}g_1(\vec{y})}.$$

Next, introduce a new unknown u by $y_2 = y_1 u$ $\left(\text{or } u = \dfrac{y_2}{y_1}\right)$. Then, (VIII.7.8) becomes

$$\text{(VIII.7.9)} \qquad y_1\frac{du}{dy_1} + u = \left(\frac{\lambda_2}{\lambda_1}\right)u + \frac{\dfrac{1}{\lambda_1 y_1}g_2(y_1\vec{u}) - \dfrac{\lambda_2 u}{\lambda_1^2 y_1}g_1(y_1\vec{u})}{1 + \dfrac{1}{\lambda_1 y_1}g_1(y_1\vec{u})},$$

where $\vec{u} = \begin{bmatrix} 1 \\ u \end{bmatrix}$. Observe that (a) $|\vec{u}| = 1$ if $|u| \leq 1$, (b) $\dfrac{\partial \vec{g}}{\partial y_2}(\vec{0}) = \lim\limits_{y_2 \to 0} \dfrac{1}{y_2}\vec{g}(0, y_2)$
$= \vec{0}$, and (c) $\dfrac{\lambda_2}{\lambda_1} - 1 < 0$.

Write system (VIII.7.9) in the form

$$\begin{cases} y_1 \dfrac{du}{dy_1} = \left(\dfrac{\lambda_2}{\lambda_1} - 1\right) u + G(y_1, u), \\[4mm] G(y_1, u) = \dfrac{\dfrac{1}{\lambda_1 y_1} g_2(y_1 \vec{u}) - \dfrac{\lambda_2 u}{\lambda_1^2 y_1} g_1(y_1 \vec{u})}{1 + \dfrac{1}{\lambda_1 y_1} g_1(y_1 \vec{u})}. \end{cases}$$

It is easy to show that $\lim\limits_{y_1 \to 0} \dfrac{\partial G}{\partial u}(y_1, u) = 0$ uniformly on $|u| \leq 1$. Hence, there exists a non-negative valued function $K(y_1)$ such that

$$|G(y_1, u) - G(y_1, v)| \leq K(y_1)|u - v|$$

whenever $|u| \leq 1$, $|v| \leq 1$ and $|y_1|$ is sufficiently small. Furthermore, $\lim\limits_{y_1 \to 0} K(y_1) = 0$.

Let $u = \psi_1(y_1)$ and $u = \psi_2(y_1)$ be two solutions of (VIII.7.9) such that
  (1) $\psi_1(y_1)$ and $\psi_2(y_1)$ are defined for $0 < y_1 < \eta$ for some sufficiently small positive number η,
  (2) $\lim\limits_{y_1 \to 0+} \psi_1(y_1) = 0$ and $\lim\limits_{y_1 \to 0+} \psi_2(y_1) = 0$.

Set $\psi(y_1) = \psi_1(y_1) - \psi_2(y_1)$. By virtue of uniqueness, assume without any loss of generality that $\psi(y_1) > 0$ for $0 < y_1 < \eta$. Then, for a sufficiently small positive η,
$y_1 \dfrac{d\psi(y_1)}{dy_1} < \gamma\psi(y_1)$ for $0 < y_1 < \eta$, where $\gamma = \dfrac{1}{2}\left[\dfrac{\lambda_2}{\lambda_1} - 1\right] < 0$. This implies that
$\dfrac{d}{dy_1}\left[y_1^{-\gamma}\psi(y_1)\right] < 0$ for $0 < y_1 < \eta$. Hence, $0 \leq y_1^{-\gamma}\psi(y_1) \leq \lim\limits_{y_1 \to 0} y_1^{-\gamma}\psi(y_1) = 0$ for $0 < y_1 < \eta$. Therefore, $\psi(y_1) = 0$ for $0 < y_1 < \eta$. $\square$
The materials of this section are also found in [CL, §§5 and 6 of Chapter 15].

VIII-8. Perturbations of a proper node

In this section, we consider a system of the form

$$\text{(VIII.8.1)} \qquad \dfrac{d\vec{y}}{dt} = \lambda\vec{y} + \vec{g}(\vec{y}), \qquad \text{where} \quad \vec{y} \in \mathbb{R}^2,$$

under the assumptions that
  (i) λ is a negative number,
  (ii) the entries of the $\mathbb{R}^2$-valued function $\vec{g}(\vec{y})$ are continuous in a neighborhood of $\vec{0}$,
  (iii) $|\vec{g}(\vec{y})| \leq K|\vec{y}|^{1+\nu}$ in a neighborhood of $\vec{0}$, where K is a non-negative number and ν is a positive number.

The main result is the following theorem.

Theorem VIII-8-1. *Under the assumption given above, the point $(0,0)$ is a stable proper node of system (VIII.8.1) as $t \to +\infty$.*

Before we prove this theorem, we illustrate the general situation by an example.

Example VIII-8-2. Look at the system

$$\text{(VIII.8.2)} \qquad \frac{d\vec{y}}{dt} = -\vec{y} + \vec{g}(\vec{y}), \qquad \vec{g}(\vec{y}) = (y_1^2 + y_2^2)\begin{bmatrix} -y_2 \\ y_1 \end{bmatrix},$$

where $\vec{y} \in \mathbb{R}^2$ with the entries y_1 and y_2. It is known that the point $(0,0)$ is a stable proper node of the linear system $\dfrac{d\vec{y}}{dt} = -\vec{y}$. To find the general solution of (VIII.8.2), change (VIII.8.2) to

$$\text{(VIII.8.3)} \qquad \frac{d\vec{z}}{dt} = e^{-2t}\vec{g}(\vec{z})$$

by the transformation $\vec{y} = e^{-t}\vec{z}$. Next, introduce a new independent variable s by $s = \dfrac{1}{2}e^{-2t}$. Then, system (VIII.8.3) becomes

$$\text{(VIII.8.4)} \qquad \frac{d\vec{z}}{ds} = -\vec{g}(\vec{z}).$$

Observe that

$$\frac{d\left(z_1^2 + z_2^2\right)}{ds} = 2\left(z_1\frac{dz_1}{ds} + z_2\frac{dz_2}{ds}\right) = 0.$$

If a solution $\vec{y}(t)$ of system (VIII.8.2) satisfies an initial condition

$$\text{(VIII.8.5)} \qquad \vec{y}(0) = \vec{\eta} = \begin{bmatrix} \eta_1 \\ \eta_2 \end{bmatrix},$$

the corresponding solution $\vec{z}(s) = e^t\vec{y}(t)$ of system (VIII.8.4) is the solution of the initial-value problem $\dfrac{d\vec{z}}{ds} = r(\vec{\eta})^2\begin{bmatrix} z_2 \\ -z_1 \end{bmatrix}$, $\vec{z}\left(\dfrac{1}{2}\right) = \vec{\eta}$, where

$$\text{(VIII.8.6)} \qquad r(\vec{\eta}) = \sqrt{(\eta_1)^2 + (\eta_2)^2}.$$

Hence,

$$z_1(s) = r(\vec{\eta})\sin(r(\vec{\eta})^2 s + \theta(\vec{\eta})), \qquad z_2(s) = r(\vec{\eta})\cos(r(\vec{\eta})^2 s + \theta(\vec{\eta})),$$

where

$$\begin{cases} z_1(0) = r(\vec{\eta})\sin\left(\theta(\vec{\eta})\right), \\ z_2(0) = r(\vec{\eta})\cos\left(\theta(\vec{\eta})\right), \end{cases} \qquad \begin{cases} \eta_1 = r(\vec{\eta})\sin\left(\dfrac{r(\vec{\eta})^2}{2} + \theta(\vec{\eta})\right), \\ \eta_2 = r(\vec{\eta})\cos\left(\dfrac{r(\vec{\eta})^2}{2} + \theta(\vec{\eta})\right) \end{cases}$$

(cf. Figure 21).

The general solution of (VIII.8.2) is

$$\begin{cases} y_1(t) = e^{-t} r(\vec{\eta}) \sin\left(\dfrac{r(\vec{\eta})^2}{2} e^{-2t} + \theta(\vec{\eta}) \right), \\[2mm] y_2(t) = e^{-t} r(\vec{\eta}) \cos\left(\dfrac{r(\vec{\eta})^2}{2} e^{-2t} + \theta(\vec{\eta}) \right), \end{cases}$$

where $\vec{y}(0)$ and $r(\vec{\eta})$ are given by (VIII.8.5) and (VIII.8.6), respectively. Every orbit of (VIII.8.2) goes around the point $(0,0)$ only a finite number of times. Figure 22 shows that the point $(0,0)$ is a stable proper node as $t \to +\infty$.

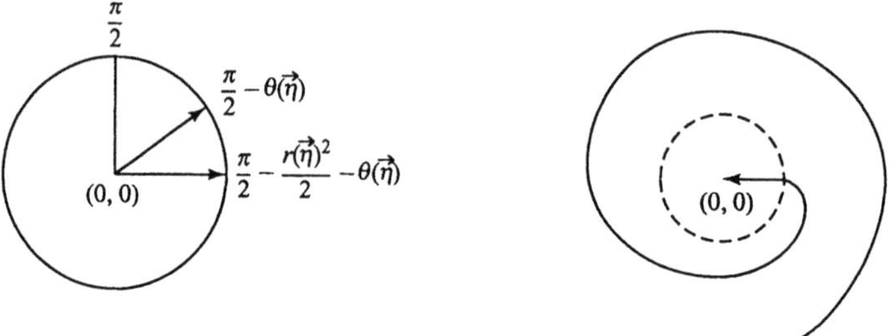

FIGURE 21.   FIGURE 22.

Remark VIII-8-3. If condition (iii) of the assumption given above is relaxed, the point $(0,0)$ may become a stable spiral point (cf. Exercise VIII-10).

Proof of Theorem VIII-8-1.

We prove Theorem VIII-8-1 in two steps. To start with, using the polar coordinates, write system (VIII.8.1) in the form

$$\text{(VIII.8.7)} \qquad \begin{cases} \dfrac{dr}{dt} = \lambda r + g_1(\vec{y})\cos(\theta) + g_2(\vec{y})\sin(\theta), \\[2mm] r\dfrac{d\theta}{dt} = -g_1(\vec{y})\sin(\theta) + g_2(\vec{y})\cos(\theta) \end{cases}$$

(cf. (VIII.7.4)). By virtue of the assumption given above, there exists a positive number r_0 such that

$$\text{(VIII.8.8)} \qquad 2\lambda r < \lambda r + g_1(\vec{y})\cos(\theta) + g_2(\vec{y})\sin(\theta) < \frac{\lambda\,r}{2} < 0$$

for $0 < r \leq r_0$.

Step 1. We show first that the point $(0,0)$ is a stable node. To do this, let $(r(t), \theta(t))$ be a solution of (VIII.8.7) satisfying the initial condition

$$\text{(VIII.8.9)} \qquad r(0) = \rho, \quad \theta(0) = \omega,$$

where ρ is a positive number and ω is a real number. This solution exists for $t \geq 0$. Furthermore, estimate (VIII.8.8) implies that the function $r(t)$ satisfies the following two conditions: (a) $r(t)$ is strictly decreasing as $t \to +\infty$ and (b) $\lim_{t \to +\infty} r(t) = 0$. Therefore, we obtain (c) $0 < r(t) \leq \rho$ for $t \geq 0$.

Since $r(t)$ is strictly decreasing for $t \geq 0$, the variable t can be regarded as a function of r, i.e., $t = t(r)$, where $t(\rho) = 0$ and $\lim_{r \to 0} t(r) = +\infty$. Use r as the independent variable. Then, (VIII.8.7) becomes

$$(\text{VIII.8.10}) \qquad \frac{d\theta}{dr} = F(r, \theta) = \frac{-g_1(\vec{y}) \sin(\theta) + g_2(\vec{y}) \cos(\theta)}{r[\lambda r + g_1(\vec{y}) \cos(\theta) + g_2(\vec{y}) \sin(\theta)]}.$$

Let $\theta = \Phi(r) = \theta(t(r))\, (0 < r \leq r_0)$ be the solution of (VIII.8.10) satisfying the initial condition $\Phi(\rho) = \omega$. Then,

$$\Phi(r) = \omega + \int_\rho^r F(s, \Phi(s))ds \qquad (0 < r \leq r_0).$$

Estimate (VIII.8.8) implies that

$$-\frac{1}{2\lambda r^2}|-g_1(\vec{y})\sin(\theta) + g_2(\vec{y})\cos(\theta)| \leq F(r, \theta) \leq -\frac{2}{\lambda r^2}\left(|g_1(\vec{y})| + |g_2(\vec{y})|\right)$$

for $0 < r \leq r_0$. Now, *by virtue of condition (iii)*, the improper integral $\int_\rho^0 F(s, \Phi(s))ds = \lim_{r \to 0} \int_\rho^r F(s, \Phi(s))ds$ exists. Observe that $\theta(t) = \Phi(r(t)) = \omega + \int_\rho^{r(t)} F(s, \Phi(s))ds$ and that $\lim_{t \to +\infty} r(t) = 0$. Therefore, $\lim_{t \to +\infty} \theta(t) = \omega + \int_\rho^0 F(s, \Phi(s))ds$. Thus, we conclude that the point $(0, 0)$ is a stable *node* of (VIII.8.1) as $t \to +\infty$.

Step 2. In this step, we prove that the point $(0, 0)$ is a proper node. To do this, for a given real number c, find a solution $(r(t), \theta(t))$ of (VIII.8.7) such that $\lim_{t \to +\infty} \theta(t) = c$ and $\lim_{t \to +\infty} r(t) = 0$. Selecting a sequence $\{\rho_m : m = 1, 2, \dots\}$ of positive numbers such that $0 < \rho_m \leq r_0$ and $\lim_{m \to +\infty} \rho_m = 0$, define a sequence of functions $\{\Phi_m(r) : m = 1, 2, \dots\}$ by the initial-value problems

$$\frac{d\theta}{dr} = F(r, \theta), \qquad \theta(\rho_m) = c \qquad (m = 1, 2, \dots),$$

respectively. Those functions are defined for $0 < r \leq r_0$, and

$$\Phi_m(r) = c + \int_{\rho_m}^r F(s, \Phi_m(s))ds \qquad (m = 1, 2, \dots).$$

Set

$$c_m = \lim_{t \to +\infty} \Phi_m(r) = c + \int_{\rho_m}^0 F(s, \Phi_m(s))ds.$$

Then, $\Phi_m(0) = c_m$, $\displaystyle\lim_{m\to+\infty} c_m = c$, and

$$\Phi_m(r) \;=\; c_m \;+\; \int_0^r F(s, \Phi_m(s))ds \qquad (m = 1, 2, \dots)$$

for $0 \le r \le r_0$. Since the sequence $\{\Phi_m(r) : m = 1, 2, \dots\}$ is bounded and equicontinuous on the interval $0 \le r \le r_0$, assume without any loss of generality that $\displaystyle\lim_{m\to+\infty} \Phi_m(r) = \Phi(r)$ exists uniformly on the interval $0 \le r \le r_0$ (otherwise choose a subsequence). It can be shown easily that

$$\Phi(r) \;=\; c \;+\; \int_0^r F(s, \Phi(s))ds \qquad (0 \le r \le r_0).$$

Therefore, $\Phi(r)$ is a solution of (VIII.8.10) such that $\displaystyle\lim_{r\to 0} \Phi(r) = c$.

Define $r(t)$ by the initial-value problem

$$\frac{dr}{dt} \;=\; \lambda r \;+\; g_1(\vec{y}(r))\cos(\Phi(r)) \;+\; g_2(\vec{y}(r))\sin(\Phi(r)), \qquad r(0) \;=\; r_0,$$

where $\vec{y}(r) = r \begin{bmatrix} \cos(\Phi(r)) \\ \sin(\Phi(r)) \end{bmatrix}$. The function $r(t)$ is defined for $t \ge 0$, is decreasing, and tends to 0 as $t \to +\infty$. Set $\theta(t) = \Phi(r(t))$. Then, it is easily shown that $(r(t), \theta(t))$ is a solution of (VIII.8.7) such that $\displaystyle\lim_{t\to+\infty} \theta(t) = \lim_{r\to 0} \Phi(r) = c$. $\square$

The materials of this section are also found in [CL, §3 of Chapter 15].

VIII-9. Perturbation of a spiral point

In this section, we show that $(0,0)$ is a stable spiral point of (VIII.7.1) as $t \to +\infty$ if $(0,0)$ is a stable spiral point of the linear system $\dfrac{d\vec{y}}{dt} = A\vec{y}$. The main result is the following theorem.

Theorem VIII-9-1. *If*
  (i) two eigenvalues of A are not real and their real parts are negative,
  (ii) the entries of the $\mathbb{R}^2$-valued function $\vec{g}(\vec{y})$ is continuous in a neighborhood of the point $(0,0)$,
  (iii) $\displaystyle\lim_{\vec{y}\to\vec{0}} \frac{\vec{g}(\vec{y})}{|\vec{y}|} = \vec{0}$,
then the point $(0,0)$ is a stable spiral point of (VIII.7.1) as $t \to +\infty$.

Proof.

Assume that

(VIII.9.1) $a_{11} = a_{22} = a < 0$ and $a_{12} = -a_{21} = b \ne 0$.

Then, from system (VIII.7.4), it follows that

$$
\begin{cases}
\dfrac{dr}{dt} = ar + g_1(\vec{y})\cos(\theta) + g_2(\vec{y})\sin(\theta), \\[2mm]
r\dfrac{d\theta}{dt} = -br - g_1(\vec{y})\sin(\theta) + g_2(\vec{y})\cos(\theta).
\end{cases}
$$

Therefore, a positive number r_0 can be fouind so that $r(t) \le r(0)\exp\left[\dfrac{at}{2}\right]$ for $t \ge 0$ if $0 \le r(0) \le r_0$. This, in turn, implies that $\lim\limits_{t\to+\infty} r(t) = 0$ and that $\dfrac{d\theta}{dt} < -\dfrac{b}{2}$ if $b > 0$, while $\dfrac{d\theta}{dt} > -\dfrac{b}{2}$ if $b < 0$, whenever $0 \le r(0) \le r_0$. Therefore, $\lim\limits_{t\to+\infty}\theta(t) = -\infty$ if $b > 0$ and $\lim\limits_{t\to+\infty}\theta(t) = +\infty$ if $b < 0$, whenever $0 \le r(0) \le r_0$. Thus, we conclude that $(0,0)$ is a stable spiral point of system (VIII.7.1) as $t \to +\infty$. $\square$

The materials of this section are also found in [CL, §3 of Chapter 15].

VIII-10. Perturbation of a center

We still consider a system (VIII.7.1) under the assumption that the entries of the $\mathbb{R}^2$-valued function $\vec{g}(\vec{y})$ is continuous in a neighborhood of $\vec{0}$ and satisfies condition (VIII.7.2). In this section, we show that if the 2×2 matrix A has two purely imaginary eigenvalues $\lambda_1 = ib$ and $\lambda_2 = -ib$, where b is a nonzero real number, then the point $(0,0)$ is either a center or a spiral point of (VIII.7.1). Assume without any loss of generality that the matrix A has the form $\begin{bmatrix} 0 & b \\ -b & 0 \end{bmatrix}$ and system (VIII.7.4) becomes

(VIII.10.1)
$$
\begin{cases}
\dfrac{dr}{dt} = g_1(\vec{y})\cos(\theta) + g_2(\vec{y})\sin(\theta), \\[2mm]
\dfrac{d\theta}{dt} = -b + \dfrac{1}{r}(-g_1(\vec{y})\sin(\theta) + g_2(\vec{y})\cos(\theta)).
\end{cases}
$$

Observe that system (VIII.10.1) can be written in the form

(VIII.10.2)
$$
\frac{dr}{d\theta} = \frac{g_1(\vec{y})\cos(\theta) + g_2(\vec{y})\sin(\theta)}{-b + \dfrac{1}{r}(-g_1(\vec{y})\sin(\theta) + g_2(\vec{y})\cos(\theta))}.
$$

If a positive number ρ is sufficiently small, the solution $r(\theta,\rho)$ of (VIII.10.2) satisfying the initial condition $r(0,\rho) = \rho$ exists for $0 \le \theta \le 2\pi$.

Case 1. If there exists a sequence $\{\rho_m : m = 1, 2, \ldots\}$ of positive numbers such that

$$
\lim_{m\to+\infty}\rho_m = 0 \quad \text{and} \quad r(2\pi, \rho_m) = \rho_m \quad (m = 1, 2, \ldots),
$$

then the point $(0,0)$ is a center of (VIII.7.1) in the sense of Bendixson [Ben2, p. 26] (cf. Figure 23).

Case 2. Assume that $b < 0$. If there exists a positive number ρ_0 such that

$$r(2\pi, \rho) > \rho \qquad (\text{respectively} \ < \rho)$$

whenever $0 < \rho \leq \rho_0$, then the point $(0,0)$ is an unstable (respectively stable) spiral point as $t \to +\infty$ (cf. Figures 24-1 and 24-2).

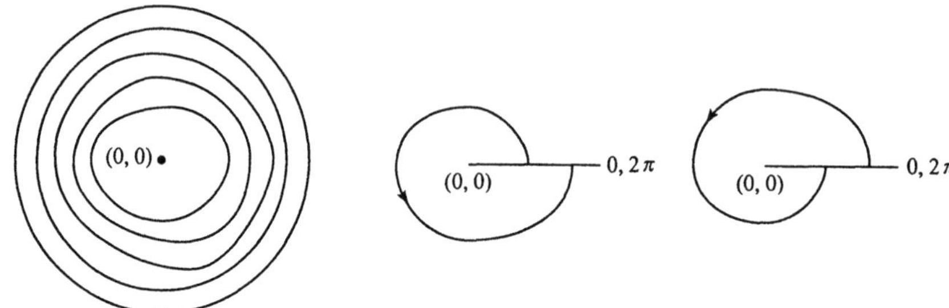

FIGURE 23.   FIGURE 24-1.   FIGURE 24-2.

Example VIII-10-1. The point $(0,0)$ is a center of the system

$$\frac{d}{dt}\begin{bmatrix} y_1 \\ y_2 \end{bmatrix} = \begin{bmatrix} y_2 \\ -y_1 \end{bmatrix} + \sqrt{y_1^2 + y_2^2}\begin{bmatrix} y_2 \\ -y_1 \end{bmatrix}$$

since $\dfrac{dr}{d\theta} = 0$.

Example VIII-10-2. The point $(0,0)$ is an unstable spiral point of the system

$$\frac{d}{dt}\begin{bmatrix} y_1 \\ y_2 \end{bmatrix} = \begin{bmatrix} y_2 \\ -y_1 \end{bmatrix} + \sqrt{y_1^2 + y_2^2}\begin{bmatrix} y_1 \\ y_2 \end{bmatrix}$$

as $t \to -\infty$, since $\dfrac{dr}{dt} = r^2$, which has the solution $r = \dfrac{1}{r_0 - t}$, $r(0) = \dfrac{1}{r_0}$ and $r \to +\infty$ when $t \to r_0$. (Note: $t < r_0$.)

Example VIII-10-3. For the system

$$\frac{dx}{dt} = -y, \qquad \frac{dy}{dt} = x + x^2 - xy + \alpha y^2,$$

the point $(0,0)$ is (a) a stable spiral point if $\alpha < -1$, (b) a center if $\alpha = -1$, and (c) an unstable spiral point if $\alpha > -1$.

Proof.

Use the polar coordinates (r, θ) for (x, y) to write the given system in the form

$$(1 + r\cos\theta(\cos^2\theta - \cos\theta\sin\theta + \alpha\sin^2\theta))\frac{dr}{d\theta} = r^2\sin\theta(\cos^2\theta - \cos\theta\sin\theta + \alpha\sin^2\theta).$$

Setting

$$r(\theta, c) \; = \; \sum_{m=1}^{+\infty} r_m(\theta)c^m, \quad \text{where } \; r(0,c) \; = \; c,$$

and comparing $r(2\pi, c)$ and c for sufficiently small positive c, it can be shown that

$$r_1(t) \; = \; 1, \qquad r_2(2\pi) \; = \; 0, \qquad r_3(2\pi) \; = \; \frac{(1+\alpha)\pi}{4}.$$

Thus, $r(2\pi, c) < c$ if $\alpha < -1$. Therefore, (a) follows. Similarly, $r(2\pi, c) > c$ if $\alpha > -1$. Therefore, (c) follows.

In the case when $\alpha = -1$, $x^2 - xy - y^2 = \left(x - \frac{(1+\sqrt{5})y}{2}\right)\left(x - \frac{(1-\sqrt{5})y}{2}\right)$. Set $\omega = -\frac{(1+\sqrt{5})}{2}$. Then, $\frac{1}{\omega} = \frac{(1-\sqrt{5})}{2}$. Therefore, changing x and y by $u = x - \frac{(1+\sqrt{5})y}{2}$ and $v = x - \frac{(1-\sqrt{5})y}{2}$, the given system is changed to

$$\frac{du}{dt} \; = \; \omega v(1 + u), \qquad \frac{dv}{dt} \; = \; -\frac{1}{\omega}u(1 + v).$$

Hence, on any solution curve, the function

$$\omega^2(v \; - \; \ln(1 + v)) \; + \; (u \; - \; \ln(1 + v)) \; = \; \frac{u^2 \; + \; \omega^2 v^2}{2} \; + \; \cdots$$

is independent of t. This implies that in the neighborhood of $(0,0)$, orbits are closed curves. This shows that $(0,0)$ is a center. $\quad\square$

Example VIII-10-4. The point $(0,0)$ is a center of the system

$$\frac{d\vec{y}}{dt} \; = \; bA\,\vec{y} \; + \; \vec{g}(\vec{y}), \qquad \vec{y} \; = \; \begin{bmatrix} y_1 \\ y_2 \end{bmatrix}$$

if (1) b is a nonzero real number, (2) $A = \begin{bmatrix} 0 & 1 \\ -1 & 0 \end{bmatrix}$, (3) the entries of the $\mathbb{R}^2$-valued function $\vec{g}(\vec{y})$ is continuous and continuously differentiable in a neighborhood of $\vec{0}$, (4) $\lim_{\vec{y}\to\vec{0}} \dfrac{\vec{g}(\vec{y})}{|\vec{y}|} = \vec{0}$, and (5) there exists a function $M(\vec{y})$ such that M is positive valued, continuous, and continuously differentiable in a neighborhood of $\vec{0}$ and that $\dfrac{\partial(Mf_1)}{\partial y_1}(\vec{y}) + \dfrac{\partial(Mf_2)}{\partial y_2}(\vec{y}) = 0$ in a neighborhood of $\vec{0}$, where $f_1(\vec{y})$ and $f_2(\vec{y})$ are the entries of the vector $bA\vec{y} + \vec{g}(\vec{y})$.

Proof.

The point $(0,0)$ is either a center or a spiral point. Look at the system

$$\frac{d\vec{y}}{dt} \; = \; bA\,\vec{y} \; + \; \vec{g}(\vec{y}) \qquad \text{and} \qquad \frac{ds}{dt} \; = \; \frac{1}{M(\vec{y})}.$$

Using s as the independent variable, change the given system to $\dfrac{d\vec{y}}{ds} = M(\vec{y})(bA\vec{y} + \vec{g}(\vec{y}))$. Upon applying Theorem VIII-1-7, we conclude that $(0,0)$ is not asymptotically stable as $t \to \pm\infty$. Hence, $(0,0)$ is a center. $\square$

The materials of this section are also found in [CL, §4 of Chapter 15] and [Sai, §21 of Chapter 3, pp. 89-100].

EXERCISES VIII

VIII-1. For each of the following five matrices A, find a phase portrait of orbits of the system $\dfrac{d\vec{x}}{dt} = A\vec{x}$:

$$A = \begin{bmatrix} 1 & 5 \\ 5 & 1 \end{bmatrix}; \quad \begin{bmatrix} 1 & 1 \\ -4 & 1 \end{bmatrix}; \quad \begin{bmatrix} 1 & 1 \\ -1 & 3 \end{bmatrix}; \quad \begin{bmatrix} 1 & 1 \\ -5 & -1 \end{bmatrix}; \quad \begin{bmatrix} 5 & 1 \\ 1 & 5 \end{bmatrix}.$$

VIII-2. Find all nontrivial solutions $(x,y) = (\phi(t), \psi(t))$, if any, of the system

$$\frac{dx}{dt} = -xy^2, \qquad \frac{dy}{dt} = -x^4 y(1+y)$$

satisfying the condition $\lim\limits_{t \to +\infty} (\phi(t), \psi(t)) = (0,0)$.

VIII-3. Let $f(x,y)$ and $g(x,y)$ be continuously differentiable functions of (x,y) such that
  (a) $f(0,0) = 0$ and $g(0,0) = 0$,
  (b) $(f(x,y), g(x,y)) \neq (0,0)$ if $(x,y) \neq (0,0)$,
  (c) f and g are homogeneous of degree m in (x,y), i.e., $f(rx, ry) = r^m f(x,y)$ and $g(rx, ry) = r^m g(x,y)$. Let (r, θ) be polar coordinates of (x,y), i.e., $x = r\cos\theta$ and $y = r\sin\theta$. Set $F(\theta) = f(\cos\theta, \sin\theta)$ and $G(\theta) = g(\cos\theta, \sin\theta)$.
  (I) Show that

$$\text{(S)} \qquad \begin{cases} \dfrac{dr}{dt} = r^m(F(\theta)\cos\theta + G(\theta)\sin\theta), \\[2mm] \dfrac{d\theta}{dt} = r^{m-1}(-F(\theta)\sin\theta + G(\theta)\cos\theta). \end{cases}$$

  (II) Using system (S), discuss the stability property of the trivial solution of each of the following three systems:

  (i) $$\frac{dx}{dt} = x^2 - y^2, \qquad \frac{dy}{dt} = 2xy;$$

  (ii) $$\frac{dx}{dt} = x^3(x^2 + y^2) - 2x(x^2 + y^2)^2, \qquad \frac{dy}{dt} = -y^3(x^2 + y^2);$$

  (iii) $$\frac{dx}{dt} = x^4 - 6x^2 y^2 + y^4, \qquad \frac{dy}{dt} = 4x^3 y - 4xy^3.$$

VIII-4. Let J be the $2n \times 2n$ matrix defined by (IV.5.2) and let H be a real constant $2n \times 2n$ symmetric matrix. Show that the trivial solution of the Hamiltonian system $\dfrac{d\vec{y}}{dt} = JH\vec{y}$ is not asymptotically stable as $t \to +\infty$.

VIII-5. Show that the trivial solution of the system

$$\frac{d\vec{y}}{dt} = A\vec{y} + \vec{g}(t,\vec{y})$$

is asymptotically stable as $t \to +\infty$ if the following conditions are satisfied:
  (i) A is a real constant $n \times n$ matrix,
  (ii) the real part of every eigenvalue of A is negative,
  (iii) the entries of the $\mathbb{R}^n$-valued function $\vec{g}(t,\vec{y})$ are continuous in the region

$$\triangle(r_0) = \mathcal{I}_0 \times \mathcal{D}(r_0) = \{(t,\vec{y}) : 0 \le t < +\infty, \; |\vec{y}| < r_0\} \text{ for some } r_0,$$

  (iv) $\vec{g}(t,\vec{y})$ satisfies the estimate

$$|\,\vec{g}(t,\vec{y})\,| \le \epsilon_0(t,\vec{y})|\vec{y}| \qquad \text{for} \quad (t,\vec{y}) \in \triangle(r_0),$$

where $\epsilon_0(t,\vec{y})$ is continuous, and positive, and $\displaystyle\lim_{t^{-1}+|\vec{y}| \to 0} \epsilon_0(t,\vec{y}) = 0$ in $\triangle(r_0)$.

VIII-6. Show that the point $(0,0)$ is a stable improper node of the system

$$\frac{d\vec{y}}{dt} = \lambda\,\vec{y} + N\vec{y} + \vec{g}(\vec{y}), \qquad \vec{y} = \begin{bmatrix} y_1 \\ y_2 \end{bmatrix},$$

where
  (1) λ is a negative number,
  (2) N is a real constant nilpotent 2×2 matrix and $N \ne O$,
  (3) the $\mathbb{R}^2$-valued function $\vec{g}(\vec{y})$ is continuous in a neighborhood of $\vec{0}$,
  (4) $|\vec{g}(\vec{y})| \le c|\vec{y}|^{1+\nu}$ for some positive numbers c and ν in a neighborhood of $\vec{0}$.

Hint. Suppose $N = \begin{bmatrix} 0 & -\dfrac{\epsilon\lambda}{2} \\ 0 & 0 \end{bmatrix}$. If we set $y_1 = r\cos\theta$ and $y_2 = r\sin\theta$, we obtain

$$\frac{dr}{dt} = r\left[\lambda - \frac{\epsilon\lambda\sin\theta\cos\theta}{2} + \frac{g_1(\vec{y})\cos\theta + g_2(\vec{y})\sin\theta}{r}\right],$$

where $\vec{g}(\vec{y}) = \begin{bmatrix} g_1(\vec{y}) \\ g_2(\vec{y}) \end{bmatrix}$. Hence, if $r(0)$ is small, $r(t)$ is bounded by $r(0)$ and $\displaystyle\lim_{t \to +\infty} r(t) = 0$ for $t \ge 0$. This implies that if $|\vec{y}(0)|$ is small, $\vec{y}(t)$ is bounded and $\displaystyle\lim_{t \to +\infty} \vec{y}(t) = \vec{0}$. Look at

$$\vec{y}(t) = e^{\lambda t}e^{tN}\left[\vec{y}(0) + \int_0^t e^{-\lambda s}e^{-sN}\vec{g}(\vec{y}(s))ds\right].$$

If we set $\vec{y}(t) = e^{\lambda t} e^{tN} \vec{u}$, we obtain

$$\vec{u}(t) \;=\; \vec{y}(0) \;+\; \int_0^t e^{-\lambda s} e^{-sN} \vec{g}(\vec{y}(s)) ds.$$

Choose $\epsilon > 0$ so that $1 - \epsilon > \dfrac{1}{1+\nu}$. Then, condition (4) implies that

$$\vec{u}(+\infty) \;=\; \vec{y}(0) \;+\; \int_0^{+\infty} e^{-\lambda s} e^{-sN} \vec{g}(\vec{y}(s)) ds$$

exists.

Now,

(1) if $\vec{u}(+\infty) = \vec{0}$, then $\vec{u}(t) = \vec{0}$ identically, since, in this case,

$$\vec{u}(t) \;=\; \int_{+\infty}^t e^{-\lambda s} e^{-sN} \vec{g}(\vec{y}(s)) ds;$$

(2) if $\vec{u}(+\infty) = \begin{bmatrix} \xi \\ 0 \end{bmatrix} \neq \vec{0}$, then $\lim\limits_{t \to +\infty} e^{-\lambda t} \vec{y}(t) = \vec{u}(+\infty)$;

(3) if $\vec{u}(+\infty) = \begin{bmatrix} \xi_1 \\ \xi_2 \end{bmatrix}$ with $\xi_2 \neq 0$, then $\lim\limits_{t \to +\infty} \left(\dfrac{e^{-\lambda t}}{t} \right) \vec{y}(t) = \begin{bmatrix} -\dfrac{\epsilon \lambda \xi_2}{2} \\ 0 \end{bmatrix}$.

Hence, the point $(0,0)$ is a stable improper node of the given system.

VIII-7. Determine whether the point $(0,0)$ is a center or a spiral point of the system

$$\frac{dx}{dt} = -y, \qquad \frac{dy}{dt} = 2x + x^3 - x^2(2-x)y.$$

VIII-8. Show that the point $(0,0)$ is the center of the system

$$\frac{dx}{dt} = y + xy^3 - y^7, \qquad \frac{dy}{dt} = -x + xy^2 - y^6.$$

Hint. This system does not change even if (t,y) is replaced by $(-t,-y)$.

VIII-9. For the system

$$\frac{dx}{dt} = 3y + 5x(x^2 + y^2), \qquad \frac{dy}{dt} = 27x + 5y(x^2 + y^2),$$

find an approximation for the orbit(s) approaching $(0,0)$ as $t \to +\infty$.

VIII-10. Show that the point $(0,0)$ is a stable spiral point of the system

$$\frac{d}{dt} \begin{bmatrix} y_1 \\ y_2 \end{bmatrix} = -\begin{bmatrix} y_1 \\ y_2 \end{bmatrix} + \frac{1}{\ln \sqrt{y_1^2 + y_2^2}} \begin{bmatrix} -y_2 \\ y_1 \end{bmatrix}$$

as $t \to +\infty$.

Hint. If we set $y_1 = r\cos\theta$ and $y_2 = r\sin\theta$, the given system becomes

$$\frac{dr}{dt} = -r, \qquad \frac{d\theta}{dt} = \frac{1}{\ln r}.$$

VIII-11. Suppose that a solution $\vec{\phi}(t)$ of a system

$$\frac{d\vec{y}}{dt} = A\vec{y} + \vec{g}(\vec{y})$$

satisfies the condition

$$\lim_{t\to\tau} \vec{\phi}(t) = \vec{0}$$

for a positive number τ. Show that if
  (1) A is an $n \times n$ constant matrix,
  (2) the n-dimensional vector $\vec{g}(\vec{y})$ is continuous in a neighborhood of $\vec{0}$,
  (3) $\displaystyle\lim_{\vec{y}\to\vec{0}} \frac{\vec{g}(\vec{y})}{|\vec{y}|} = \vec{0}$,
then $\vec{\phi}(t) = \vec{0}$ for all values of t.
(Note that the uniqueness of solutions of initial-value problems is not assumed.)

VIII-12. Assume that
  (i) $\lambda_1, \ldots, \lambda_n$ are complex numbers that are in the interior of a half-plane in the complex λ-plane whose boundary contains $\lambda = 0$,
  (ii) there are no relations $\lambda_j = p_1\lambda_1 + p_2\lambda_2 + \cdots + p_n\lambda_n$ for $j = 1,\ldots,n$ and non-negative integers $p_1,\ldots,p_n$ such that $p_1 + \cdots + p_n \geq 2$,
  (iii) $\vec{f}(\vec{y}) = \displaystyle\sum_{|\rho|\geq 2} \vec{y}^{\rho} \vec{f}_{\rho}$ is a convergent power series in $\vec{y} \in \mathbb{C}^n$ with coefficients
  $\vec{f}_{\rho} \in \mathbb{C}^n$.
Show that there exists a convergent power series $\vec{Q}(\vec{u}) = \displaystyle\sum_{|\rho|\geq 2} \vec{u}^{\rho}\vec{Q}_{\rho}$ with coefficients
$\vec{Q}_{\rho} \in \mathbb{C}^n$ such that the transformation $\vec{y} = \vec{u} + \vec{Q}(\vec{u})$ changes the system $\dfrac{d\vec{y}}{dt} =$
$A\vec{y} + \vec{f}(\vec{y})$ to $\dfrac{d\vec{u}}{dt} = \Lambda\vec{u}$, where $\Lambda = \mathrm{diag}[\lambda_1, \lambda_2, \lambda_3, \ldots, \lambda_n]$.

VIII-13. Show that the trivial solution of the system

$$\frac{dx_1}{dt} = -x_1(x_1^2 + x_2^2) + x_2 e^{x_1+x_2}, \qquad \frac{dx_2}{dt} = -x_2(x_1^2 + x_2^2) - x_1 e^{x_1+x_2}$$

is aymptotically stable as $t \to +\infty$. Sketch a phase portrait of the orbits of this system.

VIII-14. In Observation VIII-6-2, it is stated that if one of the ω_m is not real, then $\vec{y}$ is a spiral point. Verify this statement.

VIII-15. Discuss the stability of the trivial solution of the system

$$\frac{dx_1}{dt} = x_2, \qquad \frac{dx_2}{dt} = -x_1^3(1 - x_1^2)$$

as $t \to +\infty$.

VIII-16. Find the general solution of the system

$$\begin{cases} \dfrac{dx_1}{dt} = x_1, \\[2mm] \dfrac{dx_2}{dt} = x_1 + x_2, \\[2mm] \dfrac{dx_3}{dt} = 2x_3 + x_1^2 + x_1 x_2 + x_2^2, \\[2mm] \dfrac{dx_4}{dt} = 3x_4 + x_1^3 + x_1 x_2^2 + x_1^2 x_2 + x_1 x_3 + x_2 x_3, \end{cases}$$

explicitly.

CHAPTER IX

AUTONOMOUS SYSTEMS

In this chapter, we explain the behavior of solutions of an autonomous system $\dfrac{d\vec{y}}{dt} = \vec{f}(\vec{y})$. We look at solution curves in the $\vec{y}$-space rather than the $(t, \vec{y})$-space. Such curves are called orbits of the given system. In general, an orbit does not tend to a limit point as $t \to +\infty$. However, a bounded orbit accumulates to a set as $t \to +\infty$. Such a set is called a limit-invariant set. In §IX-1, we explain the basic properties of limit-invariant sets. In §IX-2, using the Liapounoff functions, we explain how to locate limit-invariant sets. The main tool is a theorem due to J. LaSalle and S. Lefschetz [LaL, Chapter 2, §13, pp. 56-71] (cf. Theorem IX-2-1). The topic of §IX-4 is the Poincaré-Bendixson theorem which characterizes limit-sets in the plane (cf. Theorem IX-4-1). In §IX-3, orbital stability and orbitally asymptotic stability are explained. In §IX-5, we explain how to use the indices of the Jordan curves to the study of autonomous systems in the plane. Most of topics discussed in this chapter are also in [CL, Chapters 13 and 16], [Har2, Chapter 7], and [SC, Chapter 4, pp. 159-171].

IX-1. Limit-invariant sets

In this chapter, we explain behavior of solutions of a system of differential equations of the form

$$(\text{IX.1.1}) \qquad \frac{d\vec{y}}{dt} = \vec{f}(\vec{y}),$$

where $\vec{y} \in \mathbb{R}^n$ and the entries of the $\mathbb{R}^n$-valued function $\vec{f}(\vec{y})$ are continuous in the entire $\vec{y}$-space $\mathbb{R}^n$. We also assume that every initial-value problem

$$(\text{IX.1.2}) \qquad \frac{d\vec{y}}{dt} = \vec{f}(\vec{y}), \qquad \vec{y}(0) = \vec{\eta}$$

has a unique solution $\vec{y} = \vec{p}(t, \vec{\eta})$. System (IX.1.1) is called an *autonomous system* since the right-hand side $\vec{f}(\vec{y})$ does not depend on the independent variable t.

Observe that $\vec{p}(t + \tau, \vec{\eta})$ is also a solution of (IX.1.1) for every real number τ. Furthermore, $\vec{p}(t + \tau, \vec{\eta}) = \vec{p}(\tau, \vec{\eta})$ at $t = 0$. Hence, uniqueness of the solution of initial-value problem (IX.1.2) implies that $\vec{p}(t + \tau, \vec{\eta}) = \vec{p}(t, \vec{p}(\tau, \vec{\eta}))$ whenever both sides are defined. For each $\vec{\eta}$, let $\mathcal{I}(\vec{\eta})$ be the maximal t-interval on which the solution $\vec{p}(t, \vec{\eta})$ is defined. Set $C(\vec{\eta}) = \{\vec{y} = \vec{p}(t, \vec{\eta}) : t \in \mathcal{I}(\vec{\eta})\}$. The curve $C(\vec{\eta})$ is called the *orbit* passing through the point $\vec{\eta}$. Two orbits $C(\vec{\eta}_1)$ and $C(\vec{\eta}_2)$ do not intersect unless they are identical as a curve. In fact,

$$(\text{IX.1.3}) \qquad C(\vec{\eta}_1) = C(\vec{\eta}_2) \qquad \text{if and only if } \vec{\eta}_2 \in C(\vec{\eta}_1).$$

If $\vec{f}(\vec{\eta}) = \vec{0}$, the point $\vec{\eta}$ is called a *stationary point*. If $\vec{\eta}$ is a stationary point, the orbit $C(\vec{\eta})$ consists of a point, i.e., $C(\vec{\eta}) = \{\vec{\eta}\}$. Generalizing property (IX.1.3) of orbits, we introduce the concept of invariant sets which play a central role in the study of autonomous system (IX.1.1).

Definition IX-1-1. *A set $\mathcal{M} \subset \mathbb{R}^n$ is said to be* invariant *if $\vec{\eta} \in \mathcal{M}$ implies $C(\vec{\eta}) \subset \mathcal{M}$.*

For example, every orbit is an invariant set.

Remark IX-1-2. If $\mathcal{M}_1$ and $\mathcal{M}_2$ are invariant sets, then $\mathcal{M}_1 \cup \mathcal{M}_2$ and $\mathcal{M}_1 \cap \mathcal{M}_2$ are also invariant. For a given set $\Omega \subset \mathbb{R}^n$, let $\mathcal{M}_\lambda$ ($\lambda \in \Lambda$, an index set) be all invariant subsets of Ω, then $\displaystyle\bigcup_{\lambda \in \Lambda} \mathcal{M}_\lambda$ is the largest invariant subset of Ω.

Hereafter, assume that every solution $\vec{p}(t, \vec{\eta})$ is defined for $t \geq 0$, i.e., $\{t : t \geq 0\} \subset \mathcal{I}(\vec{\eta})$. In general, $\displaystyle\lim_{t \to +\infty} \vec{p}(t, \vec{\eta})$ may not exist. However, if $\vec{p}(t, \vec{\eta})$ is bounded for $t \geq 0$, the orbit $C(\vec{\eta})$ accumulates to a set as $t \to +\infty$. This set is very important in the study of behavior of $C(\vec{\eta})$ as $t \to +\infty$.

Definition IX-1-3. *Let $\mathcal{L}^+(\vec{\eta})$ denote the set of all $\vec{y} \in \mathbb{R}^n$ such that $\displaystyle\lim_{k \to +\infty} \vec{p}(t_k, \vec{\eta}) = \vec{y}$ for some increasing sequence $\{t_k : k = 1, 2, \ldots\}$ of real numbers such that $\displaystyle\lim_{k \to +\infty} t_k = +\infty$. The set $\mathcal{L}^+(\vec{\eta})$ is called the* limit-invariant set *for the initial point $\vec{\eta}$.*

The basic properties of $\mathcal{L}^+(\vec{\eta})$ are given in the following theorem.

Theorem IX-1-4. *If $\vec{p}(t, \vec{\eta})$ is bounded for $t \geq 0$, then $\mathcal{L}^+(\vec{\eta})$ is nonempty, bounded, closed, connected, and invariant.*

Proof.

(1) $\mathcal{L}^+(\vec{\eta})$ *is nonempty*: In fact, let $\{s_k : k = 1, 2, \ldots\}$ be an increasing sequence of real numbers such that $\displaystyle\lim_{k \to +\infty} s_k = +\infty$. Since $\vec{p}(t, \vec{\eta})$ is bounded for $t \geq 0$, there exists a subsequence $\{t_k : k = 1, 2, \ldots\}$ such that $\displaystyle\lim_{k \to +\infty} t_k = +\infty$ and that $\displaystyle\lim_{k \to +\infty} \vec{p}(t_k, \vec{\eta})$ exists. This limit belongs to $\mathcal{L}^+(\vec{\eta})$.

(2) *The boundedness of $\mathcal{L}^+(\vec{\eta})$* follows from the boundedness of $\vec{p}(t, \vec{\eta})$ immediately.

(3) $\mathcal{L}^+(\vec{\eta})$ *is closed*: To prove this, suppose that $\displaystyle\lim_{k \to +\infty} \vec{y}_k = \vec{y}$ for $\vec{y}_k \in \mathcal{L}^+(\vec{\eta})$. It must be shown that $\vec{y} \in \mathcal{L}^+(\vec{\eta})$. Since $\vec{y}_k \in \mathcal{L}^+(\vec{\eta})$, it follows that $\vec{y}_k = \displaystyle\lim_{\ell \to +\infty} \vec{p}(t_{k,\ell}, \vec{\eta})$ for some $\{t_{k,\ell} : \ell = 1, 2, \ldots\}$ such that $\displaystyle\lim_{\ell \to +\infty} t_{k,\ell} = +\infty$. Choose ℓ_k so that $\displaystyle\lim_{k \to +\infty} t_{k,\ell_k} = +\infty$ and $|\vec{y}_k - \vec{p}(t_{k,\ell_k}, \vec{\eta})| \leq \dfrac{1}{k}$. Then, since $|\vec{y} - \vec{p}(t_{k,\ell_k}, \vec{\eta})| \leq \dfrac{1}{k} + |\vec{y} - \vec{y}_k|$, we obtain $\displaystyle\lim_{k \to +\infty} \vec{p}(t_{k,\ell_k}, \vec{\eta}) = \vec{y} \in \mathcal{L}^+(\vec{\eta})$.

(4) $\mathcal{L}^+(\vec{\eta})$ *is connected*: Otherwise, there must be two nonempty, bounded, and closed sets $\mathcal{S}_1$ and $\mathcal{S}_2$ in $\mathbb{R}^n$ such that

(1) $\mathcal{S}_1 \cap \mathcal{S}_2 = \emptyset$,

(2) $\mathcal{S}_1 \cup \mathcal{S}_2 = \mathcal{L}^+(\vec{\eta})$.

Set $d = \mathrm{distance}(\mathcal{S}_1, \mathcal{S}_2) = \min\{|\vec{y}_1 - \vec{y}_2| : \vec{y}_1 \in \mathcal{S}_1, \vec{y}_2 \in \mathcal{S}_2\}$. Note that $d > 0$. Set also $\mathcal{S}_0 = \left\{\vec{y} : \mathrm{distance}(\vec{y}, \mathcal{S}_1) = \dfrac{d}{2}\right\}$. Then, $\mathcal{S}_0$ is not empty, bounded, and closed. Furthermore, $\mathcal{S}_0 \cap \mathcal{L}^+(\vec{\eta}) = \emptyset$. Choose two points $\vec{y}_1 \in \mathcal{S}_1$ and $\vec{y}_2 \in \mathcal{S}_2$ and two sequences $\{t_k : k = 1, 2, \ldots\}$ and $\{s_k : k = 1, 2, \ldots\}$ of real numbers so that $t_k < s_k \, (k = 1, 2, \ldots)$, $\lim\limits_{k \to +\infty} t_k = +\infty$, $\lim\limits_{k \to +\infty} s_k = +\infty$, $\lim\limits_{k \to +\infty} \vec{p}(t_k, \vec{\eta}) = \vec{y}_1$, and $\lim\limits_{k \to +\infty} \vec{p}(s_k, \vec{\eta}) = \vec{y}_2$. Assume that $\mathrm{distance}(\vec{p}(t_k, \vec{\eta}), \mathcal{S}_1) < \dfrac{d}{2}$ and $\mathrm{distance}(\vec{p}(s_k, \vec{\eta}), \mathcal{S}_1) > \dfrac{d}{2}$. Then, there exists a τ_k for each k such that $t_k < \tau_k < s_k$ and $\vec{p}(\tau_k, \vec{\eta}) \in \mathcal{S}_0$. Choose a subsequence $\{\sigma_\ell : \ell = 1, 2, \ldots\}$ of $\{\tau_k : k = 1, 2, \ldots\}$ so that $\lim\limits_{\ell \to +\infty} \sigma_\ell = +\infty$ and $\lim\limits_{\ell \to +\infty} \vec{p}(\sigma_\ell, \vec{\eta}) = \vec{y}$ exist. Then, $\vec{y} \in \mathcal{S}_0 \cap \mathcal{L}^+(\vec{\eta}) = \emptyset$. This is a contradiction.

(5) $\mathcal{L}^+(\vec{\eta})$ *is invariant*: It must be shown that if $\vec{y} \in \mathcal{L}^+(\vec{\eta})$, then $\vec{p}(t, \vec{y}) \in \mathcal{L}^+(\vec{\eta})$ for all $t \in \mathcal{I}(\vec{y})$. In fact, there exists a sequence $\{t_k : k = 1, 2, \ldots\}$ of real numbers such that $\lim\limits_{k \to +\infty} t_k = +\infty$ and $\lim\limits_{k \to +\infty} \vec{p}(t_k, \vec{\eta}) = \vec{y}$. From (IX.1.1) and the continuity of $\vec{p}(t, \vec{y})$ of $\vec{y}$, it follows that for each fixed t,

$$\lim_{k \to +\infty} \vec{p}(t + t_k, \vec{\eta}) = \lim_{k \to +\infty} \vec{p}(t, \vec{p}(t_k, \vec{\eta})) = \vec{p}(t, \vec{y}) \in \mathcal{L}^+(\vec{\eta}). \qquad \square$$

The materials of this section are also found in [CL, Chapter 16, §1, pp. 389-391] and [Har2, Chapter VII, §1, pp. 144-146].

IX-2. Liapounoff's direct method

In order to find the behavior of $\vec{p}(t, \vec{\eta})$ as $t \to +\infty$, it is important to locate $\mathcal{L}^+(\vec{\eta})$. As a matter of fact, if $\vec{p}(t, \vec{\eta})$ is bounded for $t \geq 0$ and if a set $\mathcal{M}$ contains $\mathcal{L}^+(\vec{\eta})$, then $\vec{p}(t, \vec{\eta})$ tends to $\mathcal{M}$ as $t \to +\infty$, i.e., $\lim\limits_{t \to +\infty} \inf(|\vec{p}(t, \vec{\eta}) - \vec{y}| : \vec{y} \in \mathcal{M}) = 0$. Otherwise, there must be a positive number ϵ_0 and a sequence $\{t_k : k = 1, 2, \ldots\}$ of real numbers such that $\lim\limits_{k \to +\infty} t_k = +\infty$, $\lim\limits_{k \to +\infty} \vec{p}(t_k, \vec{\eta})$ exists, and $|\vec{p}(t_k, \vec{\eta}) - \vec{y}| \geq \epsilon_0$ for all $\vec{y} \in \mathcal{M}$. Hence, $\lim\limits_{k \to +\infty} \vec{p}(t_k, \vec{\eta}) \notin \mathcal{M}$. This is a contradiction. Keeping this fact in mind, let us prove the following theorem (cf. [LaL, Chapter 2, §13, pp. 56-71]).

Theorem IX-2-1. *Let $V(\vec{y})$ be a real-valued, continuous, and continuously differentiable function for $|\vec{y}| < r_0$, where $0 < r_0 \leq +\infty$. Set*

$$\begin{cases} \mathcal{D}_\ell = \{\vec{y} : V(\vec{y}) < \ell\}, & \mathcal{S}_\ell = \{\vec{y} \in \mathcal{D}_\ell : V_{\vec{y}}(\vec{y}) \cdot \vec{f}(\vec{y}) = 0\}, \\ \mathcal{M}_\ell = & \text{the largest invariant set in } \mathcal{S}_\ell, \end{cases}$$

where $V_{\vec{y}}(\vec{y}) \cdot \vec{f}(\vec{y}) = \sum\limits_{j=1}^{n} \dfrac{\partial V}{\partial y_j} f_j(\vec{y})$, $\vec{y} = \begin{bmatrix} y_1 \\ \vdots \\ y_n \end{bmatrix}$, *and* $\vec{f}(\vec{y}) = \begin{bmatrix} f_1(\vec{y}) \\ \vdots \\ f_n(\vec{y}) \end{bmatrix}$. *Suppose* *that there exist a real number ℓ and a positive number r such that $0 < r < r_0$, $\mathcal{D}_\ell \subset \{\vec{y} : |\vec{y}| \leq r\}$, and $V_{\vec{y}}(\vec{y}) \cdot \vec{f}(\vec{y}) \leq 0$ on $\mathcal{D}_\ell$. Then, $\mathcal{L}^+(\vec{\eta}) \subset \mathcal{M}_\ell$ for all $\vec{\eta} \in \mathcal{D}_\ell$.*

Proof.

Set $u(t) = V(\vec{p}(t, \vec{\eta}))$. Then,

$$\frac{du(t)}{dt} = V_{\vec{y}}(\vec{p}(t, \vec{\eta})) \cdot \frac{d}{dt}\vec{p}(t, \vec{\eta}) = V_{\vec{y}}(\vec{p}(t, \vec{\eta})) \cdot \vec{f}(\vec{p}(t, \vec{\eta})).$$

Therefore, $\dfrac{du(t)}{dt} \leq 0$ as long as $\vec{p}(t, \vec{\eta}) \in \mathcal{D}_\ell$. This implies that $u(t) \leq V(\vec{\eta}) < \ell$ for $\vec{\eta} \in \mathcal{D}_\ell$ as long as $\vec{p}(t, \vec{\eta}) \in \mathcal{D}_\ell$. Hence, $\vec{p}(t, \vec{\eta}) \in \mathcal{D}_\ell$ for $t \geq 0$ if $\vec{\eta} \in \mathcal{D}_\ell$. Consequently, $|\vec{p}(t, \vec{\eta})| \leq r$ for $t \geq 0$. It is known that $\mathcal{L}^+(\vec{\eta})$ is nonempty, bounded, closed, connected, and invariant (cf. Theorem IX-1-4). Furthermore, $\mathcal{L}^+(\vec{\eta}) \subset \mathcal{D}_\ell$.

Let ρ_0 be the minimum of $V(\vec{\eta})$ for $|\vec{y}| \leq r$. Then, $\rho_0 \leq u(t)$ and $\dfrac{du(t)}{dt} \leq 0$ if $\vec{\eta} \in \mathcal{D}_\ell$. Hence, $\lim\limits_{t \to +\infty} u(t) = u_0$ exists and $u_0 \geq \rho_0$. This implies that $V(\vec{y}) = u_0$ for all $\vec{y} \in \mathcal{L}^+(\vec{\eta})$. Since $\mathcal{L}^+(\vec{\eta})$ is invariant, $V(\vec{p}(t, \vec{y})) = u_0$ for all $\vec{y} \in \mathcal{L}^+(\vec{\eta})$ and $t \geq 0$. Therefore, $V_{\vec{y}}(\vec{p}(t, \vec{y})) \cdot \vec{f}(\vec{p}(t, \vec{y})) = 0$ for all $\vec{y} \in \mathcal{L}^+(\vec{\eta})$ and $t \geq 0$. Setting $t = 0$, we obtain $V_{\vec{y}}(\vec{y}) \cdot \vec{f}(\vec{y}) = 0$ if $\vec{y} \in \mathcal{L}^+(\vec{\eta})$, i.e., $\mathcal{L}^+(\vec{\eta}) \subset \mathcal{S}_\ell$ if $\vec{\eta} \in \mathcal{D}_\ell$. Hence, $\mathcal{L}^+(\vec{\eta}) \subset \mathcal{M}_\ell$ for $\vec{\eta} \in \mathcal{D}_\ell$. $\square$

The following theorem is useful in many situations and it can be proved in a way similar to the proof of Theorem IX-2-1.

Theorem IX-2-2. *Let $V(\vec{y})$ be a real-valued and continuously differentiable function for all $\vec{y} \in \mathbb{R}^n$ such that $V_{\vec{y}}(\vec{y}) \cdot \vec{f}(\vec{y}) \leq 0$ for all $\vec{y} \in \mathbb{R}^n$. Assume also that the orbit $\vec{p}(t, \vec{\eta})$ is bounded for $t \geq 0$. Set $\mathcal{S} = \{\vec{y} : V_{\vec{y}}(\vec{y}) \cdot \vec{f}(\vec{y}) = 0\}$ and let $\mathcal{M}$ be the largest invariant set in $\mathcal{S}$. Then, $\mathcal{L}^+(\vec{\eta}) \subset \mathcal{M}$.*

The proof of this theorem is left to the reader as an exercise.

In order to use Theorem IX-2-2, the boundedness of $\vec{p}(t, \vec{\eta})$ must be shown in advance. To do this, the following theorem is useful.

Theorem IX-2-3. *If $V(\vec{y})$ is a real-valued and continuously differentiable function for all $\vec{y} \in \mathbb{R}^n$ such that $V_{\vec{y}}(\vec{y}) \cdot \vec{f}(\vec{y}) \leq 0$ for $|\vec{y}| \geq r_0$, where r_0 is a positive number, and that $\lim\limits_{|\vec{y}| \to +\infty} V(\vec{y}) = +\infty$, then all solutions $\vec{p}(t, \vec{\eta})$ are bounded for $t \geq 0$.*

Proof.

It suffices to consider the case when $\vec{p}(t_0, \vec{\eta}) > r_0$ for some $t_0 > 0$. There are two possibilities:

(1) $|\vec{p}(t, \vec{\eta})| > r_0$ for $t \geq t_0$,

(2) $|\vec{p}(t, \vec{\eta})| > r_0$ for $t_0 \leq t < t_1$ and $|\vec{p}(t_1, \vec{\eta})| = r_0$ for some $t_1 > t_0$.

Case (1). In this case, $\dfrac{d}{dt}V(\vec{p}(t, \vec{\eta})) = V_{\vec{y}}(\vec{p}(t, \vec{\eta})) \cdot \vec{f}(\vec{p}(t, \vec{\eta})) \leq 0$ for $t \geq 0$. Therefore, $V(\vec{p}(t, \vec{\eta})) \leq V(\vec{p}(t_0, \vec{\eta}))$ for $t \geq t_0$. Hence, $\vec{p}(t, \vec{\eta})$ is bounded for $t \geq 0$.

Case (2). In this case, it can be shown that

$$V(\vec{p}(t, \vec{\eta})) \leq \max\left[V(\vec{p}(t_0, \vec{\eta})), \ \max(V(\vec{y}) : |\vec{y}| \leq r_0)\right]$$

in a way similar to Case (1). The details are left to the reader as an exercise.   □

IX-3. Orbital stability

In this section, we introduce a concept of stability (respectively asymptotic stability) in a sense different from the stability (respectively the asymptotic stability) of Chapter VIII (cf. Definitions VIII-1-1 and VIII-1-3). We denote again by $C(\vec{\eta})$ the orbit passing through $\vec{\eta}$. Also, assume that every solution $\vec{p}(t, \vec{\eta})$ is defined for $t \geq 0$.

Definition IX-3-1. *An orbit $C(\vec{\eta}_0)$ is said to be* orbitally stable *as $t \to +\infty$ if, for any given positive number ϵ, there exists another positive number $\delta(\epsilon)$ such that*

$$\text{distance}(\vec{p}(t, \vec{\eta}), C(\vec{\eta}_0)) \leq \epsilon \quad \text{for} \quad t \geq 0 \quad \text{whenever} \quad |\vec{\eta} - \vec{\eta}_0| \leq \delta(\epsilon).$$

This definition is independent of the choice of the point $\vec{\eta}_0$ on the orbit $C(\vec{\eta}_0)$. If $C(\vec{\eta}_0) = C(\vec{\eta}_1)$ and if the condition of this definition is satisfied in terms of $\vec{\eta}_0$, then the same condition is satisfied also in terms of $\vec{\eta}_1$ with a different choice of $\delta(\epsilon)$. The following example illustrates these new concepts.

Example IX-3-2. The orbit of the system

$$\frac{d}{dt}\begin{bmatrix} y_1 \\ y_2 \end{bmatrix} = \sqrt{y_1^2 + y_2^2}\begin{bmatrix} -y_2 \\ y_1 \end{bmatrix}$$

passing through the point $\vec{\eta} = \begin{bmatrix} \eta_1 \\ \eta_2 \end{bmatrix}$ is given by

$$y_1 = r(\vec{\eta})\cos\{r(\vec{\eta})t + \theta(\vec{\eta})\}, \qquad y_2 = r(\vec{\eta})\sin\{r(\vec{\eta})t + \theta(\vec{\eta})\},$$

where $\vec{\eta} = \begin{bmatrix} r(\vec{\eta})\cos\theta(\vec{\eta}) \\ r(\vec{\eta})\sin\theta(\vec{\eta}) \end{bmatrix}$. It is easy to show that if $r(\vec{\eta}_1) - r(\vec{\eta}_2) = \delta > 0$ and $\theta(\vec{\eta}_1) = \theta(\vec{\eta}_2)$, then $\text{distance}(\vec{p}(t, \vec{\eta}_2), C(\vec{\eta}_1)) = \delta$. Hence, the orbit $C(\vec{\eta})$ is orbitally stable as $t \to +\infty$ for every $\vec{\eta}$. However, $\left|\vec{p}\left(\frac{\pi}{\delta}, \vec{\eta}_1\right) - \vec{p}\left(\frac{\pi}{\delta}, \vec{\eta}_2\right)\right| = r(\vec{\eta}_1) + r(\vec{\eta}_2)$ since $[r(\vec{\eta}_1)t + \theta(\vec{\eta}_1)] - [r(\vec{\eta}_2)t + \theta(\vec{\eta}_2)] = \delta t$ (cf. Figure 1). Therefore, every nontrivial solution $\vec{p}(t, \vec{\eta})$ is not stable as $t \to +\infty$ in the sense of Definition VIII-1-1.

Definition IX-3-3. *An orbit $C(\vec{\eta}_0)$ is said to be* orbitally asymptotically stable *as $t \to +\infty$ if*
  (i) $C(\vec{\eta}_0)$ is orbitally stable as $t \to +\infty$,
  (ii) there exists a positive number δ_0 such that $\lim\limits_{t \to +\infty} \text{distance}(\vec{p}(t, \vec{\eta}), C(\vec{\eta}_0)) = 0$
      whenever $|\vec{\eta} - \vec{\eta}_0| \leq \delta_0$.

This definition is independent of the choice of the point $\vec{\eta}_0$ on the orbit. However, the choice of δ_0 depends on $\vec{\eta}_0$.

Consider a system of differential equations

$$(\text{IX.3.1}) \qquad\qquad \frac{d\vec{y}}{dt} = \vec{f}(\vec{y}),$$

where the entries of the $\mathbb{R}^n$-valued function $\vec{f}$ is continuously differentiable on the entire $\vec{y}$-space $\mathbb{R}^n$. Assume also that the solution $\vec{p}(t,\vec{\eta}_0)$ is periodic in t of period 1 (i.e., $\vec{p}(t+1,\vec{\eta}_0) = \vec{p}(t,\vec{\eta}_0)$ for $-\infty < t < +\infty$) and that $\vec{f}(\vec{p}(t,\vec{\eta}_0)) \neq \vec{0}$. The system of linear differential equations

$$\text{(IX.3.2)} \qquad\qquad \frac{d\vec{u}}{dt} \;=\; \frac{\partial \vec{f}}{\partial \vec{y}}(\vec{p}(t,\vec{\eta}_0))\vec{u}$$

is called the *first variation* of system (IX.3.1) with respect to the solution $\vec{p}(t,\eta_0)$. The coefficients matrix of (IX.3.2) is periodic in t of period 1.

Let $\rho_1,\rho_2,\ldots,\rho_n$ be the multipliers of (IX.3.2) (cf. Definition IV-4-5). Since $\frac{d^2}{dt^2}\vec{p}(t,\vec{\eta}_0) = \frac{\partial \vec{f}}{\partial \vec{y}}(\vec{p}(t,\vec{\eta}_0))\frac{d}{dt}\vec{p}(t,\vec{\eta}_0)$, linear system (IX.3.2) has a nontrivial periodic solution $\frac{d}{dt}\vec{p}(t,\vec{\eta}_0)$. This implies that one of the multipliers must be 1. Set $\rho_1 = 1$. The following theorem gives a basic sufficient condition for orbitally asymptotic stability.

Theorem IX-3-4. *If $n-1$ multipliers $\rho_2,\ldots,\rho_n$ satisfy the condition $|\rho_j| < 1$ $(j = 2,\ldots,n)$, then the periodic orbit $C(\vec{\eta}_0)$ is orbitally asymptotically stable as $t \to +\infty$.*

Proof.

We prove this theorem in five steps. It suffices to find an $(n-1)$-dimensional manifold $\mathcal{M}$ in a neighborhood of the point $\vec{\eta}_0$ so that
 (1) $\mathcal{M}$ is transversal to the orbit $C(\vec{\eta}_0)$ at $\vec{\eta}_0$; i.e., the tangent of $C(\vec{\eta}_0)$ is not in the tangent space of $\mathcal{M}$ at $\vec{\eta}_0$,
 (2) there exist two positive numbers K and σ such that

$$|\vec{p}(t,\vec{\eta}) \;-\; \vec{p}(t,\vec{\eta}_0)| \;\leq\; K|\vec{\eta} \;-\; \vec{\eta}_0|e^{-\sigma t} \qquad \text{for} \quad \vec{\eta} \in \mathcal{M} \quad \text{and} \quad t \geq 0$$

(cf. Figure 2).

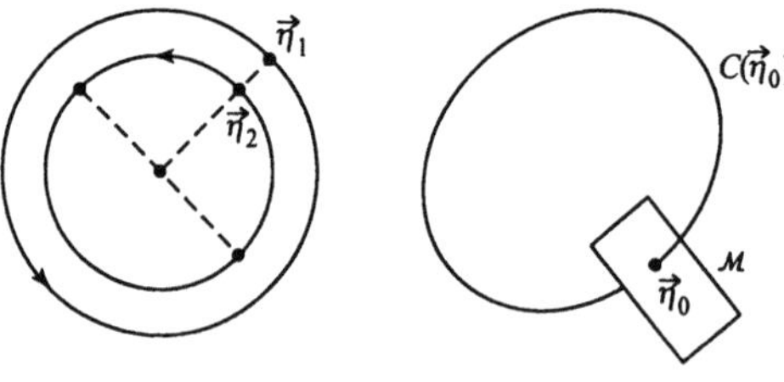

FIGURE 1. 　　　　　　　　 FIGURE 2.

Step 1. There exists an invertible $n \times n$ matrix $P(t)$ whose entries are real-valued, continuously differentiable, and periodic in t of period 1 (or 2) such that the transformation

$$\text{(IX.3.3)} \qquad\qquad \vec{u} \;=\; P(t)\vec{v}$$

changes (IX.3.2) to

(IX.3.4) $$\frac{d\vec{v}}{dt} = A\vec{v}, \qquad A = \begin{bmatrix} 0 & 0 \\ 0 & B \end{bmatrix},$$

where B is a constant $(n-1) \times (n-1)$ matrix (cf. Theorems IV-4-1 and IV-4-3). Since the absolute values of $n-1$ multipliers are less than 1, there exist two positive numbers C_0 and σ such that

(IX.3.5) $$|\exp[tB]| \leq C_0 e^{-2\sigma t} \qquad \text{for} \quad t \geq 0.$$

A fundamental matrix solution of (IX.3.2) is given by

$$\Psi(t) = P(t)\exp[tA], \qquad \exp[tA] = \begin{bmatrix} 1 & 0 \\ 0 & \exp[tB] \end{bmatrix}.$$

This implies that the first column vector of $P(t)$ is a periodic solution of (IX.3.2) and all the other columns of $\Psi(t)$ are not periodic. Hence, assume that

(IX.3.6) $$P(t) = \left[\frac{d}{dt}\vec{p}(t,\vec{\eta}_0), \ Q(t) \right],$$

where $Q(t)$ is an $n \times (n-1)$ matrix. Note that $\det[P(t)] \neq 0$.

Step 2. Change (IX.3.1) by

(IX.3.7) $$\vec{y} = \vec{z} + \vec{p}(t,\vec{\eta}_0)$$

to

(IX.3.8) $$\frac{d\vec{z}}{dt} = \frac{\partial \vec{f}}{\partial \vec{y}}(\vec{p}(t,\vec{\eta}_0))\vec{z} + \vec{h}(t,\vec{z}),$$

where

$$\vec{h}(t,\vec{z}) = \vec{f}(\vec{z} + \vec{p}(t,\vec{\eta}_0)) - \vec{f}(\vec{p}(t,\vec{\eta}_0)) - \frac{\partial \vec{f}}{\partial \vec{y}}(\vec{p}(t,\vec{\eta}_0))\vec{z}.$$

It is easy to show that $\vec{h}(t,\vec{0}) = \vec{0}$ and $\dfrac{\partial \vec{h}}{\partial \vec{z}}(t,\vec{0}) = O$.

Step 3. Change (IX.3.8) by the transformation $\vec{z} = P(t)\vec{w}$. Since $A = P(t)^{-1}\left[\dfrac{\partial \vec{f}}{\partial \vec{y}}(\vec{p}(t,\vec{\eta}_0))P(t) - \dfrac{dP(t)}{dt} \right]$, it follows that

(IX.3.9) $$\frac{d\vec{w}}{dt} = A\vec{w} + \vec{g}(t,\vec{w}),$$

where $\vec{g}(t,\vec{w}) = P(t)^{-1}\vec{h}(t,P(t)\vec{w})$. Note that $\vec{g}(t,\vec{0}) = \vec{0}$ and $\dfrac{\partial \vec{g}}{\partial \vec{w}}(t,\vec{0}) = O$. This implies that for any given positive number r, there exists another positive number $K(r)$ such that

(IX.3.10) $$|\vec{g}(t,\vec{w}_1) - \vec{g}(t,\vec{w}_2)| \leq K(r)|\vec{w}_1 - \vec{w}_2| \qquad \text{for} \quad t \geq 0$$

whenever $|\vec{w}_1| \leq r$ and $|\vec{w}_2| \leq r$ and that

(IX.3.11) $$\lim_{r \to 0} K(r) = 0.$$

Set $\vec{w}_1 = \vec{w}$ and $\vec{w}_2 = \vec{0}$. Then, (IX.3.10) becomes

(IX.3.12) $$|\vec{g}(t,\vec{w})| \leq K(r)|\vec{w}| \qquad \text{for} \quad t \geq 0.$$

Step 4. Let us write (IX.3.9) in the form

$$(\text{IX.3.13}) \qquad \frac{du}{dt} = g_1(t, \vec{w}), \qquad \frac{d\vec{v}}{dt} = B\vec{v} + \vec{g}_2(t, \vec{w}),$$

where $\vec{v}$ and $\vec{g}_2$ are $(n-1)$-dimensional vectors $\vec{w} = \begin{bmatrix} u \\ \vec{v} \end{bmatrix}$ and $\vec{g} = \begin{bmatrix} g_1 \\ \vec{g}_2 \end{bmatrix}$. System (IX.3.13) can be changed to the system of integral equations

$$(\text{IX.3.14}) \qquad \begin{cases} u(t, \vec{\xi}) = \displaystyle\int_{+\infty}^{t} g_1(s, \vec{w}(s, \vec{\xi}))ds, \\[2mm] \vec{v}(t, \vec{\xi}) = \exp[tB]\,\vec{\xi} + \displaystyle\int_0^t \exp[(t-s)B]\,\vec{g}_2(s, \vec{w}(s, \vec{\xi}))ds, \end{cases}$$

where $\vec{\xi}$ is an $(n-1)$-dimensional arbitrary constant vector. In this step, a solution $\vec{w}(t, \vec{\xi})$ of (IX.3.14) will be constructed in such a way that

$$(\text{IX.3.15}) \qquad\qquad |\vec{w}(t, \vec{\xi})| \leq k_0\,|\vec{\xi}|e^{-\sigma t}$$

for

$$(\text{IX.3.16}) \qquad\qquad t \geq 0 \quad \text{and} \quad |\vec{\xi}| \leq \delta_0,$$

where k_0 and δ_0 are suitable positive numbers.

First fix three positive numbers k_0, r_0, and δ_0 so that $k_0 > 2$, $k_0\delta_0 < r_0$, and $\dfrac{K(r)}{\sigma} < \dfrac{1}{2}$ for $0 < r \leq r_0$. Then, from (IX.3.5), (IX.3.10), (IX.3.11), (IX.3.14), and (IX.3.15), it follows that

$$(\text{I}) \qquad \left| \int_{+\infty}^t g_1(s, \vec{w}(s, \vec{\xi}))ds \right| \leq K(k_0|\vec{\xi}|)k_0|\vec{\xi}| \int_t^{+\infty} e^{-\sigma s}ds$$

$$= \frac{K(k_0|\vec{\xi}|)}{\sigma}k_0|\vec{\xi}|e^{-\sigma t} \leq k_0|\vec{\xi}|e^{-\sigma t}$$

and

$$(\text{II}) \qquad \left| \exp[tB]\,\vec{\xi} + \int_0^t \exp[(t-s)B]\,\vec{g}_2(s, \vec{w}(s, \vec{\xi}))ds \right|$$

$$\leq e^{-2\sigma t}|\vec{\xi}| + e^{-2\sigma t}K(k_0|\vec{\xi}|)k_0|\vec{\xi}| \int_0^t e^{\sigma s}ds$$

$$\leq e^{-\sigma t}|\vec{\xi}|\left(1 + \frac{K(k_0|\vec{\xi}|)}{\sigma}k_0\right) \leq e^{-\sigma t}|\vec{\xi}|\left(1 + \frac{k_0}{2}\right) \leq k_0|\vec{\xi}|e^{-\sigma t}.$$

Next, set $\|\vec{\psi}\|(\vec{\xi}) = \sup\left(e^{\sigma t}|\vec{\psi}(t, \vec{\xi})| : t \geq 0\right)$ if the entries of an $\mathbb{R}^n$-valued function $\vec{\psi}(t, \vec{\xi})$ are continuous for (IX.3.16) and $|\vec{\psi}(t, \vec{\xi})| \leq k_0|\vec{\xi}|e^{-\sigma t}$ for (IX.3.16). Then,

$$(\text{III}) \qquad \left| \int_{+\infty}^t g_1(s, \vec{\psi}_1(s, \vec{\xi}))ds - \int_{+\infty}^t g_1(s, \vec{\psi}_2(s, \vec{\xi}))ds \right|$$

$$\leq \frac{K(k_0|\vec{\xi}|)}{\sigma}\|\vec{\psi}_1 - \vec{\psi}_2\|(\vec{\xi})e^{-\sigma t} \leq \frac{1}{2}\|\vec{\psi}_1 - \vec{\psi}_2\|(\vec{\xi})e^{-\sigma t}$$

and

$$\text{(IV)} \quad \left| \int_0^t \exp\left[(t-s)B\right] \vec{g}_2(s, \vec{\psi}_1(s, \vec{\xi}))ds - \int_{+\infty}^t \exp\left[(t-s)B\right] \vec{g}_2(s, \vec{\psi}_2(s, \vec{\xi}))ds \right|$$

$$\leq \frac{K(k_0|\vec{\xi}|)}{\sigma} \|\vec{\psi}_1 - \vec{\psi}_2\|(\vec{\xi})e^{-\sigma t} \leq \frac{1}{2}\|\vec{\psi}_1 - \vec{\psi}_2\|(\vec{\xi})e^{-\sigma t}$$

for (IX.3.16) if the entries of $\mathbb{R}^n$-valued functions $\vec{\psi}_1(t, \vec{\xi})$ and $\vec{\psi}_2(t, \vec{\xi})$ are continuous for (IX.3.16) and that $|\vec{\psi}_j(t, \vec{\xi})| \leq k_0|\vec{\xi}|e^{-\sigma t}$ $(j = 1, 2)$ for (IX.3.16).

Let us define successive approximations as follows:

$$\vec{\psi}_m(t, \vec{\xi}) = \begin{bmatrix} u_m(t, \vec{\xi}) \\ \vec{v}_m(t, \vec{\xi}) \end{bmatrix} \qquad (m = 0, 1, 2, \ldots),$$

where $\vec{\psi}_0(t, \vec{\xi}) = \vec{0}$ and

$$\begin{cases} u_{m+1}(t, \vec{\xi}) = \displaystyle\int_{+\infty}^t g_1(s, \vec{\psi}_m(s, \vec{\xi}))ds, \\[3mm] \vec{v}_{m+1}(t, \vec{\xi}) = \exp\left[tB\right]\vec{\xi} + \displaystyle\int_0^t \exp\left[(t-s)B\right] \vec{g}_2(s, \vec{\psi}_m(s, \vec{\xi}))ds. \end{cases}$$

Then, it can be shown without any difficulties that $\displaystyle\lim_{m\to+\infty} \vec{\psi}_m(t, \vec{\xi}) = \vec{\psi}(t, \vec{\xi})$ exists uniformly for (IX.3.16) and the limit $\vec{\psi}(t, \vec{\xi})$ is a solution of integral equations (IX.3.14) satisfying condition (IX.3.15) for (IX.3.16). Note that $\vec{\psi}(0, \vec{\xi}) = \begin{bmatrix} a(\vec{\xi}) \\ \vec{\xi} \end{bmatrix}$, where $a(\vec{\xi}) = \displaystyle\int_{+\infty}^0 g_1(s, \vec{\psi}(s, \vec{\xi}))ds$. This implies that

$$\text{(IX.3.17)} \qquad |a(\vec{\xi})| \leq \frac{K(k_0|\vec{\xi}|)}{\sigma} k_0|\vec{\xi}| \qquad \text{for} \quad |\vec{\xi}| \leq \delta_0.$$

Step 5. Set
$$\vec{\phi}(t, \vec{\xi}) = P(t)\vec{\psi}(t, \vec{\xi}) + \vec{p}(t, \vec{\eta}_0).$$

Then, $\vec{\phi}$ is a solution of system (IX.3.1) such that

$$\text{(IX.3.18)} \qquad |\vec{\phi}(t, \vec{\xi}) - \vec{p}(t, \vec{\eta}_0)| \leq K_0|\vec{\xi}|e^{-\sigma t} \qquad \text{for} \quad \text{(IX.3.16)},$$

where K_0 is a positive constant. Define an $(n-1)$-dimensional manifold $\mathcal{M}$ by

$$\mathcal{M} = \{\vec{y} = \vec{\phi}(0, \vec{\xi}) : |\vec{\xi}| \leq \delta_0\}.$$

Using (IX.3.6), (IX.3.17), (IX.3.18), and (IX.3.11), we obtain

$$\begin{cases} \vec{\phi}(0, \vec{\xi}) = P(0)\vec{\psi}(0, \vec{\xi}) + \vec{\eta}_0 = a(\vec{\xi})\dfrac{d\vec{p}}{dt}(0, \vec{\eta}_0) + Q(0)\,\vec{\xi} + \vec{\eta}_0, \\[3mm] a(\vec{\xi}) = o(|\vec{\xi}|), \qquad \det\left[\dfrac{d\vec{p}}{dt}(0, \vec{\eta}_0),\, Q(0)\right] \neq 0. \end{cases}$$

Thus, the manifold $\mathcal{M}$ is transversal to the orbit $C(\vec{\eta}_0)$ at $\vec{\eta}_0$. Note that $\vec{\phi}(t, \vec{\xi}) = \vec{p}(t, \vec{\phi}(0, \vec{\xi}))$.

Moreover, by continuity of the solutions of (IX.3.1) with respect to the initial conditions, there exist two positive numbers δ_1 and τ_0 such that for $\vec{\eta}$ satisfying the condition

$$|\vec{\eta} - \vec{\eta}_0| \leq \delta_1,$$

there is a real number $\tau(\vec{\eta})$ such that $|\tau(\vec{\eta})| \leq \tau_0$ and $\vec{p}(\tau(\vec{\eta}), \vec{\eta}) \in \mathcal{M}$. Therefore, (IX.3.18) implies that

$$|\vec{p}(t + \tau(\vec{\eta}), \vec{\eta}) - \vec{p}(t, \vec{\eta}_0)| \leq K_0 \delta_0 e^{-\sigma t} \qquad \text{for} \quad t \geq 0.$$

Thus, $C(\vec{\eta}_0)$ is orbitally asymptotically stable as $t \to +\infty$. $\square$

Remark IX-3-5. If the entries of an $\mathbb{R}^n$-valued function $\vec{f}(t, \vec{y})$ is continuously differentiable on the entire $\vec{y}$-space $\mathbb{R}^n$ and system (IX.3.1) has a periodic solution $\vec{p}(t, \vec{\eta}_0)$ of period 1 such that $\vec{f}(\vec{p}(t, \vec{\eta}_0)) \neq \vec{0}$, there exist $n - 1$ vectors $\vec{q}_j(t) \in \mathbb{R}^n$ $(j = 2, \ldots, n)$ such that

(i) the entries of vectors $\vec{q}_j(t)$ $(j = 2, \ldots, n)$ are continuously differentiable and periodic in t of period 1,

(ii) n vectors $\dfrac{d\vec{p}(t, \vec{\eta}_0)}{dt}$, $\vec{q}_2(t)$, $\ldots$, $\vec{q}_n(t)$ form an orthogonal system.

Proof.

If $n = 2$, it is easy to find $\vec{q}_2(t)$. For $n \geq 3$, there is the recipe.

Note first that $\dfrac{d\vec{p}(t, \vec{\eta}_0)}{dt} = \vec{f}(\vec{p}(t, \vec{\eta}_0)) \neq \vec{0}$. Set $\vec{q}_1(t) = \dfrac{\vec{f}(\vec{p}(t, \vec{\eta}_0))}{\sqrt{\vec{f}(\vec{p}(t, \vec{\eta}_0)) \cdot \vec{f}(\vec{p}(t, \vec{\eta}_0))}}$,

where $\vec{a} \cdot \vec{b}$ denotes the usual dot product. Since $\vec{q}_1(t)$ is smooth, there exists a constant vector $\vec{\eta}_1$ such that $\vec{\eta}_1 \cdot \vec{\eta}_1 = 1$ and that $\vec{\eta}_1 + \vec{q}_1(t) \neq \vec{0}$ for $0 \leq t \leq 1$. Choosing an orthonormal set $\{\vec{\eta}_1, \ldots, \vec{\eta}_n\}$, where $\vec{\eta}_j$ are constant vectors, define

$$\vec{q}_h(t) = \vec{\eta}_h - \frac{\alpha_h(t)}{1 + \alpha_1(t)}(\vec{\eta}_1 + \vec{q}_1(t)) \qquad (j = 2, \ldots, n),$$

where $\alpha_h(t) = \vec{\eta}_h \cdot \vec{q}_1(t)$. $\square$

For this construction, see, for example, [U].

Example IX-3-6. The orthogonal system of Remark IX-3-5 is useful to study solutions of (IX.3.1) in a neighborhood of a periodic solution. To illustrate this application, let us look at a system

$$(S_1) \qquad\qquad \frac{d\vec{y}}{dt} = \vec{f}(\vec{y}),$$

where $\vec{f}$ is an $\mathbb{R}^3$-valued function whose entries are continuously differentiable on the entire $\vec{y}$-space $\mathbb{R}^3$. Assume that (S_1) has a periodic solution $\vec{p}(t, \vec{\eta}_0)$ of period 1 such that $\vec{f}(\vec{p}(t, \vec{\eta}_0)) \neq \vec{0}$. In this case, there exist two vectors $\vec{q}_2(t) \in \mathbb{R}^3$ and $\vec{q}_3(t) \in \mathbb{R}^3$ such that

(i) the entries of vectors $\vec{q}_j(t)$ $(j = 2, 3)$ are continuously differentiable and periodic in t of period 1,

(ii) three vectors $\dfrac{d\vec{p}(t, \vec{\eta}_0)}{dt}$, $\vec{q}_2(t)$, and $\vec{q}_3(t)$ form an orthogonal system.

Note that $\dfrac{d\vec{p}(t, \vec{\eta}_0)}{dt} = \vec{f}(\vec{p}(t, \vec{\eta}_0))$.

For any fixed real non-negative number τ, the set

$$\mathcal{P}(\tau) \;=\; \{\vec{p}(\tau, \eta_0) \;+\; u_1 \vec{q}_2(\tau) \;+\; u_2 \vec{q}_3(\tau) : (u_1, u_2) \in \mathbb{R}^2\}$$

is the plane which is perpendicular to the orbit $C(\vec{\eta}_0)$ at $\vec{p}(\tau, \eta_0)$. Letting t, u_1, and u_2 be three functions of τ to be determined, set

(IX.3.19)
$$\vec{y}(t) \;=\; \vec{p}(\tau, \eta_0) \;+\; u_1 \vec{q}_2(\tau) \;+\; u_2 \vec{q}_3(\tau).$$

From (IX.3.19) and the given system (S_1), we derive

$$\vec{f}(\vec{y}(t)) \frac{dt}{d\tau} \;=\; \vec{f}(\vec{p}(\tau, \eta_0)) \;+\; \frac{du_1}{d\tau} \vec{q}_2(\tau) \;+\; \frac{du_2}{d\tau} \vec{q}_3(\tau) \;+\; u_1 \frac{d\vec{q}_2(\tau)}{d\tau} \;+\; u_2 \frac{d\vec{q}_3(\tau)}{d\tau}.$$

This yields

(IX.3.20)
$$\begin{cases} \dfrac{du_1}{d\tau} = \vec{q}_2(\tau) \cdot \vec{f}(\vec{y}(t)) \dfrac{dt}{d\tau} - \left(\vec{q}_2(t) \cdot \dfrac{d\vec{q}_2(\tau)}{d\tau} \right) u_1 - \left(\vec{q}_2(t) \cdot \dfrac{d\vec{q}_3(\tau)}{d\tau} \right) u_2, \\[3mm] \dfrac{du_2}{d\tau} = \vec{q}_3(\tau) \cdot \vec{f}(\vec{y}(t)) \dfrac{dt}{d\tau} - \left(\vec{q}_3(t) \cdot \dfrac{d\vec{q}_2(\tau)}{d\tau} \right) u_1 - \left(\vec{q}_3(t) \cdot \dfrac{d\vec{q}_3(\tau)}{d\tau} \right) u_2 \end{cases}$$

and

(IX.3.21)
$$\begin{aligned} \frac{dt}{d\tau} &= \left[\vec{f}(\vec{p}(\tau, \eta_0)) \cdot \vec{f}(\vec{p}(\tau, \eta_0)) \;+\; \left(\vec{f}(\vec{p}(\tau, \eta_0)) \cdot \frac{d\vec{q}_2(\tau)}{d\tau} \right) u_1 \right. \\[2mm] &\quad \left. +\; \left(\vec{f}(\vec{p}(\tau, \eta_0)) \cdot \frac{d\vec{q}_3(\tau)}{d\tau} \right) u_2 \right] \times \left[\vec{f}(\vec{p}(\tau, \eta_0)) \cdot \vec{f}(\vec{y}(t)) \right]^{-1}, \end{aligned}$$

where $\vec{a} \cdot \vec{b}$ denotes the usual dot product and we assumed that $\vec{q}_j(\tau) \cdot \vec{q}_j(\tau) = 1$ $(j = 2, 3)$.

Note that

$$\vec{f}(\vec{y}(t)) \;=\; \vec{f}(\vec{p}(\tau, \eta_0)) + u_1 \frac{\partial \vec{f}}{\partial \vec{y}}(\vec{p}(\tau, \eta_0)) \vec{q}_2(\tau) + u_2 \frac{\partial \vec{f}}{\partial \vec{y}}(\vec{p}(\tau, \eta_0)) \vec{q}_3(\tau) + o(|u_1| + |u_2|).$$

Hence, from (IX.3.20) and (IX.3.21), we derive

$$\frac{dt}{d\tau} \;=\; 1 \;+\; O(|u_1| + |u_2|)$$

and

(IX.3.22)

$$
\begin{cases}
\dfrac{du_1}{d\tau} = \left[\vec{q}_2(\tau)\cdot\left(\dfrac{\partial\vec{f}}{\partial\vec{y}}(\vec{p}(\tau,\eta_0))\vec{q}_2(\tau)\right)\right]u_1 + \left[\vec{q}_2(\tau)\cdot\left(\dfrac{\partial\vec{f}}{\partial\vec{y}}(\vec{p}(\tau,\eta_0))\vec{q}_3(\tau)\right)\right]u_2 \\[3mm]
\qquad - \left(\vec{q}_2(t)\cdot\dfrac{d\vec{q}_2(\tau)}{d\tau}\right)u_1 \;-\; \left(\vec{q}_2(t)\cdot\dfrac{d\vec{q}_3(\tau)}{d\tau}\right)u_2 \;+\; o(|u_1|+|u_2|), \\[4mm]
\dfrac{du_2}{d\tau} = \left[\vec{q}_3(\tau)\cdot\left(\dfrac{\partial\vec{f}}{\partial\vec{y}}(\vec{p}(\tau,\eta_0))\vec{q}_2(\tau)\right)\right]u_1 + \left[\vec{q}_3(\tau)\cdot\left(\dfrac{\partial\vec{f}}{\partial\vec{y}}(\vec{p}(\tau,\eta_0))\vec{q}_3(\tau)\right)\right]u_2 \\[3mm]
\qquad - \left(\vec{q}_3(t)\cdot\dfrac{d\vec{q}_2(\tau)}{d\tau}\right)u_1 \;-\; \left(\vec{q}_3(t)\cdot\dfrac{d\vec{q}_3(\tau)}{d\tau}\right)u_2 \;+\; o(|u_1|+|u_2|).
\end{cases}
$$

Let $Q(t)$ be the 3×3 matrix whose column vectors are $\vec{f}(\vec{p}(t,\eta_0))$, $\vec{q}_2(t)$, and $\vec{q}_3(t)$, i.e., $Q(t) = \left[\vec{f}(\vec{p}(t,\eta_0))\;\;\vec{q}_2(t)\;\;\vec{q}_3(t)\right]$. The transformation $\vec{w} = Q(t)\vec{v}$ changes the linear system

$$(S_2)\qquad\qquad \frac{d\vec{w}}{dt} = \frac{\partial\vec{f}}{\partial\vec{y}}(\vec{p}(t,\eta_0))\vec{w}$$

to

$$\frac{d\vec{v}}{dt} = \begin{bmatrix} 0 & \beta_1(t) & \beta_2(t) \\ 0 & a_{11}(t) & a_{12}(t) \\ 0 & a_{21}(t) & a_{22}(t) \end{bmatrix}\vec{v}.$$

Using these notations, write (IX.3.22) in the form

(IX.3.23)
$$
\begin{cases}
\dfrac{du_1}{d\tau} = a_{11}(\tau)u_1 + a_{12}(\tau)u_2 + g_1(\tau,u_1,u_2), \\[3mm]
\dfrac{du_2}{d\tau} = a_{21}(\tau)u_1 + a_{22}(\tau)u_2 + g_2(\tau,u_1,u_2),
\end{cases}
$$

where

$$|g_j(\tau,u_1,u_2) - g_j(\tau,v_1,v_2)| \leq K(r)(|u_1-v_1|+|u_2-v_2|) \qquad (j=1,2)$$

with $K(r) > 0$ and $\lim_{r\to0}K(r) = 0$ for $|u_1|+|u_2| \leq r$ and $|v_1|+|v_2| \leq r$. Observe that the two multipliers of the linear system

$$\frac{d\vec{u}}{dt} = \begin{bmatrix} a_{11}(t) & a_{12}(t) \\ a_{21}(t) & a_{22}(t) \end{bmatrix}\vec{u}$$

are also multipliers of system (S_2). Therefore, using system (IX.3.23), Theorem IX-3-4 can be proven. In general, we obtain more precise information concerning the behavior of solutions in a neighborhood of a periodic solution in this way.

The materials of this section are also found in [CL, Chapter 13, §2, pp. 321-327].

IX-4. The Poincaré-Bendixson theorem

In this section, we explain the structure of $\mathcal{L}^+(\vec{\eta})$ on the plane. Consider an $\mathbb{R}^2$-valued function $\vec{f}(\vec{y})$ of $\vec{y} \in \mathbb{R}^2$ such that the entries of $\vec{f}(\vec{y})$ are continuously differentiable on the entire $\vec{y}$-plane $\mathbb{R}^2$. Denote again by $\vec{p}(t, \vec{\eta})$ the unique solution of the initial-value problem $\dfrac{d\vec{y}}{dt} = \vec{f}(\vec{y}), \quad \vec{y}(0) = \vec{\eta}$. The main result of this section is the following theorem due to H. Poincaré [Poi1] and I. Bendixson [Ben2].

Theorem IX-4-1. *Suppose that the solution $\vec{p}(t, \vec{\eta}_0)$ is bounded for $t \geq 0$ and that $\mathcal{L}^+(\vec{\eta}_0)$ contains only a finite number of stationary points. Then, there are the following three possibilities:*
   (i) $\mathcal{L}^+(\vec{\eta}_0)$ is a periodic orbit,
   (ii) $\mathcal{L}^+(\vec{\eta}_0)$ consists of a stationary point,
   (iii) $\mathcal{L}^+(\vec{\eta}_0)$ consists of a finite number of stationary points and a set of orbits each of which tends to one of these stationary points as $|t|$ tends to $+\infty$.

To prove this theorem, we need some preparation.

Definition IX-4-2. *A finite closed segment ℓ of a straight line in $\mathbb{R}^2$ is called a transversal with respect to $\vec{f}$ if $\vec{f}(\vec{y}) \neq \vec{0}$ at every point on ℓ and if the vector $\vec{f}(\vec{y})$ is not parallel to ℓ at every point on ℓ (cf. Figure 3).*

Observation IX-4-3. For every transversal ℓ and every point $\vec{\eta}$, the set $\ell \cap \mathcal{L}^+(\vec{\eta})$ contains at most one point (cf. Figures 4-1 and 4-2).

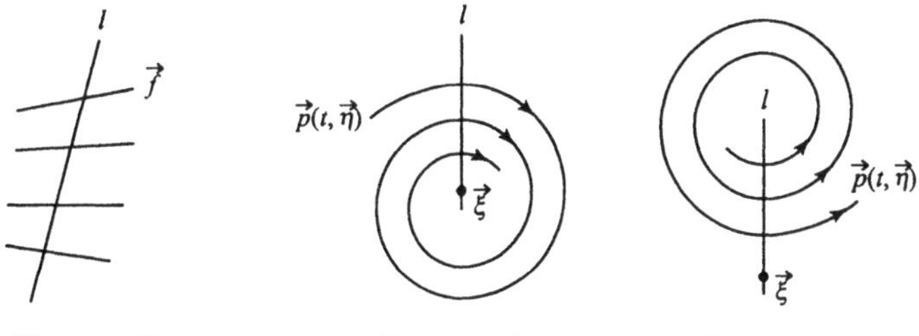

FIGURE 3.   FIGURE 4-1.   FIGURE 4-2.

The following lemma is the main part of the proof of Theorem IX-4-1.

Lemma IX-4-4. *If $\vec{p}(t, \vec{\eta}_0)$ is bounded for $t \geq 0$ and if there exists a point $\vec{\eta}_1 \in \mathcal{L}^+(\vec{\eta}_0)$ such that $\mathcal{L}^+(\vec{\eta}_1)$ contains a nonstationary point, then $\mathcal{L}^+(\vec{\eta}_0)$ is a periodic orbit.*

Proof.

We prove this lemma in three steps. Since $\mathcal{L}^+(\vec{\eta}_0)$ is invariant and $\vec{\eta}_1 \in \mathcal{L}^+(\vec{\eta}_0)$, it follows that $\vec{p}(t, \vec{\eta}_1) \in \mathcal{L}^+(\vec{\eta}_0)$ for all t. Furthermore, $\mathcal{L}^+(\vec{\eta}_1) \subset \mathcal{L}^+(\vec{\eta}_0)$ since $\mathcal{L}^+(\vec{\eta}_0)$ is closed.

Step 1. Let $\vec{\eta}$ be a nonstationary point on $\mathcal{L}^+(\vec{\eta}_1)$. Also, let ℓ be a transversal with respect to $\vec{f}$ that passes through $\vec{\eta}$. Then, $\ell \cap \mathcal{L}^+(\vec{\eta}_0) = \{\vec{\eta}\}$ since $\vec{\eta} \in \mathcal{L}^+(\vec{\eta}_1) \subset \mathcal{L}^+(\vec{\eta}_0)$.

Step 2. Since $\vec{\eta} \in \mathcal{L}^+(\vec{\eta}_1)$, there exists a sequence $\{t_k : k = 1, 2, \dots\}$ of real numbers such that $\lim\limits_{m \to +\infty} t_k = +\infty$ and $\vec{p}(t_k, \vec{\eta}_1) \in \ell \, (k = 1, 2, \dots)$. Note that $\vec{p}(t_k, \vec{\eta}_1) \in \mathcal{L}^+(\vec{\eta}_0)$. Therefore, $\vec{p}(t_k, \vec{\eta}_1) = \vec{\eta} \, (k = 1, 2, \dots)$. This implies that there exist two distinct real numbers τ_1 and τ_2 such that $\vec{\eta} = \vec{p}(\tau_1, \vec{\eta}_1) = \vec{p}(\tau_2, \vec{\eta}_1)$ and hence $\vec{p}(t, \vec{\eta}) = \vec{p}(t, \vec{p}(\tau_1, \vec{\eta}_1)) = \vec{p}(t, \vec{p}(\tau_2, \vec{\eta}_1))$. This, in turn, implies that $\vec{p}(t + \tau_1, \vec{\eta}_1) = \vec{p}(t + \tau_2, \vec{\eta}_1)$ for $t \geq 0$. Therefore, the orbit $C(\vec{\eta}_1)$ is periodic in t of period $|\tau_1 - \tau_2|$. Furthermore, $C(\vec{\eta}_1) \subset \mathcal{L}^+(\vec{\eta}_0)$. Note that there is no stationary point on $C(\vec{\eta}_1)$.

Step 3. Since $\mathcal{L}^+(\vec{\eta}_0)$ is connected, it follows that $\mathrm{distance}(C(\vec{\eta}_1), \mathcal{L}^+(\vec{\eta}_0) - C(\vec{\eta}_1)) = 0$. Hence, if $C(\vec{\eta}_1) \neq \mathcal{L}^+(\vec{\eta}_0)$, there exists a sequence $\{\vec{\xi}_k \in \mathcal{L}^+(\vec{\eta}_0) : k = 1, 2, \dots\}$ such that $\vec{\xi}_k \notin C(\vec{\eta}_1) \, (k = 1, 2, \dots)$ and $\lim\limits_{k \to +\infty} \vec{\xi}_k = \vec{\xi} \in C(\vec{\eta}_1)$. Assume that there exists a transversal ℓ such that $\vec{\xi} \in \ell$ and $\vec{\xi}_k \in \ell \, (k = 1, 2, \dots)$. Note that $\vec{\xi} \in C(\vec{\eta}_1) \subset \mathcal{L}^+(\vec{\eta}_0)$. Then, $\vec{\xi}_k = \vec{\xi} \in C(\vec{\eta}_1) \, (k = 1, 2, \dots)$. This is a contradiction. Thus, it is concluded that $\mathcal{L}^+(\eta_0) = C(\vec{\eta}_1)$. $\square$

Now, we complete the proof of Theorem IX-4-1 as follows.

Proof of Theorem IX-4-1.

If $\mathcal{L}^+(\vec{\eta}_0)$ does not contain any stationary points, then (i) follows (cf. Lemma IX-4-4). If $\mathcal{L}^+(\vec{\eta}_0)$ consists of stationary points only, we obtain (ii), since $\mathcal{L}^+(\vec{\eta}_0)$ is connected. If $\mathcal{L}^+(\vec{\eta}_0)$ contains stationary and nonstationary points, then $\mathcal{L}^+(\vec{\eta})$ for any point $\vec{\eta} \in \mathcal{L}^+(\vec{\eta}_0)$ does not contain nonstationary points (cf. Lemma IX-4-4). This is true also for $t < 0$. Hence, (iii) follows. $\square$

Observation IX-4-5. In cases (i) and (iii), the set $\mathbb{R}^2 - \mathcal{L}^+(\vec{\eta}_0)$ is not connected. Furthermore, if an orbit $C(\vec{\eta})$ is contained in $\mathcal{L}^+(\vec{\eta}_0)$, two sides of the curve $\vec{p}(t, \vec{\eta})$ belong to two different connected components of $\mathbb{R}^2 - \mathcal{L}^+(\vec{\eta}_0)$. In fact, if we consider a simple Jordan curve C which intersects with the orbit $C(\vec{\eta})$ at $\vec{\eta}$ transversally, then the curve $\vec{p}(t, \vec{\eta}_0)$ intersects with C in a neighborhood of two distinct points on C infinitely many times (cf. Figure 5).

Theorem IX-4-6. *If $C(\vec{\eta}_0) \cap \mathcal{L}^+(\vec{\eta}_0) \neq \emptyset$, then $\mathcal{L}^+(\vec{\eta}_0) = C(\vec{\eta}_0)$ and either $\vec{\eta}_0$ is a stationary point or the orbit $C(\vec{\eta}_0)$ is periodic.*

Proof.

In this case, $C(\vec{\eta}_0) \subset \mathcal{L}^+(\eta_0)$. If $\mathcal{L}^+(\eta_0)$ contains nonstationary points, the orbit $C(\vec{\eta}_0)$ consists of nonstationary points. Choose a transversal ℓ at $\vec{\eta}_0$. Then, it can be shown that $C(\vec{\eta}_0)$ is periodic in a way similar to the proof of Lemma IX-4-4, since $\vec{\eta}_0 \in \mathcal{L}^+(\eta_0)$. $\square$

Observation IX-4-7. If $C(\vec{\eta}_0) \cap \mathcal{L}^+(\eta_0) = \emptyset$, then $\lim\limits_{t \to +\infty} \mathrm{distance}(\vec{p}(t, \vec{\eta}_0)$ and $\mathcal{L}^+(\vec{\eta}_0)) = 0$. It follows that if $\mathcal{L}^+(\vec{\eta}_0)$ consists of a stationary point $\vec{\xi}$, then $\lim\limits_{t \to +\infty} \vec{p}(t, \vec{\eta}_0) = \vec{\xi}$. If $\mathcal{L}^+(\vec{\eta}_0)$ is a periodic orbit, then $\mathcal{L}^+(\vec{\eta}_0)$ is called a *limit cycle* (cf. Figures 6-1 and 6-2).

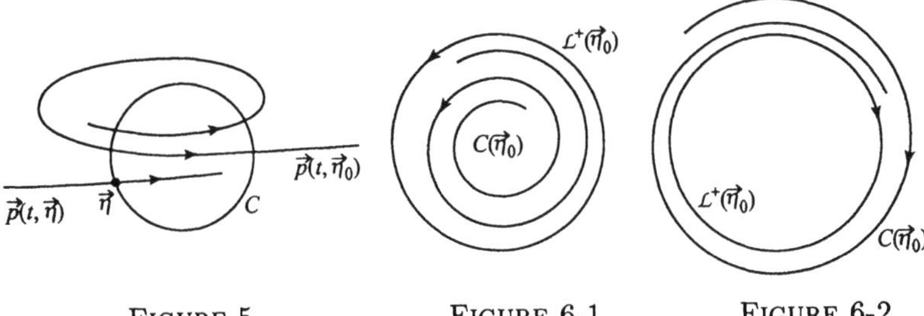

FIGURE 5.   FIGURE 6-1.   FIGURE 6-2.

The materials of this section are also found in [CL, Chapter 16, §§1 and 2, pp. 391-398] and [Har2, Chapter VII, §§4 and 5, pp. 151-158]. In [Har2], the Poincaré-Bendixson Theorem was proved without the uniqueness of solutions of initial-value problems.

IX-5. Indices of Jordan curves

In this section, we explain applications of *index* of a Jordan curve in the plane to the study of solutions of an autonomous system in $\mathbb{R}^2$. Consider again an $\mathbb{R}^2$-valued function $\vec{f}(\vec{y})$ of $\vec{y} \in \mathbb{R}^2$ whose entries are continuously differentiable on the entire $\vec{y}$-plane $\mathbb{R}^2$. Denote also by $\vec{p}(t,\vec{\eta})$ the unique solution of the initial-value problem
$$\frac{d\vec{y}}{dt} = \vec{f}(\vec{y}), \quad \vec{y}(0) = \vec{\eta}.$$

To begin with, let us introduce the concept of indices of Jordan curves. Let C be a Jordan curve $\vec{y} = \vec{\eta}(s)\,(0 \leq s \leq 1)$ with the counterclockwise orientation (cf. Figure 7). Assume that $\vec{f}(\vec{\eta}(s)) \neq \vec{0}$ for $0 \leq s \leq 1$. Set $\vec{u}(s) = \dfrac{\vec{f}(\vec{\eta}(s))}{|\vec{f}(\vec{\eta}(s))|}$ $(0 \leq s \leq 1)$. Then, $\vec{u}(s)$ is the unit vector in the direction of $\vec{f}(\vec{\eta}(s))$. There exists a real-valued continuous function $\theta(s)$ defined on the interval $0 \leq s \leq 1$ such that
$$\vec{u}(s) = \begin{bmatrix} \cos\theta(s) \\ \sin\theta(s) \end{bmatrix}.$$

Definition IX-5-1. *The* index *of the Jordan curve C with respect to the vector field $\vec{f}(\vec{y})$ is given by* $I_{\vec{f}}(C) = \dfrac{\theta(1) - \theta(0)}{2\pi}$.

This definition is independent of the choice of a parameterization $\vec{\eta}(s)$ of C and the function $\theta(s)$. Let us denote by $\mathcal{D}$ the domain bounded by C (cf. Figure 7).

Observation IX-5-2. If the domain $\mathcal{D}$ is divided into two domains $\mathcal{D}_1$ and $\mathcal{D}_2$ by a simple curve ℓ, then $I_{\vec{f}}(C) = I_{\vec{f}}(\partial\mathcal{D}_1) + I_{\vec{f}}(\partial\mathcal{D}_2)$ if $\vec{f}(\vec{\eta}) \neq \vec{0}$ on $\ell \cup C$, where $\partial\mathcal{D}_j\,(j = 1, 2)$ denote the boundaries of domains $\mathcal{D}_1$ and $\mathcal{D}_2$, respectively, as the portions of $I_{\vec{f}}(\partial\mathcal{D}_1)$ and $I_{\vec{f}}(\partial\mathcal{D}_1)$ along ℓ canceled each other (cf. Figure 8).

Observation IX-5-3. If $\vec{f}(\vec{\eta}) \neq \vec{0}$ on $C \cup \mathcal{D}$, then $I_{\vec{f}}(C) = 0$. To show this, divide $\mathcal{D}$ into sufficiently small subdomains, use the fact that the vector field $\vec{u}(s)$ has no change on a small subdomain, and apply Observation IX-5-2.

Observation IX-5-4. Assume that a point $\vec{a} \in \mathcal{D}$ is a stationary point, i.e., $\vec{f}(\vec{a}) = \vec{0}$. Assume also that $\vec{f}(\vec{\eta}) \neq \vec{0}$ on $\mathcal{C} \cup \mathcal{D}$ except at $\vec{a}$. Then, $I_{\vec{f}}(\mathcal{C})$ depends only on $\vec{a}$. Therefore, we define the *index of an isolated stationary point* $\vec{a}$ with respect to $\vec{f}$ by $I_{\vec{f}}(\vec{a}) = I_{\vec{f}}(\partial \mathcal{V})$, where $\mathcal{V}$ is a neighborhood of $\vec{a}$ such that there are no stationary points in $\mathcal{V}$ other than $\vec{a}$.

Observation IX-5-5. If $\mathcal{D}$ contains only a finite number of stationary points $\vec{a}_1, \vec{a}_2, \ldots, \vec{a}_N$, then $I_{\vec{f}}(\mathcal{C}) = \sum_{k=1}^{N} I_{\vec{f}}(\vec{a}_k)$. Proof of this result is similar to that of Observation IX-5-3.

Observation IX-5-6. If $\vec{f}(\vec{y}) \neq \vec{0}$ on $\mathcal{C}$ and if $\vec{f}(\vec{\eta}(s))$ is tangent to $\mathcal{C}$ at every point $\vec{\eta}(s)$ $(0 \leq s \leq 1)$ of $\mathcal{C}$, then $I_{\vec{f}}(\mathcal{C}) = 1$ (cf. [CL, Chapter 16, §4, Theorem 4.3]).

Proof.

It is evident that the index of $\mathcal{C}$ with respect to $\vec{f}$ and the index of $\mathcal{C}$ with respect to the tangent vector $\vec{\eta}'(\cdot) = \dfrac{d\vec{\eta}}{ds}(\cdot)$ to $\mathcal{C}$ are the same, i.e.,

$$I_{\vec{f}}(\mathcal{C}) \;=\; I_{\vec{\eta}'(\cdot)}(\mathcal{C}).$$

Assume without any loss of generalities that

$$\vec{\eta}(0) \;=\; \vec{\eta}(1) \;=\; \vec{0} \qquad \text{and} \qquad \eta_2(s) > 0 \quad \text{for} \quad 0 < s < 1,$$

where $\eta_j(s)$ $(j = 1, 2)$ are the entries of the vector $\vec{\eta}(s)$ (cf. Figure 9).

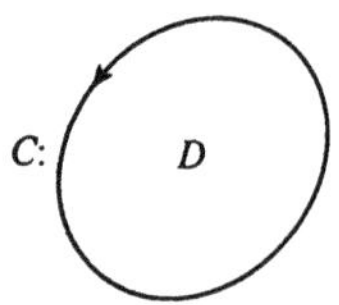

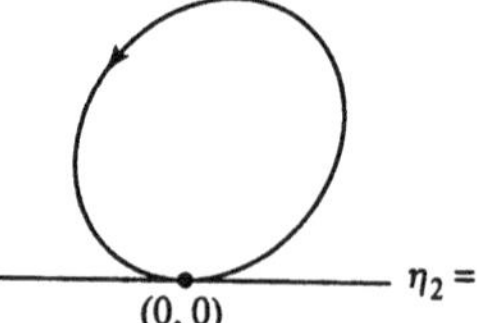

FIGURE 7.   FIGURE 8.   FIGURE 9.

For $0 \leq \tau \leq s \leq 1$, set

$$\vec{v}(\tau, s) \;=\; \begin{cases} \dfrac{\dfrac{d\vec{\eta}}{ds}(\tau)}{\left| \dfrac{d\vec{\eta}}{ds}(\tau) \right|} & \text{if} \quad \tau = s, \\[4ex] \dfrac{\vec{\eta}(s) - \vec{\eta}(\tau)}{|\,\vec{\eta}(s) - \vec{\eta}(\tau)\,|} & \text{if} \quad \tau < s \pmod 1, \\[4ex] -\dfrac{\dfrac{d\vec{\eta}}{ds}(0)}{\left| \dfrac{d\vec{\eta}}{ds}(0) \right|} & \text{if} \quad (\tau, s) = (0, 1). \end{cases}$$

Then, $\vec{v}(\tau, s)$ is continuous in (τ, s) for $0 \leq \tau \leq s \leq 1$. Hence, studying how $\vec{v}(0, s)$ changes from $s = 0$ to $s = 1$ and how $\vec{v}(\tau, 1)$ changes from $\tau = 0$ to $\tau = 1$, we obtain $I_{\vec{\eta}'(\cdot)}(\mathcal{C}) = 1$. $\square$

Observation IX-5-7. If C is a periodic orbit of the autonomous system

$$\text{(IX.5.1)} \qquad\qquad \frac{d\vec{\eta}}{dt} = \vec{f}(\vec{\eta}),$$

then $I_{\vec{f}}(C) = 1$.

Observation IX-5-8. If C is a periodic orbit of autonomous system (IX.5.1), then the domain $\mathcal{D}$ contains at least a stationary point (cf. Observations IX-5-3 and IX-5-7).

Observation IX-5-9. Suppose that $\vec{f}(\vec{\eta}(s))$ is tangent to C only at a finite number of points $\vec{\eta}(\tau_1), \vec{\eta}(\tau_2), \dots, \vec{\eta}(\tau_N)$ on C. Assume also that at each point $\vec{\eta}(\tau_k)$, either
  (1) $\vec{p}(t, \vec{\eta}(\tau_k)) \notin \mathcal{D}$ for $|t| < \delta_k$, or
  (2) $\vec{p}(t, \vec{\eta}(\tau_k)) \in \mathcal{D}$ for $0 < |t| < \delta_k$,
for some sufficiently small positive numbers $\delta_1, \dots, \delta_N$. Let us call $\vec{\eta}(\tau_k)$ an *exterior* (respectively *interior*) *contact point* in case (1) (respectively (2)) (cf. Figure 10). Let us denote by E (respectively H) the total number of interior (respectively exterior) contact points among N points $\vec{\eta}(\tau_1)$, $\vec{\eta}(\tau_2)$, $\dots$, $\vec{\eta}(\tau_N)$. Then $E+H = N$ and

$$\text{(IX.5.2)} \qquad\qquad I_{\vec{f}}(C) = \frac{E - H + 2}{2}.$$

Sketch of proof.

Let $\vec{\eta}(\tau_j)$ and $\vec{\eta}(\tau_k)$ be two consecutive contact points such that $\tau_j < \tau_k$. Then,
  (i) if both of these two points are exterior contact points, then the tangent to C changes π in angle more than the vector field $\vec{f}$ does from the point $\vec{\eta}(\tau_j)$ to the point $\vec{\eta}(\tau_k)$ (cf. Figure 11-1),
  (ii) if both of these two points are interior contact points, then the vector field $\vec{f}$ changes π in angle more than the tangent to C does from the point $\vec{\eta}(\tau_j)$ to the point $\vec{\eta}(\tau_k)$ (cf. Figure 11-2),
  (iii) if these two points are an exterior contact point and an interior contact point, then the tangent to C and the vector field $\vec{f}$ change in the same amount in angle (cf. Figures 12-1 and 12-2).
Since the total amount of change of the tangent to C in angle is 2π (cf. Observation IX-5-6), we arrive at formula (IX.5.2). $\square$

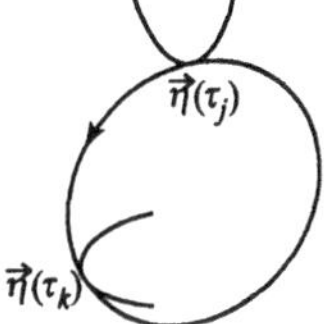

FIGURE 10.

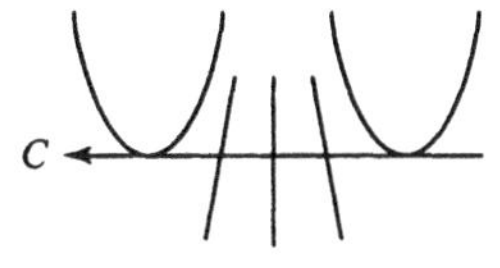

FIGURE 11-1.

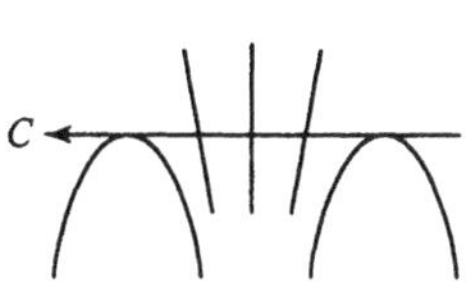

FIGURE 11-2.

Observation IX-5-10. Let $\vec{a}$ be an isolated stationary point. Assume that a neighborhood $\mathcal{V}$ of $\vec{a}$ is divided into a finite number of *sectorial regions* $\mathcal{S}_1, \mathcal{S}_2, \ldots, \mathcal{S}_N$ by a finite number of orbits $C(\vec{\eta}_1), C(\vec{\eta}_2), \ldots, C(\vec{\eta}_N)$ in such a way that

(α) $\vec{\eta}_1, \vec{\eta}_2, \ldots, \vec{\eta}_N \in \partial\mathcal{V}$,

(β) either $\vec{p}(\tau, \vec{\eta}_k) \in \mathcal{V}$ for $t > 0$ and tends to $\vec{a}$ as $t \to +\infty$, or $\vec{p}(\tau, \vec{\eta}_k) \in \mathcal{V}$ for $t < 0$ and tends to $\vec{a}$ as $t \to -\infty$ (cf. Figures 13-1 and 13-2).

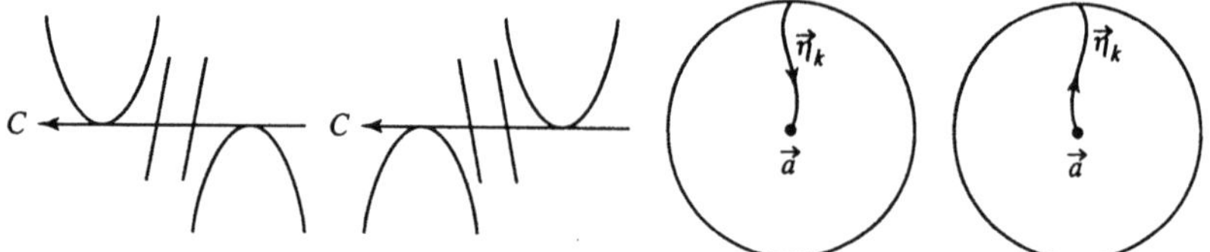

FIGURE 12-1.   FIGURE 12-2.   FIGURE 13-1.   FIGURE 13-2.

Let us assume also that each sectorial region $\mathcal{S}_k$ satisfies one of the following four conditions:

(I) if $\vec{\eta} \in \mathcal{S}_k$ and $|\vec{\eta} - \vec{a}|$ is sufficiently small, then $C(\vec{\eta}) \subset \mathcal{S}_k$ and $\vec{p}(t, \vec{\eta})$ tends to $\vec{a}$ as $|t| \to +\infty$ (cf. Figure 14),

(II) if $\vec{\eta} \in \mathcal{S}_k$ and $|\vec{\eta} - \vec{a}|$ is sufficiently small, then $\vec{p}(t, \vec{\eta}) \in \mathcal{S}_k$ only on a finite t-interval $\alpha_k < t < \beta_k$ (cf. Figure 15),

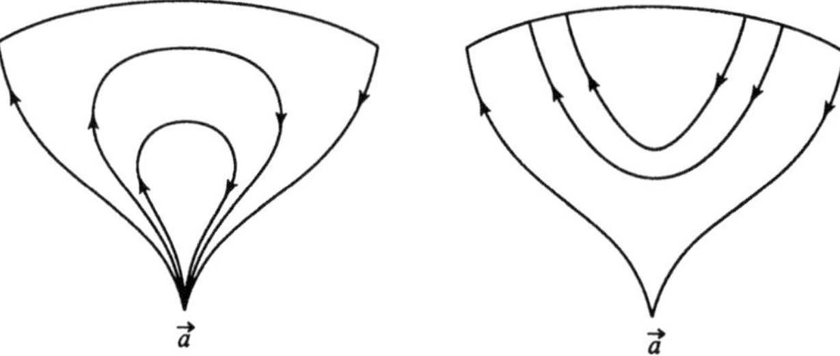

FIGURE 14.   FIGURE 15.

(III) if $\vec{\eta} \in \mathcal{S}_k$ and $|\vec{\eta} - \vec{a}|$ is sufficiently small, then $\vec{p}(t, \vec{\eta})$ tends to $\vec{a}$ in $\mathcal{S}_k$ as $t \to +\infty$, but $\vec{p}(t, \vec{\eta})$ does not tend to $\vec{a}$ in $\mathcal{S}_k$ as $t \to -\infty$ (cf. Figure 16-1),

(IV) if $\vec{\eta} \in \mathcal{S}_k$ and $|\vec{\eta} - \vec{a}|$ is sufficiently small, then $\vec{p}(t, \vec{\eta})$ tends to $\vec{a}$ in $\mathcal{S}_k$ as $t \to -\infty$, but $\vec{p}(t, \vec{\eta})$ does not tend to $\vec{a}$ in $\mathcal{S}_k$ as $t \to +\infty$ (cf. Figure 16-2).

The sectorial region $\mathcal{S}_k$ is said to be *elliptic* (respectively *hyperbolic*) in case (I) (respectively (II)). In cases (III) and (IV), the sectorial region $\mathcal{S}_k$ is said to be *parabolic*. Let us denote by E (respectively H) the total number of elliptic sectorial regions (respectively hyperbolic sectorial regions) among $\mathcal{S}_1, \ldots, \mathcal{S}_N$. Then, it is known that

(IX.5.3)
$$I_f(\vec{a}) = \frac{E - H + 2}{2}.$$

This result is similar to formula (IX.5.2) of Observation IX-5-9. For a proof in detail, see [Har2, Chapter VII, §9, pp. 166-172].

Observation IX-5-11. Suppose that the vector field $\vec{f}(\vec{y}, \epsilon)$ depends on a parameter $\epsilon \in \Delta$ continuously, where Δ is a connected set. In this case, if $\vec{f}(\vec{\eta}(s), \epsilon) \neq \vec{0}$ at every point $\vec{\eta}(s)$ on the Jordan curve C for all $\epsilon \in \Delta$, then the index $I_{\vec{f}(\cdot,\epsilon)}(C)$ is a constant independent of ϵ. In fact, in this case, the integer $I_{\vec{f}(\cdot,\epsilon)}(C)$ depends on ϵ continuously, and hence it is a constant.

Example IX-5-12. If an isolated stationary point $\vec{a}$ is a node, a spiral point, or a center, then $E = H = 0$. Hence $I_{\vec{f}}(\vec{a}) = 1$.

Example IX-5-13. If an isolated stationary point $\vec{a}$ is a saddle point, then $E = 0$ and $H = 4$. Hence $I_{\vec{f}}(\vec{a}) = -1$ (cf. Figure 17).

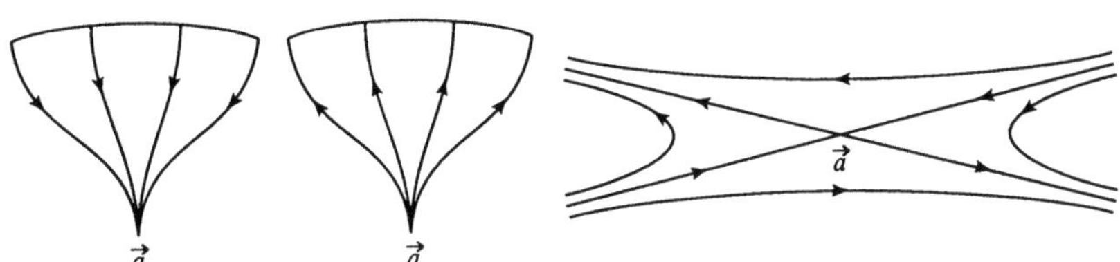

FIGURE 16-1.   FIGURE 16-2.   FIGURE 17.

Example IX-5-14. More general cases of isolated stationary points $\vec{a}$ are shown by Figures 18-1, 18-2, and 18-3. In fact,

$$\begin{cases} E = 0, \quad H = 2 \ \text{ and hence } I_{\vec{f}}(\vec{a}) = 0 & \text{in the case of Figure 18-1,} \\ E = 2, \quad H = 0 \ \text{ and hence } I_{\vec{f}}(\vec{a}) = 2 & \text{in the case of Figure 18-2,} \\ E = 1, \quad H = 1 \ \text{ and hence } I_{\vec{f}}(\vec{a}) = 1 & \text{in the case of Figure 18-3.} \end{cases}$$

Example IX-5-15. Let $\wp(z)$ be a polynomial in a complex variable z with complex coefficients. Set $y_1 = \Re[z]$ and $y_2 = \Im[z]$ (i.e., $z = y_1 + iy_2$) and regard the differential equation $\dfrac{dz}{dt} = \wp(z)$ as a system of two differential equations

$$(\text{IX.5.4}) \qquad \frac{dy_1}{dt} = \Re[\wp(y_1 + iy_2)], \qquad \frac{dy_2}{dt} = \Im[\wp(y_1 + iy_2)]$$

on the $\vec{y}$-plane. It is easy to see that, if $z_0 = \eta_1 + i\eta_2$ is a zero of $\wp(z)$ of multiplicity m, then $I_{\vec{f}}(\vec{\eta}) = m$, where $\vec{f}(\vec{y}) = \begin{bmatrix} \Re[\wp(y_1 + iy_2)] \\ \Im[\wp(y_1 + iy_2)] \end{bmatrix}$ and $\vec{\eta} = \begin{bmatrix} \eta_1 \\ \eta_2 \end{bmatrix}$. For example, if $\wp(z) = iz^2$, system (IX.5.4) becomes

$$(\text{IX.5.5}) \qquad \frac{dy_1}{dt} = -2y_1 y_2, \qquad \frac{dy_2}{dt} = y_1^2 - y_2^2.$$

The point $\vec{0}$ is an isolated stationary point and $I_{\vec{f}}(\vec{0}) = 2$.

If we define a vector field $\vec{f}(\vec{y}, \epsilon)$ by $\vec{f}(\vec{y}, \epsilon) = \begin{bmatrix} -\epsilon y_1 y_2 \\ y_1^2 - y_2^2 \end{bmatrix}$, then, $I_{\vec{f}(\cdot,\epsilon)}(\vec{0}) = 2$ for $\epsilon > 0$ (cf. Observation IX-5-11).

Example IX-5-16. Assume that $\vec{f}(\vec{\eta}) = \begin{bmatrix} f_1(y_1, y_2) \\ f_2(y_1, y_2) \end{bmatrix}$ satisfies the condition $\vec{f}(\lambda\vec{\eta})$

$= \lambda^p \vec{f}(\vec{\eta})$, where λ is a real variable and p is an integer. Set $\vec{y} = r \begin{bmatrix} \cos\theta \\ \sin\theta \end{bmatrix}$. Then,

the autonomous system $\dfrac{d\vec{\eta}}{dt} = \vec{f}(\vec{\eta})$ can be written in the form

$$\frac{dr}{dt} = r^p F_1(\theta), \qquad r\frac{d\theta}{dt} = r^p F_2(\theta),$$

where

$$\begin{cases} F_1(\theta) = f_1(\cos\theta, \sin\theta)\cos\theta + f_2(\cos\theta, \sin\theta)\sin\theta, \\ F_2(\theta) = -f_1(\cos\theta, \sin\theta)\sin\theta + f_2(\cos\theta, \sin\theta)\cos\theta \end{cases}$$

(cf. Exercise VIII-3). If a ray ℓ_θ is defined by $F_2(\theta) = 0$, then ℓ_θ is an orbit of the system $\dfrac{d\vec{\eta}}{dt} = \vec{f}(\vec{\eta})$. Thus, the entire $\vec{y}$-plane can be divided into sectorial regions by those orbits ℓ_θ determined by equation $F_2(\theta) = 0$. For example, in the case of system (IX.5.5), we obtain $F_1(\theta) = -\sin\theta$ and $F_2(\theta) = \cos\theta$. Observe that

$$\frac{dr}{dt}\begin{cases} > 0 & \text{for} \quad -\pi < \theta < 0, \\ = 0 & \text{for} \quad \theta = -\pi \text{ and } 0, \\ < 0 & \text{for} \quad 0 < \theta < \pi \end{cases} \quad \text{and} \quad \frac{d\theta}{dt}\begin{cases} > 0 & \text{for} \quad -\dfrac{\pi}{2} < \theta < \dfrac{\pi}{2}, \\ = 0 & \text{for} \quad \theta = -\dfrac{\pi}{2} \text{ and } \dfrac{\pi}{2}, \\ < 0 & \text{for} \quad \dfrac{\pi}{2} < \theta < \dfrac{3\pi}{2}. \end{cases}$$

Hence, $E = 2$ and $H = 0$ (cf. Figure 19). Therefore, $I_{\vec{f}}(\vec{0}) = 2$ (cf. Example IX-5-15).

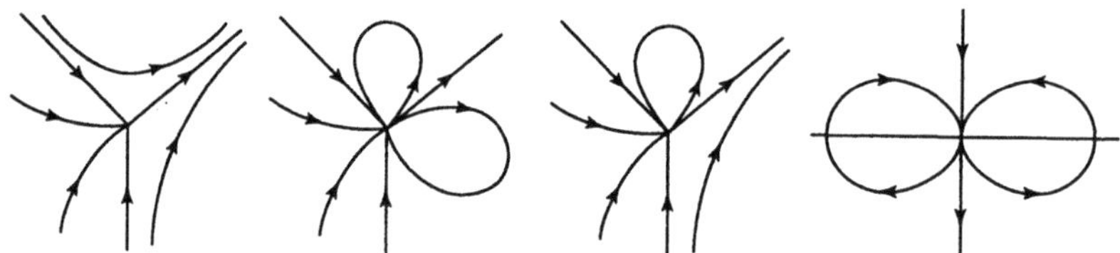

FIGURE 18-1.   FIGURE 18-2.   FIGURE 18-3.   FIGURE 19.

The materials of this section are also found in [CL, Chapter 16, §§4 and 5, pp. 398-402] and [Har2, Chapter VII, §§2 and 3, pp. 146-151, and §§8 and 9, pp. 161-174]. For a geometric and topological treatment of indices, see [Mi].

EXERCISES IX

IX-1. For each of the following three systems, using the given function $V(x, y)$, show that all orbits are bounded as $t \to +\infty$.

$$(1) \quad \frac{dx}{dt} = -xy^2 - 4y, \qquad \frac{dy}{dt} = -yx^2 + 3x, \qquad V(x,y) = 3x^2 + 4y^2;$$

$$(2) \qquad \frac{dx}{dt} = y, \qquad \frac{dy}{dt} = -x^3 - y, \qquad V(x,y) = x^4 + 2y^2;$$

$$(3) \quad \begin{cases} \dfrac{dx}{dt} = y, \quad \dfrac{dy}{dt} = -(x^5 - 3x^4 + 2x^3 + 120x^2 - 23x + 5) - (1+x^2)y, \\[2mm] V(x,y) = \dfrac{y^2}{2} + \displaystyle\int_0^x (s^5 - 3s^4 + 2s^3 + 120s^2 - 23s + 5)\,ds. \end{cases}$$

IX-2. Show that every solution $y(t)$ and its derivative $\dfrac{dy}{dt}(t)$ of the differential equation $\dfrac{d^2y}{dt^2} + \dfrac{dy}{dt} + y^3 = 0$ tend to 0 as $t \to +\infty$.

IX-3. Consider an autonomous system

$$(S_1) \qquad \frac{d}{dt}\begin{bmatrix} y_1 \\ y_2 \end{bmatrix} = \begin{bmatrix} f_1(y_1, y_2) \\ f_2(y_1, y_2) \end{bmatrix},$$

where f_1 and f_2 are continuously differentiable on the entire (y_1, y_2)-plane. Assume that
(1)

$$\begin{cases} f_1(y_1, y_2) > 0 & \text{for} \quad y_2 > 0 \quad \text{and} \quad -\infty < y_1 < +\infty, \\ f_1(y_1, y_2) < 0 & \text{for} \quad y_2 < 0 \quad \text{and} \quad -\infty < y_1 < +\infty, \end{cases}$$

(2) $f_1(y_1, 1) \geq 1$ and $f_1(y_1, -1) \leq -1$ for $-\infty < y_1 < +\infty$,
(3) $f_2(y_1, 1) = 0$ and $f_2(y_1, -1) = 0$ for $-\infty < y_1 < +\infty$,
(4)

$$\begin{cases} f_2(y_1, y_2) \leq y_1(y_2^2 - 1) & \text{for} \quad |y_2| < 1 \quad \text{and} \quad 0 < y_1 < +\infty, \\ f_2(y_1, y_2) \geq y_1(y_2^2 - 1) & \text{for} \quad |y_2| < 1 \quad \text{and} \quad -\infty < y_1 < 0, \end{cases}$$

Find $\mathcal{L}^+((\eta_1, \eta_2))$ for (η_1, η_2) such that $|\eta_2| < 1$.

Hint. There are two possibilities:

Case 1. The solution $(y_1(t), y_2(t))$ of (S_1) that satisfies the initial condition

$$(C) \qquad\qquad y_1(0) = \eta_1 \qquad \text{and} \qquad y_2(0) = \eta_2$$

is bounded as $t \to +\infty$.

Case 2. The solution $(y_1(t), y_2(t))$ of (S_1) that satisfies the initial condition (C) is unbounded as $t \to +\infty$.

In Case 1, $\mathcal{L}^+((\eta_1, \eta_2))$ is either $\{(0,0)\}$ or a periodic orbit. In Case 2, $\mathcal{L}^+((\eta_1, \eta_2))$ is $\{(x, \pm 1) : -\infty < x < +\infty\}$.

Examples.

(a) Every orbit in $|y_2| < 1$ of the system with $f_1(y_1, y_2) = y_2$ and $f_2(y_1, y_2) = y_1(y_2^2 - 1)$ is periodic.

(b) The stationary point $(0,0)$ is a stable spiral point if $f_1(y_1, y_2) = y_2(2 - \sin(y_1 y_2))$ and $f_2(y_1, y_2) = 2y_1(y_2^2 - 1)$.

(c) The stationary point $(0,0)$ is an unstable spiral point if $f_1(y_1, y_2) = y_2(2 + \sin(y_1 y_2))$ and $f_2(y_1, y_2) = 2y_1(y_2^2 - 1)$.

Verify these statements by using the function $V(y_1, y_2) = y_1^2 - \ln(1 - y_2^2)$.

IX-4. Let us consider a system

$$(S_2) \qquad \frac{d\vec{y}}{dt} = \vec{f}(\vec{y}), \qquad \vec{y} = \begin{bmatrix} y_1 \\ y_2 \end{bmatrix}, \qquad \vec{f}(\vec{y}) = \begin{bmatrix} f_1(\vec{y}) \\ f_2(\vec{y}) \end{bmatrix},$$

where $f_1(\vec{y})$ and $f_2(\vec{y})$ are continuously differentiable with respect to $\vec{y}$ in a domain $\mathcal{D}_0 \subset \mathbb{R}^2$. Assume that

(i) system (S_2) has a periodic orbit $\vec{p}(t, \vec{\eta}_0)$ of period 1 which is contained in the domain $\mathcal{D}_0$,

(ii) $\vec{f}(\vec{p}(t, \vec{\eta}_0)) \neq \vec{0}$,

(iii) the integral $\displaystyle\int_0^1 \left(\frac{\partial f_1}{\partial y_1}(\vec{p}(t, \vec{\eta}_0)) + \frac{\partial f_2}{\partial y_2}(\vec{p}(t, \vec{\eta}_0)) \right) dt$ is negative.

Show that the periodic orbit $\vec{p}(t, \vec{\eta}_0)$ is orbitally asymptotically stable as $t \to +\infty$.

Hint. This is called *Poincaré's criterion*. Look at

$$\frac{d\vec{w}}{dt} = \frac{\partial \vec{f}}{\partial \vec{y}}(\vec{p}(t, \eta_0))\vec{w}.$$

Then, $|\det Y(1)| < 1$ for the fundamental matrix solution $Y(t)$ of this system such that $Y(0) = I_2$ (cf. (4) of Remark IV-2-7). Hence, an eigenvalue ρ of $Y(1)$ must satisfy the condition $|\rho| < 1$. (The other eigenvalue of $Y(1)$ is 1.) Now, use Theorem IX-3-4.

IX-5. Assume that two functions $f(x)$ and $g(x)$ are continuously differentiable in $x \in \mathbb{R}$. Assume also that the differential equation $\dfrac{d^2 x}{dt^2} + f(x)\dfrac{dx}{dt} + g(x) = 0$ has a nontrivial periodic solution $x(t)$ of period 1 such that $\displaystyle\int_0^1 f(x(t))dt > 0$. Show that $(x(t), x'(t))$ is an orbitally asymptotically stable orbit as $t \to +\infty$ in the (x, x') phase plane.

Hint. Use Exercise IX-4.

IX-6. Show that there exists a nontrivial periodic orbit of the system

$$\frac{d\vec{y}}{dt} = \vec{f}(\vec{y}), \qquad \vec{y} = \begin{bmatrix} y_1 \\ y_2 \end{bmatrix}, \qquad \vec{f}(\vec{y}) = \begin{bmatrix} f_1(\vec{y}) \\ f_2(\vec{y}) \end{bmatrix},$$

where

(1) the entries of the $\mathbb{R}^2$-valued function $\vec{f}(\vec{y})$ is continuously differentiable on the entire $\vec{y}$-plane,

(2) $\vec{f}(\vec{0}) = \vec{0}$ and $\vec{f}(\vec{y}) \neq \vec{0}$ if $\vec{y} \neq \vec{0}$,

(3) $\dfrac{\partial f_1}{\partial y_1}(\vec{0}) = 1$, $\dfrac{\partial f_1}{\partial y_2}(\vec{0}) = 8$, $\dfrac{\partial f_2}{\partial y_1}(\vec{0}) = -2$ and $\dfrac{\partial f_2}{\partial y_2}(\vec{0}) = 1$,

(4) $\displaystyle\lim_{|y_1|+|y_2|\to+\infty}(f_1(y_1, y_2) + y_1^7)$ and $\displaystyle\lim_{|y_1|+|y_2|\to+\infty}(f_2(y_1, y_2) + y_2^7)$ exist.

IX-7. Given that

$$\vec{f}(\vec{y}, \epsilon) = \begin{bmatrix} \epsilon y_1 y_2 \\ y_1^2 - y_2^2 \end{bmatrix}, \qquad \vec{y} = \begin{bmatrix} y_1 \\ y_2 \end{bmatrix},$$

find the total number E of elliptic sectorial regions, the total number H of hyperbolic sectorial regions, and the total number of parabolic sectorial regions in the neighborhood of the isolated stationary point $\vec{0}$ of the system $\dfrac{d\vec{y}}{dt} = \vec{f}(\vec{y}, \epsilon)$ in the following two cases: (i) $\epsilon = \dfrac{1}{2}$ and (ii) $\epsilon = -\dfrac{1}{2}$. Also, find $I_{\vec{f}(\cdot,\epsilon)}(\vec{0})$ for $\epsilon \neq 0$.

IX-8. Let us consider a system

$$(S_3) \qquad\qquad\qquad \dfrac{d\vec{y}}{dt} = \vec{f}(\vec{y}),$$

where the entries of the $\mathbb{R}^3$-valued function $\vec{f}$ is continuously differentiable on the entire $\vec{y}$-space $\mathbb{R}^3$. Assume that (S_3) has a periodic orbit $\vec{p}(t, \vec{\eta}_0)$ of period 1 such that $\vec{f}(\vec{p}(t, \vec{\eta}_0)) \neq \vec{0}$. Assume also that the first variation of system (S_3) with respect to the solution $\vec{p}(t, \vec{\eta}_0)$, i.e.,

$$\dfrac{d\vec{u}}{dt} = \dfrac{\partial \vec{f}}{\partial \vec{y}}(\vec{p}(t, \vec{\eta}_0))\vec{u},$$

has three multipliers $\rho_1 = 1$, ρ_2, and ρ_3 satisfying the condition: $|\rho_2| < 1$ and $|\rho_3| > 1$, respectively. Construct the general orbits $\vec{p}(t, \vec{\eta})$ of (S_3) such that distance$(\vec{p}(t, \vec{\eta}), C(\vec{\eta}_0))$ tends to 0 as $t \to +\infty$.

IX-9. Show that the differential equation $\dfrac{d^2 x}{dt^2} + (x+3)(x+2)\dfrac{dx}{dt} + x(x+1) = 0$ does not have nontrivial periodic solutions.

Hint. Two stationary points are a node $(0,0)$ and a saddle $(-1,0)$. Furthermore, setting $\vec{f}(x_1, x_2) = \begin{bmatrix} x_2 \\ -(x_1+3)(x_1+2)x_2 - x_1(x_1+1) \end{bmatrix}$, we obtain div$\vec{f}(x_1, x_2) = -(x_1+3)(x_1+2) < 0$ if $x_1 > -1$. Also, use index in §IX-5.

IX-10. For the system

$$\dfrac{dx}{dt} = f(x, y) = x(1 - x^2 - y^2) - 3y, \qquad \dfrac{dy}{dt} = g(x, y) = y(1 - x^2 - y^2) + 3x,$$

(1) find and classify all critical points,

(2) find

$$\dfrac{dV}{dt} = f(x, y)\dfrac{\partial V}{\partial x} + g(x, y)\dfrac{\partial V}{\partial y}$$

for the function $V(x, y) = x^2 + y^2$,

(3) find the set $S = \left\{ (x, y) : \dfrac{dV}{dt} = 0 \right\}$,

(4) examine if S is an invariant set,

(5) find the phase-portrait of orbits.

IX-11. For the system

$$
\begin{cases}
\dfrac{dx}{dt} = f(x, y, z) = -x(1 - x^2 - y^2)^2 + 3xz + y, \\[2mm]
\dfrac{dy}{dt} = g(x, y, z) = -y(1 - x^2 - y^2)^2 + 3yz - x, \\[2mm]
\dfrac{dz}{dt} = h(x, y, z) = -z - 3(x^2 + y^2),
\end{cases}
$$

(1) find all critical points and determine if they are asymptotically stable,

(2) find

$$
\frac{dV}{dt} = f(x, y, z)\frac{\partial V}{\partial x} + g(x, y, z)\frac{\partial V}{\partial y} + h(x, y, z)\frac{\partial V}{\partial z}
$$

for the function $V(x, y, z) = x^2 + y^2 + z^2$,

(3) find the set $S = \left\{ (x, y, z) : \dfrac{dV}{dt} = 0 \right\}$,

(4) find the maximal invariant set $\mathcal{M}$ in S,

(5) find the phase portrait of orbits.

IX-12. Find the a phase portrait of orbits of the system

$$
\frac{dx}{dt} = y, \qquad \frac{dy}{dt} = y + (1 - x^2)x^2(4 - x^2).
$$

IX-13. Let $f(x, y)$ and $g(x, y)$ be real-valued, continuous, and continuously differentiable functions of two real variables (x, y) in an open, connected, and simply connected set $\mathcal{D}$ in the (x, y)-plane such that $\dfrac{\partial f}{\partial x}(x, y) + \dfrac{\partial f}{\partial y}(x, y) \neq 0$ for all $(x, y) \in \mathcal{D}$. Show that the system $\dfrac{dx}{dt} = f(x, y)$, $\dfrac{dy}{dx} = g(x, y)$ does not have any nontrivial periodic orbit that is contained entirely in $\mathcal{D}$.

IX-14. Let $\wp(z)$ be a polynomial in a complex variable z and $\deg \wp(z) > 0$. Set $x = \Re[z]$ and $y = \Im[z]$. Verify the following statements.

(i) In the neighborhood of each of stationary points of system

$$
\text{(A)} \qquad \frac{dx}{dt} = \Re[\wp(z)], \quad \frac{dy}{dt} = \Im[\wp(x)],
$$

there are no hyperbolic sectors. Also, system (A) does not have any isolated nontrivial periodic orbit.

(ii) In the neighborhood of each of stationary points of system

$$
\text{(B)} \qquad \frac{dx}{dt} = \Re[\wp(z)], \quad \frac{dy}{dt} = -\Im[\wp(x)],
$$

there are no elliptic sectors. Also, system (B) does not have any nontrivial periodic orbits.

IX-15. Find the phase portrait of orbits of the system

$$\frac{dx}{dt} = x^2 - y^2 - 3x + 2, \qquad \frac{dy}{dt} = -2xy + 3y.$$

Hint. See Exercise IX-14 with $\wp(z) = (z - 1)(z - 2)$.

IX-16. Find explicitly a two-dimensional system $\dfrac{dx}{dt} = f(x,y)$, $\dfrac{dy}{dt} = g(x,y)$ so that it has exactly five stationary points and all of them are centers.

IX-17. Consider a system

$$(S_4) \qquad\qquad\qquad \frac{d\vec{y}}{dt} = \vec{f}(\vec{y}),$$

where the entries of the $\mathbb{R}^n$-valued function $\vec{f}$ are analytic with respect to $\vec{y}$ in a domain $\mathcal{D}_0 \subset \mathbb{R}^n$. Assume that
  (i) system (S_2) has a periodic orbit $\vec{p}(t, \vec{\eta}_0)$ of period 1 which is contained in the domain $\mathcal{D}_0$,
 (ii) $\vec{f}(\vec{p}(t, \vec{\eta}_0)) \neq \vec{0}$,
(iii) for any open subset $\mathcal{V}$ of $\mathcal{D}_0$ which contains the periodic orbits $\vec{p}(t, \vec{\eta}_0)$, there exists an open subset $\mathcal{U}$ of $\mathcal{V}$ which also contains $\vec{p}(t, \vec{\eta}_0)$ such that for any point $\vec{\eta}$ in $\mathcal{U}$, the orbits $\vec{p}(t, \vec{\eta})$ of (S_4) is contained in $\mathcal{V}$ and periodic in t.
Show that if $\mathcal{U}$ is sufficiently small, for any point $\vec{\eta}$ in $\mathcal{U}$, we can fix a positive period $\mathcal{T}(\vec{\eta})$ of $\vec{p}(t, \vec{\eta})$ so that $\mathcal{T}(\vec{\eta})$ is bounded and analytic with respect to $\vec{\eta}$ in any simply connected bounded open subset of $\mathcal{U}$.

Hint. Apply the following observation.

Observation. *Let $\mathcal{D}_0$ be a connected, simply connected, open, and bounded set in $\mathbb{R}^k$ and let $T_j\,(j = 1, 2, \dots)$ be analytic mappings of $\mathcal{D}_0$ to $\mathbb{R}^k$. Suppose that, for any point $\vec{y} \in \mathcal{D}_0$, there exists a j such that $T_j[\vec{y}] = \vec{y}$, where j may depends on $\vec{y}$. Then, there exists a j_0 such that $T_{j_0}[\vec{y}] = \vec{y}$ for all $\vec{y} \in \mathcal{D}_0$.*

Proof.

Set

$$E_j = \{\vec{y} \in \mathcal{D}_0 : T_j[\vec{y}] = \vec{y}\}, \qquad j = 1, 2, \dots.$$

Then,
  (1) E_j is closed in $\mathcal{D}_0$,
  (2) $\mathcal{D}_0 = \bigcup_{j=1}^{+\infty} E_j$,
  (3) $\mathcal{D}_0$ is of the second category in the sense of Baire.
Hence, for some j_0, the set E_{j_0} contains a nonempty open subset (cf. Baire's Theorem). Since T_{j_0} is analytic, we obtain $T_{j_0}[\vec{y}] = \vec{y}$ for all $\vec{y} \in \mathcal{D}_0$. $\square$

For Baire's Theorem, see, for example, [Bar, pp. 91-92].

CHAPTER X

THE SECOND-ORDER DIFFERENTIAL

$$\textbf{EQUATION} \quad \frac{d^2x}{dt^2} + h(x)\frac{dx}{dt} + g(x) = 0$$

In this chapter, we explain the basic results concerning the behavior of solutions
of a system

$$\frac{d}{dt}\begin{bmatrix} y_1 \\ y_2 \end{bmatrix} = \begin{bmatrix} y_2 \\ -h(y_1)y_2 - g(y_1) \end{bmatrix}$$

as $t \to +\infty$. In §X-2, using results given in §IX-2, we show the boundedness of
solutions and apply these results to the van der Pol equation

$$(E) \qquad\qquad \frac{d^2x}{dt^2} + \epsilon(x^2 - 1)\frac{dx}{dt} + x = 0$$

(cf. Example X-2-5). The boundedness of solutions and the instability of the unique
stationary point imply that the van der Pol equation has a nontrivial periodic
solution. This is a consequence of the Poincaré-Bendixson Theorem (cf. Theorem
IX-4-1). In §X-3, we prove the uniqueness of periodic orbits in such a way that it
can be applied to equation (E). In §X-4, we show that the absolute value of one
of the two multipliers of the unique periodic solution of (E) is less than 1. The
argument in §X-4 gives another proof of the uniqueness of periodic orbit of (E). In
§X-5, we explain how to approximate the unique periodic solution of (E) in the case
when ϵ is positive and small. This is a typical problem of regular perturbations. In
§X-6, we explain how to locate the unique periodic solution of (E) geometrically as
$\epsilon \to +\infty$. In §X-8, we explain how to find an approximation of the periodic solution
of (E) analytically as $\epsilon \to +\infty$. This is a typical problem of singular perturbations.
Concerning singular perturbations, we also explain a basic result due to M. Nagumo
[Na6] in §X-7. In §X-1, we look at a boundary-value problem

$$\frac{d^2y}{dt^2} = F\left(t, y, \frac{dy}{dt}\right), \qquad y(a) = \alpha, \quad y(b) = \beta.$$

Using the Kneser Theorems (cf. Theorems III-2-4 and III-2-5), we show the existence of solutions for this problem in the case when $F(t, y, u)$ is bounded on the
entire (y, u)-space. Also, we explain a basic theorem due to M. Nagumo [Na4] (cf.
Theorem X-1-3) which we can use in more general situations including singular
perturbation problems (cf. [How]).

For more singular perturbation problems, see, for example, [Levi2], [LeL], [FL],
[HabL], [Si5], [How], [Was1], and [O'M].

X-1. Two-point boundary-value problems

In this section, first as an application of Theorems III-2-4 and III-2-5 (cf. [Kn]), we prove the following theorem concerning a boundary-value problem

$$(X.1.1) \qquad \frac{d^2y}{dt^2} = F\left(t, y, \frac{dy}{dt}\right), \qquad y(a) = \alpha, \quad y(b) = \beta.$$

Theorem X-1-1. *If the function $F(t, y_1, y_2)$ is continuous and bounded on a region $\Omega = \{(t, y_1, y_2) : a \leq t \leq b, |y_1| < +\infty, |y_2| < +\infty\}$, then problem (X.1.1) has a solution (or solutions).*

Proof.

For any positive number K, the set $\mathcal{A}_0 = \{(a, \alpha, y_2) : |y_2| \leq K\}$ is a compact and connected subset of Ω. We shall show that $\mathcal{A}_0$ satisfies Assumptions 1 and 2 of §III-2 for every positive number K. In fact, writing the second-order equation (X.1.1) as a system

$$(X.1.2) \qquad \frac{dy_1}{dt} = y_2, \qquad \frac{dy_2}{dt} = F(t, y_1, y_2),$$

we derive

$$\begin{cases} y_1(t) = y_1(a) + \displaystyle\int_a^t y_2(s)ds, \\[2mm] y_2(t) = y_2(a) + \displaystyle\int_a^t F(s, y_1(s), y_2(s))ds. \end{cases}$$

Hence, if $(a, y_1(a), y_2(a)) \in \mathcal{A}_0$, we obtain

$$\begin{cases} |y_2(t)| \leq K + M(b-a), \\ |y_1(t)| \leq |\alpha| + [K + M(b-a)](b-a), \end{cases}$$

where $|F(t, y_1, y_2)| \leq M$ on Ω. Therefore, $\mathcal{A}_0$ satisfies Assumptions 1 and 2 of §III-2. Thus, Theorem III-2-5 implies that $\mathcal{S}_c$ is also compact and connected for every c on the interval $\mathcal{I}_0 = \{t : a \leq t \leq b\}$.

We shall prove that if $K > 0$ is sufficiently large, the set $\mathcal{S}_b$ contains two points (η_1, η_2) and (ζ_1, ζ_2) such that

$$(X.1.3) \qquad\qquad \eta_1 < \beta < \zeta_1.$$

In fact, by using the Taylor series at $t = a$, write $y_1(b)$ in the form

$$y_1(b) = \alpha + y_2(a)(b-a) + \frac{1}{2}\frac{dy_2}{dt}(c)(b-a)^2,$$

where c is a certain point in the interval $\mathcal{I}_0$. Since $\left|\dfrac{dy_2}{dt}(c)\right| \leq M$, the quantity $|y_1(b)|$ can be made as large as we wish by choosing $|y_2(a)|$ sufficiently large. Thus, there are two points (η_1, η_2) and (ζ_1, ζ_2) in $\mathcal{S}_b$ such that (X.1.3) is satisfied.

Since the set $\mathcal{S}_b$ is compact and connected, there must be a point (β, ζ) in the set $\mathcal{S}_b$. This implies the existence of a solution of problem (X.1.1). $\square$

Example X-1-2. Theorem X-1-1 applies to the following two problems:

$$\text{(X.1.4)} \qquad \frac{d^2y}{dt^2} + \sin y = 0, \qquad y(a) = \alpha, \quad y(b) = \beta$$

and

$$\text{(X.1.5)} \qquad \frac{d^2y}{dt^2} + \frac{1}{y^2+1} = 0, \qquad y(a) = \alpha, \quad y(b) = \beta.$$

However, Theorem X-1-1 does not apply to

$$\text{(X.1.6)} \qquad \frac{d^2y}{dt^2} + y = 0, \qquad y(a) = \alpha, \quad y(b) = \beta.$$

For more general cases, the following theorem due to M. Nagumo [Na4] is useful.

Theorem X-1-3. *Assume that*

(i) a real-valued function $f(t,x,y)$ and its derivatives $\dfrac{\partial f}{\partial x}$ and $\dfrac{\partial f}{\partial y}$ are continuous in a region $\mathcal{D} = \{(t,x,y) : (t,x) \in \Delta, -\infty < y < +\infty\}$, where Δ is a bounded and closed set in the (t,x)-space;

(ii) in the region $\mathcal{D}$, the function f satisfies the condition

$$\text{(I)} \qquad |f(t,x,y)| \leq \phi(|y|),$$

where $\phi(u)$ is a positive-valued function on the interval $0 \leq u < +\infty$ such that

$$\text{(II)} \qquad \int_0^{+\infty} \frac{u\,du}{\phi(u)} = +\infty;$$

(iii) two real-valued functions $w_1(t)$ and $w_2(t)$ are twice continuously differentiable on an interval $a \leq t \leq b$ and satisfy the conditions

$$\text{(III)} \qquad \begin{cases} w_1(t) < w_2(t) & \text{for} \quad a \leq t \leq b, \\ \Delta_0 = \{(t,x) : a \leq t \leq b, \ w_1(t) \leq x \leq w_2(t)\} \subset \Delta, \end{cases}$$

and

$$\text{(IV)} \qquad \begin{cases} \dfrac{d^2w_1(t)}{dt^2} > f\left(t, w_1(t), \dfrac{dw_1(t)}{dt}\right), \\ \dfrac{d^2w_2(t)}{dt^2} < f\left(t, w_2(t), \dfrac{dw_2(t)}{dt}\right), & \text{for} \quad a \leq t \leq b; \end{cases}$$

(iv) two real numbers A and B satisfy the condition

$$\text{(V)} \qquad w_1(a) < A < w_2(a), \qquad \text{and} \qquad w_1(b) < B < w_2(b).$$

Then, the boundary-value problem

$$\frac{d^2x}{dt^2} = f\left(t, x, \frac{dx}{dt}\right), \qquad x(a) = A, \quad x(b) = B,$$

has a solution $x(t)$ such that $(t, x(t)) \in \Delta_0$ for $a \leq t \leq b$, i.e.,

$$w_1(t) \leq x(t) \leq w_2(t), \qquad \text{for} \quad a \leq t \leq b.$$

Proof.

The main tools are the following two lemmas.

Lemma X-1-4. *Let $x(t, t_0, \xi, \eta)$ be the solution of the initial-value problem*

$$\frac{d^2 x}{dt^2} = f\left(t, x, \frac{dx}{dt}\right), \qquad x(t_0) = \xi, \quad x'(t_0) = \eta,$$

where $a \leq t_0 \leq b$, $(t_0, \xi) \in \Delta_0$. Then, for any given positive number M, there exists a positive number $\alpha(M)$ such that $|x'(t, t_0, \xi, \eta)| < \alpha(M)$ if

(X.1.7) $|\eta| \leq M$ *and* $(\tau, x(\tau, t_0, \xi, \eta)) \in \Delta_0$ *for* $t_0 \leq \tau \leq t$ *or* $t \leq \tau \leq t_0$.

Proof.

Letting L be a positive number such that

(X.1.8) $$w_2(t) - w_1(t) \leq L \qquad \text{for} \qquad a \leq t \leq b,$$

choose $\alpha(M) > 0$ for any given positive number M in such a way that $\alpha(M) > M$ and

(X.1.9) $$\int_M^{\alpha(M)} \frac{u\,du}{\phi(u)} > L.$$

Suppose that there exist τ_1 and τ_2 such that $t_0 \leq \tau_1 < \tau_2 \leq t$ and that

$$x'(\tau_1) = M < x'(\tau) < x'(\tau_2) = \alpha(M) \qquad \text{for} \qquad \tau_1 < \tau < \tau_2,$$

where $x(\tau) = x(\tau, t_0, \xi, \eta)$. Then, since $x'(\tau) > 0$ for $\tau_1 \leq \tau \leq \tau_2$, it follows that

$$\frac{x'(\tau) x''(\tau)}{\phi(x'(\tau))} \leq x'(\tau) \qquad \text{for} \qquad \tau_1 \leq \tau \leq \tau_2.$$

Hence,

$$\int_M^{\alpha(M)} \frac{u\,du}{\phi(u)} \leq x(\tau_2) - x(\tau_1) \leq L.$$

This contradicts the choice of L by (X.1.9). Therefore, Lemma X-1-4 is true for $t_0 \leq \tau \leq t$. We can treat the case $t \leq \tau \leq t_0$ similarly, since if we change t by $-t$, the differential equation $x'' = f(t, x, x')$ becomes $x'' = f(-t, x, -x')$. $\square$

Lemma X-1-5. *Set*

$$\gamma = \max\left\{\alpha\left(\frac{L}{b-a}\right),\ w_1'(a) + \epsilon,\ -w_2'(a) - \epsilon\right\},$$

where ϵ is an arbitrarily fixed positive number. Let also $x(t, c)$ be the solution to the initial-value problem

$$\frac{d^2 x}{dt^2} = f\left(t, x, \frac{dx}{dt}\right), \qquad x(a) = A, \quad x'(a) = c.$$

Then,

(1) *two curves $x = x(t, c)$ and $x = w_2(t)$ meet for some t on the interval $a < t < b$ if $c \geq \gamma$;*

(2) *two curves $x = x(t, c)$ and $x = w_1(t)$ meet for some t on the interval $a < t < b$ if $c \leq -\gamma$.*

Proof.

For part (1), by virtue of Lemma X-1-4, $x'(\tau, c) > \dfrac{L}{b - a}$ if $(\tau, x(t, c)) \in \Delta_0$ for $a \leq \tau \leq t$. The proof of part (2) is similar. $\square$

Proof of Theorem X-1-3.

Now, let us complete the proof of Theorem X-1-3. The main point is that when two curves $x = x(t, c)$ and $x = w_2(t)$ or two curves $x = x(t, c)$ and $x = w_1(t)$ meet, they cut through each other. So look at Figure 1. $\square$

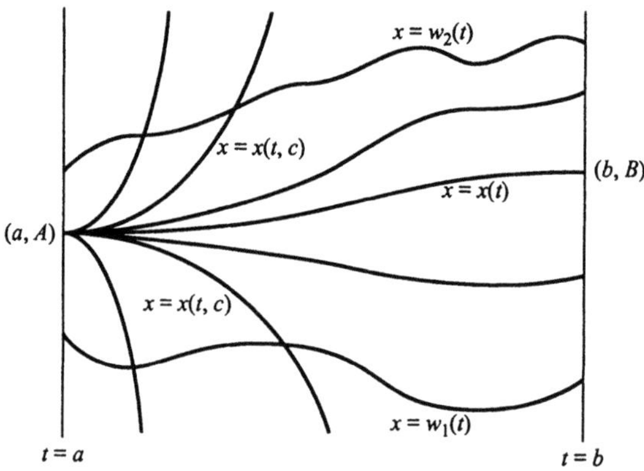

FIGURE 1.

Example X-1-6. Theorem X-1-3 applies to the boundary-value problem

$$(X.1.10) \qquad \frac{d^2 x}{dt^2} = \lambda x, \qquad x(0) = A, \quad x(1) = B$$

if λ is a positive number. In fact, assume that $\phi(u)$ is a suitable positive constant. If $w_1(t) = \sinh(\sqrt{\lambda} t) - \alpha$ and $w_2(t) = \sinh(\sqrt{\lambda} t) + \beta$ with two positive numbers α and β such that $-\alpha < A < \beta$ and $\sinh(\sqrt{\lambda}) - \alpha < B < \sinh(\sqrt{\lambda}) + \beta$, all requirements of Theorem X-1-3 are satisfied.

If λ is negative, Theorem X-1-3 does not apply to problem (X.1.10). Details are left to the reader as an exercise.

X-2. Applications of the Liapounoff functions

In this section, using the results of §IX-2, we explain the behavior of orbits of a system

$$\text{(X.2.1)} \qquad \frac{d}{dt}\begin{bmatrix} y_1 \\ y_2 \end{bmatrix} = \begin{bmatrix} y_2 \\ -h(y_1)y_2 - g(y_1) \end{bmatrix}$$

as $t \to +\infty$. Set $\vec{y} = \begin{bmatrix} y_1 \\ y_2 \end{bmatrix}$ and $\vec{f}(\vec{y}) = \begin{bmatrix} y_2 \\ -h(y_1)y_2 - g(y_1) \end{bmatrix}$. Let us assume that $h(x)$, $g(x)$, and $\dfrac{dg(x)}{dx}$ are continuous with respect to x on the entire real line $\mathbb{R}$. Also, we denote by $\vec{p}(t, \vec{\eta})$ the solution of (X.2.1) satisfying the initial condition $\vec{y}(0) = \vec{\eta}$.

First set $V(\vec{y}) = \dfrac{1}{2}y_2^2 + G(y_1)$, where $G(x) = \displaystyle\int_0^x g(s)ds$. Then,

$$\frac{\partial V}{\partial \vec{y}} = [g(y_1),\ y_2], \qquad \frac{\partial V}{\partial \vec{y}} \cdot \vec{f}(\vec{y}) = -h(y_1)\,y_2^2.$$

Set also, $\mathcal{S} = \left\{ \vec{y} : \dfrac{\partial V}{\partial \vec{y}} \cdot \vec{f}(\vec{y}) = 0 \right\}$. Then, $\vec{y} \in \mathcal{S}$ if and only if either $h(y_1) = 0$ or $y_2 = 0$.

Observation X-2-1. Denote by $\mathcal{M}$ the set of all stationary points of system (X.2.1), i.e., $\mathcal{M} = \{\vec{y} : g(y_1) = 0, y_2 = 0\}$. Then, $\mathcal{M}$ is the largest invariant set in $\mathcal{S}$ if the following three conditions are satisfied:
  (1) $h(x) \geq 0$ for $-\infty < x < +\infty$,
  (2) $h(x)$ has only isolated zeros on the entire real line $\mathbb{R}$,
  (3) $g(x)$ has only isolated zeros on the entire real line $\mathbb{R}$.
The proof of this result is left to the reader as an exercise (cf. Figure 2, where $g(\xi_1) > 0$, $g(\xi_2) < 0$ and $h(\xi_3) = 0$).

By using Theorem IX-2-2, we conclude that $\lim\limits_{t \to +\infty} \vec{p}(t, \vec{\eta}) = \vec{y} \in \mathcal{M}$ if conditions (1), (2), and (3) are satisfied and if the solution $\vec{p}(t, \vec{\eta})$ is bounded for $t \geq 0$. Note that $\mathcal{L}^+(\vec{\eta})$ is a connected subset of $\mathcal{M}$.

In Observation X-2-1, the boundedness of the solution $\vec{p}(t, \vec{\eta})$ for $t \geq 0$ was assumed. In the following three observations, we explore the boundedness of all solutions of (X.2.1). Set

$$G(x) = \int_0^x g(s)ds \qquad \text{and} \qquad H(x) = \int_0^x h(s)ds.$$

Observation X-2-2. Every solution $\vec{p}(t, \vec{\eta})$ of (X.2.1) is bounded for $t \geq 0$ if
  (i) $h(x) \geq 0$ for $-\infty < x < +\infty$ and (ii) $\lim\limits_{|x| \to +\infty} G(x) = +\infty$.
This is a simple consequence of Theorem IX-2-3. In fact, $\lim\limits_{|\vec{y}| \to +\infty} V(\vec{y}) = +\infty$.

Observation X-2-3. Every solution $\vec{p}(t, \vec{\eta})$ of (X.2.1) is bounded for $t \geq 0$, if
  (i) $h(x) \geq 0$ for $-\infty < x < +\infty$, (ii) $\lim\limits_{|x| \to +\infty} |H(x)| = +\infty$, and (iii) $xg(x) \geq 0$
for $-\infty < x < +\infty$.

Proof.

Change system (X.2.1) to

$$\text{(X.2.2)} \qquad \frac{d}{dt} \begin{bmatrix} z_1 \\ z_2 \end{bmatrix} = \begin{bmatrix} z_2 - H(z_1) \\ -g(z_1) \end{bmatrix} \quad (= \vec{F}(\vec{z})),$$

by the transformation

$$y_1 = z_1, \qquad y_2 = z_2 - H(z_1).$$

Denote by $\vec{q}(t, \vec{\zeta})$ the solution of (X.2.2) such that $\vec{z}(0) = \vec{\zeta}$. Set $V_1(\vec{z}) = \frac{1}{2} z_2^2 + G(z_1)$. Then,

$$\frac{\partial V_1}{\partial \vec{z}} = [g(z_1), \; z_2], \qquad \frac{\partial V_1}{\partial \vec{z}} \cdot \vec{F} = -g(z_1)H(z_1).$$

Note that $g(x)H(x) \geq 0$ for $-\infty < x < +\infty$ and that

$$\frac{d}{dt} V_1(\vec{q}(t, \vec{\zeta})) = \frac{\partial V_1}{\partial \vec{z}}(\vec{q}(t, \vec{\zeta})) \cdot \vec{F}(\vec{q}(t, \vec{\zeta})) \leq 0 \quad \text{for} \quad t \geq 0.$$

Hence, setting $\vec{q}(t, \vec{\zeta}) = \begin{bmatrix} z_1(t, \vec{\zeta}) \\ z_2(t, \vec{\zeta}) \end{bmatrix}$ and $V_1(\vec{\zeta}) = c \geq 0$, we obtain

$$\frac{1}{2}[z_2(t, \vec{\zeta})]^2 \leq V_1(\vec{q}(t, \vec{\zeta})) \leq c \qquad \text{for} \quad t \geq 0,$$

since $G(x) \geq 0$ for $-\infty < x < +\infty$. Therefore, Figure 3 clearly shows that $\vec{q}(t, \vec{\zeta})$ is bounded for $t \geq 0$. This implies that all solutions of (X.2.2) are also bounded for $t \geq 0$.

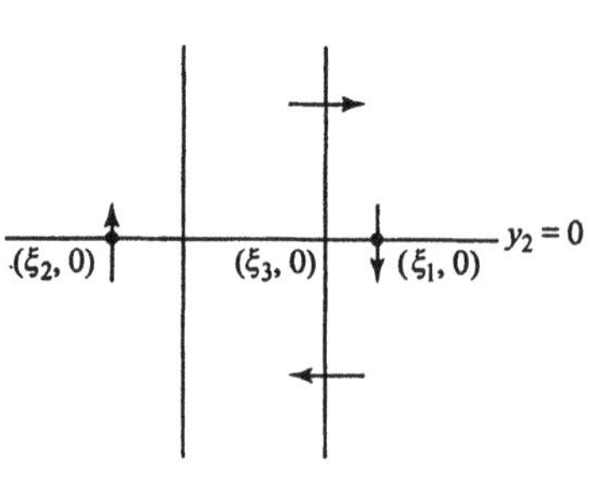

FIGURE 2.

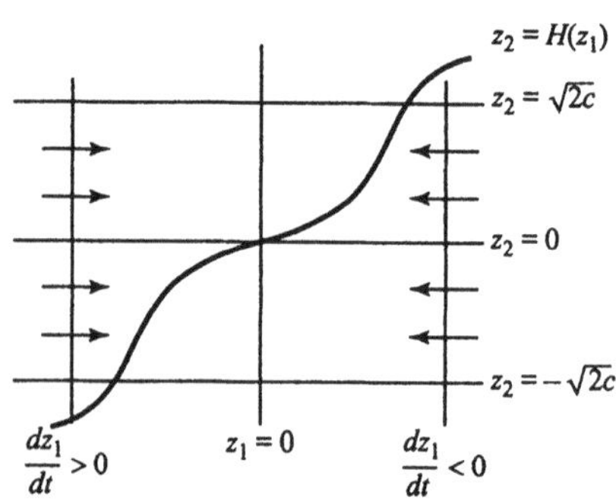

FIGURE 3.

Observation X-2-4. Every solution $\vec{p}(t, \vec{\eta})$ of (X.2.2) is bounded for $t \geq 0$ if
(i) $\lim\limits_{x \to +\infty} H(x) = +\infty$, (ii) $\lim\limits_{x \to -\infty} H(x) = -\infty$, (iii) $g(x) \geq \alpha$ for $x \geq a_0 > 0$,
and (iv) $g(x) \leq -\alpha$ for $x \leq -a_0 < 0$, where α and a_0 are some positive numbers.

Proof.

In Observation X-2-3, the *Liapounoff function*

$$V_1(\vec{z}) = \frac{1}{2}z_2^2 + G(z_1) = \frac{1}{2}[y_2 + H(y_1)]^2 + G(y_1)$$

was used. Now, let us modify V_1 to a form

$$V_2(\vec{y}) = \frac{1}{2}[y_2 + H(y_1) - k(y_1)]^2 + G(y_1).$$

Then,

$$\frac{\partial V_2}{\partial \vec{y}} = \left[[y_2 + H(y_1) - k(y_1)] \left\{ h(y_1) - \frac{dk(y_1)}{dy_1} \right\} + g(y_1), \ \ y_2 + H(y_1) - k(y_1) \right]$$

and

$$\frac{\partial V_2}{\partial \vec{y}} \cdot \vec{f} = -\frac{dk(y_1)}{dy_1} \left\{ y_2^2 + [H(y_1) - k(y_1)] y_2 \right\}$$
$$- g(y_1) [H(y_1) - k(y_1)].$$

Using (i) and (ii), three positive numbers M, a, and c can be chosen so that

$$\begin{cases} \alpha [H(x) - c] \geq M & \text{for} \quad x \geq a \geq a_0, \\ -\alpha [H(x) + c] \geq M & \text{for} \quad x \leq -a \leq -a_0. \end{cases}$$

Also choose a function $k(x)$ so that

(I) $$k(x) = \begin{cases} c & \text{for} \quad x \geq 2a, \\ -c & \text{for} \quad x \leq -2a, \end{cases}$$

(II) $$|k(x)| \leq c \quad \text{and} \quad \frac{dk(x)}{dx} \geq 0 \quad \text{for} \quad -\infty < x < +\infty,$$

and

$$\frac{dk(x)}{dx} \geq m > 0 \quad \text{for} \quad |x| \leq a$$

for some positive number m (cf. Figure 4).

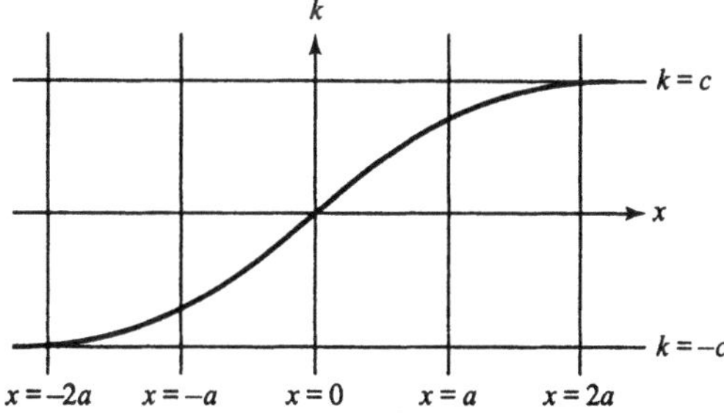

FIGURE 4.

If a positive number b is chosen sufficiently large,

$$\frac{\partial V_2}{\partial \vec{y}} \cdot \vec{f} \leq 0 \qquad \text{for} \quad |y_1| \geq 2a, \quad \text{or} \quad |y_2| \geq b.$$

In fact,

$$(\mathrm{A}) \quad -g(y_1)\,[H(y_1) - k(y_1)] \leq \begin{cases} -\alpha\,[H(y_1) - c] \leq -M < 0 & \text{for } y_1 \geq a, \\ \alpha\,[H(y_1) + c] \leq -M < 0 & \text{for } y_1 \leq -a \end{cases}$$

and

$$(\mathrm{B}) \qquad y_2^2 + [H(y_1) - k(y_1)]\,y_2 \geq 1 \qquad \text{for} \quad |y_1| \leq 2a, \ |y_2| \geq b$$

if $b > 0$ is sufficiently large. Therefore,

$$(\mathrm{i}) \quad \frac{\partial V_2}{\partial \vec{y}} \cdot \vec{f} = -g(y_1)\,[H(y_1) - k(y_1)] \leq 0 \quad \text{for} \quad |y_1| \geq 2a, \ |y_2| < +\infty,$$

$$(\mathrm{ii}) \qquad \frac{\partial V_2}{\partial \vec{y}} \cdot \vec{f} \leq -\frac{dk(y_1)}{dy_1} \leq 0 \qquad \text{for} \quad a \leq |y_1| \leq 2a, \ |y_2| \geq b,$$

and

$$(\mathrm{iii}) \qquad \frac{\partial V_2}{\partial \vec{y}} \cdot \vec{f} \leq -m\left\{ y_2^2 + [H(y_1) - k(y_1)]\,y_2 \right\}$$
$$- g(y_1)\,[H(y_1) - k(y_1)] \leq 0 \quad \text{for} \ |y_1| \leq a, \ |y_2| \geq b$$

if $b > 0$ is sufficiently large.

Since $\displaystyle \lim_{|x| \to +\infty} G(x) = +\infty$, Theorem IX-2-3 implies that every solution of (X.2.2) is bounded. $\square$

Example X-2-5. For the van der Pol equation

$$\frac{d^2 x}{dt^2} + \epsilon(x^2 - 1)\frac{dx}{dt} + x = 0,$$

where ϵ is a positive number, $h(x) = \epsilon(x^2 - 1)$ and $g(x) = x$. Hence,

$$G(x) = \frac{1}{2}x^2 \qquad \text{and} \qquad H(x) = \epsilon\left(\frac{1}{3}x^3 - x\right).$$

Therefore, conditions (i), (ii), (iii), and (iv) of Observation X-2-4 are satisfied. This implies that every solution of the van der Pol equation

$$(\mathrm{X.2.3}) \qquad \frac{d}{dt}\begin{bmatrix} y_1 \\ y_2 \end{bmatrix} = \begin{bmatrix} y_2 \\ -\epsilon(y_1^2 - 1)y_2 - y_1 \end{bmatrix}$$

is bounded for $t \geq 0$. System (X.2.3) has only one stationary point $\vec{0}$. It is easy to see that $\vec{0}$ is an unstable stationary point as $t \to +\infty$. Therefore, using the Poincaré-Bendixson Theorem (cf. Theorem IX-4-1), we conclude that there exists at least one limit cycle. In §X-3, it will be shown that system (X.2.3) has exactly one periodic solution.

Example X-2-6. For a given positive number a, the system

$$\text{(X.2.4)} \qquad \frac{d}{dt}\begin{bmatrix} y_1 \\ y_2 \end{bmatrix} = \begin{bmatrix} y_2 \\ -ay_2 - \sin(y_1) \end{bmatrix}$$

satisfies three conditions (1), (2), and (3) of Observation X-2-1. But, (X.2.4) does not satisfy conditions of Observations X-2-2, X-2-3, and X-2-4. Therefore, in order to prove the boundedness of solutions of (X.2.4), we must use some other methods. In fact, using the Liapounoff function $V(\vec{y}) = \frac{1}{2}y_2^2 - \cos(y_1)$, we obtain $\dfrac{\partial V}{\partial \vec{y}} \cdot \vec{f} = -ay_2^2 \le 0$, where $\vec{f}(\vec{y}) = \begin{bmatrix} y_2 \\ -a\,y_2 - \sin(y_1) \end{bmatrix}$. Since $V(\vec{y}) \ge -1$,

$$\text{(X.2.5)} \qquad \lim_{t\to+\infty} V(\vec{p}(t,\vec{\eta})) = \alpha$$

exists for every solution $\vec{p}(t,\vec{\eta})$ of (X.2.4). Now, observe that

(1) We must have $\alpha \le 1$. Otherwise we would have $y_2(t,\vec{\eta})^2 \ge 2(\alpha + \cos(y_1(t,\vec{\eta}))$ for $t > 0$. This implies that $\dfrac{d}{dt}V(\vec{p}(t,\vec{\eta})) \le -2a(\alpha - 1) < 0$. This contradicts (X.2.5).

(2) If $-1 \le \alpha < 1$, the solution $\vec{p}(t,\vec{\eta})$ must stay in one of connected components of the set $\{\vec{y} : V(\vec{y}) \le \alpha + \epsilon < 1\}$ for large positive t. Those connected components are bounded sets.

(3) In case $\alpha = 1$, we can show the boundedness of $\vec{p}(t,\vec{\eta})$ by investigating the behavior of solutions of (X.2.4) on the boundary of the set $\{\vec{y} : V(\vec{y}) \le 1\}$.

X-3. Existence and uniqueness of periodic orbits

In this section, we prove the following theorem (cf. [CL, p. 402, Problem 5]).

Theorem X-3-1. *Assume that*

(i) two real-valued functions $h(x)$ and $g(x)$, and $\dfrac{dg(x)}{dx}$ are continuous for $-\infty < x < +\infty$,

(ii) $g(-x) = -g(x)$ and $h(-x) = h(x)$ for $-\infty < x < +\infty$,

(iii) $g(x) > 0$ for $x > 0$,

(iv) $h(0) < 0$,

(v) $H(x) = \displaystyle\int_0^x h(s)ds$ has only one positive zero at $x = a$,

(vi) $h(x) \ge 0$ for $x \ge a$,

(vii) $H(x)$ tends to $+\infty$ as $x \to +\infty$.

Then, the system

$$\text{(X.3.1)} \qquad \frac{d}{dt}\begin{bmatrix} y_1 \\ y_2 \end{bmatrix} = \begin{bmatrix} y_2 \\ -h(y_1)y_2 - g(y_1) \end{bmatrix}$$

has exactly one nontrivial periodic orbit and all the other orbits (except for the stationary point $\vec{0}$) tend asymptotically to this periodic orbit as $t \to +\infty$.

Proof.

Change system (X.3.1) to

$$(X.3.2) \qquad \frac{d}{dt}\begin{bmatrix} z_1 \\ z_2 \end{bmatrix} = \begin{bmatrix} z_2 - H(z_1) \\ -g(z_1) \end{bmatrix}$$

by the transformation

$$\begin{bmatrix} y_1 \\ y_2 \end{bmatrix} = \begin{bmatrix} z_1 \\ z_2 - H(z_1) \end{bmatrix}.$$

Setting

$$V(\vec{z}) = \frac{z_2^2}{2} + G(z_1),$$

where

$$G(x) = \int_0^x g(s)ds \qquad \text{and} \qquad \vec{z} = \begin{bmatrix} z_1 \\ z_2 \end{bmatrix},$$

look at the way in which the function $V(\vec{z})$ changes along an orbit of (X.3.2). For example,

$$\frac{d}{dt}V(\vec{z}) = g(z_1)[z_2 - H(z_1)] - z_2 g(z_1) = -g(z_1)H(z_1)$$

along an orbit of (X.3.2). Hence,

$$\begin{cases} \dfrac{dz_2}{dz_1} = -\dfrac{g(z_1)}{z_2 - H(z_1)}, & \dfrac{dz_1}{dz_2} = -\dfrac{z_2 - H(z_1)}{g(z_1)}, \\[2mm] \dfrac{dV}{dz_1} = -\dfrac{g(z_1)H(z_1)}{z_2 - H(z_1)}, & \dfrac{dV}{dz_2} = H(z_1) \end{cases}$$

along an orbit of (X.3.2).

Observation 1. Let $\vec{z}(t,\alpha) = \begin{bmatrix} z_1(t,\alpha) \\ z_2(t,\alpha) \end{bmatrix}$ be the orbit of (X.3.2) such that $\vec{z}(0,\alpha) = \begin{bmatrix} 0 \\ \alpha \end{bmatrix}$. Then, $V(\vec{z}(0,\alpha)) = \frac{1}{2}\alpha^2$. There exists exactly one positive number α_0 such that

$$\vec{z}(\sigma_0, \alpha_0) = \begin{bmatrix} a \\ 0 \end{bmatrix} \qquad \text{and} \qquad z_2(t, \alpha_0) > 0 \quad \text{for} \quad 0 \le t < \sigma_0$$

for some positive number σ_0, where a is the unique positive zero of $H(x)$ given in condition (v) (cf. Figure 5).

Observation 2. Since $\dfrac{dz_1}{dz_2} = 0$ when $z_2 = H(z_1)$, there exist two positive numbers $\tau(\alpha)$ and $\beta(\alpha)$ such that

$$\vec{z}(\tau(\alpha), \alpha) = \begin{bmatrix} 0 \\ -\beta(\alpha) \end{bmatrix} \qquad \text{and} \qquad 0 < z_1(t, \alpha) \le a \quad \text{for} \quad 0 < t < \tau(\alpha)$$

if $0 < \alpha \le \alpha_0$. Also, since $H(x) < 0$ for $0 < x < a$, we obtain

$$\frac{d}{dt}V(\vec{z}(t,\alpha)) = -g(z_1(t,\alpha))H(z_1(t,\alpha)) > 0 \qquad \text{for} \quad 0 < t < \tau(\alpha)$$

except for $\alpha = \alpha_0$ and $t = \sigma_0$. Therefore,

(A) $V(\vec{z}(\tau(\alpha), \alpha)) - V(\vec{z}(0, \alpha)) > 0$ for $0 < \alpha \leq \alpha_0$ (cf. Figure 6).

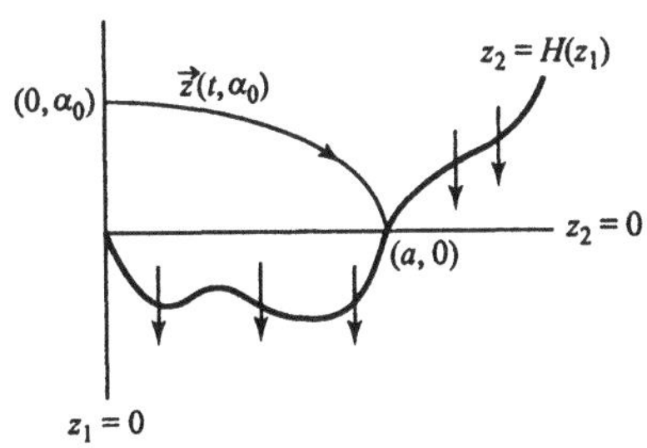

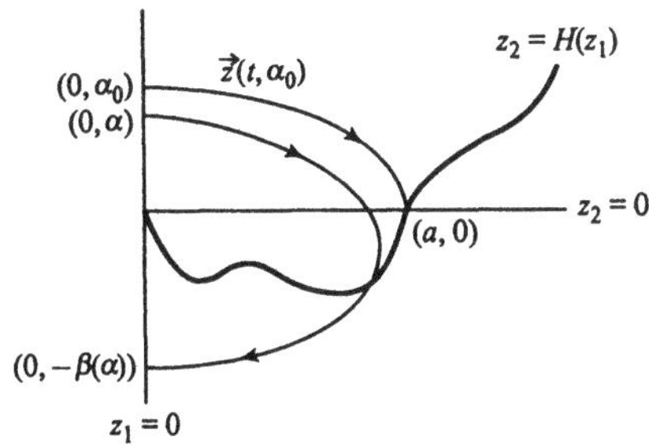

FIGURE 5.   FIGURE 6.

Observation 3. If $\alpha_0 \leq \alpha$, there exists a positive number $\tau_0(\alpha)$ such that

$$\begin{cases} z_1(\tau_0(\alpha), \alpha) = a, & 0 < z_1(t, \alpha) < a \quad \text{for} \quad 0 < t < \tau_0(\alpha), \\ 0 < z_2(t, \alpha) - H(z_1(t, \alpha)) & \text{for} \quad 0 < t < \tau_0(\alpha) \end{cases}$$

(cf. Figure 7). In particular $\tau_0(\alpha_0) = \sigma_0$ (cf. Observation 1).

If the variable t is restricted to the interval $0 \leq t \leq \tau_0(\alpha)$, the quantity $z_2(t, \alpha)$ can be regarded as a function of $z_1(t, \alpha)$, i.e.,

$$z_2(t, \alpha) = \mathcal{Z}(z_1(t, \alpha), \alpha),$$

where $\mathcal{Z}(x, \alpha)$ is a continuous function of (x, α) for $0 \leq x \leq a$ and $\alpha \geq \alpha_0$, and continuously differentiable for $0 \leq x \leq a$ and $\alpha \geq \alpha_0$ except for $x = a$ and $\alpha = \alpha_0$. Furthermore, $\mathcal{Z}(x, \alpha_1) < \mathcal{Z}(x, \alpha_2)$ for $0 \leq x \leq a$ if $\alpha_0 \leq \alpha_1 < \alpha_2$ (cf. Figure 7).

Set $\vec{\mathcal{Z}}(x, \alpha) = \begin{bmatrix} x \\ \mathcal{Z}(x, \alpha) \end{bmatrix}$. Then,

$$\frac{d}{dx} V(\vec{\mathcal{Z}}(x, \alpha)) = -\frac{g(x)H(x)}{\mathcal{Z}(x, \alpha) - H(x)} > 0 \quad \text{for} \quad 0 < x < a$$

and, hence,

$$\frac{d}{dx} V(\vec{\mathcal{Z}}(x, \alpha_1)) > \frac{d}{dx} V(\vec{\mathcal{Z}}(x, \alpha_2)) \quad \text{for} \quad 0 < x < a$$

if $\alpha_0 \leq \alpha_1 < \alpha_2$. Thus, we obtain

(I) $V(\vec{z}(\tau_0(\alpha_1), \alpha_1)) - V(\vec{z}(0, \alpha_1)) > V(\vec{z}(\tau_0(\alpha_2), \alpha_2)) - V(\vec{z}(0, \alpha_2)) > 0$

for $\alpha_0 \leq \alpha_1 < \alpha_2$.

Observation 4. If $\alpha_0 < \alpha$, there exists a positive number $\tau_1(\alpha)$ such that $\tau_1(\alpha) > \tau_0(\alpha)$, $z_1(\tau_1(\alpha), \alpha) = a$, and $z_1(t, \alpha) > a$ for $\tau_0(\alpha) < t < \tau_1(\alpha)$ (cf. Figure 8).

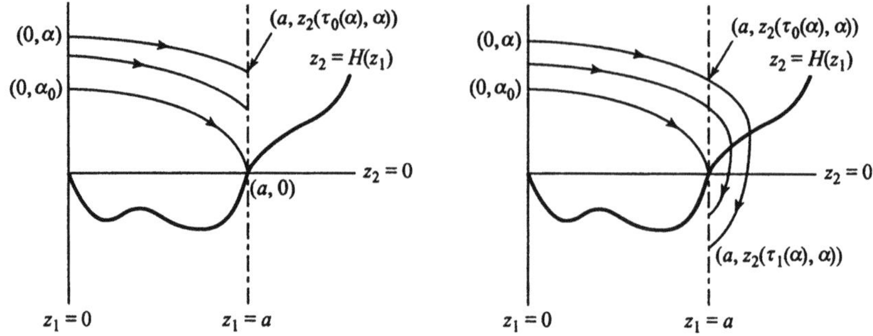

FIGURE 7.   FIGURE 8.

Note that $z_2(\tau_1(\alpha), \alpha) < 0$ and that $H(z_1(t, \alpha)) > 0$ for $\tau_0(\alpha) < t < \tau_1(\alpha)$ since $z_1(t, \alpha) > a$ on this t-interval. Regarding $z_1(t, \alpha)$ as a function of $z_2(t, \alpha)$ for $z_2(\tau_1(\alpha), \alpha) \leq z_2 \leq z_2(\tau_0(\alpha), \alpha)$, we obtain

$$(\text{II}) \quad 0 > V(\vec{z}(\tau_1(\alpha_1), \alpha_1)) - V(\vec{z}(\tau_0(\alpha_1), \alpha_1)) > V(\vec{z}(\tau_1(\alpha_2), \alpha_2)) - V(\vec{z}(\tau_0(\alpha_2), \alpha_2))$$

for $\alpha_0 < \alpha_1 < \alpha_2$ in a way similar to Observation 3 (cf. Figure 8).

Observation 5. If $\alpha_0 < \alpha$, there exists a positive number $\tau(\alpha)$ such that $\tau(\alpha) > \tau_1(\alpha)$, $z_1(\tau(\alpha), \alpha) = 0$, and $0 < z_1(t, \alpha) < a$ for $\tau_1(\alpha) < t < \tau(\alpha)$ (cf. Figure 9). Note that $z_2(t, \alpha) < H(z_1(t, \alpha)) < 0$ for $\tau_1(\alpha) < t < \tau(\alpha)$. Again, regarding $z_2(t, \alpha)$ as a function of $z_1(t, \alpha)$ in the same way as in Observation 3, we can derive

$$(\text{III}) \quad V(\vec{z}(\tau(\alpha_1), \alpha_1)) - V(\vec{z}(\tau_1(\alpha_1), \alpha_1)) > V(\vec{z}(\tau(\alpha_2), \alpha_2)) - V(\vec{z}(\tau_1(\alpha_2), \alpha_2)) > 0$$

if $\alpha_0 < \alpha_1 < \alpha_2$ (cf. Figure 9).

Observation 6. Thus, by adding (I), (II), and (III), we obtain

$$(\text{B}) \qquad V(\vec{z}(\tau(\alpha_1), \alpha_1)) - V(\vec{z}(0, \alpha_1)) > V(\vec{z}(\tau(\alpha_2), \alpha_2)) - V(\vec{z}(0, \alpha_2)) > 0$$

if $\alpha_0 \leq \alpha_1 < \alpha_2$. This implies that the function $\mathcal{G}(\alpha)$ defined by

$$\mathcal{G}(\alpha) = V(\vec{z}(\tau(\alpha), \alpha)) - V(\vec{z}(0, \alpha)) = \frac{1}{2} z_2(\tau(\alpha), \alpha)^2 - \frac{1}{2}\alpha^2$$

is strictly decreasing for $\alpha \geq \alpha_0$ as $\alpha \to +\infty$. Also, $\mathcal{G}(\alpha) > 0$ for $0 < \alpha \leq \alpha_0$ (cf. (A)).

Observation 7. Since

$$\begin{cases} \displaystyle\lim_{|z_2| \to +\infty} \frac{dV}{dz_1} = 0 & \text{uniformly} \quad \text{for} \quad 0 \leq z_1 \leq a, \\[2ex] \displaystyle\lim_{z_1 \to +\infty} \frac{dV}{dz_2} = +\infty & \text{uniformly} \quad \text{for} \quad -\infty < z_2 < +\infty, \end{cases}$$

it follows that

$$\begin{cases} \lim_{\alpha \to +\infty} [V(\vec{z}(\tau_0(\alpha),\alpha)) - V(\vec{z}(0,\alpha))] = 0, \\ \lim_{\alpha \to +\infty} [V(\vec{z}(\tau_1(\alpha),\alpha)) - V(\vec{z}(\tau_0(\alpha),\alpha))] = -\infty, \\ \lim_{\alpha \to +\infty} [V(\vec{z}(\tau(\alpha),\alpha)) - V(\vec{z}(\tau_1(\alpha),\alpha))] = 0. \end{cases}$$

Therefore,

$$\text{(C)} \qquad \lim_{\alpha \to +\infty} \mathcal{G}(\alpha) = -\infty.$$

Thus, we conclude that $\mathcal{G}$ has exactly one positive zero α_+, i.e.,

$$\text{(X.3.3)} \qquad \mathcal{G}(\alpha) = \frac{1}{2}z_2(\tau(\alpha),\alpha)^2 - \frac{1}{2}\alpha^2 \begin{cases} > 0, & 0 < \alpha < \alpha_+, \\ = 0, & \alpha = \alpha_+, \\ < 0, & \alpha > \alpha_+ \end{cases}$$

(cf. Figure 10).

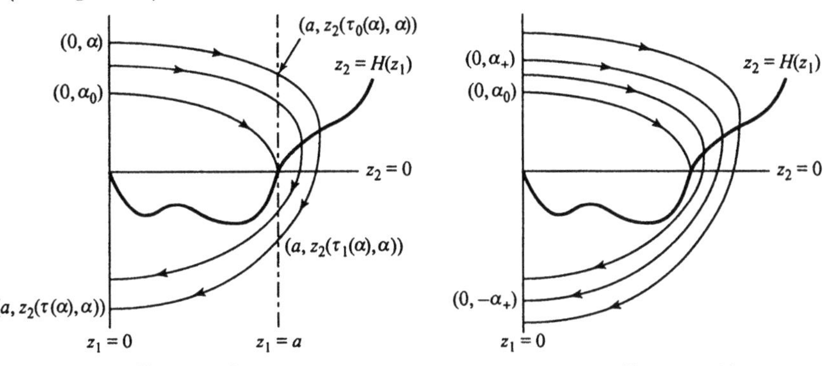

FIGURE 9.   FIGURE 10.

From (X.3.3) and symmetric properties (ii) of the functions $h(x)$ and $g(x)$, we conclude that $\vec{z}(t,\alpha_+)$ is the only periodic orbit, and all the other orbits tends to $\vec{z}(t,\alpha_+)$ asymptotically since

$$\begin{aligned} |z_2(\tau(\alpha),\alpha)| > \alpha \quad &\text{if } 0 < \alpha < \alpha_+, \\ |z_2(\tau(\alpha),\alpha)| < \alpha \quad &\text{if } \alpha > \alpha_+. \end{aligned}$$

Thus, we complete the proof of Theorem X-3-1.   □

Remark X-3-2. Condition (iv) of Theorem X-3-1 can be replaced by the following condition:

(iv′) there exists a positive number δ such that $H(x) < 0$ for $0 < x < \delta$.

X-4. Multipliers of the periodic orbit of the van der Pol equation

In Example X-2-5, we looked at the van der Pol equation

$$(X.4.1) \qquad \frac{d^2x}{dt^2} + \epsilon(x^2 - 1)\frac{dx}{dt} + x = 0,$$

where ϵ is a positive number. Using Observation X-2-4, it was shown that every orbit of the system

$$(X.4.2) \qquad \frac{d}{dt}\begin{bmatrix} y_1 \\ y_2 \end{bmatrix} = \begin{bmatrix} y_2 \\ -\epsilon(y_1^2 - 1)y_2 - y_1 \end{bmatrix}$$

is bounded for $t \geq 0$. It was also remarked that $\vec{0}$ is the only stationary point of system (X.4.2) and that the stationary point $\vec{0}$ is not stable as $t \to +\infty$. In fact, the linear part of the right-hand side of (X.4.2) at $\vec{y} = \vec{0}$ is $A\vec{y}$, where $A = \begin{bmatrix} 0 & 1 \\ -1 & \epsilon \end{bmatrix}$ and $\vec{y} = \begin{bmatrix} y_1 \\ y_2 \end{bmatrix}$. Since trace$[A]=\epsilon$ and det$[A] = 1$, the stationary point $\vec{0}$ is an unstable node for $\epsilon \geq 2$, while $\vec{0}$ is an unstable spiral point for $0 < \epsilon < 2$. Therefore, using the Poincaré-Bendixson Theorem (cf. Theorem IX-4-1), we conclude that there exists at least one limit cycle. Now, Theorem X-3-1 implies that system (X.4.2) has exactly one limit cycle and all the other orbits except for the stationary point $\vec{0}$ approach this limit cycle asymptotically as $t \to +\infty$. In fact, since $h(x) = \epsilon(x^2 - 1)$, $H(x) = \epsilon\left[\frac{x^3}{3} - x\right]$, and $g(x) = x$, the seven conditions (i) – (vii) of Theorem X-3-1 are satisfied. In particular, the positive zero of $H(x)$ is $a = \sqrt{3} > 1$.

In this section, we prove the following theorem concerning the multipliers of the unique periodic solution $x = x(t, \epsilon)$ of the van der Pol equation (X.4.1).

Theorem X-4-1. *The multipliers of the periodic solution $x(t, \epsilon)$ of (X.4.1) are 1 and ρ such that $|\rho| < 1$.*

Proof.

If we set $v = \dfrac{x^2 + y^2}{2}$, where $y = \dfrac{dx}{dt}$, then $\dfrac{dv}{dt} = -\epsilon(x^2 - 1)y^2$. This implies that $\epsilon\displaystyle\int (x(t)^2 - 1)dt = -\int \dfrac{dv}{y(t)^2}$. Now look at Figure 11.

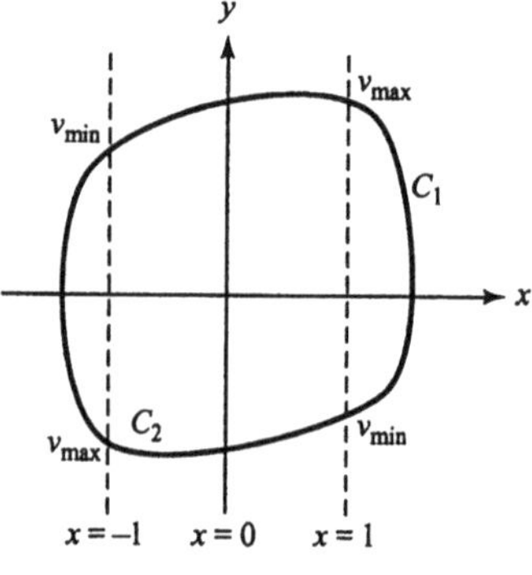

FIGURE 11.

On each of two curves C_1 and C_2, let us denote y as a function of v by $y_1(v)$ and $y_2(v)$ respectively. Note that $x^2 > 1$ on C_1, but $x^2 < 1$ on C_2. Hence, $y_1(v)^2 < y_2(v)^2$. This implies that

$$\epsilon \int_0^T (x(t)^2 - 1)dt \; > \; 0,$$

where T is a period. Therefore, we can complete the proof by using the following lemma.

Lemma X-4-2. *Assume that two functions $f(x)$ and $g(x)$ are continuously differentiable in $\mathbb{R}$. Assume also that the differential equation*

$$\frac{d^2x}{dt^2} \; + \; f(x)\frac{dx}{dt} \; + \; g(x) \; = \; 0$$

has a nontrivial periodic solution $x(t)$ of period 1 such that $\displaystyle\int_0^1 f(x(t))dt > 0$. Then, the multipliers of the periodic solution $x(t)$ are 1 and ρ such that $|\rho| < 1$.

Remark X-4-3. If system (X.4.2) has more than one periodic orbit, then at least one of them must be orbitally unstable. Therefore, the proof of Theorem X-4-1 is another proof of the uniqueness of periodic orbit of (X.4.2).

X-5. The van der Pol equation for a small $\epsilon > 0$

In this section, we explain a method to locate the unique periodic orbit of differential equation (X.4.1) (or system (X.4.2)) for a small $\epsilon > 0$.

Set $\epsilon = 0$. Then, system (X.4.2) becomes

$$(\text{X.5.1}) \qquad\qquad \frac{d}{dt}\begin{bmatrix} y_1 \\ y_2 \end{bmatrix} \; = \; \begin{bmatrix} y_2 \\ -y_1 \end{bmatrix}.$$

Every orbit of (X.5.1) is a circle and of period 2π in t. We expect that the periodic orbit of (X.4.2) must be approximated by one of those circles as $\epsilon \to 0$. The main problem is to find the radius of the circle which approximates the periodic orbit of (X.4.2) for a sufficiently small $\epsilon > 0$.

Since the periodic orbit and its period are functions of ϵ, we normalize the independent variable t by the change of independent variable

$$(\text{X.5.2}) \qquad\qquad t \; = \; (1 + \epsilon\omega)\tau$$

so that the period of the periodic orbit becomes 2π for every $\epsilon > 0$. Transformation (X.5.2) changes differential equation (X.4.1) to

$$\frac{d^2x}{d\tau^2} \; + \; \epsilon(1 + \epsilon\omega)(x^2 - 1)\frac{dx}{d\tau} \; + \; (1 + \epsilon\omega)^2 x \; = \; 0$$

that can be written in the form

$$(\text{X.5.3}) \qquad \frac{d^2x}{d\tau^2} \; + \; x \; = \; \epsilon(1 + \epsilon\omega)(1 - x^2)\frac{dx}{d\tau} \; - \; \epsilon(2\omega + \epsilon\omega^2)x.$$

Observation X-5-1. In the case when a real-valued function $f(\tau)$ is continuous and periodic of period 2π in τ, the general solution of the linear nonhomogeneous differential equation

$$(\text{X.5.4}) \qquad \frac{d^2x}{d\tau^2} + x = f(\tau)$$

is given by

$$(\text{X.5.5}) \qquad x(\tau) = K\cos(\tau + \phi) + \frac{1}{2\pi}\int_\tau^{\tau+2\pi} sf(s)\sin(\tau - s)ds$$

as it is shown with a straight forward calculation, where K and ϕ are arbitrary constants. This solution is periodic in τ of period 2π if and only if

$$(\text{X.5.6}) \qquad \int_0^{2\pi} f(s)\sin(s)ds = 0 \qquad \text{and} \qquad \int_0^{2\pi} f(s)\cos(s)ds = 0.$$

Observation X-5-2. In the case when condition (X.5.6) is not satisfied, set

$$(\text{X.5.7}) \qquad S[f] = \frac{1}{\pi}\int_0^{2\pi} f(s)\sin(s)ds \quad \text{and} \quad C[f] = \frac{1}{\pi}\int_0^{2\pi} f(s)\cos(s)ds.$$

Then,

$$(\text{X.5.8}) \qquad \begin{aligned} v(\tau) &= K\cos(\tau + \phi) \\ &+ \frac{1}{2\pi}\int_\tau^{\tau+2\pi} s\left[f(s) - S[f]\sin(s) - C[f]\cos(s)\right]\sin(\tau - s)ds \end{aligned}$$

is the general solution of the differential equation

$$\frac{d^2x}{d\tau^2} + x = f(\tau) - S[f]\sin(\tau) - C[f]\cos(\tau).$$

Furthermore, $v(\tau)$ is periodic of period 2π.

Observation X-5-3. Set

$$f\left(x, \frac{dx}{d\tau}, \omega, \epsilon\right) = (1 + \epsilon\omega)(1 - x^2)\frac{dx}{d\tau} - (2\omega + \epsilon\omega^2)x.$$

Letting K be a parameter and a function $v(\tau, K, \omega, \epsilon)$ be periodic in τ of period 2π, set

$$\begin{cases} S(K,\omega,\epsilon) = \dfrac{1}{\pi}\displaystyle\int_0^{2\pi} f\left(v(s,K,\omega,\epsilon), \frac{dv}{d\tau}(s,K,\omega,\epsilon), \omega, \epsilon\right)\sin(s)ds, \\[2ex] C(K,\omega,\epsilon) = \dfrac{1}{\pi}\displaystyle\int_0^{2\pi} f\left(v(s,K,\omega,\epsilon), \frac{dv}{d\tau}(s,K,\omega,\epsilon), \omega, \epsilon\right)\cos(s)ds. \end{cases}$$

Now, let us consider an integral equation

$$v(\tau, K, \omega, \epsilon) = K\cos(\tau) + \frac{\epsilon}{2\pi} \int_\tau^{\tau+2\pi} s\left[f\left(v, \frac{dv}{d\tau}, \omega, \epsilon\right)\right.$$

(X.5.9)

$$\left. - S(K, \omega, \epsilon)\sin(s) - C(K, \omega, \epsilon)\cos(s)\right]\sin(\tau - s)ds.$$

Solutions of (X.5.9) satisfy the differential equation

$$(\text{X.5.10}) \quad \frac{d^2v}{d\tau^2} + v = \epsilon\left[f\left(v, \frac{dv}{d\tau}, \omega, \epsilon\right) - S(K, \omega, \epsilon)\sin(\tau) - C(K, \omega, \epsilon)\cos(\tau)\right]$$

as long as v is periodic in τ of period 2π. This integral equation can be solved by using successive approximations in such a way that the solution $v(\tau, K, \omega, \epsilon)$ is a convergent power series in ϵ:

$$(\text{X.5.11}) \quad v(\tau, K, \omega, \epsilon) = \sum_{m=0}^\infty \epsilon^m v_m(\tau, K, \omega) = K\cos(\tau) + \epsilon\, v_1(\tau, K, \omega) + \cdots,$$

where $v_m(\tau, K, \omega)$ are polynomials in (K, ω) with coefficients periodic in τ of period 2π. For any given positive numbers K_0 and ω_0, there exists another positive number $\epsilon_0(K_0, \omega_0)$ such that series (X.5.11) is uniformly convergent for

$$|K| \leq K_0, \qquad |\omega| \leq \omega_0, \qquad |\epsilon| \leq \epsilon_0(K_0, \omega_0), \qquad -\infty < \tau < +\infty.$$

This implies that $S(K, \omega, \epsilon)$ and $C(K, \omega, \epsilon)$ are also convergent power series in ϵ:

$$(\text{X.5.12}) \quad S(K, \omega, \epsilon) = \sum_{m=0}^\infty \epsilon^m S_m(K, \omega) \quad \text{and} \quad C(K, \omega, \epsilon) = \sum_{m=0}^\infty \epsilon^m C_m(K, \omega),$$

where $S_m(K, \omega)$ and $C_m(K, \omega)$ are polynomials in (K, ω) with constant coefficients. These two power series also converge uniformly for

$$|K| \leq K_0, \qquad |\omega| \leq \omega_0, \qquad |\epsilon| \leq \epsilon_0(K_0, \omega_0).$$

Observation X-5-4. Inserting series (X.5.11) and (X.5.12) into system (X.5.10), we obtain

$$\frac{d^2v_1}{d\tau^2} + v_1 = (1 - K^2\cos^2(\tau))(-K\sin(\tau)) - 2\omega K\cos(\tau)$$

$$- S_0(K, \omega)\sin(\tau) - C_0(K, \omega)\cos(\tau)$$

$$= K^3\cos^2(\tau)\sin(\tau) - K\sin(\tau) - 2\omega K\cos(\tau)$$

$$- S_0(K, \omega)\sin(\tau) - C_0(K, \omega)\cos(\tau)$$

$$= \frac{1}{4}K^3\sin(3\tau) + \left[\frac{1}{4}K^3 - K - S_0(K, \omega)\right]\sin(\tau)$$

$$- [2\omega K + C_0(K, \omega)]\cos(\tau).$$

Since v_1 is periodic in τ of period 2π, we must have

$$S_0(K,\omega) = \frac{1}{4}K^3 - K \qquad \text{and} \qquad C_0(K,\omega) = -2\omega K.$$

This implies that $S(2,0,0) = 0$, $C(2,0,0) = 0$, and

$$\begin{vmatrix} \dfrac{\partial S}{\partial K}(2,0,0) & \dfrac{\partial S}{\partial \omega}(2,0,0) \\[2mm] \dfrac{\partial C}{\partial K}(2,0,0) & \dfrac{\partial C}{\partial \omega}(2,0,0) \end{vmatrix} = \begin{vmatrix} 2 & 0 \\ 0 & -4 \end{vmatrix} = -8.$$

Therefore, the system of equations $S(K,\omega,\epsilon) = 0$, $C(K,\omega,\epsilon) = 0$ has a solution $K(\epsilon) = 2 + O(\epsilon)$, $\omega(\epsilon) = O(\epsilon)$. The functions $K(\epsilon)$ and $\omega(\epsilon)$ are power series in ϵ which converge if $|\epsilon|$ is sufficiently small.

Observation X-5-5. Set

$$x(t,\epsilon) = v\left(\frac{t}{1+\epsilon\omega(\epsilon)}, K(\epsilon), \omega(\epsilon), \epsilon\right).$$

Then, $x(t,\epsilon)$ is a periodic solution of (X.5.3) and

$$x(t,\epsilon) = K(\epsilon)\cos\left(\frac{t}{1+\epsilon\omega(\epsilon)}\right) + O(\epsilon) \qquad \text{as} \quad \epsilon \to 0.$$

From the fact that

$$y_2(t,\epsilon) = \frac{dx}{dt}(t,\epsilon) = -\frac{K(\epsilon)}{1+\epsilon\omega(\epsilon)}\sin\left(\frac{t}{1+\epsilon\omega(\epsilon)}\right) + O(\epsilon) \qquad \text{as} \quad \epsilon \to 0,$$

we obtain the following conclusion.

Conclusion X-5-6. *The unique periodic orbit of (X.4.2) tends to the circle* $y_1^2 + y_2^2 = 4$ *as* $\epsilon \to 0^+$.

X-6. The van der Pol equation for a large parameter

In this section, we consider the van der Pol equation (X.4.1) for a large ϵ. Let us write (X.4.1) in the form

$$(\text{X.6.1}) \qquad \frac{d}{dt}\begin{bmatrix} z_1 \\ z_2 \end{bmatrix} = \begin{bmatrix} z_2 - \epsilon\left(\dfrac{z_1^3}{3} - z_1\right) \\ -z_1 \end{bmatrix}$$

by setting $x = z_1$ and $\dfrac{dx}{dt} = z_2 - \epsilon\left(\dfrac{z_1^3}{3} - z_1\right)$.

Set $\begin{bmatrix} z_1 \\ z_2 \end{bmatrix} = \begin{bmatrix} w_1 \\ \epsilon\, w_2 \end{bmatrix}$ and $t = \epsilon\tau$. Then, system (X.6.1) becomes

$$(\text{X.6.2}) \qquad \frac{d}{d\tau}\begin{bmatrix} \beta^2\, w_1 \\ w_2 \end{bmatrix} = \begin{bmatrix} w_2 - \left(\dfrac{w_1^3}{3} - w_1\right) \\ -w_1 \end{bmatrix},$$

where $\beta = \dfrac{1}{\epsilon}$.

Observation X-6-1. Note that, along an orbit of (X.6.2),

$$(\text{X.6.3}) \qquad \frac{dw_2}{dw_1} = -\frac{\beta^2 w_1}{w_2 - \left(\dfrac{w_1^3}{3} - w_1\right)}.$$

Therefore, if $\beta > 0$ is sufficiently small, the slope $\dfrac{dw_2}{dw_1}$ of any orbit is small at a point (w_1, w_2) far away from the curve $C : w_2 = \dfrac{w_1^3}{3} - w_1$. This implies that every orbit moves toward the curve C almost horizontally (cf. Figure 12).

Observation X-6-2. From Observation X-6-1 a rough picture of orbits of (X.6.2) is obtained (cf. Figure 13).

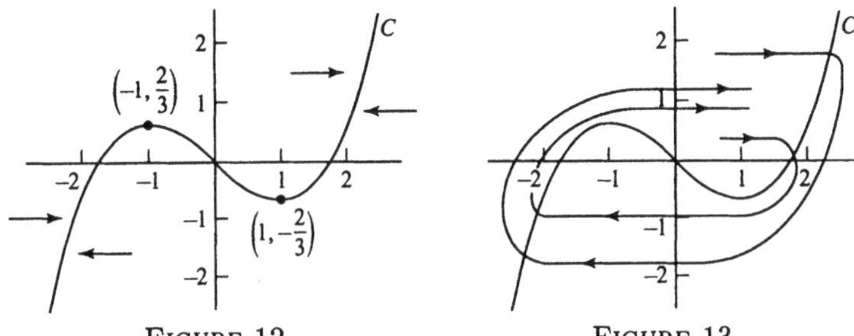

FIGURE 12.   FIGURE 13.

Actually, defining the curve C_0 by Figure 14, we prove the following theorem.

Theorem X-6-3. *For a sufficiently small positive number β, the unique periodic orbit of (X.6.2) is located in an open set $\mathcal{V}(\beta)$ such that the closure of $\mathcal{V}(\beta)$ contains the curve C_0 and shrinks to C_0 as $\beta \to 0^+$.*

Proof of Theorem X-6-3.   In eight steps, we construct an open set $\mathcal{V}(\beta)$ so that
   (1) the closure of $\mathcal{V}(\beta)$ contains the curve C_0,
   (2) the closure of $\mathcal{V}(\beta)$ shrinks to the curve C_0 as $\beta \to 0^+$,
   (3) if an orbit of (X.6.2) enters in the open set $\mathcal{V}(\beta)$ at $\tau = \tau_0$, then the orbit stays in $\mathcal{V}(\beta)$ for $\tau \geq \tau_0$.

Step 1. Fixing a number $a(\beta) > 2$, we use the line segment

$$C_1(\beta) = \left\{ \left(w_1, \frac{a(\beta)^3}{3} - a(\beta) \right) : 0 \leq w_1 \leq a(\beta) \right\}$$

as a part of the boundary $\partial \mathcal{V}(\beta)$ of $\mathcal{V}(\beta)$ (cf. Figure 15).

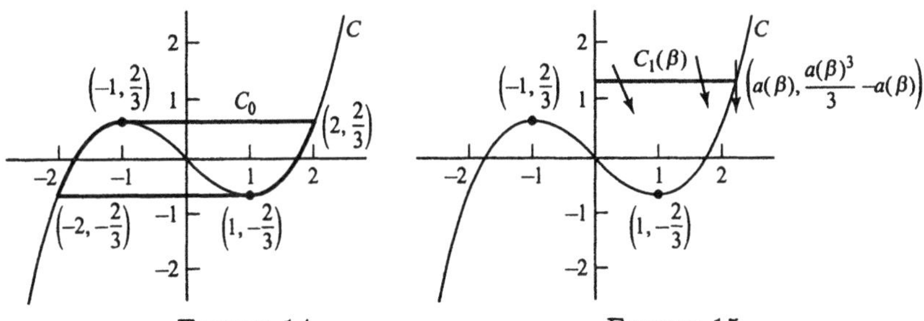

FIGURE 14.   FIGURE 15.

Note that the slope of an orbit given by (X.6.3) is negative on $C_1(\beta)$ and decreasing to $-\infty$ as w_1 increases to $a(\beta)$. Also, $\dfrac{dw_2}{dw_1}$ is 0 on $C_1(\beta)$.

Step 2. Let us consider a curve

$$C_2(\beta) \; = \; \left\{ (w_1, w_2) : \; w_2 \; = \; \frac{w_1^3}{3} - w_1 - \beta\mu(w_1, \beta), \quad 1 \le w_1 \le a(\beta) \right\},$$

where

   (i) $\mu(w_1, \beta)$ is continuous for $1 \le w_1 \le a(\beta)$ and continuously differentiable for $1 \le w_1 < a(\beta)$,

  (ii) $\mu(w_1, \beta) > 0$ for $1 \le w_1 < a(\beta)$, and (iii) $\mu(a(\beta), \beta) = 0$.

On the curve $C_2(\beta)$,

$$w_2 - \left(\frac{w_1^3}{3} - w_1 \right) \; = \; -\beta\mu(w_1, \beta)$$

and, hence,

$$-\frac{\beta^2 w_1}{w_2 - \left(\dfrac{w_1^3}{3} - w_1 \right)} \; = \; \frac{\beta w_1}{\mu(w_1, \beta)}.$$

On the other hand, the slope of the curve $C_2(\beta)$ is

$$\frac{dw_2}{dw_1} \; = \; w_1^2 - 1 - \beta \frac{d\mu(w_1, \beta)}{dw_1}.$$

This implies that if μ is fixed by the initial-value problem

(X.6.4) $$\frac{d\mu}{dw_1} \; = \; -\frac{w_1}{\mu}, \qquad \mu(a(\beta), \beta) = 0,$$

we obtain

$$-\frac{\beta^2 w_1}{w_2 - \left(\dfrac{w_1^3}{3} - w_1 \right)} \; < \; \frac{dw_2}{dw_1} \qquad \text{for} \quad 1 < w_1 < a(\beta),$$

where $\dfrac{dw_2}{dw_1}$ denotes the slope of $C_2(\beta)$. The unique solution to problem (X.6.4) is given by

$$\mu(w_1, \beta) \; = \; \sqrt{a(\beta)^2 - w_1^2}.$$

Thus, we choose the curve

$$C_2(\beta) = \left\{ (w_1, w_2) : \; w_2 = \frac{w_1^3}{3} - w_1 - \beta\sqrt{a(\beta)^2 - w_1^2}, \quad 1 \le w_1 \le a(\beta) \right\}$$

as a part of the boundary $\partial V(\beta)$ of $V(\beta)$ (cf. Figure 16). Note that on the curve $C_2(\beta)$,

$$w_2 \; = \; -\frac{2}{3} - \beta\sqrt{a(\beta)^2 - 1} \qquad \text{at} \quad w_1 = 1.$$

Step 3. We choose the curve

$$\mathcal{C}_3(\beta) = \left\{ (w_1, w_2) : \ w_2 = -\frac{2}{3} - \beta\sqrt{a(\beta)^2 - w_1^2}, \ \ 0 \leq w_1 \leq 1 \right\}$$

as a part of the boundary $\partial \mathcal{V}(\beta)$ of $\mathcal{V}(\beta)$ (cf. Figure 17).

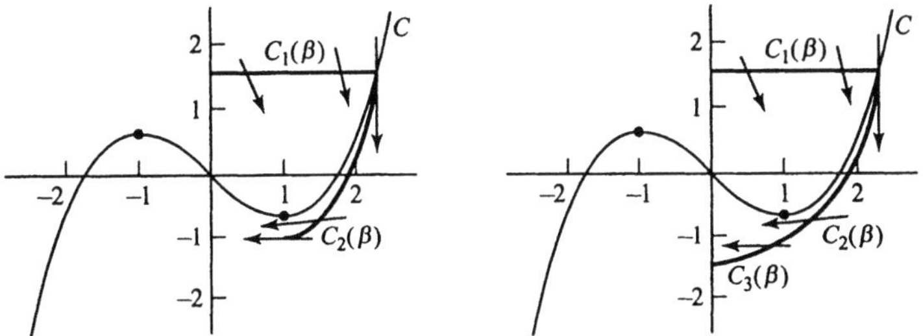

FIGURE 16.   FIGURE 17.

On the curve $\mathcal{C}_3(\beta)$,

$$w_2 - \left(\frac{w_1^3}{3} - w_1 \right) = -\frac{2}{3} - \left(\frac{w_1^3}{3} - w_1 \right) - \beta\sqrt{a(\beta)^2 - w_1^2}$$

and, hence,

$$-\frac{\beta^2 w_1}{w_2 - \left(\dfrac{w_1^3}{3} - w_1 \right)} = \frac{\beta^2 w_1}{\beta\sqrt{a(\beta)^2 - w_1^2} + \dfrac{2}{3} + \left(\dfrac{w_1^3}{3} - w_1 \right)}.$$

On the other hand, the slope of the curve $\mathcal{C}_3(\beta)$ is $\dfrac{dw_2}{dw_1} = \dfrac{\beta w_1}{\sqrt{a(\beta)^2 - w_1^2}}$. This implies that

$$-\frac{\beta^2 w_1}{w_2 - \left(\dfrac{w_1^3}{3} - w_1 \right)} < \frac{dw_2}{dw_1} \quad \text{for} \ \ 0 \leq w_1 < 1,$$

where $\dfrac{dw_2}{dw_1}$ denotes the slope of $\mathcal{C}_3(\beta)$ (cf. Figure 17). Note that on the curve $\mathcal{C}_3(\beta)$, $w_2 = -\dfrac{2}{3} - \beta a(\beta)$ at $w_1 = 0$.

Step 4. Now, let us fix the positive number $a(\beta) > 2$ by the equation

$$(\mathrm{X}.6.5) \qquad \frac{2}{3} + \beta a(\beta) = \frac{a(\beta)^3}{3} - a(\beta).$$

Denote by $\mathcal{C}_+(\beta)$ the curve consisting of $\mathcal{C}_1(\beta)$, $\mathcal{C}_2(\beta)$, and $\mathcal{C}_3(\beta)$, i.e., $\mathcal{C}_+(\beta) = \mathcal{C}_1(\beta) \cup \mathcal{C}_2(\beta) \cup \mathcal{C}_3(\beta)$. Let $\mathcal{C}_-(\beta)$ be the symmetric image of $\mathcal{C}_+(\beta)$ with respect to the point $(0,0)$. Then, $\mathcal{C}_+(\beta) \cup \mathcal{C}_-(\beta)$ is a closed curve if $a(\beta)$ satisfies condition $(\mathrm{X}.6.5)$. We use the closed curve $\mathcal{C}_+(\beta) \cup \mathcal{C}_-(\beta)$ as a part of the boundary $\partial \mathcal{V}(\beta)$ of $\mathcal{V}(\beta)$ (cf. Figure 18).

Step 5. Fixing a positive number $b(\beta) < 2$ so that $\dfrac{2}{3} = \dfrac{b(\beta)^3}{3} - b(\beta) + \beta b(\beta)$, we choose the curve

$$\Gamma_1(\beta) = \left\{ (w_1, w_2) : w_2 = \frac{b(\beta)^3}{3} - b(\beta) + \beta\sqrt{b(\beta)^2 - w_1^2}, \quad 0 \le w_1 \le b(\beta) \right\}$$

as a part of the boundary $\partial V(\beta)$ of $V(\beta)$ (cf. Figure 19). Note that, on $\Gamma_1(\beta)$, $w_2 = \dfrac{2}{3}$ at $w_1 = 0$.

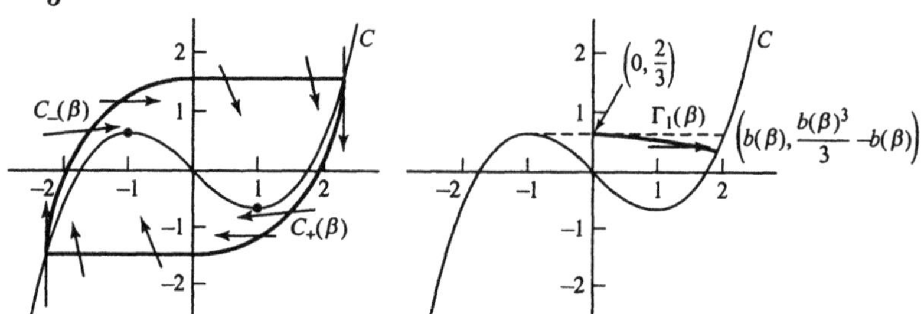

FIGURE 18.   FIGURE 19.

On the curve $\Gamma_1(\beta)$,

$$-\frac{\beta^2 w_1}{w_2 - \left(\dfrac{w_1^3}{3} - w_1\right)} = -\frac{\beta^2 w_1}{\beta\sqrt{b(\beta)^2 - w_1^2} + \dfrac{b(\beta)^3}{3} - b(\beta) - \left(\dfrac{w_1^3}{3} - w_1\right)}$$

$$> -\frac{\beta\, w_1}{\sqrt{b(\beta)^2 - w_1^2}}.$$

On the other hand, the slope of the curve $\Gamma_1(\beta)$ is $\dfrac{dw_2}{dw_1} = -\dfrac{\beta w_1}{\sqrt{b(\beta)^2 - w_1^2}}$.

Step 6. We use the two curves

$$\begin{cases} \Gamma_2(\beta) = \left\{ (w_1, w_2) : w_2 = \dfrac{w_1^3}{3} - w_1, \ 1 \le w_1 \le b(\beta) \right\}, \\[2mm] \Gamma_3(\beta) = \left\{ (w_1, w_2) : w_2 = -\dfrac{2}{3}, \ 0 \le w_1 \le 1 \right\} \end{cases}$$

as parts of the boundary $\partial V(\beta)$ of $V(\beta)$ (cf. Figure 20). Note that on $\Gamma_3(\beta)$, the slope of an orbit given by (X.6.3) is positive and $\dfrac{dw_2}{d\tau}$ is negative.

Step 7. Denote by $\Gamma_+(\beta)$ the curve consisting of $\Gamma_1(\beta)$, $\Gamma_2(\beta)$, and $\Gamma_3(\beta)$, i.e.,

$$\Gamma_+(\beta) = \Gamma_1(\beta) \cup \Gamma_2(\beta) \cup \Gamma_3(\beta).$$

Let $\Gamma_-(\beta)$ be the symmetric image of $\Gamma_+(\beta)$ with respect to the point $(0,0)$. Then, $\Gamma_+(\beta) \cup \Gamma_-(\beta)$ is a closed curve. We use the closed curve $\Gamma_+(\beta) \cup \Gamma_-(\beta)$ as a part of the boundary $\partial V(\beta)$ of $V(\beta)$ (cf. Figure 21).

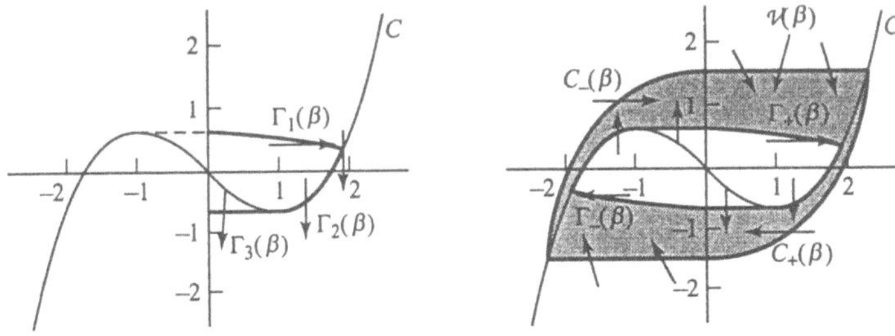

FIGURE 20.   FIGURE 21.

Step 8. Denote by $\mathcal{V}(\beta)$ the domain bounded by two closed curves $\mathcal{C}_+(\beta) \cup \mathcal{C}_-(\beta)$ and $\Gamma_+(\beta) \cup \Gamma_-(\beta)$. It is easy to show that $\lim_{\beta \to 0} a(\beta) = 2$ and $\lim_{\beta \to 0} b(\beta) = 2$. Therefore, the closure of $\mathcal{V}(\beta)$ shrinks to $\mathcal{C}_0$ as $\beta \to 0^+$ (cf. Figure 14). $\square$

X-7. A theorem due to M. Nagumo

In the next section, we shall return to the van der Pol equation (X.4.1) for a large positive parameter ϵ and look at (X.4.1) as a problem of singular perturbations. As a preparation, in this section we prove the following theorem concerning a singular perturbation problem

$$\lambda \frac{d^2 y}{dx^2} + f\left(x, y, \frac{dy}{dx}\right) = 0.$$

Theorem X-7-1 (M. Nagumo [Na6]). *Assume that*
(i) $f(x,y,z)$ and $F(x,y,z)$ are two real-valued continuous functions of three real variables (x,y,z) in a region

$$\mathcal{D} = \left\{ (x,y,z): \ 0 \leq x \leq \ell, \ |y - Y(x)| \leq \alpha(x), \right.$$

$$\left. \left| z - \frac{dY(x)}{dx} \right| \leq pe^{-cx} + b(x) \right\},$$

where ℓ, p, and c are positive constants, $Y(x)$ is a real-valued and two-times continuously differentiable function of x on the interval $0 \leq x \leq \ell$, and the two quantities $\alpha(x)$ and $b(x)$ are positive continuous functions of x on the interval $0 \leq x \leq \ell$,
(ii) in the region $\mathcal{D}$, we have

$$\begin{cases} |f(x,y,z) - F(x,y,z)| \leq \epsilon, \\ |F(x,y_2,z) - F(x,y_1,z)| \leq K|y_2 - y_1|, \\ \dfrac{F(x,y,z_2) - F(x,y,z_1)}{z_2 - z_1} \geq L, \end{cases}$$

where ϵ, K, and L are positive constants,

(iii) the function $Y(x)$ is a solution of the differential equation

$$F\left(x, Y, \frac{dY}{dx}\right) \; = \; 0,$$

and $\left|\dfrac{d^2Y(x)}{dx^2}\right| \leq M$ for a positive constant M on the interval $0 \leq x \leq \ell$. Let $y(x)$ be any solution of the differential equation

$$\lambda\frac{d^2y}{dx^2} \; + \; f\left(x, y, \frac{dy}{dx}\right) \; = \; 0 \qquad (\lambda > 0)$$

satisfying the conditions $y(0) = Y(0)$ and $|y'(0) - Y'(0)| \leq p$. Then,

$$|y(x) \; - \; Y(x)| \; < \; \left\{\frac{\epsilon}{K} \; + \; \lambda\left(\frac{p}{L} \; + \; \frac{M}{K}\right)\right\}\exp\left[\frac{Kx}{L}\right]$$

on the interval $0 \leq x \leq \ell$ if ϵ and λ are sufficiently small.

Proof.

We prove this theorem in four steps.

Step 1. Setting

$$y \; = \; u \; + \; Y(x) \qquad \text{and} \qquad \frac{dy}{dx} \; = \; v \; + \; \frac{dY(x)}{dx},$$

change the equation

$$\lambda\frac{d^2y}{dx^2} \; + \; f\left(x, y, \frac{dy}{dx}\right) \; = \; 0 \qquad (\lambda > 0)$$

to the system

$$\text{(X.7.1)} \qquad\qquad \frac{du}{dx} \; = \; v, \qquad \lambda\frac{dv}{dx} \; = \; \mathcal{F}(x, u, v, \lambda),$$

where

$$\left|F\left(x, Y(x), v + \frac{dY(x)}{dx}\right) \; + \; \mathcal{F}(x, u, v, \lambda)\right| \; \leq \; (\epsilon + \lambda M) \; + \; K|u|,$$

as long as (x, u, v) is in the region

$$\mathcal{D}_0 \; = \; \{(x, u, v) : \; 0 \; \leq \; x \; \leq \; \ell, \; |u| \; \leq \; \alpha(x), \; |v| \; \leq \; pe^{-cx} \; + \; b(x)\}.$$

Also,

$$\begin{cases} F\left(x, Y(x), v + \dfrac{dY(x)}{dx}\right) \; \geq \; Lv & \text{for} \qquad v \geq 0, \\[4mm] F\left(x, Y(x), v + \dfrac{dY(x)}{dx}\right) \; \leq \; Lv & \text{for} \qquad v \leq 0, \end{cases}$$

in $\mathcal{D}_0$. Hence, in $\mathcal{D}_0$,

$$\text{(X.7.2)} \qquad \begin{cases} \mathcal{F}(x, u, v, \lambda) \; \leq \; -Lv \; + \; K|u| \; + (\epsilon + \lambda M) & \text{for} \qquad v \geq 0, \\[2mm] \mathcal{F}(x, u, v, \lambda) \; \geq \; -Lv \; - \; K|u| \; - (\epsilon + \lambda M) & \text{for} \qquad v \leq 0. \end{cases}$$

Step 2. Suppose that two functions $w_1(x, \lambda)$ and $w_2(x, \lambda)$ satisfy the following conditions:

(X.7.3)
$$\begin{cases} 0 < w_1(x, \lambda) < \alpha(x), \qquad 0 < w_2(x, \lambda) < pe^{-cx} + b(x), \\ w_1(0, \lambda) > 0, \qquad w_2(0, \lambda) > p, \\ w_1'(x, \lambda) > w_2(x, \lambda), \qquad \lambda w_2'(x, \lambda) > -Lw_2(x, \lambda) + Kw_1(x, \lambda) + (\epsilon + \lambda M) \end{cases}$$

on the interval $0 \le x \le \ell$. Then, as long as (x, u, v) is in the region

$$\mathcal{D}_1 = \{(x, u, v) : 0 \le x \le \ell, |u| \le w_1(x, \lambda), |v| \le w_2(x, \lambda)\},$$

it holds that

$$\begin{cases} |v| < w_1'(x, \lambda), \\ \mathcal{F}(x, u, w_2(x, \lambda), \lambda) < \lambda w_2'(x, \lambda), \\ \mathcal{F}(x, u, -w_2(x, \lambda), \lambda) > -\lambda w_2'(x, \lambda). \end{cases}$$

Look at the right-hand side of (X.7.1) on the boundary of $\mathcal{D}_1$. Then, it can be easily seen that if a solution of (X.7.1) starts from $\mathcal{D}_1$, it will stay in $\mathcal{D}_1$ on the interval $0 \le x \le \ell$.

Step 3. Show that two functions

$$\begin{cases} w_1(x, \lambda) = \displaystyle\int_0^x w_2(\xi, \lambda)d\xi + \delta_2(x + 1), \\ w_2(x, \lambda) = pe^{-Lx/\lambda} + \delta_1 e^{Kx/L}, \end{cases}$$

where $\delta_1 = \dfrac{\lambda pK}{L^2} + \dfrac{\epsilon + \lambda M + K\delta_2(\ell + 1)}{L}$, satisfy the requirements (X.7.3) if ϵ, λ, and a positive constant δ_2 are sufficiently small. Observe that two roots of $\lambda X^2 + LX - K = 0$ are $-\dfrac{L}{\lambda} - \zeta_0$ and $\zeta_0 = \dfrac{K}{L} + O(\lambda)$.

Step 4. Note that

$$w_1(x, \lambda) = \left(\frac{\lambda p}{L}\right)(1 - e^{-Lx/\lambda}) + \left(\frac{\delta_1 L}{K}\right)(e^{Kx/L} - 1) + \delta_2(x + 1).$$

To complete the proof, look at $|u| < w_1(x, \lambda)$ as $\delta_2 \to 0$. $\square$

The inequality $|v| < w_2(x, \lambda)$, as $\delta_2 \to 0$, yields the following estimate of $\dfrac{dy}{dx}$:

$$\left|\frac{dy}{dx} - \frac{dY(x)}{dx}\right| \le pe^{-Lx/\lambda} + \left[\frac{\epsilon}{L} + \frac{\lambda}{L}\left(\frac{pK}{L} + M\right)\right]e^{Kx/L}.$$

Note that

$$\lim_{\lambda \to 0^+} e^{-Lx/\lambda} = 0 \qquad \text{if} \quad x > 0.$$

X-8. A singular perturbation problem

In this section, we look at behavior of solutions of the van der Pol equation (X.4.1) as $\epsilon \to +\infty$ more closely. Setting $t = \epsilon\tau$ and $\lambda = \epsilon^{-2}$, let us change (X.4.1) to

$$(\text{X.8.1}) \qquad \lambda\frac{d^2x}{d\tau^2} + (x^2 - 1)\frac{dx}{d\tau} + x = 0.$$

Set $\lambda = 0$. Then, (X.8.1) becomes

$$(\text{X.8.2}) \qquad (x^2 - 1)\frac{dx}{d\tau} + x = 0.$$

Solving (X.8.2) with an initial value $x(0) = x_0 < -1$, we obtain $x = \phi(\tau)$, where

$$\frac{\phi(\tau)^2}{2} - \ln|\phi(\tau)| = -\tau + \frac{x_0^2}{2} - \ln(-x_0).$$

Observe that

$$\frac{d\phi(\tau)}{d\tau} = -\frac{\phi(\tau)}{\phi(\tau)^2 - 1} > 0 \qquad \text{if} \qquad \phi(\tau) < -1.$$

Note also that setting $\tau_0 = \dfrac{x_0^2}{2} - \ln(-x_0) - \dfrac{1}{2} > 0$, we obtain $\phi(\tau_0) = -1$ and $\phi(\tau)^2 - 1 > 0$ for $0 \leq \tau < \tau_0$. The graph of $\phi(\tau)$ is given in Figure 22.

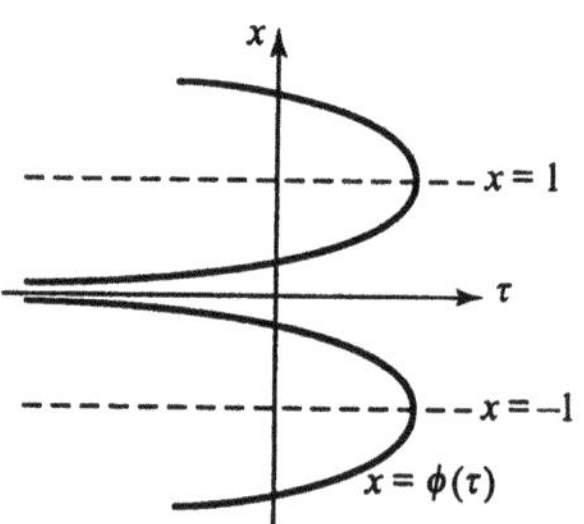

FIGURE 22.

(i) Behavior for $0 \leq \tau \leq \tau_0 - \delta_0$, where $\delta_0 > 0$: Let us denote by $x(\tau, \lambda)$ the solution to the initial-value problem

$$(\text{X.8.3}) \qquad \lambda\frac{d^2x}{d\tau^2} + (x^2 - 1)\frac{dx}{d\tau} + x = 0, \qquad x(0, \lambda) = x_0, \quad x'(0, \lambda) = \eta,$$

where the prime is $\dfrac{d}{d\tau}$ and η is a fixed constant. Using Theorem X-7-1 [Na6], we derive

$$\begin{cases} |x(\tau, \lambda) - \phi(\tau)| \leq \lambda K, \\ |x'(\tau, \lambda) - \phi'(\tau)| \leq |\eta - \phi'(0)|e^{-\mu\tau/\lambda} + \lambda K, \end{cases}$$

for $0 \leq \tau \leq \tau_0 - \delta_0$, where K and μ are suitable positive constants.

(ii) Behavior for $|x+1| \leq \sigma$: First note that $\lim\limits_{t \to \tau_0 - 0} \phi'(\tau) = +\infty$. Set $\phi'(\tau_0 - 2\delta_0) = \dfrac{1}{\rho_1} > 0$. Then, there exists $\tau_1(\lambda)$ for sufficiently small $\lambda > 0$ such that

$$
\begin{cases}
0 < \tau_1(\lambda) < \tau_0 - \delta_0, \qquad \lim\limits_{\lambda \to 0+} \tau_1(\lambda) = \tau_0 - 2\delta_0, \\[2mm]
x(\tau_1(\lambda), \lambda) = \phi(\tau_0 - 2\delta_0), \quad x'(\tau, \lambda) > \dfrac{1}{2\rho_1} \quad \text{for} \quad \tau_1(\lambda) \leq \tau \leq \tau_0 - \delta_0.
\end{cases}
$$

Set $p = \dfrac{1}{x'(\tau, \lambda)} = \dfrac{d\tau}{dx}$. Then,

$$
\text{(X.8.4)} \qquad \lambda \frac{dp}{dx} = p^2(x^2 - 1) + p^3 x,
$$

regarding $p = p(x, \lambda)$ as a function of x and λ. Note that

$$
0 < p(x, \lambda) < 2\rho_1 \quad \text{for} \quad \phi(\tau_0 - 2\delta_0) \leq x \leq x(\tau_0 - \delta_0, \lambda) < -1, \quad 0 < \lambda < \lambda_0,
$$

where λ_0 is a sufficiently small positive number. It is important to notice that if we make δ_0 small and if we make λ_0 also small accordingly, we can make ρ_1 small.
Set

$$
\text{(X.8.5)} \qquad \sigma = -1 - \phi(\tau_0 - 2\delta_0) \qquad \text{and} \qquad M_0 = \max_{|1+x| \leq \sigma} |x^2 - 1|.
$$

Then,

$$
\text{(X.8.6)} \qquad \sigma > 0, \qquad \lim_{\delta_0 \to 0+} \sigma = 0, \qquad \text{and} \qquad \lim_{\sigma \to 0+} M_0 = 0.
$$

Furthermore, $0 < \rho_1 < M_0$ since $\rho_1 = \dfrac{1 - \phi(\tau_0 - 2\delta_0)^2}{\phi(\tau_0 - 2\delta_0)}$. Therefore,

$$
0 < p(x, \lambda) < 2M_0 \quad \text{for} \quad \phi(\tau_0 - 2\delta_0) \leq x \leq x(\tau_0 - \delta_0, \lambda), \quad 0 < \lambda < \lambda_0.
$$

Now, we shall show that

$$
\text{(X.8.7)} \qquad 0 < p(x, \lambda) < 2M_0 \qquad \text{for} \qquad |1 + x| \leq \sigma, \qquad 0 < \lambda < \lambda_0.
$$

If (X.8.7) is not true, there must exist a ξ such that

$$
-1 - \sigma < \xi \leq -1 + \sigma, \quad p(\xi, \lambda) = 2M_0, \quad 0 < p(x, \lambda) < 2M_0 \ \text{for} \ -1 - \sigma \leq x < \xi.
$$

Then, this implies that

$$
\begin{aligned}
\lambda \frac{dp}{dx}(\xi, \lambda) &= p(\xi, \lambda)^2[(\xi^2 - 1) + p(\xi, \lambda)\xi] \\
&\leq p(\xi, \lambda)^2[M_0 + 2M_0\xi] = M_0 p(\xi, \lambda)^2(1 + 2\xi) < 0.
\end{aligned}
$$

This is impossible.

(iii) Behavior for $-1+\sigma \leq x \leq 2-\sigma_1$: To begin with, it should be remarked that

$$\text{(X.8.8)} \quad \begin{cases} \displaystyle\int_{-1}^{2}(\xi^2-1)d\xi = \left[\frac{\xi^3}{3}-\xi\right]_{-1}^{2} = \frac{8}{3}-2-\left(-\frac{1}{3}+1\right) = 0, \\[2ex] \displaystyle\int_{-1}^{x}i(\xi^2-1)d\xi = \left(\frac{x^3}{3}-x\right)-\frac{2}{3} < 0, \quad \text{for} \quad -1 < x < 2. \end{cases}$$

Fix a positive number σ_1 so that

$$\text{(X.8.9)} \qquad \int_{-1}^{x}(\xi^2-1)d\xi \ \leq \ -\gamma \qquad \text{for} \qquad -1+\sigma \leq \ x \ \leq \ 2-\sigma_1,$$

where γ is a sufficiently small positive constant. Note that $\gamma \to 0$ as $\delta_0 \to 0^+$ and $\sigma_1 \to 0^+$.

Integrating (X.8.4), it follows that

$$\frac{\lambda}{p(x,\lambda)} = \frac{\lambda}{p(-1,\lambda)} - \int_{-1}^{x}(\xi^2-1)d\xi - \int_{-1}^{x}p(\xi,\lambda)\xi d\xi.$$

(I) $\dfrac{\lambda}{p(x,\lambda)} > \gamma$ or $0 < p(x,\lambda) < \dfrac{\lambda}{\gamma}$ for $-1+\sigma \leq x \leq 0$. Hence, if we choose $\lambda_0 > 0$ so that

$$\frac{\lambda}{\gamma} < \frac{\gamma}{4K} \qquad \text{for} \qquad 0 < \lambda < \lambda_0, \qquad \text{where} \qquad K = \int_{0}^{2}|\xi|d\xi,$$

we obtain $0 < p(x,\lambda) < \dfrac{\gamma}{4K}$ for $-1+\sigma \leq x \leq 0, 0 < \lambda < \lambda_0$.

(II) $\dfrac{\lambda}{p(x,\lambda)} > \gamma - K \max_{0 \leq \xi \leq x} p(\xi,\lambda)$ for $0 \leq x \leq 2-\sigma_1$. Suppose that

$$0 < p(\xi,\lambda) \leq \frac{\gamma}{2K} \qquad \text{for} \qquad 0 \leq \xi \leq x, \quad 0 < \lambda < \lambda_0.$$

Then, $0 < p(x,\lambda) < \dfrac{2\lambda}{\gamma} < \dfrac{\gamma}{2K}$. Thus, we proved that

$$\text{(X.8.10)} \quad 0 < p(x,\lambda) < \frac{\gamma}{2K} \qquad \text{for} \qquad -1+\sigma \leq \ x \ \leq 2 - \sigma_1, \quad 0 < \lambda < \lambda_0.$$

(iv) Behavior for $2-\sigma_1 \leq x \leq 2+\sigma_2$: First look at

$$\frac{\lambda}{p(2-\sigma_1,\lambda)} = \frac{\lambda}{p(\phi(\tau_0-2\delta_0),\lambda)} - \int_{\phi(\tau_0-2\delta_0)}^{2-\sigma_1}[(\xi^2-1) + p(\xi,\lambda)\xi]d\xi.$$

Notice that $p(2 - \sigma_1, \lambda)$ is independent of δ_0. Then, it can be shown that

$$\frac{\lambda}{p(2 - \sigma_1, \lambda)} \to 0^+ \quad \text{as} \quad \sigma_1 \to 0^+ \quad \text{and} \quad \lambda \to 0^+.$$

For a fixed positive number σ_2, assume that

$$0 < p(x, \lambda) < 1 \quad \text{for} \quad 2 - \sigma_1 \leq x \leq 2 + \sigma_2, \quad 0 < \lambda < \lambda_0.$$

Then,

$$\frac{\lambda}{p(2 + \sigma_2, \lambda)} = \frac{\lambda}{p(2 - \sigma_1, \lambda)} - \int_{2-\sigma_1}^{2+\sigma_2} [(\xi^2 - 1) + p(\xi, \lambda)\xi] d\xi > 0.$$

Hence,

$$\int_{2-\sigma_1}^{2+\sigma_2} [(\xi^2 - 1) + p(\xi, \lambda)\xi] d\xi < \frac{\lambda}{p(2 - \sigma_1, \lambda)} \quad \text{for} \quad 2 - \sigma_1 \leq x \leq 2 + \sigma_2.$$

This is a contradiction if σ_1 and λ are small. Therefore, if $\lambda_0 > 0$ is sufficiently small, there exists an $x(\lambda)$ such that

$$2 - \sigma_1 \leq x(\lambda) \leq 2 + \sigma_2 \quad \text{and} \quad p(x(\lambda), \lambda) = 1 \quad \text{for} \quad 0 < \lambda < \lambda_0.$$

It can be shown that $\lim_{\lambda \to 0^+} x(\lambda) = 2$.

Setting $\tau(x(\lambda)) = \tau(\lambda)$, it can be shown that $\lim_{\lambda \to 0^+} \tau(\lambda) = \tau_0$. Now again, apply Theorem X-7-1 [Na6] to the initial-value problem

$$\lambda \frac{d^2 x}{d\tau^2} + (x^2 - 1)\frac{dx}{d\tau} + x = 0, \quad x(\tau(\lambda)) = x(\lambda), \quad x'(\tau(\lambda)) = 1.$$

Figure 23 shows the general behavior of $x(\tau, \lambda)$ for small positive λ.

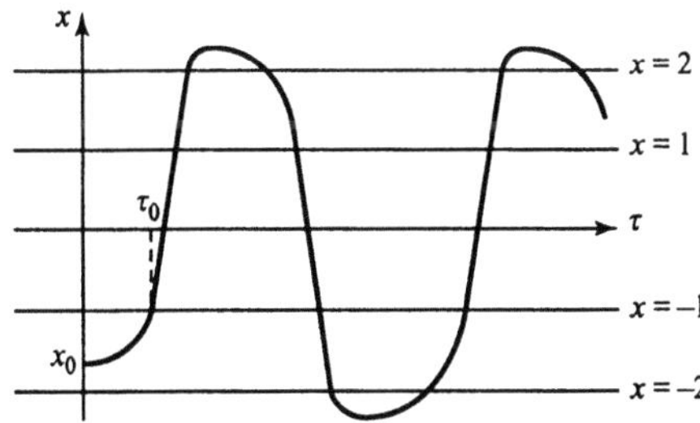

FIGURE 23.

Exercises X

X-1. (1) Show that if $F(t, y_1, y_2)$ is continuous and bounded on
$$\Omega = \{(t, y_1, y_2) : a \leq t \leq b, \ |y_1| < +\infty, \ |y_2| < +\infty\},$$
the boundary-value problem
$$\frac{d^2 y}{dt^2} = F\left(t, y, \frac{dy}{dt}\right), \qquad \frac{dy}{dt}(a) = \alpha, \quad y(b) = \beta$$
has a solution.

(2) Does the boundary-value problem
$$\frac{d^2 y}{dt^2} = F\left(t, y, \frac{dy}{dt}\right), \qquad \frac{dy}{dt}(a) = \alpha, \quad \frac{dy}{dt}(b) = \beta$$
have a solution?

(3) Show that the boundary-value problem
$$\frac{d^2 x}{dt^2} = tx + x^3, \qquad x(-2) = A, \quad x(3) = B$$
has a solution for any real numbers A and B.

Hint. A counterexample for (2) is
$$\frac{d^2 y}{dt^2} = 0, \qquad \frac{dy}{dt}(a) = 0, \quad \frac{dy}{dt}(b) = 1.$$

For (3), apply Theorem X-1-3 [Na4] with $w_1(t) = \alpha$ and $w_2(t) = \beta$, where $-\alpha$ and β are sufficiently large positive constants.

X-2. Find the global phase portrait of each of the following two differential equations:

(i)
$$\frac{d^2 x}{dt^2} + x^2(1-x)^2 \frac{dx}{dt} + (x-1)^2(x+1)x = 0;$$

(ii)
$$\frac{d^2 x}{dt^2} + \frac{dx}{dt} - \frac{1}{1+x^2} = 0.$$

X-3. Suppose that
   (i) $f(x, y, t)$ is continuously differentiable everywhere in the (x, y, t)-space (i.e., in $\mathbb{R}^3$),
   (ii) $g(x)$ is continuously differentiable in $-\infty < x < +\infty$,
   (iii) $e(t)$ is continuous on $0 \leq t < +\infty$,
   (iv) $f(x, y, t) \geq 0$ for $x^2 + y^2 \geq r^2$ and $t \geq 0$, where r is a positive number r,
   (v) $G(x) = \displaystyle\int_0^x g(\xi) d\xi \to +\infty$ as $|x| \to +\infty$,
   (vi) $\displaystyle\int_0^t |e(\tau)| d\tau$ is bounded for $0 \leq t < +\infty$.
Show that every solution $(x(t), y(t))$ of the system
$$\frac{dx}{dt} = y, \qquad \frac{dy}{dt} = -f(x, y, t)y - g(x) + e(t)$$
is well defined and bounded for $0 \leq t < +\infty$.

Hint. Set $V(x, y) = \dfrac{y^2}{2} + G(x)$. Then, $\dfrac{dV}{dt} = -f(x, y, t)y^2 + e(t)y$. There exists a positive number $r_0 \geq r$ such that $V(x, y) \geq \dfrac{y^2}{4}$ for $x^2 + y^2 \geq r_0^2$. Hence, if an orbit $(x(t), y(t))$ satisfies conditions that $x(t)^2 + y(t)^2 \geq r_0$ for $t_0 \leq t \leq t_1$, we obtain

$$\frac{d}{dt}V(x(t), y(t)) \leq |e(t)||y(t)| \leq 2|e(t)|\sqrt{V(x(t), y(t))} \qquad \text{for} \qquad t_0 \leq t \leq t_1.$$

This yields

$$\sqrt{V(x(t), y(t))} \; \leq \; \sqrt{V(x(t_0), y(t_0))} \; + \; \int_{t_0}^{t} |e(\tau)|d\tau \qquad \text{for} \qquad t_0 \leq t \leq t_1.$$

X-4. Show that the differential equation

(E.1)
$$\frac{d}{dt}\begin{bmatrix} y_1 \\ y_2 \end{bmatrix} = \begin{bmatrix} y_2 \\ -\epsilon(y_1^2 - 1)y_2 - y_1 - \epsilon y_1^3 \end{bmatrix}$$

has exactly one periodic orbit, where ϵ is a positive parameter.

X-5. Find an approximation for the unique periodic orbit of (E.1) for small $\epsilon > 0$ and for large ϵ.

X-6. Considering the system of two differential equations

(E.2)
$$\frac{d}{dt}\begin{bmatrix} y_1 \\ y_2 \end{bmatrix} = \begin{bmatrix} y_2 \\ -\epsilon(y_1^2 - 1)y_2 + y_1 - \delta y_1^3 \end{bmatrix},$$

where ϵ and δ are positive numbers,
  (a) show that every orbit is bounded for $t \geq 0$,
  (b) determine whether each stationary point is stable or unstable, examining every possibility,
  (c) using the function

$$V(\vec{y}) = \frac{1}{2}y_2^2 + \epsilon[-y_1 + \delta y_1^3]y_2 + \int_0^{y_1} [-s + \delta s^3][1 + \epsilon^2(s_1^2)]ds$$

and Theorem IX-2-2, show that if $0 < \delta \leq \dfrac{1}{3}$, every orbit of system (E.2) tends to one of the stationary points as $t \to +\infty$,
  (d) discuss the uniqueness of periodic orbits.

X-7. Show that the differential equation

$$\frac{d^2x}{dt^2} + f(x)\frac{dx}{dt} + g(x) = 0$$

has at most one nontrivial periodic solution if $f(x)$ and $g(x)$ are continuously differentiable in $\mathbb{R}$ and satisfy the following conditions

(i)
$$f(x) \begin{cases} < 0 & \text{for} \quad |x| < 1, \\ > 0 & \text{for} \quad |x| > 1, \end{cases}$$

(ii)
$$g(x) \begin{cases} < 0 & \text{for} \quad -2 < x < 0, \\ > 0 & \text{for} \quad (x + 2)x > 0, \end{cases}$$

(iii)
$$\int_{-1}^{1} g(x)dx = 0.$$

Hint. See [Sat].

The main ideas are as follows:

(1) If there are more than one nontrivial periodic orbits, then at least one of them should not be orbitally asymptotically stable.

(2) For a nontrivial periodic orbit $(x(t), x'(t))$ of period $T > 0$, the inequality
$$\int_0^T f(x(t))dt > 0$$
implies that this orbit is orbitally asymptotically stable (cf. Exercise IX-5).

(3) Set
$$\lambda(t) \; = \; V(x(t), x'(t)) \; = \; \frac{(x'(t))^2}{2} \; + \; \int_0^{x(t)} g(\xi)d\xi.$$

Then,
$$\frac{d\lambda(t)}{dt} \; = \; -f(x(t))(x'(t))^2.$$

Therefore, we obtain

(I)
$$\int f(x(t))dt \; = \; -\int \frac{d\lambda}{(x')^2}.$$

Now, investigate the quantity $(x'(t))^2$ as a function of λ along the periodic orbit $(x(t), x'(t))$. For a fixed value of λ, compare $(x'(t))^2$ for different values of $x(t)$, using the assumptions on f and g.

Step 1. First the following remarks are very important.

(a) If we set $y = x'$, the given differential equation is reduced to the system

(S)
$$\frac{dx}{dt} \; = \; y, \qquad \frac{dy}{dt} \; = \; -f(x)y \; - \; g(x).$$

(b) A careful observation shows that the index of the critical point $(0,0)$ is 1, while the index of the critical point $(-2,0)$ is -1. There are only two critical points of (S).

(c) The periodic orbit and any line $\{x = \text{a constant}\}$ intersect each other at most twice.

(d) The critical point $(-2,0)$ should not be contained in the domain bounded by the periodic orbit. To show this, use the fact that the index of any periodic orbit is 1.

These facts imply that the periodic orbit should be confined in the half-plane $x > -2$.

Step 2. A level curve of $V(x,y) = \dfrac{y^2}{2} + \displaystyle\int_0^x g(\xi)d\xi$ is a closed curve if the curve is totally confined in the strip $|x| \le 1$. In particular, $(1,0)$ and $(-1,0)$ are on the same level curve $V(x,y) = \displaystyle\int_0^1 g(\xi)d\xi = \int_0^{-1} g(\xi)d\xi$. Since $\lambda(t)$ is increasing as long as $|x(t)| < 1$, the periodic orbit cannot intersect the level curve $V(x,y) = \displaystyle\int_0^1 g(\xi)d\xi$.

Therefore, the periodic orbit must intersect the lines $x = 1$ (as well as the line $x = -1$) twice. Denote these four points by $(1, \eta(A))$, $(1, -\eta(B))$, $(-1, -\eta(C))$, and $(-1, \eta(D))$, where $\eta(A)$, $\eta(B)$, $\eta(C)$, and $\eta(D)$ are positive numbers.

Step 3. Set

$$\lambda_A = V(1, \eta(A)), \quad \lambda_B = V(1, -\eta(B)), \quad \lambda_C = V(-1, -\eta(C)), \quad \lambda_D = V(-1, \eta(D)).$$

Then,

$$\lambda_A > \lambda_B, \qquad \lambda_C > \lambda_B, \qquad \lambda_C > \lambda_D, \qquad \lambda_A > \lambda_D.$$

Set

$$\begin{aligned}
\mathcal{I}(A) &= \{\, \lambda : \lambda_A > \lambda > \max(\lambda_B, \lambda_D) \,\}, \\
\mathcal{I}(B) &= \{\, \lambda : \lambda_B < \lambda < \min(\lambda_A, \lambda_C) \,\}, \\
\mathcal{I}(C) &= \{\, \lambda : \lambda_C > \lambda > \max(\lambda_B, \lambda_D) \,\}, \\
\mathcal{I}(D) &= \{\, \lambda : \lambda_D < \lambda < \min(\lambda_A, \lambda_C) \,\}.
\end{aligned}$$

Note that the interval $\mathcal{I}_0 = \{\lambda : \min(\lambda_A, \lambda_C) > \lambda > \max(\lambda_B, \lambda_D)\}$ is contained in $\mathcal{I}(A) \cap \mathcal{I}(B) \cap \mathcal{I}(C) \cap \mathcal{I}(D)$.

Now,
(α) on the arc between $(1, \eta(A))$ and $(1, -\eta(B))$ of the periodic orbit, regard y as a function of λ and denote it by $y_1^+(\lambda)$, where $\lambda_B \le \lambda \le \lambda_A$,

(β) on the arc between $(1, -\eta(B))$ and $(-1, -\eta(C))$ of the periodic orbit, regard y as a function of λ and denote it by $y_2^-(\lambda)$, where $\lambda_B \le \lambda \le \lambda_C$,

(γ) on the arc between $(-1, -\eta(C))$ and $(-1, \eta(D))$ of the periodic orbit, regard y as a function of λ and denote it by $y_1^-(\lambda)$, where $\lambda_D \le \lambda \le \lambda_C$,

(δ) on the arc between $(-1, \eta(D))$ and $(1, \eta(A))$ of the periodic orbit, regard y as a function of λ and denote it by $y_2^+(\lambda)$, where $\lambda_D \le \lambda \le \lambda_A$.

Then, it can be shown that

$$\begin{cases}
y_1^+(\lambda)^2 < y_2^+(\lambda)^2 & \text{on} \quad \mathcal{I}(A), \\
y_1^+(\lambda)^2 < y_2^-(\lambda)^2 & \text{on} \quad \mathcal{I}(B), \\
y_1^-(\lambda)^2 < y_2^-(\lambda)^2 & \text{on} \quad \mathcal{I}(C), \\
y_1^-(\lambda)^2 < y_2^+(\lambda)^2 & \text{on} \quad \mathcal{I}(D).
\end{cases}$$

Step 4. Now, fixing a $\lambda_0 \in \mathcal{I}(A) \cap \mathcal{I}(B) \cap \mathcal{I}(C) \cap \mathcal{I}(D)$, evaluate the integral (I) as follows:

$$\int_0^T f(x(t))\,dt = -\left\{ \int_{\lambda_A}^{\lambda_B} \frac{d\lambda}{y_1^+(\lambda)^2} + \int_{\lambda_B}^{\lambda_C} \frac{d\lambda}{y_2^-(\lambda)^2} + \int_{\lambda_C}^{\lambda_D} \frac{d\lambda}{y_1^-(\lambda)^2} + \int_{\lambda_D}^{\lambda_A} \frac{d\lambda}{y_2^+(\lambda)^2} \right\},$$

where

$$\begin{cases}
\displaystyle \int_{\lambda_A}^{\lambda_B} \frac{d\lambda}{y_1^+(\lambda)^2} = -\int_{\lambda_B}^{\lambda_0} \frac{d\lambda}{y_1^+(\lambda)^2} - \int_{\lambda_0}^{\lambda_A} \frac{d\lambda}{y_1^+(\lambda)^2}, \\[2.5ex]
\displaystyle \int_{\lambda_B}^{\lambda_C} \frac{d\lambda}{y_2^-(\lambda)^2} = \int_{\lambda_B}^{\lambda_0} \frac{d\lambda}{y_2^-(\lambda)^2} + \int_{\lambda_0}^{\lambda_C} \frac{d\lambda}{y_2^-(\lambda)^2}, \\[2.5ex]
\displaystyle \int_{\lambda_C}^{\lambda_D} \frac{d\lambda}{y_1^-(\lambda)^2} = -\int_{\lambda_D}^{\lambda_0} \frac{d\lambda}{y_1^-(\lambda)^2} - \int_{\lambda_0}^{\lambda_C} \frac{d\lambda}{y_1^-(\lambda)^2}, \\[2.5ex]
\displaystyle \int_{\lambda_D}^{\lambda_A} \frac{d\lambda}{y_2^+(\lambda)^2} = \int_{\lambda_D}^{\lambda_0} \frac{d\lambda}{y_2^+(\lambda)^2} + \int_{\lambda_0}^{\lambda_A} \frac{d\lambda}{y_2^+(\lambda)^2}.
\end{cases}$$

Thus, we conclude that

$$\int_0^T f(x(t))dt \; > \; 0.$$

This implies that every periodic orbit is orbitally asymptotically stable. Hence, there exists at most one periodic orbit.

X-8. For a nonzero real number ϵ and a real-valued, continuous, and periodic function $f(t)$ of period T which is defined on $-\infty < t < +\infty$, find a unique solution $\phi(t, \epsilon)$ of the differential equation

$$\epsilon \frac{dy}{dt} = -y + f(t)$$

which is periodic of period T. Also, find the uniform limit of $\phi(t, \epsilon)$ on the interval $-\infty < t < +\infty$ as $\epsilon \to 0$.

X-9. Assuming that $X(t)$ satisfies the condition

$$(X(t)^2 - 1)\frac{dX(t)}{dt} \; + \; X(t) \; = \; 0 \qquad \text{and} \qquad |X(t)| \; > \; 1$$

on an interval $0 \leq t \leq \ell$, where ℓ is a positive constant, and that $x(t, \lambda)$ is the unique solution to the initial-value problem

$$\lambda \frac{d^2 x}{dt^2} + (x^2 - 1)\frac{dx}{dt} + x = 0, \qquad x(0, \lambda) = X(0), \quad x'(0, \lambda) = X'(0),$$

find $\displaystyle \lim_{\lambda \to 0+} \frac{d^2 x(t, \lambda)}{dt^2}$ as a function of $X(t)$ for $0 < t \leq \ell$.

X-10. Assuming that $y(t, \epsilon)$ is the solution to the initial-value problem

$$\epsilon \frac{d^2 y}{dt^2} + 2y\frac{dy}{dt} = 0, \qquad y(0) = \alpha, \quad y'(0) = \frac{\beta}{\epsilon},$$

where ϵ is a positive parameter, α is a real number, and β is a positive number, find $\displaystyle \lim_{\epsilon \to 0+} y(t, \epsilon)$ for any fixed $t > 0$.

Hint. Look at $\epsilon \dfrac{dy}{dt} = c - y^2$. Then, $c = \beta + \alpha^2 > \alpha^2$. Hence, y^2 increases to c very quickly. Then, use Theorem X-7-1 [Na6], or find the solution explicitly (cf. Exercise I-1).

X-11. Find, if any, solution(s) $y(t, \epsilon)$ of the boundary-value problem

$$\epsilon \frac{d^2 y}{dt^2} + 2y\frac{dy}{dt} = 0, \qquad y(0) = A, \; y(1) = B,$$

in the following six cases: (1) $0 < B < A$, (2) $B = A$, (3) $B > |A|$, (4) $B = -A > 0$, (5) $|B| < -A$, and (6) $B \leq 0 < A$, assuming that ϵ is a positive parameter. Also, find $\displaystyle \lim_{\epsilon \to 0+} y(t, \epsilon)$ for $0 < t < 1$ in each of the six cases.

Hint. Use explicit solutions together with the Nagumo Theorems (Theorems X-1-3 and X-7-1) on boundary-value and singular perturbation problems. See also Exercises X-10 and X-12, and [How].

X-12. Let $f(x, y, t, \epsilon)$ be a real-valued funcion of four real variables (x, y, t, ϵ). Assume that

(i) $\phi(t)$ is a real-valued, continuous, twice-continuously differentiable function on the interval $\mathcal{I}_0 = \{t : 0 \leq t \leq 1\}$ and satisfies the conditions $0 = f(\phi(t), \phi'(t), t, 0)$ and $\phi(1) = B$ on $\mathcal{I}_0$, where B is a given real number,

(ii) the function $f(x, y, t, \epsilon)$ and its partial derivatives with respect to (x, y) are continuous in (x, y, t, ϵ) on a region $\mathcal{R} = \{(x, y, t, \epsilon) : |x - \phi(t)| \leq r_1, |y| < +\infty, t \in \mathcal{I}_0, 0 \leq \epsilon \leq r_2\}$, where r_1 and r_2 are positive nmbers,

(iii) $|f(\phi(t), \phi'(t), t, \epsilon)| \leq K\epsilon$ for $t \in \mathcal{I}_0$, where K is a positive number,

(iv) there exists a positive number μ such that $\dfrac{\partial f}{\partial y}(x, y, t, \epsilon) \leq -\mu$ on $\mathcal{R}$,

(v) there is a positive-valued and continuous function $\psi(s)$ defined on the interval $0 \leq s < +\infty$ such that $\displaystyle\int^{+\infty} \frac{s\, ds}{\psi(s)} = +\infty$ and that $|f(x, y, t, \epsilon)| \leq \psi(|y|)$ on $\mathcal{R}$,

(vi) A is a given real number.

Then, there exists a positive number ϵ_0 such that for each positive ϵ not greater than ϵ_0, there exists a solution $x(t, \epsilon)$ of the boundary-value problem

$$\epsilon \frac{d^2 x}{dt^2} = f\left(x, \frac{dx}{dt}, t, \epsilon\right), \quad x(0, \epsilon) = A, \quad x(1, \epsilon) = B$$

such that

$$\begin{cases} |x(t, \epsilon) - \phi(t)| \leq |A - \phi(0)| e^{c_1 t} e^{-\mu t/\epsilon} + c_2 \epsilon & \text{on} \quad \mathcal{I}_0, \\[2mm] |x'(t, \epsilon) - \phi'(t)| \leq \dfrac{c_3 e^{-\mu t/\epsilon}}{\epsilon} + c_4 \epsilon & \text{for} \quad 0 < c_5 \epsilon \leq t \leq 1, \end{cases}$$

where c_1, c_2, c_3, c_4, and c_5 are positive numbers.

Hint. Cf. Theorem X-1-3. See also [How].

X-13. Let $a(t)$, $b(t)$, and $f(t)$ be real-valued, continuous, and continuously differentiable functions of t which are defined on the interval $0 \leq t \leq 1$. Also, let $\phi_0(t)$ be a real-valued solution of the differential equation $\dfrac{d^2 z}{dt^2} + a(t)\dfrac{dz}{dt} + b(t)z = f(t)$. For a positive number ϵ, denote by $y(t, \epsilon)$ the unique solution of the initial-value problem

$$\epsilon \frac{d^3 y}{dt^3} + \frac{d^2 y}{dt^2} + a(t)\frac{dy}{dt} + b(t)y = f(t),$$

$$y(0) = \eta_0(\epsilon), \quad y'(0) = \eta_1(\epsilon), \quad y''(0) = \eta_2(\epsilon).$$

Show that $|y(t, \epsilon) - \phi_0(t)| + |y'(t, \epsilon) - \phi_0'(t)| + |y''(t, \epsilon) - \phi_0''(t)|$ tends to zero uniformly on the interval $0 \leq t \leq 1$ if $|\eta_0(\epsilon) - \phi_0(0)| + |\eta_1(\epsilon) - \phi_0'(0)| + |\eta_2(\epsilon) - \phi_0''(0)| \to 0$ as $\epsilon \to 0^+$.

X-14. Let $\vec{x} \in \mathbb{R}^n$, $\vec{y} \in \mathbb{R}^m$, $t \in \mathbb{R}$, and $\epsilon \in \mathbb{R}$. Also, let $\vec{f}(\vec{x}, \vec{y}, t, \epsilon)$ and $\vec{g}(\vec{x}, \vec{y}, t, \epsilon)$ be respectively $\mathbb{R}^n$-valued and $\mathbb{R}^m$-valued functions of $(\vec{x}, \vec{y}, t, \epsilon)$. Assume that

  (i) an $\mathbb{R}^n$-valued function $\vec{\phi}(t)$ and an $\mathbb{R}^m$-valued function $\vec{\psi}(t)$ are continuous and continuously differentiable on an interval $\mathcal{I}_0 = \{t : a \leq t \leq b\}$ and satisfy the system of equations

$$\frac{d\vec{x}}{dt} = \vec{f}(\vec{x}, \vec{y}, t, 0), \quad \vec{0} = \vec{g}(\vec{x}, \vec{y}, t, 0) \quad \text{on} \ \mathcal{I}_0,$$

  (ii) $\vec{f}(\vec{x}, \vec{y}, t, \epsilon)$, $\vec{g}(\vec{x}, \vec{y}, t, \epsilon)$, and their partial derivatives with respect to $(\vec{x}, \vec{y})$ are continuous on a region $\mathcal{R} = \{(\vec{x}, \vec{y}, t, \epsilon) : |\vec{x} - \vec{\phi}(t)| \leq r_1, |\vec{y} - \vec{\psi}(t)| \leq r_2, t \in \mathcal{I}_0, 0 \leq \epsilon \leq r_3\}$, where r_1, r_2, and r_3 are positive numbers,

  (iii) all eigenvalues of the matrix $\dfrac{\partial \vec{g}}{\partial \vec{y}}(\vec{\phi}(t), \vec{\psi}(t), t, 0)$ are less than a negative number μ on $\mathcal{I}_0$.

Show that the initial-value problem

$$\frac{d\vec{x}}{dt} = \vec{f}(\vec{x}, \vec{y}, t, \epsilon), \quad \epsilon \frac{d\vec{y}}{dt} = \vec{g}(\vec{x}, \vec{y}, t, \epsilon), \quad \vec{x}(a) = \vec{\xi}(\epsilon), \quad \vec{y}(a) = \vec{\eta}(\epsilon)$$

has a unique solution $(\vec{x}(t, \epsilon), \vec{y}(t, \epsilon))$ for $t \in \mathcal{I}_0$ and $0 < \epsilon \leq r_4$ if $\epsilon + |\vec{\xi}(\epsilon) - \vec{\phi}(a)| + |\vec{\eta}(\epsilon) - \vec{\psi}(a)|$ is sufficiently small on $0 < \epsilon \leq r_4$, where r_4 is a positive number. Also, show that $|\vec{x}(t, \epsilon) - \vec{\phi}(t)| + |\vec{y}(t, \epsilon) - \vec{\psi}(t)| \to 0$ uniformly on $\mathcal{I}_0$ as $\epsilon + |\vec{\xi}(\epsilon) - \vec{\phi}(a)| + |\vec{\eta}(\epsilon) - \vec{\psi}(a)| \to 0$.

Hint. See [LeL].

X-15. Let $\vec{x} \in \mathbb{R}^n$, $\vec{y} \in \mathbb{R}^m$, $t \in \mathbb{R}$, and $\epsilon \in \mathbb{R}$. Also, let $\vec{f}(\vec{x}, \vec{y}, t, \epsilon)$ and $\vec{g}(\vec{x}, \vec{y}, t, \epsilon)$ be respectively $\mathbb{R}^n$-valued and $\mathbb{R}^m$-valued functions of $(\vec{x}, \vec{y}, t, \epsilon)$ which are periodic in t of period T. Assume that

  (i) an $\mathbb{R}^n$-valued function $\vec{\phi}(t)$ and an $\mathbb{R}^m$-valued function $\vec{\psi}(t)$ are continuous, continuously differentiable, and periodic of period T on the interval $\mathcal{I}_0 = \{t : -\infty < t < +\infty\}$ and satisfy the system of equations $\dfrac{d\vec{x}}{dt} = \vec{f}(\vec{x}, \vec{y}, t, 0)$, $\vec{0} = \vec{g}(\vec{x}, \vec{y}, t, 0)$ on $\mathcal{I}_0$,

  (ii) $\vec{f}(\vec{x}, \vec{y}, t, \epsilon)$, $\vec{g}(\vec{x}, \vec{y}, t, \epsilon)$, and their partial derivatives with respect to $(\vec{x}, \vec{y})$ are continuous on a region $\mathcal{R} = \{(\vec{x}, \vec{y}, t, \epsilon) : |\vec{x} - \vec{\phi}(t)| \leq r_1, |\vec{y} - \vec{\psi}(t)| \leq r_2, t \in \mathcal{I}_0, |\epsilon| \leq r_3\}$, where r_1, r_2, and r_3 are positive numbers,

  (iii) there exists a real $m \times m$ matrix $P(t)$ such that

    (iiia) the entries of $P(t)$ and $P(t)^{-1}$ are real-valued, continuous, continuously differentiable, and periodic of period T on $\mathcal{I}_0$,

    (iiib) $P(t)^{-1} \dfrac{\partial \vec{g}}{\partial \vec{y}}(\vec{\phi}(t), \vec{\psi}(t), t, 0) P(t) = \begin{bmatrix} B_1(t) & O \\ O & B_2(t) \end{bmatrix}$, where $B_1(t)$ is a real $m_1 \times m_1$ matrix with eigenvaules having negative real parts on $\mathcal{I}_0$, while $B_2(t)$ is a real $(m - m_1) \times (m - m_1)$ matrix with eigenvalues having positive real parts on $\mathcal{I}_0$,

  (iv) the system

$$\frac{d\vec{z}}{dt} = \left\{ \frac{\partial \vec{f}}{\partial \vec{x}}(\vec{\phi}(t), \vec{\psi}(t), t, 0) \right.$$

$$\left. - \frac{\partial \vec{f}}{\partial \vec{y}}(\vec{\phi}(t), \vec{\psi}(t), t, 0) \left[\frac{\partial \vec{g}}{\partial \vec{y}}(\vec{\phi}(t), \vec{\psi}(t), t, 0) \right]^{-1} \frac{\partial \vec{g}}{\partial \vec{x}}(\vec{\phi}(t), \vec{\psi}(t), t, 0) \right\} \vec{z}$$

does not have nontrivial periodic solutions of period T.

Show that the system

$$\frac{d\vec{x}}{dt} = \vec{f}(\vec{x}, \vec{y}, t, \epsilon), \quad \epsilon \frac{d\vec{y}}{dt} = \vec{g}(\vec{x}, \vec{y}, t, \epsilon)$$

has a unique periodic solution $(\vec{x}(t, \epsilon), \vec{y}(t, \epsilon))$ of period T on $\mathcal{I}_0$ if $0 < |\epsilon|$ is sufficiently small. Also, show that $|\vec{x}(t, \epsilon) - \vec{\phi}(t)| + |\vec{y}(t, \epsilon) - \vec{\psi}(t)| \to 0$ uniformly on $\mathcal{I}_0$ as $|\epsilon| \to 0$.

Hint. See [FL].

CHAPTER XI

ASYMPTOTIC EXPANSIONS

In §§V-1 and V-2, we defined formal solutions of a system of analytic differential equations. Formal solutions are not necessarily convergent. For example, as we mentioned it in Remark V-1-4, the divergent formal power series $\hat{f} = \sum_{m=0}^{\infty} (-1)^m (m!) x^{m+1}$ is a formal solution of $x^2 \dfrac{dy}{dx} + y - x = 0$. This equation has an actual solution $f(x) = e^{1/x} \displaystyle\int_0^x t^{-1} e^{-1/t} dt$ for $x > 0$. Integrating by parts, we obtain $f = \sum_{m=0}^{N} (-1)^m (m!) x^{m+1} + (-1)^{N+1} ((N+1)!) e^{1/x} \displaystyle\int_0^x t^N e^{-1/t} dt$. Since $0 <$
$$e^{1/x} \int_0^x t^N e^{-1/t} dt = x^{N+2} - (N+2) e^{1/x} \int_0^x t^{N+1} e^{-1/t} dt < x^{N+2},$$ we conclude that
$$\left| f(x) - \sum_{m=0}^{N} (-1)^m (m!) x^{m+1} \right| < ((N+1)!) x^{N+2} \quad \text{for } x > 0.$$ This is an example of an asymptotic representation of an actual solution by means of a formal solution. In this chapter, we explain the asymptotic expansions of functions in the sense of Poincaré and in the sense of the Gevrey asymptotics. In the Poincaré asymptotics, flat functions are characterized by the condition $\lim_{x \to 0} \dfrac{f(x)}{x^m} = 0$ for all positive integers m, whereas in the Gevrey asymptotic, flat functions are characterized by the condition $|f(x)| \exp(c|x|^{-k}) \leq M$ as $x \to 0$, where c, k, and M are some positive numbers. Generally speaking, the Poincaré asymptotics is too general for the study of ordinary differential equations. A motivation of the Gevrey asymptotics is also given by the Maillet Theorem (cf. Theorem V-1-5). In §XI-1, we summarize the basic properties of asymptotic expansions of functions in the sense of Poincaré. The Gevrey asymptotics is explained in §§XI-2–XI-5.

For more information concerning the Poincaré asymptotics, see, for example, [Was1]. The Gevrey asymptotics was originally introduced in [Wat] and further developed in [Ne1]. To understand the materials concerning the Gevrey asymptotics of this chapter, [Ram 1], [Ram 2], [Ram3], [Si17, Appendices], [Si18], and [Si19] are helpful.

XI-1. Asymptotic expansions in the sense of Poincaré

In this section, we explain the asymptotic expansions in the sense of Poincaré. Let $x = a$ be a point on the extended complex x-plane. Consider a formal power series

$$(XI.1.1) \qquad\qquad p(x) = \sum_{m=0}^{\infty} c_m (x - a)^m.$$

Let $\mathcal{D}$ be a sector in the x-plane with vertex at $x = a$ and $\mathcal{D}_0$ be a neighborhood of $x = a$ in $\mathcal{D}$. Assume that $f(x)$ is defined and continuous in $\mathcal{D}_0$.

Definition XI-1-1. *The formal series (XI.1.1) is said to be an* asymptotic (series) *expansion of $f(x)$ as $x \to a$ in $\mathcal{D}$ if for every non-negative integer N, there exists a constant K_N such that*

$$(\text{XI.1.2}) \qquad \left| f(x) - \sum_{m=0}^{N} c_m (x-a)^m \right| \leq K_N |x-a|^{N+1}, \qquad N = 0, 1, 2, \ldots$$

for all x in $\mathcal{D}_0$.

Such an asymptotic relation is denoted by

$$(\text{XI.1.3}) \qquad f(x) \simeq p(x) \qquad \text{as } x \to a \text{ in } \mathcal{D}.$$

This definition of an asymptotic expansion of a function was originally given by H. Poincaré [Poi2].

Before we explain some basic properties of asymptotic expansions, it is worthwhile to make the following remarks.

Remark XI-1-2. The vertex $x = a$ can be $x = \infty$. In that case, the asymptotic series is in the form $\displaystyle\sum_{m=0}^{\infty} c_m x^{-m}$.

Remark XI-1-3. Assume that $f(x,t)$ is a function defined and continuous in (x,t) for x in $\mathcal{D}$ and t a parameter in a domain Ω in the t-plane. A formal power series $\displaystyle\sum_{m=0}^{N} c_m(t)(x-a)^m$, where $c_m(t)$ is a function of t, is said to be an asymptotic (series) expansion of $f(x,t)$ as $x \to 0$ in $\mathcal{D}$ if for every non-negative integer N, there exists a function $K_N(t)$, independent of x, such that

$$\left| f(x,t) - \sum_{m=0}^{N} c_m(t)(x-a)^m \right| \leq K_N(t)|x-a|^{N+1}, \quad N = 0, 1, 2, \ldots$$

for all x in $\mathcal{D}_0$. If, moreover, $K_N(t)$ are independent of t, then the asymptotic expansion is said to be *uniform* with respect to t.

Remark XI-1-4. If $f(x)$ is holomorphic (i.e., analytic and single-valued) in a neighborhood of $x = a$, by virtue of the Taylor's Remainder Theorem, $f(x)$ admits its Taylor's series expansion as its asymptotic expansion in any sector with vertex at $x = a$.

Theorem XI-1-5. *For a continuous function $f(x)$, there is at most one asymptotic expansion as $x \to a$ in a sector with vertex at $x = a$.*

Proof.

Assume that there are two asymptotic expansions of $f(x)$ at $x = a$

$$(\text{XI.1.4}) \qquad f(x) \simeq \sum_{m=0}^{\infty} c_m (x-a)^m \qquad \text{as } x \to a \text{ in } \mathcal{D}$$

and

$$f(x) \simeq \sum_{m=0}^{\infty} \gamma_m (x-a)^m \qquad \text{as } x \to a \text{ in } \mathcal{D}.$$

Then, for every non-negative integer N, there exist two constants K_N and L_N such that

$$(\text{XI.1.5}) \qquad \left| f(x) - \sum_{m=0}^{N} c_m (x-a)^m \right| \le K_N |x-a|^{N+1}, \qquad N = 0,1,2,\ldots$$

and

$$(\text{XI.1.6}) \qquad \left| f(x) - \sum_{m=0}^{N} \gamma_m (x-a)^m \right| \le L_N |x-a|^{N+1}, \qquad N = 0,1,2,\ldots$$

for all x in a neighborhood $\mathcal{D}_0$ of $x = a$ in $\mathcal{D}$. For $N = 0$, we have $|c_0 - \gamma_0| \le (K_0 + L_0)|x-a|$ for x in $\mathcal{D}_0$. Let $x \to a$ in $\mathcal{D}$; then, we obtain $c_0 = \gamma_0$.

Now, assume that $c_k = \gamma_k$ for $k = 0,1,\ldots,N-1$. Then, from (XI.1.5) and (XI.1.6), it follows that $|c_N - \gamma_N||x-a|^N \le (K_N + L_N)|x-a|^{N+1}$ for x in $\mathcal{D}_0$. Let $x \to a$ in $\mathcal{D}$. Then, $c_N = \gamma_N$. Thus, $c_m = \gamma_m$ is true for every non-negative integer m. $\square$

Theorem XI-1-6. *Assume that $f(x)$ and $g(x)$ are two continuous functions such that*
(XI.1.7)

$$f(x) \simeq \sum_{m=0}^{\infty} c_m (x-a)^m \quad and \quad g(x) \simeq \sum_{m=0}^{\infty} \gamma_m (x-a)^m \quad as \ x \to a \ in \ \mathcal{D}.$$

Then,

$$(\text{XI.1.8}) \qquad f(x) \pm g(x) \simeq \sum_{m=0}^{\infty} (c_m \pm \gamma_m)(x-a)^m \qquad as \ x \to a \ in \ \mathcal{D}$$

and

$$(\text{XI.1.9}) \qquad f(x)g(x) \simeq \sum_{m=0}^{\infty} \left(\sum_{h+k=m} c_h \gamma_k \right) (x-a)^m \qquad as \ x \to a \ in \ \mathcal{D}.$$

Proof.

Fix a non-negative integer N and put $f(x) = \sum_{m=0}^{N} c_m (x-a)^m + E_1(x)(x-a)^{N+1}$

and $g(x) = \sum_{m=0}^{N} \gamma_m (x-a)^m + E_2(x)(x-a)^{N+1}$. Then, there exists two constants K_N and L_N such that

$$(\text{XI.1.10}) \qquad |E_1(x)| \le K_N, \qquad |E_2(x)| \le L_N$$

for x in $\mathcal{D}_0$. Therefore,

$$(XI.1.11) \qquad \left| [f(x) \pm g(x)] - \sum_{m=0}^{N} [(c_m \pm \gamma_m)(x-a)^m \right| \leq (K_N + L_N)|x-a|^{N+1}$$

for x in $\mathcal{D}_0$. Thus, (XI.1.8) holds. Also, it is easy to see that there exists a positive constant $\hat{K}$ such that

$$\left| f(x)g(x) - \sum_{m=0}^{N} \left(\sum_{h+k=m} c_h \gamma_k \right) (x-a)^m \right| \leq \hat{K}|x-a|^{N+1}$$

for x in $\mathcal{D}_0$. Thus, (XI.1.9) holds. $\square$

Theorem XI-1-7. *Suppose that $f(x)$ is a function holomorphic in $0 < |x-a| < \delta$ and admits an asymptotic expansion (XI.1.4) with $\mathcal{D} = \{0 < |x-a| < \delta\}$. Then, the asymptotic series actually converges to $f(x)$ in $0 < |x-a| < \delta$.*

Proof.

Since $\lim_{x \to a} f(x) = c_0$, $x = a$ is a removable singularity. Thus, $f(x)$ is holomorphic in $|x-a| < \delta$. Therefore, the asymptotic series agrees with the Taylor's series (cf. Remark XI-1-4 and Theorem XI-1-5). Thus, the asymptotic series converges in $|x-a| < \delta$. $\square$

For a vector $\vec{z} \in \mathbb{C}^m$ with entries $(z_1, z_2, \ldots, z_m)$ and $\wp = (p_1, p_2, \ldots, p_m)$ with non-negative integers p_j $(j = 1, 2, \ldots, m)$, we define $|\wp| = p_1 + p_2 + \cdots + p_m$, $\vec{z}^{\wp} = z_1^{p_1} z_2^{p_2} \cdots z_m^{p_m}$, and $|\vec{z}| = \max\{|z_1|, |z_2|, \ldots, |z_m|\}$.

Theorem XI-1-8. *Suppose that $F(x, \vec{z})$ is a function with power series expansion*

$$F(x, \vec{z}) = \sum_{|\wp|=0}^{\infty} F_{\wp}(x) \vec{z}^{\wp}, \quad \text{which converges uniformly for } x \in D, |\vec{z}| < \delta_0, \text{ where } F_{\wp}(x)$$

is continuous in $\mathcal{D}$ and admits an asymptotic expansion $F_{\wp}(x) \simeq \sum_{k=0}^{\infty} F_{\wp k}(x-a)^k$

as $x \to a \in \mathcal{D}$. Define the formal power series in $(x, \vec{z})$

$$(XI.1.12) \qquad \Phi(x, \vec{z}) = \sum_{|\wp|=0}^{\infty} \Phi_{\wp}(x) \vec{z}^{\wp},$$

where

$$(XI.1.13) \qquad \Phi_{\wp}(x) = \sum_{k=0}^{\infty} F_{\wp k}(x-a)^k.$$

If $\vec{f}(x)$ is a continuous $\mathbb{C}^m$-valued function with entrywise asymptotic expansions

$$\vec{f}(x) \simeq \sum_{j=0}^{\infty} \vec{f}_j(x-a)^j \equiv \vec{p}(x) \quad as \quad x \to a \quad in \quad \mathcal{D},$$

then $\Phi(x, \vec{p}(x) - \vec{f}_0)$ defines a formal power series of $(x - a)$ and

$$(XI.1.14) \qquad F(x, \vec{f}(x) - \vec{f}_0) \simeq \Phi(x, \vec{p}(x) - \vec{f}_0) \qquad as \quad x \to a \quad in \quad \mathcal{D}.$$

Proof.

Choose ρ so small that $|\vec{f}(x) - \vec{f}_0| < \delta_0$ for $|x - a| < \rho$. Fix a positive integer N and put

$$F(x, \vec{f}(x) - \vec{f}_0) = \sum_{|\wp|=0}^{N} F_\wp(x)(\vec{f}(x) - \vec{f}_0)^\wp + \sum_{|\wp|=N+1}^{\infty} F_\wp(x)(\vec{f}(x) - \vec{f}_0)^\wp.$$

Then, by virtue of Theorem XI-1-6, we obtain

$$(XI.1.15) \qquad \sum_{|\wp|=0}^{N} F_\wp(x)(\vec{f}(x) - \vec{f}_0)^\wp \simeq \sum_{|\wp|=0}^{N} \Phi_\wp(x)(\vec{p}(x) - \vec{f}_0)^\wp.$$

Also,

$$(XI.1.16) \qquad \sum_{|\wp|=N+1}^{\infty} F_\wp(x)(\vec{f}(x) - \vec{f}_0)^\wp = O(|x - a|^{N+1}).$$

Set

$$\hat{H}(x) \equiv \Phi(x, \vec{p}(x) - \vec{f}_0) = \sum_{m=0}^{\infty} H_m(x - a).$$

Then,

$$(XI.1.17) \qquad \sum_{|\wp|=0}^{N} \Phi_\wp(x)(\vec{p}(x) - \vec{f}_0)^\wp = \sum_{m=0}^{N} H_m(x - a)^m + O((x - a)^{N+1})$$

and, hence,

$$(XI.1.18) \qquad \sum_{|\wp|=0}^{N} F_\wp(x)(\vec{f}(x) - \vec{f}_0)^\wp = \sum_{m=0}^{N} H_m(x - a)^m + O(|x - a|^{N+1}).$$

Since this is true for all positive integers N, the theorem follows immediately. $\square$

Theorem XI-1-9. *Suppose that $f(x)$ is a continuous function with asymptotic expansion (XI.1.4) and $c_0 \neq 0$. Let $p(x)$ denote the formal power series (XI.1.1). Then, $\dfrac{1}{p(x)}$ defines a formal power series in $(x - a)$ and $\dfrac{1}{f(x)} \simeq \dfrac{1}{p(x)}$ as $x \to a$ in $\mathcal{D}$.*

Proof.

Consider $F(z) = \dfrac{1}{c_0 + z}$. Then, the theorem follows from Theorem XI-1-8. $\square$

Theorem XI-1-10. *Suppose that $f(x)$ is a continuous function with asymptotic expansion (XI.1.4) as $x \to a$ in $\mathcal{D}$. Let $p(x)$ denote the formal power series (XI.1.1). Then,*

$$\int_a^x f(\xi)d\xi \simeq \sum_{m=0}^{\infty} \frac{1}{m+1} c_m (x-a)^{m+1} \qquad as \ x \to a \ \ in \ \ \mathcal{D},$$

where the path of integration is the line segment joining a and x, i.e.,

$$\int_a^x f(\xi)d\xi \simeq \int_a^x p(\xi)d\xi \qquad as \ x \to a \ \ in \ \ \mathcal{D}.$$

Proof.

Set $p_N(x) = \sum_{m=0}^{N} c_m (x-a)^m$. From (XI.1.4) and (XI.1.5), it follows that

$$\left| \int_a^x [f(\xi) - p_N(\xi)]d\xi \right| \leq \int_0^{|x-a|} |f(\xi) - p_N(\xi)| \, ds$$

$$\leq K_N \int_0^{|x-a|} s^{N+1} ds = K_N \frac{|x-a|^{N+2}}{N+2}$$

for every non-negative integer N, where $s = |\xi - a|$. $\square$

Theorem XI-1-11. *Suppose that $f(x)$ is a holomorphic function with asymptotic expansion (XI.1.4). Let $p(x)$ denote the formal power series (XI.1.1). Let $\tilde{\mathcal{D}}$ be a proper subsector of $\mathcal{D}$ with vertex $x = a$. Then, $\dfrac{df}{dx} \simeq \dfrac{dp}{dx}$ as $x \to a$ in $\tilde{\mathcal{D}}$, where*

$$\frac{dp}{dx} = \sum_{m=1}^{\infty} m c_m (x-a)^{m-1}.$$

Proof.

For a non-negative integer N, put $f(x) = p_N(x) + E_N(x)$. Then, there exists a positive constant K_N such that

$$(XI.1.19) \qquad\qquad |E_N(x)| \leq K_N |x-a|^{N+1}$$

for x in a neighborhood of $x = a$ in $\mathcal{D}$ (cf. Definition XI-1-1). Let the radial boundaries of $\mathcal{D}$ be $\arg(x-a) = \Theta_{\pm}$ and those of $\tilde{\mathcal{D}}$ be $\arg(x-a) = \theta_{\pm}$ ($\Theta_- < \theta_- < \theta_+ < \Theta_+$). Let $\theta = \min\{\Theta_+ - \theta_+, \ \theta_- - \Theta_-\}$. Let x be a point of $\tilde{\mathcal{D}}$. Consider a circle $\Gamma = \left\{ \xi \ \Big| \ |\xi - x| = \frac{1}{2}|x-a|\sin\theta \right\}$. Then, Γ and its interior are contained in $\mathcal{D}$. Using Cauchy's integral formula, we obtain

$$\frac{df(x)}{dx} = \frac{1}{2\pi i} \int_\Gamma \frac{f(\xi)}{(\xi-x)^2} d\xi \ = \frac{1}{2\pi i} \int_\Gamma \frac{p_N(\xi)}{(\xi-x)^2} d\xi + \frac{1}{2\pi i} \int_\Gamma \frac{E_N(\xi)}{(\xi-x)^2} d\xi.$$

Let $\eta = \xi - x = \dfrac{e^{i\phi}}{2}|x - a|\sin\theta$. Then, $\Gamma = \left\{\eta \;\middle|\; \eta = \dfrac{e^{i\phi}}{2}|x - a|\sin\theta, 0 \le \phi \le 2\pi\right\}$ and (XI.1.19) implies

$$\frac{1}{2\pi i}\int_\Gamma \frac{E_N(\xi)}{(\xi - x)^2}d\xi = \frac{1}{2\pi i}\int_\Gamma \frac{E_N(\xi)}{\eta^2}d\eta$$
$$= \frac{1}{\pi}\frac{1}{|x - a|\sin^2\theta}\int_0^{2\pi} E_N(\xi)e^{i\phi}d\phi = O\left(|x - a|^N\right).$$

Thus, the theorem is proved. $\square$

Theorem XI-1-12. *Suppose that $\{f_n(x)|n = 1, 2, 3, \dots\}$ is a sequence of continuous functions such that they admit uniform asymptotic expansions*

$$\text{(XI.1.20)}\qquad f_n(x) \simeq \sum_{m=0}^{\infty} c_{nm}(x - a)^m \quad (n = 1, 2, 3, \dots) \ \ as \ \ x \to a \ \ in \ \ \mathcal{D}.$$

Assume that $\{f_n(x)|n = 1, 2, 3, \dots\}$ converges uniformly to a function $f(x)$ in a subsector $\mathcal{D}_0$ of $\mathcal{D}$ with vertex $x = a$. Then,

$$\text{(XI.1.21)}\qquad\qquad \lim_{n\to\infty} c_{nm} = c_m$$

and

$$\text{(XI.1.22)}\qquad f(x) \simeq \sum_{m=0}^{\infty} c_m(x - a)^m \qquad as \ \ x \to a \ \ in \ \ \mathcal{D}_0.$$

Proof.

The assumption implies that for each non-negative integer N, there exists a positive constant K_N, independent of n, such that

$$\text{(XI.1.23)}\qquad \left|f_n(x) - \sum_{m=0}^{N} c_{nm}(x - a)^m\right| \le K_N|x - a|^{N+1} \qquad (n = 1, 2, 3, \dots)$$

for x in a neighborhood of a in $\mathcal{D}$. Furthermore, for each pair of positive integers (j, k), there exists a positive constant δ_{jk} such that

$$\text{(XI.1.24)}\qquad\qquad |f_j(x) - f_k(x)| \le \delta_{jk}$$

for x in $\mathcal{D}_0$, where

$$\text{(XI.1.25)}\qquad\qquad \delta_{jk} \to 0 \qquad as \ \ j, k \to \infty.$$

Put

$$\text{(XI.1.26)}\qquad p_{jN}(x) = \sum_{m=0}^{N} c_{jm}(x - a)^m \qquad (j = 1, 2, 3, \dots).$$

Then,

$$(\text{XI.1.27}) \qquad |f_j(x) - p_{jN}(x)| \leq K_N |x - a|^{N+1} \qquad (j = 1, 2, 3, \dots).$$

Thus,

$$(\text{XI.1.28}) \qquad |p_{jN}(x) - p_{kN}(x)| \leq \delta_{jk} + 2K_N |x - a|^{N+1} \qquad (j, k = 1, 2, 3, \dots).$$

In particular,

$$(\text{XI.1.29}) \qquad |c_{j0} - c_{k0}| \leq \delta_{jk} + 2K_0 |x - a| \qquad (j, k = 1, 2, 3, \dots)$$

for $N = 0$ and x in $\mathcal{D}_0$. Therefore, (XI.1.25) and (XI.1.29) imply that $\lim_{j \to \infty} c_{j0} = c_0$ exists.

Assume that $\lim_{j \to \infty} c_{jm} = c_m$ exist for $m < N$. Then, from (XI.1.28), it follows that

$$\left| \sum_{m=0}^{N-1} (c_{jm} - c_{km})(x - a)^m + (c_{jN} - c_{kN})(x - a)^N \right|$$
$$\leq \delta_{jk} + 2K_N |x - a|^{N+1} \qquad (j, k = 1, 2, 3, \dots)$$

for x in $\mathcal{D}_0$. Thus,

$$|c_{jN} - c_{kN}|\, |x - a|^N \leq \epsilon_{jk} + \delta_{jk} + 2K_N |x - a|^{N+1} \qquad (j, k = 1, 2, 3, \dots),$$

where $\epsilon_{jk} \to 0$ as $j, k \to \infty$. Hence,

$$|c_{jN} - c_{kN}| \leq \frac{\epsilon_{jk} + \delta_{jk}}{|x - a|^N} + 2K_N |x - a| \qquad (j, k = 1, 2, 3, \dots).$$

Therefore, $\lim_{j \to \infty} c_{jN} = c_N$. Consequently, $\lim_{j \to \infty} c_{jm} = c_m$ for all m.

Furthermore, since (XI.1.27) holds independently of j, we obtain

$$\left| f(x) - \sum_{m=0}^{N} c_m (x - a)^m \right| \leq K_N |x - a|^{N+1}$$

for x in $\mathcal{D}_0$. Thus, (XI.1.22) holds. $\square$

The following basic theorem is due to E. Borel and J. F. Ritt.

Theorem XI-1-13 ([Bor] and [Ri]). *For a given formal power series in $(x - a)$*

$$(\text{XI.1.30}) \qquad p(x) = \sum_{m=0}^{\infty} c_m (x - a)^m$$

and a sector with vertex $x = a$

$$(\text{XI.1.31}) \qquad \mathcal{D} = \{x \,|\, 0 < |x - a| \leq \gamma, \ \beta_1 \leq \arg(x - a) \leq \beta_2\},$$

there exists a function $f(x)$ which is continuous in $\mathcal{D}$ and holomorphic in the interior of $\mathcal{D}$, and

$$\text{(XI.1.32)} \qquad f(x) \simeq \sum_{m=0}^{\infty} c_m (x-a)^m \qquad as \quad x \to a \quad in \quad \mathcal{D}.$$

Proof.

Without loss of generality, assume that the sector $\mathcal{D}$ contains the ray $\arg(x - a) = 0$. Construct functions $\alpha_m(x)$ for $m = 1, 2, \ldots$ such that $f(x) = c_0 + \sum_{m=1}^{\infty} c_m \alpha_m(x)(x-a)^m$ is continuous in $\mathcal{D}$, holomorphic in the interior of $\mathcal{D}$, and satisfies (XI.1.32). Let b_m and θ be positive numbers, where b_m are to be specified later, whereas θ is chosen to satisfy $0 < \theta < 1$ and

$$\text{(XI.1.33)} \qquad \theta |\arg(x-a)| < \frac{\pi}{2} \qquad \text{for} \quad x \in \mathcal{D}.$$

Consider

$$\text{(XI.1.34)} \qquad \alpha_m(x) = 1 - \exp\left[- \frac{b_m}{2^m \gamma^m (x-a)^\theta} \right], \qquad m = 1, 2, \ldots .$$

Then, $\alpha_m(x)$ are holomorphic in a sector containing $\mathcal{D}$. Also, from (XI.1.33) and the fact that

$$|1 - e^z| = \left| \int_0^z e^t dt \right| < |z| \qquad \text{for} \quad \Re z < 0,$$

it follows that

$$\text{(XI.1.35)} \qquad |\alpha_m(x)| \le \frac{b_m}{2^m \gamma^m |x-a|^\theta}, \qquad m = 1, 2, \ldots ,$$

in $\mathcal{D}$. Put

$$\text{(XI.1.36)} \qquad b_m = \begin{cases} |c_m|^{-1} & \text{if} \ c_m \ne 0, \\ 0 & \text{if} \ c_m = 0. \end{cases}$$

Then,

$$\sum_{m=1}^{\infty} |c_m \alpha_m(x)(x-a)^m| \le \sum_{m=1}^{\infty} \frac{|x-a|^{m-\theta}}{2^m \gamma^m} \qquad \text{for} \quad x \in \mathcal{D}.$$

Hence,

$$f(x) = c_0 + \sum_{m=1}^{\infty} c_m \alpha_m(x)(x-a)^m$$

converges uniformly in $\mathcal{D}$. Consequently, $f(x)$ is continuous in $\mathcal{D}$ and holomorphic in the interior of $\mathcal{D}$.

To show that $f(x)$ satisfies (XI.1.32), let N be a positive integer and observe that

$$f(x) - c_0 - \sum_{m=1}^{N+1} c_m (x-a)^m = \sum_{m=1}^{N+1} c_m \exp\left[- \frac{b_m}{2^m \gamma^m (x-a)^\theta} \right] (x-a)^m$$
$$+ (x-a)^{N+1} \left[\sum_{m=N+2}^{\infty} c_m \alpha_m(x)(x-a)^{m-N-1} \right].$$

Since (XI.1.33) implies that

$$\left| \exp\left[- \frac{b_m}{2^m \gamma^m (x-a)^\theta} \right] \right| |x-a|^{-r}, \qquad r = 0, 1, 2, \ldots, N,$$

are bounded for $x \in \mathcal{D}$, there exists a positive constant H_1 such that

$$\left| \sum_{m=1}^{N} c_m \exp\left[- \frac{b_m}{2^m \gamma^m (x-a)^\theta} \right] (x-a)^m \right| \leq H_1 |x-a|^{N+1} \quad \text{for } x \in \mathcal{D}.$$

Also, (XI.1.35) implies that there exists a positive constant H_2 such that

$$\left| \sum_{m=N+2}^{\infty} c_m \alpha_m(x)(x-a)^{m-N-1} \right| \leq \frac{1}{(2\gamma)^{N+1}|x-a|^\theta} \sum_{m=N+2}^{\infty} \frac{|x-a|^{m-N-1}}{(2\gamma)^{m-N-1}}$$
$$= \frac{|x-a|^{1-\theta}}{(2\gamma)^{N+2}} \frac{1}{1 - \dfrac{|x-a|}{2\gamma}} \leq H_2 \quad \text{for } x \in \mathcal{D}.$$

Thus, there exists a positive constant K_N such that

$$\left| f(x) - \sum_{m=0}^{N} c_m (x-a)^m \right| \leq K_N |x-a|^{N+1} \qquad \text{for } x \in \mathcal{D}.$$

Therefore, (XI.1.32) is satisfied. $\square$

Similarly to Theorem XI-1-13, we can prove the following theorem.

Theorem XI-1-14. *For a given formal series $p(x,t) = \sum_{m=0}^{\infty} c_m(t)(x-a)^m$ in powers of $(x-a)$ and a sector $\mathcal{D}$ given by (XI.1.31), where $c_m(t)$ $(m = 0, 1, \ldots)$ are holomorphic and $\dfrac{dc_m}{dt}$ are bounded in a domain Ω, there exists a function $f(x,t)$ which is holomorphic with respect to (x,t) in $\mathcal{D} \times \Omega$ and satisfies the conditions*

$$(\text{XI.1.37}) \qquad f(x,t) \simeq \sum_{m=0}^{\infty} c_m(t)(x-a)^m \qquad \text{as } x \to a \text{ in } \mathcal{D}$$

and

(XI.1.38)
$$\frac{df(x,t)}{dt} \simeq \sum_{m=0}^{\infty} \frac{dc_m(t)}{dt}(x-a)^m \qquad as \ \ x \to a \ \ in \ \ \mathcal{D}$$

uniformly for t in Ω.

Proof.

As $|c_m(t)|$ and $\left|\dfrac{dc_m(t)}{dt}\right|$ are bounded in Ω $(m = 0, 1, \ldots)$, choose

$$b_m = \begin{cases} \left\{ \left[\max_{t\in\Omega} |c_m(t)| \right] + \left[\max_{t\in\Omega} \left| \frac{dc_m(t)}{dt} \right| \right] \right\}^{-1} & \text{if } c_m(t) \not\equiv 0, \\[2mm] 0 & \text{if } c_m(t) \equiv 0. \end{cases}$$

Set

$$f(x,t) = c_0(t) + \sum_{m=1}^{\infty} c_m(t)\alpha_m(x)(x-a)^m,$$

where $\alpha_m(x)$ are given in (XI.1.34). Then, it can be shown that $f(x,t)$ is holomorphic with respect to (x,t) in $\mathcal{D} \times \Omega$ and satisfies (XI.1.37) and (XI.1.38) in a manner similar to the proof of Theorem XI-1-13. $\square$

Examples XI-1-15. The followings are two examples of functions admitting asymptotic expansions.

1. Let $\Gamma(z)$ denote the Gamma function and $\mathrm{Log}\,z$ denote the principal branch of the complex natural logarithm of z. If $\theta = \arg z$ satisfies $|\theta| < \pi$, a real number ω satisfies $|\omega| < \dfrac{\pi}{2}$, and $|\theta + \omega| < \dfrac{\pi}{2}$, then

$$\mathrm{Log}[\Gamma(z)] = \left(z - \frac{1}{2}\right)\mathrm{Log}\,z \ - \ z \ + \ \frac{\ln(2\pi)}{2} \ + \ \sum_{m=1}^{N} \frac{(-1)^{m-1}B_m}{2m(2m-1)}z^{-(2m-1)}$$

$$+ \ \Theta_N(z)\frac{B_{N+1}}{2(N+1)(2N+1)}z^{-(2N+1)},$$

where $|\Theta_N(z)| \leq \dfrac{1}{(\cos(\theta+\omega))^{2N+1}|\sin(2\omega)|}$ and the B_m are the Bernoulli numbers (see, for example, [Ol, p. 294]).

2. The exponential integral function $Ei(z) = \displaystyle\int_{-\infty}^{z} e^t t^{-1}dt$ has the following form:

$$Ei(z) = e^z \sum_{k=1}^{N}(k-1)! z^{-k} + N! \int_{-\infty}^{z} e^t t^{-N-1}dt.$$

From this observation, it follows that

$$\left| e^{-z}Ei(z) \ - \ \sum_{k=1}^{N}(k-1)! z^{-k} \right| \leq \left\{ N! + (N+1)!(\pi \ + \ (N+1)^{-1}) \right\} |z|^{-N-1}$$

for $|\arg(-z)| \leq \pi$. The asymptotic expansion is valid for $|\arg(-z)| < \dfrac{3\pi}{2}$ (see, for example, [Was1, p. 31]).

XI-2. Gevrey asymptotics

The Gevrey asymptotics is based on the following two definitions.

Definition XI-2-1. *Let s be a non-negative number. A formal power series $p = \sum_{m=0}^{+\infty} a_m x^m \in \mathbb{C}[[x]]$ is said to be of Gevrey order s if there exist two non-negative numbers C and A such that*

$$(XI.2.1) \qquad |a_m| \le C(m!)^s A^m \qquad for \quad m = 0, 1, 2, \dots .$$

We denote by $\mathbb{C}[[x]]_s$ the set of all power series of Gevrey order s.

Definition XI-2-2. *A function ϕ of x is said to admit an* asymptotic expansion *of Gevrey order s $(s \ge 0)$ as $x \to 0$ on a sector*

$$\mathcal{D}(r, a, b) = \{x : \ 0 < |x| < r, \ \ a < \arg(x) < b\},$$

where a and b are two real numbers such that $a < b$ and r is a positive number if

(i) ϕ is holomorphic on $\mathcal{D}(r, a, b)$,

(ii) there exists a formal power series $p = \sum_{m=0}^{+\infty} a_m x^m \in \mathbb{C}[[x]]_s$ such that an in-

equality

$$(XI.2.2) \qquad \left| \phi(x) - \sum_{m=0}^{N-1} a_m x^m \right| \le K_{\rho,\alpha,\beta}(N!)^s (B_{\rho,\alpha,\beta})^N |x|^N$$

holds on $\mathcal{D}(\rho, \alpha, \beta)$ for every positive integer N and every (ρ, α, β) satisfying the inequalities $0 < \rho < r$ and $a < \alpha < \beta < b$, where $K_{\rho,\alpha,\beta}$ and $B_{\rho,\alpha,\beta}$ are non-negative numbers determined by (ρ, α, β).

We denote by $\mathcal{A}_s(r, a, b)$ the set of all functions admitting asymptotic expansions of Gevrey order s as $x \to 0$ on the sector $\mathcal{D}(r, a, b)$. We also set $J(\phi) = p$ for $\phi \in \mathcal{A}_s(r, a, b)$. In §XI-3, we shall explain the basic properties of $\mathbb{C}[[x]]_s$, $\mathcal{A}_s(r, a, b)$, and the map $J : \mathcal{A}_s(r, a, b) \to \mathbb{C}[[x]]_s$.

In the example given at the beginning of this chapter, the formal solution $p = \sum_{m=0}^{\infty} (-1)^m (m!) x^{m+1}$ of $x^2 \dfrac{dy}{dx} + y - x = 0$ is of Gevrey order 1 and the solution $f(x) = e^{1/x} \int_0^x t^{-1} e^{-1/t} dt$ admits an asymptotic expansion of Gevrey order 1. Furthermore, $J(f) = p$. Also, the Maillet Theorem (cf. Theorem V-1-5) states that any formal solutions of an algebraic differential equation belong to $\mathbb{C}[[x]]_s$ for some s, where s depends on each solution.

It is well known that if a complex-valued function $\phi(x)$ is holomorphic and bounded on a domain $0 < |x| < r$, where r is a positive number, then ϕ is represented by a convergent power series in x. The Gevrey asymptotic expansions arise in a similar but more general situation. To explain such a situation, let us consider N sectors

$$S_\ell = \{x : \ 0 < |x| < r, \ \ a_\ell < \arg(x) < b_\ell\} \qquad (\ell = 1, 2, \dots, N)$$

which satisfy the condition $\bigcup\limits_{\ell=1}^{N} S_\ell = \{x : \ 0 < |x| < r\}$. The set $\{S_1, S_2, \dots, S_N\}$ is called a *covering at* $x = 0$. Also, a covering $\{S_1, S_2, \dots, S_N\}$ at $x = 0$ is said to be *good* if

(i) $a_\ell < a_{\ell+1} \ (\ell = 1, 2, \dots, N)$, where $a_{N+1} = a_1 + 2\pi$,

(ii) $b_\ell - a_\ell < \pi \ (\ell = 1, 2, \dots, N)$,

(iii) $S_\ell \cap S_{\ell+1} \neq \emptyset \ (\ell = 1, 2, \dots, N)$ and $S_\ell \cap S_k = \emptyset$ otherwise if $\ell \neq k$, where $S_{N+1} = S_1$.

The following theorem is the most basic fact in the Gevrey asymptotics.

Theorem XI-2-3. *Assume that a covering* $\{S_1, S_2, \dots, S_N\}$ *at* $x = 0$ *is good and that* N *functions* $\phi_1(x), \phi_2(x), \dots, \phi_N(x)$ *satisfy the conditions*

(1) $\phi_\ell(x)$ *is holomorphic on* S_ℓ,

(2) $\phi_\ell(x)$ *is bounded on* S_ℓ,

(3)   *it holds that*

$$(\text{XI.2.3}) \qquad |\phi_\ell(x) - \phi_{\ell+1}(x)| \ \leq \ \gamma \exp\left[-\frac{\lambda}{|x|^k}\right] \qquad on \quad S_\ell \cap S_{\ell+1},$$

where $\gamma \geq 0$, $\lambda > 0$, *and* $k > 0$ *are suitable numbers independent of* ℓ.
Set

$$(\text{XI.2.4}) \qquad\qquad\qquad\qquad s \ = \ \frac{1}{k}.$$

Then, there exists a formal power series $p = \sum\limits_{m=0}^{+\infty} a_m x^m \in \mathbb{C}[[x]]_s$ *such that* $\phi_\ell \in$
$\mathcal{A}_s\left(r, a_\ell, b_\ell\right)$ *and* $J(\phi_\ell) = p$ *for each* ℓ.

There are various situations in which Gevrey asymptotic expansions arise. To illustrate such a situation, let $\phi(x)$ be a convergent power series in x with coefficients in $\mathbb{C}$. For two positive numbers r and k, set

$$\ell_{r,k}(\phi) \ = \ \frac{k}{x^k} \int_0^r \phi(t) e^{-(t/x)^k} t^{k-1} dt.$$

This integral is called an *incomplete Leroy transform* of ϕ of order k.

Theorem XI-2-4. *For every* $\phi = \sum\limits_{m=0}^{+\infty} c_m x^m \in \mathbb{C}\{x\}$, *it holds that*

$$\ell_{r,k}(\phi) \ \in \ \mathcal{A}_{1/k}\left(\rho, -\frac{\pi}{2k}, \frac{\pi}{2k}\right)$$

and

$$J(\ell_{r,k}(\phi)) = \sum_{m=0}^{+\infty} \Gamma\left(1 + \frac{m}{k}\right) c_m x^m,$$

where k and ρ are any positive numbers and r is any positive number smaller than the radius of convergence of ϕ.

Proof.

The following proof is suggested by B. L. J. Braaksma. For an arbitrary small positive number ϵ, let $\{\mathcal{S}_1(\epsilon), \mathcal{S}_2(\epsilon), \ldots, \mathcal{S}_N(\epsilon)\}$ be a good covering at $x = 0$, where

$$\mathcal{S}_\ell(\epsilon) = \left\{ x: \ |\arg x \ - \ d_\ell| \ < \ \frac{\pi}{2k} - \epsilon, \ \ 0 \ < \ |x| \ < \ r \right\},$$

with real numbers d_ℓ such that $-\pi < d_1 < \cdots < d_N \leq \pi$, and set

$$\begin{cases} \psi_\ell(\xi) = \phi(\xi e^{id_\ell}), \\ \phi_\ell(x) = \ell_{r,k}(\psi_\ell)(xe^{-id_\ell}) \end{cases} \qquad (\ell = 1, 2, \ldots, N).$$

In particular, choose $d_\ell = 0$ for some ℓ. Then, $\{\mathcal{S}_1(\epsilon), \mathcal{S}_2(\epsilon), \ldots, \mathcal{S}_N(\epsilon)\}$ and $\{\phi_1, \phi_2, \ldots, \phi_N\}$ satisfy conditions (1), (2), and (3) of Theorem XI-2-3. In fact, (1) and (2) are evident. To prove (3), note that

$$\phi_\ell(x) = \frac{k}{x^k} \int_0^{re^{id_\ell}} \phi(\sigma)e^{-(\sigma/x)^k} \sigma^{k-1} d\sigma,$$

where the path of integration is the line segment connecting 0 to re^{id_ℓ}. Therefore,

$$\phi_\ell(x) - \phi_{\ell+1}(x) = \frac{k}{x^k} \int_{\gamma_\ell} \phi(\sigma)e^{-(\sigma/x)^k} \sigma^{k-1} d\sigma,$$

where the path γ_ℓ of integration is the circular arc connecting $re^{id_{\ell+1}}$ to re^{id_ℓ}. Statement (3) follows, since

$$\left| e^{-(\sigma/x)^k} \right| = \exp\left[-\left(\frac{r}{|x|}\right)^k \cos[k(\arg \sigma \ - \ \arg x)] \right]$$

and $k|\arg \sigma - \arg x| \leq \dfrac{\pi}{2} - k\epsilon$ for $\sigma \in \gamma_\ell$ and $x \in \mathcal{S}_\ell(\epsilon) \cap \mathcal{S}_{\ell+1}(\epsilon)$. Since ϵ is arbitrary, it follows that

$$\phi_\ell \in \mathcal{A}_{1/k}\left(\rho, d_\ell - \frac{\pi}{2k}, d_\ell + \frac{\pi}{2k}\right) \qquad (\ell = 1, 2, \ldots, N).$$

Furthermore, $J(\phi_\ell)$ can be computed easily (cf. Exercise XI-13). $\square$

Observe that a power series $p = \displaystyle\sum_{m=0}^{+\infty} a_m x^m \in \mathbb{C}[[x]]$ belongs to $\mathbb{C}[[x]]_s$ if and only if $\phi(x) = \displaystyle\sum_{m=0}^{+\infty} \frac{a_m}{\Gamma(1 + sm)} x^m$ belongs to $\mathbb{C}\{x\}$. Therefore, we obtain an important corollary of Theorem XI-2-4.

Corollary XI-2-5. *For any $p \in \mathbb{C}[[x]]_s$ and any real number d, there exists a function $\phi(x) \in \mathcal{A}_s\left(r, d - \dfrac{s\pi}{2}, d + \dfrac{s\pi}{2}\right)$ such that $J(\phi) = p$.*

This corollary corresponds to the Borel-Ritt Theorem (cf. Theorem XI-1-13) of the Poincaré asymptotics. Also, this corollary implies that the map J : $\mathcal{A}_s\left(r, d - \dfrac{s\pi}{2}, d + \dfrac{s\pi}{2}\right) \to \mathbb{C}[[x]]_s$ is onto.

Theorem XI-2-3 is a corollary of the following lemma.

Lemma XI-2-6. *Assume that a covering $\{\mathcal{S}_\ell : \ell = 1, 2, \ldots, N\}$ at $x = 0$ is good and that N functions $\delta_1(x), \delta_2(x), \ldots, \delta_N(x)$ satisfy the following conditions:*
  (i) δ_ℓ is holomorphic on $\mathcal{S}_\ell \cap \mathcal{S}_{\ell+1}$,
  (ii) $|\delta_\ell(x)| \leq \gamma \exp[-\lambda |x|^{-k}]$ on $\mathcal{S}_\ell \cap \mathcal{S}_{\ell+1}$, where $\gamma \geq 0$, $\lambda > 0$, and $k > 0$ are suitable numbers independent of ℓ.
Define s by (XI.2.4). Then, there exist N functions $\psi_1(x), \psi_2(x), \ldots, \psi_N(x)$ and a formal power series $p = \displaystyle\sum_{m=0}^{+\infty} a_m x^m \in \mathbb{C}[[x]]_s$ such that
  (a) $\psi_\ell \in \mathcal{A}_s(r, a_\ell, b_\ell)$ and $J(\psi_\ell) = p$, where $\mathcal{S}_\ell = \{x : 0 < |x| < r,\ a_\ell < \arg(x) < b_\ell\}$ $(\ell = 1, 2, \ldots, N)$,
  (b) $\delta_\ell(x) = \psi_\ell(x) - \psi_{\ell+1}(x)$ on $\mathcal{S}_\ell \cap \mathcal{S}_{\ell+1}$.

Let us prove Theorem XI-2-3 by using Lemma XI-2-6.

Proof.

Set
$$\delta_\ell(x) \ = \ \phi_\ell(x) \ - \ \phi_{\ell+1}(x) \qquad (\ell = 1, 2, \ldots, N).$$
Then, there exist N functions $\psi_1(x), \psi_2(x), \ldots, \psi_N(x)$ satisfying conditions (a) and (b) of Lemma XI-2-6. In particular, (b) implies that
$$\phi_\ell(x) \ - \ \phi_{\ell+1}(x) \ = \ \psi_\ell(x) \ - \ \psi_{\ell+1}(x) \qquad (\ell = 1, 2, \ldots, N)$$
on $\mathcal{S}_\ell \cap \mathcal{S}_{\ell+1}$. This, in turn, implies that
$$\phi_\ell(x) \ - \ \psi_\ell(x) \ = \ \phi_{\ell+1}(x) \ - \ \psi_{\ell+1}(x) \qquad (\ell = 1, 2, \ldots, N)$$
on $\mathcal{S}_\ell \cap \mathcal{S}_{\ell+1}$. Define a function ϕ by
$$\phi(x) \ = \ \phi_\ell(x) \ - \ \psi_\ell(x) \qquad \text{on} \quad \mathcal{S}_\ell \quad (\ell = 1, 2, \ldots, N).$$

Then, ϕ is holomorphic and bounded for $0 < |x| < r$. Therefore, ϕ is represented by a convergent power series. Since $\phi_\ell = \psi_\ell + \phi$, Theorem XI-2-3 follows immediately. $\square$

We shall prove Lemma XI-2-6 in §XI-5.

Because the Gevrey asymptotics of functions containing parameters will be used later, we state the following two definitions.

Definition XI-2-7. *Let s be a positive number. A formal power series $\displaystyle\sum_{m=0}^{\infty} a_m(\vec{u})\epsilon^m$
is said to be of* Gevrey order s *uniformly on a domain $\mathcal{D}$ in the $\vec{u}$-space if there
exist two non-negative numbers C_0 and C_1 such that*

$$\text{(XI.2.5)} \qquad\qquad |a_m(\vec{u})| \leq C_0(m!)^s C_1^m$$

for $\vec{u} \in \mathcal{D}$ and $m = 0,\ 1,\ 2,\ \ldots$.

Set $V = \mathcal{D}(\delta_0, \alpha_0, \beta_0) = \{\epsilon : 0 < |\epsilon| < \delta_0,\ \alpha_0 < \arg \epsilon < \beta_0\}$ and $W = \mathcal{D}(\delta, \alpha, \beta)$.

Definition XI-2-8. *Let s be a positive number. A function $f(\vec{u}, \epsilon)$ is said to admit
an asymptotic expansion $\displaystyle\sum_{m=0}^{\infty} a_m(\vec{u})\epsilon^m$ of Gevrey order s as $\epsilon \to 0$ in V uniformly
on $\mathcal{D}$ if*

(i) $\displaystyle\sum_{m=0}^{\infty} a_m(\vec{u})\epsilon^m$ is of Gevrey order s uniformly on $\mathcal{D}$,

*(ii) for each W such that $\alpha_0 < \alpha < \beta < \beta_0$ and $0 < \delta < \delta_0$, there exist two
non-negative numbers K_W and L_W such that*

$$\text{(XI.2.6)} \qquad \left| f(\vec{u}, \epsilon) - \sum_{m=0}^{N} a_m(\vec{u})\epsilon^m \right| \leq K_W[(N+1)!]^s L_W^{N+1} |\epsilon|^{N+1}$$

for $\vec{u} \in \mathcal{D}$, $\epsilon \in W$ and $N = 1,\ 2,\ \ldots$.

Theorem XI-2-3, Theorem XI-2-4, Corollary XI-2-5, and Lemma XI-2-6 can be
extended in a natural way so that we can use them for functions containing parameters. We leave such details to the reader as an exercise.

The materials of this section are also found in [Ram 1], [Ram 2], [Si17, Appen-
dices], and [Si18].

XI-3. Flat functions in the Gevrey asymptotics

In the next section, we shall show that $\mathbb{C}[[x]]_s$ and $\mathcal{A}_s(r, a, b)$ are differential
algebras over $\mathbb{C}$ and the map $J : \mathcal{A}_s(r, a, b) \to \mathbb{C}[[x]]_s$ is a homomorphism of
differential algebras over $\mathbb{C}$. In §XI-2, it was shown that the map J is onto if
$b - a \leq s\pi$ (cf. Corollary XI-2-5). In this section, we explain the basic results
concerning the nullspace of J. To begin with, we introduce the following definition.

Definition XI-3-1. *A function $f(\vec{u}, \epsilon)$ is said to be* flat of Gevrey order s *as $\epsilon \to 0$
in a sector*

$$V = \mathcal{D}(r, a, b) = \{\epsilon : 0 < |\epsilon| < r,\ a < \arg(\epsilon) < b\}$$

*uniformly on a domain $\mathcal{D}$ in the $\vec{u}$-space if $f(\vec{u}, \epsilon)$ admits an asymptotic expansion
$p = \displaystyle\sum_{m=0}^{+\infty} a_m(\vec{u})\epsilon^m$ of Gevrey order s as $\epsilon \to 0$ in V uniformly on $\mathcal{D}$ and the expansion
of f is 0, i.e., all the coefficients $a_m(\vec{u})$ of p are equal to zero.*

The following theorem characterizes flat functions of Gevrey order s.

Theorem XI-3-2. *A function $f(\vec{u}, \epsilon)$ is flat of Gevrey order s as $\epsilon \to 0$ in a sector $\mathcal{D}(r, a, b)$ uniformly on a domain $\mathcal{D}$ in the $\vec{u}$-space if and only if for each $W = \mathcal{D}(\rho, \alpha, \beta)$ such that $0 < \rho < r$ and $a < \alpha < \beta < b$, there exist non-negative numbers K_W and λ_W such that*

$$(\text{XI.3.1}) \qquad |f(\vec{u}, \epsilon)| \le K_W \exp\left[-\frac{\lambda_W}{|\epsilon|^k}\right] \quad \text{for} \ \ (\vec{u}, \epsilon) \in \mathcal{D} \times W,$$

where

$$(\text{XI.3.2}) \qquad k = \frac{1}{s}.$$

Proof.

Suppose that for a fixed W, there exist two non-negative numbers C and A such that

$$|f(\vec{u}, \epsilon)| \ \le \ C(m!)^s A^m |\epsilon|^m \qquad \text{for} \qquad (\vec{u}, \epsilon) \ \in \ \mathcal{D} \times W \quad \text{and} \quad m = 0, 1, 2, \ldots.$$

Then, by virtue of Stirling's formula

$$(\text{XI.3.3}) \qquad \begin{cases} \Gamma\left(\dfrac{m}{k}\right) = \sqrt{\dfrac{2\pi k}{m}} \left(\dfrac{m}{k}\right)^{m/k} e^{-m/k} \left[1 + O\left(\dfrac{k}{m}\right)\right], \\[2em] m! = \sqrt{2\pi m}\, m^m e^{-m} \left[1 + O\left(\dfrac{1}{m}\right)\right], \end{cases}$$

there exist two positive numbers C_0 and A_0 such that

$$|f(\vec{u}, \epsilon)| \ \le \ C_0 m^{m/k} A_0^m |\epsilon|^m \qquad \text{for} \qquad (\vec{u}, \epsilon) \in \mathcal{D} \times W \quad \text{and} \quad m = 0, 1, 2, \ldots.$$

For a given $\epsilon \in W$, choose a non-negative integer m so that

$$me \ < \ \left(\frac{1}{A_0 |\epsilon|}\right)^k \ \le \ (m+1)e.$$

Then,

$$|f(\vec{u}, \epsilon) \ \le \ C_0 m^{m/k} \left(\frac{1}{me}\right)^{m/k} \ = \ C_0 e^{-m/k} \ = \ C_0 \exp\left[\frac{1}{ke}\{e - (m+1)e\}\right]$$

$$\le \ C_0 \exp\left[\frac{1}{k} - \frac{1}{ke}\left(\frac{1}{A_0 |\epsilon|}\right)^k\right].$$

Hence, $|f(\vec{u}, \epsilon)| \ \le \ K \exp[-\lambda |\epsilon|^{-k}]$ for every $(\vec{u}, \epsilon) \in \mathcal{D} \times W$, where $K = C_0 \exp[s]$ and $\lambda = \dfrac{s}{e A_0^k}$. The converse is evident. $\quad\square$

Remark XI-3-3. It is known that a function $f(\vec{u}, \epsilon)$ is identically zero on $\mathcal{D} \times V$ if

(i) f is holomorphic in ϵ on V for each fixed $\vec{u}$ in $\mathcal{D}$,

(ii) f is flat of Gevrey order s as $\epsilon \to 0$ in V uniformly on $\mathcal{D}$,

This implies that the homomorphism $J : \mathcal{A}_s(r, a, b) \to \mathbb{C}[[x]]_s$ is one-to-one if $b - a > s\pi$. Remark XI-3-3 is a consequence of the following lemma.

Lemma XI-3-4 ([Wat]). *Let* $V = \mathcal{D}\left(\rho, -\dfrac{\pi}{2k}, \dfrac{\pi}{2k}\right)$, *where* ρ *is a positive number. If* $f(x)$ *is holomorphic in* V *and* $|f(x)| \le C \exp[-B|x|^{-k}]$ *for* $x \in V$ *for some positive numbers* k, B, *and* C, *then* $f = 0$ *identically in* V.

Proof.

For $a > k$, set

$$F(x, a) \;=\; f(x) \exp\left[\frac{B}{\cos(\frac{k\pi}{2a})} x^{-k}\right].$$

Then, for each fixed value of a, F is holomorphic in V and

$$|F(x, a)| \;\le\; C \exp\left[-B\left(1 - \frac{\cos(k\theta)}{\cos}\left(\frac{k\pi}{2a}\right)\right)|x|^{-k}\right] \qquad \text{for} \qquad x \in V,$$

where $\theta = \arg x$. In particular,

$$\begin{cases} |F(x, a)| \le C & \text{for} \qquad \arg x = \pm\dfrac{\pi}{2a}, \;\; x \in V, \\[2mm] |F(x, a)| \le C \exp\left[B\left(\dfrac{1}{\cos(\frac{k\pi}{2a})} - 1\right) r^{-k}\right] & \text{for} \quad |x| = r < \rho \text{ and } |\arg x| \le \dfrac{\pi}{2a}. \end{cases}$$

Therefore, by virtue of the Phragmén-Lindelöf Theorem (cf. Lemma XI-3-5), we conclude that

$$|F(x, a)| \;\le\; C \exp\left[B\left(\frac{1}{\cos(\frac{k\pi}{2a})} - 1\right)\rho^{-k}\right] \qquad \text{for} \qquad |\arg x| \le \frac{\pi}{2a}, \;\; x \in V.$$

In particular,

$$|f(x)| \;\le\; C \exp\left[B\left(\frac{1}{\cos(\frac{k\pi}{2a})} - 1\right)\rho^{-k} - \frac{B}{\cos(\frac{k\pi}{2a})} x^{-k}\right] \qquad \text{for} \qquad \arg x = 0, \; x \in V.$$

Letting $a \to k$, we derive $f(x) = 0$ if $\arg x = 0$ and $x < \rho$. This completes the proof of Lemma III-3-5. $\quad\square$

Since use was made of the Phragmén-Lindelöf Theorem, we shall state and prove the theorem precisely (cf. [Ne2, pp. 43–44]).

Lemma XI-3-5 (Phragmén-Lindelöf).   *Let W be a closed sector in $\mathbb{C}$ given by*

$$W = \{x : \alpha \leq \arg x \leq \beta, \ 0 < |x| \leq \rho(\arg x)\},$$

where $\rho(\omega)$ is a positive-valued and continuous function on the interval $\alpha \leq \omega \leq \beta$. Assume that

(1) $\beta - \alpha < \dfrac{\pi}{k}$, where k is a positive number,

(2) f is continuous on W,

(3) f is holomorphic in $\overset{\circ}{W}$, where $\overset{\circ}{W}$ is the interior of W,

(4) $|f(x)| \leq K \exp[A|x|^{-k}]$ for $x \in W$, where A and K are positive numbers,

(5) $|f(x)| \leq M$ for x on the boundary of W except for $x = 0$, where M is a positive number.

Then,

$$(\text{XI.3.4}) \qquad\qquad |f(x)| \leq M \qquad for \qquad x \in W.$$

Proof.

Choose $\ell > k$ so that $\beta - \alpha < \dfrac{\pi}{\ell} < \dfrac{\pi}{k}$. Set $g(x) = f(x)\exp[-\epsilon(e^{-i\theta}x)^{-\ell}]$, where $\theta = \dfrac{\alpha + \beta}{2}$ and $\epsilon > 0$. Then,

$$\begin{aligned}
|g(x)| &= |f(x)|\exp[-\epsilon\cos(\ell(\arg x - \theta))|x|^{-\ell}] \\
&\leq K\exp[-|x|^{-\ell}(\epsilon\cos(\ell(\arg x - \theta)) - A|x|^{\ell-k})]
\end{aligned}$$

for $x \in W$. Note that $\ell|\arg x - \theta| \leq \ell\left(\dfrac{\beta - \alpha}{2}\right) < \dfrac{\pi}{2}$ for $x \in W$. Therefore,

$\lim\limits_{|x|\to 0} g(x) = 0$ and $|g(x)| \leq |f(x)|$ for $x \in W$. Since g is holomorphic in $\overset{\circ}{W}$, it follows that $|g(x)| \leq M$ for $x \in W$. Therefore,

$$|f(x)| \leq M\left|\exp(\epsilon(e^{-i\theta}x)^{-\ell})\right| \qquad for \qquad x \in W .$$

Letting $\epsilon \to 0$, we derive (XI.3.4). $\square$

We can also prove the following theorem without any complication.

Theorem XI-3-6. *If f is holomorphic in ϵ on $\mathcal{D}(r,a,b)$ for each fixed $\vec{u}$ in $\mathcal{D}$ and f is flat of Gevrey order s as $\epsilon \to 0$ in $\mathcal{D}(r,a,b)$ uniformly on a domain $\mathcal{D}$ in the $\vec{u}$-space, then $\displaystyle\int_0^\epsilon f(\vec{u},\sigma)d\sigma$ and $\dfrac{\partial f(\vec{u},\epsilon)}{\partial\epsilon}$ are also flat of Gevrey order s as $\epsilon \to 0$ in $\mathcal{D}(r,a,b)$ uniformly on $\mathcal{D}$.*

The proof of this theorem is left to the reader as an exercise.

For more information see [Ram1], [Ram2] and [Si17, Appendices].

XI-4. Basic properties of Gevrey asymptotic expansions

In this section, we summarize the basic properties of the Gevrey asymptotic expansions. First we prove the following theorem.

Theorem XI-4-1. *Assume that*

(i) *a $\mathbb{C}$-valued function $F(u_1, \dots, u_n, \epsilon)$ is holomorphic in $(u_1, \dots, u_n, \epsilon)$ in a domain $\mathcal{D}(r_1, \dots, r_n) \times \mathcal{D}(r, a, b)$, where $\mathcal{D}(r_1, \dots, r_n) = \{(u_1, \dots, u_n) : |u_j| < r_j, j = 1, \dots, n\}$,*

(ii) *$F(u_1, \dots, u_n, \epsilon)$ admits an asymptotic expansion $p(u_1, \dots, u_n, \epsilon) = \sum_{m=0}^{+\infty} a_m(u_1, \dots, u_n)\epsilon^m$ of Gevrey order s as $\epsilon \to 0$ in $\mathcal{D}(r, a, b)$ uniformly on $\mathcal{D}(r_1, \dots, r_n)$, where the coefficients $a_m(u_1, \dots, u_n)$ are holomorphic in $\mathcal{D}(r_1, \dots, r_n)$,*

(iii) *n $\mathbb{C}$-valued functions $\phi_1(\vec{v}, \epsilon), \dots, \phi_n(\vec{v}, \epsilon)$ are holomorphic in a domain $\mathcal{D} \times \mathcal{D}(r, a, b)$, where $\mathcal{D}$ is a domain in the $\vec{v}$-space,*

(iv) *for each j, the function $\phi_j(\vec{v}, \epsilon)$ admits an asymptotic expansion $p_j(\vec{v}, \epsilon) = \sum_{m=1}^{+\infty} b_{j,m}(\vec{v})\epsilon^m$ of Gevrey order s_j as $\epsilon \to 0$ in $\mathcal{D}(r, a, b)$ uniformly on $\mathcal{D}$, where the coefficients $b_{j,m}(\vec{v})$ are holomorphic in $\mathcal{D}$.*

Regarding the coefficients $a_m(u_1, \dots, u_n)$ as power series in $(u_1, \dots, u_n)$, define a formal power series in ϵ by

$$P(\vec{v}, \epsilon) = p(p_1(\vec{v}, \epsilon), \dots, p_n(\vec{v}, \epsilon), \epsilon) = \sum_{m=0}^{+\infty} P_m(\vec{v})\epsilon^m,$$

where the coefficients $P_m(\vec{v})$ are holomorphic in $\mathcal{D}$. Then, for any (α, β) satisfying the condition $a < \alpha < \beta < b$, there exists a positive number $\rho(\alpha, \beta)$ such that $0 < \rho(\alpha, \beta) < r$ and that the function $F(\phi_1(\vec{v}, \epsilon), \dots, \phi_n(\vec{v}, \epsilon), \epsilon)$ is holomorphic in $\mathcal{D} \times \mathcal{D}(\rho(\alpha, \beta), \alpha, \beta)$ and admits the formal power series $P(\vec{v}, \epsilon)$ as its asymptotic expansion of Gevrey order $\max(s, s_1, \dots, s_n)$ as $\epsilon \to 0$ in $\mathcal{D}(\rho(\alpha, \beta), \alpha, \beta)$ uniformly on the domain $\mathcal{D}$.

Remark XI-4-2. As a consequence of this theorem, we conclude that under the same assumptions as in Theorem XI-4-1, the formal power series $P(\vec{v}, \epsilon)$ in ϵ is of Gevrey order $\max(s, s_1, \dots, s_n)$ uniformly on $\mathcal{D}$.

Proof of Theorem XI-4-1.

Fixing a pair (α, β) satisfying the condition $a < \alpha < \beta < b$, choose a suitable good covering $\{\mathcal{S}_1, \dots, \mathcal{S}_N\}$ at $\epsilon = 0$ and $N(n+1)$ functions $F_\ell(u_1, \dots, u_n, \epsilon)$, $\phi_{1,\ell}(\vec{v}, \epsilon), \dots, \phi_{n,\ell}(\vec{v}, \epsilon)\,(\ell = 1, \dots, N)$ such that

(1) for each ℓ, the function $F_\ell(u_1, \dots, u_n, \epsilon)$ is holomorphic in the domain $\mathcal{D}(r_1, \dots, r_n) \times \mathcal{S}_\ell$ and admits the formal power series $p(u_1, \dots, u_n, \epsilon)$ as its asymptotic expansion of Gevrey order s as $\epsilon \to 0$ in $\mathcal{S}_\ell$ uniformly on $\mathcal{D}(r_1, \dots, r_n)$,

(2) for each (j, ℓ), the function $\phi_{j,\ell}(\vec{v}, \epsilon)$ is holomorphic in the domain $\mathcal{D} \times \mathcal{S}_\ell$ and admits the formal power series $p_j(\vec{u}, \epsilon)$ as its asymptotic expansion of Gevrey order s_j as $\epsilon \to 0$ in $\mathcal{S}_\ell$ uniformly on $\mathcal{D}$,

(3) there exist two positive numbers K and λ such that

$$|F_\ell(u_1, \dots, u_n, \epsilon) - F_{\ell+1}(u_1, \dots, u_n, \epsilon)| \leq K \exp\left[-\frac{\lambda}{|\epsilon|^k}\right]$$

on $\mathcal{D}(r_1,\dots,r_n) \times (\mathcal{S}_\ell \cap \mathcal{S}_{\ell+1})$, where $k = \dfrac{1}{s}$,

(4) for each $j = 1,\dots,n$, it holds that

$$\left|\phi_{j,\ell}(\vec{v},\epsilon) - \phi_{j,\ell+1}(\vec{v},\epsilon)\right| \leq K \exp\left[-\frac{\lambda}{|\epsilon|^{k_j}}\right] \qquad \text{on} \qquad \mathcal{D} \times (\mathcal{S}_\ell \cap \mathcal{S}_{\ell+1}),$$

where $k_j = \dfrac{1}{s_j}$,

(5) $\mathcal{S}_{\ell_0} = \mathcal{D}(\rho,\alpha,\beta)$, $F_{\ell_0} = F$, and $\phi_{j,\ell_0} = \phi_j$ $(j = 1,\dots,n)$ for some ℓ_0.

We can accomplish this by virtue of Corollary XI-2-5 and Theorem XI-3-2.

Observe that

$$
\begin{aligned}
&\left|F_\ell(\phi_{1,\ell}(\vec{v},\epsilon),\dots,\phi_{n,\ell}(\vec{v},\epsilon),\epsilon) \;-\; F_{\ell+1}(\phi_{1,\ell+1}(\vec{v},\epsilon),\dots,\phi_{n,\ell+1}(\vec{v},\epsilon),\epsilon)\right| \\
&\leq \left|F_\ell(\phi_{1,\ell}(\vec{v},\epsilon),\dots,\phi_{n,\ell}(\vec{v},\epsilon),\epsilon) \;-\; F_\ell(\phi_{1,\ell+1}(\vec{v},\epsilon),\dots,\phi_{n,\ell+1}(\vec{v},\epsilon),\epsilon)\right| \\
&\quad+ \left|F_\ell(\phi_{1,\ell+1}(\vec{v},\epsilon),\dots,\phi_{n,\ell+1}(\vec{v},\epsilon),\epsilon) \;-\; F_{\ell+1}(\phi_{1,\ell+1}(\vec{v},\epsilon),\dots,\phi_{n,\ell+1}(\vec{v},\epsilon),\epsilon)\right| \\
&\leq L \sum_{j=1}^{n} K \exp\left[-\frac{\lambda}{|\epsilon|^{k_j}}\right] \;+\; K \exp\left[-\frac{\lambda}{|\epsilon|^{k}}\right],
\end{aligned}
$$

for $(\vec{v},\epsilon) \in \mathcal{D} \times (\mathcal{S}_\ell \cap \mathcal{S}_{\ell+1})$, where L, K, and λ are suitable positive numbers. Therefore, using Theorem XI-2-3, we complete the proof of Theorem XI-4-1. $\square$

As a corollary of Theorem XI-4-1, the following result is obtained without any complication.

Theorem XI-4-3. $\mathbb{C}[[x]]_s$ *and* $\mathcal{A}_s(r,a,b)$ *are commutative differential algebras over* $\mathbb{C}$, *i.e.,*

(i) $f + g \in \mathbb{C}[[x]]_s$ *(respectively* $\mathcal{A}_s(r,a,b)$*) if* f *and* $g \in \mathbb{C}[[x]]_s$ *(respectively* $\mathcal{A}_s(r,a,b)$*),*

(ii) $fg \in \mathbb{C}[[x]]_s$ *(respectively* $\mathcal{A}_s(r,a,b)$*) if* f *and* $g \in \mathbb{C}[[x]]_s$ *(respectively* $\mathcal{A}_s(r,a,b)$*),*

(iii) $cf \in \mathbb{C}[[x]]_s$ *(respectively* $\mathcal{A}_s(r,a,b)$*) if* $c \in \mathbb{C}$ *and* $f \in \mathbb{C}[[x]]_s$ *(respectively* $\mathcal{A}_s(r,a,b)$*),*

(iv) $\dfrac{df}{dx} \in \mathbb{C}[[x]]_s$ *(respectively* $\mathcal{A}_s(r,a,b)$*) if* $f \in \mathbb{C}[[x]]_s$ *(respectively* $\mathcal{A}_s(r,a,b)$*),*

(v) $\displaystyle\int_0^x f\,dx \in \mathbb{C}[[x]]_s$ *(respectively* $\mathcal{A}_s(r,a,b)$*) if* $f \in \mathbb{C}[[x]]_s$ *(respectively* $\mathcal{A}_s(r,a,b)$*) .*

Furthermore, the map $J : \mathcal{A}_s(r,a,b) \to \mathbb{C}[[x]]_s$ *is a homomorphism of differential algebras over* $\mathbb{C}$.

Remark XI-4-4. Using Theorem XI-3-6, we can prove (iv) and (v) of Theorem XI-4-3 in a way similar to the proof of Theorem XI-4-1. Also, it can be shown that if $f(x) \in \mathcal{A}_s(r,a,b)$ with $J(f) = \displaystyle\sum_{m=0}^{+\infty} a_m x^m$ and $a_0 \neq 0$, we obtain $\dfrac{1}{f} \in$

$\mathcal{A}_s\left(\rho(\alpha,\beta),\alpha,\beta\right)$ and $J\left(\dfrac{1}{f}\right) = \dfrac{1}{J(f)}$, where $a < \alpha < \beta < b$ and $\rho(\alpha,\beta)$ is a suitable positive number depending on (α,β).

The materials in this section are also found in [Ram1], [Ram2], and [Si17, Appendices].

XI-5. Proof of Lemma XI-2-6

In order to prove Lemma XI-2-6, choose N line segments $\mathcal{C}_1$, $\mathcal{C}_2$, ..., $\mathcal{C}_N$ so that $\mathcal{C}_\ell \in \mathcal{S}_\ell \cap \mathcal{S}_{\ell+1}$ $(\ell = 1, 2, \ldots, N)$, i.e., $\mathcal{C}_\ell : z = te^{i\theta_\ell} (0 < t < r)$, where for each ℓ, θ_ℓ is a fixed number satisfying the condition $a_{\ell+1} < \theta_\ell < b_\ell$. These N line segments divide the open punctured disk $\mathcal{D} = \{x : 0 < |z| < r\}$ into N open sectors $\hat{\mathcal{S}}_1, \hat{\mathcal{S}}_2, \ldots, \hat{\mathcal{S}}_N$, where

$$\hat{\mathcal{S}}_\ell = \{x : \theta_{\ell-1} < \arg(x) < \theta_\ell, \ 0 < |x| < r\}.$$

Set

$$\psi_\ell(x) = \frac{-1}{2\pi i} \sum_{h=1}^{N} \int_{\mathcal{C}_h} \frac{\delta_h(\xi)}{\xi - x} d\xi$$

for $x \in \hat{\mathcal{S}}_\ell$, $\ell = 1, 2, \ldots, N$. The functions ψ_ℓ can be continued analytically onto $\mathcal{S}_\ell$ by deforming $\mathcal{C}_{\ell-1}$ and $\mathcal{C}_\ell$ without moving any of their endpoints. In doing this, we do not change other line segments $\mathcal{C}_h$ $(h \neq \ell - 1, \ell)$. Thus, the function ψ_ℓ is holomorphic on $\mathcal{S}_\ell$, $\ell = 1, 2, \ldots, N$, respectively.

Now, assuming that $x \in \mathcal{S}_\ell \cap \mathcal{S}_{\ell+1}$, compute $\psi_\ell(x) - \psi_{\ell+1}(x)$. To do this, write ψ_ℓ and $\psi_{\ell+1}$ in the following forms respectively:

$$\begin{cases} \psi_\ell(x) = \dfrac{(-1)}{2\pi i} \displaystyle\int_{\hat{\mathcal{C}}_\ell} \frac{\delta_\ell(\xi)}{\xi - x} d\xi + \dfrac{(-1)}{2\pi i} \sum_{h \neq \ell} \int_{\mathcal{C}_h} \frac{\delta_h(\xi)}{\xi - x} d\xi, \\[3mm] \psi_{\ell+1}(x) = \dfrac{(-1)}{2\pi i} \displaystyle\int_{\hat{\mathcal{C}}_{\ell+1}} \frac{\delta_\ell(\xi)}{\xi - x} d\xi + \dfrac{(-1)}{2\pi i} \sum_{h \neq \ell} \int_{\mathcal{C}_h} \frac{\delta_h(\xi)}{\xi - x} d\xi, \end{cases}$$

where the paths $\hat{\mathcal{C}}_\ell$ and $\hat{\mathcal{C}}_{\ell+1}$ of integration are obtained by deforming $\mathcal{C}_\ell$ without moving either of its endpoints so that

(1) $\hat{\mathcal{C}}_\ell \subset \mathcal{S}_\ell \cap \mathcal{S}_{\ell+1}$ and $\hat{\mathcal{C}}_{\ell+1} \subset \mathcal{S}_\ell \cap \mathcal{S}_{\ell+1}$,

(2) $-\hat{\mathcal{C}}_\ell + \hat{\mathcal{C}}_{\ell+1}$ is a simple closed curve whose interior contains x.

Thus, (b) follows, i.e.,

$$\psi_\ell(x) - \psi_{\ell+1}(x) = \frac{1}{2\pi i} \int_{-\hat{\mathcal{C}}_\ell + \hat{\mathcal{C}}_{\ell+1}} \frac{\delta_\ell(\xi)}{\xi - x} d\xi = \delta_\ell(x) \quad \text{on} \quad \mathcal{S}_\ell \cap \mathcal{S}_{\ell+1}.$$

Let us derive asymptotic properties of ψ_ℓ. To do this, fix an ℓ and a closed sector $\mathcal{W}$ contained in $\mathcal{S}_\ell$. Let $\tilde{\mathcal{C}}_\ell$ (respectively $\tilde{\mathcal{C}}_{\ell-1}$) be a path obtained by deforming $\mathcal{C}_\ell$ (respectively $\mathcal{C}_{\ell-1}$) without moving the endpoints so that $\mathcal{W}$ is contained in the interior of the simple closed curve $\tilde{C}_{\ell-1} + \gamma_\ell - \tilde{C}_\ell$, where γ_ℓ is a circular arc joining two points $re^{i\theta_{\ell-1}}$ and $re^{i\theta_\ell}$. For $x \in \mathcal{W}$, it holds that

$$\psi_\ell(x) = \frac{(-1)}{2\pi i} \int_{\tilde{C}_\ell} \frac{\delta_\ell(\xi)}{\xi - x} d\xi + \frac{(-1)}{2\pi i} \int_{\tilde{C}_{\ell-1}} \frac{\delta_{\ell-1}(\xi)}{\xi - x} d\xi + \frac{(-1)}{2\pi i} \sum_{h \neq \ell, \ell-1} \int_{C_h} \frac{\delta_h(\xi)}{\xi - x} d\xi.$$

It may be assumed that the path $\tilde{C}_\ell$ is given as a union of a line segment L_ℓ and a curve Γ_ℓ defined by

$$\begin{cases} L_\ell : x = te^{i\omega_\ell} & (0 < t \leq r_1 < r), \\ \Gamma_\ell : x = \mu_\ell(\tau) & (0 \leq \tau < 1), \end{cases}$$

where ω_ℓ is a constant such that $\theta_\ell < \omega_\ell < b_\ell$, and

$$\mu_\ell(0) = r_1 e^{i\omega_\ell}, \quad \mu_\ell(1) = re^{i\theta_\ell}, \quad r_1 \leq |\mu_\ell(\tau)| < r \quad (0 \leq \tau < 1).$$

It can be also assumed that there exists a positive number $\sigma < 1$ such that $|x| \leq \sigma r_1$ for $x \in \mathcal{W}$. Then, $\dfrac{1}{2\pi i} \displaystyle\int_{\Gamma_\ell} \frac{\delta_\ell(\xi)}{\xi - x} d\xi$ is represented by a convergent power series in x whose radius of convergence is not less than r_1.

Let us estimate the integral $\dfrac{1}{2\pi i} \displaystyle\int_{L_\ell} \frac{\delta_\ell(\xi)}{\xi - x} d\xi$. Since

$$\frac{1}{\xi - x} = \frac{1}{\xi\left(1 - \dfrac{x}{\xi}\right)} = \frac{1}{\xi} \sum_{m=0}^{M} \left(\frac{x}{\xi}\right)^m + \left(\frac{x}{\xi}\right)^{M+1} \frac{1}{\xi - x},$$

we obtain

$$\frac{1}{2\pi i} \int_{L_\ell} \frac{\delta_\ell(\xi)}{\xi - x} d\xi = \sum_{m=0}^{M} \alpha_m x^m + x^{M+1} E_{M+1}(x),$$

where

$$\alpha_m = \frac{1}{2\pi i} \int_{L_\ell} \frac{\delta_\ell(\xi)}{\xi^{m+1}} d\xi \quad (m \geq 0) \quad \text{and} \quad E_{M+1}(x) = \frac{1}{2\pi i} \int_{L_\ell} \frac{\delta_\ell(\xi)}{\xi^{M+1}(\xi - x)} d\xi.$$

Now, estimate the α_m as follows:

$$|\alpha_m| = \frac{1}{2\pi} \left| \int_0^{r_1} \frac{\delta_\ell(\tau e^{i\omega_\ell}) e^{i\omega_\ell}}{e^{i(m+1)\omega_\ell} \tau^{m+1}} d\tau \right| \leq \frac{\gamma}{2\pi} \int_0^{r_1} \frac{e^{-\lambda\tau^{-k}}}{\tau^{m+1}} d\tau$$

$$< \frac{\gamma}{2\pi} \int_0^{+\infty} \tau^{-m-1} e^{-\lambda\tau^{-k}} d\tau = \frac{\gamma}{2k\pi} \left(\lambda^{-1/k}\right)^m \Gamma\left(\frac{m}{k}\right)$$

$$< \frac{\gamma}{2k\pi} \left(\lambda^{-1/k}\right)^m \left(\sqrt{2\pi}\right)^{1-k} \left(\sqrt{k}\right)^{-m/k} (m!)^{1/k},$$

where the Stirling's formula (XI.3.3) was used as $m \to +\infty$.

To estimate $E_{N+1}(x)$, note that there exists a positive number θ depending on $\mathcal{W}$ such that

$$|\xi - x| \geq |\xi| \sin(\theta) \qquad \text{for} \quad \xi \in L_\ell, \ x \in \mathcal{W}.$$

Hence,

$$|E_{N+1}(x)| \leq \frac{1}{2\pi} \int_0^{r_1} \frac{\gamma e^{-\lambda \tau^{-k}}}{\tau^{N+2} \sin(\theta)} d\tau \ < \ \frac{\gamma \lambda^{-(N+1)/k}}{2k\pi \sin(\theta)} \Gamma\left(\frac{N+1}{k}\right)$$

$$< \ \frac{\gamma}{k \sin(\theta)} \left(\frac{N+1}{\lambda k e}\right)^{(N+1)/k} .$$

We can estimate

$$\frac{-1}{2\pi i} \int_{\tilde{C}_{\ell-1}} \frac{\delta_{\ell-1}(\xi)}{\xi - x} d\xi \qquad \text{and} \qquad \frac{-1}{2\pi i} \int_{C_h} \frac{\delta_h(\xi)}{\xi - x} d\xi \qquad (h \neq \ell, \ell-1)$$

in a similar manner. Thus, the proof of Lemma XI-2-6 is completed. $\quad\square$

The material of this section is also found in [Si18].

EXERCISES XI

XI-1. Let $\mathcal{S}$ be an open sector $\mathcal{D}(r_0, a, b)$, i.e.,

$$\mathcal{S} = \{x : 0 < |x| < r_0, \ a < \arg x < b\},$$

where a and b are real numbers and r_0 is a positive number. Denote by
  (a) $\mathcal{A}(\mathcal{S})$ the set of all functions which are holomorphic in $\mathcal{S}$ and admit asymptotic expansions in powers of x as $x \to 0$ in every open subsector

$$\mathcal{W} = \{x : 0 < |x| < r < r_0, \ a < \alpha < \arg x < \beta < b\}$$

of $\mathcal{S}$,
  (b) $J[f]$ the asymptotic expansion for $f \in \mathcal{A}(\mathcal{S})$,
  (c) $\mathcal{A}_0(\mathcal{S})$ the set of all those functions $f(x)$ in $\mathcal{A}(\mathcal{S})$ such that $f(x) \simeq 0$ as $x \to 0$ in every $\mathcal{W}$ (i.e. $\mathcal{A}_0(\mathcal{S}) = \{f \in \mathcal{A}(\mathcal{S}) : J[f] = 0\}$).
Show that $\mathbb{C}[[x]]$, $\mathcal{A}(\mathcal{S})$, and $\mathcal{A}_0(\mathcal{S})$ are differential algebras over the field $\mathbb{C}$ and that the map $J : \mathcal{A}(\mathcal{S}) \to \mathbb{C}[[x]]$ is a homomorphism of differential algebras over the field $\mathbb{C}$.

XI-2. Using the same notations as in Exercise XI-1 and considering a linear differential operator $P = \sum_{k=0}^{N} a_k(x) D^k$ with coefficients $a_k(x)$ which are holomorphic in a disk $|x| < r$ containing $\mathcal{S}$, where $D = \dfrac{d}{dx}$, show that

(i) P defines three homomorphisms

$$P_1 = P : \mathcal{A}_0(\mathcal{S}) \to \mathcal{A}_0(\mathcal{S}), \quad P_2 = P : \mathcal{A}(\mathcal{S}) \to \mathcal{A}(\mathcal{S}), \quad P_3 = P : \mathbb{C}[[x]] \to \mathbb{C}[[x]]$$

of vector spaces over the field $\mathbb{C}$,

(ii) two vector spaces $\mathcal{A}(\mathcal{S})/\mathcal{A}_0(\mathcal{S})$ and $\mathbb{C}[[x]]$ are isomorphic,

(iii) dimensions of three vector spaces Nullspace ofP_1, Nullspace ofP_2, and Nullspace ofP_3 over $\mathbb{C}$ are not greater than the order of P,

(iv) two vector spaces $\{$Nullspace of $P_2\}/\{$Nullspace of $P_1\}$ and $J[$Nullspace of $P_2]$ are isomorphic,

(v) there exists a homomorphism $T :$ Nullspace of $P_3 \to \mathcal{A}_0(\mathcal{S})/P_1[\mathcal{A}_0(\mathcal{S})]$ of vector spaces over $\mathbb{C}$ such that Nullspace of $T = J[$Nullspace of $P_2]$ and $T[$Nullspace of $P_3] = \{P_2[\mathcal{A}(\mathcal{S})] \cap \mathcal{A}_0(\mathcal{S})\}/P_1[\mathcal{A}_0(\mathcal{S})]$.

Hint.

(ii) J is onto (cf. Theorem XI-1-13) and $\mathcal{A}_0(\mathcal{S})$ is Nullspace of J.

(iii) Use Theorem IV-2-1 for Nullspace of P_1 and Nullspace of P_2; use an idea similar to the proof of Theorem V-I-3 for Nullspace of P_3.

(iv) Nullspace of the homomorphism $J :$ Nullspace of $P_2 \to J[$Nullspace of $P_2]$ is Nullspace of P_1. Also, this homomorphism is onto.

(v) Assume that $P_3[\hat{\phi}] = 0$ in $\mathbb{C}[[x]]$. There exists a $\phi \in \mathcal{A}(\mathcal{S})$ such that $J[\phi] = \hat{\phi}$ (cf. Theorem XI-1-13). Such ϕ is not unique. Actually, $\phi + \psi$ also satisfies the condition $J[\phi + \psi] = \hat{\phi}$ if and only if $\psi \in \mathcal{A}_0(\mathcal{S})$. So define T by $T[\hat{\phi}] = P_2[\phi]$ (mod $P_1[\mathcal{A}_0(\mathcal{S})]$). Note that $J[P_2[\phi]] = P_3[\hat{\phi}] = 0$. This implies that $P_2[\phi] \in \mathcal{A}_0(\mathcal{S})$.

XI-3. For each of the following three power series, find s such that $f \in \mathbb{C}[[x]]_s$.

(a) $f(x) = \displaystyle\sum_{m=0}^{\infty} (mx)^m$, (b) $f(x) = \displaystyle\sum_{m=0}^{\infty} \frac{\Gamma\left(1 + \frac{m}{2}\right)}{\Gamma\left(1 + \frac{m}{5}\right)} x^m$,

(c) $f(x) = \displaystyle\sum_{m=0}^{\infty} x^m \prod_{h=m}^{2m} (5 + \sqrt{5h})$.

XI-4. For two real numbers $s \geq 0$ and $A > 0$ and for a formal power series $f = \displaystyle\sum_{m=0}^{+\infty} a_m x^m \in \mathbb{C}[[x]]$, set $\|f\|_{s,A} = \displaystyle\sum_{m=0}^{+\infty} \frac{|a_m|}{(m!)^s A^m}$. Denote by $\mathcal{E}(s, A)$ the set of all $f \in \mathbb{C}[[x]]$ satisfying the condition $\|f\|_{s,A} < +\infty$. Show that

(i) $\mathcal{E}(s, A)$ is a Banach space with the norm $\|f\|_{s,A}$,

(ii) $\|fg\|_{s,A} \leq \|f\|_{s,A} \|g\|_{s,A}$ if $f \in \mathcal{E}(s, A)$ and $g \in \mathcal{E}(s, A)$,

(iii) if $B \geq A$, then $\mathcal{E}(s, A) \subset \mathcal{E}(s, B)$, and $\|f\|_{s,B} \leq \|f\|_{s,A}$ for $f \in \mathcal{E}(s, A)$,

(iv) if $B \geq A$, then $\|f\|_{s,B} - |f(0)| \leq \dfrac{A}{B} [\|f\|_{s,A} - |f(0)|]$ for $f \in \mathcal{E}(s, A)$, where $f(0) = a_0$ if $f = \sum_{m \geq 0} a_m x^m$,

(v) for $f \in \mathcal{E}(s, A)$ and an integer q such that $sq \geq 1$, it holds that

$$\left\| x^{q+1} \frac{df}{dx} \right\|_{s,A} \leq \frac{1}{A^q} \|f\|_{s,A}.$$

Comment. The formal power series $f \in \mathbb{C}[[x]]$ is of Gevrey order s if and only if $f \in \mathcal{E}(s, A)$ for some positive number A.

XI-5. Show that $f(\epsilon) = 0$ identically in a sector $\mathcal{S}$ if f is holomorphic and flat of Gevrey order 0 as $\epsilon \to 0$ in $\mathcal{S}$.

Hint. $|f(\epsilon)| \leq K A^N |\epsilon|^N$ for $\epsilon \in \mathcal{S}$ and $N = 0, 1, 2, \ldots$, where K and A are positive numbers independent of N.

XI-6. Show that $\mathbb{C}[[x]]_0 = \mathbb{C}\{x\}$.

Hint. $f = \sum_{m=0}^{\infty} c_m x^m \in \mathbb{C}[[x]]_0$ if and only if $|c_m| \leq K A^m |\epsilon|^m$ $(m = 0, 1, 2, \ldots)$, where K and A are positive numbers independent of m.

XI-7. Consider a system of differential equations

$$\text{(S)} \qquad x^{q+1} \frac{d\vec{y}}{dx} = \vec{f}_0 + \Phi \vec{y} + \sum_{|\wp| \geq 0} \vec{y}^{\wp} \vec{f}_{\wp}, \qquad \vec{y} = \begin{bmatrix} y_1 \\ \vdots \\ y_n \end{bmatrix},$$

where q is a positive integer, $\vec{f}_0$ and $\vec{f}_{\wp} \in \mathbb{C}[[x]]^n$, Φ is an $n \times n$ matrix whose entries are formal power series in x, $\wp = (p_1, \ldots, p_n)$ with n non-negative integers $p_1, \ldots, p_n$, $|\wp| = p_1 + \cdots + p_n$, and $\vec{y}^{\wp} = y_1^{p_1} \cdots y_n^{p_n}$. Assume that $sq \geq 1$ and that

(i) $\vec{f}_0$ and $\vec{f}_{\wp} \in \mathcal{E}(s, A)^n$, and the entries of Φ belongs to $\mathcal{E}(s, A)$ for some $A > 0$,

(ii) the power series $\displaystyle\sum_{|\wp| \geq 2} \|\vec{f}_{\wp}\|_{s,A} \vec{y}^{\wp}$ is a convergent power series in $\vec{y}$, where
$$\|\vec{y}\|_{s,A} = \max\{\|y_j\|_{s,A} : 1 \leq j \leq n\},$$

(iii) $\vec{f}_0(0) = \vec{0}$,

(iv) $\Phi(0)$ is invertible.

Show that there exists a unique power series $\vec{\phi} \in \mathbb{C}[[x]]^n$ satisfying system (S) and the condition $\vec{\phi}(0) = \vec{0}$. Furthermore, $\vec{\phi} \in (\mathbb{C}[[x]]_s)^n$.

Hint. Write system (S) in a form $\vec{y} = \vec{g}_0 + x^{q+1} \Psi \dfrac{d\vec{y}}{dx} + \displaystyle\sum_{|\wp| \geq 2} \vec{y}^{\wp} \vec{g}_{\wp}$ and use the

Banach fixed-point theorem in terms of the norm $\| \ \|_{s,A}$.

XI-8. Let $\mathcal{D}(r) = \{z : |z| < r\}$, $\mathcal{S}(r, \rho) = \{z : 0 < |z| < r, |\arg z| < \rho\}$, and a power series $f(z, \epsilon) = \displaystyle\sum_{n=0}^{+\infty} f_n(\epsilon) z^n$ is convergent uniformly in a domain $\mathcal{D}(r_0) \times \mathcal{S}(r, \rho)$, where r_0, r, and ρ are positive numbers. Assume that the coefficients $f_n(\epsilon)$ of the series f are holomorphic in $\mathcal{S}(r, \rho)$ and admit asymptotic expansions in powers of ϵ as $\epsilon \to 0$ in $\mathcal{S}(r, \rho)$. Suppose also that $\lim_{\epsilon \to 0} f_0(\epsilon) = 0$ and $\lim_{\epsilon \to 0} f_1(\epsilon) \neq 0$. Show that there exist a positive number r_1 and a function $\phi(\epsilon)$ such that (1) $\phi(\epsilon)$ is holomorphic in $\mathcal{S}(r_1, \rho)$ and admits an asymptotic expansion in powers of ϵ as $\epsilon \to 0$ in $\mathcal{S}(r_1, \rho)$, (2) $\lim_{\epsilon \to 0} \phi(\epsilon) = 0$, and (3) $f(\phi(\epsilon), \epsilon) = 0$ identically in $\mathcal{S}(r_1, \rho)$.

XI-9. Consider a formal power series

$$\text{(P)} \qquad\qquad \vec{F}(x, \vec{y}, \vec{z}) = \sum_{|\wp_1| + |\wp_2| \geq 0} \vec{y}^{\wp_1} \vec{z}^{\wp_2} \vec{f}_{\wp_1, \wp_2},$$

where $\vec{y} = \begin{bmatrix} y_1 \\ \vdots \\ y_n \end{bmatrix}$, $\vec{z} = \begin{bmatrix} z_1 \\ \vdots \\ z_m \end{bmatrix}$, $\vec{f}_{\wp_1, \wp_2} \in \mathbb{C}[[x]]^n$, $\wp_1 = (p_{11}, \ldots, p_{1n})$, and $\wp_2 =$

$(p_{21}, \ldots, p_{2m})$ with non-negative integers p_{1k} and p_{2k}. Denote by $\vec{F}_{\vec{y}}(x, \vec{y}, \vec{z})$ the Jacobian matrix of $\vec{F}$ with respect to $\vec{y}$. Assume that power series (P) satisfies the conditions (1) $\vec{F}(0, \vec{0}, \vec{0}) = 0$, (2) $\vec{F}_{\vec{y}}(0, \vec{0}, \vec{0})$ is invertible, (3) $\vec{f}_{\wp_1, \wp_2} \in \mathcal{E}(s, A)^n$ for some $s \geq 0$ and $A > 0$, and (4) the power series $\displaystyle\sum_{|\wp_1| + |\wp_2| \geq 0} \|\vec{f}_{\wp_1, \wp_2}\|_{s,A} \vec{y}^{\wp_1} \vec{z}^{\wp_2}$ is a convergent power series in $\vec{y}$ and $\vec{z}$. Show that there exists a unique power series $\vec{\phi}(x, \vec{z}) = \displaystyle\sum_{|\wp_2| \geq 0} \vec{z}^{\wp_2} \vec{\phi}_{\wp_2}$ in $(x, \vec{z})$ such that (i) $\vec{\phi}_{\wp_2} \in \mathbb{C}[[x]]^n$, (ii) $\vec{\phi}(0, \vec{0}) = \vec{0}$, and (iii) $\vec{F}(x, \vec{\phi}(x, \vec{z}), \vec{z}) = \vec{0}$. Also, show that $\vec{\phi}_{\wp_2} \in \mathcal{E}(s, B)^n$ for some $B > 0$ and the power series $\displaystyle\sum_{|\wp_2| \geq 0} \|\vec{\phi}_{\wp_2}\|_{s,B} \vec{z}^{\wp_2}$ is a convergent power series in $\vec{z}$.

Hint. This is an implicit function theorem. Write $\vec{F}$ in the form $\vec{F} = \vec{f}_0 + \Phi \vec{y} + \sum' \vec{y}^{\wp_1} \vec{z}^{\wp_2} \vec{f}_{\wp_1, \wp_2}$, where $\vec{f}_0$ and $\vec{f}_{\wp_1, \wp_2}$ are in $\mathcal{E}(s, A)^n$, $\Phi \in \mathcal{M}_n(\mathcal{E}(s, A))$, and $\sum'$ is the sum over all $(\wp_1, \wp_2)$ such that either $|\wp_1| = 0$ and $|\wp_2| = 1$ or $|\wp_1| + |\wp_2| \geq 2$. Here, $\mathcal{M}_n(R)$ denotes the set of all $n \times n$ matrices with entries in a ring R. Assumption (1) implies that $\vec{f}_0(0) = \vec{0}$. Assumption (2) implies that $\Phi(0)$ is invertible. Therefore, Φ^{-1} exists in $\mathcal{M}_n(\mathbb{C}[[x]]_s)$, and hence $\|\Phi^{-1}\|_{s,\tilde{A}} < +\infty$ for some $\tilde{A} > 0$, where $\|\Phi^{-1}\|_{s,\tilde{A}} = \sup\{\|\Phi^{-1}\vec{f}\|_{s,\tilde{A}} : \vec{f} \in \mathcal{E}(s, \tilde{A})^n, \|\vec{f}\|_{s,\tilde{A}} = 1\}$. Since $\|\vec{f}\|_{s,B} \leq \|\vec{f}\|_{s,A}$ for $\vec{f} \in (\mathbb{C}[[x]]_s)^n$ if $B \geq A$, assume without any loss of generality that $\tilde{A} = A$. Then, $\Phi^{-1}\vec{F} = \vec{g}_0 + \vec{y} + \sum' \vec{y}^{\wp_1} \vec{z}^{\wp_2} \vec{g}_{\wp_1, \wp_2}$, where $\vec{g}_0 \in \mathcal{E}(s, A)^n$, $\vec{g}_{\wp_1, \wp_2} \in \mathcal{E}(s, A)^n$, $\vec{g}_0(0) = \vec{0}$, and $\sum' \|\vec{g}_{\wp_1, \wp_2}\|_{s,A} \vec{y}^{\wp_1} \vec{z}^{\wp_2}$ is a convergent power series in $\vec{y}$ and $\vec{z}$. Consider the equation $\vec{0} = \vec{u} + \vec{y} + \sum' \vec{y}^{\wp_1} \vec{z}^{\wp_2} \vec{g}_{\wp_1, \wp_2}$, $\vec{u} = \begin{bmatrix} u_1 \\ \vdots \\ u_n \end{bmatrix}$. Then, there exists a unique power series $\vec{\alpha}(x, \vec{u}, \vec{z}) = \displaystyle\sum_{|\wp_1| + |\wp_2| \geq 1} \vec{u}^{\wp_1} \vec{z}^{\wp_2} \vec{\alpha}_{\wp_1, \wp_2}$ such that $\vec{\alpha}_{\wp_1, \wp_2} \in \mathcal{E}(s, A)^n$ and $\vec{0} = \vec{u} + \vec{\alpha} + \sum' (\vec{\alpha})^{\wp_1} \vec{z}^{\wp_2} \vec{g}_{\wp_1, \wp_2}$ identically. The unique power series satisfying conditions (i), (ii), and (iii) is given by

$$\text{(S)} \qquad\qquad \vec{\phi}(x, \vec{z}) = \vec{\alpha}(x, \vec{g}_0, \vec{z}).$$

Now, consider the equation

$$\text{(R)} \quad \vec{y} = \vec{u} + \sum' \vec{y}^{\wp_1} \vec{z}^{\wp_2} \vec{G}_{\wp_1, \wp_2} \qquad \text{where} \qquad \vec{G}_{\wp_1, \wp_2} = \begin{bmatrix} \|\vec{g}_{\wp_1, \wp_2}\|_{s,A} \\ \vdots \\ \|\vec{g}_{\wp_1, \wp_2}\|_{s,A} \end{bmatrix} \in \mathbb{R}^n.$$

Equation (R) has a unique solution $\vec{y} = \displaystyle\sum_{|\wp_1|+|\wp_2|\geq 1} \vec{u}^{\wp_1} \vec{z}^{\wp_2} \vec{\beta}_{\wp_1,\wp_2}$ such that $\vec{\beta}_{\wp_1,\wp_2} \in$ $\mathbb{R}^n$, the entries of $\vec{\beta}_{\wp_1,\wp_2}$ are non-negative, and the series is a convergent power series in $\vec{u}$ and $\vec{z}$. Use this series as a majorant to show that $\vec{\phi}$ defined by (S) satisfies conditions (iv) and (v).

XI-10. Assume that a covering $\{\mathcal{S}_1, \mathcal{S}_2, \ldots, \mathcal{S}_N\}$ at $x = 0$ is good and that N functions $\phi_1(x), \phi_2(x), \ldots, \phi_N(x)$ satisfy the conditions:
  (1) $\phi_\ell(x)$ is holomorphic in $\mathcal{S}_\ell$,
  (2) $\phi_\ell(x) \simeq 0$ as $x \to 0$ in $\mathcal{S}_\ell$,
  (3) $|\phi_\ell(x) - \phi_{\ell+1}(x)| \leq \gamma \exp[-\lambda|x|^{-k}]$ on $\mathcal{S}_\ell \cap \mathcal{S}_{\ell+1}$, where $\gamma \geq 0$, $\lambda > 0$ and
      $k > 0$ are suitable numbers independent of ℓ.
Show that there exists a positive number H such that

$$|\phi_\ell(x)| \leq H \exp[-\lambda|x|^{-k}] \quad \text{in} \quad \mathcal{S}_\ell.$$

Hint. See [Si15].

XI-11. Assume that a covering $\{\mathcal{S}_1, \mathcal{S}_2, \ldots, \mathcal{S}_N\}$ at $x = 0$ is good and that N functions $\phi_1(x), \phi_2(x), \ldots, \phi_N(x)$ satisfy the conditions
  (1) $\phi_\ell(x)$ is holomorphic on $\mathcal{S}_\ell$,
  (2) $\phi_\ell(x)$ is bounded on $\mathcal{S}_\ell$,
  (3) we have

$$|\phi_\ell(x) - \phi_{\ell+1}(x)| \leq K_n|x|^n \quad (n = 1, 2, \ldots) \quad \text{on} \quad \mathcal{S}_\ell \cap \mathcal{S}_{\ell+1},$$

  where K_n are positive numbers.
Show that there exists a formal power series $p = \displaystyle\sum_{m=0}^{+\infty} a_m x^m \in \mathbb{C}[[x]]$ such that for each ℓ, we obtain $\phi_\ell \in \mathcal{A}(\mathcal{S}_\ell)$ and $J(\phi_\ell) = p$, where the notations $\mathcal{A}(\mathcal{S})$ and J are defined in Exercise XI-1.

XI-12. Assume that a covering $\{\mathcal{S}_1, \mathcal{S}_2, \ldots, \mathcal{S}_N\}$ at $x = 0$ is good and that $n \times n$ matrices $\Phi_1(x), \Phi_2(x), \ldots, \Phi_N(x)$ satisfy the following conditions: (1) the entries of $\Phi_\ell(x)$ belong to $\mathcal{A}(\mathcal{S}_\ell \cap \mathcal{S}_{\ell+1})$ and (2) $J(\Phi_\ell) = I_n$. Show that there exists a formal power series $Q = I_n + \displaystyle\sum_{m=1}^{+\infty} x^m Q_m$ having constants $n \times n$ matrices Q_m as coefficients, and $n \times n$ matrices $P_1(x), P_2(x), \ldots, P_N(x)$ such that
  (i) for each ℓ, the entries of $P_\ell(x)$ and $P_\ell(x)^{-1}$ belong to $\mathcal{A}(\mathcal{S}_\ell)$,
  (ii) $J(P_\ell) = Q \ (\ell = 1, 2, \ldots, N)$,
  (iii) $\Phi_\ell(x) = P_\ell(x)^{-1} P_{\ell+1}(x) \ (\ell = 1, 2, \ldots, N)$,
where the notation $\mathcal{A}(\mathcal{S})$ and J are defined in Exercise XI-1. Also, show that if the entries of $\Phi_\ell(x)$ belong to $\mathcal{A}_s(\mathcal{S}_\ell)$, respectively, then the entries of $P_\ell(x)$ and $P_\ell^{-1}(x)$ also belong to $\mathcal{A}_s(\mathcal{S}_\ell)$, respectively.

Hint. See [Si17, Theorem 6.4.1 on p. 150, its proof on pp. 152-161, and §A.2.4 on pp. 207–208].

XI-13. Prove the formula

$$J\left(\ell_{r,k}(\phi)\right) \ = \ \sum_{m=0}^{\infty} \Gamma\left(1 + \frac{m}{k}\right) c_m x^m,$$

which is given in Therorem XI-2-4.

Hint. Assume that $|\arg x| \leq \dfrac{\pi}{2k} - \delta$, where δ is a small positive number. Then,

$$\frac{k}{x^k} \int_0^r t^m e^{-(t/x)^k} t^{k-1} dt \ = \ x^m \int_0^{(r/x)^k} \sigma^{m/k} e^{-\sigma} d\sigma$$

$$= \ \Gamma\left(1 + \frac{m}{k}\right) x^m \ - \ x^m \int_{(t/x)^k}^{\infty} \sigma^{m/k} e^{-\sigma} d\sigma.$$

Hence,

$$\ell_{r,k}(\phi) \ = \ \sum_{m=0}^{N} \Gamma\left(1 + \frac{m}{k}\right) c_m x^m \ - \ \sum_{m=0}^{N} c_m x^m \int_{(t/k)^k}^{\infty} \sigma^{m/k} e^{-\sigma} d\sigma$$

$$+ \ \frac{k}{x^k} \int_0^r \left(\sum_{m=N+1}^{+\infty} c_m t^m\right) e^{-(t/x)^k} t^{k-1} dt.$$

XI-14. Let α be a positive number larger than 1. Also, let

$$\mathcal{S}_j = \{x : \ a_j < \arg x < b_j, \ 0 < |x| < \rho\} \quad (j = 1, 2, \ldots, \nu)$$

be a good covering of the sector $\mathcal{S} = \left\{x : |\arg x| < \dfrac{\pi}{2\alpha}, 0 < |x| < \rho\right\}$, where $a_1 = -\dfrac{\pi}{2\alpha}$, $b_\nu = \dfrac{\pi}{2\alpha}$, and ρ is a positive number. Assume that ν functions $\phi_1(x)$, $\phi_2(x)$, $\ldots$, $\phi_\nu(x)$ satisfy the following conditions:

  (i) $\phi_j(x)$ is holomorphic in $\mathcal{S}_j$ and continuous on the closed sector
$$\bar{\mathcal{S}}_j = \{x : a_j \leq \arg x \leq b_j, 0 < |x| \leq \rho\},$$

  (ii) $|\phi_j(x)| \leq A \exp[c|x|^{-1}]$ in $\mathcal{S}_j$ for some positive numbers A and c,

 (iii) $|\phi_{j+1}(x) - \phi_j(x)| \leq M_0$ in $\mathcal{S}_j \cap \mathcal{S}_{j+1}$ for some positive number M_0,

 (iv) $|\phi_1(x)| \leq M_0$ on the line segment $\arg x = a_1$, $0 < |x| < \rho$,

  (v) $|\phi_\nu(x)| \leq M_0$ on the line segment $\arg x = b_\nu$, $0 < |x| < \rho$.

Show that there exists a positive number M such that

$$|\phi_j(x)| \ \leq \ M \quad \text{in} \quad \mathcal{S}_j \quad (j = 1, 2, \ldots, \nu).$$

Hint. This is a generalization of the Phragmén-Lindelöf Theorem (Lemma XI-3-5). See [Lin].

XI-15. Let k be a positive number. A formal power series $f \in \mathbb{C}[[x]]$ is said to be *k-summable* in a direction $\arg x = \theta$ if

$$f \in \text{Image} \left\{ J : \mathcal{A}_{1/k} \left(\rho_0, \theta - \frac{\pi}{2k} - \epsilon, \theta + \frac{\pi}{2k} + \epsilon \right) \to \mathbb{C}[[x]]_{1/k} \right\}$$

for some positive numbers ρ_0 and ϵ. Show that if a formal power series $f \in \mathbb{C}[[x]]$ is k-summable in a direction $\arg x = \theta$, there exists one and only one function $F \in \mathcal{A}_{1/k} \left(\rho_0, \theta - \frac{\pi}{2k} - \epsilon, \theta + \frac{\pi}{2k} + \epsilon \right)$ such that $J[F] = f$.

Hint. Use Remark XI-3-3. For more informations concerning summability, see, for example, [Bal3], [Ram2], [Ram3], [Si17, Appendices], and [Si19].

XI-16. Consider the integral $f(x) = \displaystyle\int_\gamma \frac{e^{-\xi^{-2}}}{\xi - x} d\xi$, where γ is a line segment $0 < x < 1$ in the sectorial domain $\mathcal{D} \left(1, -\frac{\pi}{4}, 2\pi + \frac{\pi}{4} \right)$. Using the argument given in §XI-5, show that

(i) $f(x)$ admits an asymptotic expansion $J[f] \in \mathbb{C}[[x]]_{1/2}$ as $x \to 0$ in
$$\mathcal{D}\left(1, -\frac{\pi}{4}, 2\pi + \frac{\pi}{4} \right),$$

(ii) $J[f]$ is 2-summable in the direction $\arg x = \theta$ if $0 < \theta < 2\pi$.

CHAPTER XII

ASYMPTOTIC SOLUTIONS IN A PARAMETER

In this chapter, we explain asymptotic solutions of a system of differential equations $\epsilon^\sigma \dfrac{d\vec{y}}{dx} = \vec{f}(x, \vec{y}, \epsilon)$ as $\epsilon \to 0$. In §§XII-1, XII-2, and XII-3, existence of such asymptotic solutions in the sense of Poincaré is proved in detail. In §XII-4, this result is used to prove a block-diagonalization theorem of a linear system $\epsilon^\sigma \dfrac{d\vec{y}}{dx} = A(x, \epsilon)\vec{y}$.

In §XII-5, we explain similar results in the Gevrey asymptotics. In §XII-6, we explain how much we can simplify a linear system by means of a linear transformation with a coefficient matrix whose entries are convergent power series in the parameter. This result is given in [Hs1] and similar to a theorem due to G. D. Birkhoff [Bi] concerning singularity with respect to the independent variable x (cf. Theorem XII-6-1 and [Si17, Chapter III]). The materials in §§XII-1–XII-5 are also found in [Was1], [Si3], [Si7], and [Si22].

XII-1. An existence theorem

In §§XII-1, XII-2, and XII-3, we consider a system of differential equations

$$\text{(XII.1.1)} \qquad \epsilon^\sigma \frac{dv_j}{dx} = f_j(x, v_1, v_2, \ldots, v_n, \epsilon) \qquad (j = 1, 2, \ldots, n),$$

where σ is a positive integer and $f_j(x, v_1, v_2, \ldots, v_n, \epsilon)$ are holomorphic with respect to complex variables $(x, v_1, v_2, \ldots, v_n, \epsilon)$ in a domain

$$\text{(XII.1.2)} \quad |x| < \delta_0, \quad 0 < |\epsilon| < \rho_0, \quad |\arg \epsilon| < \alpha_0, \quad |v_j| < \gamma_0 \quad (j = 1, 2, \ldots, n),$$

δ_0, ρ_0, α_0, and γ_0 being positive constants. Set

$$\text{(XII.1.3)} \qquad f_j(x, \vec{v}, \epsilon) = f_{j0}(x, \epsilon) + \sum_{h=1}^{n} a_{jh}(x, \epsilon)v_h + \sum_{|\wp| \geq 2} f_{j\wp}(x, \epsilon)\vec{v}^{\wp},$$

where $\vec{v} \in \mathbb{C}^n$ with the entries $(v_1, v_2, \ldots, v_n)$.

We look at (XII.1.1) under the following three assumptions.

Assumption I. *Each function* $f_j(x, \vec{v}, \epsilon)\, (j = 1, 2, \ldots, n)$ *has an asymptotic expansion*

$$\text{(XII.1.4)} \qquad f_j(x, \vec{v}, \epsilon) \simeq \sum_{\nu=0}^{\infty} \hat{f}_{j\nu}(x, \vec{v})\epsilon^\nu$$

in the sense of Poincaré as $\epsilon \to 0$ *in the sector*

$$\text{(XII.1.5)} \qquad 0 < |\epsilon| < \rho_0, \quad |\arg \epsilon| < \alpha_0,$$

372

where coefficients $\hat{f}_{j\nu}(x, \vec{v})$ are holomorphic in the domain

$$(XII.1.6) \qquad |x| < \delta_0, \quad |\vec{v}| < \gamma_0.$$

Furthermore, we assume that

$$(XII.1.7) \qquad \hat{f}_{j0}(x, \vec{0}) = 0 \qquad (j = 1, 2, \ldots, n) \quad \text{for} \quad |x| < \delta_0.$$

Observation XII-1-1. Under Assumption I, $f_{j0}(x, \epsilon)$, $a_{jh}(x, \epsilon)$, and $f_{j\wp}(x, \epsilon)$ admit asymptotic expansions

$$(XII.1.8) \qquad f_{j0}(x, \epsilon) \simeq \sum_{\nu=1}^{\infty} f_{j0\nu}(x)\epsilon^\nu, \qquad f_{j\wp}(x, \epsilon) \simeq \sum_{\nu=0}^{\infty} f_{j\wp\nu}(x)\epsilon^\nu,$$

and

$$(XII.1.9) \qquad a_{jh}(x, \epsilon) \simeq \sum_{\nu=0}^{\infty} a_{jh\nu}(x)\epsilon^\nu \qquad (j, h = 1, 2, \ldots, n)$$

as $\epsilon \to 0$ in sector (XII.1.5) with coefficients holomorphic in the domain

$$(XII.1.10) \qquad |x| < \delta_0.$$

Let $A(x, \epsilon)$ be the $n \times n$ matrix whose (j, k)-entry is $a_{jk}(x, \epsilon)$, respectively (i.e., $A(x, \epsilon) = (a_{jk}(x, \epsilon))$). Then, $A(x, \epsilon)$ admits an asymptotic expansion

$$(XII.1.11) \qquad A(x, \epsilon) \simeq \sum_{\nu=0}^{\infty} \epsilon^\nu A_\nu(x)$$

as $\epsilon \to 0$ in sector (XII.1.5), where the entries of coefficient matrices $A_\nu(x) = (a_{jk\nu}(x))$ are holomorphic in domain (XII.1.10). The following second assumption is technical and we do not lose any generality with it.

Assumption II. *The matrix $A_0(0)$ has the following S-N decomposition*

$$A_0(0) = \text{diag}\,[\mu_1, \mu_2, \ldots, \mu_n] \, + \, \mathcal{N},$$

where $\mu_1, \mu_2, \ldots, \mu_n$ are eigenvalues of $A_0(0)$ and $\mathcal{N}$ is a lower-triangular nilpotent matrix.

Note that $|\mathcal{N}|$ can be made as small as we wish (cf. Lemma VII-3-3).

The following third assumption plays a key roll.

Assumption III. *The matrix $A_0(0)$ is invertible, i.e.,*

$$\mu_j \neq 0 \qquad (j = 1, 2, \ldots, n).$$

In §§XII-2 and XII-3, we shall prove the following theorem.

Theorem XII-1-2. *Under Assumptions I, II, and III, system (XII.1.1) has a solution*

$$(XII.1.12) \qquad\qquad v_j = p_j(x, \epsilon) \qquad (j = 1, 2, \dots, n)$$

such that
 (i) $p_j(x, \epsilon)$ are holomorphic in a domain

$$(XII.1.13) \qquad\qquad |x| < \delta, \quad 0 < |\epsilon| < \rho, \quad |\arg \epsilon| < \alpha,$$

where δ, ρ, and α are suitable positive constants such that $0 < \delta \leq \delta_0$, $0 < \rho \leq \rho_0$, and $0 < \alpha \leq \alpha_0$,
 (ii) $p_j(x, \epsilon)$ admit asymptotic expansions

$$(XII.1.14) \qquad\qquad p_j(x, \epsilon) \simeq \sum_{\nu=1}^{\infty} p_{j\nu}(x)\epsilon^{\nu} \qquad (j = 1, 2, \dots, n)$$

as $\epsilon \to 0$ in the sector

$$(XII.1.15) \qquad\qquad 0 < |\epsilon| < \rho, \quad |\arg \epsilon| < \alpha,$$

where coefficients $p_{j\nu}(x)$ of (XII.1.14) are holomorphic with respect to x in the domain $\{x : |x| < \delta\}$.

XII-2. Basic estimates

In order to prove Theorem XII-1-2, let us change system (XII-1-1) to a system of integral equations.

Observation XII-2-1. Expansion (XII.1.14) of the solution $p_j(x, \epsilon)$

$$(XII.2.1) \qquad\qquad v_j = \sum_{\nu=1}^{\infty} p_{j\nu}(x)\epsilon^{\nu} \qquad (j = 1, 2, \dots, n)$$

must be a formal solution of system (XII.1.1). The existence of such a formal solution (XII.2.1) of system (XII.1.1) follows immediately from Assumptions I and III. The proof of this fact is left to the reader as an exercise.

Observation XII-2-2. For each $j = 1, 2, \dots, n$, using Theorem XI-1-14, let us construct a function $q_j(x, \epsilon)$ such that
 (i) q_j is holomorphic in a domain

$$(XII.2.2) \qquad\qquad |x| < \delta', \quad 0 < |\epsilon| < \rho', \quad |\arg \epsilon| < \alpha',$$

where $0 < \delta' < \delta_0$, $0 < \rho' < \rho_0$, and $0 < \alpha' < \alpha_0$,

(ii) q_j and $\dfrac{dq_j}{dx}$ admit asymptotic expansions

$$(\text{XII.2.3}) \qquad q_j(x,\epsilon) \simeq \sum_{\nu=1}^{\infty} p_{j\nu}(x)\epsilon^{\nu} \quad \text{and} \quad \frac{dq_j(x,\epsilon)}{dx} \simeq \sum_{\nu=1}^{\infty} \frac{dp_{j\nu}(x)}{dx}\epsilon^{\nu}$$

as $\epsilon \to 0$ in the sector

$$(\text{XII.2.4}) \qquad 0 < |\epsilon| < \rho', \quad |\arg \epsilon| < \alpha'.$$

Consider the change of variables

$$v_j = u_j + q_j(x,\epsilon) \qquad (j = 1,2,\dots,n).$$

Denote $(q_1, q_2, \dots, q_n)$ and $(u_1, u_2, \dots, u_n)$ by $\vec{q}$ and $\vec{u}$, respectively. Then, $\vec{u}$ satisfies the system of differential equations

$$(\text{XII.2.5}) \qquad \epsilon^{\sigma}\frac{du_j}{dx} = g_j(x,\vec{u},\epsilon) \qquad (j = 1,2,\dots,n),$$

where

$$g_j(x,\vec{u},\epsilon) = f_j(x,\vec{u}+\vec{q},\epsilon) - \epsilon^{\sigma}\frac{dq_j(x,\epsilon)}{dx} \qquad (j = 1,2,\dots,n).$$

Set

$$(\text{XII.2.6}) \qquad g_j(x,\vec{u},\epsilon) = g_{j0}(x,\epsilon) + \sum_{k=1}^{n} b_{jk}(x,\epsilon)u_k + \sum_{|\wp|\geq 2}^{\infty} b_{j\wp}(x,\epsilon)\vec{u}^{\wp}$$
$$(j = 1,2,\dots,n).$$

In particular,

$$(\text{XII.2.7}) \qquad g_{j0}(x,\epsilon) = f_j(x,\vec{q},\epsilon) - \epsilon^{\sigma}\frac{dq_j(x,\epsilon)}{dx} \simeq 0 \qquad (j = 1,2,\dots,n)$$

and

$$b_{jk}(x,\epsilon) - a_{jk}(x,\epsilon) = O(\epsilon) \qquad (j,k = 1,2,\dots,n)$$

as $\epsilon \to 0$ in sector (XII.2.4). Therefore,

$$(\text{XII.2.8}) \qquad b_{jk}(x,\epsilon) = a_{jk0}(x) + O(\epsilon) \qquad (j,k = 1,2,\dots,n)$$

as $\epsilon \to 0$ in sector (XII.2.4).

Set

$$(\text{XII.2.9}) \qquad g_j(x,\vec{u},\epsilon) = \mu_j u_j + R_j(x,\vec{u},\epsilon) \qquad (j = 1,2,\dots,n).$$

Then, for sufficiently small positive numbers δ, ρ, and γ, there exists a positive constant $\hat{c}$, independent of j, such that for every positive integer N, the estimates

$$(\text{XII.2.10}) \qquad |R_j(x,\vec{u},\epsilon)| \leq \hat{c}|\vec{u}| + B_N|\epsilon|^{N} \qquad (j = 1,2,\dots,n)$$

and

$$(\text{XII.2.11}) \qquad |R_j(x,\vec{u},\epsilon) - R_j(x,\vec{u}',\epsilon)| \leq \hat{c}|\vec{u} - \vec{u}'| \qquad (j = 1, 2, \dots, n)$$

hold whenever $(x, \vec{u}, \epsilon)$ and $(x, \vec{u}', \epsilon)$ are in the domain

$$(\text{XII.2.12}) \quad |x| < \delta, \quad 0 < |\epsilon| < \rho, \quad |\arg \epsilon| < \alpha', \quad |u_j| < \gamma \quad (j = 1, 2, \dots, n).$$

Here, B_N is a positive constant depending on N and $|\vec{u}|$.

From (XII.2.5) and (XII.2.9), it follows that

$$(\text{XII.2.13}) \qquad \epsilon^\sigma \frac{du_j}{dx} = \mu_j u_j + R_j(x, \vec{u}, \epsilon) \qquad (j = 1, 2, \dots, n).$$

Change system (XII.2.13) to the system of integral equations

$$(\text{XII.2.14}) \qquad u_j = \frac{1}{\epsilon^\sigma} \int_{x_j}^{x} \exp\left[\frac{-\mu_j}{\epsilon^\sigma}(t - x)\right] R_j(t, \vec{u}, \epsilon) dt$$
$$(j = 1, 2, \dots, n),$$

where the paths of integration must be chosen carefully so that uniformly convergent successive approximations can be defined.

Hereafter in this section, we explain how to choose paths of integration on the right-hand side of (XII.2.14).

Observation XII-2-3. Set

$$(\text{XII.2.15}) \qquad \omega_j = \arg \mu_j \qquad (j = 1, 2, \dots, n)$$

and suppose that

$$(\text{XII.2.16}) \qquad -\frac{3}{2}\pi < \omega_j \leq \frac{1}{2}\pi \qquad (j = 1, 2, \dots, n).$$

Then, there exists a positive number Θ less than $\frac{1}{2}\pi$ such that

$$-\frac{3}{2}\pi < \omega_j - \Theta < \frac{1}{2}\pi \quad \text{and} \quad \omega_j - \Theta \neq -\frac{1}{2}\pi \qquad (j = 1, 2, \dots, n).$$

Without loss of generality, suppose that n real numbers ω_j are divided into the following two groups:

$$(\text{I}) \qquad -\frac{1}{2}\pi < \omega_j - \Theta < \frac{1}{2}\pi \qquad (j = 1, 2, \dots, m')$$

and

$$(\text{II}) \qquad -\frac{3}{2}\pi < \omega_j - \Theta < -\frac{1}{2}\pi \qquad (j = m' + 1, \dots, n).$$

Choose two positive numbers α and β sufficiently small such that

$$\begin{cases} -\dfrac{1}{2}\pi + \sigma\alpha + \beta < \omega_j - \Theta < \dfrac{1}{2}\pi - (\sigma\alpha + \beta) & (j = 1, 2, \ldots, m'), \\[2mm] -\dfrac{3}{2}\pi + \sigma\alpha + \beta < \omega_j - \Theta < -\dfrac{1}{2}\pi - (\sigma\alpha + \beta) & (j = m' + 1, \ldots, n). \end{cases}$$

Set

$$x^{(1)} = \delta e^{-i\Theta}, \quad x^{(2)} = ix^{(1)}\tan\beta, \quad x^{(3)} = -x^{(1)}, \quad \text{and} \quad x^{(4)} = -x^{(2)},$$

where δ is a sufficiently small positive number. Then, a rhombus is defined by its four vertices $x^{(j)}$ $(j = 1, 2, 3, 4)$ (cf. Figure 1). Note that the angle at $x^{(1)}$ is 2β.

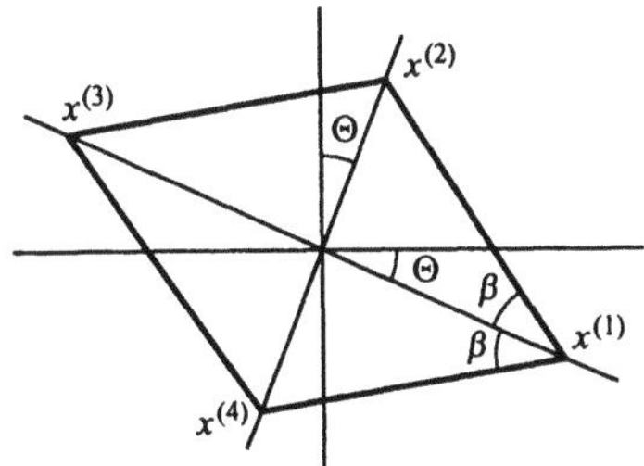

FIGURE 1.

Denote the interior of this rhombus by $\mathcal{D}(\delta)$. It is noteworthy that the domain $\mathcal{D}(\delta)$ contains a small open neighborhood of $x = 0$ and is contained in the domain $\{x : |x| < \delta\}$.

The basic estimates for the proof of Theorem XII-1-2 are given by the following lemma.

Lemma XII-2-4. *For each $j = 1, \ldots, n$, consider the function*

(XII.2.17)
$$U_j(x, \epsilon) = \int_{x_j}^{x} \exp\left[-\frac{\mu_j(t - x_j)}{\epsilon^\sigma}\right] dt,$$

where the path of integration is the straight line $\overline{x_j x}$ and

$$x_j = \begin{cases} x^{(1)} & (j = 1, 2, \ldots, m'), \\ x^{(3)} & (j = m' + 1, \ldots, m). \end{cases}$$

Then, there exists a positive number c such that

(XII.2.18)
$$|U_j(x, \epsilon)| \le c|\epsilon|^\sigma \left|\exp\left[-\frac{\mu_j(x - x_j)}{\epsilon^\sigma}\right]\right|$$

for

(XII.2.19)
$$x \in \mathcal{D}(\delta), \qquad |\arg \epsilon| < \alpha, \qquad |\epsilon| > 0.$$

Proof.

We prove this lemma for the cases $j = 1, 2, \ldots, m'$. The cases $j = m'+1, \ldots, m$ can be treated in a similar manner. Set $\xi = |t - x^{(1)}|$ and $\theta = \arg(t - x^{(1)})$ for $t \in \mathcal{D}(\delta)$ and set $\omega = \arg \epsilon$. Then,

(XII.2.20)
$$|U_j(x, \epsilon)| \leq \int_0^{|x-x^{(1)}|} \exp\left[-\frac{|\mu_j|}{|\epsilon|^\sigma} \xi \cos(\omega_j + \theta - \sigma\omega)\right] d\xi \qquad (j = 1, 2, \ldots, m').$$

From the definition of $\mathcal{D}(\delta)$, it follows that $\pi - \Theta - \beta < \theta < \pi - \Theta + \beta$ for $x \in \mathcal{D}(\delta)$. Thus, if ϵ is in the sector $|\arg \epsilon| < \alpha$, the inequalities

$$\frac{1}{2}\pi < \omega_j + \pi - \Theta - (\sigma\alpha + \beta) < \omega_j + \theta - \sigma\omega < \omega_j + \pi - \Theta + (\sigma\alpha + \beta) < \frac{3}{2}\pi$$

hold for $x \in \mathcal{D}(\delta)$ and $j = 1, 2, \ldots, m'$. Therefore, there exists a positive constant c' such that

(XII.2.21)
$$-\cos(\omega_j + \theta - \sigma\omega) \geq c' \qquad (j = 1, 2, \ldots, m')$$

for (x, ϵ) in (XII.2.19). By virtue of (XII.2.20) and (XII.2.21), we obtain

$$|U_j(x, \epsilon)| \leq \frac{|\epsilon|^\sigma}{c'|\mu_j|} \left|\exp\left[-\frac{\mu_j(x - x_j)}{\epsilon^\sigma}\right]\right| \qquad (j = 1, 2, \ldots, m')$$

for (x, ϵ) in (XII.2.19). Thus, Lemma XII-2-4 is proved. $\square$

In the next section, we shall consider system (XII-2-14) of integral equations assuming that $x \in \mathcal{D}(\delta)$ and the paths of integration are chosen in the same way as in Lemma XII-2-4.

XII-3. Proof of Theorem XII-1-2

Let us construct a solution $u_j = \phi_j(x, \epsilon)$ of (XII.2.14) so that

(XII.3.1)
$$\phi_j(x, \epsilon) \simeq 0 \qquad (j = 1, 2, \ldots, n)$$

as $\epsilon \to 0$ in (XII.1.15).

Now, by virtue of Assumption II, three positive quantities δ, ρ, and $|\mathcal{N}|$ can be chosen so small that $\hat{c}$ in (XII.2.10) and (XII.2.11) satisfies the condition

(XII.3.2)
$$c\hat{c} < 1,$$

where c is the constant given in (XII.2.18).

Define successive approximations of a solution of (XII.2.14) in the following way:

(XII.3.3)
$$\begin{cases} u_j^{(0)}(x, \epsilon) \equiv 0, \\ u_j^{(h)}(x, \epsilon) = \dfrac{1}{\epsilon^\sigma} \displaystyle\int_{x_j}^x \exp\left[\dfrac{-\mu_j}{\epsilon^\sigma}(t - x)\right] R_j(t, \vec{u}^{(h-1)}(t, \epsilon), \epsilon) dt \\ \qquad\qquad\qquad (j = 1, 2, \ldots, n; \quad h = 1, 2, \ldots), \end{cases}$$

where $x \in \mathcal{D}(\delta)$ and the integration is taken over the straight line $\overline{x_j x}$. For a given positive integer N, it will be shown that

  (i) for each j, the sequence $\{u_j^{(h)}(x, \epsilon) : h = 0, 1, \dots\}$ is well defined,

  (ii) for the given integer N, there exist positive constants K_N and ρ_N $(0 < \rho_N \leq \rho)$ such that

$$(\text{XII.3.4}) \qquad |u_j^{(h)}(x, \epsilon)| \leq K_N |\epsilon|^N \qquad (j = 1, 2, \dots, n; \quad h = 0, 1, \dots),$$

uniformly for (x, ϵ) in the domain

$$(\text{XII.3.5}) \qquad x \in \mathcal{D}(\delta), \quad 0 < |\epsilon| < \rho_N, \quad |\arg \epsilon| < \alpha',$$

  (iii) the sequence $\{\vec{u}^{(h)}(x, \epsilon) : h = 0, 1, \dots\}$ converges uniformly to $\vec{\phi}(x, \epsilon) = (\phi_1(x, \epsilon), \phi_2(x, \epsilon), \dots, \phi_n(x, \epsilon))$ in (XII.3.5), where $\vec{u}^{(h)}(x, \epsilon)$ is the $\mathbb{C}^n$-valued function with the entries $(u_1^{(h)}(x, \epsilon), \dots, u_n^{(h)}(x, \epsilon))$.

The limit function $\vec{\phi}(x, \epsilon)$ is independent of N since the successive approximations are independent of N.

If (i), (ii), and (iii) are proved, it follows that

$$(\text{XII.3.6}) \qquad \phi_j(x, \epsilon) = \lim_{h \to \infty} u_j^{(h)}(x, \epsilon) \simeq 0 \qquad (j = 1, 2, \dots, n)$$

as $\epsilon \to 0$ in the sector $\mathcal{S} = \left\{ \epsilon : 0 < |\epsilon| < \sup_N(\rho_N), \ |\arg \epsilon| < \alpha' \right\}$, and the functions $\phi_j(x, \epsilon)$ are holomorphic in the domain $\mathcal{D}(\delta) \times \mathcal{S}$ and satisfy (XII.2.14). Setting $p_j(x, \epsilon) = \phi_j(x, \epsilon) + q_j(x, \epsilon)$ $(j = 1, \dots, n)$, we obtain a solution $v_j = p_j(x, \epsilon)$ $(j = 1, 2, \dots, n)$ of (XII.1.1) which satisfies all of the requirements of Theorem XII-1-2. Thus, the proof of Theorem XII-1-2 will be completed.

To show (i) and (ii), choose two constants K_N and ρ_N so that $K_N \geq \dfrac{cB_N}{1 - c\hat{c}}$ and $\rho_N{}^N K_N \leq \gamma$. This is possible since condition (XII.3.2) is satisfied. Now, assuming that (i) and (ii) are true for $u_j^{(h-1)}(x, \epsilon)$ in (XII.3.5), let us prove that $u_j^{(h)}(x, \epsilon)$ also satisfy conditions (i) and (ii). First from (XII.3.3), it follows that

$$(\text{XII.3.7}) \qquad \begin{aligned} u_j^{(h)}(x, \epsilon) &= \frac{1}{\epsilon^\sigma} \exp\left[\frac{\mu_j}{\epsilon^\sigma}(x - x_j) \right] \\ &\quad \times \int_{x_j}^{x} \exp\left[\frac{-\mu_j}{\epsilon^\sigma}(t - x_j) \right] R_j(t, \vec{u}^{(h-1)}, \epsilon)\, dt \quad (j = 1, 2, \dots, n). \end{aligned}$$

Then, from Lemma XII-2-4, (XII.2.10), (XII.3.4), and the inductive assumption, we conclude that

$$|u_j^{(h)}(x, \epsilon)| \leq c\{\hat{c} K_N + B_N\}|\epsilon|^N \leq K_N |\epsilon|^N$$

for (x, ϵ) in (XII.3.5). Thus, (i) and (ii) are true for $u_j^{(h)}(x, \epsilon)$ $(j = 1, 2, \dots, n)$.

To show (iii), from (XII.2.11), (XII.3.3), and Lemma XII-2-4, we obtain

$$|\vec{u}^{(h+1)}(x,\epsilon) - \vec{u}^{(h)}(x,\epsilon)| \leq c\hat{c} \sup_{t\in\mathcal{D}(\delta)} |\vec{u}^{(h)}(t,\epsilon) - \vec{u}^{(h-1)}(t,\epsilon)|$$

in (XII.3.5) for $h = 1, 2, 3, \ldots$. Hence, the sequence $\{\vec{u}^{(h)}(x,\epsilon) : h = 0, 1, 2, \ldots\}$ is convergent to $\vec{\phi}(x,\epsilon)$ uniformly in (XII.3.5). Furthermore, $u_j = \phi_j(x,\epsilon)$ $(j = 1, 2, \ldots, n)$ satisfy system (XII.2.14). Thus, the proof of Theorem XII-1-2 is completed. $\square$

XII-4. A block-diagonalization theorem

Consider a system of linear differential equations

$$(\text{XII.4.1}) \qquad \epsilon^\sigma \frac{d\vec{y}}{dx} = A(x,\epsilon)\vec{y},$$

where σ is a positive integer, $\vec{y} \in \mathbb{C}^n$, and $A(x,\epsilon)$ is an $n \times n$ matrix. The entries of $A(x,\epsilon)$ are holomorphic with respect to a complex variable x and a complex parameter ϵ in a domain

$$(\text{XII.4.2}) \qquad |x| < \delta_0, \quad 0 < |\epsilon| < \rho_0, \quad |\arg\epsilon| < \alpha_0,$$

where δ_0, ρ_0, and α_0 are positive numbers. Assume that the matrix $A(x,\epsilon)$ admits a uniform asymptotic expansion in the sense of Poincaré,

$$(\text{XII.4.3}) \qquad A(x,\epsilon) \simeq \sum_{\nu=0}^{\infty} \epsilon^\nu A_\nu(x),$$

in domain (XII.4.2) as $\epsilon \to 0$ in the sector

$$(\text{XII.4.4}) \qquad 0 < |\epsilon| < \rho_0, \quad |\arg\epsilon| < \alpha_0,$$

where the entries of coefficients $A_\nu(x)$ are holomorphic with respect to x in the domain

$$(\text{XII.4.5}) \qquad |x| < \delta_0.$$

Suppose that $A_0(0)$ has ℓ distinct eigenvalues $\lambda_1, \lambda_2, \ldots, \lambda_\ell$ with multiplicities $n_1, n_2, \ldots, n_\ell$, respectively $(n_1 + n_2 + \cdots + n_\ell = n)$. Without loss of generality, assume that $A_0(0)$ is in a block-diagonal form

$$(\text{XII.4.6}) \qquad A_0(0) = \text{diag}\left[\mathring{A}_1, \mathring{A}_2, \ldots, \mathring{A}_\ell\right],$$

where $\mathring{A}_j$ are $n_j \times n_j$ matrices in the form

$$(\text{XII.4.7}) \qquad \mathring{A}_j = \lambda_j I_{n_j} + \mathcal{N}_j \quad (j = 1, 2, \ldots, \ell).$$

Here, I_{n_j} is the $n_j \times n_j$ identity matrix and $\mathcal{N}_j$ is an $n_j \times n_j$ lower-triangular nilpotent matrix. The main result of this section is the following theorem.

Theorem XII-4-1 ([Si7]).   *Under assumptions (XII.4.3) and (XII.4.6), there exists an $n \times n$ matrix $P(x, \epsilon)$ such that*
  (i) the entries of $P(x, \epsilon)$ are holomorphic with respect to (x, ϵ) in a domain

$$(XII.4.8) \qquad\qquad |x| < \delta, \quad 0 < |\epsilon| < \rho, \quad |\arg \epsilon| < \alpha,$$

where δ, ρ, and α are positive numbers such that $0 < \delta \leq \delta_0$, $0 < \rho < \rho_0$ and $0 < \alpha \leq \alpha_0$,
  (ii) $P(x, \epsilon)$ admits a uniform asymptotic expansion

$$(XII.4.9) \qquad\qquad P(x, \epsilon) \simeq \sum_{\nu=0}^{\infty} \epsilon^{\nu} P_{\nu}(x) \qquad (P_0(0) = I_n)$$

in domain (XII.4.8) as $\epsilon \to 0$ in the sector

$$(XII.4.10) \qquad\qquad 0 < |\epsilon| < \rho, \quad |\arg \epsilon| < \alpha,$$

where the entries of coefficients $P_{\nu}(x)$ are holomorphic with respect to x in the domain

$$(XII.4.11) \qquad\qquad |x| < \delta,$$

  (iii) the transformation

$$(XII.4.12) \qquad\qquad \vec{y} = P(x, \epsilon)\vec{z}$$

reduces system (XII.4.1) to a system

$$(XII.4.13) \qquad\qquad \epsilon^{\sigma} \frac{d\vec{z}}{dx} = B(x, \epsilon)\vec{z}$$

with the coefficient matrix $B(x, \epsilon)$ in a block-diagonal form

$$(XII.4.14) \qquad B(x, \epsilon) = \mathrm{diag}[B_1(x, \epsilon), B_2(x, \epsilon), \dots, B_{\ell}(x, \epsilon)],$$

where $B_j(x, \epsilon)$ is an $n_j \times n_j$ matrix $(j = 1, 2, \dots, \ell)$,
  (iv) the matrix $B_j(x, \epsilon)$ admits a uniform asymptotic expansion

$$(XII.4.15) \qquad\qquad B_j(x, \epsilon) \simeq \sum_{\nu=0}^{\infty} \epsilon^{\nu} B_{j\nu}(x)$$

in domain (XII.4.8) as $\epsilon \to 0$ in sector (XII.4.10), where the entries of coefficients $B_{j\nu}(x)$ are holomorphic with respect to x in domain (XII.4.11).

Remark XII-4-2.

(a) Set

$$(\text{XII.4.16}) \qquad B_\nu(x) = \text{diag}\,[B_{1\nu}(x), B_{2\nu}(x), \dots, B_{\ell\nu}(x)].$$

Then, the coefficient matrix $B(x,\epsilon)$ of system (XII.4.13) admits a uniform asymptotic expansion $B(x,\epsilon) \simeq \sum_{\nu=0}^{\infty} \epsilon^\nu B_\nu(x)$.

(b) When a fundamental matrix solution $Z(x,\epsilon)$ of (XII.4.13) is known, a fundamental matrix solution $Y(x,\epsilon)$ of (XII.4.1) is given by $Y(x,\epsilon) = P(x,\epsilon)Z(x,\epsilon)$.

(c) In the case when the matrix $A_0(0)$ has n distinct eigenvalues, by Theorem XII-4-1, we can diagonalize system (XII.4.1).

(d) In the case when eigenvalues of the matrix $A_0(0)$ are not completely distinct, the point $x = 0$ is, in general, a so-called transition point. In order to study behavior of solutions in the neighborhood of a transition point, we need a much deeper analysis of solutions of system (XII.4.1). For these informations, see, for example, [Was1], [Was2], and [Si12].

Proof of Theorem XII-4-1.

We prove this theorem in two steps. The proof is similar to that of Theorem VII-3-1.

Step 1. We show that there exist a positive number δ $(\le \delta_0)$ and an $n \times n$ matrix $P_0(x)$ such that

(i) the entries of $P_0(x)$ and $P_0(x)^{-1}$ are holomorphic in the domain $\{x : |x| < \delta\}$ and $P_0(0) = I_n$,

(ii) the matrix $C_0(x) = P_0(x)^{-1}A_0(x)P_0(x)$ is in a block-diagonal form

$$(\text{XII.4.17}) \qquad C_0(x) = \text{diag}\left[C_0^{(1)}(x), C_0^{(2)}(x), \dots, C_0^{(\ell)}(x)\right],$$

where $C_0^{(j)}(x)$ is an $n_j \times n_j$ matrix such that

$$C_0^{(j)}(0) \;=\; \mathring{A}_j = \lambda_j I_{n_j} + \mathcal{N}_j \qquad (j = 1, 2, \dots, \ell).$$

In fact, two matrices $P_0(x)$ and $C_0(x)$ must be determined by the equation

$$(\text{XII.4.18}) \qquad A_0(x)P_0(x) - P_0(x)C_0(x) = O.$$

Set

$$A_0(x) = \begin{bmatrix} A_0^{(11)}(x) & A_0^{(12)}(x) & A_0^{(13)}(x) & \cdots & A_0^{(1\ell)}(x) \\ A_0^{(21)}(x) & A_0^{(22)}(x) & A_0^{(23)}(x) & \cdots & A_0^{(2\ell)}(x) \\ \vdots & \vdots & \vdots & \vdots & \vdots \\ A_0^{(\ell 1)}(x) & A_0^{(\ell 2)}(x) & A_0^{(\ell 3)}(x) & \cdots & A_0^{(\ell\ell)}(x) \end{bmatrix},$$

and

$$P_0(x) = \begin{bmatrix} P_0^{(11)}(x) & P_0^{(12)}(x) & P_0^{(13)}(x) & \cdots & P_0^{(1\ell)}(x) \\ P_0^{(21)}(x) & P_0^{(22)}(x) & P_0^{(23)}(x) & \cdots & P_0^{(2\ell)}(x) \\ \vdots & \vdots & \vdots & \vdots & \vdots \\ P_0^{(\ell 1)}(x) & P_0^{(\ell 2)}(x) & P_0^{(\ell 3)}(x) & \cdots & P_0^{(\ell\ell)}(x) \end{bmatrix},$$

where $A_0^{(jk)}(x)$ and $P_0^{(jk)}(x)$ are $n_j \times n_k$ matrices. Furthermore, set $P_0^{(jj)}(x) = I_{n_j}$ $(j = 1, 2, \ldots, \ell)$. Then,

$$(\text{XII.4.19}) \qquad C_0^{(j)}(x) = A_0^{(jj)}(x) + \sum_{h \neq j} A_0^{(jh)}(x) P_0^{(hj)}(x) \quad (j = 1, 2, \ldots, \ell)$$

and

$$
\begin{aligned}
(\text{XII.4.20}) \qquad & A_0^{(jj)}(x) P_0^{(jk)}(x) - P_0^{(jk)}(x) C_0^{(k)}(x) \\
& + \sum_{h \neq j,k} A_0^{(jh)}(x) P_0^{(hk)}(x) + A_0^{(jk)}(x) = O \qquad (j \neq k)
\end{aligned}
$$

from (XII.4.18). Combining (XII.4.19) and (XII.4.20), we obtain

$$
\begin{aligned}
(\text{XII.4.21}) \qquad & A_0^{(jj)}(x) P_0^{(jk)}(x) - P_0^{(jk)}(x)\Big(A_0^{(kk)}(x) + \sum_{h \neq k} A_0^{(kh)}(x) P_0^{(hk)}(x)\Big) \\
& + \sum_{h \neq j,k} A_0^{(jh)}(x) P_0^{(hk)}(x) + A_0^{(jk)}(x) = O \qquad (j \neq k).
\end{aligned}
$$

Upon applying the implicit function theorem to (XII.4.21), matrices $P_0^{(jk)}(x)$ $(j \neq k)$ can be constructed. Then, $C_0(x)$ is given by (XII.4.19) and (XII.4.17).

Step 2. Now, assume without loss of generality that $A_0(x)$ is in a block-diagonal form

$$(\text{XII.4.22}) \qquad A_0(x) = \text{diag}\left[A_0^{(1)}(x), A_0^{(2)}(x), \ldots, A_0^{(\ell)}(x)\right].$$

To prove Theorem XII-4-1, it suffices to solve the differential equation

$$(\text{XII.4.23}) \qquad \epsilon^\sigma \frac{dP(x,\epsilon)}{dx} = A(x,\epsilon)P(x,\epsilon) - P(x,\epsilon)B(x,\epsilon),$$

where

$$
\begin{aligned}
(\text{XII.4.24}) \qquad & A(x,\epsilon) = A_0(x) + \epsilon\hat{A}(x,\epsilon), \quad P(x,\epsilon) = I_n + \epsilon\hat{P}(x,\epsilon) \quad \text{and} \\
& B(x,\epsilon) = A_0(x) + \epsilon\hat{B}(x,\epsilon).
\end{aligned}
$$

Set

$$
\hat{A}(x,\epsilon) = \begin{bmatrix}
\hat{A}^{(11)}(x,\epsilon) & \hat{A}^{(12)}(x,\epsilon) & \cdots & \hat{A}^{(1\ell)}(x,\epsilon) \\
\hat{A}^{(21)}(x,\epsilon) & \hat{A}^{(22)}(x,\epsilon) & \cdots & \hat{A}^{(2\ell)}(x,\epsilon) \\
\vdots & \vdots & \vdots & \vdots \\
\hat{A}^{(\ell 1)}(x,\epsilon) & \hat{A}^{(\ell 2)}(x,\epsilon) & \cdots & \hat{A}^{(\ell\ell)}(x,\epsilon)
\end{bmatrix},
$$

$$
\hat{P}(x,\epsilon) = \begin{bmatrix}
\hat{P}^{(11)}(x,\epsilon) & \hat{P}^{(12)}(x,\epsilon) & \cdots & \hat{P}^{(1\ell)}(x,\epsilon) \\
\hat{P}^{(21)}(x,\epsilon) & \hat{P}^{(22)}(x,\epsilon) & \cdots & \hat{P}^{(2\ell)}(x,\epsilon) \\
\vdots & \vdots & \vdots & \vdots \\
\hat{P}^{(\ell 1)}(x,\epsilon) & \hat{P}^{(\ell 2)}(x,\epsilon) & \cdots & \hat{P}^{(\ell\ell)}(x,\epsilon)
\end{bmatrix},
$$

and

$$\hat{B}(x,\epsilon) = \operatorname{diag}\left[\hat{B}^{(1)}(x,\epsilon), \hat{B}^{(2)}(x,\epsilon), \ldots, \hat{B}^{(\ell)}(x,\epsilon)\right],$$

where $\hat{A}^{(jk)}(x)$ and $\hat{P}^{(jk)}(x)$ are $n_j \times n_k$ matrices and $\hat{B}_j(x,\epsilon)$ is an $n_j \times n_j$ matrix $(j, k = 1, \ldots, \ell)$. Furthermore, set

$$(\text{XII.4.25}) \qquad \hat{P}^{(jj)}(x,\epsilon) = O \qquad (j = 1, 2, \ldots, \ell).$$

Then, equation (XII.4.23) becomes

$$(\text{XII.4.26}) \qquad \hat{B}^{(j)}(x,\epsilon) = \hat{A}^{(jj)}(x,\epsilon) + \epsilon \sum_{h=1}^{\ell} \hat{A}^{(jh)}(x,\epsilon)\hat{P}^{(hj)}(x,\epsilon)$$
$$(j = 1, 2, \ldots, \ell)$$

and
(XII.4.27)
$$\epsilon^\sigma \frac{d\hat{P}^{(jk)}(x,\epsilon)}{dx} = A_0^{(j)}(x)\hat{P}^{(jk)}(x,\epsilon) - \hat{P}^{(jk)}(x,\epsilon)A_0^{(k)}(x) + \hat{A}^{(jk)}(x,\epsilon)$$
$$+ \epsilon \left\{ \sum_{h \neq k} \hat{A}^{(jh)}(x,\epsilon)\hat{P}^{(hk)}(x,\epsilon) - \hat{P}^{(jk)}(x,\epsilon)\hat{B}^{(k)}(x,\epsilon) \right\} \quad (j \neq k).$$

Combining (XII.4.26) and (XII.4.27), we obtain

$$\epsilon^\sigma \frac{d\hat{P}^{(jk)}(x,\epsilon)}{dx} = A_0^{(j)}(x)\hat{P}^{(jk)}(x,\epsilon) - \hat{P}^{(jk)}(x,\epsilon)A_0^{(k)}(x) + \hat{A}^{(jk)}(x,\epsilon)$$
$$+ \epsilon \sum_{h=1}^{\ell} \hat{A}^{(jh)}(x,\epsilon)\hat{P}^{(hk)}(x,\epsilon) - \epsilon\hat{P}^{(jk)}(x,\epsilon)\hat{A}^{(kk)}(x,\epsilon)$$
$$- \epsilon^2 \hat{P}^{(jk)}(x,\epsilon) \sum_{h=1}^{\ell} \hat{A}^{(kh)}(x,\epsilon)\hat{P}^{(hk)}(x,\epsilon) \qquad (j \neq k).$$

Replacing $\hat{P}^{(jk)}(x,\epsilon)$ by an $n_j \times n_k$ matrix $X^{(jk)}$, we derive a system of nonlinear differential equations

$$\epsilon^\sigma \frac{dX^{(jk)}}{dx} = A_0^{(j)}(x)X^{(jk)} - X^{(jk)}A_0^{(k)}(x) + \hat{A}^{(jk)}(x,\epsilon)$$
$$(\text{XII.4.28}) \qquad + \epsilon \sum_{h=1}^{\ell} \hat{A}^{(jh)}(x,\epsilon)X^{(hk)} - \epsilon X^{(jk)}\hat{A}^{(kk)}(x,\epsilon)$$
$$- \epsilon^2 X^{(jk)} \sum_{h=1}^{\ell} \hat{A}^{(kh)}(x,\epsilon)X^{(hk)} \qquad (j \neq k).$$

Consider the entries of $X^{(jk)}$ $(j, k = 1, 2, \ldots, \ell; \ j \neq k)$ altogether to form a system of nonlinear differential equations. Upon applying Theorem XII-1-2, we construct an analytic solution of this nonlinear system in a domain (XII.4.8) admitting an asymptotic expansion in ϵ as described in Theorem XII-4-1. This completes the proof of Theorem XII-4-1. $\square$

XII-5. Gevrey asymptotic solutions in a parameter

In this section, using Theorems XI-2-3 and the proof of Theorem XII-1-2, we construct Gevrey asymptotic solutions of a system of differential equations

$$(\text{XII.5.1}) \qquad \epsilon^\sigma \frac{d\vec{y}}{dx} = \vec{f}(x, \vec{y}, \epsilon),$$

where x is a complex variable, $\vec{y} \in \mathbb{C}^n$, ϵ is a complex parameter, σ is a positive integer, and $\vec{f}(x, \vec{y}, \epsilon)$ is a $\mathbb{C}^n$-valued function of $(x, \vec{y}, \epsilon)$. Define three domains by

$$(\text{XII.5.2}) \qquad \begin{cases} \Delta(\delta_0) = \{x \in \mathbb{C} : |x| < \delta_0\}, \\ \Omega(\rho_0) = \{\vec{y} \in \mathbb{C}^n : |\vec{y}| < \rho_0\}, \\ \mathcal{S}(r_0, \alpha_0) = \{\epsilon \in \mathbb{C} : |\arg \epsilon| < \alpha_0, \ 0 < |\epsilon| < r_0\}. \end{cases}$$

Also, define two matrices $A(x, \epsilon)$ and $A_0(x)$ by

$$(\text{XII.5.3}) \qquad A(x, \epsilon) = \frac{\partial \vec{f}}{\partial \vec{y}}(x, \vec{0}, \epsilon) = \begin{bmatrix} \dfrac{\partial f_1}{\partial y_1}(x, \vec{0}, \epsilon) & \cdots & \dfrac{\partial f_1}{\partial y_n}(x, \vec{0}, \epsilon) \\ \vdots & \vdots & \vdots \\ \dfrac{\partial f_n}{\partial y_1}(x, \vec{0}, \epsilon) & \cdots & \dfrac{\partial f_n}{\partial y_n}(x, \vec{0}, \epsilon) \end{bmatrix},$$

and

$$(\text{XII.5.4}) \qquad A_0(x) = \lim_{\epsilon \to 0} A(x, \epsilon),$$

respectively.

We first prove the following lemma.

Lemma XII-5-1. *Assume that*
  (i) *$\vec{f}(x, \vec{y}, \epsilon)$ is holomorphic with respect to $(x, \vec{y}, \epsilon)$ in a domain $\Delta(\delta_0) \times \Omega(\rho_0) \times \mathcal{S}(r_0, \alpha_0)$, where δ_0, ρ_0, r_0, and α_0 are positive numbers,*
  (ii) *$\vec{f}(x, \vec{y}, \epsilon)$ is bounded on $\Delta(\delta_0) \times \Omega(\rho_0) \times \mathcal{S}(r_0, \alpha_0)$,*
  (iii) *the matrix $A_0(x)$ defined by (XII.5.4) exists as $\epsilon \to 0$ in $\mathcal{S}(r_0, \alpha_0)$ uniformly in $x \in \Delta(\delta_0)$ and $A_0(0)$ is invertible,*
  (iv) *$\vec{f}(x, \vec{0}, \epsilon)$ is flat of Gevrey order τ as $\epsilon \to 0$ in $\mathcal{S}(r_0, \alpha_0)$ uniformly in $x \in \Delta(\delta_0)$, where τ is a non-negative number.*
Then, there exist three positive numbers δ, r, and α such that system (XII.5.1) has a solution $\vec{\phi}(x, \epsilon)$ which is holomorphic in $(x, \epsilon) \in \Delta(\delta) \times \mathcal{S}(r, \alpha)$ and that $\vec{\phi}(x, \epsilon)$ is flat of Gevrey order τ as $\epsilon \to 0$ in $\mathcal{S}(r, \alpha)$ uniformly in $x \in \Delta(\delta)$.

Proof.

Note first that

$$\text{(XII.5.5)} \qquad |\exp[-c\epsilon^{-k}]| \;=\; \exp[-c|\epsilon|^{-k}\cos(k(\arg\epsilon))]$$

for any positive numbers c and k. Note next that assumptions (i) and (iv) imply that in the case when $\tau = 0$, $\vec{f}(x,\vec{0},\epsilon)$ is identically equal to $\vec{0}$ for $(x,\epsilon) \in \Delta(\delta) \times \mathcal{S}(r,\alpha)$. Hence, in this case, $\vec{y} = \vec{0}$ is a solution of (XII.5.1). In the case when $\tau > 0$, it holds that

$$\text{(XII.5.6)} \qquad |\vec{f}(x,\vec{0},\epsilon)| \;\leq\; K\exp[-2c|\epsilon|^{-k}]$$

for some positive numbers K and c if $(x,\epsilon) \in \Delta(\delta) \times \mathcal{S}(r,\alpha)$ for sufficiently small positive numbers δ, r, and α and if

$$\text{(XII.5.7)} \qquad k \;=\; \frac{1}{\tau}$$

(cf. Theorem XI-3-2). Therefore, (XII.5.5) and (XII.5.6) imply that

$$\text{(XII.5.8)} \qquad \begin{aligned} |\exp[c\epsilon^{-k}]|\,|\vec{f}(x,\vec{0},\epsilon)| &= |\vec{f}(x,\vec{0},\epsilon)|\exp[c|\epsilon|^{-k}\cos(k(\arg\epsilon))]\\ &\leq K\,\exp[-c|\epsilon|^{-k}] \end{aligned}$$

for $(x,\epsilon) \in \Delta(\delta) \times \mathcal{S}(r,\alpha)$. Note also that

$$\text{(XII.5.9)} \qquad \cos(k(\arg\epsilon)) \;\geq\; \cos(k\alpha) \;>\; 0 \quad \text{for}\quad \epsilon \in \mathcal{S}(r,\alpha) \quad \text{if}\quad k\alpha < \frac{\pi}{2}.$$

Let us change $\vec{y}$ in (XII.5.1) by

$$\text{(XII.5.10)} \qquad \vec{y} \;=\; \exp[-c\epsilon^{-k}]\vec{u}.$$

Then, (XII.5.1) is reduced to

$$\text{(XII.5.11)} \qquad \epsilon^{\sigma}\frac{d\vec{u}}{dx} \;=\; \exp[c\epsilon^{-k}]\vec{f}(x,\exp[-c\epsilon^{-k}]\vec{u},\epsilon).$$

Set

$$\vec{f}(x,\vec{y},\epsilon) \;=\; \vec{f}(x,\vec{0},\epsilon) \;+\; A(x,\epsilon)\vec{y} \;+\; \sum_{|\wp|\geq 2} \vec{y}^{\wp}\vec{f}_{\wp}(x,\epsilon).$$

Then,

$$\exp[c\epsilon^{-k}]\vec{f}(x,\exp[-c\epsilon^{-k}]\vec{u},\epsilon)$$
$$= \exp[c\epsilon^{-k}]\vec{f}(x,\vec{0},\epsilon) \;+\; A(x,\epsilon)\vec{u} \;+\; \sum_{|\wp|\geq 2} \exp[c(1-|\wp|)\epsilon^{-k}]\vec{u}^{\wp}\,\vec{f}_{\wp}(x,\epsilon).$$

Using a method similar to the proof of Theorem XII-1-2, we construct a bounded solution $\vec{u} = \vec{\psi}(x,\epsilon)$ of (XII.5.11). Therefore, system (XII.5.1) has a solution of the form $\vec{y} = \vec{\phi}(x,\epsilon) = \exp[-c\epsilon^{-k}]\vec{\psi}(x,\epsilon)$. This completes the proof of Lemma XII.5.1. $\square$

The main result of this section is the following theorem, which was originally conjectured by J.-P. Ramis.

Theorem XII-5-2. *Assume that*

(i) $\vec{f}(x, \vec{y}, \epsilon)$ *is holomorphic in* $(x, \vec{y}, \epsilon)$ *on a domain* $\Delta(\delta_0) \times \Omega(\rho_0) \times S(r_0, \alpha_0)$, *where* δ_0, ρ_0, r_0, *and* α_0 *are positive numbers*,

(ii) $\vec{f}(x, \vec{y}, \epsilon)$ *admits an asymptotic expansion* $\vec{F}(x, \vec{y}, \epsilon)$ *of Gevrey order* s *as* $\epsilon \to 0$ *in* $S(r_0, \alpha_0)$ *uniformly in* $(x, \vec{y}) \in \Delta(\delta_0) \times \Omega(\rho_0)$, *where* s *is a non-negative number*,

(iii) *the matrix* $A_0(x)$ *given by (XII.5.4) is invertible on* $\Delta(\delta_0)$,

(iv) *it holds that*

$$\text{(XII.5.12)} \qquad \lim_{\epsilon \to 0} \vec{f}(x, \vec{0}, \epsilon) = \vec{0} \qquad on \;\; \Delta(\delta_0).$$

Then,

(1) (XII.5.1) has a unique formal solution

$$\text{(XII.5.13)} \qquad \vec{p}(x, \epsilon) = \sum_{\ell=1}^{+\infty} \epsilon^{\ell} \vec{p}_{\ell}(x)$$

with coefficients $\vec{p}_{\ell}(x)$ *which are holomorphic on* $\Delta(\delta_0)$,

(2) there exist three positive numbers δ, r, *and* α *such that (XII.5.1) has an actual solution* $\vec{\phi}(x, \epsilon)$ *which is holomorphic in* $(x, \epsilon) \in \Delta(\delta) \times S(r, \alpha)$ *and that* $\vec{\phi}(x, \epsilon)$ *admits the formal solution* $\vec{p}(x, \epsilon)$ *as its asymptotic expansion of Gevrey order* $\max\left\{\dfrac{1}{\sigma}, s\right\}$ *as* $\epsilon \to 0$ *in* $S(r, \alpha)$ *uniformly in* $x \in \Delta(\delta)$.

Proof.

If a positive number $\tilde{\alpha}$ is sufficiently small, for every real number θ, there exists a $\mathbb{C}^n$-valued function $\vec{f}_{\theta}(x, \vec{y}, \epsilon)$ such that

(a) $\vec{f}_{\theta}(x, \vec{y}, \epsilon)$ is holomorphic in $(x, \vec{y}, \epsilon)$ on a domain $\Delta(\delta_0) \times \Omega(\rho_0) \times S_{\theta}(r_0, \tilde{\alpha})$, where

$$\text{(XII.5.14)} \qquad S_{\theta}(r_0, \tilde{\alpha}) = \{\epsilon : |\arg \epsilon - \theta| < \tilde{\alpha}, \; 0 < |\epsilon| < r_0\},$$

(b) $\vec{f}_{\theta}(x, \vec{y}, \epsilon)$ admits an asymptotic expansion $\vec{F}(x, \vec{y}, \epsilon)$ of Gevrey order s as $\epsilon \to 0$ in $S_{\theta}(r_0, \tilde{\alpha})$ uniformly in $(x, \vec{y}) \in \Delta(\delta_0) \times \Omega(\rho_0)$, where s is the non-negative number given in Theorem XII.5.2.

Such a function $\vec{f}_{\theta}(x, \vec{y}, \epsilon)$ exists if $\tilde{\alpha}$ is sufficiently small (cf. Corollary XI-2-5). In particular, set

$$\text{(XII.5.15)} \qquad \vec{f}_0(x, \vec{y}, \epsilon) = \vec{f}(x, \vec{y}, \epsilon).$$

Let $\vec{y} = \vec{\phi}_{\theta}(x, \epsilon)$ be a solution of the system

$$\text{(XII.5.16)} \qquad \epsilon^{\sigma} \frac{d\vec{y}}{dx} = \vec{f}_{\theta}(x, \vec{y}, \epsilon)$$

such that $\vec{\phi}_{\theta}(x, \epsilon)$ is holomorphic and bounded in $(x, \epsilon) \in \Delta(\delta_1) \times S_{\theta}(r_1, \alpha_1)$ for suitable positive numbers δ_1, r_1, and α_1. Using Theorem XII-1-2, it can be shown that such a solution of (XII.5.16) exists.

Suppose that $\mathcal{S}_{\theta_1}(r_1,\alpha_1) \cap \mathcal{S}_{\theta_2}(r_1,\alpha_1) \neq \emptyset$. Set

$$(\text{XII.5.17}) \qquad \vec{\psi}(x,\epsilon) \;=\; \vec{\phi}_{\theta_1}(x,\epsilon) \;-\; \vec{\phi}_{\theta_2}(x,\epsilon)$$

for $(x,\epsilon) \in \Delta(\delta_1) \times \{\mathcal{S}_{\theta_1}(r_1,\alpha_1) \cap \mathcal{S}_{\theta_2}(r_1,\alpha_1)\}$. Then, $\vec{\psi}$ satisfies the following linear system

$$(\text{XII.5.18}) \qquad \epsilon^\sigma \frac{d\vec{\psi}}{dx} \;=\; B(x,\epsilon)\vec{\psi} \;+\; \vec{b}(x,\epsilon),$$

where

$$(\text{XII.5.19}) \qquad \begin{cases} B(x,\epsilon)\vec{\psi}(x,\epsilon) \;=\; \vec{f}_{\theta_1}(x,\vec{\phi}_{\theta_1}(x,\epsilon),\epsilon) \;-\; \vec{f}_{\theta_1}(x,\vec{\phi}_{\theta_2}(x,\epsilon),\epsilon), \\ \vec{b}(x,\epsilon) \;=\; \vec{f}_{\theta_1}(x,\vec{\phi}_{\theta_2}(x,\epsilon),\epsilon) \;-\; \vec{f}_{\theta_2}(x,\vec{\phi}_{\theta_2}(x,\epsilon),\epsilon). \end{cases}$$

Note that

$$(\text{XII.5.20}) \qquad B(x,\epsilon) \;=\; \int_0^1 \frac{\partial \vec{f}_{\theta_1}}{\partial \vec{y}}(x,t\vec{\phi}_{\theta_1}(x,\epsilon) + (1-t)\vec{\phi}_{\theta_2}(x,\epsilon),\epsilon)dt$$

and that, if $s > 0$, then

$$(\text{XII.5.21}) \qquad |\vec{b}(x,\epsilon)| \leq \mu\exp\{-\nu|\epsilon|^{-k}\}$$

for $(x,\epsilon) \in \Delta(\delta_1) \times \{\mathcal{S}_{\theta_1}(r_1,\alpha_1) \cap \mathcal{S}_{\theta_2}(r_1,\alpha_1)\}$, where

$$(\text{XII.5.22}) \qquad k \;=\; \frac{1}{s}$$

and μ and ν are suitable positive numbers (cf. Theorem XI-3-2). From (XII.5.20), it follows that

$$(\text{XII.5.23}) \qquad \lim_{\epsilon \to 0} B(x,\epsilon) \;=\; A_0(x),$$

where the matrix $A_0(x)$ is given by (XII.5.4). If $s = 0$, the power series $\vec{F}$ is convergent in ϵ (cf. Exercise XI-6). Hence, $\vec{b}(x,\epsilon) = \vec{0}$.

If

$$(\text{XII.5.24}) \qquad \mathcal{S}_{\theta}(r_2,\alpha_2) \;\subset\; \mathcal{S}_{\theta_1}(r_1,\alpha_1) \cap \mathcal{S}_{\theta_2}(r_1,\alpha_1)$$

and if positive numbers δ_2, r_2, and α_2 are sufficiently small, applying Lemma XII-5-1 to (XII.5.18), the functions $\vec{\psi}(x,\epsilon)$ can be written in the form

$$(\text{XII.5.25}) \qquad \vec{\psi}(x,\epsilon) \;=\; \Phi(x,\epsilon)\vec{\gamma}(\epsilon) \;+\; \vec{w}(x,\epsilon),$$

where

$$(\text{XII.5.26}) \qquad |\vec{w}(x,\epsilon)| \begin{cases} = 0 & \text{if} \quad s = 0, \\ \leq K\exp\{-c|\epsilon|^{-k}\} & \text{if} \quad s > 0 \end{cases}$$

for $(x, \epsilon) \in \Delta(\delta_2) \times \mathcal{S}_\theta(r_2, \alpha_2)$, K and c are suitable positive numbers, $\Phi(x, \epsilon)$ is a fundamental matrix solution of the homogeneous system

$$(\text{XII.5.27}) \qquad \epsilon^\sigma \frac{d\vec{v}}{dx} = B(x, \epsilon)\vec{v},$$

and $\vec{\gamma}(\epsilon)$ is a $\mathbb{C}^n$-valued function of ϵ independent of x.

Assuming that $A_0(0)$ has distinct eigenvalues $\lambda_1, \lambda_2, \ldots, \lambda_\ell$ with multiplicities $n_1, n_2, \ldots, n_\ell$, respectively $(n_1 + n_2 + \cdots + n_\ell = n)$ and that $A_0(0)$ is in a block-diagonal form (XII.4.6) with (XII.4.7), where N_j $(j = 1, 2, \ldots, \ell)$ are $n_j \times n_j$ nilpotent matrices, respectively, apply Theorem XII-4-1 to block-diagonalize (XII.5.27). More precisely speaking, there exist three positive numbers δ_3, r_3, α_3, and an $n \times n$ matrix $P(x, \epsilon)$ which is holomorphic in $(x, \epsilon) \in \Delta(\delta_3) \times \mathcal{S}_\theta(r_3, \alpha_3)$ such that

(i) $P(x, \epsilon)$ admits an asymptotic expansion (XII.4.9) in the sense of Poincaré as $\epsilon \to 0$ in $\mathcal{S}(r_3, \alpha_3)$ uniformly in $x \in \Delta(\delta_3)$,

(ii) the transformation

$$(\text{XII.5.28}) \qquad \vec{v} = P(x, \epsilon)\vec{z}$$

reduces (XII.5.27) to

$$(\text{XII.5.29}) \qquad \epsilon^\sigma \frac{d\vec{z}}{dx} = E(x, \epsilon)\vec{z},$$

where $E(x, \epsilon)$ is in the block-diagonal form

$$(\text{XII.5.30}) \qquad E(x, \epsilon) = \mathrm{diag}\left[E_1(x, \epsilon), E_2(x, \epsilon), \ldots, E_\ell(x, \epsilon)\right],$$

with an $n_j \times n_j$ matrix $E_j(x, \epsilon)$ for each $j = 1, 2, \ldots, \ell$, and $E(x, \epsilon)$ admits an asymptotic expansion $\sum_{\nu=0}^\infty \epsilon^\nu E_\nu(x)$ in the sense of Poincaré as $\epsilon \to 0$ in $\mathcal{S}_\theta(r_3, \alpha_3)$ uniformly in $x \in \Delta(\delta_3)$, where $E_0(0) = A_0(0)$.

Let us construct a rhombus $\mathcal{D}(\delta_4)$ with vertices $x^{(1)}$, $x^{(2)}$, $x^{(3)}$, and $x^{(4)}$ as it is defined in §XII-2, in such a way that

(a) $\mathcal{D}(\delta_4) \subset \Delta(\delta_3)$,

(b) it holds that

$$(\text{XII.5.31}) \qquad \begin{cases} \Re\left(\dfrac{\lambda_j(x - x^{(1)})}{\epsilon^\sigma}\right) < -\dfrac{\mu|\lambda_j||x - x^{(1)}|}{|\epsilon|^\sigma} & \text{for } j = 1, \ldots, m', \\[2ex] \Re\left(\dfrac{\lambda_j(x - x^{(3)})}{\epsilon^\sigma}\right) < -\dfrac{\mu|\lambda_j||x - x^{(3)}|}{|\epsilon|^\sigma} & \text{for } j = m' + 1, \ldots, \ell \end{cases}$$

on the domain $\mathcal{D}(\delta_4) \times \mathcal{S}_\theta(r_4, \alpha_4)$, where δ_4, r_4, and α_4 are suitable positive numbers.

Write $\vec{z}$ in the form $\vec{z} = \begin{bmatrix} \vec{z}_1 \\ \vdots \\ \vec{z}_\ell \end{bmatrix}$, where $\vec{z}_j \in \mathbb{C}^{n_j}$ $(j = 1, 2, \ldots, \ell)$. Then, (XII.5.29) can be written in the form

$$(\text{XII.5.32}) \qquad \epsilon^\sigma \frac{d\vec{z}_j}{dx} = E_j(x, \epsilon)\vec{z}_j \qquad (j = 1, 2 \ldots, \ell).$$

For each j, let $\Phi_j(x, \epsilon)$ be a fundamental matrix solution of $\epsilon^\sigma \dfrac{d\vec{z}_j}{dx} = E_j(x, \epsilon)\vec{z}_j$. Then, the general solution of (XII.5.29) is given by

$$\text{(XII.5.33)} \qquad \vec{z}_j(x, \epsilon) \; = \; \Phi_j(x, \epsilon)\Phi_j(x_j, \epsilon)^{-1}\vec{z}_j(x_j, \epsilon),$$

where

$$\text{(XII.5.34)} \qquad x_j \; = \; \begin{cases} x_1 & \text{for} \quad j = 1, \ldots, m', \\ x_3 & \text{for} \quad j = m'+1, \ldots, \ell. \end{cases}$$

In this way, we can prove that $\vec{\psi}(x, \epsilon) - \vec{w}(x, \epsilon)$ is flat of Gevrey order $\dfrac{1}{\sigma}$ as $\epsilon \to 0$ in $\mathcal{S}_{\theta_1}(r_1, \alpha_1) \cap \mathcal{S}_{\theta_2}(r_1, \alpha_1)$ uniformly for $x \in \Delta(\delta)$ if a positive number δ is sufficiently small (cf. [Si6; Proof of Lemma 1 on pp. 377-379]). Therefore, by using Theorem XI-2-3, we can complete the proof of Theorem XII-5-2. $\square$

Materials of this section are also found in [Si22].

XII-6. Analytic simplification in a parameter

For a system of linear differential equations of the form

$$\text{(E)} \qquad \frac{d\vec{y}}{dx} = x^k A\!\left(\frac{1}{x}\right)\vec{y},$$

where k is an integer, $\vec{y} \in \mathbb{C}^n$, and $A(t)$ is an $n \times n$ matrix whose entries are holomorphic in a domain

$$\text{(D)} \qquad \left\{t : |t| < \frac{1}{R_0}\right\}, \quad \left(t = \frac{1}{x}; \; R_0 : \text{positive constant}\right),$$

G. D. Birkhoff proved the following theorem.

Theorem XII-6-1 ([Bi]). *There exists an $n \times n$ matrix $P(t)$ such that the entries of $P(t)$ and $P(t)^{-1}$ are holomorphic in domain (D) and that the transformation*

$$\text{(T)} \qquad \vec{y} = P\!\left(\frac{1}{x}\right)\vec{v}$$

reduces system (E) to a system of the form

$$\text{(E}_0\text{)} \qquad \frac{d\vec{v}}{dx} = x^k B\!\left(\frac{1}{x}\right)\vec{v},$$

where $B(t)$ is an $n \times n$ matrix whose entries are polynomials in t. Moreover, the matrix $P(t)$ can be chosen so that $x = 0$ is, at worst, a regular singular point of (E).

Observe that if we set $t = x^{-1}$, (E) becomes

$$\text{(E}_1\text{)} \qquad \frac{d\vec{y}}{dt} = -t^{-k-2}A(t)\vec{y}.$$

If $k \leq -2$, (E_1) has the coefficient holomorphic in (D). Hence, there is a fundamental matrix solution $V(t)$ of (E_1) whose entries are holomorphic in (D). Then, the transformation $\vec{v} = V\left(\dfrac{1}{x}\right)\vec{z}$ reduces (E) to the system $\dfrac{d\vec{z}}{dx} = \vec{0}$. Therefore, the main claim of Theorem XII-6-1 concerns the case when $k \geq -1$. In this theorem of G. D. Birkhoff, the entries of the matrix $B(t)$ are polynomials in t. However, even though we can choose $P(t)$ so that $x = 0$ is, at worst, a regular singular point of (E_0), the degree of $B(t)$ with respect to t may be very large. In order that the degree of $B(t)$ with respect to t is at most $k + 1$ so that $x = 0$ is a singularity of the first kind of system (E_0), we must impose a certain condition on $A(t)$ (cf. Exercises XII-8, XII-9, and XII-10). For interesting discussions on this matter, see, for example, [Tu2], [JLP], [Bal1], and [Bal2]. A complete proof of Theorem XII-6-1 is found, for example, in [Si17, Chapter 3].

In this section, we prove a result similar to Theorem XII-6-1 for a system of linear differential equations

$$(XII.6.1) \qquad \epsilon^{\sigma}\frac{d\vec{y}}{dx} = A(x,\epsilon)\vec{y},$$

under the assumption that σ is a positive integer, $\vec{y} \in \mathbb{C}^n$, and $A(x,\epsilon)$ is an $n \times n$ matrix whose entries are holomorphic with respect to complex variables (x,ϵ) in a domain

$$(XII.6.2) \qquad x \in \mathcal{D}_0, \qquad |\epsilon| < \delta_0,$$

where $\mathcal{D}_0$ is a domain in the x-plane containing $x = 0$, and δ_0 is a positive constant. Let

$$(XII.6.3) \qquad A(x,\epsilon) = \sum_{k=0}^{\infty} \epsilon^k A_k(x)$$

be the expansion of $A(x,\epsilon)$ in powers of ϵ, where the entries of coefficients $A_k(x)$ are holomorphic in $\mathcal{D}_0$. We assume that the series $\sum_{k=0}^{\infty} \delta_0^k |A_k(x)|$ is convergent uniformly in $\mathcal{D}_0$. The main result of this section is the following theorem, which was originally proved in [Hs1].

Theorem XII-6-2 ([Hs1]). *For each non-negative integer m, there is an $n \times n$ matrix $P(x,\epsilon)$ satisfying the following conditions:*
(i) the entries of $P(x,\epsilon)$ are holomorphic in (x,ϵ) in a domain

$$(XII.6.4) \qquad x \in \mathcal{D}_1, \qquad |\epsilon| < \delta_0,$$

where $\mathcal{D}_1$ is a subdomain of $\mathcal{D}_0$ containing $x = 0$,
(ii) $P(x,0) = I_n$ for $x \in \mathcal{D}_1$ and $P(0,\epsilon) = I_n$ for $|\epsilon| < \delta_0$,
(iii) the system (XII.6.1) is reduced to a system of the form

$$(XII.6.5) \qquad \epsilon^{\sigma}\frac{d\vec{u}}{dx} = \left\{\sum_{k=0}^{m} \epsilon^k A_k(x) + \epsilon^{m+1}\sum_{k=0}^{\sigma-1} \epsilon^k B_k(x)\right\}\vec{u}$$

by the transformation

$$(XII.6.6) \qquad\qquad \vec{y} = P(x, \epsilon)\vec{u},$$

where $B_k(x)$ $(k = 1, 2, \ldots, \sigma - 1)$ are $n \times n$ matrices whose entries are holomorphic for $x \in \mathcal{D}_1$.

Remark XII-6-3. In case when $\sigma = 1$, (XII.6.5) becomes

$$\epsilon \frac{d\vec{u}}{dx} = \left\{ \sum_{k=0}^{m} \epsilon^k A_k(x) + \epsilon^{m+1} B_0(x) \right\} \vec{u}.$$

In particular, if $\sigma = 1$ and $m = 0$, (XII.6.5) has the form

$$\epsilon \frac{d\vec{u}}{dx} = \left\{ A_0(x) + \epsilon B_0(x) \right\} \vec{u}.$$

It should be noticed that in Theorem XII-6-2, if we put $m = 0$, then the right-hand side of (XII.6.5) is a polynomial in ϵ of degree at most σ without any restrictions on $A(x, \epsilon)$. We prove Theorem XII-6-2 by a direct method based on the theory of ordinary differential equations in a Banach space. (See also [Si8].)

Proof of Theorem XII-6-2.

The proof is given in three steps.

Step 1. Put

$$(XII.6.7) \qquad P(x, \epsilon) = I_n + \epsilon^{m+1} \sum_{k=0}^{\infty} \epsilon^k P_k(x) \quad \text{and} \quad B(x, \epsilon) = \sum_{k=0}^{m+\sigma} \epsilon^k \hat{B}_k(x),$$

where

$$(XII.6.8) \qquad \hat{B}_k(x) = \begin{cases} A_k(x) & (k = 0, 1, \ldots, m), \\ B_{k-m-1}(x) & (k = m+1, m+2, \ldots, m+\sigma). \end{cases}$$

From (XII.6.1), (XII.6.5), and (XII.6.6), it follows that the matrices $P(x, \epsilon)$ and $B(x, \epsilon)$ must satisfy the equation

$$(XII.6.9) \qquad\qquad \epsilon^\sigma \frac{dP}{dx} = A(x, \epsilon)P - PB(x, \epsilon).$$

From (XII.6.3), (XII.6.7), (XII.6.8), and (XII.6.9), we obtain

$$(XII.6.10) \qquad O = A_{m+1+k}(x) - \hat{B}_{m+1+k}(x) + \sum_{h=0}^{k} \{A_{k-h}(x)P_h(x) - P_h(x)\hat{B}_{k-h}(x)\}$$

$$(k = 0, 1, \ldots, \sigma - 1)$$

and

$$\frac{dP_k(x)}{dx} = A_{m+1+\sigma+k}(x) + \sum_{h=0}^{\sigma+k} A_{\sigma+k-h}(x)P_h(x)$$

(XII.6.11)

$$- \sum_{h=k-m}^{\sigma+k} P_h(x)\hat{B}_{\sigma+k-h}(x) \qquad (k = 0, 1, 2, \dots),$$

where

(XII.6.12) $$P_h(x) \equiv O \qquad \text{if} \qquad h < 0.$$

It should be noted that the formal power series P and B that satisfy the equation (XII.6.9) are not convergent in general. In order to construct P as a convergent power series in ϵ, we must choose a suitable B. To do this, first solve equation (XII.6.10) for $\hat{B}_{m+1+k}(x)$ to derive

(XII.6.13)
$$\hat{B}_{m+1+k}(x) = A_{m+1+k}(x) + H_{m+1+k}(x; P_0, P_1, \dots, P_k)$$
$$(k = 0, 1, \dots, \sigma - 1),$$

where H_j are defined by
(XII.6.14)

$$\left\{ \begin{array}{l} H_j = O, \qquad (k = 0, 1, \dots, m), \\[2mm] H_{m+1+k}(x; P_0, P_1, \dots, P_k) = \sum_{h=0}^{k}\{A_{k-h}(x)P_h - P_h A_{k-h}(x)\} - \sum_{h=0}^{k} P_h H_{k-h}, \\[2mm] \qquad\qquad\qquad\qquad\qquad (k = 0, 1, \dots, \sigma - 1). \end{array} \right.$$

Denote by $\mathcal{P}$ an infinite-dimensional vector $\{P_k : k = 0, 1, 2, \dots\}$. Then, by substituting (XII.6.13) into (XII.6.11), we obtain

(XII.6.15) $$\frac{dP_k(x)}{dx} = f_k(x; \mathcal{P}) \qquad (k = 0, 1, 2, \dots),$$

where

$$f_k(x; \mathcal{P}) = A_{m+1+\sigma+k}(x) + \sum_{h=0}^{\sigma+k} A_{\sigma+k-h}(x)P_h - \sum_{h=k-m}^{\sigma+k} P_h A_{\sigma+k-h}(x)$$

(XII.6.16)

$$- \sum_{h=k-m}^{\sigma+k} P_h H_{\sigma+k-h}(x; \mathcal{P}) \qquad (k = 0, 1, 2, \dots).$$

Denote by $\mathcal{F}(x; \mathcal{P})$ the infinite-dimensional vector $\{f_k(x; \mathcal{P}) : k = 0, 1, 2, \dots\}$. Then, equation (XII.6.15) can be written in the form

(XII.6.17) $$\frac{d\mathcal{P}}{dx} = \mathcal{F}(x; \mathcal{P}).$$

Solve this differential equation in a suitable Banach space with the initial condition $\mathcal{P}(0) = O$. Actually, we solve the integral equation

$$(\text{XII.6.18}) \qquad \mathcal{P}(x) = \int_0^x \mathcal{F}(\xi; \mathcal{P}(\xi))d\xi.$$

This is equivalent to the system

$$(\text{XII.6.19}) \qquad P_k(x) = \int_0^x f_k(\xi; \mathcal{P}(\xi))d\xi \qquad (k = 0, 1, 2, \dots).$$

If $\mathcal{P}$ is determined, then the matrices B_k are determined by (XII.6.13) and (XII.6.14).

Step 2. We still assume that $A(x, \epsilon)$ is holomorphic in domain (XII.6.2). Denote by $\mathcal{B}$ the set of all infinite-dimensional vectors $\mathcal{P} = \{P_k : k = 0, 1, 2, \dots\}$ such that

(i) P_k are $n \times n$ matrices of complex entries,

(ii) $\displaystyle\sum_{k=0}^{\infty}\delta_0^k\|P_k\| < \infty$, where $\|P_k\|$ is the sum of the absolute values of entries

of P_k.

For each $\mathcal{P}$, define a norm $\|\mathcal{P}\|$ by

$$(\text{XII.6.20}) \qquad \|\mathcal{P}\| = \sum_{k=0}^{\infty} \delta_0^k \, \|P_k\|.$$

Then, we can regard $\mathcal{B}$ as a Banach space over the field of complex numbers.

Now, we can establish the following lemma.

Lemma XII-6-4. *Let* $\mathcal{F}(x; \mathcal{P})$ *be the infinite-dimensional vector whose entries* $f_k(x, \mathcal{P})$ *are given by (XII.6.16). Then, for each positive numbers R, there exist two positive numbers $G(R)$ and $K(R)$ such that*

$$(\text{XII.6.21}) \qquad \|\mathcal{F}(x; \mathcal{P})\| \le G(R) \qquad for \quad \|\mathcal{P}\| \le R$$

and

$$(\text{XII.6.22}) \quad \|\mathcal{F}(x; \mathcal{P}) - \mathcal{F}(x; \hat{\mathcal{P}})\| \le K(R)\|\mathcal{P} - \hat{\mathcal{P}}\| \quad for \quad \|\mathcal{P}\| \le R, \ \|\hat{\mathcal{P}}\| \le R.$$

In order to prove this lemma, consider a formal power series of ϵ which is defined by

$$(\text{XII.6.23}) \qquad \mathcal{F}(x, \mathcal{P}, \epsilon) = \sum_{k=0}^{\infty} \epsilon^k f_k(x; \mathcal{P}).$$

Then, using (XII.6.16), we obtain
(XII.6.24)

$$\mathcal{F}(x, \mathcal{P}, \epsilon) = \sum_{k=0}^{\infty} \epsilon^k A_{m+1+\sigma+k}(x) + \frac{1}{\epsilon^\sigma}\left\{\left(\sum_{k=0}^{\infty} \epsilon^k A_k(x)\right)\left(\sum_{k=0}^{\infty} \epsilon^k P_k\right)\right.$$

$$\left. - \sum_{k=0}^{\sigma-1} \epsilon^k \sum_{h=0}^{k} A_{k-h}P_h\right\} - \frac{1}{\epsilon^\sigma}\left\{\left(\sum_{k=0}^{\infty} \epsilon^k P_k\right)\left(\sum_{k=0}^{m+\sigma} \epsilon^k [A_k(x) + H_k(x; \mathcal{P})]\right)\right.$$

$$\left. - \sum_{k=0}^{\sigma-1} \epsilon^k \sum_{h=0}^{k} P_h[A_{k-h}(x) + H_{k-h}(x; \mathcal{P})]\right\}.$$

Hence, Lemma XII-6-4 follows immediately.

Step 3. We construct the matrix $P(x, \epsilon)$, solving the integral equation (XII.6.18) by the method of successive approximations similar to that given in Chapter I. By virtue of Lemma XII-6-4, we can construct a solution $\mathcal{P}(x)$ in a subdomain D_1 of D_0 containing $x = 0$ in its interior. Since (XII.6.18) is equivalent to differential equation (XII.6.15) with the initial condition $\mathcal{P}(0) = O$, the solution $\mathcal{P}(x)$ gives the desired $P(x, \epsilon)$. The matrix $B(x, \epsilon)$ is given by (XII.6.13) and (XII.6.14). $\square$

EXERCISES XII

XII-1. Find a formal power series solution $\vec{y} = \displaystyle\sum_{m=0}^{\infty} \epsilon^m \vec{a}_m(x)$ of the system of differential equations

$$\epsilon \frac{d\vec{y}}{dx} = A\vec{y} + \sum_{m=0}^{\infty} \epsilon^m \vec{b}_m(x),$$

where $\vec{y} \in \mathbb{C}^n$, A is an invertible constant $n \times n$ matrix, and $\vec{a}_m(x)$ and $\vec{b}_m(x)$ are $\mathbb{C}^n$-valued functions whose entries are holomorphic in x in a neighborhood of $x = 0$.

XII-2. Using Theorem XII-4-1, diagonalize the system

$$\epsilon \frac{d\vec{y}}{dx} = \begin{bmatrix} 0 & 1+x \\ 1-x & \epsilon x \end{bmatrix} \vec{y}, \qquad \text{where} \qquad \vec{y} = \begin{bmatrix} y_1 \\ y_2 \end{bmatrix}.$$

XII-3. Using Theorem XII-4-1, find two linearly independent formal solutions of each of the following two differential equations which do not involve any fractional powers of ϵ.

$$(1) \; \epsilon^2 \frac{d^2 y}{dx^2} + y = \epsilon q(x)y, \qquad (2) \; \epsilon \frac{d^2 y}{dx^2} + y = \epsilon q(x)y,$$

where $q(x)$ is holomorphic in x for small $|x|$.

Hint. If we set $\beta^2 = \epsilon$, differential equation (2) has two linearly independent solutions $e^{ix/\beta}\phi(x, \beta)$ and $e^{-ix/\beta}\phi(x, -\beta)$. The two solutions

$$\begin{cases} \phi_1(x, \epsilon) = e^{ix/\beta}\phi(x, \beta) + e^{-ix/\beta}\phi(x, -\beta), \\ \phi_2(x, \epsilon) = \beta\left[e^{ix/\beta}\phi(x, \beta) - e^{-ix/\beta}\phi(x, -\beta)\right] \end{cases}$$

do not involve any fractional powers of ϵ.

XII-4. Let x be a complex independent variable, $\vec{y} \in \mathbb{C}^n$, $\vec{z} \in \mathbb{C}^m$, ϵ be a complex parameter, $A(x, \vec{y}, \vec{z}, \epsilon)$ be an $n \times n$ matrix whose entries are holomorphic with respect to $(x, \vec{y}, \vec{z}, \epsilon)$ in a domain $\mathcal{D}_0 = \{(x, \vec{y}, \vec{z}, \epsilon) : |x| < r_0, |\vec{y}| < \rho_1, |\vec{z}| < \rho_2, 0 < |\epsilon| < \alpha_0, |\arg \epsilon| < \beta_0\}$, $\vec{f}(x, \vec{y}, \vec{z}, \epsilon)$ be a $\mathbb{C}^m$-valued function whose entries are holomorphic with respect to $(x, \vec{y}, \vec{z}, \epsilon)$ in $\mathcal{D}_0$, and $\vec{g}(x, \vec{z}, \epsilon)$ be a $\mathbb{C}^n$-valued function whose entries are holomorphic with respect to $(x, \vec{z}, \epsilon)$ in the domain $\mathcal{U}_0 = \{(x, \vec{z}, \epsilon) : |x| < r_0, |\vec{z}| < \rho_2, 0 < |\epsilon| < \alpha_0, |\arg \epsilon| < \beta_0\}$. Assume that the entries of the matrix

$A(x, \vec{y}, \vec{z}, \epsilon)$, the functions $\vec{f}(x, \vec{y}, \vec{z}, \epsilon)$, and $\vec{g}(x, \vec{z}, \epsilon)$ admit asymptotic expansions as $\epsilon \to 0$ in the sector $S_0 = \{\epsilon : 0 < |\epsilon| < \alpha_0, |\arg \epsilon| < \beta_0\}$ uniformly in $(x, \vec{y}, \vec{z})$ in the domain $\Delta_0 = \{(x, \vec{y}, \vec{z}) : |x| < r_0, |\vec{y}| < \rho_1, |\vec{z}| < \rho_2\}$ or $(x, \vec{z})$ in $\mathcal{V}_0 = \{(x, \vec{z}) : |x| < r_0, |\vec{z}| < \rho_2\}$. The coefficients of those expansions are holomorphic with respect to $(x, \vec{y}, \vec{z})$ in Δ_0 or $(x, \vec{z})$ in $\mathcal{V}_0$. Assume also that $\det A(0, \vec{0}, \vec{0}, 0) \neq 0$. Show that the system

$$\epsilon \frac{d\vec{y}}{dx} = A(x, \vec{y}, \vec{z}, \epsilon)\vec{y} + \epsilon \vec{g}(x, \vec{z}, \epsilon), \qquad \frac{d\vec{z}}{dx} = \epsilon \vec{f}(x, \vec{y}, \vec{z}, \epsilon)$$

has one and only one solution $(\vec{y}, \vec{z}) = (\vec{\phi}(x, \epsilon), \vec{\psi}(x, \epsilon))$ satisfying the following conditions:

(a) the entries of functions $\vec{\phi}(x, \epsilon)$ and $\vec{\psi}(x, \epsilon)$ are holomorphic with respect to (x, ϵ) in a domain $\{(x, \epsilon) : |x| < r, 0 < |\epsilon| < \alpha, |\arg \epsilon| < \beta\}$ for some positive numbers r, α, and β such that $r < r_0$, $\alpha < \alpha_0$ and $\beta < \beta_0$,

(b) the entries of functions $\vec{\phi}(x, \epsilon)$ and $\vec{\psi}(x, \epsilon)$ admit asymptotic expansions in powers of ϵ as $\epsilon \to 0$ in the sector $\{\epsilon : 0 < |\epsilon| < \alpha, |\arg \epsilon| < \beta\}$ uniformly in x in the disk $\{x : |x| < r\}$, where coefficients of these expansions are holomorphic with respect to x in the disk $\{x : |x| < r\}$,

(c) $\vec{\psi}(0, \epsilon) = \vec{0}$ for $\epsilon \in \{\epsilon : 0 < |\epsilon| < \alpha, |\arg \epsilon| < \beta\}$.

Hint. Use a method similar to that of §§XII-2 and XII-3.

XII-5. Find the following limits: (1) $\displaystyle \lim_{\epsilon \to 0} \int_0^1 f(t) \sin\left(\frac{\alpha t}{\epsilon}\right) dt$ and (2) $\displaystyle \lim_{\epsilon \to 0} \int_0^1 f(t) \sin^2\left(\frac{\alpha t}{\epsilon}\right) dt$, where α is a nonzero real number and $f(t)$ is continuous and continuously differentiable on the interval $0 \leq t \leq 1$.

XII-6. Discuss the behavior of real-valued solutions of the system

$$\epsilon \frac{d\vec{y}}{dx} = \begin{bmatrix} \epsilon x & -1 + \epsilon x \\ \epsilon & \epsilon x \end{bmatrix} \vec{y} \quad \text{as} \quad \epsilon \to 0, \qquad \text{where} \quad \vec{y} = \begin{bmatrix} y_1 \\ y_2 \end{bmatrix}.$$

XII-7. Discuss the behavior of real-valued solutions of the following two differential equations as $\epsilon \to 0^+$: (1) $\epsilon^2 \dfrac{d^3 y}{dx^3} + \dfrac{dy}{dx} + xy = 0$, (2) $\epsilon^3 \dfrac{d^3 y}{dx^3} + (1 + x)y = 0$.

XII-8. Assume that

(i) the entries of a $\mathbb{C}^n$-valued function $\vec{f}(x, \vec{y}, \epsilon)$ are holomorphic with respect to $(x, \vec{y}, \epsilon)$ in a domain $\Delta_1(\delta_0) \times \Omega(\rho_0) \times \Delta_2(r_0)$, where δ_0, ρ_0, and r_0 are positive numbers and

$$\Delta_1(\delta_0) = \{ x \in \mathbb{C} : |x| < \delta_0 \}, \quad \Omega(\rho_0) = \{\vec{y} \in \mathbb{C}^n : |\vec{y}| < \rho_0\},$$
$$\Delta_2(r_0) = \{\epsilon \in \mathbb{C} : |\epsilon| < r_0\},$$

(ii) the matrix $A_0(x) = \dfrac{\partial \vec{f}}{\partial \vec{y}}(x, \vec{0}, 0)$ is invertible on $\Delta(\delta_0)$,

(iii) $\vec{f}(x, \vec{0}, 0) = \vec{0}$ on $\Delta(\delta_0)$,

(iv) σ is a positive integer.

Denote by $\mathcal{S}(r, \alpha, \theta)$ the sector $\{\epsilon \in \mathbb{C} : 0 < |\epsilon| < r, |\arg \epsilon - \theta| < \alpha\}$. Show that

(1) the system (S) $\epsilon^\sigma \dfrac{d\vec{y}}{dx} = \vec{f}(x, \vec{y}, \epsilon)$ has a unique formal solution $\vec{p}(x, \epsilon) = \displaystyle\sum_{\ell=1}^{+\infty} \epsilon^\ell \vec{p}_\ell(x)$, with coefficients $\vec{p}_\ell(x)$, which are holomorphic in $\Delta(\delta_0)$,

(2) for any real number θ, there exist three positive numbers δ, r, and α such that (S) has an actual solution $\vec{\phi}(x, \epsilon)$, which is holomorphic in $(x, \epsilon) \in \Delta(\delta) \times \mathcal{S}(r, \alpha, \theta)$, and that $\vec{\phi}(x, \epsilon)$ has the formal solution $\vec{p}(x, \epsilon)$ as its asymptotic expansion of Gevrey order $\dfrac{1}{\sigma}$ as $\epsilon \to 0$ in $\mathcal{S}(r, \alpha, \theta)$ uniformly in $x \in \Delta(\delta)$.

Hint. This is a special case of Theorem XII-5-2.

XII-9. Assume that $\vec{p}(x, \epsilon) = \displaystyle\sum_{m=1}^{+\infty} \epsilon^m \vec{p}_m(x)$ is a formal solution of a system (S)

$\epsilon^\sigma \dfrac{d\vec{y}}{dx} = \vec{f}(x, \vec{y}, \epsilon)$, where σ is a positive integer, $\vec{y} \in \mathbb{C}^n$, $\vec{f}(x, \vec{y}, \epsilon)$ is a $\mathbb{C}^n$-valued function whose entries are holomorphic in a neighborhood of $(x, \vec{y}, \epsilon) = (0, \vec{0}, 0)$, and the entries of $\vec{p}_m(x)\, (m = 0, 1, \dots)$ are holomorphic in a disk $|x| < r_0$, where r_0 is a positive nmber. Assume also that $\vec{p}(x, \epsilon) \in \{\mathbb{C}[[\epsilon]]_s\}^n$ uniformly for $|x| < r_0$ and that $0 \leq s < \dfrac{1}{\sigma}$. Show that if $\mathcal{S}$ is an open sector in the ϵ-plane with vertex at $\epsilon = 0$ and whose opening is smaller than $s\pi$, there exists two positive numbers r_1 and r_2 and a solution $\vec{\phi}(x, \epsilon)$ of (S) such that the entries of $\vec{\phi}(x, \epsilon)$ are holomorphic in (x, ϵ) for $|x| < r_1$, $|\epsilon| < r_2, \epsilon \in \mathcal{S}$, and that $\vec{\phi}(x, \epsilon)$ admits the formal solution $\vec{p}(x, \epsilon)$ as the asymptotic expansion of Gevrey order s as $\epsilon \to 0$ in $\mathcal{S} \cap \{|\epsilon| < r_2\}$ uniformly for $|x| < r_1$.

Note. No additional conditions on the linear part of $\vec{f}(x, \vec{y}, \epsilon)$ are assumed.

XII-10. Assume that $\vec{p}(x, \epsilon) = \displaystyle\sum_{m=0}^{+\infty} \epsilon^m \vec{p}_m(x)$ is a formal solution of a system

$\epsilon \dfrac{d\vec{y}}{dx} = \vec{f}(x, \vec{y}, \epsilon)$, where $\vec{y} \in \mathbb{C}^n$, $\vec{f}(x, \vec{y}, \epsilon)$ is a $\mathbb{C}^n$-valued function whose entries are holomorphic in a neighborhood of $(x, \vec{y}, \epsilon) = (0, \vec{0}, 0)$, $\vec{p}_m \in \mathbb{C}[[x]]^n \ (m = 0, 1, \dots)$, $\vec{p}_0(0) = \vec{0}$, and $\vec{p}(0, \epsilon)$ is convergent. Show that $\vec{p}(x, \epsilon)$ is convegent for $|x| < r$ and $|\epsilon| < \rho$ for some positive numbers r and ρ.

Hint. Regard $\vec{p}(\epsilon\tau, \epsilon)$ as the solution of the initial-value problem $\dfrac{d\vec{y}}{d\tau} = \vec{f}(\epsilon\tau, \vec{y}, \epsilon)$, $\vec{y}(0) = \vec{p}(0, \epsilon)$.

XII-11. Let t and ϵ be complex variables, $\vec{y} \in \mathbb{C}^n$, and the entries of a $\mathbb{C}^n$-valued function $\vec{f}(t, \vec{y}, \epsilon)$ be holomorphic with respect to $(t, \vec{y}, \epsilon)$ in a domain $\mathcal{D}_0 = \{(t, \vec{y}, \epsilon) : |\Im t| < d_0, -\infty < \Re t < \infty, |\vec{y}| < \rho_0, 0 < |\epsilon| < r_0, |\arg \epsilon| < \alpha_0\}$. Assume that $\vec{f}(t, \vec{y}, \epsilon) \simeq \displaystyle\sum_{m=0}^{\infty} \epsilon^m \vec{f}_m(t, \vec{y})$ as $\epsilon \to 0$ in the sector $\mathcal{S}_0 = \{\epsilon : 0 < |\epsilon| < r_0, |\arg \epsilon| < \alpha_0\}$ uniformly with respect to $(t, \vec{y})$ in the domain $\Delta_0 = \{(t, \vec{y}) : |\Im t| < d_0, -\infty < \Re t < +\infty, |\vec{y}| < \rho_0\}$, where the coefficients $\vec{f}_m(t, \vec{y})$ are holomorphic in

Δ_0. Assume also that $\vec{f}(t, \vec{y}, \epsilon)$ and the coefficients $\vec{f}_m(t, \vec{y})$ are periodic in t of period 1. Show that the differential equation $\dfrac{d\vec{y}}{dt} = 2\pi i \vec{y} + \epsilon[\vec{y} + \epsilon \vec{f}(t, \vec{y}, \epsilon)]$ has a periodic solution $\vec{y} = \epsilon \vec{\phi}(t, \epsilon)$ of period 1 such that

(a) the entries of $\vec{\phi}(t, \epsilon)$ are holomorphic with respect to (t, ϵ) in a domain $\mathcal{D} = \{(t, \epsilon) : |\Im t| < d, -\infty < \Re t < +\infty, 0 < |\epsilon| < r, |\arg \epsilon| < \alpha\}$ for some (d, r, α) such that $0 < d < d_0$, $0 < r < r_0$, and $0 < \alpha < \alpha_0$,

(b) $\vec{\phi}(t, \epsilon) \simeq \displaystyle\sum_{m=0}^{\infty} \epsilon^m \vec{\phi}_m(t)$ as $\epsilon \to 0$ in the sector $\mathcal{S} = \{\epsilon : 0 < |\epsilon| < r, |\arg \epsilon| < \alpha\}$, where coefficients $\vec{\phi}_m(t)$ are holomorphic and periodic of period 1 in the domain $\mathcal{D}' = \{t : |\Im t| < d, -\infty < \Re t < +\infty\}$.

Hint.

Step 1. Construct the solution $\vec{\psi}(t, \vec{c}, \epsilon)$ to the initial-value problem

$$\text{(IP)} \qquad \frac{d\vec{y}}{dt} = 2\pi i \vec{y} + \epsilon[\vec{y} + \epsilon \vec{f}(t, \vec{y}, \epsilon)], \qquad \vec{y}(0) = \vec{c}.$$

Substep 1. First, construct a formal solution

$$\vec{\psi}(t, \vec{c}, \epsilon) = \sum_{m=0}^{+\infty} \epsilon^m \vec{\psi}_m(t, \vec{c}) = e^{2\pi i t}\vec{c} + \epsilon t e^{2\pi i t}\vec{c} + O(\epsilon^2).$$

The coefficients $\vec{\psi}_m(t, \vec{c})$ are holomorphic in a domain

$$\Omega_0 = \{(t, \vec{c}) : |\Im t| < d_0, -\infty < \Re t < +\infty, |\vec{c}| < \gamma_0\},$$

where γ_0 is a positive number. To prove this, set $\vec{y} = e^{2\pi i t}(\vec{c} + \epsilon \vec{z})$ to change (IP) to

$$\text{(IP}')\qquad \frac{d\vec{z}}{dt} = \vec{c} + \epsilon[\vec{z} + e^{-2\pi i t}\vec{f}(t, e^{2\pi i t}(\vec{c} + \epsilon \vec{z}), \epsilon)], \qquad \vec{z}(0) = \vec{0}.$$

The function $\vec{f}(t, e^{2\pi i t}(\vec{c} + \epsilon \vec{z}), \epsilon)$ admits an asymptotic expansion in powers of ϵ (cf. Proof of Theorem XI-1-8) whose coefficients are polynomials in the entries of the vector $\vec{z}$. Coefficients $\vec{\psi}_m(t, \vec{c})$ can be calculated successively.

Substep 2. Show that $\vec{\psi}(t, \vec{c}, \epsilon) \simeq \vec{\psi}(t, \vec{c}, \epsilon)$ as $\epsilon \to 0$ in a sector $\mathcal{S} = \{\epsilon : 0 < |\epsilon| < r, |\arg \epsilon| < \alpha\}$ uniformly in a domain $\mathcal{D}' = \{(t, c) : |\Im t| < d, |\Re t| < T, |\vec{c}| < \gamma\}$, where T is an arbitrary positive number and $\gamma > 0$ depends on T. In this argument, the Gronwall's inequality (Lemma I-1-5) is useful.

Step 2. Solve the equation $\vec{\psi}(1, \vec{c}, \epsilon) = \vec{c}$. This equation has the form

$$\text{(E)} \qquad\qquad \vec{c} = \epsilon \vec{h}(\vec{c}, \epsilon),$$

where $\vec{h}(\vec{c}, \epsilon)$ admits an asymptotic expansion as $\epsilon \to 0$ in the sector $\mathcal{S}$ uniformly for $|\vec{c}| < \gamma$. To solve this, we can use successive approximations as follows:

$$\vec{c}_0(\epsilon) = \vec{0}, \qquad \vec{c}_h(\epsilon) = \epsilon \vec{h}(\vec{c}_{h-1}(\epsilon), \epsilon).$$

Then, the approximations $\vec{c}_h(\epsilon)$ converge to a limit $\vec{c}(\epsilon)$. Using Theorem XI-1-12, we can conclude that $\vec{c}(\epsilon)$ admits an asymptotic expansion in powers of ϵ.

Step 3. The particular solution $\vec{\phi}(t,\epsilon) = \vec{\psi}(t,\vec{c}(\epsilon),\epsilon)$ is the periodic solution satisfying all of the requirements.

XII-12. Consider a system of differential equations

$$(s) \qquad\qquad \frac{dy}{dx} = x^k a(x,\epsilon)y,$$

where x is a complex independent variable, ϵ is a complex parameter, k is a nonnegative integer, and y is an unknown element in a Banach algebra $\mathcal{U}$ over the field $\mathbb{C}$ of complex numbers with a unit element I . Assume that

$$(a) \qquad\qquad a(x,\epsilon) = \sum_{m=0}^{+\infty} x^{-m} a_m(\epsilon),$$

where $a_m(\epsilon) \in \mathcal{U}$ and these quantities are holomorphic and bounded with respect to ϵ in a sector $S = \{\epsilon : 0 < |\epsilon| < \delta_0, \ |\arg \epsilon| < \delta_1\}$, and the series (a) is convergent in norm for $|x| > R_0$ uniformly for ϵ in S. Also, assume that $a(x,\epsilon)$ admits an asymptotic expansion in powers of ϵ uniformly for x in $|x| > R_0$ as $\epsilon \to 0$ in S. Show that if a positive integer N is sufficiently large, there exist elements $p(x,\epsilon)$, $b_0(\epsilon), \ldots, b_M(\epsilon)$ of $\mathcal{U}$ such that

  (i) $p(x,\epsilon)$ is holomorphic and bounded in S and large $|x|$,
 (ii) $p(x,\epsilon)$ admits an asymptotic expansion in powers of ϵ uniformly for large $|x|$ as $\epsilon \to 0$ in S,
(iii) M is a positive integer and the quantities $b_0(\epsilon), \ldots, b_M(\epsilon)$ are holomorphic in S and admit asymptotic expansions in powers of ϵ as $\epsilon \to 0$ in S,
 (iv) $b_m(\epsilon) \simeq 0$ as $\epsilon \to 0$ in S for $m \geq k+1$,
  (v) the transformation $y = [I + x^{-(N+1)}p(x,\epsilon)]u$ changes (s) to

$$(s') \qquad \frac{du}{dx} = x^k \left[\sum_{m=0}^{N} x^{-m} a_m(\epsilon) + x^{-(N+1)} \sum_{m=0}^{M} x^{-m} b_m(\epsilon) \right] u.$$

Comment and Hint. See [Si11; in particular Theorem 2 on p. 157]. The point $x = 0$ is not necessarily a regular singular point of (s') as in Theorem XII-6-1.

Step 1. The main idea is, assuming that $N > k$, to solve equations of the following forms:

$$(I) \quad \begin{aligned} (N+1)x^{-(k+1)}Q(x) &- x^{-k}\frac{dQ(x)}{dx} + \alpha(x)Q(x) - Q(x)\alpha(x) - B(x) \\ &+ x^{-(N+1)}[a(x)Q(x) - Q(x)B(x)] = F(x) \end{aligned}$$

and

$$(II) \quad \begin{aligned} (N+1)x^{-(k+1)}Q(x) &- x^{-k}\frac{dQ(x)}{dx} + \alpha(x)Q(x) - Q(x)\alpha(x) - B(x) \\ &+ x^{-(N+1)}[a(x)Q(x) - Q(x)\beta(x) - \gamma(x)B(x)] = F(x), \end{aligned}$$

where

(1) $Q(x)$ is an unknown quantity which should be a convergent power series in x^{-1},

(2) $B(x)$ is an unknown quantity which should be a polynomial in x^{-1} of degree k,

(3) $\alpha(x)$ is a given polynomial in x^{-1} of degree N,

(4) $a(x)$ is a given convergent power series in x^{-1},

(5) β is a given polynomial in x^{-1} of degree k,

(6) $\gamma(x)$ is a given convergent power series in x^{-1},

(7) $F(x)$ is a given convergent power series in x^{-1}.

To solve these equations, first express B in terms of Q. In fact, setting

$$Q(x) = \sum_{m=0}^{+\infty} x^{-m} Q_m, \quad B(x) = \sum_{m=0}^{k} x^{-m} B_m, \quad \alpha(x) = \sum_{m=0}^{N} x^{-m} \alpha_m,$$

we obtain

$$B_m = \sum_{h=1}^{m} [\alpha_{m-h} Q_h - Q_h \alpha_{m-h}], \qquad m = 0, 1, \ldots, k.$$

Then, we can write (I) and (II) in the form

(III) $\qquad Q_m = \dfrac{1}{N+1+m} \mathcal{G}_m(Q_n; 0 \le n < +\infty) \qquad (m \ge 0),$

where $\mathcal{G}_m$ is either quadratic or linear in Q_n $(n \ge 0)$.

Step 2. If $N > 0$ is sufficiently large, we can solve (III) by defining a norm for a convergent power series $P(x) = \sum_{m=0}^{+\infty} x^{-m} P_m$ by $\|P\| = \sum_{m=0}^{+\infty} \rho^m \|P_m\|$, where ρ is a sufficiently small positive number.

Step 3. Using Steps 1 and 2, we can construct a formal power series $\hat{q}(x, \epsilon) = \sum_{\ell=0}^{+\infty} \epsilon^\ell q_\ell(x)$ such that the coefficients $q_\ell(x)$ are holomorphic and bounded in a disk $\Delta = \{x : |x| > R > 0\}$ and that the formal transformation $y = [I + x^{-(N+1)} \hat{q}(x, \epsilon)] u$ changes (s) to (s').

Step 4. Find a function $q(x, \epsilon)$ such that q is holomorphic and bounded in $\Delta \times \mathcal{S}$ and that $q(x, \epsilon) \simeq \hat{q}(x, \epsilon)$ as $\epsilon \to 0$ in $\mathcal{S}$ uniformly for $x \in \Delta$. Then, transformation $y = [I + x^{-(N+1)} q(x, \epsilon)] v$ changes (s) to

(s'')

$$\frac{dv}{dx} = x^k \left[\sum_{m=0}^{N} x^{-m} a_m(\epsilon) + x^{-(N+1)} \left(\sum_{m=0}^{k} x^{-m} b_m(\epsilon) + \sum_{m=k+1}^{+\infty} x^{-m} \delta_m(\epsilon) \right) \right] v,$$

where

$$b_m(\epsilon) \simeq \hat{b}_m(\epsilon) \qquad \text{as} \quad \epsilon \to 0 \text{ in } \mathcal{S}$$

and

$$\sum_{m=k+1}^{+\infty} x^{-m} \delta_m(\epsilon) \simeq 0$$

as $\epsilon \to 0$ in $\mathcal{S}$ uniformly for $x \in \Delta$.

Step 5. Using again Steps 1 and 2, find a function $r(x,\epsilon)$ such that
  (a) $r(x,\epsilon)$ is holomorphic and bounded for $|x| > R' > 0$ and $\epsilon \in \mathcal{S}$,
  (b) $r(x,\epsilon) \simeq 0$ as $\epsilon \to 0$ in $\mathcal{S}$ uniformly for $|x| > R'$,
  (c) for a sufficiently large $M' > 0$, the transformation $v = [I + x^{-(M'+1)}r(x,\epsilon)]w$
  changes (s'') to

$$\frac{dw}{dx} = x^k \left[\sum_{m=0}^{N} x^{-m} a_m(\epsilon) + x^{-(N+1)} \sum_{m=0}^{M} x^m b_m(\epsilon) \right] w,$$

where $b_m(\epsilon) \simeq 0$ as $\epsilon \to 0$ in $\mathcal{S}$ for $m \geq k+1$.

XII-13. Let $A(t)$ be a 2×2 matrix whose entries are holomorphic in a disk $\{t : |t| < \rho_0\}$ such that the matrix

(M)
$$P\left(\frac{1}{x}\right)^{-1} \left\{ x^k A\left(\frac{1}{x}\right) P\left(\frac{1}{x}\right) - \frac{d}{dx} P\left(\frac{1}{x}\right) \right\}$$

is not triangular for any 2×2 matrix $P(t)$ such that the entries of $P(t)$ and $P(t)^{-1}$ are holomorphic in the disk $\mathcal{D} = \{t : |t| < \rho_0\}$. Show that there exists such a 2×2 matrix $P(x)$ for which matrix (M) has the form $x^k B\left(\dfrac{1}{x}\right)$ with a 2×2 matrix $B(t)$ whose entries are polynomials in t and whose degree in t is at most $k+1$.

Hint. See [JLP]. This result can be generalized to the case when $A(t)$ is a 3×3 matrix (cf. [Bal1] and [Bal2]).

XII-14. Let $P(t)$ be a 2×2 matrix such that the entries of $P(t)$ and $P(t)^{-1}$ are holomorphic in a disk $\mathcal{D} = \{t : |t| < \rho_0\}$ and that the transformation $\vec{y} = P\left(\dfrac{1}{x}\right)\vec{u}$ changes the system

$$\frac{d\vec{y}}{dx} = \begin{bmatrix} 1 + x^{-1} & 0 \\ x^{-3} & 1 - x^{-1} \end{bmatrix} \vec{y}, \qquad \vec{y} = \begin{bmatrix} y_1 \\ y_2 \end{bmatrix}$$

to

$$\frac{d\vec{u}}{dr} = B\left(\frac{1}{r}\right)\vec{u}, \qquad \vec{u} = \begin{bmatrix} u_1 \\ u_2 \end{bmatrix}$$

with a 2×2 matrix $B(t)$ whose entries are polynomials in t. Show that the degree of the polynomial $B(t)$ is not less than 2. Also, show that there exists $P(t)$ such that the degree of the polynomial $B(t)$ is equal to 2.

Hint. To prove that there exists $P(t)$ such that the degree of the polynomial $B(t)$ is equal to 2, apply Theorem V-5-1 to the system $t\dfrac{d\vec{v}}{dt} = \begin{bmatrix} -1 + \alpha t & \beta t \\ \gamma t & 1 + \delta t \end{bmatrix} \vec{v}$ with suitable constants α, β, γ, and δ. To show that the degree of the polynomial $B(t)$ is not less than 2, assuming that the degree of the polynomial $B(t)$ is less than 2, derive a contradiction from the following fact:

The transformation $\vec{y} = \begin{bmatrix} x^2 & 0 \\ 0 & 1 \end{bmatrix} \vec{u}$ changes the system

$$\frac{d\vec{y}}{dx} = \begin{bmatrix} 1 + x^{-1} & 0 \\ x^{-3} & 1 - x^{-1} \end{bmatrix} \vec{y}, \qquad \vec{y} = \begin{bmatrix} y_1 \\ y_2 \end{bmatrix}$$

to

$$\frac{d\vec{u}}{dx} = \begin{bmatrix} 1 - x^{-1} & 0 \\ x^{-1} & 1 - x^{-1} \end{bmatrix} \vec{u}, \qquad \vec{u} = \begin{bmatrix} u_1 \\ u_2 \end{bmatrix}.$$

XII-15. Let $A(t)$ be a 2×2 triangular matrix whose entries are holomorphic in a disk $\{t : |t| < \rho_0\}$. Show that there exists a 2×2 matrix $Q(x)$ such that

(i) the entries of $Q(x)$ and $Q(x)^{-1}$ are holomorphic for $|x| > \dfrac{1}{r_0}$ and meromorphic at $x = \infty$,

(ii) the transformation $\vec{y} = Q(x)\vec{u}$ changes the system $\dfrac{d\vec{y}}{dx} = x^k A\left(\dfrac{1}{x}\right)\vec{y}$ to $\dfrac{d\vec{u}}{dx} = x^k B\left(\dfrac{1}{x}\right)\vec{u}$ with a 2×2 matrix $B(t)$ whose entries are polynomials in t and whose degree in t is at most $k + 1$.

Hint. See [JLP]. This result can be generalized to the case when $A(t)$ is a 3×3 matrix (cf. [Bal1] and [Bal2]).

CHAPTER XIII

SINGULARITIES OF THE SECOND KIND

In this chapter, we explain the structure of asymptotic solutions of a system of differential equations at a singular point of the second kind. In §§XIII-1, XIII-2, and XIII-3, a basic existence theorem of asymptotic solutions in the sense of Poincaré is proved in detail. In §XII-4, this result is used to prove a block-diagonalization theorem of a linear system. The materials in §§XIII-1–XIII-4 are also found in [Si7]. The main topic of §XIII-5 is the equivalence between a system of linear differential equations and an n-th-order linear differential equation. The equivalence is based on the existence of a cyclic vector for a linear differential operator. The existence of cyclic vectors was originally proved in [Del]. In §XIII-6, we explain a basic theorem concerning the structure of solutions of a linear system at a singular point of the second kind. This theorem was proved independently in [Huk4] and [Tu1]. In §XIII-7, the Newton polygon of a linear differential operator is defined. This polygon is useful when we calculate formal solutions of an n-th-order linear differential equation (cf. [St]). In §XIII-8, we explain asymptotic solutions in the Gevrey asymptotics. To understand materials in §XIII-8, the expository paper [Ram3] is very helpful. In §§XIII-1–XIII-4, the singularity is at $x = \infty$, but from §XIII-5 through §XIII-8, the singularity is at $x = 0$. Any singularity at $x = \infty$ is changed to a singularity of the same kind at $\xi = 0$ by the transformation $x = \dfrac{1}{\xi}$.

XIII-1. An existence theorem

In §§XIII-1, XIII-2, and XIII-3, we consider a system of differential equations

$$(XIII.1.1) \qquad \frac{dv_j}{dx} = x^r f_j(x, v_1, v_2, \ldots, v_n) \qquad (j = 1, 2, \ldots, n),$$

where r is a non-negative integer and $f_j(x, v_1, v_2, \ldots, v_n)$ are holomorphic with respect to complex variables $(x, v_1, v_2, \ldots, v_n)$ in a domain

$$(XIII.1.2) \qquad |x| > N_0, \quad |\arg x| < \alpha_0, \quad |v_j| < \delta_0 \quad (j = 1, 2, \ldots, n),$$

N_0, α_0, and δ_0 being positive constants. Set

$$(XIII.1.3) \qquad f_j(x, \vec{v}) = f_{j0}(x) + \sum_{h=1}^{n} a_{jh}(x) v_h + \sum_{|\wp| \geq 2} f_{j\wp}(x) \vec{v}^{\wp},$$

where $\vec{v} \in \mathbb{C}^n$ with the entries $(v_1, v_2, \ldots, v_n)$.

We look at (XIII.1.1) under the following three assumptions.

403

Assumption I. *Each function $f_j(x, \vec{v})$ $(j = 1, 2, \ldots, n)$ admits a uniform asymptotic expansion*

$$(XIII.1.4) \qquad f_j(x, \vec{v}) \simeq \sum_{k=0}^{\infty} \hat{f}_{jk}(\vec{v}) x^{-k}$$

in the sense of Poincaré as $x \to \infty$ in the sector

$$(XIII.1.5) \qquad |x| > N_0, \quad |\arg x| < \alpha_0,$$

where coefficients $\hat{f}_{jk}(\vec{v})$ are holomorphic in the domain

$$(XIII.1.6) \qquad |\vec{v}| < \delta_0.$$

Furthermore, we assume that

$$(XIII.1.7) \qquad \hat{f}_{j0}(\vec{0}) = 0 \qquad (j = 1, 2, \ldots, n).$$

Observation XIII-1-1. Under Assumption I, $f_{j0}(x)$, $a_{jh}(x)$, and $f_{j\wp}(x)$ admit asymptotic expansions

$$(XIII.1.8) \qquad f_{j0}(x) \simeq \sum_{k=1}^{\infty} f_{j0k} x^{-k}, \qquad f_{j\wp}(x) \simeq \sum_{k=0}^{\infty} f_{j\wp k} x^{-k},$$

and

$$(XIII.1.9) \qquad a_{jh}(x) \simeq \sum_{k=0}^{\infty} a_{jhk} x^{-k} \qquad (j, h = 1, 2, \ldots, n)$$

as $x \to \infty$ in sector (XIII.1.5) with coefficients in $\mathbb{C}$. Let $A(x)$ be the $n \times n$ matrix whose (j, k)-entry is $a_{jk}(x)$, respectively (i.e., $A(x) = (a_{jh}(x))$). Then, $A(x)$ admits an asymptotic expansion

$$(XIII.1.10) \qquad A(x) \simeq \sum_{k=0}^{\infty} x^{-k} A_k$$

as $x \to \infty$ in sector (XIII.1.5), where $A_k = (a_{jhk})$. The following assumption is technical and we do not lose any generality with it.

Assumption II. *The matrix A_0 has the following S-N decomposition:*

$$(XIII.1.11) \qquad A_0 = \mathrm{diag}\,[\mu_1, \mu_2, \ldots, \mu_n] + \mathcal{N},$$

where $\mu_1, \mu_2, \ldots, \mu_n$ are eigenvalues of A_0 and $\mathcal{N}$ is a lower-triangular nilpotent matrix.

Note that $|\mathcal{N}|$ can be made as small as we wish (cf. Lemma VII-3-3).

The following assumption plays a key roll.

Assumption III. *The matrix A_0 is invertible, i.e.,*

$$(\text{XIII.1.12}) \qquad \mu_j \neq 0 \qquad (j = 1, 2, \ldots, n).$$

Set

$$(\text{XIII.1.13}) \qquad \omega = \arg x, \qquad \omega_j = \arg \mu_j \qquad (j = 1, 2, \ldots, n),$$

and denote by $\mathcal{D}_j(N, \gamma, q)$ the domain defined by

$$(\text{XIII.1.14}) \qquad \mathcal{D}_j(N, \gamma, q) = \left\{ x : \; |x| > N, \; |\omega| < \alpha_0, \right.$$
$$\left. \left(2q - \frac{3}{2}\right)\pi + \gamma < \omega_j + (r+1)\omega < \left(2q + \frac{3}{2}\right)\pi - \gamma \right\},$$

where q is an integer, N is a sufficiently large positive constant, and γ is a sufficiently small positive constant. For each j, there exists at least one integer q_j such that the real half-line defined by $x > N$ is contained in the interior of $\mathcal{D}_j(N, \gamma, q_j)$. Set

$$(\text{XIII.1.15}) \qquad \mathcal{D}(N, \gamma) = \bigcap_{j=1}^{n} \mathcal{D}_j(N, \gamma, q_j).$$

In §§XIII-2 and XIII-3, we shall prove the following theorem.

Theorem XIII-1-2. *If N^{-1} and γ are sufficiently small positive numbers, then under Assumptions I, II, and III, system (XIII.1.1) has a solution*

$$(\text{XIII.1.16}) \qquad v_j = p_j(x) \qquad (j = 1, 2, \ldots, m),$$

such that
  (i) $p_j(x)$ are holomorphic in $\mathcal{D}(N, \gamma)$,
  (ii) $p_j(x)$ admit asymptotic expansions

$$(\text{XIII.1.17}) \qquad p_j(x) \simeq \sum_{k=1}^{\infty} p_{jk} x^{-k} \qquad (j = 1, 2, \ldots, m)$$

as $x \to \infty$ in $\mathcal{D}(N, \gamma)$, where $p_{jk} \in \mathbb{C}$.

To illustrate Theorem XIII-1-2, we prove the following corollary.

Corollary XIII-1-3. *Let $A(x)$ be an $n \times n$ matrix whose entries are holomorphic and bounded in a domain $\Delta_0 = \{x : |x| > R_0\}$ and let $\vec{f}(x)$ be a $\mathbb{C}^n$-valued function whose entries are holomorphic and bounded in the domain Δ_0. Also, let $\lambda_1, \lambda_2, \ldots, \lambda_n$ be eigenvalues of $A(\infty)$. Assume that $A(\infty)$ is invertible. Assume also that none of the quantities $\lambda_j e^{ik\theta}$ $(j = 1, 2, \ldots, n)$ are real and negative for a real number θ and a positive integer k. Then, there exist a positive number ϵ and a solution $\vec{y} = \vec{\phi}(x)$ of the system $\dfrac{d\vec{y}}{dx} = x^{k-1} A(x)\vec{y} + x^{-1}\vec{f}(x)$ such that the entries of*

$\vec{\phi}(x)$ *are holomorphic and admit asymptotic expansions in powers of* x^{-1} *as* $x \to \infty$ *in the sector* $\mathcal{S} = \left\{ x : |x| > R_0, \ |\arg x - \theta| < \dfrac{\pi}{2k} + \epsilon \right\}$.

Proof.

The main claim of this corollary is that the asymptotic expansion of the solution is valid in a sector $|\arg x - \theta| < \dfrac{\pi}{2k} + \epsilon$ *whose opening is greater than* $\dfrac{\pi}{k}$. So, we look at the sector $\mathcal{D}(N, \gamma)$ of Theorem XIII-1-2. In the given case, $\alpha_0 = +\infty$, $r = k - 1$, and $\mu_j = \lambda_j$ $(j = 1, \dots, n)$. The assumptions given in the corollary imply that

$$\omega_j + k\theta \neq (2p+1)\pi, \qquad \text{where} \quad \omega_j = \arg \lambda_j \quad (j = 1, \dots, n),$$

for any integer p. Therefore, for each j, there exists an integer q_j such that

$$\text{either} \quad -\pi < \omega_j - 2q_j\pi + k\theta \leq 0 \quad \text{or} \quad 0 \leq \omega_j - 2q_j\pi + k\theta < \pi.$$

Therefore, either

$$-\frac{3\pi}{2} < \omega_j - 2q_j\pi + k\theta - \frac{\pi}{2} < \omega_j - 2q_j\pi + k\theta + \frac{\pi}{2} \leq \frac{\pi}{2}$$

or

$$-\frac{\pi}{2} \leq \omega_j - 2q_j\pi + k\theta - \frac{\pi}{2} < \omega_j - 2q_j\pi + k\theta + \frac{\pi}{2} < \frac{3\pi}{2}.$$

This proves that a sector $\mathcal{S} = \left\{ x : |x| > R_0, \ |\arg x - \theta| < \dfrac{\pi}{2k} + \epsilon \right\}$ is in $\mathcal{D}(N, \gamma)$ for a sufficiently large $R_0 > 0$ and a sufficiently small $\epsilon > 0$ if we use $xe^{-i\theta}$ as the independent variable instead of x. $\quad \square$

XIII-2. Basic estimates

In order to prove Theorem XIII-1-2, let us change system (XIII-1-1) to a system of integral equations.

Observation XIII-2-1. Expansion (XIII.1.17) of the solution $p_j(x)$

$$(\text{XIII.2.1}) \qquad v_j = \sum_{k=1}^{\infty} p_{jk} x^{-k} \qquad (j = 1, 2, \dots, n)$$

must be a formal solution of system (XIII.1.1). The existence of such a formal solution (XIII.2.1) of system (XIII.1.1) follows immediately from Assumptions I and III. The proof of this fact is left to the reader as an exercise.

Observation XIII-2-2. For each $j = 1, 2, \ldots, n$, using Theorem XI-1-14, let us construct a function $z_j(x)$ such that

(i) $z_j(x)$ is holomorphic in a sector

$$(\text{XIII.2.2}) \qquad |x| > \tilde{N}_0, \quad |\arg x| < \alpha_0,$$

where $\tilde{N}_0$ is a positive number not smaller than N_0,

(ii) $z_j(x)$ and $\dfrac{dz_j(x)}{dx}$ admit asymptotic expansions

$$(\text{XIII.2.3}) \qquad z_j(x) \simeq \sum_{k=1}^{\infty} p_{jk} x^{-k} \quad \text{and} \quad \frac{dz_j(x)}{dx} \simeq \sum_{k=1}^{\infty} (-k) p_{jk} x^{-k-1}$$

as $x \to \infty$ in sector (XIII.2.2), respectively.

Consider the change of variables

$$(\text{XIII.2.4}) \qquad v_j = u_j + z_j(x) \qquad (j = 1, 2, \ldots, n).$$

Denote $(z_1, z_2, \ldots, z_n)$ and $(u_1, u_2, \ldots, u_n)$ by $\vec{z}$ and $\vec{u}$, respectively. Then, $\vec{u}$ satisfies the system of differential equations

$$(\text{XIII.2.5}) \qquad \frac{du_j}{dx} = x^r g_j(x, \vec{u}) \qquad (j = 1, 2, \ldots, n),$$

where

$$g_j(x, \vec{u}) = f_j(x, \vec{u} + \vec{z}) - x^{-r} \frac{dz_j(x)}{dx} \qquad (j = 1, 2, \ldots, n).$$

Set

$$(\text{XIII.2.6}) \quad g_j(x, \vec{u}) = g_{j0}(x) + \sum_{k=1}^{m} b_{jk}(x) u_k + \sum_{|\wp| \geq 2}^{\infty} b_{j\wp}(x) \vec{u}^{\wp} \quad (j = 1, 2, \ldots, n).$$

In particular,

$$(\text{XIII.2.7}) \qquad g_{j0}(x) = f_j(x, \vec{z}) - x^{-r} \frac{dz_j(x)}{dx} \simeq 0 \qquad (j = 1, 2, \ldots, n)$$

and

$$b_{jk}(x) - a_{jk}(x) = O(|x|^{-1}) \qquad (j, k = 1, 2, \ldots, n)$$

as $x \to \infty$ in sector (XIII.2.2). Thus,

$$(\text{XIII.2.8}) \qquad b_{jk}(x) = a_{jk}(\infty) + O(|x|^{-1}) \qquad (j, k = 1, 2, \ldots, n)$$

as $x \to \infty$ in sector (XIII.2.2).

Observation XIII-2-3. Set

$$(\text{XIII.2.9}) \qquad g_j(x, \vec{u}) = \mu_j u_j + R_j(x, \vec{u}) \qquad (j = 1, 2, \dots, n).$$

Then, by virtue of Assumption III, (XIII.2.7), and (XIII.2.8), for sufficiently small positive numbers $\tilde{N}_0^{-1}$ and δ, there exists a positive constant c, independent of j, such that for any positive integer h, the estimates

$$(\text{XIII.2.10}) \qquad |R_j(x, \vec{u})| \leq c|\vec{u}| + b_h|x|^{-(h+1)} \qquad (j = 1, 2, \dots, n)$$

and

$$(\text{XIII.2.11}) \qquad |R_j(x, \vec{u}) - R_j(x, \vec{u}')| \leq c|\vec{u} - \vec{u}'| \qquad (j = 1, 2, \dots, n)$$

hold whenever $(x, \vec{u})$ and $(x, \vec{u}')$ are in the domain

$$(\text{XIII.2.12}) \qquad |x| > \tilde{N}_0, \quad |\arg x| < \alpha_0, \quad |u_j| < \delta \quad (j = 1, 2, \dots, n).$$

Here, b_h is a positive constant depending on h. Furthermore, by virtue of Assumption II, it can be assumed without loss of generality that the constant c satisfies the condition

$$(\text{XIII.2.13}) \qquad c < \frac{r+1}{H},$$

where H is a positive constant to be specified later.

From (XIII.2.5) and (XIII.2.9), it follows that

$$(\text{XIII.2.14}) \qquad \frac{du_j}{dx} = x^r[\mu u_j + R_j(x, \vec{u})].$$

Change system (XIII.2.14) to a system of integral equations

$$(\text{XIII.2.15}) \quad u_j = \int_{L_{jx}} t^r R_j(t, \vec{u}) \exp\left[\frac{-\mu_j}{r+1}(t^{r+1} - x^{r+1})\right] dt \quad (j = 1, 2, \dots, n),$$

where the paths of integration L_{jx} start from x. The paths of integration must be chosen carefully so that uniformly convergent successive approximations can be defined in such a way that the limit is a solution $u_j(x)$ of (XIII.2.15) which satisfies the conditions (i) $u_j(x)\,(j = 1, 2, \dots, n)$ are holomorphic in $\mathcal{D}(N, \gamma)$ and (ii) $u_j(x) \simeq 0 \ (j = 1, 2, \dots, n)$ as $x \to \infty$ in $\mathcal{D}(N, \gamma)$ for suitable positive numbers N and γ.

Hereafter in this section, we explain how to choose paths of integration on the right-hand side of (XIII.2.15).

Observation XIII-2-4. Since for each j, the domain $\mathcal{D}_j(N, \gamma, q_j)$ contains the real half-line defined by $x > N$ in its interior, their common part $\mathcal{D}(N, \gamma)$ is given by the inequalities

$$(\text{XIII.2.16}) \qquad |x| > N, \quad -\ell < \arg x < \ell',$$

where ℓ and ℓ' are positive constants. It is noteworthy that if $x \in \mathcal{D}(N,\gamma)$, then x satisfies the inequalities

(XIII.2.17)
$$|\arg x| < \alpha_0, \quad -\frac{3}{2}\pi + \gamma < \omega_j + (r+1)\arg x < \frac{3}{2}\pi - \gamma$$
$$(j = 1, 2, \ldots, n),$$

provided that $\omega_j = \arg \mu_j$ are chosen suitably. Therefore,

(XIII.2.18)
$$-\frac{3}{2}\pi + \gamma \le \omega_j - (r+1)\ell < \omega_j + (r+1)\arg x$$
$$< \omega_j + (r+1)\ell' \le \frac{3}{2}\pi - \gamma \quad (j = 1, 2, \ldots, n).$$

Moreover, since $\mathcal{D}(N,\gamma)$ is the common part of $\mathcal{D}_j(N,\gamma,q_j)$, the equalities

(XIII.2.19)
$$-\frac{3}{2}\pi + \gamma = \omega_j - (r+1)\ell$$

and

(XIII.2.20)
$$\frac{3}{2}\pi - \gamma = \omega_h + (r+1)\ell'$$

hold for some j and h. This implies that ℓ and ℓ' depend on γ. Thus, the quantity γ can be chosen so that

(XIII.2.21)
$$\omega_j - (r+1)\ell \ne \pm\frac{1}{2}\pi$$

and

(XIII.2.22)
$$\omega_j + (r+1)\ell' \ne \pm\frac{1}{2}\pi$$

for all j.

Observation XIII-2-5. Define the paths of integration in each of the following two cases.

Case 1. Consider first the case when

(XIII.2.23)
$$(r+1)(\ell + \ell') < \pi.$$

In this case, the set of indices $J = \{j : j = 1, 2, \ldots, n\}$ is divided into four groups:

(XIII.2.24)
$$G_1 = \left\{ j : \frac{1}{2}\pi < \omega_j - (r+1)\ell < \omega_j + (r+1)\ell' < \frac{3}{2}\pi \right\},$$

(XIII.2.25)
$$G_2 = \left\{ j : -\frac{3}{2}\pi < \omega_j - (r+1)\ell < \omega_j + (r+1)\ell' < -\frac{1}{2}\pi \right\},$$

$$(\text{XIII.2.26}) \qquad G_3 = \left\{ j : \ -\frac{1}{2}\pi < \omega_j + (r+1)\ell' < \frac{1}{2}\pi \right\},$$

and

$$(\text{XIII.2.27}) \qquad G_4 = J - \{G_1 \cup G_2 \cup G_3\}.$$

If γ' is a sufficiently small positive number, then it holds that

$$(\text{XIII.2.28}) \qquad \frac{1}{2}\pi + \gamma' < \omega_j - (r+1)\ell < \omega_j + (r+1)\ell' < \frac{3}{2}\pi - \gamma' \quad (j \in G_1),$$

$$(\text{XIII.2.29}) \qquad -\frac{3}{2}\pi + \gamma' < \omega_j - (r+1)\ell < \omega_j + (r+1)\ell' < -\frac{1}{2}\pi - \gamma' \quad (j \in G_2),$$

$$(\text{XIII.2.30}) \qquad -\frac{1}{2}\pi + \gamma' < \omega_j + (r+1)\ell' < \frac{1}{2}\pi - \gamma' \quad (j \in G_3),$$

and

$$(\text{XIII.2.31}) \qquad -\frac{1}{2}\pi + \gamma' < \omega_j - (r+1)\ell < \frac{1}{2}\pi - \gamma' \quad (j \in G_4).$$

Set

$$(\text{XIII.2.32}) \qquad \xi = x^{r+1}.$$

The domain $\mathcal{D}(N)$ in the ξ-plane that is defined by

$$(\text{XIII.2.33}) \qquad |\xi| > N^{r+1}, \quad -(r+1)\ell < \arg\xi < (r+1)\ell'$$

corresponds to domain (XIII.2.16) in the x-plane. Letting

$$(\text{XIII.2.34}) \qquad \xi_0 = N^{r+1} \exp\left[i\frac{1}{2}(r+1)(\ell' - \ell) \right],$$

define a domain $\tilde{\mathcal{D}}(N)$ in the ξ-plane by

$$(\text{XIII.2.35}) \qquad |\xi| \geq N^{r+1}, \quad -(r+1)\ell - \frac{1}{2}\gamma' < \arg(\xi - \xi_0) < (r+1)\ell' + \frac{1}{2}\gamma',$$

and set $\hat{\mathcal{D}}(N) = \mathcal{D}(N) \cap \tilde{\mathcal{D}}(N)$. Denote by $\mathcal{D}_0(N)$ the domain in the x-plane which corresponds to the domain $\hat{\mathcal{D}}(N)$ in the ξ-plane (cf. Figure 1).

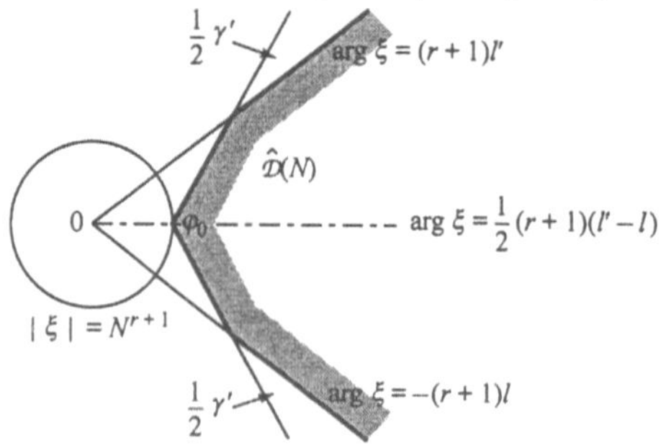

FIGURE 1.

For every point $\xi \in \hat{\mathcal{D}}(N)$, the lines $\hat{L}_{j\xi}$ in $\hat{\mathcal{D}}(N)$ (except possibly for their starting points) are defined by

$$(\text{XIII.2.36}) \quad \begin{cases} s = \xi_0 + t\exp[i\arg(\xi - \xi_0)], & 0 < t < |\xi - \xi_0| & \text{for} \quad j \in G_1 \cup G_2, \\ s = \xi + t\exp[i(r+1)\ell'], & 0 < t < \infty & \text{for} \quad j \in G_3, \\ s = \xi + t\exp[-i(r+1)\ell], & 0 < t < \infty & \text{for} \quad j \in G_4. \end{cases}$$

Note that from (XIII.2.28) and (XIII.2.29), it follows that

$$\begin{cases} \dfrac{1}{2}\pi + \dfrac{\gamma'}{2} < \omega_j + \arg(\xi - \xi_0) < \dfrac{3}{2}\pi - \dfrac{\gamma'}{2} & (j \in G_1), \\[2mm] -\dfrac{3}{2}\pi + \dfrac{\gamma'}{2} < \omega_j + \arg(\xi - \xi_0) < -\dfrac{1}{2}\pi - \dfrac{\gamma'}{2} & (j \in G_2). \end{cases}$$

Case 2. For the case when

$$(\text{XIII.2.37}) \quad (r+1)(\ell + \ell') \geq \pi,$$

the set of indices J is divided into three groups:

$$(\text{XIII.2.38.1}) \quad \hat{G}_1 = \left\{ j : \begin{array}{l} -\dfrac{3}{2}\pi + \gamma' < \omega_j - (r+1)\ell < -\dfrac{1}{2}\pi - \gamma', \\[2mm] -\dfrac{1}{2}\pi + \gamma' < \omega_j + (r+1)\ell' < \dfrac{1}{2}\pi - \gamma' \end{array} \right\},$$

$$(\text{XIII.2.38.2}) \quad \hat{G}_2 = \left\{ j : \begin{array}{l} -\dfrac{1}{2}\pi + \gamma' < \omega_j - (r+1)\ell < \dfrac{1}{2}\pi - \gamma', \\[2mm] \dfrac{1}{2}\pi + \gamma' < \omega_j + (r+1)\ell' < \dfrac{3}{2}\pi - \gamma' \end{array} \right\},$$

$$(\text{XIII.2.38.3}) \quad \hat{G}_3 = \left\{ j : \begin{array}{l} -\dfrac{3}{2}\pi + \gamma' < \omega_j - (r+1)\ell < -\dfrac{1}{2}\pi - \gamma', \\[2mm] \dfrac{1}{2}\pi + \gamma' < \omega_j + (r+1)\ell' < \dfrac{3}{2}\pi - \gamma' \end{array} \right\},$$

where γ' is a sufficiently small positive constant such that $\pi - 2\gamma' > 0$. Note that $(r+1)(\ell + \ell') > \pi - \gamma'$. These inequalities imply that $\dfrac{1}{2}\pi - \dfrac{1}{2}\gamma' > 0$ and

$$-(r+1)\ell < -(r+1)\ell - \frac{1}{2}\gamma' + \frac{1}{2}\pi < (r+1)\ell' + \frac{1}{2}\gamma' - \frac{1}{2}\pi < (r+1)\ell'.$$

In $\mathcal{D}(N)$, let

$$\begin{cases} \xi_1 = N^{r+1}\exp\left[i\left\{(r+1)\ell' + \dfrac{1}{2}\gamma' - \dfrac{1}{2}\pi\right\}\right], \\[3mm] \xi_2 = N^{r+1}\exp\left[i\left\{-(r+1)\ell - \dfrac{1}{2}\gamma' + \dfrac{1}{2}\pi\right\}\right], \end{cases}$$

and denote by T_1 and T_2 the tangents of the circle $|\xi| = N^{r+1}$ at the points ξ_1 and ξ_2, respectively. The tangent T_1 intersects the line $\arg \xi = (r+1)\ell'$ at an angle $\frac{1}{2}\gamma'$, while the tangent T_2 intersects the line $\arg \xi = -(r+1)\ell$ at an angle $\frac{1}{2}\gamma'$. Let $\xi^{(1)}$ and $\xi^{(2)}$ be those two points of intersection and denote by $\hat{D}(N)$ the subdomain of $D(N)$ whose boundary curve consists of the following arcs (cf. Figure 2 and Figure 3):

$$(\text{A-1}) \qquad s = \xi^{(1)} + t \exp[i(r+1)\ell'] \qquad\qquad (0 < t < +\infty),$$

$$(\text{A-2}) \qquad s = \xi^{(1)} + t \exp[i \arg(\xi_1 - \xi^{(1)})] \qquad\qquad (0 < t < |\xi_1 - \xi^{(1)}|),$$

$$(\text{A-3}) \qquad s = N^{r+1} e^{it} \qquad\qquad (\arg \xi_2 < t < \arg \xi_1),$$

$$(\text{A-4}) \qquad s = \xi^{(2)} + t \exp[i \arg(\xi_1 - \xi^{(1)})] \qquad\qquad (0 < t < |\xi_2 - \xi^{(2)}|),$$

and

$$(\text{A-5}) \qquad s = \xi^{(2)} + t \exp[-i(r+1)\ell] \qquad\qquad (0 < t < +\infty).$$

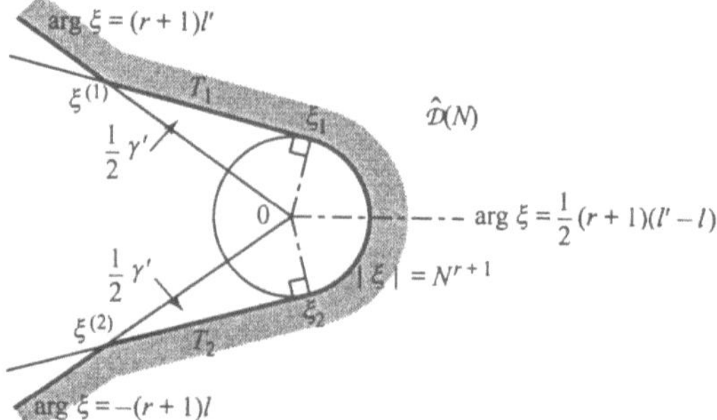

FIGURE 2. $(r+1)(\ell + \ell') < 2\pi$.

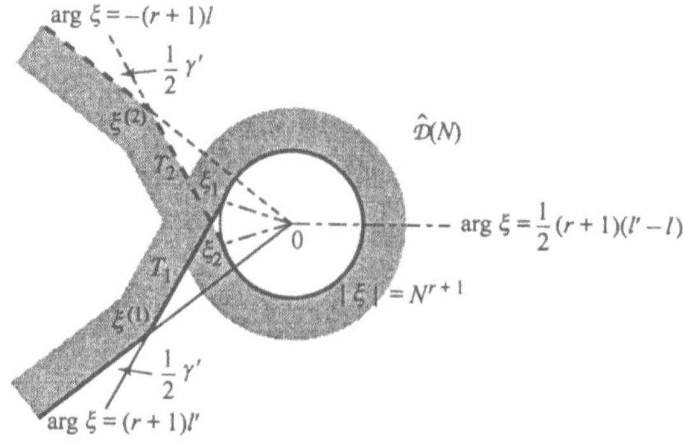

FIGURE 3. $(r+1)(\ell + \ell') > 2\pi$.

For every $\xi \in \hat{\mathcal{D}}(N)$, the paths of integration in $\hat{\mathcal{D}}(N)$ are defined in the following manner:

(a) For $j \in \hat{G}_1$, from (XIII.2.37) and (XIII.2.38.1), it follows that $-\frac{1}{2}\pi + \frac{1}{2}\gamma' < \omega_j - (r+1)\ell - \frac{1}{2}\gamma' + \pi$ and $\omega_j - (r+1)\ell - \frac{1}{2}\gamma' + \pi < \omega_j + (r+1)\ell' < \frac{1}{2}\pi - \gamma'$ and, hence,

$$-\frac{1}{2}\pi + \frac{1}{2}\gamma' < \omega_j - (r+1)\ell - \frac{1}{2}\gamma' + \pi < \omega_j + (r+1)\ell' < \frac{1}{2}\pi - \gamma'.$$

The lines $\hat{L}_{j\xi}$ in $\hat{\mathcal{D}}(N)$ are defined by

$$\text{(XIII.2.39)} \qquad\qquad s = \xi + t\exp(i(\theta(\xi))) \qquad\qquad (0 < t < \infty),$$

where $\theta(\xi)$ is a real-valued and continuous function of ξ such that

$$-(r+1)\ell - \frac{1}{2}\gamma' + \pi \leq \theta(\xi) \leq (r+1)\ell'$$

in $\hat{\mathcal{D}}(N)$. This implies that $-\frac{\pi}{2} + \frac{\gamma'}{2} < \omega_j + \theta(\xi) < \frac{\pi}{2} - \frac{\gamma'}{2}$. Precisely speaking, the function $\theta(\xi)$ is defined in the following way. Note that (XIII.2.37) implies

$$-(r+1)\ell - \frac{1}{2}\gamma' + \frac{1}{2}\pi < (r+1)\ell' - \frac{1}{2}\pi < (r+1)\ell' - \frac{1}{2}\pi + \frac{1}{2}\gamma'.$$

Let

$$\xi_0 = N^{r+1}\exp\left[i\left\{(r+1)\ell' - \frac{1}{2}\pi\right\}\right].$$

Then, ξ_0 is on arc (A-3). Let ξ' be a point on arc (A-3) such that $\arg \xi_2 < \arg \xi' < \arg \xi_0$. Then, the tangent $T_{\xi'}$ of the circle $|\xi| = N^{r+1}$ at the point ξ' is given by $s = \xi' + t\exp(i\phi(\xi'))\,(-\infty < t < +\infty)$, where $\phi(\xi')$ is continuous in ξ' and $-(r+1)\ell - \frac{1}{2}\gamma' + \pi < \phi(\xi') < (r+1)\ell'$. Moreover, the part of $T_{\xi'}$ in $\hat{\mathcal{D}}(N)$ is given by

$$\text{(XIII.2.40)} \qquad\qquad s = \xi' + t\exp(i\phi(\xi')) \qquad (0 < t < +\infty).$$

If a point ξ in $\hat{\mathcal{D}}(N)$ is on such a tangent (XIII.2.40), set $\theta(\xi) = \phi(\xi')$. Note that the tangent of the circle $|\xi| = N^{r+1}$ at ξ_0 is parallel to the line (A-1). If a point ξ is in a domain between these two lines, then set $\theta(\xi) = (r+1)\ell'$. For all points other than those given above, set $\theta(\xi) = -(r+1)\ell - \frac{1}{2}\gamma' + \pi$.

(b) For $j \in \hat{G}_2$, (XIII.2.37) and (XIII.2.38.2) imply that

$$-\frac{1}{2}\pi + \gamma' < \omega_j - (r+1)\ell < \omega_j + (r+1)\ell' + \frac{1}{2}\gamma' - \pi < \frac{1}{2}\pi - \frac{1}{2}\gamma'.$$

The lines $\hat{L}_{j\xi}$ is defined by (XIII.2.39) with $\theta(\xi)$ such that

$$-(r+1)\ell \leq \theta(\xi) \leq (r+1)\ell' + \frac{1}{2}\gamma' - \pi$$

in $\hat{D}(N)$. This implies that $-\dfrac{\pi}{2} + \dfrac{\gamma'}{2} < \omega_j + \theta(\xi) < \dfrac{\pi}{2} - \dfrac{\gamma'}{2}$. The function $\theta(\xi)$ is defined in a way similar to the previous case.

(c) For $j \in \hat{G}_3$, from (XIII.2.38.3), it follows that

$$\begin{cases} -\dfrac{1}{2}\pi + \dfrac{1}{2}\gamma' < \omega_j - (r+1)\ell - \dfrac{1}{2}\gamma' + \pi < \dfrac{1}{2}\pi - \dfrac{3}{2}\gamma', \\[2mm] -\dfrac{1}{2}\pi + \dfrac{3}{2}\gamma' < \omega_j + (r+1)\ell' + \dfrac{1}{2}\gamma' - \pi < \dfrac{1}{2}\pi - \dfrac{1}{2}\gamma'. \end{cases}$$

Let

$$\xi_0 = N^{r+1} \exp\left[i\left\{ (r+1)\ell' + \dfrac{1}{2}\gamma' - \dfrac{3}{2}\pi \right\} \right].$$

In the case when $(r+1)(\ell + \ell') \geq 2\pi - \gamma'$, the inequality

(XIII.2.41) $$\arg \xi_0 \geq \arg \xi_2,$$

holds, whereas in the case when $(r+1)(\ell + \ell') < 2\pi - \gamma'$, it holds that

(XIII.2.42) $$\arg \xi_0 < \arg \xi_2.$$

In the case when (XIII.2.41) holds, the lines $\hat{L}_{j\xi}$ are defined by (XIII.2.39) with

$$-(r+1)\ell - \dfrac{1}{2}\gamma' + \pi \leq \theta(\xi) \leq (r+1)\ell' + \dfrac{1}{2}\gamma' - \pi$$

in $\hat{D}(N)$. This implies that $-\dfrac{\pi}{2} + \dfrac{\gamma'}{2} < \omega_j + \theta(\xi) < \dfrac{\pi}{2} - \dfrac{\gamma'}{2}$. Since ξ_0 is on arc (A-3), $\theta(\xi)$ is defined as given above. In the case when (XIII.2.42) holds, the lines $\hat{L}_{j\xi}$ are defined by (XIII.2.39) with

$$(r+1)\ell' + \dfrac{1}{2}\gamma' - \pi \leq \theta(\xi) \leq -(r+1)\ell - \dfrac{1}{2}\gamma' + \pi$$

in $\hat{D}(N)$. This implies that $-\dfrac{\pi}{2} + \dfrac{3\gamma'}{2} < \omega_j + \theta(\xi) < \dfrac{\pi}{2} - \dfrac{3\gamma'}{2}$. In this case, ξ_0 is not on arc (A-3), but $\theta(\xi)$ is still defined in a way similar to the previous cases.

For all of the cases considered above, we prove the following lemma.

Lemma XIII-2-6. *There exists a positive constant H independent of h such that the inequalities*

(XIII.2.43) $$\int_{\hat{L}_{j\xi}} |s|^{-h} \left| \exp\left[-\dfrac{\mu_j}{r+1} s \right] \right| |ds| < H |\xi|^{-h} \left| \exp\left[-\dfrac{\mu_j}{r+1} \xi \right] \right|$$
$$(j = 1, 2, \dots, n)$$

hold in a domain $\hat{D}(N_h)$ for any positive number h, where N_h is a sufficiently large positive constant depending only on h.

Proof.

First let $j \in G_1 \cup G_2$. Then,

$$
\begin{aligned}
\left| \exp\left[-\frac{\mu_j}{r+1} s \right] \right| &= \left| \exp\left[-\frac{\mu_j}{r+1} \{ \xi_0 + t \exp(i \arg(\xi - \xi_0)) \} \right] \right| \\
&= \left| \exp\left[-\frac{\mu_j}{r+1} \xi_0 \right] \right| \left| \exp\left[-\frac{\mu_j t}{r+1} \exp(i \arg(\xi - \xi_0)) \right] \right| \\
&= \left| \exp\left[-\frac{\mu_j}{r+1} \xi_0 \right] \right| \exp\left[-\frac{|\mu_j| t}{r+1} \cos(\omega_j + \arg(\xi - \xi_0)) \right]
\end{aligned}
$$

for $\xi \in \hat{\mathcal{D}}(N)$. By virtue of (XIII.2.28), (XIII.2.29), and (XIII.2.35), it holds that

$$
\frac{1}{2}\pi + \frac{1}{2}\gamma' < \omega_j + \arg(\xi - \xi_0)) < \frac{3}{2}\pi - \frac{1}{2}\gamma' \qquad (j \in G_1),
$$

$$
-\frac{3}{2}\pi + \frac{1}{2}\gamma' < \omega_j + \arg(\xi - \xi_0)) < -\frac{1}{2}\pi - \frac{1}{2}\gamma' \qquad (j \in G_2).
$$

Therefore, $-\cos(\omega_j + \arg(\xi - \xi_0)) > \sin\left(\frac{1}{2}\gamma'\right)$ for $\xi \in \hat{\mathcal{D}}(N_h)$ and $j \in G_1 \cup G_2$. Moreover, (XIII.2.34) implies

$$
|\arg(\xi - \xi_0) - \arg \xi_0| < \frac{1}{2}(r+1)(\ell + \ell') + \frac{1}{2}\gamma',
$$

and (XIII.2.28) and (XIII.2.29) imply $(r+1)(\ell + \ell') < \pi - \gamma'$. Hence,

$$
(\text{XIII.2.44}) \qquad\qquad |\arg(\xi - \xi_0) - \arg \xi_0| < \frac{1}{2}\pi - \frac{1}{2}\gamma'.
$$

Observe that $|\xi_0| = N_h^{r+1}$ implies $|s|^2 = M^2 + t^2 + 2\hat{c}Mt$, where $M = N_h^{r+1}$ and $\hat{c} = -\cos(\pi - \theta)$ with $\theta = \arg(\xi - \xi_0) - \arg \xi_0$ (cf. Figure 4).

Let $b = \sin\left(\frac{1}{2}\gamma'\right)$. Then,

$$
(\text{XIII.2.45}) \qquad\qquad 0 < b < \hat{c} < 1
$$

(cf. (XIII.2.44)). Set

$$
Y(\tau) = (M^2 + \tau^2 + 2\hat{c}M\tau)^{h/2} \exp(M_j \tau) \int_0^\tau \frac{\exp(-M_j t)}{(M^2 + t^2 + 2\hat{c}Mt)^{h/2}} dt,
$$

where $M_j = \dfrac{|\mu_j| \cos(\omega_j + \arg(\xi - \xi_0))}{r+1}$. Then,

$$
(\text{XIII.2.46}) \qquad\qquad \frac{dY}{d\tau} = \left\{ M_j + h\frac{\tau + \hat{c}M}{M^2 + \tau^2 + 2\hat{c}M\tau} \right\} Y + 1
$$

and $M_j < -\dfrac{|\mu_j|b}{r+1} < 0$. From (XIII.2.45), it follows that $\dfrac{\tau + \hat{c}M}{M^2 + \tau^2 + 2\hat{c}M\tau} <$ $\dfrac{1}{Mb}$ $(\tau \geq 0)$. If M satisfies an inequality $M > \dfrac{2h(r+1)}{|\mu_j|b^2}$, then $\dfrac{dY}{d\tau} < -\dfrac{|\mu_j|b}{2(r+1)}Y +$ 1. Since $Y = 0$ for $\tau = 0$, we obtain $Y(\tau) < \dfrac{2(r+1)}{|\mu_j|b}$ $(\tau \geq 0)$. This implies that

$$
Y(|\xi - \xi_0|) = |\xi|^h \left| \exp\left\{\frac{\mu_j(\xi - \xi_0)}{r+1}\right\} \right|
$$
$$
\times \int_{L_{j\xi}} |s|^{-h} \left| \exp\left[-\frac{\mu_j t}{r+1} \exp\{i\arg(\xi - \xi_0)\} \right] \right| |ds| < \frac{2(r+1)}{|\mu_j|b}.
$$

Therefore, (XIII.2.43) follows. In this case, $H = \dfrac{2(r+1)}{\mu b}$, where $\mu = \min\limits_{j}\{\mu_j\}$.

In other cases (i.e., $G_3, G_4, \hat{G}_1, \hat{G}_2$, and $\hat{G}_3$), the lines $\hat{L}_{j\xi}$ are given by (XIII.2.39), where $\theta(\xi)$ satisfies $-\dfrac{1}{2}\pi + \dfrac{1}{2}\gamma' < \omega_j + \theta(\xi) < \dfrac{1}{2}\pi - \dfrac{1}{2}\gamma'$ in $\hat{\mathcal{D}}(N_h)$. Therefore,

$$
\text{(XIII.2.47)} \qquad \cos(\omega_j + \theta(\xi)) > \sin\left(\frac{1}{2}\gamma'\right) = b
$$

for $\xi \in \hat{\mathcal{D}}(N_h)$.

(i) Consider the case when $|\theta(\xi) - \arg \xi| < \dfrac{1}{2}\pi$ for a given point $\xi \in \hat{\mathcal{D}}(N_h)$. Then,

$$
|s|^2 = |\xi|^2 + t^2 - 2|\xi|t\cos(\pi - |\theta(\xi) - \arg\xi|) \geq |\xi|^2.
$$

Therefore,

$$
\int_{\hat{L}_{j\xi}} |s|^{-h} \left| \exp\left[-\frac{\mu_j s}{r+1} \right] \right| |ds| < |\xi|^{-h} \int_0^\infty \left| \exp\left[-\frac{\mu_j}{r+1}\{\xi + t\exp(i\theta(\xi))\} \right] \right| dt
$$
$$
= |\xi|^{-h} \left| \exp\left[-\frac{\mu_j \xi}{r+1} \right] \right| \left| \int_0^\infty \exp\left[-\frac{\mu_j t}{r+1}\cos(\omega_j + \theta(\xi)) \right] dt \right|
$$
$$
< |\xi|^{-h} \left| \exp\left[-\frac{\mu_j \xi}{r+1} \right] \right| \frac{(r+1)}{|\mu_j|b}.
$$

This implies (XIII.2.43).

(ii) Consider the case when $\dfrac{1}{2}\pi < |\theta(\xi) - \arg\xi| < \pi$ for a given point $\xi \in \hat{\mathcal{D}}(N_h)$. Since the lines $\hat{L}_{j\xi}$ are in $\hat{\mathcal{D}}(N_h)$, the distance D from the origin to $\hat{L}_{j\xi}$ is not less than $M = N_h^{r+1}$. Then, $|s|^2$ can be written in the form $|s|^2 = D^2 + \sigma^2$ (cf. Figure 5).

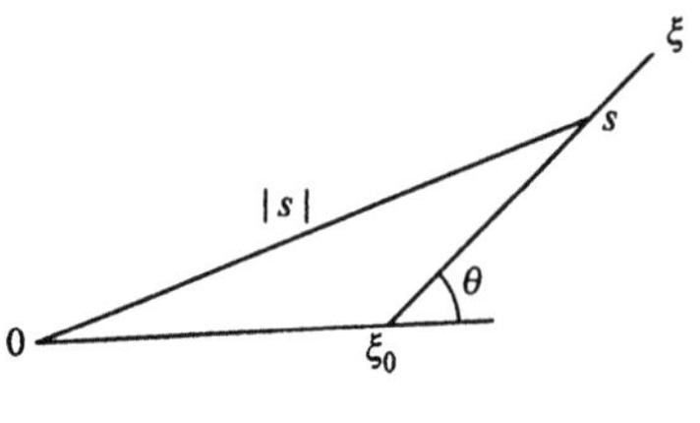

FIGURE 4.

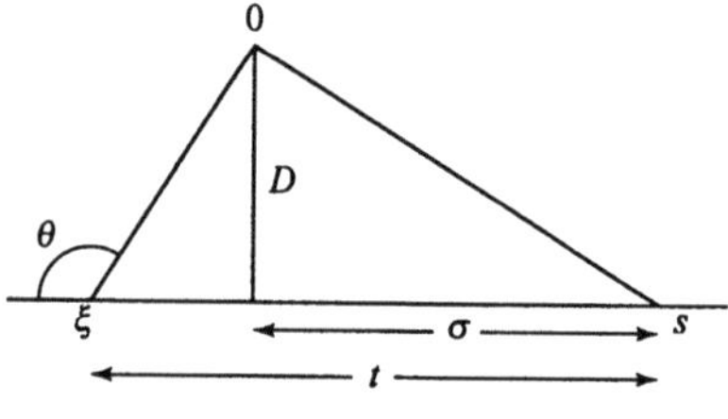

FIGURE 5.

Consider a function $Y(\tau) = (D^2 + \tau^2)^{h/2} \exp(M_j \tau) \int_\tau^\infty \dfrac{\exp(-M_j \sigma)}{(D^2 + \sigma^2)^{h/2}} d\sigma$, where

$M_j = \dfrac{|\mu_j| \cos(\omega_j + \theta(\xi))}{r+1}$. Note that $b > 0$. Hence, for $\tau \geq 0$, (XIII.2.47) implies

that $Y(\tau) < \dfrac{1}{M_j}$. On the other hand, $Y(\tau)$ satisfies the differential equation $\dfrac{dY}{d\tau} =$

$\left\{ M_j + h \dfrac{\tau}{D^2 + \tau^2} \right\} Y - 1$. Since $\dfrac{\tau}{D^2 + \tau^2} < \dfrac{1}{2D}$, it follows that $\dfrac{dY}{d\tau} > \dfrac{|\mu_j| b}{2(r+1)} Y - 1$

if $M > \dfrac{h(r+1)}{|\mu_j| b}$. Here, a use was made of the inequality $M \leq D$. Since $Y(0) <$

$\dfrac{1}{M_j} < \dfrac{(r+1)}{|\mu_j| b}$, we obtain $Y(\tau) < \dfrac{2(r+1)}{|\mu_j| b}$ $(\tau < 0)$. Therefore, (XIII.2.43) follows.

This completes the proof of Lemma XIII-2-6. In this case, again, $H = \dfrac{2(r+1)}{\mu b}$,

where $\mu = \min_j \{\mu_j\}$. $\quad\square$

We fix the paths of integration on the right-hand sides of (XIII.2.15) as explained above. The constant H in (XIII.2.13) is chosen to be equal to the constant H given in Lemma XIII-2-6. Note that H is independent of h.

XIII-3. Proof of Theorem XIII-1-2

Let us construct a solution $u_j = \phi_j(x)$ of (XIII.2.15) so that

$$(XIII.3.1) \qquad\qquad \phi_j(x) \simeq 0 \qquad (j = 1, 2, \dots, n)$$

as $x \to \infty$ in (XIII.1.15).

Observation XIII-3-1. As in the previous section, denote by $\mathcal{D}_0(N_h)$ the domain in the x-plane which corresponds to the domain $\hat{\mathcal{D}}(N_h)$ in the ξ-plane. The domain $\mathcal{D}_0(N_h)$ depends on h. Set $\xi_0^{(h)} = N_h^{r+1} \exp\left[i \dfrac{1}{2}(r+1)(\ell + \ell') \right]$, and denote by $x_0^{(h)}$ the corresponding point in $\mathcal{D}_0(N_h)$. Also, setting $h' = \dfrac{h+1}{r+1}$ $(h = 1, 2, 3, \dots)$, consider the domain $\mathcal{D}_0(N_{h'})$. As mentioned in the previous section, the constant H in (XIII.2.13) is chosen to be equal to the constant H in Lemma XIII-2-6. Choose a positive constant $\hat{\delta}$ so that $\dfrac{H}{r+1}\left\{ c\hat{\delta} + \dfrac{b_1}{N_{h'}^2} \right\} \leq \hat{\delta}$. Furthermore, by virtue of (XIII.2.13), $N_{h'}$ can be chosen so large that $N_{h'} \geq \tilde{N}_0$ and $\hat{\delta} \leq \delta$ (cf. (XIII.2.12)). Fix $N_{h'}$ in this way. Then, by a method similar to that in §XII-3 for each j $(j = 1, 2, \dots, n)$, successive approximations can be defined to construct a solution $u_j = \phi_j(x)$ of (XIII.2.15) which is holomorphic in $\mathcal{D}_0(N_{h'})$ and $|\phi_j(x)| \leq \hat{\delta}|\xi|^{-1}$ $(j = 1, 2, \dots, n)$, where $\xi = x^{r+1}$. In this way, the existence of a bounded solution $u_j = \phi_j(x)$ $(j = 1, 2, \dots, n)$ of system (XIII.2.15) is proved.

Observation XIII-3-2. In order to prove that $\phi_j(x) \simeq 0 \, (j = 1, 2, \ldots, n)$ as $x \to \infty$ in $\mathcal{D}_0(N_{h'})$, use the fact that, for each positive integer k, the functions $\phi_j(x) \, (j = 1, 2, \ldots, n)$ also satisfy the integral equations

$$(\text{XIII.3.2}) \qquad u_j(x) = \begin{cases} \phi_j(x_0^{(k)}) \exp\left[\dfrac{\mu_j}{r+1}(x^{r+1} - (x_0^{(k)})^{r+1})\right] \\ \qquad + \displaystyle\int_{L_{jx}} t^r R_j(t, \vec{u}(t)) \exp\left[\dfrac{-\mu_j}{r+1}(t^{r+1} - x^{r+1})\right] dt \\ \qquad\qquad\qquad\qquad\qquad\qquad (j \in G_1 \cup G_2), \\[4pt] \displaystyle\int_{L_{jx}} t^r R_j(t, \vec{u}(t)) \exp\left[\dfrac{-\mu_j}{r+1}(t^{r+1} - x^{r+1})\right] dt \\ \qquad\qquad\qquad\qquad\qquad\qquad (j \notin G_1 \cup G_2) \end{cases}$$

in the domain $\mathcal{D}_0(N_k)$. Here, $\vec{u}(x)$ denotes the $\mathbb{C}^n$-valued function whose entries are $(u_1(x), u_2(x), \ldots, u_n(x))$. Note that we can assume $\mathcal{D}_0(N_k) \subset \mathcal{D}_0(N_{h'})$ without loss of generality.

Upon applying successive approximations similar to those of §XII-3 together with Lemma XIII-2-6 to (XIII.3.2), it can be proved that (XIII.3.2) has a solution $u_j(x) = \psi_j(x) \, (j = 1, 2, \ldots, n)$ such that $|\psi_j(x)| \leq C_k|x|^{-k} \, (j = 1, 2, \ldots, n)$ in $\mathcal{D}_0(N_k)$, where C_k is a suitable positive number. Also, using Lemma XIII-2-6 and (XIII.2.13), it can be shown that $\phi_j(x) = \psi_j(x) \, (j = 1, 2, \ldots, n)$ in $\mathcal{D}_0(N_k)$, since

$$\phi_j(x) - \psi_j(x) = \int_{L_{jx}} t^r [R_j(t, \vec{\phi}(t)) - R_j(t, \vec{\psi}(t))] \exp\left[\frac{-\mu_j}{r+1}(t^{r+1} - x^{r+1})\right] dt,$$

where $j = 1, 2, \ldots, n$, and $\vec{\phi}(x)$ and $\vec{\psi}(x)$ denote $\mathbb{C}^n$-valued functions whose entries are $(\phi_1(x), \phi_2(x), \ldots, \phi_n(x))$ and $(\psi_1(x), \psi_(x), \ldots, \psi_n(x))$, respectively. Thus, we conclude that $\phi_j(x) \simeq 0 \, (j = 1, 2, \ldots, n)$ as $x \to \infty$ in $\mathcal{D}_0(N_{h'})$.

Now, let us return to Observation XIII-2-2. If we set $p_j(x) = \phi_j(x) + z_j(x)$, the solution $v_j = p_j(x) \, (j = 1, 2, \ldots, n)$ of (XIII.1.1) satisfies all of the requirements of Theorem XIII-1-2. $\square$

Remark XIII-3-3. Let $\vec{f}(x, \vec{y}, \vec{\mu}, \epsilon)$ be a $\mathbb{C}^n$-valued function of a variable $(x, \vec{y}, \vec{\mu}, \epsilon) \in \mathbb{C} \times \mathbb{C}^n \times \mathbb{C}^m \times \mathbb{C}$ such that
 (a) the entries of $\vec{f}$ are bounded and holomorphic in a domain

$$\mathcal{D}_0 = \{\, (x, \vec{y}, \vec{\mu}, \epsilon) : |x| > R_0, |\vec{y}| < \delta_0, |\mu| < \mu_0, |\epsilon| < \epsilon_0, |\arg \epsilon| < \rho_0 \},$$

where R_0, δ_0, μ_0, ϵ_0, and ρ_0 are some positive constants,
 (b) $\vec{f}$ admits a uniform asymptotic expansion in $\mathcal{D}_0$

$$\vec{f}(x, \vec{y}, \vec{\mu}, \epsilon) \simeq \sum_{h=0}^{+\infty} \epsilon^h \vec{f}_h(x, \vec{y}, \vec{\mu}) \qquad \text{as} \quad \epsilon \to 0,$$

where coefficients $\vec{f}_h(x, \vec{y}, \vec{\mu})$ are single-valued, bounded, and holomorphic in the domain

$$\Delta_0 = \{(x, \vec{y}, \vec{\mu}) : |x| > R_0, \quad |\vec{y}| < \delta_0, \quad |\mu| < \mu_0\},$$

(c) if $\vec{f}(x, \vec{y}, \vec{\mu}, \epsilon) = \vec{F}_0(x, \vec{\mu}, \epsilon) + A(x, \vec{\mu}, \epsilon)\vec{y} + O(|\vec{y}|^2)$, then

$$\vec{F}_0(\infty, \vec{\mu}, \epsilon) = \vec{0} \quad \text{and} \quad \det A(\infty, \vec{0}, 0) \neq 0.$$

Let $\lambda_1, \lambda_2, \ldots, \lambda_n$ be n eigenvalues of $A(\infty, \vec{0}, 0)$ and set $\omega_j = \arg \lambda_j$ $(j = 1, 2, \ldots, n)$. Note that the ω_j are not unique. Set also $\mathcal{S}(\rho_1, \rho_2) = \{x : \rho_1 < \arg x < \rho_2\}$. Let σ and r be two non-negative integers. Assume that the sector $\mathcal{S}(\rho_1, \rho_2)$ is contained in the sector

$$-\frac{3\pi}{2} - \min_j\{\omega_j\} + \sigma\rho_0 < (r+1)\arg x < \frac{3\pi}{2} - \max_j\{\omega_j\} - \sigma\rho_0$$

for a suitable choice of $\{\omega_j : j = 1, 2, \ldots, n\}$.

Under these assumptions, we can prove the following theorem in a way similar to the proof of Theorem XIII-1-2.

Theorem XIII-3-4. *The system* $\epsilon^\sigma \dfrac{d\vec{y}}{dx} = x^r \vec{f}(x, \vec{y}, \vec{\mu}, \epsilon)$ *has two solutions* $\vec{\phi}(x, \vec{\mu}, \epsilon)$ *and* $\vec{\psi}(x, \vec{\mu}, \epsilon)$ *such that*

(i) these two solutions are bounded and holomorphic in the domain

$$\mathcal{S}_0 = \{(x, \vec{\mu}, \epsilon) : |x| > N, \ x \in \mathcal{S}(\rho_1, \rho_2), \ |\vec{\mu}| < \mu_1, \ 0 < |\epsilon| < \epsilon_1, |\arg \epsilon| < \rho_0\},$$

if $\dfrac{1}{N}$, μ_1 *and* ϵ_1 *are sufficiently small,*

(ii) the solution $\vec{\phi}(x, \vec{\mu}, \epsilon)$ *admits an uniform asymptotic expansion*

$$\vec{\phi}(x, \vec{\mu}, \epsilon) \simeq \sum_{k=1}^{+\infty} x^{-k} \vec{\phi}_k(\vec{\mu}, \epsilon) \qquad \text{as} \quad x \to \infty$$

in $\mathcal{S}_0$, *where coefficients* $\vec{\phi}_k(\vec{\mu}, \epsilon)$ *are bounded, holomorphic, and admit uniform asymptotic expansions in powers of* ϵ *as* $\epsilon \to 0$ *in the domain* $\{(\vec{\mu}, \epsilon) : |\vec{\mu}| < \mu_1, \ 0 < |\epsilon| < \epsilon_1, \ |\arg \epsilon| < \rho_0\}$,

(iii) the solution $\vec{\psi}(x, \vec{\mu}, \epsilon)$ *admits an uniform asymptotic expansion:*

$$\vec{\psi}(x, \vec{\mu}, \epsilon) \simeq \sum_{h=0}^{+\infty} \epsilon^h \vec{\psi}_h(x, \vec{\mu}) \qquad \text{as} \quad \epsilon \to 0$$

in $\mathcal{S}_0$, *where coefficients* $\vec{\psi}_h(x, \vec{\mu})$ *are bounded, holomorphic, and admit uniform asymptotic expansions in powers of* x^{-1} *as* $x \to \infty$ *in the domain* $\{(x, \vec{\mu}) : |x| > N, \ x \in \mathcal{S}(\rho_1, \rho_2), \ |\vec{\mu}| < \mu_1\}$.

Furthermore, there exist functions $\vec{\phi}_\ell(x, \vec{\mu}, \epsilon)$ *for* $\ell = 0, 1, \ldots$ *such that*

(a) these functions $\vec{\phi}_\ell$ *satisfy conditions (i) and (ii) given above,*

(b)

$$\vec{\psi}(x, \vec{\mu}, \epsilon) = \sum_{h=0}^{\ell} \epsilon^h \vec{\psi}_h(x, \vec{\mu}) + \epsilon^{\ell+1} \vec{\phi}_\ell(x, \vec{\mu}, \epsilon).$$

A complete proof of Theorem XIII-3-4 is found in [Si7] and [Si10].

Remark XIII-3-5. In the proof of Theorem XIII-1-2, we used the assumption that the matrix A_0 on the right-hand side of (XIII.1.10) is invertible (cf. Assumption III of §XIII-1). Without such an assumption, we can prove the following theorem.

Theorem XIII-3-6. *Let $F(x, y_0, y_1, \ldots, y_n)$ be a nonzero polynomial in $y_0, y_1, \ldots,$ y_n whose coefficients are convergent power series in x^{-1}, and let $p(x) = \displaystyle\sum_{m=0}^{\infty} a_m x^{-m}$ $\in \mathbb{C}[[x^{-1}]]$ be a formal solution of the differential equation*

$$(\text{XIII.3.3}) \qquad F\left(x, y, \frac{dy}{dx}, \ldots, \frac{d^n y}{dx^n}\right) = 0.$$

Then, for any given direction $\arg x = \theta$, there exist two positive numbers δ and α and a function $\phi(x)$ such that

(i) $\phi(x)$ is holomorphic in the sector $\mathcal{S} = \{x \in \mathbb{C} : 0 < |x| < \delta, \ |\arg x - \theta| < \alpha\}$,

(ii) $\phi(x)$ admits the formal solution $p(x)$ as its asymptotic expansion as $x \to \infty$ in $\mathcal{S}$,

(iii) $\phi(x)$ satisfies differential equation (XIII.3.3) in $\mathcal{S}$.

The main idea of the proof is similar to the proof of Theorem XIII-1-2. However, we need the thorough knowledge of the structure of solutions of a linear system $x\dfrac{d\vec{y}}{dx} = A(x)\vec{y}$ that we shall explain in §XIII-6. Also, it is more difficult to define the paths of integration (cf. [I] and [RS1]). A complete proof of Theorem XIII-3-6 is found in [RS1]. See, also, §XIII-8.

XIII-4. A block-diagonalization theorem

Consider a system of linear differential equations

$$(\text{XIII.4.1}) \qquad \frac{d\vec{y}}{dx} = x^r A(x)\vec{y},$$

where r is a positive integer, $\vec{y} \in \mathbb{C}^n$, and $A(x)$ is an $n \times n$ matrix. The entries of $A(x)$ are holomorphic with respect to a complex variable x in a sector

$$(\text{XIII.4.2}) \qquad |x| > N_0, \qquad |\arg x| < \alpha_0,$$

where N_0 and α_0 are positive numbers. Assume that the matrix $A(x)$ admits an asymptotic expansion in the sense of Poincaré

$$(\text{XIII.4.3}) \qquad A(x) \simeq \sum_{\nu=0}^{\infty} x^{-\nu} A_\nu$$

as $x \to \infty$ in sector (XIII.4.2), where coefficients A_ν are constant $n \times n$ matrices. Suppose also that $A_0 = A(\infty)$ has ℓ distinct eigenvalues $\lambda_1, \lambda_2, \ldots, \lambda_\ell$ with multiplicities $n_1, n_2, \ldots, n_\ell$, respectively ($n_1 + n_2 + \cdots + n_\ell = n$). Without loss of generality, assume that A_0 is in a block-diagonal form:

$$(\text{XIII.4.4}) \qquad A_0 = \mathrm{diag}\left[\mathring{A}_1, \mathring{A}_2, \ldots, \mathring{A}_\ell\right],$$

where $\mathring{A}_j$ are $n_j \times n_j$ matrices in the form

$$(\text{XIII.4.5}) \qquad \mathring{A}_j = \lambda_j I_{n_j} + \mathcal{N}_j \qquad (j = 1, 2, \dots, \ell).$$

Here, $\mathcal{N}_j$ is a lower-triangular nilpotent $n_j \times n_j$ matrix. The main result of this section is the following theorem (cf. [Si7]).

Theorem XIII-4-1. *Under the assumption (XIII.4.3) and (XIII.4.4), there exists an $n \times n$ matrix $P(x)$ whose entries are holomorphic in a sector*

$$(\text{XIII.4.6}) \qquad |x| > N_1, \qquad |\arg x| < \alpha_1,$$

where N_1^{-1} and α_1 are sufficiently small positive numbers, such that
(i) $P(x)$ admits an asymptotic expansion

$$(\text{XIII.4.7}) \qquad P(x) \simeq \sum_{\nu=0}^{\infty} x^{-\nu} P_\nu \qquad (P_0 = I_n),$$

as $x \to \infty$ in sector (XIII.4.6), where coefficients P_ν are constant $n \times n$ matrices,
(ii) the transformation

$$(\text{XIII.4.8}) \qquad \vec{y} = P(x)\vec{z}$$

reduces system (XIII.4.1) to

$$(\text{XIII.4.9}) \qquad \frac{d\vec{z}}{dx} = x^r B(x)\vec{z},$$

where $B(x)$ is in a block-diagonal form

$$(\text{XIII.4.10}) \qquad B(x) = \mathrm{diag}\,[B_1(x), B_2(x), \dots, B_\ell(x)].$$

Here, $B_j(x)$ are $n_j \times n_j$ matrices and admit asymptotic expansions

$$(\text{XIII.4.11}) \qquad B_j(x) \simeq \sum_{\nu=0}^{\infty} x^{-\nu} B_{j\nu}$$

as $x \to \infty$ in (XIII.4.6), where coefficients $B_{j\nu}$ are constant $n_j \times n_j$ matrices.

Proof.

From (XIII.4.1), (XIII.4.8), and (XIII.4.9), we derive the equation

$$(\text{XIII.4.12}) \qquad \frac{dP}{dx} = x^r [A(x)P - PB(x)]$$

that determines the matrices $P(x)$ and $B(x)$. Set

$$(\text{XIII.4.13}) \quad A(x) = \mathring{A}_0 + E(x), \qquad B(x) = \mathring{A}_0 + F(x), \qquad P(x) = I_n + Q(x).$$

Then, $E(x) = O(x^{-1})$, $F(x) = O(x^{-1})$, and $Q(x) = O(x^{-1})$. Furthermore, (XIII.4.12) becomes

$$\text{(XIII.4.14)} \qquad \frac{dQ}{dx} = x^r[A_0 Q - Q A_0 + E - F + EQ - QF].$$

Write each of three matrices $E(x)$, $F(x)$, and $Q(x)$ in a block-matrix form according to that of A_0 in (XIII.4.4), i.e.,

(XIII.4.15)

$$\left\{ \begin{array}{l} F(x) = \operatorname{diag}\left[F_1, F_2, \ldots, F_\ell\right], \\[2ex] E(x) = \begin{bmatrix} E_{11} & E_{12} & \cdots & E_{1\ell} \\ E_{21} & E_{22} & \cdots & E_{2\ell} \\ \vdots & \vdots & \vdots & \vdots \\ E_{\ell 1} & E_{\ell 2} & \cdots & E_{\ell\ell} \end{bmatrix}, \qquad Q(x) = \begin{bmatrix} Q_{11} & Q_{12} & \cdots & Q_{1\ell} \\ Q_{21} & Q_{22} & \cdots & Q_{2\ell} \\ \vdots & \vdots & \vdots & \vdots \\ Q_{\ell 1} & Q_{2\ell} & \cdots & Q_{\ell\ell} \end{bmatrix}, \end{array} \right.$$

where E_{jk} and Q_{jk} are $n_j \times n_k$ matrices and F_j are $n_j \times n_j$ matrices. Set

$$\text{(XIII.4.16)} \qquad Q_{jj} = O \qquad (j = 1, 2, \ldots, \ell).$$

From (XIII.4.4), (XIII.4.14), (XIII.4.15), and (XIII.4.16), it follows that

$$\text{(XIII.4.17)} \qquad F_j = E_{jj} + \sum_{h \neq j} E_{jh} Q_{hj} \qquad (j = 1, 2, \ldots, \ell)$$

and

$$\text{(XIII.4.18)} \quad \frac{dQ_{jk}}{dx} = x^r\left[\mathring{A}_j Q_{jk} - Q_{jk}\mathring{A}_k + E_{jk} + \sum_{h \neq k} E_{jh} Q_{hk} - Q_{jk} F_k\right] \quad (j \neq k).$$

Substituting (XIII.4.17) into (XIII.4.18), a system of nonlinear differential equations

(XIII.4.19)

$$\begin{aligned} \frac{dQ_{jk}}{dx} = x^r\Bigg[&\mathring{A}_j Q_{jk} - Q_{jk}\mathring{A}_k + \sum_{h \neq k} E_{jh} Q_{hk} \\ &- Q_{jk}\Big(E_{kk} + \sum_{h \neq k} E_{kh} Q_{hk}\Big) + E_{jk}\Bigg] \qquad (j \neq k) \end{aligned}$$

is obtained. Since it is assumed that $\lambda_1, \ldots, \lambda_\ell$ are distinct eigenvalues of A_0 and that A_0 is in the block-diagonal form (XIII.4.5), upon applying Theorem XIII-1-2 to (XIII.4.19) we can construct a desired holomorphic solution $Q_{jk}(x)$ of (XIII.4.19) which admit an asymptotic expansion $Q_{jk}(x) \simeq \sum_{\nu=1}^{\infty} x^{-\nu} Q_{jk\nu}$ $(j, k = 1, 2, \ldots, s; j \neq k)$, where $Q_{jk\nu}$ are constant n_j by n_k matrices. Defining F_j by (XIII.4.17) and then $B(x)$ by (XIII.4.13), the proof of Theorem XIII-4-1 is completed. $\square$

Theorem XIII-4-1 concerns the behavior of solutions of system (XIII.4.1) near $x = \infty$. Since it is useful to give a similar result concerning behavior of solutions near $x = 0$, we consider, hereafter in this section, a system of differential equations

$$\text{(XIII.4.20)} \qquad x^{d+1}\frac{d\vec{y}}{dx} = A(x)\vec{y},$$

where d is a positive integer and the entries of $n \times n$ matrix $A(x)$ are holomorphic in a neighborhood of $x = 0$. Also, assume that $A(0)$ is in a block-diagonal form

$$(\text{XIII.4.21}) \qquad A(0) \;=\; \text{diag}\,[\lambda_1 I_{n_1} + \mathcal{N}_1, \lambda_2 I_{n_2} + \mathcal{N}_2, \dots, \lambda_\ell I_{n_\ell} + \mathcal{N}_\ell]\,,$$

where $\lambda_1, \dots, \lambda_\ell$ are distinct eigenvalues of $A(0)$ with multiplicities $n_1, n_2, \dots, n_\ell$, respectively $(n_1 + n_2 + \cdots + n_\ell = n)$, and, for each j, $\mathcal{N}_j$ is a lower-triangular and nilpotent $n_j \times n_j$ matrix.

Comparing the present situation with that of Theorem XIII-4-1, we notice the following two differences:

(a) singularity is at $x = 0$ in the present situation, while singularity is at $x = \infty$ in Theorem XIII-4-1,

(b) The power series expansion of $A(x)$ is convergent in the present situation, while $A(x)$ in Theorem XIII-4-1 admits only an asymptotic expansion in a sector containing the direction $\arg x = 0$.

We can change any singularity at $x = 0$ to a singularity at $x = \infty$ by changing the independent variable x by $\dfrac{1}{x}$. Also, any direction $\arg x = \theta$ can be changed to the direction $\arg x = 0$ by rotating the independent variable x. Furthermore, the asymptotic expansion $\hat{P}$ of $P_\theta(x)$ and the expansion $\hat{B}$ of B_θ are formal power series satisfying the equation $x^{d+1}\dfrac{d\hat{P}}{dx} = A\hat{P} - \hat{P}\hat{B}$. This implies that two matrices $\hat{P}$ and $\hat{B}$ are independent of θ. Hence, using Corollary XIII-1-3, the following result is obtained.

Theorem XIII-4-2. *Let $A(x)$ be an $n \times n$ matrix whose entries are holomorphic in a neighborhood of $x = 0$. Also, let d be a positive integer. Assume that the matrix $A(0)$ is in block-diagonal form (XIII.4.21), where $\lambda_1, \dots, \lambda_\ell$ are distinct eigenvalues of $A(0)$ with multiplicities $n_1, n_2, \dots, n_\ell$, respectively $(n_1 + n_2 + \cdots + n_\ell = n)$, and for each j, $\mathcal{N}_j$ is a lower-triangular and nilpotent $n_j \times n_j$ matrix. Fix a real number θ so that $(\lambda_j - \lambda_k)e^{-id\theta} \notin \mathbb{R}$ for $j \neq k$. Then, there exists two positive numbers δ_θ and ϵ_θ and an $n \times n$ matrix $P_\theta(x)$ such that*

(a) the entries of $P_\theta(x)$ are holomorphic and admit asymptotic expansions in powers of x as $x \to 0$ in the sector $\mathcal{S}_\theta = \left\{ x : 0 < |x| < \delta_\theta, \, |\arg x - \theta| < \dfrac{\pi}{2k} + \epsilon_\theta \right\}$,

(b) if $\displaystyle\sum_{m=0}^{\infty} x^m P_m$ is the asymptotic expansion of $P_\theta(x)$, then this expansion is independent of θ and $P_0 = I_n$,

(c) the transformation $\vec{y} = P_\theta(x)\vec{u}$ changes system (XIII.4.20) to a system

$$(\text{XIII.4.22}) \qquad\qquad x^{d+1}\dfrac{d\vec{u}}{dx} \;=\; B_\theta(x)\vec{u},$$

where the matrix $B_\theta(x)$ is in a block-diagonal form

$$(\text{XIII.4.23}) \qquad\qquad B_\theta(x) \;=\; \text{diag}\,[B_{1\theta}(x), B_{2\theta}(x), \dots, B_{\ell\theta}(x)]\,.$$

For each j, $B_{j\theta}$ is an $n_j \times n_j$ matrix which admits also an asymptotic expansion in x as $x \to 0$ in $\mathcal{S}_\theta$.

The main claims of this theorem are

(i) the asymptotic expansion of $P_\theta(x)$ is independent of θ,

(ii) the opening of the sector $\mathcal{S}_\theta$ is larger than $\dfrac{\pi}{k}$.

Proof of Theorem XIII-4-2 is left to the reader as an exercises.

Remark XIII-4-3. Using Theorem XIII-3-4, we can generalize Theorem XIII-4-1 to the system $\epsilon^\sigma \dfrac{d\vec{y}}{dx} = x^r A(x, \vec{\mu}, \epsilon)\vec{y}$, where r and σ are non-negative integers, $\vec{y} \in \mathbb{C}^n$, and $A(x, \vec{\mu}, \epsilon)$ is an $n \times n$ matrix with the entries that are bounded and holomorphic with respect to the variable $(x, \vec{\mu}, \epsilon) \in \mathbb{C} \times \mathbb{C}^m \times \mathbb{C}$ in a domain $\Delta_0 = \{(x, \vec{\mu}, \epsilon) : |x| > N,\ |\vec{\mu}| < \mu_0,\ 0 < |\epsilon| < \epsilon_0,\ 0 < |\arg \epsilon| < \rho_0\}$. Also, assume that A admits a uniform asymptotic expansion $A(x, \vec{\mu}, \epsilon) \simeq \sum_{h=0}^{\infty} \epsilon^h A_h(x, \vec{\mu})$ in Δ_0 as $\epsilon \to 0$, where coefficients $A_h(x, \vec{\mu})$ are bounded and holomorphic in the domain $\{(x, \vec{\mu}) : |x| > N,\ |\vec{\mu}| < \mu_0\}$. A complete proof of this result is found in [Si7] and [Hs2].

XIII-5. Cyclic vectors (A lemma of P. Deligne)

In the study of singularities, a single n-th-order differential equation is, in many cases, easier to treat than a system of differential equations. In this section, we explain equivalence between a system of linear differential equations and a single n-th-order linear differential equation.

Let us denote by $\mathcal{K}$ the field of fractions of the ring $\mathbb{C}[[x]]$ of formal power series in x, i.e.,

$$\mathcal{K} = \left\{ \frac{p}{q} :\ p \in \mathbb{C}[[x]],\ q \in \mathbb{C}[[x]],\ q \neq 0 \right\}.$$

Also, denote by $\mathcal{V}$ the set of all *row vectors* $(c_1(x),\ c_2(x),\ \ldots,\ c_n(x))$, where the entries are in the field $\mathcal{K}$. The set $\mathcal{V}$ is an n-dimensional vector space over the field $\mathcal{K}$.

Define a linear differential operator $\mathcal{L} : \mathcal{V} \to \mathcal{V}$ by $\mathcal{L}[\vec{v}] = \delta\vec{v} + \vec{v}\Omega(x)\ (\vec{v} \in \mathcal{V})$, where $\delta = x\dfrac{d}{dx}$ and $\Omega(x)$ is an $n \times n$ matrix whose entries are in the field $\mathcal{K}$. We first prove the following lemma.

Lemma XIII-5-1 (P. Deligne [Del]). *There exists an element $\vec{v}_0 \in \mathcal{V}$ such that $\{\vec{v}_0, \mathcal{L}\vec{v}_0, \mathcal{L}^2\vec{v}_0, \ldots, \mathcal{L}^{n-1}\vec{v}_0\}$ is a basis for $\mathcal{V}$ as a vector space over $\mathcal{K}$.*

Proof.

For each nonzero element $\vec{v}$ of $\mathcal{V}$, denote by $\mu(\vec{v})$ the largest integer ℓ such that $\{\vec{v}, \mathcal{L}\vec{v}, \mathcal{L}^2\vec{v}, \ldots, \mathcal{L}^\ell\vec{v}\}$ is linearly independent over $\mathcal{K}$. In two steps, we shall derive a contradiction from the assumption that $\max\{\mu(\vec{v}) : \vec{v} \in \mathcal{V}\} < n - 1$.

Step 1. First, we introduce a criterion for linear dependence of a set of elements of $\mathcal{V}$. Consider a set $\{\vec{v}_1,\dots,\vec{v}_m\} \subset \mathcal{V}$, where m is a positive integer not greater than $n = \dim_{\mathcal{K}} \mathcal{V}$. Let $\vec{v}_j = (c_{j1}, c_{j2}, \dots, c_{jn})$ $(j = 1, 2, \dots, m)$. Set $\mathcal{J} = \{(j_1,\dots,j_m) : 1 \le j_1 < j_2 < \cdots < j_m \le n\}$, and introduce a linear order $\mathcal{J} \to \{1, 2, \dots, \binom{n}{m}\}$ in the set $\mathcal{J}$. Let us now define a map

$$\mathcal{V}^m = \{(\vec{v}_1, \vec{v}_2, \dots, \vec{v}_m) : \vec{v}_j \in \mathcal{V} \ (j = 1, \dots, m)\} \ \to \ \mathcal{K}^{\binom{n}{m}}$$

by $(\vec{v}_1, \dots, \vec{v}_m) \to \vec{v}_1 \wedge \vec{v}_2 \wedge \cdots \wedge \vec{v}_m$, where

$$\vec{v}_1 \wedge \vec{v}_2 \wedge \cdots \wedge \vec{v}_m$$
$$= \left(\det \begin{bmatrix} c_{1j_1} & c_{1j_2} & \cdots & c_{1j_m} \\ \vdots & \vdots & \vdots & \vdots \\ c_{mj_1} & c_{mj_2} & \cdots & c_{mj_m} \end{bmatrix} : (j_1, \dots, j_m) \in \mathcal{J} \right).$$

It is easy to verify the following properties of $\vec{v}_1 \wedge \vec{v}_2 \wedge \cdots \wedge \vec{v}_m$:

$$(1) \qquad \begin{aligned} &\vec{v}_1 \wedge \cdots \wedge \vec{v}_{k-1} \wedge (\vec{v}_k + \vec{w}_k) \wedge \vec{v}_{k+1} \wedge \cdots \wedge \vec{v}_{m-1} \wedge \vec{v}_m \\ &= \vec{v}_1 \wedge \cdots \wedge \vec{v}_{k-1} \wedge \vec{v}_k \wedge \vec{v}_{k+1} \wedge \cdots \wedge \vec{v}_{m-1} \wedge \vec{v}_m \\ &\quad + \vec{v}_1 \wedge \cdots \wedge \vec{v}_{k-1} \wedge \vec{w}_k \wedge \vec{v}_{k+1} \wedge \cdots \wedge \vec{v}_{m-1} \wedge \vec{v}_m, \end{aligned}$$

$$(2) \qquad \begin{aligned} &\vec{v}_1 \wedge \cdots \wedge \vec{v}_{k-1} \wedge (\alpha\,\vec{v}_k) \wedge \vec{v}_{k+1} \wedge \cdots \wedge \vec{v}_{m-1} \wedge \vec{v}_m \\ &= \alpha\,(\vec{v}_1 \wedge \cdots \wedge \vec{v}_{k-1} \wedge \vec{v}_k \wedge \vec{v}_{k+1} \wedge \cdots \wedge \vec{v}_{m-1} \wedge \vec{v}_m), \end{aligned}$$

for all $\alpha \in \mathcal{K}$,

$$(3) \qquad \begin{aligned} &\vec{v}_1 \wedge \cdots \wedge \vec{v}_k \wedge \cdots \wedge \cdots \wedge \vec{v}_j \wedge \cdots \wedge \vec{v}_m \\ &= -(\vec{v}_1 \wedge \cdots \wedge \vec{v}_j \wedge \cdots \wedge \cdots \wedge \vec{v}_k \wedge \cdots \wedge \vec{v}_m), \end{aligned}$$

(4) a set $\{\vec{v}_1, \vec{v}_2, \dots, \vec{v}_m\} \in \mathcal{V}$ is linearly dependent if and only if $\vec{v}_1 \wedge \vec{v}_2 \wedge \cdots \wedge \vec{v}_m = \vec{0}$ in $\mathcal{K}^{\binom{n}{m}}$.

Step 2. Fix an element $\vec{v}_0$ of $\mathcal{V}$ such that $\mu(\vec{v}_0) = \max\{\mu(\vec{v}) : \vec{v} \in \mathcal{V}\} < n-1$. Since $\mu(\vec{v}_0) < n-1$, another element $\vec{w}$ of $\mathcal{V}$ can be chosen so that $\{\vec{v}_0, \mathcal{L}\vec{v}_0, \mathcal{L}^2\vec{v}_0, \dots, \mathcal{L}^{n_0}\vec{v}, \vec{w}\}$ is linearly independent, where $n_0 = \max\{\mu(\vec{v}) : \vec{v} \in \mathcal{V}\}$. Set $\vec{v} = \vec{v}_0 + \lambda x^m \vec{w} \in \mathcal{V}$, where $\lambda \in \mathbb{C}$ and m is an integer. Then, $\mathcal{L}^\ell \vec{v} = \mathcal{L}^\ell \vec{v}_0 + \mathcal{L}^\ell(\lambda x^m \vec{w}) = \mathcal{L}^\ell \vec{v}_0 + \lambda x^m (\mathcal{L} + m)^\ell \vec{w}$. Since $\{\vec{v}, \mathcal{L}\vec{v}, \dots, \mathcal{L}^{n_0+1}\vec{v}\}$ is linearly dependent, it follows that $\vec{v} \wedge \mathcal{L}\vec{v} \wedge \cdots \wedge \mathcal{L}^{n_0}\vec{v} \wedge \mathcal{L}^{n_0+1}\vec{v} = \vec{0}$ for all $\lambda \in \mathbb{C}$ and all integers m. Note that $\vec{v} \wedge \mathcal{L}\vec{v} \wedge \cdots \wedge \mathcal{L}^{n_0}\vec{v} \wedge \mathcal{L}^{n_0+1}\vec{v}$ is a polynomial in λ. Since this polynomial is identically zero, each coefficients must be zero. For example, the constant term of this polynomial is $\vec{v}_0 \wedge \mathcal{L}\vec{v}_0 \wedge \cdots \wedge \mathcal{L}^{n_0+1}\vec{v}_0$. This is zero since $\{\vec{v}_0, \mathcal{L}\vec{v}_0, \dots, \mathcal{L}^{n_0+1}\vec{v}_0\}$ is linearly dependent. Compute the coefficient of the linear term in λ of the polynomial. Then,

$$\vec{w} \wedge \mathcal{L}\vec{v}_0 \wedge \cdots \wedge \mathcal{L}^{n_0+1}\vec{v}_0 + \vec{v}_0 \wedge \cdots \wedge \mathcal{L}^{n_0}\vec{v}_0 \wedge (\mathcal{L}+m)^{n_0+1}\vec{w}$$

$$+ \sum_{j=1}^{n_0} \vec{v}_0 \wedge \cdots \wedge \mathcal{L}^{j-1}\vec{v}_0 \wedge (\mathcal{L}+m)^j \vec{w} \wedge \mathcal{L}^{j+1}\vec{v}_0 \wedge \cdots \wedge \mathcal{L}^{n_0+1}\vec{v}_0 = \vec{0}$$

identically for all integers m. The left-hand side of this identity is a polynomial in m of degree $n_0 + 1$. Hence, each coefficient of this polynomial must be zero. In particular, computing the coefficient of m^{n_0+1}, we obtain $\vec{v}_0 \wedge \mathcal{L}\vec{v}_0 \wedge \cdots \wedge \mathcal{L}^{n_0}\vec{v}_0 \wedge \vec{w} = \vec{0}$. This is a contradiction, since $\{\vec{v}_0, \mathcal{L}\vec{v}_0, \ldots, \mathcal{L}^{n_0}\vec{v}_0, \vec{w}\}$ is linearly independent. This completes the proof of Lemma XIII-5-1. $\square$

Definition XIII-5-2. *An element $\vec{v}_0 \in \mathcal{V}$ is called a* cyclic vector *of $\mathcal{L}$ if $\{\vec{v}_0, \mathcal{L}\vec{v}_0, \mathcal{L}^2\vec{v}_0, \ldots, \mathcal{L}^{n-1}\vec{v}_0\}$ is a basis for $\mathcal{V}$ as a vector space over $\mathcal{K}$.*

Observation XIII-5-3. Let $\vec{v}_0$ be a cyclic vector of $\mathcal{L}$ and let $P(x)$ be the $n \times n$ matrix whose row vectors are $\{\vec{v}_0, \mathcal{L}\vec{v}_0, \ldots, \mathcal{L}^{n-1}\vec{v}_0\}$, i.e., $P(x) = \begin{bmatrix} \vec{v}_0 \\ \mathcal{L}\vec{v}_0 \\ \vdots \\ \mathcal{L}^{n-1}\vec{v}_0 \end{bmatrix}$. Then,

$\mathcal{L}[P(x)] = \begin{bmatrix} \mathcal{L}\vec{v}_0 \\ \mathcal{L}^2\vec{v}_0 \\ \vdots \\ \mathcal{L}^n\vec{v}_0 \end{bmatrix}$ and, hence, setting $A(x) = \mathcal{L}[P(x)]P(x)^{-1} = \delta P(x)P(x)^{-1} + P(x)\Omega(x)P(x)^{-1}$, we obtain

$$A = \begin{bmatrix} 0 & 1 & 0 & 0 & \cdots & 0 \\ 0 & 0 & 1 & 0 & \cdots & 0 \\ \vdots & \vdots & \vdots & \vdots & \vdots & \vdots \\ 0 & 0 & 0 & 0 & \cdots & 1 \\ \alpha_0 & \alpha_1 & \alpha_2 & \alpha_3 & \cdots & \alpha_{n-1} \end{bmatrix}$$

with the entries $\alpha_j \in \mathcal{K}$. Thus, we proved the following theorem, which is the main result of this section.

Theorem XIII-5-4. *The system of differential equations*

$$(\text{XIII.5.1}) \qquad \delta\vec{y} = \Omega(x)\vec{y}, \quad where \quad \vec{y} = \begin{bmatrix} y_1 \\ \vdots \\ y_n \end{bmatrix}$$

becomes

$$(\text{XIII.5.2}) \qquad \delta\vec{u} = A(x)\vec{u},$$

if $\vec{y}$ is changed by $\vec{u} = P(x)\vec{y}$. System (XIII.5.2) is equivalent to the n-th-order differential equation

$$(\text{XIII.5.3}) \qquad \delta^n\eta - \sum_{\ell=0}^{n-1} \alpha_\ell \delta^\ell \eta = 0, \quad where \quad \eta = y_1.$$

Example XIII-5-5.

(I) Let us consider the system

(a) $$\delta\vec{y} = \vec{0}, \qquad \text{where} \qquad \vec{y} = \begin{bmatrix} y_1 \\ \vdots \\ y_n \end{bmatrix}.$$

The transformation

(T) $$\vec{u} = \text{diag}\left[1, x, \ldots, x^{n-1}\right] \vec{y}$$

changes system (a) to

(E) $$\delta\vec{u} = \text{diag}\left[0, 1, \ldots, n-1\right] \vec{u}.$$

Further, the transformation

(T') $$\vec{w} = \begin{bmatrix} 1 & 1 & 1 & \cdots & 1 \\ 0 & 1 & 2 & \cdots & n-1 \\ 0 & 1 & 2^2 & \cdots & (n-1)^2 \\ \vdots & \vdots & \vdots & \vdots & \vdots \\ 0 & 1 & 2^{n-1} & \cdots & (n-1)^{n-1} \end{bmatrix} \vec{u}$$

changes (E) to the form

(E') $$\delta\vec{w} = \begin{bmatrix} 0 & 1 & 0 & 0 & \cdots & 0 \\ 0 & 0 & 1 & 0 & \cdots & 0 \\ \vdots & \vdots & \vdots & \vdots & \vdots & \vdots \\ 0 & 0 & 0 & 0 & \cdots & 1 \\ \alpha_0 & \alpha_1 & \alpha_2 & \alpha_3 & \cdots & \alpha_{n-1} \end{bmatrix} \vec{w},$$

where $\alpha_0, \alpha_1, \ldots, \alpha_{n-1}$ are integers. Hence, the transformation

$$\vec{w} = \begin{bmatrix} 1 & 1 & 1 & \cdots & 1 \\ 0 & 1 & 2 & \cdots & n-1 \\ 0 & 1 & 2^2 & \cdots & (n-1)^2 \\ \vdots & \vdots & \vdots & \vdots & \vdots \\ 0 & 1 & 2^{n-1} & \cdots & (n-1)^{n-1} \end{bmatrix} \text{diag}\left[1, x, \ldots, x^{n-1}\right] \vec{y}$$

changes (a) to (E'). This implies that $\vec{v}_0 = (1, x, x^2, \ldots, x^{n-1})$ is a cyclic vector in this case.

(II) Next, consider the system

(b) $$\delta\vec{y} = \Lambda\vec{y},$$

where Λ is a constant diagonal $n \times n$ matrix. Choose a transformation similar to (T) of (a) to change system (b) to

(E'') $$\delta\vec{u} = \Lambda'\vec{u}$$

so that Λ' is a diagonal matrix with n distinct diagonal entries. Then, a transformation similar to (T$'$) can be found so that (E$''$) is changed to (E$'$) with suitable constants $\alpha_0, \alpha_1, \dots, \alpha_{n-1}$.

XIII-6. The Hukuhara-Turrittin theorem

In this section, we explain a theorem due to M. Hukuhara and H. L. Turrittin that clearly shows the structure of solutions of a system of linear differential equations of the form

$$\text{(XIII.6.1)} \qquad x\frac{d\vec{y}}{dx} = A\vec{y},$$

where the entries of the $n \times n$ matrix A are in $\mathcal{K}$ (cf. §XIII-5). In order to state this theorem, we must introduce a field extension $\mathcal{L}$ of $\mathcal{K}$. To define $\mathcal{L}$, we first set

$$\mathcal{K}_\nu = \left\{ \sum_{m=M}^{+\infty} a_m x^{m/\nu} : a_m \in \mathbb{C} \quad \text{and} \quad M \in \mathbb{Z} \right\},$$

where $\mathbb{Z}$ is the set of all integers. For any element $a = \sum_{m=M}^{+\infty} a_m x^{m/\nu}$ of $\mathcal{K}_\nu$, we define $x\dfrac{da}{dx}$ by $x\dfrac{da}{dx} = \sum_{m=M}^{+\infty} \left(\dfrac{m}{\nu}\right) a_m x^{m/\nu}$. Then, $\mathcal{K}_\nu$ is a differential field. The field $\mathcal{L}$ is given by $\mathcal{L} = \bigcup_{\nu=1}^{+\infty} \mathcal{K}_\nu$ which is also a differential field containing $\mathcal{K}$ as a subfield. Furthermore, $\mathcal{L}$ is algebraically closed. The Hukuhara-Turrittin theorem is given as follows.

Theorem XIII-6-1 ([Huk4] and [Tu1]). *There exists a transformation*

$$\text{(XIII.6.2)} \qquad \vec{y} = U\vec{z}$$

such that
  (i) the entries of the matrix U are in $\mathcal{L}$ and $\det U \neq 0$,
  (ii) transformation (XIII.6.2) changes system (XIII.6.1) to

$$\text{(XIII.6.3)} \qquad x\frac{d\vec{z}}{dx} = B\vec{z},$$

  where B is an $n \times n$ matrix in the Jordan canonical form

$$\text{(XIII.6.4)} \qquad \begin{cases} B = \operatorname{diag}[B_1, B_2, \dots, B_p], \quad B_j = \operatorname{diag}[B_{j1}, B_{j2}, \dots, B_{jm_j}], \\ B_{jk} = \lambda_j I_{n_{jk}} + J_{n_{jk}}. \end{cases}$$

Here, $I_{n_{jk}}$ is the $n_{jk} \times n_{jk}$ identity matrix, $J_{n_{jk}}$ is an $n_{jk} \times n_{jk}$ nilpotent matrix of the form

$$(\text{XIII.6.5}) \qquad J_{n_{jk}} = \begin{bmatrix} 0 & 1 & 0 & \cdots & 0 \\ 0 & 0 & 1 & \cdots & 0 \\ \vdots & \vdots & \vdots & \vdots & \vdots \\ 0 & 0 & 0 & \cdots & 1 \\ 0 & 0 & 0 & \cdots & 0 \end{bmatrix},$$

and the λ_j are polynomials in $x^{-1/s}$ for some positive integer s, i.e.,

$$(\text{XIII.6.6}) \quad \lambda_j = \sum_{\ell=0}^{d_j} \lambda_{j\ell}\, x^{-\ell/s} \qquad where \quad \lambda_{j\ell} \in \mathbb{C} \qquad (j = 1, 2, \ldots, p)$$

and
(XIII.6.7)
$$\lambda_{j d_j} \neq 0 \quad if \quad d_j > 0 \qquad and \qquad \lambda_j - \lambda_i \ \text{are not integers} \ \ if\ i \neq j.$$

Proof.

Without loss of generality, assume that the matrix A of system (XIII.6.1) has the form

$$A = \begin{bmatrix} 0 & 1 & 0 & 0 & \cdots & 0 \\ 0 & 0 & 1 & 0 & \cdots & 0 \\ \vdots & \vdots & \vdots & \vdots & \vdots & \vdots \\ 0 & 0 & 0 & 0 & \cdots & 1 \\ \alpha_0 & \alpha_1 & \alpha_2 & \alpha_3 & \cdots & \alpha_{n-1} \end{bmatrix},$$

where $\alpha_k \in \mathcal{K}$ $(k = 0, 1, \ldots, n-1)$ (cf. Theorem XIII-5-4). Set $E = A - \dfrac{\alpha_{n-1}}{n} I_n$. Then,

$$(\text{XIII.6.8}) \qquad\qquad \text{trace}\,[E] = 0.$$

Consider the system

$$(\text{XIII.6.9}) \qquad\qquad x\frac{d\vec{w}}{dx} = E\vec{w}.$$

Case 1. If there exists an $n \times n$ matrix $\tilde{S}$ with the entries in $\mathcal{K}$ such that $\det \tilde{S} \neq 0$ and the transformation $\vec{w} = \tilde{S}\vec{u}$ changes (XIII.6.9) to $x\dfrac{d\vec{u}}{dx} = \tilde{A}(x)\vec{u}$, where the entries of $\tilde{A}$ are in $\mathbb{C}[[x]]$, then there exists another $n \times n$ matrix S with the entries in $\mathcal{K}$ such that $\det S \neq 0$ and the transformation

$$(\text{XIII.6.10}) \qquad\qquad \vec{w} = S\vec{u}$$

changes (XIII.6.9) to $x\dfrac{d\vec{u}}{dx} = A_0\vec{u}$, where the entries of the matrix A_0 are in $\mathbb{C}$. Furthermore, any two distinct eigenvalues of A_0 do not differ by an integer (cf. Theorem V-5-4). Hence, in this case, system (XIII.6.1) is changed by transformation (XIII.6.10) to

$$x\frac{d\vec{u}}{dx} = \left[\frac{\alpha_{n-1}}{n} I_n + A_0\right]\vec{u}.$$

This proved Theorem XIII-6-1 in this case.

Case 2. Assume that there is no $n \times n$ matrix $\tilde{S}$ with entries in $\mathcal{K}$ such that $\det \tilde{S} \neq 0$ and the transformation $\vec{w} = \tilde{S}\vec{u}$ changes (XIII.6.9) to $x\dfrac{d\vec{u}}{dx} = \tilde{A}(x)\vec{u}$, where the entries of $\tilde{A}$ are in $\mathbb{C}[[x]]$. Since

$$
E = \begin{bmatrix}
-\dfrac{1}{n}\alpha_{n-1} & 1 & 0 & 0 & \cdots & \cdots & 0 \\[2mm]
0 & -\dfrac{1}{n}\alpha_{n-1} & 1 & 0 & \cdots & \cdots & 0 \\[2mm]
\vdots & \vdots & \vdots & \vdots & \vdots & \vdots & \vdots \\[2mm]
0 & 0 & 0 & \cdots & \cdots & -\dfrac{1}{n}\alpha_{n-1} & 1 \\[2mm]
\alpha_0 & \alpha_1 & \alpha_2 & \alpha_3 & \cdots & \alpha_{n-2} & \alpha_{n-1} - \dfrac{1}{n}\alpha_{n-1}
\end{bmatrix},
$$

a cyclic vector can be found for system (XIII.6.9) by using the matrix W defined by

$$
\begin{cases}
W = \begin{bmatrix} w_{11} & \cdots & w_{1n} \\ \vdots & \vdots & \vdots \\ w_{n1} & \cdots & w_{nn} \end{bmatrix}, & \text{with} \\[4mm]
[w_{11} \ \cdots \ w_{1n}] = [1 \ 0 \ \cdots \ 0], \\[2mm]
[w_{j1} \ \cdots \ w_{jn}] = \nabla^{j-1}[1 \ 0 \ \cdots \ 0], & (j = 2, 3, \ldots, n),
\end{cases}
$$

where $\nabla([c_1 \ \cdots \ c_n]) = x\dfrac{d}{dx}[c_1 \ \cdots \ c_n] + [c_1 \ \cdots \ c_n]E$. The matrix W is lower-triangular and the diagonal entries are $\{1, \ldots, 1\}$, i.e.,

$$
\text{(XIII.6.11)} \qquad W = \begin{bmatrix}
1 & 0 & 0 & \cdots & 0 \\
\cdots & 1 & 0 & \cdots & 0 \\
\vdots & \vdots & \vdots & \vdots & \vdots \\
\cdots & \cdots & \cdots & \cdots & 1
\end{bmatrix}.
$$

If (XIII.6.9) is changed by the transformation $\vec{v} = W\vec{w}$, then

$$
\text{(XIII.6.12)} \qquad x\frac{d\vec{v}}{dx} = \left[x\frac{dW}{dx} + WE \right] W^{-1}\vec{v}.
$$

It follows from (XIII.6.8) and (XIII.6.11) that

$$
\text{(XIII.6.13)} \qquad \text{trace}\left[x\frac{dW}{dx} + WE \right] W^{-1} = 0.
$$

Also,

$$
\left[x\frac{dW}{dx} + WE \right] W^{-1} = \nabla[W]W^{-1} = \begin{bmatrix}
0 & 1 & 0 & 0 & \cdots & 0 \\
0 & 0 & 1 & 0 & \cdots & 0 \\
\vdots & \vdots & \vdots & \vdots & \vdots & \vdots \\
0 & 0 & 0 & 0 & \cdots & 1 \\
\beta_0 & \beta_1 & \beta_2 & \beta_3 & \cdots & \beta_{n-1}
\end{bmatrix},
$$

where $\beta_k \in \mathcal{K}$ $(k = 0, 1, \ldots, n-1)$. In particular, from (XIII.6.13), it follows that $\beta_{n-1} = 0$. Under our assumption, not all β_ℓ are in $\mathbb{C}[[x]]$. Set $\mathcal{J} = \{\ell : \beta_\ell \notin \mathbb{C}[[x]]\}$ and set Also, $\beta_\ell = x^{-\mu_\ell} \sum_{m=0}^{+\infty} \beta_{\ell m} x^m$ $(\ell \in \mathcal{J})$, where, for each ℓ, the quantity μ_ℓ is a positive integer, $\sum_{m=0}^{+\infty} \beta_{\ell m} x^m \in \mathbb{C}[[x]]$ and $\beta_{\ell 0} \neq 0$. Set

$$k = \max\left\{\frac{\mu_\ell}{n-\ell} : \ell \in \mathcal{J}\right\}.$$

Then, $\mu_\ell \leq k(n-\ell)$ for every $\ell \in \mathcal{J}$ and $\mu_\ell = k(n-\ell)$ for some $\ell \in \mathcal{J}$. This implies that

$$\text{(I)} \qquad \beta_\ell = \sum_{m \geq -k(n-\ell)} \beta_{\ell,m} x^m \qquad (\ell = 0, 1, \ldots, n-1)$$

and

$$\text{(II)} \qquad \beta_\ell = x^{-k(n-\ell)}(c_\ell + x q_\ell)$$

for some ℓ such that $k(n-\ell)$ is a positive integer, c_ℓ is a nonzero number in $\mathbb{C}$, $q_\ell \in \mathbb{C}[[x]]$, and $\beta_{\ell,m} \in \mathbb{C}$. We may assume without any loss of generality that $k = \dfrac{h}{q}$ for some positive integers h and q.

Let us change system (XIII.6.12) by the transformation

$$\vec{v} = \operatorname{diag}\left[1, x^{-k}, \ldots, x^{-(n-1)k}\right] \vec{u}.$$

Then,

$$\text{(XIII.6.14)} \qquad x\frac{d\vec{u}}{dx} = x^{-k} F \vec{u},$$

where

$$\text{(XIII.6.15)} \qquad F = \begin{bmatrix} 0 & 1 & 0 & 0 & \cdots & 0 \\ 0 & 0 & 1 & 0 & \cdots & 0 \\ \vdots & \vdots & \vdots & \vdots & \vdots & \vdots \\ 0 & 0 & 0 & 0 & \cdots & 1 \\ \gamma_0 & \gamma_1 & \gamma_2 & \gamma_3 & \cdots & \gamma_{n-1} \end{bmatrix} + kx^k \operatorname{diag}[0, 1, \ldots, n-1]$$

and

$$\gamma_\ell = x^{k(n-\ell)} \beta_\ell = \sum_{m \geq -k(n-\ell)} \beta_{\ell,m} x^{m+k(n-\ell)} \in \mathbb{C}[[x^{1/q}]] \qquad (0 \leq \ell \leq n-1).$$

In particular, $\gamma_{n-1} = 0$.

Setting $F = \sum_{m=0}^{+\infty} x^{m/q} F_m$, where the entries of F_m are in $\mathbb{C}$, we obtain

$$F_0 = \begin{bmatrix} 0 & 1 & 0 & 0 & \cdots & 0 & 0 \\ 0 & 0 & 1 & 0 & \cdots & 0 & 0 \\ \vdots & \vdots & \vdots & \vdots & \vdots & \vdots & \vdots \\ 0 & 0 & 0 & 0 & \cdots & 0 & 1 \\ c_0 & c_1 & c_2 & c_3 & \cdots & c_{n-2} & 0 \end{bmatrix} \qquad (c_\ell \in \mathbb{C},\ \ell = 0, 1, \ldots, n-2),$$

where the constants $c_0, c_1 \ldots, c_{n-2}$ are not all zero. This implies that the matrix F_0 must have at least two distinct eigenvalues. Hence, there exists an $n \times n$ matrix T such that

(1) $T = \sum_{m=0}^{+\infty} x^{m/q} T_m$, where the entries of the matrices T_m are in $\mathbb{C}$ and T_0 is invertible,

(2) the transformation

(XIII.6.16) $$\vec{y} = T\vec{z}$$

changes system (XIII.6.14) to a system $x\dfrac{d\vec{z}}{dx} = x^{-k} G\vec{z}$ with a matrix G in a block-diagonal form $G = \begin{bmatrix} G_1 & O \\ O & G_2 \end{bmatrix}$, where G_1 and G_2 are respectively $n_1 \times n_1$ and $n_2 \times n_2$ matrices with entries in $\mathbb{C}[[x^{1/q}]]$ and that $n_1 + n_2 = n$ and $n_j > 0$ $(j = 1, 2)$ (cf. §XIII-5). Therefore, the proof of Theorem XIII-6-1 can be completed recursively on n. $\square$

Observation XIII-6-2. In order to find a fundamental matrix solution of (XIII.6.1), let us construct a fundamental matrix solution of (XIII.6.3) in the following way:

Step 1. For each (j, k), set

$$\Phi_{jk} = x^{\lambda_{j0}} \exp[\Lambda_j(x)] \exp[(\log x) J_{n_{jk}}],$$

where

$$\Lambda_j(x) = \begin{cases} 0 & \text{if} \quad d_j = 0, \\ -\sum_{\ell=1}^{d_j} \dfrac{s}{\ell} \lambda_{j\ell}\, x^{-\ell/s} & \text{if} \quad d_j > 0. \end{cases}$$

Step 2. For each j, set

$$\Phi_j = \operatorname{diag}\left[\Phi_{j1}, \Phi_{j2}, \ldots, \Phi_{jm_j}\right] = x^{\lambda_{j0}} \exp[\Lambda_j(x)] \exp[(\log x) J_j],$$

where $J_j = \operatorname{diag}\left[J_{n_{j1}}, J_{n_{j2}}, \ldots, J_{n_{jm_j}}\right]$.

Step 3. Set

$$\begin{cases} \Lambda = \operatorname{diag}\left[\Lambda_1(x)I_{n_1}, \Lambda_2(x)I_{n_2}, \dots, \Lambda_p(x)I_{n_p}\right], \\ C = \operatorname{diag}\left[\lambda_{10}I_{n_1} + J_1, \lambda_{20}I_{n_2} + J_2, \dots, \lambda_{p0}I_{n_p} + J_p\right]. \end{cases}$$

Then,

$$(\text{XIII.6.17}) \qquad \Phi = \operatorname{diag}\left[\Phi_1, \Phi_2, \dots, \Phi_p\right] = x^C \exp[\Lambda]$$

is a fundamental matrix solution of (XIII.6.3), where

$$x^C = \operatorname{diag}\left[x^{\lambda_{10}} x^{J_1}, x^{\lambda_{20}} x^{J_2}, \dots, x^{\lambda_{p0}} x^{J_p}\right], \qquad x^{J_j} = \exp[(\log x)J_j].$$

The matrix

$$(\text{XIII.6.18}) \qquad U\Phi = U x^C \exp[\Lambda]$$

is a formal fundamental matrix solution of system (XIII.6.1), where U is the matrix of transformation (XIII.6.2) of Theorem XIII-6-1. The two matrices $\exp[\Lambda]$ and x^C commute.

Observation XIII-6-3. Theorem XIII-6-1 is given totally in terms of formal power series. However, even if the matrix $A(x)$ of system (XIII.6.1) is given analytically, the entries of U of transformation (XIII.6.2) are, in general, formal power series in $x^{1/s}$, since the entries of the matrix $T(x)$ of transformation (XIII.6.16) are formal power series in general. Transformation (XIII.6.16) changes system (XIII.6.14) to a block-diagonal form. Therefore, in a situation to which Theorem XIII-4-1 applies, transformation (XIII.6.2) can be justified analytically. The following theorem gives such a result.

Theorem XIII-6-4. *Assume that the entries of an $n \times n$ matrix $A(x)$ are holomorphic in a sector $S_0 = \{x \in \mathbb{C} : 0 < |x| < r_0, |\arg x| < \alpha_0\}$ and admit asymptotic expansions in powers of x as $x \to 0$ in S_0, where r_0 and α_0 are positive numbers. Assume also that d is a positive integer and $\vec{y} \in \mathbb{C}^n$. Let S be a subsector of S_0 whose opening is sufficiently small. Then, Theorem XIII-6-1 applies to the system*

$$(\text{XIII.6.19}) \qquad x^{d+1}\frac{d\vec{y}}{dx} = A(x)\vec{y}$$

with transformation (XIII.6.2) such that the entries of the matrix U of (XIII.6.2) are holomorphic in S and each of them is in a form $x^\rho \phi(x)$ where ρ is a rational number and $\phi(x)$ admits an asymptotic expansion in powers of $x^{1/s}$ as $x \to 0$ in S, where s is a positive integer.

Observation XIII-6-5. In the case when the entries of the matrix $A(x)$ on the right-hand side of (XIII.6.19) are in $\mathbb{C}[[x]]$ and $A(0)$ has n distinct eigenvalues, the matrix $A(x)$ also has n distinct eigenvalues $\lambda_1(x), \lambda_2(x), \dots, \lambda_n(x)$ which are in $\mathbb{C}[[x]]$. Furthermore, the corresponding eigenvectors $\vec{p}_1(x), \vec{p}_2(x), \dots, \vec{p}_n(x)$ can be constructed in such a way that their entries are in $\mathbb{C}[[x]]$ and that $\vec{p}_1(0), \vec{p}_2(0), \dots,$

$\vec{p}_n(0)$ are n eigenvectors of $A(0)$. Denote by $P(x)$ the $n \times n$ matrix whose column vectors are $\vec{p}_1(x), \vec{p}_2(x), \ldots, \vec{p}_n(x)$. Then, $\det P(0) \neq 0$ and $P(x)^{-1}A(x)P(x) = \text{diag}[\lambda_1(x), \lambda_2(x), \ldots, \lambda_n(x)]$. This implies that the transformation $\vec{y} = P(x)\vec{u}$ changes system (XIII.6.19) to

$$\text{(XIII.6.20)} \quad x^{d+1}\frac{d\vec{u}}{dx} = \left\{ \text{diag}\left[\lambda_1(x), \lambda_2(x), \ldots, \lambda_n(x)\right] - x^{d+1}P(x)^{-1}\frac{dP(x)}{dx} \right\}\vec{u}.$$

It is easy to construct another $n \times n$ matrix $Q(x)$ so that (a) the entries of $Q(x)$ are in $\mathbb{C}[[x]]$, (b) $Q(0) = I_n$, and (c) the transformation $\vec{u} = Q(x)\vec{v}$ changes system (XIII.6.20) to

$$\text{(XIII.6.21)} \qquad\qquad x^{d+1}\frac{d\vec{v}}{dx} = \text{diag}\left[\mu_1(x), \mu_2(x), \ldots, \mu_n(x)\right]\vec{v},$$

where $\mu_1(x), \mu_2(x), \ldots, \mu_n(x)$ are polynomials in x of degree at most d such that $\lambda_j(x) = \mu_j(x) + O(x^{d+1})$ $(j = 1, 2, \ldots, n)$. Therefore, in this case, the entries of the matrix U of transformation (XIII.6.2) are in $\mathcal{K}$.

Observation XIII-6-6. Assume that the entries of $A(x)$ of (XIII.6.19) are in $\mathbb{C}[[x]]$. Assume also that $A(0)$ is invertible. Then, upon applying Theorem XIII-6-1 to system (XIII.6.19), we obtain $\dfrac{d_j}{s} = d$ for all j. Using this fact, we can prove the following theorem.

Theorem XIII-6-7. *Let $\vec{\phi}_1(x)$ and $\vec{\phi}_2(x)$ be two solutions of a system*

$$\text{(XIII.6.22)} \qquad\qquad x^{d+1}\frac{d\vec{y}}{dx} = A(x)\vec{y} + x\vec{f}(x),$$

where d is a positive integer, the entries of the $n \times n$ matrix $A(x)$ and the $\mathbb{C}^n$-valued function $\vec{f}(x)$ are holomorphic in a neighborhood of $x = 0$, and $A(0)$ is invertible. Assume that for each $j = 1, 2$, the solution $\vec{\phi}_j(x)$ admits an asymptotic expansion in powers of x as $x \to 0$ in a sector $\mathcal{S}_j = \{x \in \mathbb{C} : |x| < r_0, a_j < \arg x < b_j\}$, where r_0 is a positive number, while a_j and b_j are real numbers. Suppose also that $\mathcal{S}_1 \cap \mathcal{S}_2 \neq \emptyset$. Then, there exist positive numbers K and λ and a closed subsector $\mathcal{S} = \{x : |x| \leq R, a \leq \arg x \leq b\}$ of $\mathcal{S}_1 \cap \mathcal{S}_2$ such that $|\vec{\phi}_1(x) - \vec{\phi}_2(x)| \leq K \exp[-\lambda|x|^{-d}]$ in $\mathcal{S}$.

Proof.

Since the matrix $A(0)$ is invertible, the asymptotic expansions of $\vec{\phi}_1(x)$ and $\vec{\phi}_2(x)$ are identical. Set $\vec{\psi}(x) = \vec{\phi}_1(x) - \vec{\phi}_2(x)$. Then, the $\mathbb{C}^n$-valued function $\vec{\psi}(x)$ satisfies system (XIII.6.19) in $\mathcal{S}_1 \cap \mathcal{S}_2$ and $\vec{\psi}(x) \simeq \vec{0}$ as $x \to 0$ in $\mathcal{S}_1 \cap \mathcal{S}_2$. By virtue of Theorem XIII-6-4, a constant vector $\vec{c} \in \mathbb{C}^n$ can be found so that $\vec{\psi}(x) = U\Phi(x)\vec{c}$, where $\Phi(x)$ is given by (XIII.6.17). Now, using Observation XIII-6-6, we can complete the proof of Theorem XIII-6-7. $\square$

Observation XIII-6-8. The matrix $\Lambda = \mathrm{diag}\left[\Lambda_1 I_{n_1}, \Lambda_2 I_{n_2}, \ldots, \Lambda_p I_{n_p}\right]$ on the right-hand side of (XIII.6.18) is unique in the following sense. Assume that another formal fundamental matrix solution $\tilde{U}x^{\tilde{C}}\exp[\tilde{\Lambda}]$ of system (XIII.6.1) is constructed with three matrices $\tilde{U}$, $\tilde{C}$, and $\tilde{\Lambda}$ similar to U, C, and Λ. Since the matrices $Ux^C\exp[\Lambda]$ and $\tilde{U}x^{\tilde{C}}\exp[\tilde{\Lambda}]$ are two formal fundamental matrices of system (XIII.6.1), there exists a constant $n \times n$ matrix $\Gamma \in \mathrm{GL}(n, \mathbb{C})$ such that $\tilde{U}x^{\tilde{C}}\exp[\tilde{\Lambda}] = Ux^C\exp[\Lambda]\Gamma$ (cf. Remark IV-2-7(1)). Hence, $\exp[\Lambda]\Gamma\exp[-\tilde{\Lambda}] = x^{-C}U^{-1}\tilde{U}x^{\tilde{C}}$. Using the fact that Γ is invertible, it can be easily shown that $\tilde{\Lambda} = \Lambda$ if the diagonal entries of $\tilde{\Lambda}$ are arranged suitably. For more information concerning the uniqueness of the Jordan form (XIII.6.3) and transformation (XIII.6.2), see, for example, [BJL], [Ju], and [Leve].

Observation XIII-6-9. The quantities $\Lambda_j(x)$ are polynomials in $x^{1/s}$. Set $\omega = \exp\left[\dfrac{2\pi i}{s}\right]$ and $\hat{x}^{1/s} = \omega x^{1/s}$. Then, $\hat{x} = x$. Therefore, if $x^{1/s}$ in $Ux^C\exp[\Lambda]$ is replaced by $\hat{x}^{1/s}$, then another formal fundamental matrix of (XIII.6.1) is obtained. This implies that the two sets $\{\Lambda_j(\hat{x}) : j = 1, 2, \ldots, p\}$ and $\{\Lambda_j(x) : j = 1, 2, \ldots, p\}$ are identical by virtue of Observation XIII-6-8.

Observation XIII-6-10. A power series $p(x)$ in $x^{1/s}$ can be written in a form
$$p(x) = \sum_{h=0}^{s-1} x^{h/s} q_h(x), \quad \text{where } q_j(x) \in \mathbb{C}[[x]] \ (j = 0, 1, \ldots, s - 1).$$
Using this fact and Observation XIII-6-9, we can derive the following result from Theorem XIII-6-1.

Theorem XIII-6-11. *There exist an integer q and an $n \times n$ matrix $T(x)$ whose entries are in $\mathbb{C}[[x]]$ such that*

(a) $\det T(x) \neq 0$ as a formal power series in x,

(b) the transformation $\vec{y} = T(x)\vec{u}$ changes system (XIII.6.1) to $x\dfrac{d\vec{u}}{dx} = E(x)\vec{u}$ with an $n \times n$ matrix $E(x)$ such that entries of $x^q E(x)$ are polynomials in x.

The main issue here is to construct, starting from Theorem XIII-6-1, a formal transformation whose matrix does not involve any fractional powers of x in such a way that the given system is reduced to another system with a matrix as simple as possible. A proof of Theorem XIII-6-11 is found in [BJL].

Changing the independent variable x by x^{-1}, we can apply Theorem XIII-6-1 to singularities at $x = \infty$. The following example illustrates such a case.

Example XIII-6-12. A system of the form
$$\frac{d\vec{y}}{dx} = \begin{bmatrix} 0 & 1 \\ P(x) & 0 \end{bmatrix}\vec{y}, \qquad \vec{y} = \begin{bmatrix} y_1 \\ y_2 \end{bmatrix},$$

where $P(x) = x^m + \sum_{h=1}^{m} a_h x^{m-h}$, m is a positive odd integer, and the a_h are complex numbers, has a formal fundamental matrix solution of the form
$$x^{-m/4}F(x)\begin{bmatrix} 1 & 0 \\ 0 & x^{-1/2} \end{bmatrix}\begin{bmatrix} 1 & 1 \\ 1 & -1 \end{bmatrix}\begin{bmatrix} e^{-E(x,a)} & 0 \\ 0 & e^{E(x,a)} \end{bmatrix},$$

where

$$\begin{cases} \left[1 + \displaystyle\sum_{k=1}^{m} \frac{a_k}{x^k}\right]^{1/2} = 1 + \displaystyle\sum_{k=1}^{+\infty} \frac{b_k(a)}{x^k}, \\[3ex] E(x,a) = \left(\dfrac{2}{m+2}\right) x^{(m+2)/2} + \displaystyle\sum_{1\le h<(m+2)/2} \left(\dfrac{2}{m+2-2h}\right) b_h(a)x^{(m+2-2h)/2}, \end{cases}$$

and $F(x) = x^q \displaystyle\sum_{h=0}^{+\infty} x^{-h} F_h$ with an integer q and 2×2 constant matrices F_h such that

$\det\left[\displaystyle\sum_{h=0}^{+\infty} x^{-h} F_h\right] \neq 0$ as a formal power series in x^{-1}. For details of construction,

see [HsS] and [Si13].

XIII-7. An n-th-order linear differential equation at a singular point of the second kind

Let us look at the formal fundamental matrix solution (XIII.6.18) of system (XIII.6.1). First, notice that if we set $k = \max\left\{\dfrac{d_j}{s} : j = 1,2,\dots,p\right\}$, then k is the order of singularity of system (XIII.6.1) at $x = 0$ (cf. Definition V-7-8). Let

(XIII.7.1) $$0 \le k_1 < k_2 < \cdots < k_{q-1} < k_q$$

be the distinct values among p non-negative rational numbers $\dfrac{d_1}{s}, \dfrac{d_2}{s}, \dots, \dfrac{d_p}{s}$. Also, set $\ell_h = \displaystyle\sum_{d_\nu/s=k_h} n_\nu$ $(h = 1,2,\dots,q)$. It is easy to see that $\ell_h > 0$ $(h = 1,2,\dots,q)$

and $\displaystyle\sum_{h=1}^{q} \ell_h = \sum_{j=1}^{p} n_j = n$.

Observation XIII-7-1. System (XIII.6.1) has ℓ_h linearly independent formal solutions of the form

(XIII.7.2) $$\vec{\psi}_{h,\nu}(x) = x^{\gamma_{h,\nu}} \exp[Q_{h,\nu}(x)]\vec{\phi}_{h,\nu}(x) \quad (\nu = 1,\dots,\ell_h; h = 1,\dots,q),$$

where $\gamma_{h,\nu} \in \mathbb{C}$, $Q_{h,\nu}(x)$ is either equal to 0 or a polynomial in $x^{-1/s}$ of the form

$$Q_{h,\nu}(x) = \mu_{h,\nu}x^{-k_h}(1 + O(x^{1/s})) \quad (\mu_{h,\nu} \in \mathbb{C} \quad \text{and} \quad \mu_{h,\nu} \neq 0),$$

and the entries of $\vec{\phi}_{h,\nu}(x)$ are polynomials in $\log x$ with coefficients in $\mathbb{C}[[x^{\frac{1}{s}}]]$.

Define $q + 1$ points (X_h, Y_h) $(h = 0,1,\dots,q)$ recursively by

$$\begin{cases} (X_0,Y_0) = (0,0), \\ (X_h,Y_h) = (X_{h-1} + \ell_h,\ Y_{h-1} + k_h\ell_h) \quad (h = 1,2,\dots,q). \end{cases}$$

Let us denote by $\mathcal{N}$ the polygon whose vertices are $q + 1$ points (X_h, Y_h) $(h = 0,1,\dots,q)$. The polygon $\mathcal{N}$ has q distinct slopes k_h given in (XIII.7.1).

Definition XIII-7-2. *The polygon $\mathcal{N}$ is called the* Newton polygon *of system* *(XIII.6.1) at $x = 0$.*

Observation XIII-7-3. In §XIII-5, it was shown that system (XIII.6.1) is equivalent to an n-th-order linear differential equation

$$(\text{XIII.7.3}) \qquad \alpha_n \delta^n \eta + \sum_{\ell=0}^{n-1} \alpha_\ell \delta^\ell \eta = 0,$$

where $\delta = x\dfrac{d}{dx}$, $\alpha_\ell \in \mathbb{C}[[x]]$, and $\alpha_n \neq 0$ (cf. Theorem XIII-5-4). For the differential operator

$$(\text{XIII.7.4}) \qquad \mathcal{L}[\eta] = \alpha_n \delta^n \eta + \sum_{\ell=0}^{n-1} \alpha_\ell \delta^\ell \eta,$$

the Newton polygon $\mathcal{N}(\mathcal{L})$ is defined in the following way.

If a power series $\alpha = \sum_{m=0}^{+\infty} c_m x^m \in \mathbb{C}[[x]]$ is not 0, we set $\nu(\alpha) = \min\{m : c_m \neq 0\}$. If $\alpha = 0$, set $\nu(0) = +\infty$. For operator (XIII.7.4), consider $n + 1$ points $(\ell, \nu(\alpha_\ell))$ $(\ell = 0, 1, \ldots, n)$ on an (X, Y)-plane. Set

$$\mathcal{P}_\ell = \{(X, Y) : 0 \leq X \leq \ell,\ Y \geq \nu(\alpha_\ell)\}, \qquad \text{and} \qquad \mathcal{P} = \bigcup_{\ell=0}^{n} \mathcal{P}_\ell.$$

Definition XIII-7-4. *The boundary curve $\mathcal{C}$ of the smallest convex set containing $\mathcal{P}$ is called the* Newton polygon *of the operator $\mathcal{L}$ at $x = 0$.*

Denote by $\mathcal{N}(\mathcal{L})$ the Newton polygon of $\mathcal{L}$ at $x = 0$.

Definition XIII-7-5. *Two Newton polygons are said to be identical if the two polygons become the same by moving one or the other upward in the direction of the Y-axis.*

Now, we prove the following theorem.

Theorem XIII-7-6. *If system (XIII.6.1) and differential equation (XIII.7.3) are equivalent in the sense of Theorem XIII-5-4, then the two Newton polygons $\mathcal{N}$ and $\mathcal{N}(\mathcal{L})$ are identical.*

Proof.

The proof of this theorem will be completed if the following three statements are verified:

(a) If $\mathcal{N}(\mathcal{L})$ has only one nonvertical side with slope k, then differential equation (XIII.7.3) is equivalent to a system $x^{k+1}\dfrac{d\vec{y}}{dx} = A(x)\vec{y}$ with a matrix $A(x)$ whose entries are power series in $x^{1/s}$, and $A(0)$ is invertible if $k > 0$, where s is a positive integer such that sk is an integer.

(b) If (XIII.7.1) gives slopes of all nonvertical sides of $\mathcal{N}(\mathcal{L})$, then the operator $\mathcal{L}$ is factored in the following way:

$$(XIII.7.5) \qquad \mathcal{L} = \mathcal{L}_1 \mathcal{L}_2 \cdots \mathcal{L}_q,$$

where, for each h, $\mathcal{N}(\mathcal{L}_h)$ has only one nonvertical side with slope k_h.

(c) If $\mathcal{L}$ is factored as in (b) and each differential equation $\mathcal{L}_h[\eta_h] = 0$ is equivalent to a system $x\dfrac{d\vec{u}_h}{dx} = A_h(x)\vec{u}_h$, then the differential equation (XIII.7.3) is equivalent to $x\dfrac{d\vec{y}}{dx} = \mathrm{diag}\,[A_1(x), A_2(x), \ldots, A_q(x)]\,\vec{y}$.

Statement (a) can be proved by an idea similar to the argument which is used to reduce system (XIII.6.12) to (XIII.6.14) in Case 2 of the proof of Theorem XIII-6-1. A proof of Statement (b) is found in [Mal1], [Si16] and [Si17, Appendix 1]. Look at the system

$$\mathcal{L}_q[u] = v, \qquad \mathcal{L}_1 \cdots \mathcal{L}_{q-1}[v] = 0.$$

Then, Statement (c) can be verified recursively on q without any complication. The proof of Theorem XIII-7-6 in detail is left to the reader as an exercise. $\square$

Combining Observation XIII-7-1 and Theorem XIII-7-6, we obtain the following theorem.

Theorem XIII-7-7. *If the distinct slopes of the nonvertical sides of $\mathcal{N}(\mathcal{L})$ are given by (XIII.7.1), then the n-th-order differential equation (XIII.7.3) has, at $x = 0$, n linearly independent formal solutions of the form*

$$(XIII.7.6) \qquad \eta_{h,\nu}(x) = x^{\gamma_{h,\nu}} \exp[Q_{h,\nu}(x)]\phi_{h,\nu}(x) \quad (\nu = 1, \ldots, \ell_h, \quad h = 1, \ldots, q),$$

where
(i) if

$$\{(X, Y_{h-1} + k_h(X - X_{h-1})) : X_{h-1} \leq X \leq X_h\}$$

is the nonvertical side of $\mathcal{N}(\mathcal{L})$ of slope k_h, then

$$\ell_h = X_h - X_{h-1} \quad (h = 1, 2, \ldots, q),$$

(ii) $\gamma_{h,\nu} \in \mathbb{C}$, $Q_{h,\nu}(x)$ is either equal to 0 or a polynomial in $x^{-1/s}$ of the form

$$Q_{h,\nu}(x) = \mu_{h,\nu} x^{-k_h}(1 + O(x^{1/s})) \quad (\mu_{h,\nu} \in \mathbb{C} \quad and \quad \mu_{h,\nu} \neq 0),$$

and the quantities $\phi_{h,\nu}(x)$ are polynomials in $\log x$ with coefficients in $\mathbb{C}[[x^{1/s}]]$. Here, s is a positive integer such that sk_h $(h = 1, \ldots, q)$ are integers, and $\ell_1 + \ell_2 + \cdots + \ell_q = n$.

A complete proof of Theorem XIII-7-7 is found in [St].

We can construct formal solutions (XIII.7.6), using an effective method with the Newton polygon $\mathcal{N}(\mathcal{L})$. The following example illustrates such a method.

Example XIII-7-8. Consider the differential operator $\mathcal{L} = x\delta^3 - x\delta^2 - \delta - 1$ or the third-order differential equation $\mathcal{L}[\eta] = 0$. The Newton polygon $\mathcal{N}(\mathcal{L})$ is given by Figure 6. In this case, $q = 2$, $k_1 = 0$, and $k_2 = \dfrac{1}{2}$.

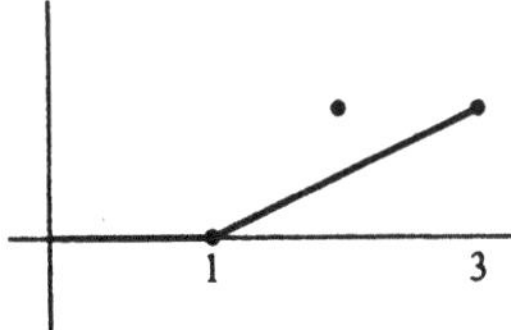

FIGURE 6.

(i) For $k_1 = 0$, set $\eta = \displaystyle\sum_{m=0}^{+\infty} c_m x^{\lambda+m}$ $(c_0 \neq 0)$. Then,

$$\mathcal{L}[\eta] = \sum_{m=0}^{+\infty} \left\{(\lambda+m)^3 - (\lambda+m)^2\right\} c_m x^{\lambda+m+1}$$
$$- \sum_{m=0}^{+\infty} (\lambda+m+1)c_m x^{\lambda+m} = 0.$$

Hence,

$$\begin{cases} (\lambda+1)c_0 = 0, \\ (\lambda+m+1)c_m = \left\{(\lambda+m-1)^3 - (\lambda+m-1)^2\right\} c_{m-1} & \text{for } m \geq 1. \end{cases}$$

Therefore,

$$\lambda = -1 \quad \text{and} \quad c_m = \frac{(m-2)^2(m-3)}{m} c_{m-1} \quad \text{for } m \geq 1.$$

Thus,

$$\lambda = -1, \quad c_1 = -2c_0, \quad c_m = 0 \ \text{ for } \ m \geq 2.$$

This implies that $\eta = c_0\left(x^{-1} - 2\right)$ is a solution of $\mathcal{L}[\eta] = 0$.

(ii) For $k_2 = \dfrac{1}{2}$, set $\eta = \exp[\lambda x^{-1/2}]\zeta$. Then,

$$\delta^n[\eta] = \exp[\lambda x^{-1/2}]\left(\delta - \frac{\lambda}{2}x^{-1/2}\right)^n [\zeta] \qquad (n = 0, 1, 2, \dots).$$

Therefore, $\mathcal{L}[\eta] = 0$ is equivalent to

$$\left\{ x\left(\delta - \frac{\lambda}{2}x^{-1/2}\right)^3 - x\left(\delta - \frac{\lambda}{2}x^{-1/2}\right)^2 - \left(\delta - \frac{\lambda}{2}x^{-1/2}\right) - 1 \right\}[\zeta] = 0.$$

Since

$$\begin{cases} \left(\delta - \dfrac{\lambda}{2}x^{-1/2}\right)^2 = \delta^2 - \lambda x^{-1/2}\delta + \dfrac{\lambda}{4}x^{-1/2} + \dfrac{\lambda^2}{4}x^{-1}, \\[2mm] \left(\delta - \dfrac{\lambda}{2}x^{-1/2}\right)^3 = \delta^3 - \dfrac{3}{2}\lambda x^{-1/2}\delta^2 + \left(\dfrac{3}{4}\lambda x^{-1/2} + \dfrac{3}{4}\lambda^2 x^{-1}\right)\delta \\[2mm] \qquad\qquad\qquad - \left(\dfrac{1}{8}\lambda x^{-1/2} + \dfrac{3}{8}\lambda^2 x^{-1} + \dfrac{1}{8}\lambda^3 x^{-3/2}\right), \end{cases}$$

it follows that

$$x\left(\delta - \dfrac{\lambda}{2}x^{-1/2}\right)^3 - x\left(\delta - \dfrac{\lambda}{2}x^{-1/2}\right)^2 - \left(\delta - \dfrac{\lambda}{2}x^{-1/2}\right) - 1$$

$$= x\delta^3 - \left(\dfrac{3\lambda}{2}x^{1/2} + x\right)\delta^2 + \left(\dfrac{3}{4}\lambda^2 - 1 + \dfrac{7\lambda}{4}x^{1/2}\right)\delta$$

$$+ \left(\dfrac{\lambda}{2} - \dfrac{\lambda^3}{8}\right)x^{-1/2} - \left(1 + \dfrac{\lambda^2}{8}\right) + \dfrac{\lambda}{8}x^{1/2}.$$

The Newton polygon of this operator has only one nonvertical side of slope $\dfrac{1}{2}$ for arbitrary λ (cf. Figure 7-1). However, if λ is determined by the equation

$$\dfrac{\lambda}{2} - \dfrac{\lambda^3}{8} = 0,$$

then the Newton polygon has a horizontal side (cf. Figure 7-2).

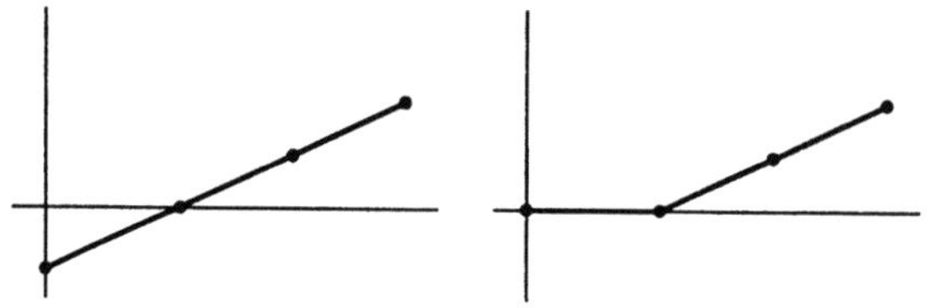

FIGURE 7-1.   FIGURE 7-2.

Hence, we can find two formal solutions of equation $\mathcal{L}[\eta] = 0$:

$$\begin{cases} \eta = \exp\left[2x^{-1/2}\right]x^{\rho_1}[1 + x^{1/2}f_1], \\[2mm] \eta = \exp\left[-2x^{-1/2}\right]x^{\rho_2}[1 + x^{1/2}f_2], \end{cases}$$

where ρ_1 and ρ_2 are constants and f_1 and f_2 are formal power series in $\mathbb{C}[[x^{1/2}]]$.

XIII-8. Gevrey property of asymptotic solutions at an irregular singular point

In this section we prove a result which is more precise than Theorem V-1-5 ([Mai]). In §XIII-3, we stated an existence theorem of asymptotic solutions for a given formal solution of an algebraic differential equation (cf. Theorem XIII-3-6). If differential equation (XIII.3.3) has a formal solution, we can transform (XIII.3.3) to the form

$$(\text{XIII.8.1}) \qquad \mathcal{L}[y] \ = \ x^M G(x, y, \delta y, \dots, \delta^{n-1} y) \qquad \left(\delta \ = \ x\frac{d}{dx} \right),$$

where

$$(\text{XIII.8.2}) \qquad \mathcal{L} \ = \ \sum_{h=0}^{n} a_h(x)\delta^h,$$

and $G(x, y_0, y_1, \dots, y_{n-1})$ is a convergent power series in $(x, y_0, y_1 \dots, y_{n-1})$ (cf. [SS3] and [Mal2]). Here, it can be assumed without any loss of generality that

(i) a_h $(h = 0, 1, \dots, n)$ are convergent power series in x and $a_n \neq 0$,

(ii) differential equation (XIII.8.1) has a formal solution $p(x) \ = \ \sum_{m=0}^{\infty} a_m x^m \in \mathbb{C}[[x]]$,

(iii) M is an integer such that for any differential operator $\mathcal{K}$ of order not greater than n, the two Newton polygons $\mathcal{N}(\mathcal{L} - x^M \mathcal{K})$ and $\mathcal{N}(\mathcal{L})$ are identical (cf. Definitions XIII-7-4 and XIII-7-5).

Using Theorem XIII-3-6, we can find

(a) a good covering $\{\mathcal{S}_1, \mathcal{S}_2, \dots, \mathcal{S}_N\}$ at $x = 0$,

(b) N solutions $\phi_1(x), \phi_2(x), \dots, \phi_N(x)$ of (XIII.8.1) in $\mathcal{S}_1, \mathcal{S}_2, \dots, \mathcal{S}_N$, respectively such that ϕ_ℓ are holomorphic and admit the formal solution $p(x)$ as their asymptotic expansions as $x \to 0$ in $\mathcal{S}_\ell$, respectively.

Set $u_\ell = \phi_\ell - \phi_{\ell+1}$ on $\mathcal{S}_\ell \cap \mathcal{S}_{j\ell+1}$. Then, u_ℓ are flat in the sense of Poincaré in sectors $\mathcal{S}_\ell \cap \mathcal{S}_{\ell+1}$, respectively, where $\mathcal{S}_{N+1} = \mathcal{S}_1$. Furthermore, if we define differential operators $\mathcal{K}_\ell$ by

$$\mathcal{K}_\ell = \sum_{0 \leq h \leq n-1} \left[\int_0^1 \frac{\partial G}{\partial y_h}(x, \dots, \delta^h[t\phi_\ell + (1-t)\phi_{\ell+1}], \dots)dt \right] \delta^h,$$

then

$$(\mathcal{L} - x^M \mathcal{K}_\ell)[u_\ell] = 0 \qquad \text{on} \qquad \mathcal{S}_\ell \cap \mathcal{S}_{\ell+1} \qquad (\ell = 1, 2, \dots, N).$$

If the Newton polygon $\mathcal{N}(\mathcal{L})$ has only one nonvertical side of slope 0, then $x = 0$ is a regular singular point of $\mathcal{L}[\eta] = 0$. Therefore, in this case the formal solution $p(x)$ is convergent (cf. Theorem V-2-7). Let us assume that $\mathcal{N}(\mathcal{L})$ has at least one side of positive slope. In such a case, let

$$(\text{XIII.8.3}) \qquad 0 \ < \ k_1 \ < \ k_2 \ < \ \cdots \ < \ k_q \ < \ +\infty$$

be all of the positive slopes of the Newton polygon $\mathcal{N}(\mathcal{L})$. Then, since $\mathcal{N}(\mathcal{L}-x^M\mathcal{K}_\ell)$ and $\mathcal{N}(\mathcal{L})$ are identical, we must have

$$|u_\ell(x)| \leq \gamma \exp[-\lambda \, |x|^{-k}] \qquad \text{on} \qquad \mathcal{S}_\ell \cap \mathcal{S}_{\ell+1}$$

for a non-negative number γ and a positive number λ, where $k \in \{k_1, k_2 \ldots, k_q\}$ (cf. Theorem XIII-7-7). Now, by virtue of Theorem XI-2-3, we obtain the following theorem.

Theorem XIII-8-1. *Under the assumptions given above, the formal solution $p(x)$ is a formal power series of Gevrey order $\dfrac{1}{k}$ and, for each ℓ, the solution ϕ_ℓ admits $p(x)$ as its asymptotic expansion of Gevrey order $\dfrac{1}{k}$ as $x \to 0$ in $\mathcal{S}_\ell$, where $k \in \{k_1, k_2 \ldots, k_q\}$.*

This theorem was originally prove in [Ram1] for a linear system. For nonlinear cases, see, for example, [Si17, §A.2.4, pp. 207-211].

Remark XIII-8-2. In Exercise XI-14, we gave the definition of a k-summable power series. As stated in Exercise XI-14, if a formal power series $f \in \mathbb{C}[[x]]$ is k-summable in a direction $\arg x = \theta$, there exists one and only one function $F \in \mathcal{A}_{1/k}\left(\rho_0, \theta - \dfrac{\pi}{2k} - \epsilon, \theta + \dfrac{\pi}{2k} + \epsilon\right)$ such that $J[F] = f$, where ρ_0 and ϵ are positive unmbers. This function F is called the *sum* of f in the direction $\arg x = \theta$. If we use the idea of Corollary XIII-1-3 and Theorem XIII-6-7, we can prove the following theorem concerning the k-summability of a formal solution of a nonlinear system

$$(XIII.8.4) \qquad x^{k+1}\frac{d\vec{y}}{dx} \; = \; A(x)\vec{y} \; + \; x\vec{b}(x,\vec{y}).$$

Theorem XIII-8-3. *Under the assumptions*
  (i) k is a positive integer,
  (ii) $A(x)$ is an $n \times n$ matrix whose entries are holomorphic in a neighborhood of $x = 0$ and $\vec{b}(x,\vec{y})$ is a $\mathbb{C}^n$-valued function whose entries are holomorphic in a neighborhood of $(x,\vec{y}) = (0,\vec{0})$,
  (iii) $A(0)$ is invertible,

system (XIII.8.4) has one and only one formal solution $\vec{y} = \vec{p}(x) = \displaystyle\sum_{m=1}^{\infty} x^m\vec{p}_m$ and $\vec{p}(x)$ is k-summable in any direction $\arg x = \theta$ except a finite number of values of θ. Furthermore, the sum of $p(x)$ in the direction $\arg x = \theta$ is a solution of (XIII.8.4).

To prove this theorem, it suffices to choose the good covering $\{\mathcal{S}_1, \mathcal{S}_2, \ldots, \mathcal{S}_N\}$ at $x = 0$ in the proof of Theorem XIII-8-1 so that opening of each of sectors $\{\mathcal{S}_1, \mathcal{S}_2, \ldots, \mathcal{S}_N\}$ is larger than $\dfrac{\pi}{k}$. We can prove a more general theorem.

Theorem XIII.8.4. *Let a linear differential operator* $\mathcal{L} = \delta - A(x)$ *be given, where* $\delta = \dfrac{d}{dx}$, *and* $A(x)$ *is an* $n \times n$ *matrix whose entries are meromorphic in a neighborhood of* $x = 0$. *Also, let (XIII.8.3) be all the positive slopes of the Newton polygon* $\mathcal{N}(\mathcal{L})$ *of the operator* $\mathcal{L}$. *Assume that*

(1) $k_1 \geq \dfrac{1}{2}$,

(2) $\mathcal{L}[\vec{f}]$ *is meromorphic at* $x = 0$ *for a* $\vec{f}(x) \in \mathbb{C}[[x]]^n$.
Then, there exist a finite number of directions $\arg x = \theta_\ell$ ($\ell = 1, 2, \ldots, p$) *such that if* $\theta \neq \theta_\ell$ *for* $\ell = 1, 2, \ldots, p$, *there exist* q *formal power series* $\vec{f}_\nu$ ($\nu = 1, 2, \ldots, q$) *satisfying the following conditions:*

(a) for each ν, *the power series* $\vec{f}_\nu$ *is* k_ν-*summable in the direction* $\arg x = \theta$,

(b) $\vec{f} = \vec{f}_1 + \vec{f}_2 + \cdots + \vec{f}_q$.

A complete proof of this theorem is found in [BBRS]. In this case, $\vec{f}$ is said to be $\{k_1, k_2 \ldots, k_q\}$-multisummable in the direction $\arg x = \theta$. We can also prove multisummability of formal solutions of a nonlinear system. For those informations, see, for example, [Ram3], [Br], [RS2], and [Bal3].

EXERCISES XIII

XIII-1. Using Observation XIII-6-5, diagonalize the following system:

$$\frac{d\vec{y}}{dx} = \begin{bmatrix} x+5 & x+8 \\ 3 & -x+1 \end{bmatrix} \vec{y}, \qquad \text{where} \qquad \vec{y} = \begin{bmatrix} y_1 \\ y_2 \end{bmatrix}$$

for large $|x|$.

XIII-2. Show that there is no rational function $f(x)$ in x such that

$$(\delta^2 f)(x) + x^2 f(x) = x \qquad \left(\delta = x\frac{d}{dx} \right).$$

Hint. One method is to show that the given equation has a unique power series solution in x^{-1} which is divergent at $x = \infty$. Another method is to observe that any solution of this equation has no singularity in $|x| < +\infty$ except possibly at $x = 0$. Furthermore, if $p(x)$ is a rational solution, then some inspection shows that $p(x)$ does not have any pole at $x = 0$. This implies that $p(x)$ must be a polynomial. But, we can easily see that this equation does not have any polynomial solution (cf. [Si20]).

XIII-3. Find a cyclic vector for the differential operator $\mathcal{L}[\vec{v}] = x\dfrac{d\vec{v}}{dx} + \vec{v}A$, with a constant $n \times n$ matrix A of the form $A = \begin{bmatrix} A_1 & O \\ O & A_2 \end{bmatrix}$, where for each $j = 1, 2$, the quantity A_j is an $n_j \times n_j$ matrix of the form

$$A_j = \begin{bmatrix} 0 & 1 & 0 & 0 & \cdots & 0 \\ 0 & 0 & 1 & 0 & \cdots & 0 \\ \vdots & \vdots & \vdots & \vdots & \vdots & \vdots \\ 0 & 0 & 0 & 0 & \cdots & 1 \\ \alpha_{0,j} & \alpha_{1,j} & \alpha_{2,j} & \alpha_{3,j} & \cdots & \alpha_{n_j-1,j} \end{bmatrix}$$

with complex constants $\alpha_{h,j}$.

XIII-4. Find the Newton polygon and a complete set of linearly independent formal solutions for each of the following three differential equations:

(a) $x\delta^2 y + 4\delta y - y = 0$, (b) $x^2\delta^2 y + x\delta y - y = 0$, (c) $x\delta^2 y + x\delta y - y = 0$,

where $\delta = x\dfrac{d}{dx}$.

XIII-5. Find the Newton polygon of the following system:

$$x^3 \frac{d\vec{y}}{dx} = \begin{bmatrix} 1 & 1 & 0 \\ 0 & x & x \\ 0 & 0 & x^2 \end{bmatrix} \vec{y}, \qquad \vec{y} = \begin{bmatrix} y_1 \\ y_2 \\ y_3 \end{bmatrix}.$$

Hint. For the given system, a cyclic vector is $(1,0,0)$. This implies that the transformation

(T) $$\vec{u} = \begin{bmatrix} 1 & 0 & 0 \\ x^{-2} & x^{-2} & 0 \\ x^{-4} - 2x^{-2} & x^{-4} + x^{-3} - 2x^{-2} & x^{-3} \end{bmatrix} \vec{y}$$

changes the given system to

(E) $$(\delta - 1)(x\delta - 1)(x^2\delta - 1)u_1 = 0, \qquad \text{where} \quad \delta = x\frac{d}{dx}.$$

Note that transformation (T) is equivalent to $u_1 = y_1$, $u_2 = \delta y_1$, $u_3 = \delta^2 y_1$, and the given system can be written in the form

$$y_2 = (x^2\delta - 1)y_1, \qquad y_3 = (x\delta - 1)y_2, \qquad (\delta - 1)y_3 = 0.$$

The Newton polygon of (E) has three sides with slopes 0, 1, and 2, respectively.

On the other hand, The standard form of the given system in the sense of Theorem XIII-6-1 is

$$x^3 \frac{d\vec{u}}{dx} = \begin{bmatrix} a(x) & 0 & 0 \\ 0 & b(x) & 0 \\ 0 & 0 & x^2 \end{bmatrix} \vec{u},$$

where

$$a(x) = 1 + O(x), \qquad b(x) = x + O(x^2).$$

Hence, this also shows that the Newton polygon of the given system has three sides with slopes 0, 1 and 2, respectively.

XIII-6. Let $A(x)$ be an $n \times n$ matrix whose entries are holomorphic and bounded in a domain $\Delta_0 = \{x : |x| < r_0\}$ and let $\vec{f}(x)$ be a $\mathbb{C}^n$-valued function whose entries are holomorphic and bounded in the domain Δ_0. Also, let $\lambda_1, \lambda_2, \ldots, \lambda_n$ be eigenvalues of $A(0)$. Assume that $\det A(0) \neq 0$. Assume further that two real numbers θ_1 and θ_2 satisfy the following conditions:

(1) $\theta_1 < \theta_2$,
(2) none of the quantities $\lambda_j e^{-ik\theta}$ $(j = 1, 2, \ldots, n)$ are real and positive for a positive integer k if $\theta_1 < \theta < \theta_2$,
(3) $\lambda_p e^{-ik\theta_1}$ and $\lambda_q e^{-ik\theta_2}$ are real and positive for some p and q.

Show that there exist one and only one solution $\vec{y} = \vec{\phi}(x)$ of the system $x^{k+1} \dfrac{d\vec{y}}{dx} = A(x)\vec{y} + x\vec{f}(x)$ such that the entries of $\vec{\phi}(x)$ are holomorphic and admit asymptotic expansions in powers of x as $x \to 0$ in the sector $\mathcal{S} = \left\{ x : 0 < |x| < r_0, \ \theta_1 - \dfrac{\pi}{2k} < \arg x < \theta_2 + \dfrac{\pi}{2k} \right\}$.

Hint. If we use Corollary XIII-1-3 at $x = 0$, it can be shown that for each θ in the interval $\theta_1 - \dfrac{\pi}{2k} < \theta < \theta_2 + \dfrac{\pi}{2k}$, a solution $\vec{y} = \vec{\phi}(x; \theta)$ is found so that the entries of $\vec{\phi}(x; \theta)$ are holomorphic and admit asymptotic expansions in powers of x as $x \to 0$ in the sectorial domain $\mathcal{S}_\theta = \left\{ x : 0 < |x| < r_0, \ |\arg x - \theta| < \dfrac{\pi}{2k} + \epsilon_\theta \right\}$, where ϵ_θ is a sufficiently small positive constant depending on θ. If $|\theta - \theta'|$ is sufficiently small, then $\vec{\phi}(x; \theta) = \vec{\phi}(x; \theta')$.

XIII-7.

(a) Show that $\hat{f}(x) = \displaystyle\sum_{m=0}^{+\infty} (-1)^m (m!) x^{m+1}$ is a formal solution of

$$(E) \qquad\qquad x^2 \dfrac{dy}{dx} + y - x = 0$$

and that $\hat{f}$ is not convergent except at $x = 0$.

(b) For a given direction θ, find a solution $f_\theta(x)$ of (E) such that $f_\theta(x) \simeq \hat{f}(x)$ as $x \to 0$ in the direction $\arg x = \theta$.

(c) Calculate $f_{\theta_1}(x) - f_{\theta_2}(x)$ for two given directions θ_1 and θ_2.

Hint. For Part (b), use the following steps:

Step 1. Apply Theorem XIII-1-2 to the given differential equation. To do this, we must change $x = 0$ to $t = \infty$ by $x = \dfrac{1}{t}$. Then, the differential equation becomes $\dfrac{dy}{dt} = y - \dfrac{1}{t}$. In this case, $n = 1$, $r = 0$, and the eigenvalue is $\mu = 1$. Set $\arg \mu = 0$. Then, the domain $\mathcal{D}(N, \gamma)$ (cf. (XIII.1.15)) is

$$\mathcal{D}(N, \gamma) = \left\{ t : |t| \geq N, \ |\arg t - 2q\pi| \leq \dfrac{3\pi}{2} - \gamma \right\},$$

where q is an integer, N is a sufficiently large positive number, and γ is a sufficiently small positive number. In terms of x, the domain $\mathcal{D}(N, \gamma)$ becomes

$$S(N, \gamma) \;=\; \left\{ x : 0 < |x| \leq \frac{1}{N}, \; |\arg x - 2p\pi| \leq \frac{3\pi}{2} - \gamma \right\},$$

where $p = -q$. Since there is no singular point of the given differential equation in the domain $\Omega = \{x : 0 < |x| < \infty\}$, we can conclude that, for each fixed integer p, there exists a solution $\phi_p(x)$ of the given differential equation such that

   (i) $\phi_p(x)$ is analytic (but not single-valued) in Ω,

   (ii) $\phi_p(x) \simeq \hat{f}(x)$ as $x \to 0$ in the domain

$$\mathcal{D}_p = \left\{ x : 0 < |x| < \infty, \; |\arg x - 2p\pi| \leq \frac{3\pi}{2} - \gamma \right\}.$$

Step 2. It is easy to see that $f(x) = e^{1/x} \displaystyle\int_0^x t^{-1} e^{-1/t} dt \;\; (x > 0)$ satisfies the given differential equation and has the asymptotic property $f(x) \simeq \hat{f}(x)$ as $x \to 0$. Since $\phi_p(x) - f(x)$ is a solution of the homogeneous differential equation $x^2 \dfrac{dy}{dx} + y = 0$, it follows that $\phi_p(x) = f(x) + c e^{1/x}$, where c is a constant. From this, $\phi_p(x) = e^{1/x} \displaystyle\int_0^x t^{-1} e^{-1/t} dt$ follows for $\arg x = 2p\pi$.

Step 3. Using an argument similar to that of Step 2,

$$f_\theta(x) \;=\; \begin{cases} \phi_p(x) & \text{if} \quad |\theta - 2p\pi| \leq \dfrac{\pi}{2}, \\[2ex] \phi_p(x) + c_\theta e^{1/x} & \text{if} \quad \dfrac{\pi}{2} < \theta - 2p\pi < \dfrac{3\pi}{2}, \end{cases}$$

is obtained, where c_θ is an arbitrary constant.

Step 4. Note that

$$\int_C t^{-1} e^{-1/t} dt \;=\; 2\pi i,$$

where C is a counterclockwise oriented circle with the center at $x = 0$. Hence, using analytic continuation of $f(x)$, we obtain

$$\phi_0(x) \;=\; \phi_p(x) + 2p\pi i e^{1/x} \quad \text{for} \quad \arg x = 2p\pi.$$

XIII-8. Show that the following differential equation has a nontrivial convergent power series solution:

$$x^2 \frac{d^3 y}{dx^3} + \frac{d^2 y}{dx^2} + y = 0.$$

Hint.
Step 1: The given differential equation has three linearly independent formal solutions

$$\hat{\phi}_1(x) = e^{1/x}\left[1 + \sum_{m=1}^{+\infty} a_m x^m\right], \quad \hat{\phi}_2(x) = 1 + \sum_{m=2}^{+\infty} b_m x^m, \quad \text{and} \quad \hat{\phi}_3(x) = x + \sum_{m=3}^{+\infty} a_m x^m.$$

In fact, $\hat{\phi}_1$ can be found through some calculation with the Newton polygon. The other two can be found by solving the equations

$$\begin{cases} \alpha_0 + 2\alpha_2 = 0, \quad \alpha_1 + 6\alpha_3 = 0, \\ \alpha_m + (m+1)m(m-1)\alpha_{m+1} + (m+2)(m+1)\alpha_{m+2} = 0 \quad \text{for} \quad m \geq 2 \end{cases}$$

for a formal solution $\displaystyle\sum_{m=0}^{+\infty} \alpha_m x^m$.

Step 2: The given differential equation has three linearly independent actual solutions such that $e^{-1/x}\phi_1(x) \simeq e^{-1/x}\hat{\phi}_1(x)$ as $x \to 0$ in the sector $|\arg x + \pi| < \dfrac{3\pi}{2}$, and $\phi_2(x) \simeq \hat{\phi}_2(x)$ and $\phi_3(x) \simeq \hat{\phi}_3(x)$ as $x \to 0$ in the sector $|\arg x| < \dfrac{3\pi}{2}$.

Step 3: In the direction $\arg x = -\pi$,

$$\phi_2(x) - \phi_2(xe^{2\pi i}) = c_2\phi_1(x) \quad \text{and} \quad \phi_3(x) - \phi_3(xe^{2\pi i}) = c_3\phi_1(x)$$

for some constants c_2 and c_3. Then, $c_3\hat{\phi}_2(x) - c_2\hat{\phi}_3(x)$ is a convergent power series solution of the given differential equation.

Remark:
  (1) See [HI].
  (2) This result was originally proved for a more general case in [Per1].
  (3) There is another proof based on Exercises V-18 and V-19 (cf. [HSW]).

XIII-9. Consider a system of differential equations

$$\text{(E)} \qquad\qquad\qquad x^{p+1}\frac{d\vec{u}}{dx} = \vec{F}(x, y, \vec{u}),$$

where p is a positive integer, x is a complex independent variable, y is a complex parameter, $\vec{u}$ and $\vec{F}$ are n-dimensional vectors (i.e., $\in \mathbb{C}^n$), and entries of $\vec{F}(x, y, \vec{u})$ are holomorphic with respect to $(x, y, \vec{u})$ in a neighborhood of $(x, y, \vec{u}) = (0, 0, \vec{0})$. Assume that there exists a formal solution of system (E)

$$\vec{u} = \vec{\psi}(x, y) = \sum_{h=0}^{\infty} y^h \vec{\psi}_h(x),$$

where coefficients $\vec{\psi}_h(x)$ are $\mathbb{R}^n$-valued functions whose entries are holomorphic in a neighborhood of $x = 0$. Assume also that $\vec{\psi}_0(0) = \vec{0}$ and $\det\left[\dfrac{\partial \vec{F}}{\partial \vec{u}}(0, 0, \vec{0})\right] \neq 0$. Show that $\vec{\psi}(x, y)$ is convergent in a neighborhood of $(x, y) = (0, 0)$.

Hint. See [Si14] and [Si21].

XIII-10. Consider a Pfaffian system

$$\text{(S)} \qquad \begin{cases} x^{p+1}\dfrac{\partial \vec{u}}{\partial x} = \vec{F}(x,y,\vec{u}), \\[2mm] y^{q+1}\dfrac{\partial \vec{u}}{\partial y} = \vec{G}(x,y,\vec{u}), \end{cases}$$

where p and q are positive integers, x and y are two complex independent variables, $\vec{u}$, $\vec{F}$, and $\vec{G}$ are n-dimensional vectors (i.e., $\in \mathbb{C}^n$), the entries of $\vec{F}$ and $\vec{G}$ are holomorphic with respect to $(x,y,\vec{u})$ in a neighborhood of $(x,y,\vec{u}) = (0,0,\vec{0})$ and system (S) is completely integrable, i.e., $\vec{F}$ and $\vec{G}$ satisfy the condition

$$y^{q+1}\frac{\partial \vec{F}}{\partial y}(x,y,\vec{u}) + \frac{\partial \vec{F}}{\partial \vec{u}}(x,y,\vec{u})\vec{G}(x,y,\vec{u}) = x^{p+1}\frac{\partial \vec{G}}{\partial x}(x,y,\vec{u}) + \frac{\partial \vec{G}}{\partial \vec{u}}(x,y,\vec{u})\vec{F}(x,y,\vec{u}).$$

Assume that $\vec{F}(0,0,\vec{0}) = \vec{0}$, $\vec{G}(0,0,\vec{0}) = \vec{0}$, $\det\left[\dfrac{\partial \vec{F}}{\partial \vec{u}}(0,0,\vec{0})\right] \neq 0$, and $\det\left[\dfrac{\partial \vec{G}}{\partial \vec{u}}(0,0,\vec{0})\right] \neq 0$. Show that system (S) has one and only one solution $\vec{u} = \vec{\psi}(x,y)$ such that $\vec{\psi}(0,0) = \vec{0}$ and that entries of $\vec{\psi}(x,y)$ are holomorphic with respect to (x,y) in a neighborhood of $(x,y) = (0,0)$.

Hint. This is an application of Exercise XIII-9.

Step 1. Construct a formal power series solution

$$\text{(FS)} \qquad \vec{u} = \vec{\psi}(x,y) = \sum_{h=0}^{+\infty} y^h \vec{\psi}_h(x)$$

of the system

$$y^{q+1}\frac{\partial \vec{u}}{\partial y} = \vec{G}(x,y,\vec{u}),$$

in such a way that coefficients $\vec{\psi}_h(x)$ are holomorphic in a neighborhood of $x = 0$.

Step 2. For $\vec{u} = \vec{\psi}(x,y)$, note that

$$y^{1+q}\frac{\partial}{\partial y}\left(x^{1+p}\frac{\partial \vec{u}}{\partial x} - \vec{F}(x,y,\vec{u})\right)$$

$$= y^{1+q}x^{1+p}\frac{\partial^2 \vec{u}}{\partial y \partial x} - y^{1+q}\frac{\partial \vec{F}}{\partial y} - \frac{\partial \vec{F}}{\partial \vec{u}}y^{1+q}\frac{\partial \vec{u}}{\partial y}$$

$$= y^{1+q}x^{1+p}\frac{\partial^2 \vec{u}}{\partial y \partial x} - y^{1+q}\frac{\partial \vec{F}}{\partial y} - \frac{\partial \vec{F}}{\partial \vec{u}}\vec{G}(x,y,\vec{u})$$

$$= x^{1+p}y^{1+q}\frac{\partial^2 \vec{u}}{\partial y \partial x} - x^{1+p}\frac{\partial \vec{G}}{\partial x} - \frac{\partial \vec{G}}{\partial \vec{u}}\vec{F}(x,y,\vec{u})$$

$$= x^{1+p}\left(\frac{\partial \vec{G}}{\partial x} + \frac{\partial \vec{G}}{\partial \vec{u}}\frac{\partial \vec{u}}{\partial x}\right) - x^{1+p}\frac{\partial \vec{G}}{\partial x} - \frac{\partial \vec{G}}{\partial \vec{u}}\vec{F}(x,y,\vec{u})$$

$$= \frac{\partial G}{\partial \vec{u}}\left(x^{1+p}\frac{\partial \vec{u}}{\partial x} - \vec{F}(x,y,\vec{u})\right).$$

Using this result, it can be shown that (FS) also satisfies the system

(E)
$$x^{p+1}\frac{\partial \vec{u}}{\partial x} \;=\; \vec{F}(x,y,\vec{u}).$$

Step 3. Upon applying Exercise XIII-9 to (E), the convergence of $\vec{\psi}(x,y)$ is proved.

XIII-11. Complete the proof of Theorem XIII-7-6 by verifying rigorously statements (a), (b), and (c) in the proof.

XIII-12. Show that the series
(FS)
$$y = p(x) = x^{-1/4}\left[1 + \sum_{h=1}^{+\infty}(-1)^h \left(\frac{3}{2}\right)^h \frac{\Gamma(3h+\frac{1}{2})}{54^h h!\,\Gamma(h+\frac{1}{2})}x^{-3h/2}\right]\exp\left[-\frac{2}{3}x^{3/2}\right]$$

is a formal solution of the differential equaton $\dfrac{d^2y}{dx^2} - xy = 0$, where Γ is the Gamma-function.

XIII-13. Show that the differential equation $\dfrac{d^2y}{dx^2} - xy = 0$ has a unique solution $\phi(x)$ such that (1) $\phi(x)$ is entire in x and (2) $\phi(x)\exp\left[\dfrac{2}{3}x^{3/2}\right]$ admits the formal series $p(x)\exp\left[\dfrac{2}{3}x^{3/2}\right]$ as its asymptotic exansion as $x \to \infty$ in the sector $|\arg x| < \pi$, where $p(x)$ is given by (FS).

Remark. $\mathrm{Ai}(x) = \dfrac{\phi(x)}{2\sqrt{\pi}}$ is called the *Airy function* (cf. [AS, p. 446], [Was1, pp. 124-126], and [Ol, pp. 392-394]).

XIII-14. Using the same notations in Exercises XIII-12 and XIII-13, show that if $\omega = \exp\left[\dfrac{2\pi i}{3}\right]$, then $\phi(\omega^{-1}x)$ and $\phi(\omega x)$ are two solutions of equation (S). Also,

(i) derive asymptotic expansions of $\phi(\omega^{-1}x)$ and $\phi(\omega x)$,
(ii) show that $\{\phi(x),\phi(\omega^{-1}x)\}$, $\{\phi(\omega^{-1}x),\phi(\omega x)\}$, and $\{\phi(\omega x),\phi(x)\}$ are three fundamental sets of solutions of (S),
(iii) show that if we set $\phi(x) = c_1\phi(\omega^{-1}x)+c_2\phi(\omega x)$, then $c_2 = -\omega$ and $\begin{bmatrix} c_1 & 0 \\ -\omega & 1 \end{bmatrix}^3$
   is equal to the 2×2 identity matrix,
(iv) using (iii), show that $c_1 = -\omega^{-1}$.

XIII-15. Show that if $\phi(x,\lambda)$ is an eigenfunction of the eigenvalue problem

(EP)
$$\frac{d^2y}{dx^2} - x^2y = \lambda y, \qquad \int_{-\infty}^{+\infty} y(x)^2 dx < +\infty,$$

then
(i) $\phi(x)$ is entire in x and $\phi(x)\exp\left[\dfrac{x^2}{2}\right]$ is a polynomial,

(ii) all negative odd integers are eigenvalues of (EP) and there is no other eigenvalue,

(iii) for every non-negative integer n, $H_n(x) = (-1)^n e^{x^2} \dfrac{d^n}{dx^n}(e^{-x^2})$ is a polynomial, and $\phi_n(x) = H_n(x)\exp\left[-\dfrac{x^2}{2}\right]$ is an eigenfunction of (EP) for the
eigenvalue $-(2n+1)$,

(iv) $\displaystyle\int_{-\infty}^{+\infty} \phi_n(x)\phi_m(x)dx = 0$ if $n \neq m$, and $\displaystyle\int_{-\infty}^{+\infty} \phi_n(x)^2 dx = 2^n n!\sqrt{\pi}$.

Remark. The polynomials $H_n(x)$ are called the *Hermit polynomials* (cf. [AS, p. 775] and [Ol, p. 49]).

XIII-16. Construct Green's function of the boundary-value problem
$$\frac{d^2 y}{dx^2} - x^2 y = f(x), \quad \int_{-\infty}^{+\infty} y(x)^2 dx < +\infty. \text{ Show also that}$$

(i) $\displaystyle\int_{-\infty}^{+\infty}\int_{-\infty}^{+\infty} G(x,\xi)^2 dx d\xi < +\infty,$

(ii) if $f(x)$ is real-valued, $f(x)$, $f'(x)$, and $f''(x)$ are continuous, $\displaystyle\int_{-\infty}^{+\infty} f(x)^2 dx <$
$+\infty$, and $\displaystyle\int_{-\infty}^{+\infty} \{f''(x) - x^2 f(x)\}^2 dx < +\infty$, then the series $\displaystyle\sum_{n=0}^{+\infty} \frac{1}{2^n n!\sqrt{\pi}}(f,\phi_n)$
$\times \phi_n(x)$ converges to $f(x)$ uniformly on the interval $-\infty < x < +\infty$, where
$\phi_n(x)$ are defined in Exercise XIII-15, and $(f,g) = \displaystyle\int_{-\infty}^{+\infty} f(x)g(x)dx$.

Hint. See §VI-4.

XIII-17. Consider a differential operator $\mathcal{L}[y] = \dfrac{d^n y}{dx^n} + \displaystyle\sum_{h=0}^{n-1} a_{n-h}(x)\dfrac{d^h y}{dx^h}$, where
$a_j(x) = \displaystyle\sum_{m=-m_j}^{+\infty} a_{jm}x^{-m} \in \mathbb{C}[[x^{-1}]]$. Also, assume that $a_{j,-m_j} \neq 0$ if $a_j(x)$ is not
equal to zero identically, while $m_j = -\infty$ if $a_j(x)$ is zero identically. In the (X,Y)-
plane, consider the points $P_j = (j, m_j)\,(j = 1, \dots, n)$. Construct a convex polygon
Π whose vertices are $(0,0)$, $\wp_1$, $\wp_2$, $\dots$, $\wp_s$ such that each $\wp_k$ is one of those points
P_j, and that all other points P_j are situated below the polygon. Set $\wp_0 = (0,0)$
and $\wp_k = (\alpha_k, \beta_k)\,(k = 1, \dots, s)$, where $\alpha_n = n$. Denote by ρ_k the slope of the
segment $\overline{\wp_k \wp_{k+1}}$ $(k = 0, \dots, s-1)$. Then, $\rho_0 > \rho_1 > \rho_2 > \cdots > \rho_{s-1}$. Assume
that $\rho_k > -1$ $(k < \nu_0)$ and $\rho_k \leq -1$ $(k \geq \nu_0)$. Show that
  (i) the differential equation $\mathcal{L}[y] = 0$ has $n - \alpha_{\nu_0}$ linearly independent formal so-
  lutions of the form $y_\ell(x) = x^{b_\ell} \displaystyle\sum_{q=0}^{M_\ell} R_{\ell q}(x)(\log x)^q$ $(\ell = \alpha_{\nu_0}+1, \dots, n)$, where
  $R_{\ell q}(x) \in \mathbb{C}[[x^{-1}]]$, $b_\ell \in \mathbb{C}$, and the M_ℓ are non-negative integers,
  (ii) if $k < \nu_0$, then the differential equation $\mathcal{L}[y] = 0$ has $\alpha_{k+1} - \alpha_k$ linearly
  independent formal solutions of the form $y_\ell = e^{\Lambda_\ell(x)} x^{b_\ell} \displaystyle\sum_{q=0}^{M_\ell} R_{\ell q}(x)(\log x)^q$ $(\ell =$

$1 + \alpha_k, \ldots, \alpha_{k+1})$, where $\Lambda_\ell(x) = \lambda_\ell x^{1+\rho_k} +$ terms of lower degree, $\lambda_\ell \in \mathbb{C}$, $\lambda_\ell \neq 0$, $b_\ell \in \mathbb{C}$, M_ℓ are non-negative integers, and $R_{\ell q} \in \mathbb{C}[[x^{-1/p}]]$, p being a positive integer.

Hint. Compare the polygon in this problem with the Newton polygon at x=0 defined in §XIII-7.

XIII-18. Consider a differential operator $\mathcal{L}[\vec{y}] = x^{q+1} \dfrac{d\vec{y}}{dx} + \Omega(x)\vec{y}$, where q is a positive integer and $\Omega(x) = \displaystyle\sum_{\ell=0}^{\infty} x^\ell \Omega_\ell$ with Ω_ℓ being $n \times n$ constant matrices. As in §V-4, the operator $\mathcal{L}$ can be represented by the matrix

$$A = \begin{bmatrix} \Omega_0 & O_1 \\ \tilde{\Omega} & J_\infty + A \end{bmatrix}, \quad \text{where} \quad \tilde{\Omega} = \begin{bmatrix} \Omega_1 \\ \Omega_2 \\ \vdots \\ \Omega_m \\ \vdots \end{bmatrix} \quad \text{and} \quad J_\infty = \begin{bmatrix} O_q \\ I_\infty \end{bmatrix}.$$

Here, O_r is the $nr \times \infty$ zero matrix and I_∞ is the $\infty \times \infty$ identity matrix. Let

$$\Omega_0 = \mathcal{S}_0 + \mathcal{N}_0 \quad \text{and} \quad A = \mathcal{S} + \mathcal{N}$$

be the S-N decomposition of Ω_0 and that of A in the sense of §V-3. Also, let $P_0, P_1, P_2, \ldots, P_m, \ldots$ be $n \times n$ constants matrices and let Λ_0 be an $n \times n$ diagonal matrix such that $\det P_0 \neq 0$ and that

$$(1) \qquad \mathcal{S}_0 P_0 = P_0 \Lambda_0, \quad \Lambda_0 = \begin{bmatrix} \lambda_1 & 0 & 0 & \cdots & 0 & 0 \\ 0 & \lambda_2 & 0 & \cdots & 0 & 0 \\ 0 & 0 & \lambda_3 & \cdots & 0 & 0 \\ \vdots & \vdots & \vdots & \vdots & \vdots & \vdots \\ 0 & 0 & 0 & \cdots & 0 & \lambda_n \end{bmatrix},$$

and

$$(2) \qquad \mathcal{S}\mathcal{P} = \mathcal{P}\Lambda_0,$$

where

$$\mathcal{P} = \begin{bmatrix} P_0 \\ P_1 \\ P_2 \\ \vdots \\ P_m \\ \vdots \end{bmatrix}.$$

Note that λ_ℓ are eigenvalues of $\mathcal{S}_0$. Hence, λ_ℓ are eigenvalues of Ω_0. Show that
  (a) the matrix $\mathcal{S}$ represents a multiplication operation: $\vec{y} \to \sigma(x)\vec{y}$ for an $n \times n$ matrix $\sigma(x) = \mathcal{S}_0 + O(x)$ whose entries are formal power series in x,

(b) the matrix $\mathcal{N}$ represents a differential operator $\mathcal{L}_0[\vec{y}] = x^{q+1}\dfrac{d\vec{y}}{dx} + \nu(x)\vec{y}$ for an $n \times n$ matrix $\nu(x) = \mathcal{N}_0 + O(x)$ whose entries are formal power series in x,

(c) $\mathcal{L}[\vec{y}] = \mathcal{L}_0[\vec{y}] + \sigma(x)\vec{y}$ and $\mathcal{L}_0[\sigma(x)\vec{y}] = \sigma(x)\mathcal{L}_0[\vec{y}]$,

(d) if we set $P(x) = \displaystyle\sum_{m=0}^{+\infty} x^m P_m$, then $P(x)^{-1}\sigma(x)P(x) = \Lambda_0$,

(e) if we set $\mathcal{K}[\vec{u}] = P(x)^{-1}\mathcal{L}[P(x)\vec{u}]$ and $\mathcal{K}_0[\vec{u}] = P(x)^{-1}\mathcal{L}_0[P(x)\vec{u}]$, then $\mathcal{K}[\vec{u}] = \mathcal{K}_0[\vec{u}] + \Lambda_0\vec{u}$,

(f) if we set

$$\mathcal{K}_0[\vec{u}] = x^{q+1}\frac{d\vec{u}}{dx} + \nu_0(x)\vec{u}, \quad \nu_0(x) = \begin{bmatrix} \nu_{11} & \cdots & \nu_{1n} \\ \vdots & \vdots & \vdots \\ \nu_{n1} & \cdots & \nu_{nn} \end{bmatrix},$$

then

$$\nu_{jk}(x) = 0 \quad \text{if} \quad \lambda_j \neq \lambda_k$$

(cf. [HKS]).

REFERENCES

[AS] M. Abramowitz and I. A. Stegun, *Handbooks of Mathematical Functions*, Dover, New York, 1965.

[Ar] C. Arzelà, Sulle funzioni di linee, *Mem. R. Accad. Bologna* (5), 5(1895), 225–244.

[As] G. Ascoli, Le curve limiti di una varietà data di curve, *Mem. R. Accad. Lincei* (3), 18(1883/4), 521–586.

[AM] M. F. Atiyah and I. M. MacDonald, *Introduction to Commutative Algebra*, Addison-Wesley, Reading, MA, 1969.

[Bal1] W. Balser, Meromorphic transformation to Birkhoff standard form in dimension three, *J. Fac. Sci. Univ. Tokyo*, Sect. IA, Math. 36(1989), 233–246.

[Bal2] W. Balser, Analytic transformation to Birkhoff standard form in dimension three, *Funk. Ekv.*, 33(1990), 59–67.

[Bal3] W. Balser, *From Divergent Power Series to Analytic Functions*, Lecture Notes in Math. No. 1582, Springer-Verlag, 1994.

[BBRS] W. Balser, B. L. J. Braaksma, J.-P. Ramis, and Y. Sibuya, Multisummability of formal power series solutions of linear ordinary differential equations, *Asymptotic Anal.*, 5(1991), 27–45.

[BJL] W. Balser, W. B. Jurkat, and D. A. Lutz, A general theory of invariants for meromorphic differential equations, *Funk. Ekv.*, 22(1979), 197–221.

[Bar] R. G. Bartle, *The Elements of Real Analysis*, John Wiley, New York, 1964.

[Bel1] R. Bellman, On the asymptotic behavior of solutions of $u'' - (1 + f(t))u = 0$, *Ann. Mat. Pura Appl.*, 31(1950), 83–91.

[Bel2] R. Bellman, *Stability Theory of Differential Equations*, McGraw-Hill, New York, 1953.

[Bel3] R. Bellman, *Introduction to Matrix Analysis*, SIAM, Philadelphia, 1995.

[Ben1] I. Bendixson, Démonstration de l'existence de l'integrale d'une équation aux derivées partielles linéaire, *Bull. Soc. Math. France*, 24(1896), 220–225.

[Ben2] I. Bendixson, Sur les courbes définiés par des équations différentielles, *Acta Math.*, 24(1901), 1–88.

[Bi] G. D. Birkhoff, Equivalent singular points of ordinary differential equations, *Math. Ann.*, 74(1913), 134–139; also in *Collected Mathematical Papers*, Vol. 1, American Mathematical Society, Providence, RI, 1950, pp. 252–257.

[Bor] E. Borel, *Leçons sur les séries divergentes*, Gauthier-Villars, Paris, 1928.

[Bou] N. Bourbaki, *Algèbre* II, Hermann, Paris, 1952.

[Br] B. L. J. Braaksma, Multisummability of formal power series solutions of nonlinear meromorphic differential equations, *Ann. Inst. Fourier, Grenoble*, 42–43(1992), 517–540.

[Ca] C. Carathéodory, *Vorlesungen über reelle Funktionen*, Teubner, Leipzig, 1927; reprinted, Chelsea, New York, 1948.

[Ce] L. Cesari, *Asymptotic Behavior and Stability Problems in Ordinary Differential Equations*, Academic Press, New York, 1963.

[CK] K. Chiba and T. Kimura, On the asymptotic behavior of solutions of a system of linear ordinary differential equations, *Comment. Math. Univ. St. Pauli*, 18(1970), 61–80.

[Cod] E. A. Coddington, *An Introduction to Ordinary Differential Equations*, Prentice-Hall, Englewood Cliffs, NJ, 1961.

[CL] E. A. Coddington and N. Levinson, *Theory of Ordinary Differential Equations*, McGraw-Hill, New York, 1955.

[Cop1] W. A. Coppel, *Stability and Asymptotic Behavior of Differential Equations*, Heath, Boston, 1965.

[Cop2] W. A. Coppel, *Disconjugacy*, Lecture Notes in Math. No. 220, Springer-Verlag, New York, 1971.

[Cu] C. G. Cullen, *Linear Algebra and Differential Equations*, 2nd ed., PWS–Kent, Boston, 1991.

[Del] P. Deligne, *Equations différentielles à points singuliers réguliers*, Lecture Notes in Math. No. 163, Springer-Verlag, New York, 1970.

[Dev] A. Devinatz, The asymptotic nature of the solutions of certain systems of differential equations, *Pacific J. Math.*, 15(1965), 75–83.

[DK] A. Devinatz and J. Kaplan, Asymptotic estimates for solutions of linear systems of ordinary differential equations having multiple characteristic roots, *Indiana Univ. Math. J.*, 22(1972), 355–366.

[Du] H. Dulac, Points singuliers des équations différentielles, *Mémor. Sci. Math.*, Fasc. 61, Gauthier-Villars, Paris, 1934.

[Dw] B. Dwork, Bessel functions as p-adic functions of the argument, *Duke Math. J.*, 41(1974) 711–738.

[E1] M. S. P. Eastham, The asymptotic solution of linear differential systems, *Mathematika*, 32(1985), 131–138.

[E2] M. S. P. Eastham, *The Asymptotic Solution of Linear Differential Systems*, Oxford Science Publications, Oxford, 1989.

[FL] L. Flatto and N. Levinson, Periodic solutions of singularly perturbed systems, *J. Rat. Mech. Anal.*, 4(1955), 943–950.

[Fl] G. Floquet, Sur les équations différentielles linéaires à coefficients périodiques, *Ann. École Norm.*, Ser. 2, 12(1883), 47–89.

[Fu] L. Fuchs, Zur Theorie der linearen Differentialgleichungen mit veränderlichen Koeffizienten, *J. Math.*, 66(1866), 121–160; 68(1868) 354–385.

[GelL] I. M. Gel'fand and B. M. Levitan, On the determination of a differential equation from its spectral function, *AMS Transl.*, Ser. 2.1, Vol. 1, 1955, 253–304.

[GérL] R. Gérard and A. H. M. Levelt, Sur les connexion à singularités régulières dans le cas de plusieurs variables, *Funk. Ekv.*, 19 (1976), 149–173.

[GérT] R. Gérard and H. Tahara, Singular nonlinear partial differential equations, *Aspects of Mathematics*, Vol. E 28, Vieweg & Sohn Verlagsgesellschaft mbH, Braunschweig/Wiesbaden, 1996.

[Gi] H. Gingold, Almost diagonal systems in asymptotic integration, *Proc. Edin. Math. Soc.*, 28(1985), 143–158.

[GH] H. Gingold and P. F. Hsieh, Globally analytic triangularization of a matrix function, *Lin. Alg. Appl.*, 169(1992), 75–101.

[GHS] H. Gingold, P. F. Hsieh and Y. Sibuya, Globally analytic simplification and the Levinson theorem, *J. Math. Anal. Appl.*, 128 (1994), 269–286.

[Gr] T. H. Gronwall, Note on the derivatives with respect to a parameter of the solutions of a system of differential equations, *Ann. Math.* (2), 20(1919), 292–296.

[Haa] A. Haar, Zur characteristikentheorie, *Acta Sci. Math Szeged*, 4(1928), 103–114.

[HabL] S. Haber and N. Levinson, A boundary value problem for a singularly perturbed differential equation, *Proc. Am. Math. Soc.*, 6(1955), 866–872.

[HarL1] W. A. Harris, Jr. and D. A. Lutz, On the asymptotic integration of linear differential system, *J. Math. Anal. Appl.*, 48(1974), 1–16.

[HarL2] W. A. Harris, Jr. and D. A. Lutz, Asymptotic integration of adiabatic oscillators, *J. Math. Anal. Appl.*, 51(1975), 76–93.

[HarL3] W. A. Harris, Jr. and D. A. Lutz, A unified theory of asymptotic integration, *J. Math. Anal. Appl.*, 57(1977), 571–586.

[HaS1] W. A. Harris, Jr. and Y. Sibuya, The reciprocals of solutions of linear ordinary differential equations, *Adv. Math.*, 58(1985), 119–132.

[HaS2] W. A. Harris, Jr. and Y. Sibuya, The n-th roots of solutions of linear ordinary differential equations, *Proc. Am. Math. Soc.*, 97(1986), 207–211.

[HaS3] W. A. Harris, Jr. and Y. Sibuya, Asymptotic behaviors of solutions of a system of linear ordinary differential equations as $t \to \infty$, in *Delay Differential Equations and Dynamical Systems*, Lecture Notes in Math. No. 1475, S. Busenberg and M. Martelli eds., Springer-Verlag, New York, 1991, pp. 210–217,

[HSW] W. A. Harris, Jr., Y. Sibuya and L. Weinberg, Holomorphic solutions of linear differential systems at singular points, *Arch. Rat. Mech. Anal.*, 35(1969), 245–248.

[Har1] P. Hartman, Unrestricted solution fields of almost separable differential equations, *Trans. Am. Math. Soc.*, 63(1948), 560–580.

[Har2] P. Hartman, *Ordinary Differential Equations*, John Wiley, New York, 1964.

[HW1] P. Hartman and A. Wintner, On nonoscillatory linear differential equations, *Am. J. Math.*, 75(1953), 717–730.

[HW2] P. Hartman and A. Wintner, Asymptotic integration of linear ordinary differential equations, *Am. J. Math.*, 77(1955), 45–87.

[HilP] E. Hille and R. S. Phillips, *Functional Analysis and Semi-Groups*, Colloquium Pblications XXXI, American Mathematical Society, Providence, RI, 1957.

[HirS] M. W. Hirsch and S. Smale, *Differential Equations, Dynamical Systems and Linear Algebra*, Academic Press, New York, 1974.

[How] F. A. Howes, Effective characterization of the asymptotic behavior of solutions of singularly perturbed boundary value problems, *SIAM J. Appl. Math.*, 30(1976), 296–306.

[Hs1] P. F. Hsieh, On an analytic simplification of a system of linear ordinary differential equations containing a parameter, *Proc. Am. Math. Soc.* 10(1968), 1201–1206.

[Hs2] P. F. Hsieh, Regular perturbation for a turning point problem, *Funk. Ekv.*, 12(1969), 155–179.

[HKS] P. F. Hsieh, M. Kohno and Y. Sibuya, Construction of a fundamental matrix solution at a singular point of the first kind by means of SN decomposition of matrices, *Lin. Alg. Appl.*, 239(1996), 29–76.

[HsS] P. F. Hsieh and Y. Sibuya, On the asymptotic integration of second order linear ordinary differential equations with polynomial coefficients, *J. Math. Anal. Appl.*, 16(1966), 84–103.

[HX1] P. F. Hsieh and F. Xie, Asymptotic diagonalization of a linear ordinary differential system, *Kumamoto J. Math.*, 7(1994), 27–50.

[HX2] P. F. Hsieh and F. Xie, Asymptotic diagonalization of a system of linear ordinary differential equations, *Dynamics of Continu. Discre. Impuls. Syst.*, 2(1996), 51–74.

[HX3] P. F. Hsieh and F. Xie, On asymptotic diagonalization of linear ordinary differential equations, *Dynamics of Continu. Discre. Impuls. Syst.*, 4(1998), 351–377.

[Huk1] M. Hukuhara, Sur les systèmes des équation différentielles ordinaires, *Proc. Imp. Acad. Japan*, 4(1928), 448–449.

[Huk2] M. Hukuhara, Sur l'ensemble des courbes intégrals d'un système d'équations différentielles ordinaires, *Proc. Imp. Acad. Tokyo*, 6(1930), 360–362; also [Huk8, 11–13].

[Huk3] M. Hukuhara, Sur les points singuliers des équations différentielles linéaires, Domaine réel, *J. Fac. Sci. Hokkaido Imp. Univ.*, 2(1934), 13–88; also [Huk8, 52–130].

[Huk4] M. Hukuhara, Sur les points singuliers des équations différentielles linéaires, II, *J. Fac. Sci. Hokkaido Imp. Univ.*, 5(1937), 123–166; also [Huk8, 131–175].

[Huk5] M. Hukuhara, Sur un système de deux équations differentielles ordinaires non linéaires à coefficients réels, *Fac. Sci. Univ. Tokyo*, 6(1951), 295–317.

[Huk6] M. Hukuhara, Sur la théorie des équations différentielles ordinaires, *J. Fac. Sci. Univ. Tokyo*, Sec. I, 7(1958), 483-510.

[Huk7] M. Hukuhara, Théorèmes fondamentaux de la théorie des équations différentielles ordinaires dans l'espace vectoriel topologique, *J. Fac. Sci. Univ. Tokyo*, Sec. I, 8(1959), 111-138.

[Huk8] M. Hukuhara, *Selected Papers*, Kyoritsu Shuppan Co., Ltd., Tokyo, 1997.

[HI] M. Hukuhara and M. Iwano, Étude de la convergence des solutions formelles d'un système différentielles ordinaires linéaires, *Funk. Ekv.*, 2(1959), 1–15; also [Huk8, 423–437].

[HN1] M. Hukuhara and M. Nagumo, On a condition of stability for a differential equation, *Proc. Imp. Acad. Tokyo*, 6(1930), 131–132; also [Na7, 136–137].

[HN2] M. Hukuhara and M. Nagumo, Sur la stabilité des intégrals d'un systèm d'équations différentielles, *Proc. Imp. Acad. Tokyo*, 6(1930), 357–359; also [Na7, 138–140].

[HN3] M. Hukuhara and M. Nagumo, Un théorème relatif à l'ensemble des courbes intégrales d'un système différentielles ordinaires, *Proc. Phys.-Math. Soc. Japan*, (3) 12(1930), 233–239; also [Na7, 141–147].

[Hum] J. E. Humphreys, *Introduction to Lie Algebras and Representation Theory*, Springer-Verlag, New York, 1972.

[I] M. Iwano, Bounded solutions and stable domains of nonlinear ordinary differential equations, *Analytic Theorey of Differential Equations*, Lecture Notes in Math. No. 183, P. F. Hsieh and A. W. J. Stoddart, eds., 59–127, Springer-Verlag, New York, 1971.

[IKSY] K. Iwasaki, H. Kimura, S. Shimomura and M. Yoshida, *From Gauss to Painlevé, Aspects of Mathematics*, Vol. E16, Vieweg & Sohn Verlagsgesellschaft mbH, Braunschweig/Wiesbaden, 1991.

[Ja] N. Jacobson, *Basic Algebra I*, Freeman and Co., New York, 1985.

[Ju] W. B. Jurkat, *Meromorphe Differentialgleichungen*, Lecture Notes in Math., No. 637, Springer-Verlag, New York, 1978.

[JL] W. B. Jurkat and D. A. Lutz, On the order of solutions of analytic linear differential equations, *Proc. London Math. Soc.*, 22(1971) 465–482.

[JLP] W. B. Jurkat, D. A. Lutz and A. Peyerimhoff, Birkhoff invariants and effective calculations for meromorphic linear differential equations, Part I, *J. Math. Anal. Appl.*, 53(1976), 438–470; Part II, *Houston J. Math.*, 2(1976), 207–238.

[KS] J. Kato and Y. Sibuya, Catastrophic deformation of a flow and non-existence of almost periodic solutions, *J. Fac. Sci. Univ. Tokyo*, 24(1977), 267–279.

[Ka] T. Kato, *Perturbation Theory for Linear Operators*, Springer-Verlag, New York, 1980.

[Kn] H. Kneser, Über die Lösungen eines Systems gewöhnlicher Differentialgleichungen das der Lipschitzschen Bedingung nicht genügt, *S.-B. Preuss. Akad. Wiss. Phys.-Math. Kl.* (1923), 171–174.

[Ko] H. Komatsu, On the index of ordinary differential operators, *J. Fac. Sci. Univ. Tokyo*, Sec. IA, 18(1971), 379–398.

[LaL] J. LaSalle and S. Lefschetz, *Stability by Liapunov's Direct Method with Applications*, Academic Press, New York, 1961.

[LM1] E. B. Lee and L. Markus, Optimal Control for Nonlinear Processes, *Arch. Rat. Mech. Anal.*, 8(1961), 36-58.

[LM2] E. B. Lee and L. Markus, *Foundations of Optimal Control Theory*, John Wiley, New York, 1967.

[Let] F. Lettenmeyer, Über die an einer Unbestimmtheitsstelle regulären Lösungen eines Systems homogener linearen Differentialgleichungen, *S.-B. Bayer. Akad. Wiss. München Math.-nat. Abt.*, (1926), 287–307.

[LeL] J. J. Levin and N. Levinson, Singular perturbations of non-linear systems of differential equations and an associated boundary layer equation, *J. Rat. Mech. Anal.*, 3(1954), 247–270.

[Levi1] N. Levinson, The asymptotic nature of solutions of linear differential equations, *Duke Math. J.*, 15(1948), 111–126.

[Levi2] N. Levinson, Perturbations of discontinuous solutions of nonlinear systems of differential equations, *Acta Math.*, 82(1950), 71 105.

[Leve] A. H. M. Levelt, Jordan decomposition for a class of singular differential operators, *Ark. Mat.*, 13(1975), 1–27.

[Lia] A. M. Liapounoff, Problème général de la stabilité du mouvement, *Ann. Fac. Sci. Univ. Toulouse*, 9(1909), 203–475; also *Ann. Math. Studies*, 17, Princeton, 1947.

[Lin] C.-H. Lin, Phragmén-Lindelöf theorem in a cohomological form, *Proc. Am. Math. Soc.*, 89(1983), 589–597.

[Lind1] E. Lindelöf, Sur l'application des méthodes d'approximations successives à l'étude des intégrales réeles des équations différentielles, *J. Math.*, Ser. 4, 10(1894), 117–128.

[Lind2] E. Lindelöf, Démonstration de quelques théorèms sur les équations différentielles, *J. Math.*, Ser. 5, 6(1900), 423–441.

[Lip] R. Lipschitz, Sur la posibilité d'intégrer complétement un système donné d'équations différentielles, *Bull. Sci. Math. Astron.*, 10(1876), 149–159.

[MW] W. Magnus and S. Winkler, *Hill's Equation*, Interscience, New York, 1966.

[Mah] K. Mahler, *Lectures on Transcendental Numbers*, Lecture Notes in Math. No. 546, Springer-Verlag, New York, 1976.

[Mai] E. Maillet, Sur les séries divergentes et les équations différentielles, *Ann. École Norm.*, 20(1903), 487–517.

[Mal1] B. Malgrange, *Sur la réduction formelle des équations différentielles à singularités irrégulières*, Preprint, Grenoble, 1979.

[Mal2] B. Malgrange, Sur le théorèm de Maillet, *Asymptotic Anal.*, 2(1989), 1–4.

[Mar] L. Markus, Exponentials in algebraic matrix groups, *Adv. Math.*, 11(1973), 351–367.

[Mi] J. W. Milnor, *Topology from the Differentiable Viewpoint*, University of Virginia, Charlottsville, 1976.

[Mo] J. Moser, The order of a singularity in Fuchs' theory, *Math. Zeitschr.*, 72(1960), 379–398.

[Mu] F. E. Mullin, On the regular perturbation of the subdominant solution to second order linear ordinary differential equations with polynomial coefficients, *Funk. Ekv.*, 11(1968), 1–38.

[Na1] M. Nagumo, Eine hinreichende Bedingung für die Unität der Lösung von Differentialgleichingen erster Ordung, *Japan J. Math.* 3, (1926), 107–112; also [Na7, 1–6].

[Na2] M. Nagumo, Eine hinreichende Bedingung für die Unität der Lösung von gewöhnlichen Differentialgleichungen n-ter Ordnuug, *Japan. J. Math.*, 4(1927), 307–309; also [Na7, 33–35].

[Na3] M. Nagumo, Über das Verfahren der sukzessive Approximationen zur Integration gewöhnlicher Differentialgleichung und die Eindeutigkeit ihrer Integrale, *Japan. J. Math.*, 7(1930) 143–160; also [Na7, 157–174].

[Na4] M. Nagumo, Über die Differentialgleichung $y'' = f(x, y, y')$, *Proc. Phys.- Math. Soc. Japan*, 19(1937), 861–866; also [Na7, 222–227].

[Na5] M. Nagumo, Über die Ungleichung $\dfrac{\partial u}{\partial y} > f\left(x, y, u, \dfrac{\partial u}{\partial x}\right)$, *Japan. J. Math.*, 15(1938), 51–56; also [Na7, 228–233].

[Na6] M. Nagumo, Über das Verhalten der Integrale von $\lambda y'' + f(x, y, y', \lambda) = 0$ für $\lambda \to 0$, *Proc. Phys.-Math. Soc. Japan*, 21(1939), 529–534; also [Na7, 234–239].

[Na7] M. Nagumo, *Mitio Nagumo Collected Papers*, Springer-Verlag, New York, 1993.

[Ne1] F. Nevanlinna, Zur Theorie der Asymptotischen Potenzreihen, *Ann. Acad. Sci. Fennicë Sarja*, Ser. A, Nid. Tom. 12(1919), 1–81.

[Ne2] R. Nevanlinna, *Analytic Functions*, Springer-Verlag, New York, 1970.

[NU] A. F. Nikiforov and V. B. Uvarov, *Special Functions of Mathematical Physics*, Birkhäuser, Boston, 1988.

[Ol] F. W. J. Olver, *Asymptotics and Special Functions*, Academic Press, New York, 1974.

[O'M] R. E. O'Malley, *Introduction to Singular Perturbations*, Academic Press, New York, 1974.

[Os] W. F. Osgood, Bewise der Existenz einer Lösung einer Differentialgleichung $dy/dx = f(x,y)$ ohne Heinzunahme der Cauchy-Lipschizschen Bedingung, *Monat. Math.-Phys.*, 9(1898), 331–345.

[Pa] B. P. Palka, *An Introduction to Complex Function Theory*, Springer-Verlag, New York, 1990.

[Pea1] G. Peano, Démonstration de l'intégrabilité des équations différentielles ordinaires, *Math. Ann.*, 37(1890), 182–228.

[Pea2] G. Peano, Generalità sulle equazioni differenziali ordinare, *Atti R. Accad. Sci. Trino*, 33(1897), 9–18.

[Per1] O. Perron, Über diejenigen Integrale linearer Differentialgleichungen, welche sich an einer Unbestimmtheitsstelle bestimmt verhalten, *Math. Ann.*, 70(1911), 1–32.

[Per2] O. Perron, Ein neue existenzbeweis für die Integrale der Differentialgleichung $y' = f(x,y)$, *Math. Ann.*, 76(1915), 471–484.

[Per3] O. Perron, Über ein vermeintliches Stabilitätskriterium, *Gött. Nachr.* (1930), 128–129.

[Pi] É. Picard, Mémoire sur la théorie des équations aux dérivées partielles et la méthode des approximations successives, *J. Math.*, ser 4, 6(1890), 145–210.

[Poi1] H. Poincaré, *Sur les courbes définies par les équations differentielles*, Oeuvres de Henri Poincaré, Vol. 1, Gauthier-Villars, Paris, 1928.

[Poi2] H. Poincaré, *Les méthodes nouvelles de la mécanique céleste*, Vol. II, Gauthier-Villars, Paris, 1893.

[Pop] J. Popken, Über arithmetische Eigenschaften analytischer Funktionen, Dissertation, Groningen, 1935.

[Rab] A. L. Rabenstein, *Elementary Differential Equations with Linear Algebra*, 4th ed., Harcourt Brace Javanovich, New York, 1991.

[Ram1] J.-P. Ramis, Dévissage Gevrey, *Astérisque*, 59/60(1978), 173–204.

[Ram2] J.-P. Ramis, Les séries k-sommables et leurs applications, *Complex Analysis, Microlocal Calculus and Relativistic Quantum Theory*, Lecture Notes in Physics No. 126, Springer-Verlag, New York, 1980, pp. 178–199.

[Ram3] J.-P. Ramis, Séries divergentes et théories asymptotiques, *Soc. Math. France, Panoramas Synthèses*, 121(1993), 1–74.

[RS1] J.-P. Ramis and Y. Sibuya, Hukuhara domains and fundamental existence and uniqueness theorems for asymptotic solutions of Gevrey type, *Asymptotic Anal.*, 2(1989), 39–94.

[RS2] J.-P. Ramis and Y. Sibuya, A new proof of multisummability of formal solutions of nonlinear meromorphic differential equations, *Ann. Inst. Fourier, Grenoble*, 44(1994), 811–848.

[Ri] J. F. Ritt, On the derivatives of a function at a point, *Ann. Math.*, 18(1916), 18–23.

[SC] G. Sansone and R. Conti, *Nonlinear Differential Equations*, MacMillan, New York, 1961.

[Sai] T. Saito, *Introduction to Analytical Mechanics*, Shibundo, Tokyo, 1964 (in Japanese).

[Sat] Y. Satô, Periodic solutions of the equation $x'' + f(x)x' + g(x) = 0$ without the hypothesis $xg(x) > 0$ for $|x| > 0$, *Comment. Math. Univ. St. Pauli*, 20(1971), 127–145.

[Sc] L. Schwartz, *Théorie des Distributions*, 3rd ed., Hermann, Paris, 1966.

[Si1] Y. Sibuya, Sur un système d'équations différentielles ordinaires linéaires à coefficients périodiques et contenants des paramètres, *J. Fac. Sci. Univ. Tokyo*, I, 7(1954), 229–241.

[Si2] Y. Sibuya, Nonlinear ordinary differential equations with periodic coefficients, *Funk. Ekv.*, 1(1958), 77–132.

[Si3] Y. Sibuya, Sur réduction analytique d'un système d'èquations différentielles ordinaires linéaires contenant un paramètre, *J. Fac. Sci. Univ. Tokyo*, 7(1958), 527–540.

[Si4] Y. Sibuya, Notes on real matrices and linear dynamical systems with periodic coefficients, *J. Math. Anal. Appl.*, 1(1960), 363–372.

[Si5] Y. Sibuya, On perturbations of discontinuous solutions of ordinary differential equations, *Natural Sci. Rep. Ochanomizu Univ.*, 11(1960), 1–18.

[Si6] Y. Sibuya, On nonlinear ordinary differential equations containing a parameter, *J. Math. Mech.*, 9(1960), 369–398.

[Si7] Y. Sibuya, Simplification of a system of linear ordinary differential equations about a singular point, *Funk. Ekv.*, 4(1962), 29–56.

[Si8] Y. Sibuya, On the convergence of formal solutions of systems of linear ordinary differential equations containing a parameter, MRC Tech. Sum. Report, No. 511, 1964.

[Si9] Y. Sibuya, A block-diagonalization theorem for systems of linear ordinary differential equations and its applications, *J. SIAM Anal. Math.*, 14(1966), 468–475.

[Si10] Y. Sibuya, Perturbation of linear ordinary differential equations at irregular singular points, *Funk. Ekv.*, 11(1968), 235–246.

[Si11] Y. Sibuya, Perturbation at an irregular singular point, *Japan–U. S. Seminar on Ordinary Differential and Functional Equations*, Lecture Notes in Math. No. 243, M. Urabe, ed., Springer-Verlag, New York, 1971, pp. 148–168.

[Si12] Y. Sibuya, *Uniform Simplification in a Full Neighborhood of a Transition Point*, Memoirs of American Mathematical Society, No. 149, Providence, RI, 1974.

[Si13] Y. Sibuya, *Global Theory of a Second Order Linear Ordinary Differential Equation with a Polynomial Coefficient*, Math. Studies 18, North-Holland, Amsterdam, 1975.

[Si14] Y. Sibuya, Convergence of formal power series solutions of a system of nonlinear differential equations at an irregular singular point, Lecture Notes in Math. No. 810, Springer-Verlag, New York, 1980, pp. 135–142.

[Si15] Y. Sibuya, A theorem concerning uniform simplification at a transition point and the problem of resonance, *SIAM J. Math. Anal.*, 12 (1981), 653–668.

[Si16] Y. Sibuya, Differential modules and theorem of Hukuhara-Turrittin, *Proc. of International Conf. on Nonlinear Phenomena in Math. Sci.*, Academic Press, New York,1982, pp. 927–937.

[Si17] Y. Sibuya, *Linear Differential Equations in the Complex Domain: Problems of Analytic Continuation*, Translations of Math. Monographs 82, American Mathematical Society, Providence, RI, 1990.

[Si18] Y. Sibuya, Gevrey asymptotics, *Ordinary and Delay Differential Equations*, J. Wiener and J. K. Hale, ed., Pitman Research Notes in Math. Series 272, Pitman, London, 1991, pp. 207–217.

[Si19] Y. Sibuya, The Gevrey asypmtotics and exact asymptotics, *Trends and Developments in Ordinary Differential Equations*, Y. Alavi and P. F. Hsieh, eds., World Scientific, Singapore, 1993, pp. 325–335.

[Si20] Y. Sibuya, A remark on Bessel functions, *Differential Equations, Dynamical Systems and Control Science*, W. N. Everitt and B. Lee, eds., Lecture Notes in Pure and Applied Math. No. 152, Marcel Dekker, New York, 1994, pp. 301–306.

[Si21] Y. Sibuya, Convergence of formal solutions of meromorphic differential equations containing parameters, *Funk. Ekv.*, 37(1994), 395–400.

[Si22] Y. Sibuya, The Gevrey asymptotics in case of singular perturbations, Preprint, Univ. of Minnesota, 1997.

[SS1] Y. Sibuya and S. Sperber, Some new results on power series solutions of algebraic differential equations, *Proc. of Advanced Seminar on Singular Perturbations and Asymptotics*, May 1980, MRC, Univ. of Wisconsin–Madison, Academic Press, New York, 1980, pp. 379–404.

[SS2] Y. Sibuya and S. Sperber, Arithmetic properties of power series solutions of algebraic differential equations, *Ann. Math.*, 113(1981), 111–157.

[SS3] Y. Sibuya and S. Sperber, *Power Series Solutions of Algebraic Differential Equations*, Publication I.R.M.A. et Département de Mathématique, Univ. Louis Pasteur, Strasbourg, France, 1982.

[St] W. Sternberg, Über die asymptotische Integration von Differentialgleichungen, *Math. Ann.*, 81(1920), 119–186.

[Sz] G. Szegö, *Orthogonal Polynomials*, Colloquium Publications XXIII, American Mathematical Society, Providence, RI, 1939.

[TD] S. Tanaka and E. Date, *KdV Equation: An Introduction to Nonlinear Mathematical Physics*, Kinokuniya, Tokyo, 1979 (in Japanese).

[Tu1] H. L. Turrittin, Convergent solutions of ordinary linear homogeneous differential equations in the neighborhood of an irregular singular point, *Acta Math.*, 93(1955), 27–66.

[Tu2] H. L. Turrittin, Reduction of ordinary differential equations to the Birkhoff canonical form, *Trans. Am. Math. Soc.*, 107(1963), 485–507.

[U] M. Urabe, Geometric study of nonlinear autonomous oscillations, *Funk. Ekv.*, 1(1958), 1–83.

[Was1] W. Wasow, *Asymptotic Expansions for Ordinary Differential Equations*, Dover, New York, 1987. (Also John Wlley, New York, 1965.)

[Was2] W. Wasow, *Linear Turning Point Theory*, Applied Math. Sci. 54, Springer-Verlag, New York, 1985.

[Wat] G. N. Watson, A theory of asymptotic series, *Phil. Trans. Roy. Soc. of London* (Ser. A), 211(1911), 279–313.

[We] H. Weyl, *The Classical Groups*, Princeton University Press, Princeton, NJ, 1946.

[WhW] E. T. Whittaker and G. N. Watson, *A Course of Modern Analysis*, Cambridge University Press, Cambridge, 1927.

[Z] A. Zygmund, *Trigonometric Series*, Vol. 1, 2nd ed., Cambridge University Press, Cambridge, 1959.